TISSUE REGENERATION – FROM BASIC BIOLOGY TO CLINICAL APPLICATION

Edited by **Jamie Davies**

Tissue Regeneration – From Basic Biology to Clinical Application
Edited by Jamie Davies

Published by InTech
Janeza Trdine 9, 51000 Rijeka, Croatia

Publishing Process Manager Dejan Grgur

Technical Editor Teodora Smiljanic

Cover Designer InTech Design Team

Additional hard copies can be obtained from orders@intechopen.com

Tissue Regeneration – From Basic Biology to Clinical Application, Edited by Jamie Davies
p. cm.
ISBN 978-953-51-0387-5

For EU product safety concerns:
IN TECH d.o.o., Prolaz Marije Krucifikse Kozulić 3, 51000 Rijeka, Croatia,
info@intechopen.com or visit our website at intechopen.com.

Meet the editor

Dr Jamie Davies obtained his PhD from the University of Cambridge and is currently Professor of Experimental Anatomy at the University of Edinburgh. His research investigates how organs develop in a growing human fetus, particularly the processes that enable cells to organize themselves into tissues without needing guidance from outside agencies. His laboratory is also active in developing ways to combine this basic knowledge with the principles of engineering, with the aim of building new tissues for people who need them. Dr Davies is Editor-in-Chief of the research journal Organogenesis.

Contents

Introductory Chapter

Tissue Regeneration – A Clinical Science Whose Time Has Come

Jamie Davies
University of Edinburgh
UK

1. Introduction

Tissue engineering is the application of knowledge gained in the study of basic developmental cell biology to the construction and repair of human bodies.

The surgically-focused side of the field has a long history, resting mainly on experience with wound healing and *ad-hoc* attempts to improve it. A well-known and long-standing example is modulation of bone healing by the application of physical force that gives rise to the image of a patient in traction, so common on humorous 'get well soon' cards.

The more cell biological side of the field is younger because its development had to await the gaining of significant amounts of basic knowledge in molecular cell biology, a field that is only a few decades old. The coming together of cell biology and experimental surgery to drive forward the development of tissue engineering is therefore a relatively recent phenomenon and only in this century has tissue engineering really taken off as a major area of research (Fig 1).

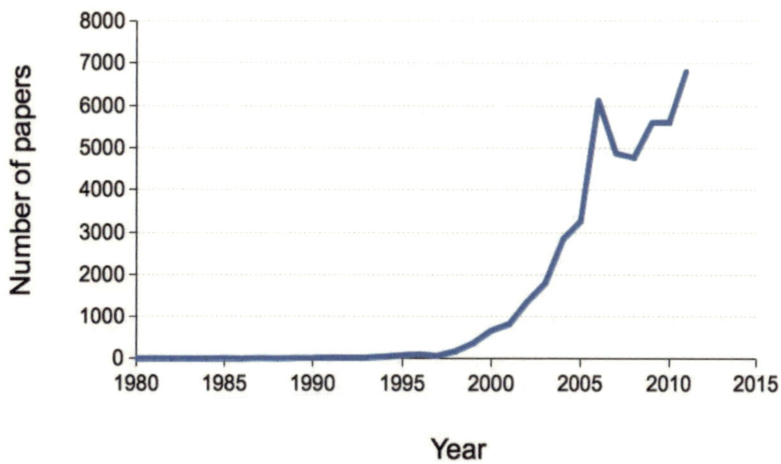

Fig. 1. Rapid growth of tissue engineering as a 21st century discipline. The graph shows the number of publications returned by a Pubmed search for ' *"tissue engineering"* <year>'.

Unlike many other young sciences, tissue engineering is growing very much as a global enterprise, perhaps because of the ubiquity of surgery and therefore the visibility of obvious need. It is noticeable, for example, that the contribution of China to research into tissue engineering is currently approximately equal to that of the European Union (judged by numbers of publications on a simple search: Figure 2).

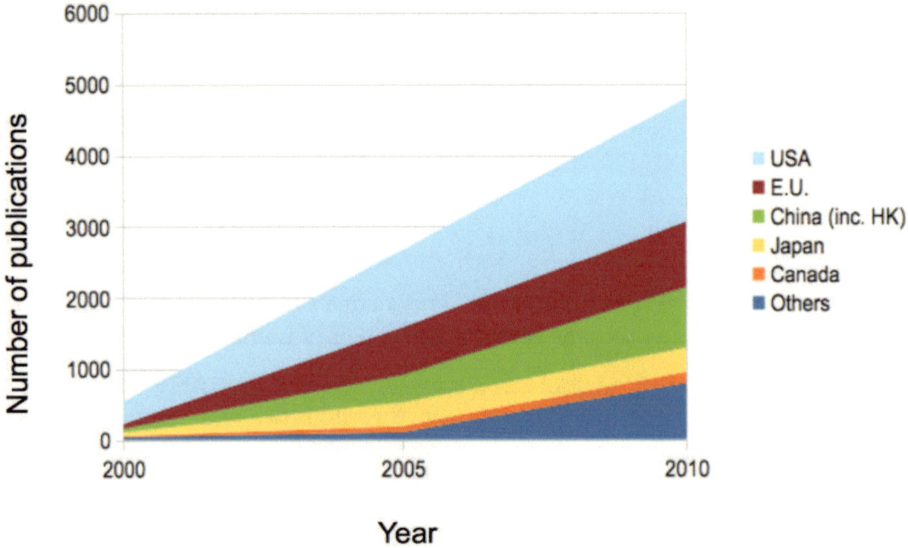

Fig. 2. Growth of tissue engineering output by country. This graph was produced by searching PubMed for '<year> *"Tissue Engineering" xxxx*', where <year> was '2000', '2005' or '2010' and 'xxxx' was the name of a country, or a Boolean expression combining, with a logical OR, a list of countries such as constituents of the European Union.

This global spread of research effort stands in marked contrast to the pattern seen in other new fields such as synthetic biology, which has grown more-or-less in parallel to that of tissue engineering and which is again very much of the twenty-first century. A comparison of pie charts of the national origins of papers in the two young sciences shows the difference immediately, about a third of research in tissue engineering coming from outside the USA and the European Union while only around fifteen percent does so in synthetic biology. The active engagement of so many countries and cultures in problems and applications of tissue regeneration ought to be a great strength for the field, encouraging the development of techniques suited to a wide range of problems and also to a wide range of health care economies.

Wherever it is done, research into tissue regeneration can be divided into three complementary sub-fields, and this book is organized around them. First, there is research that aims to understand and manipulate the endogenous healing processes in human tissues. This is the oldest part of the field. Second, there is the application of stem cell science to the regeneration of tissues (or to their *de novo* generation). Third, there is the construction of engineered scaffolds to guide normal healing processes and the behaviour of stem cells either in culture or *in vivo*. These different aspects of tissue regeneration link and overlap but, for convenience of organization, they will be considered in different sections of this book.

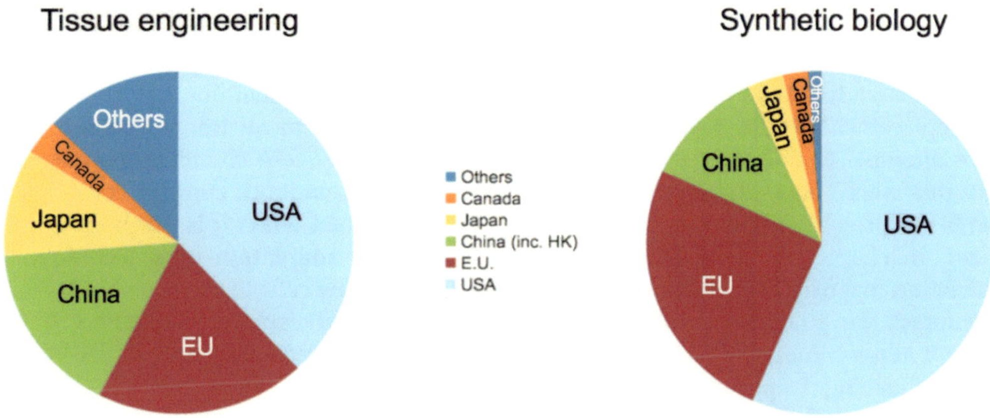

Fig. 3. Research in the young field of tissue engineering is a more global enterprise than research in synthetic biology, a field chosen for comparison because it is about the same age, is taking off about as quickly and is also an application of molecular cell biological knowledge. As with figures 1 and 2, data plotted were obtained by conducting PubMed searches for the name of the field AND the name of the country, or a Boolean expression combining several countries as appropriate.

2. Understanding and manipulating endogenous healing

There are two broad methods for assisting endogenous healing; physical and biological. Physical forces are relevant to healing because many cell types sense and respond to mechanical forces using sensory systems in their cell-cell junctions, cell-matrix junctions and cytoskeleton. The sensing systems, though currently the subject of intense research, remain poorly understood (Katoh et al. 2008). Their detection of appropriate mechanical forces can nevertheless be critically important and the development of many cells and tissues, including heart muscles (Vandenburgh et al. 1996), heart valves (Jacot et al. 2010), and blood vessels (Poelmann et al. 2008), is modulated or determined by mechanical force.

Three chapters in this book illustrate how purely physical interventions can be used to assist healing. Monici and Cialdai (chapter 1) set the scene by reviewing in detail the ways in which living cells sense and respond to physical forces. These include mechanical and gravitational forces (which are, in terms of how they interact with living things, effectively mechanical) and also electric fields, which modulate various types of cell signalling and can be used to stimulate cell motility. Electrical forces are relatively easy to apply from outside the body and therefore offer exciting possibilities for safe, long-term treatment to encourage the regeneration of damaged nerves amongst other things (Hamid & Hayek 2008). High-frequency electromagnetic radiation, in the form of microwaves, infra-red and light, has a long history of medical use (Kenkre et al. 1996, McCulley & Petroll 2008, Gravas et al. 2003, Owens et al. 2003, Simon et al. 2005, Goldberg & Sand 2000). Chapter 1 presents examples of laser light being used to promote regeneration of muscle and bone. In Chapter 2, Li illustrates how a very specific type of mechanical force, pressure waves from ultrasound, can be used to promote the repair of damaged nerves. Again, this technology holds particular promise because it can be applied from outside the body.

One of the major problems faced by surgeons and their patients is the replacement of bones that have to be removed because they harbour neoplasia. Amputation of a limb is clearly a major loss for a patient, so surgeons attempt to conserve whenever possible. If a large section of bone has to be removed, there is a problem finding something to replace it. The ideal replacement would be the excised bone itself, freed from all traces of the tumour. Crude attempts to kill tumour cells, by autoclaving or by severe chemical treatment, do result in a sterile bone but destroy so many of the normal guidance cues for cells that the bone is not properly integrated into the body and maintained when it is put back into the patient. In chapter 3, Diehl and colleagues describe an adaptation of a technique for sterilization that uses high hydrostatic pressures. This kills any cells in the bone but seems to leave intact the guidance cues necessary for the bone's efficient recolonization. It may therefore allow tumour-infested bones to be removed, sterilized safely and replaced, with the aim of leaving the patient with a normal limb.

Biochemical manipulation of cell behaviour is as ancient as medicine itself and the herbal remedies of the first physicians worked - when they worked at all - in this way. An improved understanding of the natural communications during healing and regeneration is allowing researchers to develop biochemical techniques for encouraging regeneration and discouraging damaging scar formation. An approach that is simple in principle even if not straightforward in practice, is to alter wound chemistry to make naturally-produced growth factors more effective. One way of doing this is to alter the extent to which wound matrix sequesters signalling molecules that would, if able to interact with cells, promote healing. In Chapter 4, van Neck and colleagues take this approach and discuss the use of synthetic, glycanase-resistant heparan sulphate mimics in mobilizing endogenous, regeneration-promoting signalling molecules. The chapter presents interesting data that suggest a beneficially-altered balance in favour of repair and against scar formation. Healthy tissue needs a good blood supply and, in chapter 5, de Mendonça describes how angiogenesis is controlled and draws attention to potential targets that would allow the process to be manipulated pharmacologically. Platelets are small cell fragments that travel in blood and that play a critical role in limiting the immediate consequences of wounding by formating a clot when a blood vessel is damaged. As well as performing this mechanical function, platelets are a rich source of growth factors and cytokines. In chapter 6, Ohkohichi and colleagues present data that argues that platelets can significantly promote the regeneration of damaged liver and possibly promote the survival of small grafts as well.

A well-known, and frustrating, fact about natural regeneration is that so-called 'lower' animals are a lot better at it than are humans. Have humans completely lost an ancient regenerative response still present in fish and amphibians, or is the difference a matter of degree, humans still retaining elements of the repair pathway but not activating it strongly enough? The difference is important, because the second of these hypotheses allows for the possibility of developing a clinical intervention that builds on an existing pathway but pushes it harder to give humans an amphibian-like power of rebuilding. Eleonora Grigoryan, in chapter 7, analyses the early cellular events that follow damage to the retina in amphibians and mammals. Although the final outcome of the type of damage is very different in these different organisms (amphibians making a new retina, mammals failing to do so), the early responses of the cells are the same but the conversion of non-neuronal cells to neurons and glia follows them only in amphibia. It is still not clear why the eventual fates of the cells diverge so much between species, but the work described in the chapter narrows down the difference to a specific stage in the time-course of the response and should help researchers to discover exactly why the cells of the different organisms end up behaving so differently from the same initial pathway.

Compared to the research in stem cell biology that is catching so much public attention, research on simple physical and biochemical methods to promote regeneration may strike some readers as pedestrian. From the point of view of the patient, though, the important requirement is not that a researcher keeps publishing in the trendiest stem cell journals, but rather that they invent something that actually works. With the honourable exception of bone marrow transplantation, most stem cell treatments are very new and most clinical trials have been small and have not been running long. It is therefore too early to say, in most cases, whether stem cells will prove as effective as is hoped (Trounson et al. 2011). If they are not, it may well be that, in the short-to-medium term at least, real patients will benefit more, and in larger numbers, from the development of 'conventional' physical and chemical based therapies.

3. Application of stem cells

Stem cells are cells that can maintain their own population and give rise to daughters that are committed to differentiate. Depending on where they come from, they can have an ability to differentiate into any cell type of the body (for example, embryonic stem cells) or only a restricted range of cell types (for example, haematopoietic stem cells). Because of their potential to make new cells, stem cells have for some years been seen as a very promising source of new tissue. Initial experiments were generally designed on the assumption that a useful stem cell therapy would work by the stem cells themselves making new tissue to replace that lost to injury. In recent years, though, the field has been made more complicated by the realization that some stem cell types, in particular mesenchymal stem cells, exert significant beneficial effects by secreting factors that modulate inflammation and healing by the host tissue, and that much new growth is from the cells that were already there rather than being from the stem cells. The dual mode of stem cell action, which places different emphasis on each mode according to the precise situation, makes analysis of stem cell-mediated repair complicated. On the positive side, the realization that secreted factors may be of significant, or even primary, importance in the mechanism of this repair is exciting from a clinical point of view because it may be possible to apply the factors directly without any need for patients to undergo potentially dangerous stem cell transplantation. This is not yet possible, though, and at the moment attempts at this type of therapy requires the use of stem cells themselves.

Development of effective stem cell therapies can be divided into two sub-problems; that of finding the most suitable type or source of stem cells to use, and that of applying them to the injured body in the most effective way possible. For convenience, section 2 of this book has been divided along precisely these lines, chapters 8-11 covering sources of stem cells and chapters 12-16 covering their application.

Arguably the most convenient source of stem cells for therapy is the patient himself. Bone marrow is a rich source of stem cells. Many attempts to use them have involved mechanical recovery of marrow followed by purification of stem cells, perhaps with additional steps of proliferation and reprogramming, followed by injection into systemic blood or directly into a site of damage. In chapter 8, Christian Drapeau and colleagues discuss an alternative approach that involves much less invasive manipulation Their strategy is to use the fact that endogenous bone marrow stem cells can be mobilized by cytokines, and they describe experiments that involve injecting pure cytokines into animals that have suffered cardiac infarction, in an effort to encourage mobilization of the animals' own stem cells to effect a repair. The authors also describe preliminary studies of this approach in humans, to treat stroke and kidney failure.

A novel source of pluripotent stem cells, capable of making any body cell, is the testis. Spermatogonial cells are the stem cells that naturally maintain production of sperm. Within the testis, they are constrained by their environment to have only the simple choice between self-renewal and spermatogenesis. Taken outside that environment, though, the cells can differentiate into a large range of cell types, making them effectively pluripotent. Liz Simon and colleagues analyze the abilities of these stem cells in chapter 9, and evaluate their potential for therapeutic use compared to the potential of other stem cell types. While spermatogonial cells can be obtained only from men, pregnant women can be a source of fetal membrane-derived mesenchymal stem cells. Shin Ishikane and colleagues describe the isolation and properties of these cells in chapter 10, and summarize their ability to modulate immune activity and to become a useful tool in regenerative medicine. The theme of mesenchymal stem cells is explored further by Arshak Alexanian in chapter 11, with a special emphasis on their ability to improve central nervous system repair.

Wherever and however they are obtained, stem cells have to be applied in a way that optimizes their ability to effect repair. Five chapters in this book focus on the application of stem cells to different clinical problems in the circulatory and the musculoskeletal systems. José Lamas and colleagues (chapter 12) address the potential of mesenchymal stem cells to treat osteoarthritis, a common and debilitating disease of joints, and outline current knowledge and future prospects for this important field. Namath Hussain and colleagues address another common and important chronic problem in chapter 13; that of debilitating lower back pain caused by damage to intervertebral discs. They explain the basic pathology of the disease and then review the results from the (very small) studies that have so far been conducted into the efficacy of stem cell treatment for disc damage in humans. Marianna Karagianni and colleagues consider a different problem in orthopaedics; the healing of defective bone. In chapter 14, they also examine the regulatory frameworks that govern the use of 'advanced therapy medicinal products' and consider how these frameworks shape research and development. Regeneration of bone is also addressed by Dilawhare Khan, Arnaldo Santos and their colleagues (chapters 15 and 16) . Khan *et al.* use an unusual source of stem cells - teeth - and having demonstrated that these show promise, they make the suggestion that milk tooth stem cells could perhaps be banked for a patient's later use. Santos *et al.* provide a wide-ranging review of different approaches to the regeneration of bone including, but not restricted to, the use of stem cells. Their chapter could have appeared in almost any section of this book: it was included as a final chapter of the stem cells section to highlight the need to compare stem cell approaches with the best of other techniques, because there is arguably a tendency, at the moment, to place undue emphasis on some ways of effecting repair perhaps to the detriment of developing others that may even show more promise in existing clinical trials.

One theme that emerges from a large number of these chapters is that studies of the effects of stem cell treatment in humans are few, and tend to be small. Perhaps because of their small size and consequent low statistical power, these studies frequently produce contradictory results. This is not helped by the lack of standardization in how experiments are performed and assessed, which makes meta-analyses problematic. Overall, it is clear that, in most areas of application, it is still far too early to decide whether stem cell treatments really are a means of effective cure and repair, or whether other approaches, such as those described in section 3 below, will actually prove more useful.

4. Construction of scaffolds

Where there are large-scale defects in tissues, caused either by injury or by congenital abnormality, simple stem cell treatments - however well they can be made to work - are unlikely to be able to make a proper repair. In terms of directly producing new tissues, stem cells are expected to work by recapitulating the processes of natural development or tissue maintenance. Embryonic development tends to take place at small scale and tissues then grow; when an embryo first makes a trachea, for example, it is less than a millimetre long, not the many centimetres it is in adult life. Also, many embryonic events depend on signals from other embryonic tissues that move or disappear by birth. There are therefore good reasons that a stem cell, even in a state that corresponds perfectly to the cells that would make a tissue in an embryo, would not be able to make it in an adult. Similarly, the stem cells concerned with tissue maintenance are regulated by the environment of the tissue in which they are situated and there is no reason to assume that they can rebuilt tissues across a large gap or scar, in which this environment is missing. In the case of genetic abnormality, the cells may be incapable of making the body part normally anyway. For all of these reasons, there is a strong argument that large scale regeneration requires the construction of scaffolds to bridge gaps and to control cell behaviour.

Bone tissue, which in its mature form is mostly inorganic matrix, lends itself to a scaffold-based approach. In chapter 17, Magdalena Cieślik and colleagues compare the ability of different matrices, based on the natural structure of bone with additional components such as bioglasses, to promote effective bone repair. On a related topic, Peter Emans and colleagues propose, in chapter 18, the use of scaffolds designed with the normal developmental process of endochondral ossification in mind. The chapter includes a critical review of clinical trials (which, as with much regenerative medicine, are less effective than original experiments gave hope to believe).

Tissue engineering of soft tissues involves different considerations, such tissues generally being much more flexible and much more cellular in terms of the ratio of cells to surrounding matrix. Chao Feng and Yue-min Xu illustrate this in chapter 19, where they explore techniques for reconstructing the lower urinary tract. They compare different scaffold materials, such as fleeces, sponges and advanced materials that include signalling molecules, and consider techniques for populating them with cells before their use. The chapter includes a review of clinical data on reconstruction of human bladder and urethra, with an encouraging rate of success.

In the last chapter in this section, chapter 20, Abir El-Sadik connects the rapidly developing field of tissue engineering with another 'hot topic': nanotechnology. Nanotechnology is young and still raises significant safety concerns so there is little clinical data, but the chapter explains the ways in which, in principle, nano-materials can modulate the function of stem cells and other tissue cells. It also illustrates, using nerve regeneration as an example, how scaffolds that incorporate nano-materials can promote useful neuronal growth and decrease glial scarring in experimental systems.

5. Assessing tissue regeneration

Having ideas about how to improve tissue regeneration is all very well, but it is essential both to the process of research that evaluates these ideas, and also to the proper care of a patient undergoing regeneration, that progress can be monitored, evaluated and perhaps even predicted. The two chapters in the last section of this book address this specific issue.

Magnetic resonance imaging is an excellent non-invasive technique that can provide high-resolution images of any part of the body. In chapter 21, Miyata illustrates how the ability of advanced MRI systems to provide quantitative information, particularly on the state of water and whether it is a free liquid or mainly bound to glycosaminoglycans, can be used to monitor the state of cartilage. This offers both researchers and clinicians an opportunity to monitor the progress of regenerative treatments over a long time-course, optimizing care and leading the way to patient-specific treatment regimes. It is possible that this approach will be extended in future to monitoring events in soft-tissue repair.

In the final chapter, Vermolen and van Rijn take an approach very different to most other authors in this book: they describe mathematical models of the processes involved in the 'healing' (by scar formation) of accidental wounds, particularly burns. The interest in the biophysics of scar formation is not simply academic and 'basic science' because, as the authors point out, creating tools that can *predict* the natural outcome of a patient's specific wound could be very useful to the design of a patient-specific programme of treatment designed to resolve that wound with as little aesthetic impact as possible. In particular, it can help physicians apply the right boost to regeneration (by whatever method) for each part of the wound, navigating successfully between the Scylla of overgrowth and the Charybdis of under-regeneration and consequent scarring, neither of which would be aesthetically satisfactory.

6. Using the chapters of this book

The whole philosophy of this multi-author book, and of its publisher, has placed unusual demands on both the authors and the editor. Unlike a conventional volume, which can only be bought or borrowed in its entirety, this book can be viewed two ways. The first is the traditional one - purchase of a single bound volume containing everything in order. The second is downloading of individual chapters from the Internet, using the Open Access model. There is much to be said for this, not least by clinicians and researchers who do not work in rich institutions that are blessed with a large library.

The knowledge that the authors are writing both for a traditional, whole-book-owning readership and for readers who may view just one chapter has, however, presented the Editor with an unusual problem: that of judging how much introductory material is necessary for a chapter to be readable in its own right, and how much can be left to the authors of other chapters. I hope that a sensible compromise has been struck, but am aware that there has had to be some repetition of some introductory material from chapter to chapter. In a book that could only be obtained as a whole, this repetition would have been removed. The editor and authors trust that readers will understand why some has had to remain, given the necessity for each chapter to stand alone.

The chapters in this book were written only a short time before publication and represent a very up-to-date overview of the field. Regenerative medicine is moving so quickly, however, that the authors expect some details of their material - especially reviews of human trials - to become out of date in only a few years. Depressing as a work's rapid obsolescence might seem to be for an author, really we rejoice in the fact, for it is the mark of a vibrant and fast-moving field that promises to have a significant impact on twenty-first century medicine. It is our hope that this book may inspire some of the new work that will one day make it obsolete.

7. References

Goldberg, R.P. & Sand, P.K., 2000. Electromagnetic pelvic floor stimulation: applications for the gynecologist. *Obstetrical & Gynecological Survey*, 55(11), pp.715-720.

Gravas, S., Laguna, M.P. & De La Rosette, J.J.M.C.H., 2003. Application of external microwave thermotherapy in urology: past, present, and future. *Journal of Endourology / Endourological Society*, 17(8), pp.659-666.

Hamid, S. & Hayek, R., 2008. Role of electrical stimulation for rehabilitation and regeneration after spinal cord injury: an overview. *European Spine Journal: Official Publication of the European Spine Society, the European Spinal Deformity Society, and the European Section of the Cervical Spine Research Society*, 17(9), pp.1256-1269.

Jacot, J.G. et al., 2010. Cardiac myocyte force development during differentiation and maturation. *Annals of the New York Academy of Sciences*, 1188, pp.121-127.

Katoh, K., Kano, Y. & Ookawara, S., 2008. Role of stress fibers and focal adhesions as a mediator for mechano-signal transduction in endothelial cells in situ. *Vascular Health and Risk Management*, 4(6), pp.1273-1282.

Kenkre, J.E. et al., 1996. A randomized controlled trial of electromagnetic therapy in the primary care management of venous leg ulceration. *Family Practice*, 13(3), pp.236-241.

McCulley, J.P. & Petroll, W.M., 2008. Quantitative assessment of corneal wound healing following IntraLASIK using in vivo confocal microscopy. *Transactions of the American Ophthalmological Society*, 106, pp.84-90; discussion 90-92.

Owens, B.D., Stickles, B.J. & Busconi, B.D., 2003. Radiofrequency energy: applications and basic science. *American Journal of Orthopedics (Belle Mead, N.J.)*, 32(3), pp.117-120; discussion 120-121.

Poelmann, R.E., Gittenberger-de Groot, A.C. & Hierck, B.P., 2008. The development of the heart and microcirculation: role of shear stress. *Medical & Biological Engineering & Computing*, 46(5), pp.479-484.

Simon, C.J., Dupuy, D.E. & Mayo-Smith, W.W., 2005. Microwave ablation: principles and applications. *Radiographics: A Review Publication of the Radiological Society of North America, Inc*, 25 Suppl 1, pp.S69-83.

Trounson, A. et al., 2011. Clinical trials for stem cell therapies. *BMC Medicine*, 9, p.52.

Vandenburgh, H.H. et al., 1996. Mechanical stimulation of organogenic cardiomyocyte growth in vitro. *The American Journal of Physiology*, 270(5 Pt 1), pp.C1284-1292

Part 1

Understanding and Manipulating Endogeneous Healing of Tissues

1

The Role of Physical Factors in Cell Differentiation, Tissue Repair and Regeneration

Monica Monici and Francesca Cialdai
ASAcampus Joint Laboratory, ASA Res. Div.
Dept. of Clinical Physiopathology, University of Florence
Italy

1. Introduction

Physical factors may induce significant biological effects, therefore they can be applied in biomedical and biotechnological fields in order to drive and modulate biological processes. It is well known that both humoral and physical factors (in particular, but not limited to the mechanical ones) are necessary for maintaining tissue homeostasis. Both biochemical and physical factors can induce the cells to reprogram their functions to adapt dynamically to the environmental conditions.

It is evident, therefore, that the only way to approach functional tissue regeneration and repair is to supply combined humoral and physical stimuli in a dose- and time-dependent manner. For example, in vitro studies have shown that a biomimetic environment simulating pulsatile flow is an indispensable condition for the tissue engineering of functional trileaflet heart valves from human marrow stromal cells. Static controls show morphological alterations and weaker mechanical properties (Hoerstrup et al., 2002).

Studies on the role of physical factors in tissue repair and regeneration cover a very broad field that extends from investigations aimed at deepening our understanding of the physiological mechanisms of tissue repair and regeneration to biotech advances in tissue engineering, such as development of biocompatible scaffolds, 3D cell culture systems and bioreactors, which in the future must integrate the delivery of biochemical factors with the provision of physical stimuli that are equally necessary. In this chapter, far from providing a comprehensive overview of this field of studies, we introduce some issues concerning the application of physical factors in biomedicine and biotechnology and report the results of our research on the application of various physical stimuli (gravitational and mechanical stresses, laser radiation, electromagnetic fields (EMF)) for modulating cell commitment and differentiation, cell adhesion/migration, production and assembly of extracellular matrix (ECM) components, with the final aim of understanding when and how physical stimuli can be useful for promoting tissue repair and formation of functional tissue constructs. We also briefly mention how, in past centuries, the role of physical factors in biological processes has been understood and physical stimuli have been applied for therapeutic purposes.

2. Mechanical stresses

The importance of gravitational and mechanical factors in modulating biological processes has been known for a long time: from Galileo Galilei onwards, studies on functional adaptation of the skeleton demonstrated that bone loss or gain is related to the magnitude, direction and frequency of the stress acting upon the skeleton during application of loads (Galilei, 1632; Wolff, 1985; Rubin, 1985; Frost, 1988; Rubin, 1984; Ingber, 1998).

Within the body, cells are subject to mechanical stimulation, caused by blood circulation, ambulation, respiration, etc., which give rise to a variety of biochemical responses. It has been demonstrated that changes in inertial conditions, shear stress, stretching, etc. can strongly affect cell machinery. Cells may sense mechanical stresses through changes in the balance of forces that are transmitted across transmembrane adhesion receptors that link the cytoskeleton to the ECM and to other cells. These changes, in turn, alter the ECM mechanics, cell shape and cytoskeletal organization (Ingber, 1998, 1999). A great deal of information has revealed that the ECM is a highly dynamic and elastic structure which undergoes continuous remodelling, in particular during development, angiogenesis, wound healing and other tissue repair processes. The ECM interacts with cells to provide relevant microenvironmental information, biochemically through the release of stored soluble and insoluble factors, and physically through imposition of structural and mechanical constraints (Carson, 2004). On the other hand, mechanical stimuli modulate ECM homeostasis: mechanical forces strictly regulate the production of ECM proteins indirectly, by stimulating the release of paracrine growth factors, or directly, by triggering intracellular signalling pathways leading to the activation of genes involved in ECM turnover (Chiquet, 2003).

Mechanical stimuli affect cells through poorly understood mechanotransductive pathways that lead to changes in morphology and orientation, modulation of gene expression, reorganization of cell structures and intercellular communication through both secretion of soluble factors and direct intercellular contact (Maul et al., 2011; Kang et al., 2011; Park et al., 2006; Papachroni et al., 2009; Wall & Banes, 2005; Bacabac et al., 2010; Hughes-Fulford & Boonstra, 2010). Over the past decade, in vitro studies have indicated that the transduction of physical stimuli involves the ECM-integrin-cytoskeleton network and also calcium channels, guanosine triphosphatases (GTPases), adenylate cyclase, phospholipase C (PLC) and mitogen-activated protein kinases (MAPKs), all of which play important roles in early signaling (Rubin et al., 2006; Hoberg et al., 2005; Adachi et al., 2033; Mobasheri et al., 2005; Chiquet et al., 2009; Bacabac et al., 2010; Hughes-Fulford & Boonstra, 2010). It has been demonstrated that, in endothelial cells, different genetic programs leading to growth, differentiation and apoptosis can be mechanically switched. Cells grow when they spread, die when fully retracted, and differentiate into capillary tubes if maintained at a moderate degree of extension (Chen, 1997).

The in vitro application of mechanical stretch, simulating the mechanical load to whom heart cells are exposed in vivo, initiated in adherent cultures of neonatal cardiomyocytes morphological alterations similar to those occurring during in vivo heart growth (Vandenburg, 1996). Stem cell commitment, the process by which a cell chooses its fate, and differentiation, the resulting development of lineage-specific characteristics, have been shown to be affected by cell shape (Roskelley, 1994; McBeath, 2044; Watt 1988; Spiegelman, 1983). Internal and external forces regulate cell shape and studies have shown that cell shape can control apoptosis, gene expression, and protein synthesis, in addition to stem cell fate (Chen, 1997; Thomas, 2002).

The cells belonging to tissues that resist the effects of gravity are particularly sensitive to mechanical and gravitational stimuli, which play a key role in the development and homeostasis of these tissues. Lack of gravitational and mechanical stresses leads to the formation of impaired tissues with lower mechanical properties and reduced function.

It is well established that bone adapts its mass and architecture in accordance with the external mechanical loads applied and osteocytes, terminally differentiated cells of the osteoblastic lineage, may be considered "mechanosensory cells" (Vatsa et al., 2010). They are sensitive to both stretching and fluid flow. Mechanical stimulation of osteocytes induces intercellular signaling which results in the modulation of osteoblast and osteoclast activity (Chow et al., 1998; Turner et al., 1997). Interestingly, it has been shown that the stimulation of a single osteocyte activates many surrounding cells (Vatsa et al., 2007).

Osteoblastic differentiation can be induced by applying mechanical stress, for example by stretching the surface on which the cells are attached (Cavalcanti-Adam et al., 2002). Many studies revealed that the micromotions at the interface between bone and artificial scaffolds play a key role in scaffold integration: they can promote tissue differentiation or induce bone resorption (Prendergast et al., 1997; Carter et al., 1998; Buchler et al., 2003; Stadelmann et al., 2008; Jasty et al.,1997). In a recent paper on biomechanics of scaffolds for bone tissue engineering applications, it has been stated that in the development of a scaffold it is important to take in account not only the structural integrity but also the load transmitted to the cells via the scaffold deformation (Pioletti, 2011).

A recent review of studies which investigated the importance of loading in maintaining the balance of matrix turnover in the intervertebral disk, reported about the possible role of overloading in the initiation and progression of disc degeneration and proposed a physiological/beneficial loading range as a basis on which to design loading regimes for testing tissue constructs or favouring differentiation of stem cells towards "discogenic" cells for tissue engineering (Chan et al., 2011)

An overview of studies on the role of mechanical stimuli in chondrogenesis showed that uniaxial loading induces the upregulation of genes associated with a chondrogenic phenotype while multiaxial loading results in a broader pattern of chondrogenic gene upregulation, revealing that not only intensity, but also direction and other parameters which characterize the stimulation are relevant for the achievement of the final effect. The physiological multiaxial pattern of loading within articulating joints is so complex that currently, even with the most sophisticated bioreactors, it would be impossible to simulate the in vivo situation. Therefore, it has been suggested to use the body as an "in vivo bioreactor"(Grad et al., 2011).

Conditions of gravitational unloading, both real and modeled by a Random Positioning Machine (RPM), negatively affect cellular organization and ECM production in cartilage constructs, even if at different extent. (Stamenkovic et al., 2010).

Our group is conducting for several years research on the role of gravitational and mechanical stimuli in cell differentiation, tissue repair and regeneration, with particular attention to the remodelling phase.

Our studies demonstrated that gravitational unloading favours the differentiation of osteoclastic precursors (FLG 29.1 cells). After 72 hours exposure to conditions of microgravity, modelled by a RPM (angular velocity of rotation 60°/s), the cells showed a dramatic increase in apoptosis, but the viable ones showed osteoclastic-like morphology, cytoskeletal reorganization, significant changes in gene expression profile. The expression of the major osteoclastic markers Receptor Activator of Nuclear Factor Kappa-B (RANK) and Receptor Activator of Nuclear Factor Kappa-B Ligand (RANKL) strongly increased and cells showed the ability to resorb bone (Fig. 1) (Monici et al., 2006).

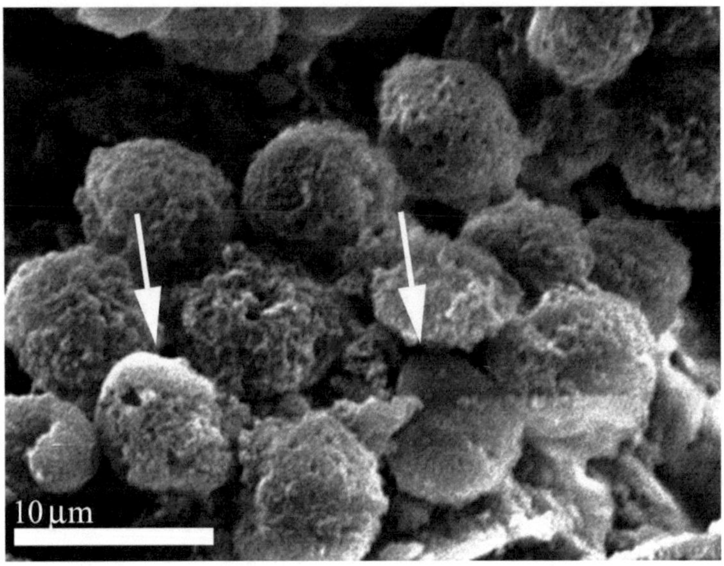

Fig. 1. Scanning electron microscopy of a bone slice exposed to FLG 29.1 cells cultured in modelled microgravity. Adherent cells on the bone surface can be observed. Arrows indicate the sealing zone.

Analysing the gene expression profile of human mesenchymal stem cells (HMSC) in loading conditions (3 hours exposure to 10x*g* in hyperfuge), we found overexpression of genes involved in osteoblastogenesis (*GLI1, NF1, MEN1*) and downregulation of genes involved in adipogenesis (*PPARγ, FABP4*) (Tab. 1) (Monici et al., 2008a).

Gene	Control	RPM	10 x g	Nd:YAG
FABP4	**27**	**304**	**3**	**9**
PPARG	**12**	**587**	**11**	**14**
GLI1	**47**	**45**	**789**	**547**
NF1	**25**	**9**	**241**	**258**
MEN1	**48**	**25**	**158**	**258**

Table 1. Gene expression profile in HMSCs.

These results, in agreement with those of other authors (Kaneuji et al., 2011; Wang et al., 2010; Searby et al., 2005) reveal that mechanical/gravitational stresses induce osteoblastic differentiation while gravitational unloading and loss of mechanical stress favour adipogenesis, osteoclastogenesis and bone resorption.

In cultures of fibroblasts exposed for 3 hours to hypergravity (10x*g*), we observed enhanced expression of collagen I and fibronectin (20% and 30% more than control, respectively), while chondrocytes exposed to the same treatment showed a marked increase in collagen II, aggrecan and Sox 9, a transcription factor which plays a key role in chondrogenesis. Therefore, after definition of optimal range of intensity and force direction, loading can be used to stimulate ECM production by cells of the connective tissues and to favour chondrogenesis.

A series of experiments we carried out with the aim of studying the effect of gravitational unloading on processes involved in tissue remodelling demonstrated that the loss of mechanical stress causes a disregulation in laminin and fibronectin (FN) production by fibroblasts and endothelial cells (Fig. 2B) (Monici et al., 2011). In particular, FN forms a disordered and intricate network, reproducing the typical condition of fibrous scars. We hypothesized that the altered FN fibrillogenesis could be a cause of impaired ECM rebuilding and altered cell adhesion/migration and could contribute to the impairment of wound healing observed in microgravity (Midura & Androjna, 2006; Delp, 2008).

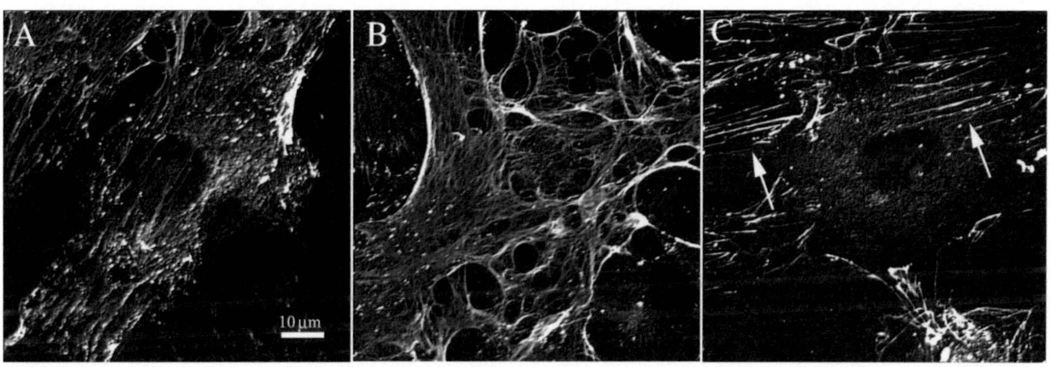

Fig. 2. FN expression in CVECs (analysed by immunofluorescence microscopy): A) control, B) exposed for 72 h to modelled microgravity and C) treated with pulsed Nd:YAG laser (1064 nm). In figure B a tight network of FN fibrils appears while in figure C the FN fibrils are parallel and ordered (see arrows).

Studying the behaviour of aortic endothelial cells cultured in micro- and hypergravity we found that the exposure to simulated microgravity conditions for 72 hours (angular velocity of rotation 60°/s) caused a reduction in coronary venular endothelial cell (CVEC) number. Genomic analysis revealed that proapoptotic signals increased, while antiapoptotic and proliferation/survival genes were downregulated by the absence of gravity. Activation of apoptosis was accompanied by morphological changes, with mitochondrial disassembly and organelles/cytoplasmic NAD(P)H redistribution, as evidenced by autofluorescence analysis. Moreover, cells were not able to respond to angiogenic stimuli in terms of migration and proliferation (Morbidelli et al., 2005)

In contrast, after exposure to hypergravity (10x*g*), no significant changes were observed in cell morphology and energy metabolism. Cells remained adherent to the substrate, but integrin distribution was modified. Accordingly, the cytoskeletal network reorganized, documenting cell activation. There was a reduction in expression of genes controlling vasoconstriction and inflammation. Proapoptotic signals were downregulated. Overall, the results documented that hypergravity exposure maintained endothelial cell survival and function by activation of adaptive mechanisms. The behavior of cells derived from microcirculation was somewhat different, because the above described effects were associated with increased anaerobic metabolism and cell detachment from the substrate (Morbidelli et al., 2009). These findings demonstrate that gravitational/mechanical stress can strongly affect endothelial function and neoangiogenesis and the biological response could also depend on the different vascular districts.

3. Electromagnetic fields

It is said that in the first century AD an "electric fish" was used to cure headache. Paracelsus (1493-1542) studied the medical use of lodestone and, in the sixteenth century, Sir Kenelm Digby described the magnetic cure of wounds. At the end of the seventeenth century, Galvani, with his famous experiments on bioelectricity, opened the way for modern studies on physiological EMFs and the effects of external EMFs on the body.

Over the past forty years, important advances have been made in research on bioelectricity: differences in electrical potentials of plants, animals and humans have been measured (Burr, 1972), changes in voltage gradients have been correlated with morphogenetic events in plants and animals (McCaig et al., 2005), physiological currents have been found to be signals for key processes in development (Levin, 2007).

An extensive discussion on electromagnetic effects from cell biology to medicine is presented in a recent review written by Funk et al. (Funk et al., 2009), where the coupling between physical mechanisms and cell biology is discussed in depth. In a nutshell, EMFs can cause polarization of bound charges, orientation of permanent dipoles (which results in topographycal changes in molecules), drift and diffusion of conduction charges, ion bound or release from proteins, ion-channel or receptor redistribution, conformational changes of voltage-sensitive enzymes, modulation of binding kinetics, reorientation of membrane phospholipids and changes in activation kinetics of ion channels (Funk et al., 2009).

Endogenous EMFs in living tissues are generated by physiological activities, for example movements of the musculoskeletal system structures. Vibrations of human muscles induce mechanical strains and currents have been measured both during postural muscle activity (5–30 Hz) and walking (<10 Hz) (Antonsson & Mann, 1985). Muscle contractions induce in the underlying bone tissue EMFs which are important for maintaining bone mass. Bone cells are selectively sensitive to low frequencies, in particular those ranging from 15 to 30 Hz. In this narrow range of frequencies, fields as low as 0.01mV/cm affect the remodelling activity (McLeod & Rubin, 1993). It has been found that EM current densities produced by mechanical loading (e.g. 1Hz during walking) in bone lie in the range 0.1–1.0 mA/cm^2 (Lisi et al., 2006). Generally, physiological EMFs are characterized by extremely low frequencies (ELF), from 0 to 300 Hz, and have low intensity.

EMFs are widely used to treat musculoskeletal diseases and many studies indicated that the most effective devices use pulsed EMFs with frequencies from 1 to 100 Hz, which induce EF of the order of μV/cm (Pilla, 2002). Therefore, physiological effects may be induced by EMFs characterized by low frequencies (optimal range 8-60 Hz) and amplitudes ≤ 1 G (Funk et al., 2009).

It has been demonstrated that pulsed EMFs can increase osteoblastic differentiation and activity and, on the other hand, inhibit osteoclastogenesis, thus shifting the balance towards osteogenesis (Otter et al., 1998; Hartig et al., 2000; Chang et al., 2004).

Studies aimed at evaluating the possibility to apply EMFs to favour ligament healing and repair demonstrated that, after exposure to pulsed EMF, fibroblasts from calf anterior cruciate ligament increased migration speed and showed enhanced collagen I expression. On the contrary, static EMF had an inhibitory effect on wound healing, which was reversed by pulsed EMF (Chao et al., 2007).

EMFs can modulate cell proliferation. The literature indicates that both intensity and frequency of the EMF are important in determining the final effect. Kwee and Raskmark (Kwee & Raskmark, 1995) have found an increase in the proliferation of human fibroblasts exposed to 0.08 mT, while Kula and Drozdz (Kula & Drozd, 1996) have shown inhibition of cell growth in murine fibroblasts exposed to 20 mT. Even trials carried out by exposing cultures of human lymphocytes have given different effects (increase, decrease or no effect in the proliferation) depending on the intensity of the applied EMF (Paile et al., 1995; Scarfì et al., 1999).

As regards frequency, many authors reported increases in proliferation of different cell types at 50 Hz frequency (Scarfì et al., 1991; Cossarizza et al., 1993).

Numerous studies have addressed the interaction between EMFs and calcium fluxes, because calcium is a principal regulator of several cellular processes. It is an activator of cyclic AMP, key molecule in triggering intracellular metabolic processes. It has been observed that the exposure to EMFs can modulate calcium concentration in a way which depends on cell type and field intensity. (Farndale, 1987; Walleczek, 1990).

The effects of EMFs on cell differentiation have been studied too. A progressive inhibition of enzyme activity and differentiation in MC-3T3 osteoblast-like cells, after exposure to 30 Hz EMF, was described by McLeod and Collazo (McLeod & Collazo, 2000). In HMSCs exposed to EMFs during chondrogenic differentiation, increase in collagen II and glycosaminoglycan (GAG)/DNA content was observed (Mayer-Wagner et al., 2010). Therefore EMFs might be a way to stimulate and maintain chondrogenesis of HMSCs and provide a new step in regenerative medicine regarding tissue engineering of cartilage.

In recent experiments aimed at studying the effects of EMFs on neuroblasts and understanding whether these effects can be useful in promoting tissue regeneration, we found that in neuroblasts (SHSY5Y human cell line derived from neuroblastoma) exposed to low frequency EMF (50 Hz; 2 mT, 3 hours) synaptophysin and TAU (microtubule-associated proteins) were overexpressed while Microtubule-Associated Protein 2 (MAP2) was downregulated. Synaptophysin participates in the formation of the channel for neurotransmitter release. TAU is associated with the protofilaments in neurites and MAP2 is a microtubule-associated protein found predominantly in the cell body. MAP2 function is not required when the cell disassembles microtubules in the cell body to give rise to the formation of neurites, while TAU is required to add new subunits to microtubules which are forming in the neurites. Moreover, in the treated neuroblasts, we observed rearrangement of microtubules and actin microfilaments, with formation of cones and cytoplasmic extensions (Fig.3), and increase of neurofilaments, a marker of neurogenic differentiation (not yet published data). Therefore we hypothesize that EMFs can favour differentiation. The expression of synaptophysin, TAU and MAP2 returned to control values 24 h after exposure. Instead, the formation of neurites continued to progress even after 24 h, with the appearance of branched extensions. This means that the changes in protein expression are part of a complex biological response that, once triggered by exposure to the EMF, proceeds even after the cessation of the stimulus (Cerrato et al., in press)

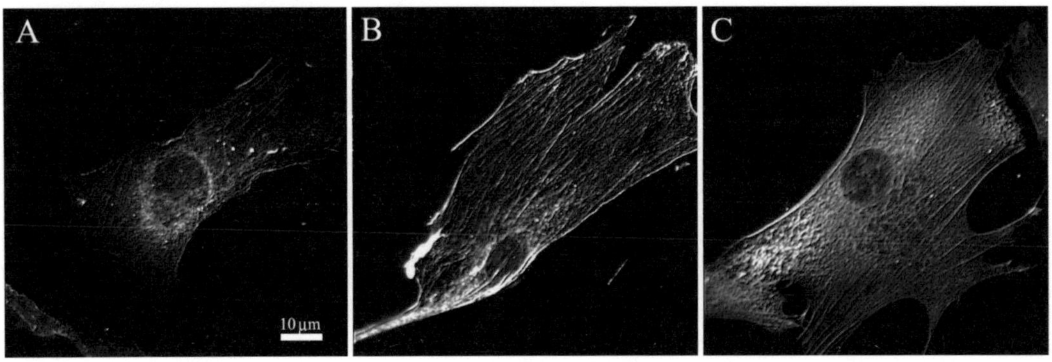

Fig. 3. Actin expression in SHSY5Y cells (analysed by immunofluorescence microscopy): control (A) and cells exposed to EMF (50 Hz, 2 mT, 3h) analysed immediately (B) and 24h (C) after the treatment. The formation of neurites can be observed in figures B and C.

Preliminar experiments on the effect of pulsed EMF on fibroblast behaviour in wound healing models showed, in agreement with other authors (Sunkari et al., 2011), that EMF can accelerate or slow the migration of fibroblasts, depending on the properties of the applied field (data not yet published). The possibility of modulating fibroblast migration during wound healing could be very interesting: in fact it might be useful to enhance the migration of fibroblasts to promote wound healing in chronic ulcers and, in general, in cases where healing is slow, while to inhibit the migration would be beneficial to prevent the formation of fibrous scars.

4. Light

Ancient civilizations had learned that light can have effects on the tissues of the body: both Romans and Greeks widely used the exposure to sun for therapeutic purposes.

In ancient China, a ritual to attain immortality in use during the Tang dynasty (fifth century AD) prescribed exposure to the sun holding in the right hand a piece of green paper with the character representing the sun in red. Subsequently, the paper previously exposed to the sun should be soaked in water and eaten for "trapping" in the body the essence of the sun. At the turn of the nineteenth century and beginning of the twentieth, it was discovered the lethal effect of ultraviolet (UV) component of sunlight on microorganisms and the efficacy of red light in preventing suppuration and scarring in patients with smallpox: the basis for the modern phototherapy were laid (Barnard & Morgan, 1903; Finsen, 1901; McDonagh, 2001). The extensive application of UV, visible, infrared (IR) radiation in biological and medical fields led to the development of suitable light sources. Actually, lasers are the latest and most advanced type of light source.

The advantages of lasers, compared to other sources, are the high intensity of radiation emitted, the directionality (which allows efficient coupling to optical fibers and focus), the monochromaticity (if needed) and, with pulsed lasers, the possibility to transfer large amounts of energy controlling the thermal effects.

Following the widespread clinical use, many studies have been conducted to investigate at the cellular and molecular level the mechanisms underlying the systemic effects produced by exposure to laser radiation. Propagation and absorption of radiation in a biological sample or in a tissue may produce photochemical, photothermal and photomechanical effects, which are able to induce a biological response (Jacques, 1992).

The actual laser systems, because of their versatility, are particularly suitable for application in biomedicine and biotechnology and the use of lasers to modulate biological and biotechnological processes has been proposed.

Hsu (Hsu et al., 2010) proved that endothelial cells pre-exposed to red-emitting laser (632 nm) and then seeded on a biomaterial surface (biomedical grade poly (carbonate) urethane) increase matrix secretion and are more resistant to flushing (greater cell retention on the graft) in comparison with non laser-exposed controls.

An overview on the state of the art in the photoengineering of bone repair showed that infrared (IR) radiation increases osteoblastic proliferation, collagen deposition and bone neoformation (Pinheiro & Gerbi, 2006)

A recent review in which the photoengineering of tissue repair in skeletal and cardiac muscles is discussed, reported that exposure to lasers with red/near infrared (NIR) emission is effective in favouring muscle repair: the application of He-Ne laser irradiation significantly enhanced muscle regeneration in rats, while Ga-Al-As laser radiation reduced muscle degeneration in the ischemia/reperfusion injury in skeletal leg muscle. Photoexposure also favoured proliferation of myogenic satellite cells. In mouse, rat, dog and pig ischemic heart models, phototherapy significantly reduced (50% - 70%) the formation of scar tissue after induction of myocardial infarction. Ventricular dilation was also reduced and ATP in the infarcted area increased (Oron, 2006).

In order to evaluate the effectiveness of light emitted by lasers and other sources in enhancing cell proliferation, Alghamdi and colleagues (Alghamdi et al., 2011) reviewed the literature in this specific field from 1923 to 2010. They concluded that light with wavelength ranging from 600 to 700 nm is helpful in enhancing the proliferation rate of various cell types, stem cells included. The increase in proliferation was generally associated with increased synthesis of ATP, RNA and DNA. The reviewed data indicated that the optimum value of energy density was between 0.5 and 0.4 J/cm^2. The possibility to develop phototreatments aimed at favouring cell proliferation could be very useful in the production of vaccines and hybrid cell lines as well as in tissue engineering and regenerative medicine.

In a review concerning the literature from 1960 to 2008 on the use of laser treatments for the improvement of tissue repair, the authors stated that, despite the difficulty in comparing results obtained with different laser sources, treatment protocols and experimental models, the majority of the reviewed reports clearly indicated that laser irradiation (the most frequently used is red/NIR radiation) speeds up tissue repair (da Silva et al., 2010).

In order to evaluate the possibility to use laser treatments as a tool to stimulate cell differentiation processes and cell functions involved in tissue repair, we studied the effect of NIR-emitting lasers on HMSCs, endothelial cells and cells of connective tissues. We found that in HMSCs treated with NIR pulses emitted by a high power Nd: YAG laser (1064 nm wavelength, 200 μs pulse duration, 10 Hz repetition rate, 458.65 mJ/cm2 energy fluence, 73 sec exposure) genes involved in osteoblastogenesis (*GLI1, NF1, MEN1*) appeared upregulated while *PPARγ*, which is a major marker of adipogenesis, and *FABP4* were downregulated, suggesting that the treatment can favour osteoblastogenesis and inhibit adipogenesis (Tab. 1). Interestingly the results obtained with laser treatment are very close to those obtained by exposure of HMSCs to hypergravity (10x*g*) (Monici et al., 2008b).

Moreover, the Nd:YAG laser pulses resulted effective in enhancing the production of ECM molecules, such as collagen I, collagen II, aggrecan and FN in cultures of connective tissue cells. A similar increase in ECM molecules was found when the cells were exposed to hypergravity (10x*g*) (Monici et al., 2008b, Basile et al., 2009).

The results of experiments in which we compared the effects of gravitational loading (10x*g*) and laser pulses on cells belonging to tissues with antigravitational function are consistent with the hypothesis that, using suitable laser pulses, it is possible to induce transient ECM rearrangements in the cell microenvironment (cell niche) which, in turn, act as indirect "photomechanical" stimuli on the cells (Rossi et al., 2010).

Experiments carried out on fibroblasts and endothelial cells, which are responsible for ECM production and neoangiogenesis in the remodelling phase of wound healing, demonstrated that NIR pulses not only increase (30%), FN production, but also favour the ordered assembly of FN fibrils in fibrillogenesis (Fig. 2C) (Monici et al., 2011). This effect is interesting because it could improve the quality of the neoformed ECM: in fact, FN fibrils act as a template for the formation of collagen fibers and strongly affect ECM properties (Shi et al., 2010). Moreover, we observed that NIR laser pulses can also favour the formation of ordered monolayer of endothelial cells (Monici et al., 2011). This effect could be of consequence in neoangiogenesis. Finally, data obtained with preliminary experiments carried out in our laboratory show that pulsed NIR radiation enhances the production of inflammation cytochines (not yet published data). The treatment could thus have the effect of accelerating the transition from inflammatory to the remodelling phase in tissue repair.

Advanced laser systems allow to apply more complex treatment protocols, to try to potentiate or to exploit synergistically the effects produced by emissions with different characteristics.

After exposure of myoblasts to a Multiwave locked System (MLS) laser, emitting at 808 and 905 nm (continuous/interrupted and pulsed mode, respectively) we observed increased activity of enzymes involved in cellular energy metabolism and enhanced expression of MyoD (Vignali et al., 2011), an early marker of muscle differentiation. These results provide an interesting premise for the future application of Multiwave systems to favour muscle tissue repair.

Physical stimulus	Parameters	Experimental model	Effects	Author
		MECHANICAL STRESS		
Mechanical stress lengthening	Lengthening of the substratum on which cells adhered (5 mm/day resulting in a 94-110% increase in 4 days	Neonatal rat cardiomyocytes	Cardiomyocytes organized into parallel arrays, increased binucleation and hypertrophy.	Vandenburg, 1996
Mechanical stress topogra-phycal stimulation	Micropatterned substrates with ECM-coated adhesive islands	Human and bovine capillary endothelial cells	Modulation of growth, differentiation and apoptosis	Chen, 1997
Mechanical stress loading	150N, 1 Hz	Female rats	Increased bone formation	Chow et al., 1998
Mechanical stress stretching	2% biaxial strain applied cyclically, 0.25 Hz.	MC3T3-E1 osteoblast-like cells	Reorganization of the cytoskeleton and increased expression of alpha(v)- beta3 integrin receptor	Cavalcanti-Adam et al., 2002
Mechanical stress	Mechanical perturbation using a glass microneedle	MC3T3-E1 osteoblastic cell	Changes in intracellular Ca^{2+} concentration and actin microfilament organization	Adachi et al., 2003
Mechanical stress stretching	Cyclic stretching, 5% to 15% elongation, 0.3 to 1 Hz	Fibroblasts	Production of ECM components (in particular tenascin-C)	Chiquet et al., 2003
Mechanical stress stretching	30 cycles of uniaxial stretching, 1 Hz, 4000 $\mu\varepsilon$	Human osteoblast-like osteosarcoma cell line MG-63 and primary human osteoblasts	Increased proliferation rates and activation of PLC beta 2 expression	Hoberg et al., 2005
Gravitational stress microgravity	Microgravity modelled by RPM, angular velocity of rotation 60°/s, 72 h	Coronary venular endothelial cells	Decreased cell number, increased proapoptotic signals and down regulation of antiapoptotic and proliferation/survival genes	Morbidelli et al., 2005
Mechanical stress	Different mechanical stresses fluid flow, strain, shear and combinations of them	Connective tissue cells	Activation of mechanotransduction pathways, Ca2+ signalling, stretch-activated channels, voltage-activated channels	Wall and Banes., 2005 **Review** on Early responses to mechanical loading in connective tissue cells

Physical stimulus	Parameters	Experimental model	Effects	Author
Gravitational stress microgravity	Microgravity modelled by RPM, angular velocity of rotation 60°/s, 72 h exposure	Osteoclastic precursors	Increased apoptosis, osteoclast-like morphology, cytoskeleton reorganization, overexpression of osteoclastic markers	Monici et al., 2006
Mechanical stress stretching	Cyclic stretching 0.5 Hz , magnitude 8% (8 % deformation of cell-seeded silicone substrate)	Ligament fibroblasts	Enhanced cell proliferation and collagen production	Park et al., 2006;
Mechanical stress	Mechanical load, strain, pressure, different parameters	Osteoblasts, osteoclasts, osteocytes and cells of the vasculature	Many effects are reported	Rubin et al., 2006 **Review** on Molecular pathways mediating mechanical signalling in bone
Mechanical stress	Mechanical stimulation by microneedles	Osteocytes	Upregulation of NO production in the stimulated cell and in the sourrading osteocytes	Vatsa et al., 2007
Gravitational stress hypergravity	5 periods of 10 min at 10 x g spaced with 10 min recovery periods at 1 x g	Human mesenchymal stem cells	Overexpression of genes involved in osteoblastogenesis	Monici et al., 2008a
Mechanical stress compression	Compression 0.5 Mpa, sinusoidal micromotion 100 µm, 1 Hz	Bone implant	Activation of bone resorption	Stadelmann et al., 2008
Gravitational stress hypergravity	5 periods of 10 min at 10 x g spaced with 10 min recovery periods at 1 x g	Human fetal fibroblast and human chondrocytes	Increased production of ECM molecules	Basile et al., 2009
Gravitational stress hypergravity	5 periods of 10 min at 10 x g spaced with 10 min recovery periods at 1 x g	Coronary venular endothelial cells	Activation of adaptive mechanisms, increased anaerobic metabolism, detachment from the substrate	Morbidelli et al.,2009
Mechanical stress compression	Compressive strain of 5% to 20% elongation, 0.15 to1 Hz, 1 to 12 h/day hydrostatic pressure 0.1 to 10 Mpa, 0.25 to 1 Hz	Intervertebral disc (IVD) and stem cells	Possible role of loading to favour differentiation of stem cells toward "discogenic" phenotype	Chan et al., 2011 **Review** on The effects of loading on IVD and stem cells

Physical stimulus	Parameters	Experimental model	Effects	Author
Mechanical stress loading	Uniaxial and multiaxial, hydrostatic pressure 7 to 10 Mpa, average tension, 3.8% radial and 2.1% circumferential tensile strains, compression 0.5 to 7.7 MPa	Chondrogenic cells	Upregulation of genes normally associated with a chondrogenic phenotype with uniaxial loading, upregulation of a broader pattern of chondrogenic genes with multiaxial loading	Grad et al., 2011 **Review** on Chondrogenic cell response to mechanical stimulation in vitro
Mechanical stress	Cyclic strain, 1Hz, 10 % strain of 3D culture + ultrasound 1.0 MHz and 30 mW/cm2	MC3T3-E1 pre-osteoblasts	Acceleration of matrix maturation	Kang et al., 2011
Mechanical stress	Cyclic stretch 5%, 1Hz; cyclic pressure 120/80mmHg, 1Hz; shear stress 10 dynes/cm2.	Mesenchymal stem cells	Increased smooth muscle cells expression with cyclic stretch and endothelial cells expression with cyclic pressure, and laminar shear stress	Maul et al., 2011
Gravitational stress microgravity	Microgravity modelled by RPM, angular velocity of rotation 60°/s, 72 h	Fibroblast and endothelial cells	Disregulation in laminin and fibronectin production	Monici et al., 2011
ELECTROMAGNETIC FIELDS				
EMF	Many different parameters and treatment protocols	Many different experimental models	Many effects are reported	Funk., 2009 **Review** on Electromagnetic effects from cell biology to medicine
ELFEMF	15 to 30 Hz, 0.01 mV/cm	Bone cells	Affect remodelling activity	McLeod & Rubin, 1993
Pulsed EF	100V external voltage, 16 Hz, EF across cell membrane 6 kV/m (estimated by computer simulation)	Osteoblast-like primary cells	Increased proliferation, enhancement of alkaline phosphatase activity, enhanced synthesis and secretion of ECM proteins	Hartig et al., 2000
Pulsed EMF	15 Hz, 0.1 mT, EF 2 mV/cm	Osteoblast-like primary cells	Increased proliferation, OPG upregulation and RANKL downregulation	Chang et al., 2004

Physical stimulus	Parameters	Experimental model	Effects	Author
EF	Static and pulsing direct current (DC) EFs	Calf anterior cruciate ligament (ACL) fibroblasts	Increased migration speed and enhanced collagen I expression with pulsing direct current (DC) EF, inhibition in wound healing with static direct current (DC) EF	Chao et al., 2007
EMF	50 Hz, 25 to 180 μT	Human fibroblasts	Increased proliferation	Kwee & Raskmark, 1995
ELFMF	50 Hz, 0.020 T	Murine fibroblasts	Inhibition of cell growth	Kula & Drozd, 1996
EMF	50 Hz sinusoidal MFs intensities: 30 μT, 300 μT, and 1 mT	Human lymphocytes	No effect on proliferation	Paile et al., 1995
EMF	50 Hz sinusoidal MF intensities: 1.0, 0.75, 0.5, 0.25, 0.05 mT exposure 72 h	Human lymphocytes	Slight decrease of cell proliferation at the intensities tested	Scarfi et al., 1999
MF	Sinusoidal 60 Hz, 44 μT	Rat thymic lymphocytes	Modulation of calcium concentration	Walleczek, 1990
EMF	30 Hz, 1.8-mT	MC-3T3 osteoblast-like cells	Progressive inhibition of enzyme activity and differentiation	McLeod & Collazo, 2000
ELFEMF	15Hz, 5mT	Human mesenchymal stem cells (hMSCs)	Increase of collagen II and glycosaminoglycan (GAG)/DNA content during chondrogenic differentiation	Mayer-Wagner et al., 2010
EMF	50 Hz; 2 mT, 3 hours exposure	SHSY5Y neuroblast model	Promotion of neurogenic differentiation	Cerrato et al., 2011
EMF	1 GHz, power density of exposure area 5 nW/cm^2	Human fibroblasts	Activation of fibroblast migration	Sunkari et al., 2011
LIGHT				
Light He-Ne and Ga-Al-As laser	Various treatment protocols and instrumental parameters	Injured muscles in rat Ischemic leg muscles Ischemic heart model of mouse, rat, dog and pig Myogenic satellite cells (SC)	Enhanced muscle regeneration; reduced muscle degeneration; reduction in scar tissue formation after induction of myocardial infarction (MI) and in ventricular dilatation. Increment of ATP in the infarcted area; MAPK/ERK activation	Oron, 2006 **Review** on Photoengineering of tissue repair in skeletal and cardiac muscles

Physical stimulus	Parameters	Experimental model	Effects	Author
Light IR radiation	Various protocols and parameters	Bone cells and tissues	Increment of osteoblastic proliferation, collagen deposition and bone neoformation	Pinheiro & Gerbi, 2006 **review** state of the art on photoengineering of bone repair using laser therapy
Light High power Nd: YAG laser	λ 1064 nm, 200 µs pulse duration, 10 Hz repetition rate, 458.65 mJ/cm² energy fluence, 73 sec exposure	Human mesenchymal stem cells	Upregulation of genes involved in osteoblastogenesis, downregulation of genes involved in adipogenesis	Monici et al., 2008a
Light High power Nd: YAG laser	λ 1064 nm, 200 µs pulse duration, 10 Hz repetition rate, 458.65 mJ/cm² energy fluence, 73 sec exposure	Human fetal fibroblasts and human chondrocytes	Increased production of ECM molecules	Monici et al., 2008b
Light	Different laser sources and treatment protocols	Different experimental models	Promotion of tissue repair	Da Silva et al., 2010 **Review** on Lasertherapy in tissue repair processes
Light He-Ne laser	λ 632.8 nm, average energy on sample 1.18 J/cm²	Endothelial cells pre-exposed to laser and then seeded on a biomaterial surface	Increase in ECM secretion and increased resistance to flushing	Hsu et al, 2010
Light Red radiation	λ from 600 to 700 nm energy density 0.4 - 0.5 J/cm²	Various cell types	Increase in proliferation associated with enhanced synthesis of ATP, RNA and DNA	Alghamdi et al., 2011 **Review** on The use LLLT for enhancing cell proliferation
Light High power Nd: YAG laser	λ 1064 nm, 200 µs pulse duration, 10 Hz repetition rate, 458.65 mJ/cm² energy fluence, 73 sec exposure	Fibroblasts and endothelial cells	Increased production of fibronectin and ordered assembly of FN fibrils in fibrillogenesis	Monici et al., 2011
Light MLS laser	λ 808 and 905 nm, 1500 Hz	Myoblasts	Enhanced expression of MyoD and increased activity of enzymes involved in cellular energy metabolism	Vignali et al., 2011

Table 1. Based on the studies cited in this chapter, the table lists physical factors, treatment parameters applied, experimental models used and observed effects.

5. Conclusion

Over the past twenty years, studies on molecular and cellular mechanisms that underlie biological responses evoked by physical stimuli have made great progress. However, in this fascinating field of study, many problems still remain and their solution will require further advances in our knowledge.

The results of our studies are a further, albeit modest, contribution to a large body of literature that shows how physical stimuli can be effective in modulating cellular functions and the production of ECM. It is obvious that the development and standardization of technologies for delivering appropriate physical stimuli, strictly controlled with regard to the intensity, frequency and timing of exposure, is a prerequisite for making progress in tissue engineering.

6. References

Adachi, T.; Sato, K., & Tomita, Y. (2003). Directional dependence of osteoblastic calcium response to mechanical stimuli. *Biomech Model Mechanobiol.* Vol. 2, No.2, pp. 73-82.

Alghamdi, KM.; Kumar, A., & Moussa, NA. (2011) Low-level laser therapy: a useful technique for enhancing the proliferation of various cultured cells. *Lasers Med Sci.* Jan 28. DOI: 10.1007/s10103-011-0885-2

Antonsson, EK., & Mann, RW. (1985) The frequency content of gait. *J Biomech* Vol.18 pp. 39-47.

Bacabac, RG.; Mizuno, D., & Koenderink, GH. (2010) Mechanical properties of living cells: On mechanosensing and microgravity, In: *Cell Mechanochemistry. Biological System and factors Inducing Mechanical Stress, Such as light, Pressure and Gravity,* Monici, M., & van Loon, J, pp. 23-54, Transworld Research Network, ISBN: 978-81-7895-458-5, Kerala, India.

Barnard, JE., & Morgan, H. (1903) Upon bactericidial action of some ultraviolet radiations as produce by a continuous current arc. *P.R. Soc. London*. Vol. 72, pp. 126-128.

Basile, V.; Romano, G.; Fusi, F. & Monici M. (2009). Comparison between the effects of hypergravity and photomechanical stress on cells producing ECM. *Microgravity Sci. Technol* ., Vol. 21, pp. 151-157. doi: 10.1007/s12217-008 -9068-6.

Büchler, P.; Pioletti, DP., & Rakotomanana, LR. (2003) Biphasic constitutive laws for biological interface evolution. *Biomech Model Mechanobiol.* Vol. 1, No.4, pp. 239-49.

Burr, HS. (1972). *Blueprint for immortality: electric patterns of life discovered in scientific break-through.* C W Daniel Co. Ltd

Carson, DD. (2004). Extracellular matrix: forum introduction. *Reproductive Biology and Endocrinology,* Vol. 2, pp. 1-5.

Carter, DR.; Beaupré, GS.; Giori, NJ., & Helms JA. (1998). Mechanobiology of skeletal regeneration. *Clin Orthop Relat Res.* (355 Suppl) pp. S41-55.

Cavalcanti-Adam, EA.; Shapiro, IM.; Composto, RJ.; Macarak, EJ., & Adams CS. (2002) RGD peptides immobilized on a mechanically deformable surface promote osteoblast differentiation. *J Bone Miner Res.,* Vol. 17, No. 12, pp. 2130-40.

Cerrato, C.; Cialdai, F., & Monici, M. (2011). Effects of Low Frequency Electromagnetic Fields on SHSY5Y cells - a neuroblast model. *Energy for Health,* in press.

Chan, SC.; Ferguson, SJ., & Gantenbein-Ritter, B. (2011). The effects of dynamic loading on the intervertebral disc. *Eur Spine J.*, Vol. 20, No. 11, pp.1796-812

Chang, WH.; Chen, LT.; Sun, JS., & Lin, FH. (2004). Effect of pulse-burst electromagnetic field stimulation on osteoblast cell activities. *Bioelectromagnetics,* Vol. 25, pp. 457–65.

Chao, PH.; Lu, HH.; Hung, CT.; Nicoll, SB., & Bulinski, JC. (2007). Effects of applied DC electric field on ligament fibroblast migration and wound healing. *Connect Tissue Res,* Vol. 48, pp. 188–97.

Chen, CS.; Mrksich, M.; Huang, S.; Whitesides, G., & Ingber, DE. (1997). Geometric control of cell life and death. *Science,* Vol. 276, No. 5317, pp.1425-8.

Chiquet, M.; Sarasa-Renedo, A.; Huber, F., & Flück M. (2003). How do fibroblast translate mechanical signals into changes in extracellular matrix production ?. *Matrix Biology,* Vol. 22, pp. 73-80.

Chiquet, M.; Gelman, L.; Lutz, R., & Maier, S. (2009). From mechanotransduction to extracellular matrix gene expression in fibroblasts. *Biochim Biophys Acta.* Vol. 1793, No. 5, pp. 911-20.

Chow, JW.; Fox, SW.; Lean, JM., & Chambers, TJ. (1998). Role of nitric oxide and prostaglandins in mechanically induced bone formation. *J Bone Miner Res.,* Vol. 13, No. 6, pp. 1039-44.

Cossarizza, A.; Capri, M.; Salvioli, S.; Monti, D.; Franceschi, C.; Bersani, F.; Cadossi, R.; Scarfi', MR.; Angioni, S., & Petraglia, F. (1993). Electromagnetic fields affects cell proliferation and citokine production in human cells. In: *Electricity and Magnetism in Biology and Medicine,* M.Blank Ed., pp. 640-642, San Francisco Press.

da Silva, JP.; da Silva, MA.; Almeida, AP.; Lombardi Junior, I., & Matos, AP. (2010). Laser therapy in the tissue repair process: a literature review. *Photomed Laser Surg.,* Vol. 28, No. 1, pp. 17-21.

Delp, M.D. (2008). Unraveling the complex web of impaired wound healing with mechanical unloading and physical deconditioning. *J. Appl. Physiol.* Vol. 104, pp. 1262–1263

Farndale, RW.; Maroudas, A., & Marsland, TP. (1987). Effects of low-amplitude pulsed magnetic fields on cellular ion transport. *Bioelectromagnetics,* Vol. 8, pp. 119-34.

Finsen, NR. (1901). *Phototherapy,* Edward Arnold, London.

Frost, HM. (1988). Vital biomechanics: proposed general concepts for skeletal adaptations to mechanical usage. *Calcif. Tissue Int.,* Vol. 42, pp. 145-156.

Funk, RH.; Monsees, T., & Ozkucur, N. (2009). Electromagnetic effects - From cell biology to medicine. *Prog Histochem Cytochem.* Vol. 43, No. 4, pp. 177-264.

Galilei G. (1974, translation). In: The Second Day, translated by Drake S. The University of Winsconsin Press, 1974 pp. 109-146.

Grad, S.; Eglin, D.; Alini, M., & Stoddart, MJ. (2011). Physical Stimulation of Chondrogenic Cells In Vitro: A Review. *Clin Orthop Relat Res.* Vol. 469, No.10, pp.2764-72

Hartig, M.; Joos, U., & Wiesmann, HP. (2000). Capacitively coupled electric fields accelerate proliferation of osteoblast-like primary cells and increase bone extracellular matrix formation in vitro. *Eur Biophys J,* Vol. 29, pp. 499-506.

Hoberg, M.; Gratz, HH.; Noll, M., & Jones, DB. (2005). Mechanosensitivity of human osteosarcoma cells and phospholipase C beta2 expression. *Biochem Biophys Res Commun.* Vol. 22, No. 333(1), pp. 142-9.

Hoerstrup, SP.; Kadner, A.; Melnitchouk, S.; Trojan, A:, Eid, K.; Tracy, J.; Sodian, R.; Visjager, JF.; Kolb, SA.; Grunenfelder, J.; Zund, G.; & Turina, MI. (2002). Tissue engineering of functional trileaflet heart valves from human marrow stromal cells. *Circulation,* Vol. 106, pp. 143-150.

Hsu, SH.; Fu, KY.; Huang, TB.; Tsai, IJ., & Yen, CY. (2010). The effect of laser preexposure on seeding endothelial cells to a biomaterial surface. *Photomed Laser Surg.* Vol. 28 Suppl 2, pp. S37-42.

Hughes-Fulford, M., & Boonstra, J. (2010). Cell mechanotrasduction: Cytoskeleton and related signalling pathways, In: *Cell Mechanochemistry. Biological System and factors Inducing Mechanical Stress, Such as light, Pressure and Gravity,* Monici, M., & van Loon, J, pp. 75-95, Transworld Research Network, ISBN: 978-81-7895-458-5, Kerala, India.

Ingber, DE. (1998). The architecture of life. *Sci. Am.,* Vol. 278, pp. 48-57.

Ingber, DE. (1999). How cells (might) sense microgravity. *The FASEB J.,* Vol. 13, pp. 3-15.

Jacques, SL. (1992). Laser-tissue interactions. Photochemical, photothermal, and photomechanical. *Surg Clin North Am.,* Vol. 72, No. 3, pp. 531-58.

Jasty, M.; Bragdon, C.; Burke, D.; O'Connor, D.; Lowenstein, J., & Harris, WH. (1997). In vivo skeletal responses to porous-surfaced implants subjected to small induced motions. *J Bone Joint Surg Am.,* Vol. 79, No. 5, pp. 707-14.

Kaneuji, T.; Ariyoshi, W.; Okinaga, T.; Toshinaga, A.; Takahashi, T., & Nishihara, T. (2011). Mechanisms involved in regulation of osteoclastic differentiation by mechanical stress-loaded osteoblasts. *Biochem Biophys Res Commun.,* Vol. 408, No.1, pp. 103-9.

Kang, KS.; Lee, SJ.; Lee, HS.; Moon, WK., & Cho, DW. (2011). Effects of combined mechanical stimulation on the proliferation and differentiation of pre-osteoblasts. *Exp Mol Med.,* Vol. 43, No.6, pp. 367-73.

Kula, B., & Drozdz, M. (1996). A study of magnetic field effects on fibroblast cultures Part 1. The evaluation of effects of static and extremely low frequency (ELF) magnetic fields on vital functions of fibroblasts. *Bioelectrochem. & Bioenerg.* Vol. 39, pp. 21-26.

Kwee, S., & Raskmark, P. (1995). Changes in cell proliferation due to environmental non-ionizing radiation ELF electromagnetic fields. *Bioelectrochem. & Bioenerg.,* Vol. 36, pp. 109-114.

Levin, M. (2007). Large-scale biophysics: ion flows and regeneration. *Trends Cell Biol* , Vol. 17, pp. 261-70.

Lisi, A.; Foletti, A.; Ledda, M.; Rosola, E.; Giuliani, L.; D'Emilia, E., & Grimaldi, S. (2006). Extremely low frequency 7Hz 100 microT electromagnetic radiation promotes differentiation in the human epithelial cell line HaCaT. *Electromagn Biol Med,* Vol. 25, pp. 269-80.

Maul, TM.; Chew, DW.; Nieponice, A., & Vorp, DA. (2011). Mechanical stimuli differentially control stem cell behavior: morphology, proliferation, and differentiation. *Biomech Model Mechanobiol.* Vol. 21.

Mayer-Wagner, S.; Passberger, A.; Sievers, B.; Aigner, J.; Summer, B.; Schiergens, TS.; Jansson, V., & Müller, PE. (2010). Effects of low frequency electromagnetic fields on the chondrogenic differentiation of human mesenchymal stem cells. *Bioelectromagnetics.*, Vol.22.

McBeath, R., et al. (2004). Cell Shape, Cytoskeletal Tension, and RhoA Regulate Stem Cell Lineage Commitment. *Developmental Cell*, Vol. 6, pp. 483-495.

McCaig, CD.; Rajnicek, AM.; Song, B., & Zhao, M. (2005). Controlling cell behavior electrically: current views and future potential. *Physiol Rev*, Vol. 85, pp. 943–78.

McDonagh, AF. (2001). Phototherapy: from ancient Egypt to the new millennium. *Journal of Perinatology*, Vol.21, pp. S7-S12.

McLeod, KJ., & Rubin, CT. (1993) Observations from mechanically and electrically induced bone remodeling. In: *Electricity and magnetism in biology and medicine,* Blank M, editor, p. 98–700, San Francisco Press, San Francisco.

McLeod, KJ., & Collazo, L. (2000). Suppression of a differentiation response in MC-3T3-E1 osteoblast-like cells by sustained, low-level, 30 Hz magnetic-field exposure. *Radiat Res.*, Vol. 153(5 Pt 2), pp. 706-14.

Midura, RJ.; Su, X., & Androjna, C. (2006). A simulated weightlessness state diminishes cortical bone healing responses. *J. Musculoskelet. Neuronal. Interact.* Vol. 6, No.4, pp. 327–328.

Mobasheri, A.; Barrett-Jolley, R.; Carter, SD.; Martín-Vasallo, P.; Schulze-Tanzil, G., & Shakibaei, M. (2005). Functional roles of mechanosensitive ion channels, β1 integrins and kinase cascades in chondrocyte mechanotransduction. In: *Mechanosensitivity in Cells and Tissues*, Kamkin, A., & Kiseleva, I, editors, Academia, Moscow.

Monici, M.; Fusi, F.; Paglierani, M.; Marziliano, N.; Cogoli, A.; Pratesi, R., & Bernabei, PA. (2006). Modeled gravitational unloading triggers differentiation and apoptosis in preosteoclastic cells. *Journal of Cellular Biochemistry*, Vol. 98, No. 1, pp. 65-80.

Monici, M.; Romano, G.; Cialdai, F.; Fusi, F.; Marziliano, N.; Benvenuti, S.; Cellai, I.; Egli, M., & Cogoli, A. (2008a). Gravitational/mechanical factors affect gene expression profile and phenotypic specification of human mesenchymal stem cells. *J Gravit. Physiol.*, Vol. 15, No. 1, pp. 191-192.

Monici, M.; Basile, V.; Cialdai, F.; Romano, G.; Fusi, F., & Conti, A. (2008b). Irradiation by pulsed Nd:YAG laser induces the production of extracellular matrix molecules by cells of the connective tissues. A tool for tissue repair. In: *Biophotonics: Photonic solutions for Better Health Care*, Popp, J.; W. Drexler, W.; Turchin VV:, & Matthews, DL. Eds., Proc. of SPIE vol. 6991, pp. 69912K1-10. doi: 10.1117/ 12.782865.

Monici, M.; Cialdai, F.; Romano, G.; Fusi, F.; Egli, M.; Pezzatini, S., & Morbidelli, L. (2011). An *in vitro* study on tissue repair: impact of unloading on cells involved in the remodelling phase. *Microgravity Sci. Technol*, DOI 10.1007/s12217-011-9259-4

Morbidelli, L.; Monici, M.; Marziliano, N.; Cogoli, A.; Fusi, F.; Waltenberger, J., & Ziche, M. (2005). Simulated hypogravity impairs the angiogenic response of endothelium by up-regulating apoptotic signals. *Biochem. Biophys. Res. Com.*, Vol. 334, pp. 491-499.

Morbidelli, L.; Marziliano, N.; Basile, V.; Pezzatini, S.; Romano, G.; Conti, A., & Monici, M. (2009). Effect of hypergravity on endothelial cell function and gene expression. *Microgravity Sci. Technol .*, Vol. 21, pp. 135-140. Doi: 10.1007/s12217-008-9067-7.

Oron, U. (2006). Photoengineering of tissue repair in skeletal and cardiac muscles. *Photomed Laser Surg.* Vol. 24, No.2, pp. 111-20.

Otter, MW.; McLeod, KJ., & Rubin, CT. (1998). Effects of electromagnetic fields in experimental fracture repair. *Clin Orthop Relat Res*, S90–104.

Paile, W.; Jokela, K.; Koivistoinen, A., & Salomaa, S. (1995). Effects of 50 Hz sinusoidal magnetic fields and spark discharges on human lymphocytes in vitro. *Bioelectrochem. & Bioenerg.* Vol. 36, pp. 15-22.

Papachroni, KK.; Karatzas, DN.; Papavassiliou, KA.; Basdra, EK., & Papavassiliou, AG. (2009). Mechanotransduction in osteoblast regulation and bone disease. *Trends Mol Med.*, Vol. 15, No. 5, pp. 208-16.

Park, SA.; Kim, IA.; Lee, YJ.; Shin, JW.; Kim, CR.; Kim, JK.; Yang, YI., & Shin, JW. (2006). Biological responses of ligament fibroblasts and gene expression profiling on micropatterned silicone substrates subjected to mechanical stimuli. *J Biosci Bioeng.* Vol. 102, No. 5, pp. 402-12.

Pilla, AA. (2002). Low-intensity electromagnetic and mechanical modulation of bone growth and repair: are they equivalent? *J Orthop Sci*, Vol. 7, pp. 420–8.

Pinheiro, AL., & Gerbi, ME. (2006). Photoengineering of bone repair processes. *Photomed Laser Surg.* Vol. 24, No. 2, pp.169-78.

Pioletti, DP. (1997). Biomechanics and tissue engineering. *Osteoporos Int.*, Vol. 22, No.6, pp. 2027-31.

Prendergast, PJ.; Huiskes, R., & Soballe, K. (1997) ESB Research Award 1996. Biophysical stimuli on cells during tissue differentiation at implant interfaces. *J Biomech*, Vol. 30, No. 6, pp. 539-548

Roskelley, C.; Desprez, P., & Bissell, M. (1994). Extracellular matrix-dependent tissue-specific gene expression in mammary epithelial cells requires both physical and biochemical signal transduction. *Proc. Natl. Acad. Sci USA* Vol. 91, pp. 12378-12382.

Rossi, F.; Pini, R., & Monici, M. (2010). Direct and indirect photomechanical effects in cells and tissues. Perspectives of application in biotechnology and medicine. In: *Cell Mechanochemistry. Biological systems and factors inducing mechanical stress, such as light, pressure and gravity*, Monici, M., & van Loon J. eds., pp. 285-301, published by Research Signpost / Transword Research Network, Trivandrum, India.

Rubin, CT.; Lanyon, LE. (1984). Regulation of bone formation by applied dynamic loads. *J. Bone Joint Surg.*, Vol. 66, pp. 397-402.

Rubin, CT. (1984). Skeletal strain and the functional significance of bone architecture. *Calcif. Tissue Int.*, Vol. 36S, pp. 11-18.

Rubin, CT., & Lanyon, LE. (1985). Regulation of bone mass by mechanical strain magnitude. *Calcif. Tissue Int.*, Vol. 37, pp. 411-417.

Rubin, J.; Rubin, C., & Jacobs, CR. (2006). Molecular pathways mediating mechanical signaling in bone. *Gene.*, Vol. 15, No. 367, pp.1-16.

Scarfi', MR.; Bersani, F.; Cossarizza, A.; Monti, D.; Franceschetti, G., & Franceschi, C. (1991). Spontaneous and mitomycin C-induced micronuclei in human lymphocytes exposed to extremely low frequency pulsed electromagnetic fields. *Biochem. Biophys. Res. Comm.*, Vol. 176, pp. 194-200.

Scarfi', MR.; Lioi, MB.; Zeni, O.; Della Noce, M.; Franceschi, C., & Bersani, F. (1999). Micronucleus frequency and cell proliferation in human lymphocytes exposed to 50 Hz sinusoidal magnetic fields. *Health Physics*, Vol. 76, No. 3, pp. 244-250.

Searby, ND.; Steele, CR.; Globus, RK. (2005). Influence of increased mechanical loading by hypergravity on the microtubule cytoskeleton and prostaglandin E2 release in primary osteoblasts. *Am J Physiol Cell Physiol.*, Vol. 289, No. 1, pp. C148-58.

Shi, F.; Harman, J.; Fujiwara, K., & Sottile, J. (2010). Collagen I matrix turnover is regulated by fibronectin polymerization. *Am J Physiol Cell Physiol.*, Vol. 298, No. 5, pp. C1265-75.

Spiegelman, BM., & Ginty, CA. (1983). Fibronectin modulation of cell shape and lipogenic gene expression in 3T3-adipocytes. *Cell*, Vol. 35, pp. 657-66.

Stadelmann, VA.; Terrier, A., & Pioletti, DP. (2008). Microstimulation at the bone-implant interface upregulates osteoclast activation pathways. *Bone.*, Vol. 42, No. 2, pp. 358-64.

Stamenkovic, I., & Yu, Q. (2010). Merlin, a "magic" linker between extracellular cues and intracellular signaling pathways that regulate cell motility, proliferation, and survival. *Curr Protein Pept Sci.*, Vol. 11, No. 6, pp. 471-84.

Sunkari, VG.; Aranovitch, B.; Portwood, N., & Nikoshkov, A. (2011). Effects of a low-intensity electromagnetic field on fibroblast migration and proliferation. *Electromagn Biol Med.*, Vol. 30, No. 2, pp. 80-5.

Thomas, CH., et al. (2002). Engineering gene expression and protein synthesis by modulation of nuclear shape. *Proc Natl Acad Sci U S A*, Vol. 99, pp. 1972-1977.

Turner, CH.; Owan, I.; Jacob, DS.; McClintock, R., & Peacock, M. (1997). Effects of nitric oxide synthase inhibitors on bone formation in rats. *Bone.*, Vol. 21, No. 6, pp. 487-90.

Vandenburg, HH.; Solerssi, R.; Shansky, J.; Adams, JW., & Henderson, SA. (1996). Mechanical stimulation of organogenic cardiomyocyte growth in vitro. *Am. J. Physiol.*, Vol. 270(5 Pt 1), pp. C1284-92.

Vatsa, A.; Smit, TH., & Klein-Nulend, J. (2007). Extracellular NO signalling from a mechanically stimulated osteocyte. *J Biomech.*, Vol. 40 Suppl 1, pp. S89-95.

Vatsa, A.; Smit, TH., & Klein-Nulend. (2010). Mechanobiology of bone: From cell to organ, In: *Cell Mechanochemistry. Biological System and factors Inducing Mechanical Stress, Such as light, Pressure and Gravity,* Monici, M., & van Loon, J, pp. 131-152, Transworld Research Network, ISBN: 978-81-7895-458-5, Kerala, India.

Vignali, L.; Cialdai, F., & Monici, M. (2011). Effects of MLS laser on myoblast cell line C2C12. *Energy for Health,* Vol. 7, pp. 12-18.

Wall, ME., & Banes, AJ. (2005). Early responses to mechanical load in tendon: role for calcium signaling, gap junctions and intercellular communication. *J Musculoskelet Neuronal Interact.* Vol. 5, No. 1, pp. 70-84.

Walleczek, J., & Liburdy, RP. (1990). Nonthermal 60 Hz sinusoidal magnetic-field exposure enhances 45Ca2+ uptake in rat thymocytes: dependence on mitogen activation. *FEBS Lett.*, Vol. 271, pp. 157-60.

Wang, L:, Zhang, X.; Guo, Y.; Chen, X.; Li, R.; Liu, L.; Shi, C.; Guo, C., & Zhang, Y. Involvement of BMPs/Smad signaling pathway in mechanical response in osteoblasts. *Cell Physiol Biochem.* Vol. 26, No.6, pp.1093-102.

Watt, FM.; Jordan, PW., & O'Neill, CH. (1988). Cell shape controls terminal differentiation of human epidermal keratinocytes. *Proc. Natl. Acad. Sci USA*, Vol. 85, pp. 5576-5580.

Wolff, JD. (1892). Das Gezetz der Transformation der Knochen, A. Hirschwald pub., Berlin.

2

Effect of Low-Intensity Pulsed Ultrasound on Nerve Repair

Jiamou Li[1], Hua Zhang[1] and Cong Ren[2]
[1]*Beijing Tiantan Hospital, Capital Medical University*
[2]*The 2nd Affiliated Hospital, Harbin Medical University*
China

1. Introduction

Low-intensity pulsed ultrasound (LIPUS) is a medical technology, generally utilizing 1-1.5 MHz frequency pulses, with a pulse width of 200 μs, repeated at 1-1.5 kHz, at an intensity of 10- 30 mW/cm^2, 20 minutes/day. There are two main types of ultrasound effects: thermal and nonthermal. Both types are thought to first "injure" the cells, resulting in their growth retardation, and then to initiate a cellular recovery response characterized by an increase in protein production(Johns 2002). Compared to high-intensity continuous ultrasound, LIPUS is much lower in intensity and has unique characteristics such as pulsed waves, which are regarded as nonthermogenic and non-destructive (Mukai, Ito et al. 2005).

Applications of LIPUS include: promoting bone fracture healing; treating orthodontically induced root resorption; regrow missing teeth; enhancing mandibular growth in children with hemifacial microsomia; promoting healing in various soft tissues such as cartilage, intervertebral disc, etc.; improving muscle healing after laceration injury. Researchers at the University of Alberta have used LIPUS to gently massage teeth roots and jawbones to cause growth or regrowth, and have grown new teeth in rabbits after lower jaw surgical lengthening (Distraction osteogenesis). As of June 2006, a device has been licensed by the Food and Drug Administration (FDA) and Health Canada for use by orthopedic surgeons. It has not yet been approved by either Canadian or American regulatory bodies and a market-ready model is currently being prepared. LIPUS is expected to be commercially available before the end of 2012. According to Dr. Chen from the University of Alberta, LIPUS may also have medical/cosmetic benefits in allowing people to grow taller by stimulating bone growth.

In recent years, data on the therapeutic effects of LIPUS have been accumulating. So far, it has been reported that LIPUS enhances cell proliferation and alters protein production in various kinds of cells such as endothelial cells, osteoblasts, chondrocytes, and fibroblasts (Ikeda, Takayama et al. 2006; Hiyama, Mochida et al. 2007; Takeuchi, Ryo et al. 2008), but there is little information on the response of Schwann cell and neurons to LIPUS irradiation. Some studies have indicated that LIPUS has positive effects on axonal regeneration during in vivo peripheral nerve injury trials (Crisci and Ferreira 2002; Chang, Hsu et al. 2005) and that its stimuli on the injured sciatic nerve can increase the number of nerve fibers compared to that of untreated injured nerves in rats (Raso, Barbieri et al. 2005). Thus, treatment with

LIPUS is likely to assist the regeneration of neuronal axons. However, the mechanism of such events is unknown.

2. Schwann cells that were subjected to LIPUS consistently demonstrated an increase in cell proliferation

Ultrasound is commonly used for diagnostic imaging and physiotherapy and can exert biological effects through either thermal or mechanical mechanisms in living tissue (Choi, Pernot et al. 2007; Nahirnyak, Mast et al. 2007). In contrast to high-intensity continuous ultrasound, LIPUS (<100 mW/cm^2) has much lower intensities, which are regarded as nonthermogenic and nondestructive (Ikeda, Takayama et al. 2006). Mechanical strains received by cells may result in biochemical events and increase membrane permeability (Danialou, Comtois et al. 2002). Despite the wide use of LIPUS for improving peripheral nerve tissue regeneration in animal models (Crisci and Ferreira 2002; Chang, Hsu et al. 2005; Raso, Barbieri et al. 2005), very little is known about its effects on the glial cells of peripheral nerves. It has been reported that Schwann cells respond somehow to LIPUS stimulation (Chang, Hsu et al. 2005; Raso, Barbieri et al. 2005). However, the results of previous investigations were somewhat inconclusive, particularly regarding the precise mechanism.

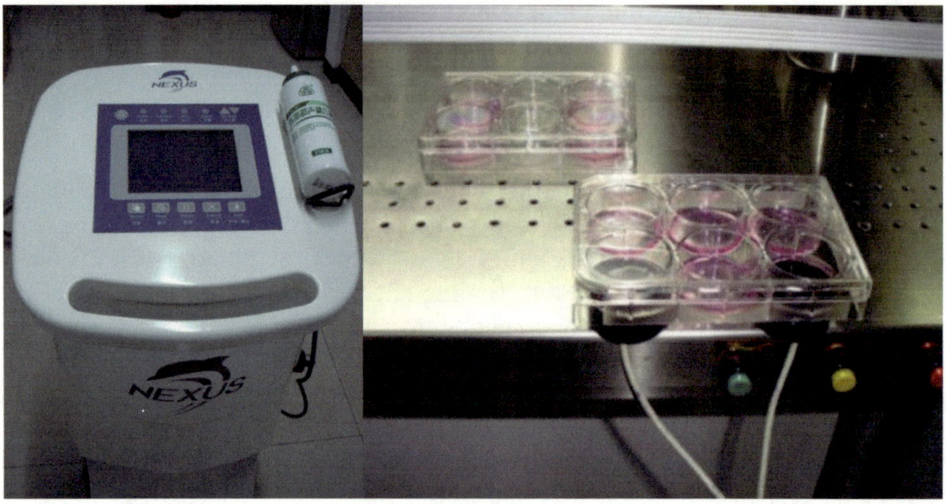

Fig. 1. Experimental apparatus for applying low- intensity pulsed ultrasound (LIPUS) which generated LIPUS with a SATA intensity of 10 mW/cm^2, pulse width of 200 microseconds, repetition rate of 1.5 KHz, and an operation frequency of 1MHz. LIPUS irradiated neurons with two probes 24 h after in culture. A six-well plate was placed on the probes. LIPUS was transmitted to culture plate via an interposed ultrasound gel. Two transducers for the control group (sham-LIPUS; LIPUS not turned on) and two probes for the LIPUS group.

Previous studies have shown that LIPUS in the cultured cells induces significant cellular responses in nucleus pulposus cells, endothelial cells, osteoblasts, chondrocytes, and fibroblasts, (Parvizi, Wu et al. 1999; Zhou, Schmelz et al. 2004; Hill, Fenwick et al. 2005; Sena, Leven et al. 2005; Hiyama, Mochida et al. 2007) but little is known about Schwann cell response to direct LIPUS stimulation. The previous work with peripheral nerve injury have demonstrated that the magnitude and duration of LIPUS has a direct effect on whether the stimulus has a positive or negative effect on nerve regeneration (Chang, Hsu et al. 2005; Raso, Barbieri et al. 2005). Yet, there are currently no data about the actual types and levels of the ultrasound for the cells within the peripheral nerve. Although using similar values from studies on chondrocytes, endothelial cells, and osteoblasts would allow direct comparison between different cell lines, these values may not have any significance to neural tissue based upon the different physiologic demands of each tissue type. Thus, we chose the magnitude and duration of stimulation for these experiments based on previous work that demonstrated that ultrasound induced a biological response in Schwann cells (Chang, Hsu et al. 2005)

The purpose of this study was to evaluate how sustained LIPUS directly affects Schwann cell function. By evaluating for the expression of the pan-specific Schwann cell marker S-100 with immunohistochemistry, we determined whether Schwann cells de-differentiated after LIPUS stimulation. Schwann cell proliferation was explored using BrdU uptake assays to ascertain if direct LIPUS stimulation is mitogenic for Schwann cells in culture plates.

2.1 Material and methods

Schwann cells culture and low-intensity pulsed ultrasound treatment

Schwann cells were prepared using a method previously described with some modifications (Cai, Campana et al. 1999). Briefly, sciatic nerves were dissected from Wistar rats (n=30) at postnatal day 1–3. The epineural sheath was removed. Thereafter, the sciatic nerves were chopped into 200-μm pieces and enzymatically digested (collagenase/trypsin, 1 mg/ml, 1 hour, 37°C). The resulting cell suspensions were plated onto a six-well plate and cultured in Schwann cell medium (DMEM/10% heat-inactivated FCS/2 mM glutamine/pen/strep). Two different cell densities were prepared for subsequent experiments, a cell density of 5,000 cells/1.77 cm^2 for proliferation assay and immunohistochemistry assays, and a cell density of 100,000 cells/1.77 cm^2 for semiquantitative RT-PCR. The fibroblasts were eliminated by 10 μM cytosine arabinoside and complement-mediated cytolysis with the fibroblast- specific antibody Thy1.1 in conjunction with baby rabbit complement (Cedarlane, Burlington, NC). The medium was changed every other day by adding 2 μM forskolin and glial growth factor (100μg/ml) for expansion of Schwann cells for up to 14 days of cells culture. The purity of cultures was monitored by immunostaining using the Schwann cell marker S-100 and the fibroblast marker Thy-1.1.

Schwann cells were cultured and subjected to LIPUS with modifications as previously described (Takayama, Suzuki et al. 2007). This device (Nexson-The-P41, Nexus Biomedical Devices, Hangzhou, China) generated LIPUS with a pulse width of 200 microseconds, repetition rate of 1.5 KHz, operation frequency of 1 MHz, spatial average temporal average of 100 mW/cm^2, 5 minutes/day. The LIPUS treatment was started 24 hours after initiation of cells culture and repeated for 14 consecutive days. In the experimental group, LIPUS was transmitted from 35- mm diameter LIPUS transducers to the bottom of the cell culture plate via a coupling gel (Smith & Nephew, Oklahoma, CA) and was administered in an incubator (see Fig. 1). In the control group, plates were placed on the same transducers for the same duration, but the LIPUS was not administered.

Schwann cells plated at a cell density of 5,000 cells/1.77cm^2 were used for the immunocytochemistry assays. At day 14 after LIPUS treatment, a total of 18 plates (nine experimental plates and nine control plates) were analyzed for S-100, NT-3, and BDNF immunostaining, respectively. The cells were fixed in plates for 10 minutes with 4% paraformaldehyde solution and then blocked in 4% goat serum with 0.25% triton in PBS. Then, the cells were incubated with either mouse anti-S100 protein monoclonal antibody (Sigma, Saint Louis, MO), mouse anti-neurotrophin-3 monoclonal antibody (Santa Cruz, Santa Cruz, CA), or mouse antibrain-derived neurotrophic factor primary antibodies (Santa Cruz, Sant Cruz, CA), then subsequently with goat anti-mouse IgG FITC (Sigma, Saint Louis, MO) or IgG TRITC (Sigma, Saint Louis, MO) for 1 hour and counterstained with DAPI (Sigma, Saint Louis, MO). The percentage of fluorescently labeled cells/DAPI-stained nuclei was counted using a fluorescent microscope-computer interface (Zeiss, Jena, Germany).

Proliferation assay with 5-Bromo-2-deoxy-uridine

The percent of proliferation was determined by the ratio of total BrdU-positive nuclei to total number of cells (DAPI-stained nuclei) as described previously (Funk, Fricke et al. 2007). Schwann cells plated at a cell density of 5,000 cells/1.77 cm^2 were used for counting of each individual Schwann cell. A total of 12 plates per time point (six experimental plates and six control plates, respectively) were analyzed. At days 4, 7, 10, and 14 after LIPUS treatment, the cells in plates were treated by Brdu (Sigma, Saint Louis, MO) for 2 hours. The cells were then fixed in methanol for 10 minutes at 48°C and treated with 1.25% proteinase K in PBS (pH 7.5) for 5 minutes. Thereafter, the cells were treated with mouse anti-BrdU monoclonal antibody (Sigma, Saint Louis, MO) for 1 hour and then with goat anti-mouse IgG FITC (Sigma, Saint Louis, MO) for 1 hour. DAPI (Sigma, Saint Louis, MO) and cover slips were added. The average proliferation percentage of the plate was counted. The average proliferation percentage was counted by examining four random images within per plate using a fluorescent microscope-computer interface (Zeiss, Jena, Germany).

2.2 LIPUS stimulus may directly trigger Schwann cell proliferation in the early phase

The Schwann cells that were subjected to LIPUS consistently demonstrated an increase in cell proliferation. Fig. 2 shows a fluorescent microscope picture demonstrating the difference in the percentage of BrdU-positive cells in both experimental and control cells at day 7. Table 1 shows the percentage of BrdU-positive cells were significantly higher in experimental groups than in control groups on day 4 ($P < 0.01$), day 7 ($P < 0.01$) and day 10 ($P < 0.01$), respectively. However, this difference between experimental and control disappeared on day 14. The percentage increase in proliferation varied depending upon the control proliferation levels of BrdU uptake. For example, when the control level of proliferation was 23.2% at day 7, there was a 100% increase in the proliferation of experimental cells. While there was a 48.6% control proliferation level at day 10, the experimental cells exhibited a 43% increase in proliferation. The variation of proliferation between both groups is not secondary to changes in media or culture media additive. Moreover, the data suggest that Schwann cells are more responsive to LIPUS at different times of their cell cycle.

In this study, we observed that LIPUS increased Schwann cell proliferation indicating that LIPUS is mitogenic for Schwann cells in vitro (See Fig.3). The LIPUS treatment could effectively improve Schwann cell proliferation at an early stage (day 4, day 7 and day 10), while at later stages (day 14) self-renewal ability of these cells reached to a much higher level but there was no obvious difference between experimental and control groups. This increase in proliferation confirmed results with previous in vitro data, which proposes that cultured cells may be mitogenic in response to LIPUS stimulation (Mukai, Ito et al. 2005; Iwashina, Mochida et al. 2006) Furthermore, these data lend credence to the possibility that the LIPUS stimulus may directly trigger Schwann cell proliferation in the early phase.

		Day 4	Day 7	Day 10	Day 14	F	P
group	Control group	13.3± 1.67	30.53±4.98	51.93±11.56	76.70±9.67	147	0.00
	experiment group	16.33±2.68	40.73±7.45	71.07±9.03	80.20±11.68		
Comparison at same time point	LSD-t	2.35	2.788	3.20	0.06		
	P	0.04	0.02	0.01	0.58		

The mitotic cells were identified by metabolic BrdU labeling. The results showed that experimental groups display a higher proliferation rate. LSD-t, Least Significant Difference t test; F, F value of One-Way ANOVA; p, the p-value is the probability that the null hypothesis is true.

Table 1. the effect of LIPUS on the cell proliferation rate of cultured Schwann cells.

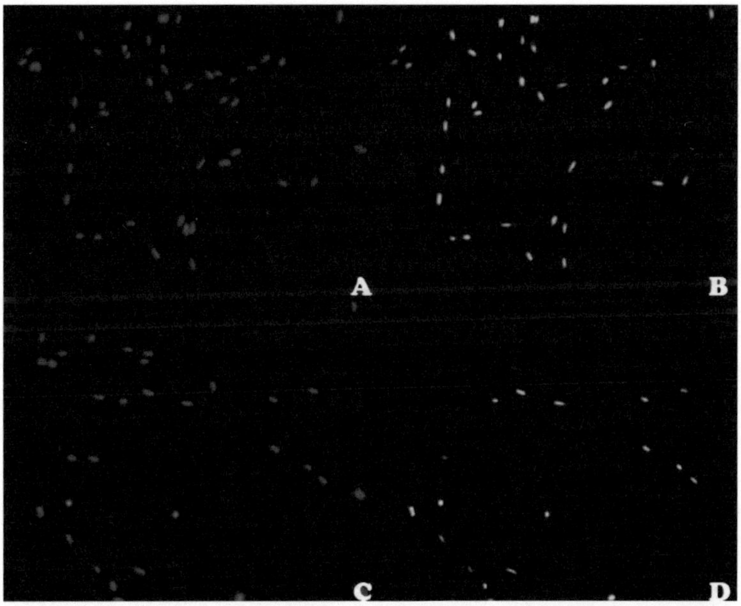

Fig. 2. LIPUS induces increased mitogenesis of in vitro cultured Schwann cells. Fluorescent images depicting the increase in the ratio of the number of BrdU-stained nuclei (green) to DAPI-stained nuclei (blue) in experimental or control cells. Note the increased number of BrdU-positive cells in experimental cells versus the control cells. (experimental A, B; control C, D) at day 4.

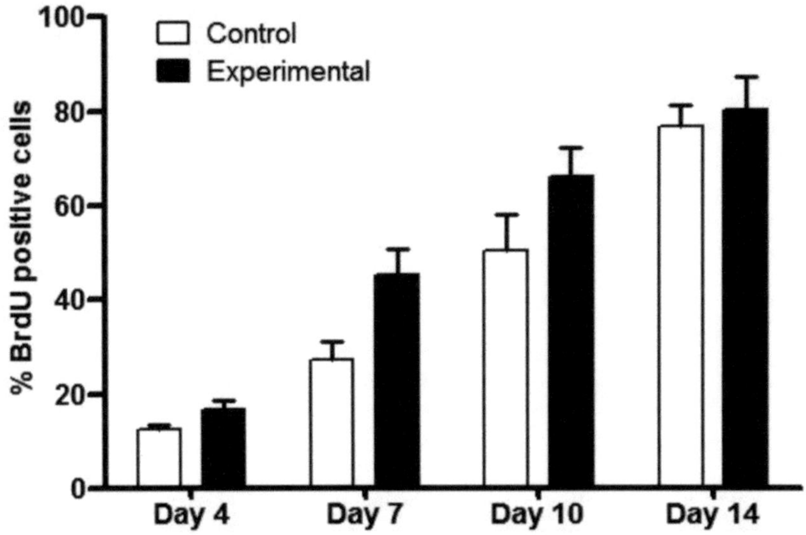

Fig. 3. Increase in proliferation of cultured Schwann cells in response to LIPUS at day 4, day 7 and day 10. No significant difference at day 14.

2.3 LIPUS treatment does not change the phenotype of the Schwann cell

In addition to the increased cell proliferation, LIPUS stimulation of cell cultures has previously been demonstrated to induce an alteration of cellular phenotype,(Ikeda, Takayama et al. 2006; Schumann, Kujat et al. 2006), but little is known about the effects of LIPUS stimulation on a Schwann cell phenotype. Thus, the initial experiments were used to determine whether LIPUS would induce a phenotype alteration to the Schwann cell or not. S-100 immunostaining results showed that Schwann cells do not de-differentiate into another cell type following LIPUS stimulation.

The immunohistochemistry study showed that more than 98% of Schwann cells were positive for the pan-specific Schwann cell marker S-100 at day 14, with or without LIPUS treatment. Moreover, immunostaining for NT-3 and BDNF shows that Schwann cells were positive in more than 98% of the evaluated cells in both the experimental and control cells at day 14. Additionally, the distribution of the positively stained cells was uniform for both the inner and outer areas of the circular plated region. These results further demonstrated that LIPUS treatment does not change the phenotype of the Schwann cell.

3. Effect of LIPUS on the expression of neurotrophin-3 and brain derived neurotrophic factor in cultured Schwann cells

Both neurotrophin-3 (NT-3) and brain-derived neurotrophic factor (BDNF) are two of key neurotrophins constituents in peripheral nervous system, NT-3 is an important regulator of neural survival, development, function, and neuronal differentiation (McAllister, Lo et al. 1995; McAllister, Katz et al. 1999). Hess (Hess, Scott et al. 2007) et al observed that NT-3 expression may modulate the number of Schwann cells at neuromuscular synapses. Otherwise, neurotrophin-3 is an important autocrine factor supporting Schwann cell survival and differentiation in the absence of axons (McAllister, Katz et al. 1999), Schwann cells also contribute to the sources of BDNF during nerve regeneration and the deprivation of endogenous BDNF results in impairment in regeneration and myelination of regenerating axons (Zhang, Luo et al. 2000), BDNF also plays a role in activity-dependent neuronal plasticity (Schmidhammer, Hausner et al. 2007). The exogenous administration of these factors has protective properties for injured neurons and stimulates axonal regeneration (Lykissas, Batistatou et al. 2007). Based on these properties, these molecules may be used as therapeutic agents for treating degenerative diseases and traumatic injuries of both the central and peripheral nervous system.

We therefore measured how LIPUS affects Schwann cells neurotrophic function by evaluated the mRNA expression of NT-3 and BDNF, two members of the neurotrophic factor family of the Schwann cells.

3.1 Semiquantitative RT-PCR for detecting the BDNF and NT-3 mRNA expression

Schwann cells plated at a cell density of 5,000 cells/ 1.77 cm^2 were used for the RT-PCR assays. At day 14 after LIPUS treatment, a total of 12 plates (six experimental plates and six control plates) were analyzed for NT-3 and BDNF mRNA expression, respectively. The cells were then incubated for 12 hours at 37°C to allow for gene transcription. Cells were then trypsinized, collected as pooled samples, and RNA was extracted using the RNeasy Mini Kit (Qiagen, Valencia, CA) following the protocol. The cDNA was prepared for experimental and control samples using 3μg of RNA with SuperScript II RNase reverse transcriptase (Invitrogen, Oklahoma, CA) and specific primers for NT-3 (forward: 5′- CTTATCTCCG TGGCATCCAAGG-3′, reverse: 5′- TCTGAAGTCAGTGCTCGGACGT-3′), BDNF (forward: 5′-ATGGGACTCTGGAGAGCGTGAA-3′, reverse: 5′-CGCCAGCCAATTCTCTTTTTGC-3′), and b-actin (forward: 5′-CCCAGAGCAAGAGAGGCATC-3′, reverse: 5′-CTCAGGAGGAG CAATGATCT-3′) (Hatami, Oryan et al. 2007).

The PCR reaction conditions were consisted of one cycle of 94°C for 5 minutes, followed by 30 cycles of thermal cycling 30 seconds at 94°C, 30 seconds at T_0°C, and 1 minute at 72°C. The T_0 was 60°C for BDNF, 64°C for NT-3, and 58°C for b-actin. The final cycle was followed by a 5-minute extension at 72°C. Ten microliters of PCR product was then differentiated on a 1.5% agarose gel and the gel image was taken with a digital camera. ImagQuant analysis software (Stratagene Company, La Jolla, CA) was used to determine the densities of the NT-3 and BDNF bands when compared with the b-actin control for both experimental and control samples.

3.2 Effect of LIPUS on the expression of NT-3 and BDNF mRNA in Schwann cell

Schwann cells that were subjected to sustained LIPUS exhibited an increase in NT-3 mRNA expression, and a decrease in BDNF mRNA expression (Fig. 4). The NT-3/β-actin ratio of RT-PCR products in the experimental group was 0.56±0.13 and 0.41±0.09 in the control group. However, the BDNF/β-actin ratio of RT-PCR products in the experimental group was 0.51±0.05 and 0.60±0.08 in the control group. The differences in NT-3 and BDNF products for experimental and control groups were found to be statistically significant ($p<0.01$ and $p<0.05$, respectively). Reverse transcriptase controls with no reverse transcriptase enzyme confirmed that there was no genomic DNA contamination.

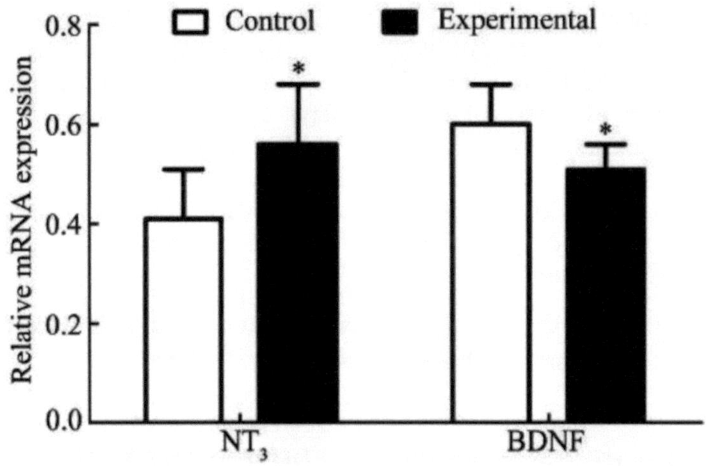

Fig. 4. Results from RT-PCR analysis of NT-3 and BDNF are expressed relative to β-actin mRNA expression 14 days after the LIPUS stimulation. There was significantly upregulated in experimental groups compared with the control in NT-3 mRNA expression ($t=2.324$, $P<0.05$), and significantly downregulated in BDNF mRNA expression ($t=2.337$, $p<0.05$).

4. LIPUS enhances elongation of neurites in rat cortical neurons through inhibition of GSK-3β

Intracellular mechanisms that enhance neurite outgrowth evidently require the reorganization of the neurite cytoskeletons including the microtubules and actin filaments (Dent and Gertler 2003). Recently, a cytoskeletal-related signaling pathway: PI 3-kinase/Atk/glycogen synthase kinase (GSK-3)/collapsin response mediator protein (CRMP-2) was reported to be important for the outgrowth of neurite, with GSK-3 being a central regulator (Jiang, Guo et al. 2005; Yoshimura, Kawano et al. 2005). GSK-3 is a multifunctional serine/threonine kinase found ubiquitously in eukaryotes (Jiang, Guo et al. 2005) and it plays key roles for various biological processes, such as the canonical Wnt signaling pathway, microtubule dynamics, and astrocyte migration (Doble and Woodgett 2003; Etienne-Manneville and Hall 2003). GSK-3 phosphorylates at least have four types of microtubule-associated proteins (MAPs), CRMP-2(Yoshimura, Kawano et al. 2005), tau (Jiang, Guo et al. 2005), adenomatous polyposis coil gene product (APC)(Frame and Cohen 2001; Grimes and Jope 2001) and MAP1B (Lucas, Goold et al. 1998; Goold, Owen et al. 1999). It modulates axial orientation during the development, differentiation, and neurite outgrowth in neurons through phosphorylation of these MAPs (Jiang, Guo et al. 2005; Yoshimura, Kawano et al. 2005; Chen, Yu et al. 2007; Conde and Caceres 2009). Some research have proved that the local inhibition of GSK-3 effectively enhances neurite/axon elongation (Kim, Zhou et al. 2006), whereas overexpression of GSK-3 could impair neurite/axon elongation (Munoz-Montano, Lim et al. 1999). During peripheral nerve regeneration, some factors such as BDNF, NT3, and laminin, locally activate the PI3-kinase/Akt/GSK-3 pathway and inhibit GSK-3, which favors neurite elongation (Kim, Zhou et al. 2006).

We measured the length of neurites to examine whether LIPUS is effective for the elongation of the neuronal processes. Then we examined the change in the activity and the mRNA expression of GSK-3β to determine the intracellular mechanism of neurite outgrowth following irradiation by LIPUS. It is concluded that LIPUS can enhance elongation of neurites and it is possible through the decreased expression of GSK-3β (Ren, Li et al. 2010).

4.1 Effect of LIPUS treatment on neurite outgrowth

4.1.1 Materials and methods: Cell culture and ultrasound treatment

Cortical neurons isolated from the brain of Wistar rats were bought from ScienCell Research Laboratories (San Diego, USA). These cortical neurons were subcultured with a density of 20 000 cells/1.6 cm^2 in poly-L-lysine coated 6-well plates (Costa, USA) for immunoblot and semi-quantitative RT-PCR analysis, and a density of 100 cells/0.32 cm^2 in poly-L-lysine coated 96-well plates (Costa, USA) for the measurement of neurite length. The cells were cultured in neuronal medium (3 mL medium per well in the 6-well plates and 0.1mL per well in the 96-well plate; ScienCell Research Laboratories, San Diego, CA, USA) in a humidified atmosphere of 5% CO_2 in air at 37 ℃. The medium was refreshed every 3 d.

A LIPUS-therapeutic apparatus, Nexson-The- P41 was constructed according to instructions from Nexus Biomedical Devices (Hangzhou, China). There were two LIPUS probes in the apparatus, both of which generated LIPUS with a SATA intensity of 10 mW/cm^2, pulse width of 200 microseconds, repetition rate of 1.5 kHz, and an operation frequency of 1 MHz. The LIPUS was applied to the cultured cortical neurons after 24 h in culture through the bottom of the 6-well plates via a coupling gel (Smith & Nephew, Oklahoma, CA, USA) and was administered for 5 min every day during the span of this experiment (Fig. 1). Ultrasound signals from this generator were detected by a hydrophone system (Model OS-111; Hewlett-Packard, Japan), and the wave amplitudes of the signals passing through the tube wall were more than 90%, which resulted in more than 85% energy irradiated. Control samples were prepared in the same manner with the exception of no LIPUS treatment.

Neurite length measurement protocol

Cultured cortical neurons in 96-well plates were randomly divided into two groups: the LIPUS-treated group and the control group. After being subcultured for 24 h, the LIPUS treatment began and was administered for 5 min every day. On the third day, both the LIPUS-treated and control groups were photographed 2 h after the treatment. A Nikon Diaphot inverted microscope with a Nikon Plan 20× objective (Nikon, Tokyo, Japan) coupled to a video camera was used to obtain cell images (Carl Zeiss, Germany). Images of at least 200 neurons for each group were obtained. For each neuron, we measured its longest neurite with the software Image-Pro Plus 6.0 (Media Cybernetics, USA).

4.1.2 Neurites in LIPUS-treated group were significant longer

There are no significant difference in morphology between LIPUS-treated group and control group except the length of neurites. In both LIPUS-treated group (Fig. 5a) and control group (Fig.5b), there were many neurons with 2-7 processes; some were thick fibers, or some were thin fibers with varicosities. We measured the length of 200 neurites in each group and most neurite measured have a length between 50 μm to 80 μm. Data showed that compared with control group, neurites in LIPUS-treated group were significant longer [(73.14 ±8.32) μm vs.

(68.18 ±8.96) μm, P<0.01](Fig. 5c). We attempted to investigate how processes of the cultured neurons were extended under the influence of LIPUS. Morphological changes revealed that LIPUS could effectively enhance elongation of neurites after three days of treatment compared to the control group. However, we failed to measure the length of neurites on the seventh or tenth day because after the fifth day, most neurites reached another neurite and, consequently, the growth of those neurites stopped. Although the mechanism by which LIPUS affects the neuronal processes is likely to be complex, the regulation of the cytoskeleton is crucial for the proper growth cone motility (Dent and Kalil 2001). To clarify the intracellular mechanism of this effect, we examined the proteins related to the cytoskeletal-related signaling pathway to determine whether the proteins in the cultured neurons were changed following the LIPUS treatment.

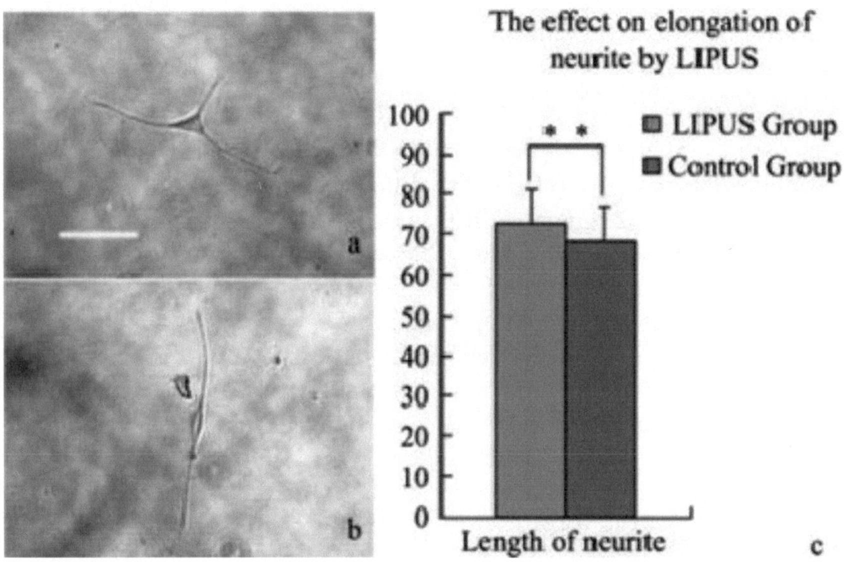

Fig. 5. On the third day, both the LIPUS-treated and control groups were photographed with a Nikon Plan 20× objective coupled to a video camera. In both LIPUS-treated group (Fig. 5a) and control group (Fig. 5b), there were many neurons with 2-7 processes; some were thick fibers, and some were thin fibers with varicosities. Bar=50 μm. Images of at least 200 neurons for each group were obtained. For each neuron, we measured its longest neurite with the software Image-Pro plus 6.0. Most neurite measured have a length between 50 μm to 80 μm. Data showed that neurites in LIPUS-treated group were significant longer than that in control group [(73.14 ±8.32) μm vs. (68.18 ±8.96) μm, P<0.01] (Fig. 5c).

4.2 Changes in protein activity related to the Cytoskeletal-signaling pathway caused by LIPUS treatment

To investigate changes in protein activity related to cytoskeletal-signaling pathway caused by LIPUS, total proteins were extracted on the third, seventh, and tenth days following daily LIPUS treatment and their acticity were examined using Western blot analysis (Fig. 6).

To measure the length of neurites to examine whether LIPUS is effective for the elongation of the neuronal processes, we examined the change in the activity and the mRNA expression of GSK-3beta to determine the intracellular mechanism of neurite outgrowth following irradiation by LIPUS (Ren, Li et al. 2010).

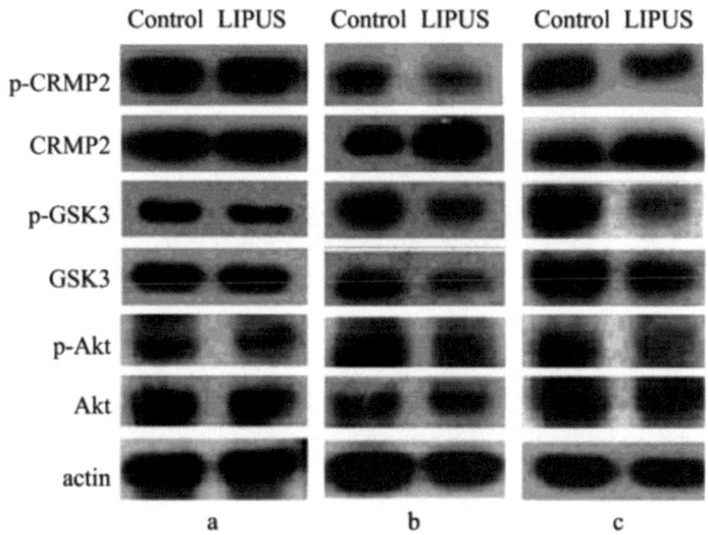

Fig. 6. Total proteins were extracted on the third, seventh, and tenth days following daily LIPUS treatment and their activity were examined using Western blot analysis. (a) On the third day, there was no significant difference in protein levels between the control and LIPUS groups. (b) On the seventh day, the levels of p-Akt, GSK-3beta, p-GSK-3beta, and p-CRMP-2 were decreased in the LIPUS group compared to the controls. (c) On the tenth day, a remarkable decrease of p-Akt, p-GSK-3beta, and p-CRMP-2 were observed while it appeared that GSK-3beta was slightly decreased. The beta-actin in each lane served as an internal control.

4.2.1 Materials and methods: Western blot analysis

For Western blot analysis, the treated and untreated cultured cells were harvested at third, seventh, and tenth days. LIPUS group cells were harvested 2 h after the last LIPUS treatment. Whole cell extracts were prepared by boiling the cells in lysis buffer (2% SDS; 10% glycerol; 10 mmol/L Tris, pH 6.8; 100 mmol/L DTT) for 10 min. Proteins were separated by electrophoresis on 4%-12% Bis-Tris gels (Novex; Invitrogen, Carlsbad, CA, USA). Separated proteins were then transferred onto PVDF membranes. The membranes were blocked with 5% nonfat dry milk in PBS, pH 7.4, and 0.1% Tween 20 (PBS-Tween) for 1 h at room temperature. The membranes were incubated with primary antibodies diluted in 5% BSA overnight at 4 ℃. The blots were washed in PBS-Tween and then incubated with diluted secondary antibodies (HRP, 1:10 000; Santa Cruz Biotechnology, Santa Cruz, CA, USA) for 1 h at room temperature. Reactive proteins were visualized with SuperSignal West Pico chemiluminescence's reagent (Pierce Biotechnology, Rockford, IL, USA) followed by exposure to x-ray film.

The primary antibodies used for the Western blot analysis were as follows: rabbit anti-GSK-3beta antibody (21001-1; Signalway Antibody, Pearland, TX, USA), rabbit anti-phospho GSK-3beta (Ser 9) antibody (11002-1; Signalway Antibody), rabbit anti-CRMP-2 antibody (SC-30228, Santa Cruz Biotechnology, Santa Cruz, CA, USA), rabbit anti-phospho CRMP-2 (Thr 514) antibody (9397, Cell Signaling Technology, Beverly, MA, USA), goat anti-Akt antibody (SC-1618; Santa Cruz Biotechnology, Santa Cruz, CA, USA), and rabbit anti-phospho-Akt (Ser 473) antibody (SC-101629; Santa Cruz Biotechnology, Santa Cruz, CA, USA). All of these primary antibodies were polyclonal and used at a dilution of 1:500. Mouse anti-beta actin polyclonal antibody (SC-81178, 1:1000, Santa Cruz, CA, USA) was used at a dilution of 1:1 000. As secondary antibodies, HRP-conjugated goat anti-rabbit (SC-2004; Santa Cruz Biotechnology, Santa Cruz, CA, USA), donkey anti-goat (SC-2020; Santa Cruz Biotechnology, Santa Cruz, CA, USA), goat anti-mouse (SC-2005; Santa Cruz Biotechnology, Santa Cruz, CA, USA) was used at a dilution of 1:10 000.

Semi-quantitative RT-PCR analysis

For RT-PCR analysis of GSK-3β gene expression, neurons were cultured in two 6-well plates. One of the plates was irradiated by LIPUS for 7 d (5 min/day; 10 mW/cm^2); the other was the control group without LIPUS treatment. Cultured cells were harvested 2 h after the last irradiation. Total RNA was extracted using Trizol reagent (Invitrogen, Carlsbad, CA, USA) according to the instruction manual, then resuspended in diethylpyrocarbonate (DEPC)-treated water. The extracted RNA was used to synthesize first strand cDNA with the PrimeScript™ RT-PCR Kit (Takara Biotechnology, Dalian, China) according to the kit's manual. Aliquots of synthesized cDNA were added to PCR mixtures containing sense and antisense primers (0.1 μmol/L each) for GSK-3beta, dNTP mixture (0.2 mmol/L of each dNTP), 1.5 mmol/L MgCl2, and rTaq DNA polymerase (1 unit) (Takara Biotechn ology, Dalian, China). The primers for GSK-3β were 5' -AGCCAGTGCAGCAGCCTTCAG C-3' for the sense strand and 5' -TCTCCTCGGACCA GCTGCT TTG-3' for the antisense strand. The primers for β-actin were 5' -GAGCTACGAGCTGCC TGACG-3' for the sense strand and 5' -CCTAGAA GCATTTGC GGTGG-3' for the antisense strand. The PCR products were electrophoretically separated in a 2% agarose gel and then visualized and photographed with an imager (Alhpa-imagerTM 2200; Alpha Innotech Corporation, San Leandro, CA, USA).

4.2.2 LIPUS enhances neurite outgrowth through the down-regulation of GSK-3β activity

On the third day, there was no significant difference in protein levels between the control and LIPUS groups. However, on the seventh and tenth days after irradiation by LIPUS, the levels of p-Akt, GSK-3beta, p-GSK-3beta, and p-CRMP-2 were decreased in the LIPUS group compared to the controls. On the tenth day, a remarkable decrease of p-Akt, p-GSK-3β, and p-CRMP-2 were observed while it appeared that GSK-3beta were slightly decreased.

During nerve regeneration, GSK-3β is locally inhibited by some factors at the growth cone through the PI3-kinase/Akt/GSK-3β signaling pathway which favors neurite outgrowth (Chen, Yu et al. 2007). The overexpression of active GSK-3beta blocks neurite growth in cultured neurons (Munoz-Montano, Lim et al. 1999). In the PI3-kinase/Akt/GSK-3beta/CRMP-2 pathway, active Akt inhibits GSK-3beta through phosphorylation at Ser 9 and GSK-3beta inhibits CRMP-2 though phosphorylation at Thr 5. If LIPUS enhances neurite elongation though this pathway, the phosphorylation of GSK-3beta should be up-regulated and the activity of GSK-3beta should be inhibited. However, in this research, the activity of GSK-3beta was inhibited by LIPUS and the phosphorylation of GSK-3beta by Akt was inhibited, too. This conflict of results revealed that LIPUS enhances neurite outgrowth through the down-regulation of GSK-3β activity but not through the PI3-kinase/Akt/GSK-3beta pathway. Therefore, we employed semi-quantitative RT-PCR to examine the mRNA of GSK-3beta. The results of the semi-quantitative RT-PCR revealed that the expression of GSK-3beta mRNA decreased after LIPUS irradiation on the seventh day. From these findings, we postulate that when neurons are irradiated by LIPUS, an unknown intracellular mechanism may be activated as a response to this "injury" and, consequently, neurons reduce the mRNA expression of GSK-3β. The decrease of GSK-3beta activity comes from reduced expression, but not through the PI3-kinase/Akt/GSK-3β signaling pathway (Ren, Li et al. 2010).

4.2.3 The expression of GSK-3β mRNA decreased after LIPUS irradiation

The mRNA expression of GSK-3beta in the cultured neurons following LIPUS treatment was examined using a semi-quantitative RT-PCR. For this analysis, the LIPUS-treated cultured neurons on the seventh day were selected as they showed a significant decrease in their mRNA levels compared to the control. Data from analysis of the imager indicated mRNA of GSK-3β decreased about 4 folds [(1.001 ±0.017) vs. (0.627 ±0.037), P<0.001] (Fig. 7). As a result, mRNA expression of GSK-3beta was also decreased on the seventh days compared to the control.

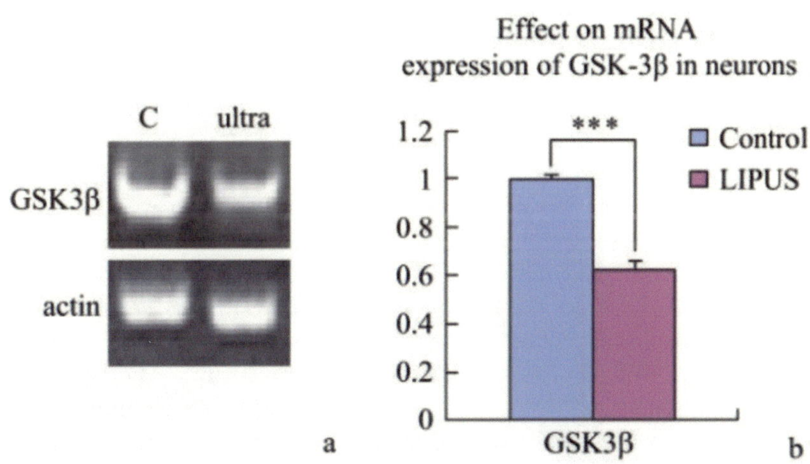

Fig. 7. Expression of GSK-3β in neurons was evaluated by semi quantitative RT-PCR. Neurons were irradiated for 7 d and harvested 2 h after the last LIPUS irradiation. The right lane represents the experimental mRNA expression, and the left lane corresponds to the control mRNA expression (Fig. 7a). Data showed that expression of GSK-3β decreased about 4 folds [(1.001 ±0.017) vs. (0.627 ±0.037), P<0.001] (Fig.7b). The β-actin in each lane served as an internal control.

The reduced expression is a kind of global inhibition of GSK-3beta that has a complex effect on neurite elongation. It favors neurite elongation at a low level of inhibition whereas it impairs neurite elongation at a high level of inhibition (Munoz-Montano, Lim et al. 1999). Strong global GSK-3beta inhibition results in excessive microtubule stability all along the neurite shaft due to the inhibition of MAP1B, which eliminates dynamic microtubules, and the abnormal distribution of APC that stabilizes microtubules all along the neurite shaft. In this case, there was no pool of dynamic microtubules at the growth cone, which are necessary for growth cone advancement, and no localization of APC to microtubule plus ends (Kim, Zhou et al. 2006).

In our research, significant morphological changes were found on the third day whereas significant changes in the activity of GSK-3beta were found on the seventh and tenth days. We postulate that daily treatment of LIPUS would result in neurons' response accumulation. Morphological changes were observed on the third day when the inhibition of GSK-3beta is not significant enough to be found. Since overly strong global inhibition of GSK-3beta impairs neurite elongation, whether LIPUS could impair neurite elongation needs further study.

5. Conclusion

5.1 LIPUS stimulation induced an alteration in Schwann cell function as demonstrated by promoted cell proliferation and NT-3 gene expression

LIPUS is one of the physical agents that is known to accelerate bone and tissue regeneration following injury (Heckman, Ryaby et al. 1994; Lu, Qin et al. 2006). Consequently, it has been accepted as an effective therapy for nonunion fractures and fresh fracture healing through an easy and non-invasive application (Azuma, Ito et al. 2001; Schortinghuis, Bronckers et al. 2005). Previous studies indicate that LIPUS has positive effects on axonal regeneration by in vivo peripheral nerve injury trials (Crisci and Ferreira 2002; Chang, Hsu et al. 2005). Raso (Raso, Barbieri et al. 2005) et al have demonstrated that the locally applied ultrasound stimuli on the injured sciatic nerve rather than the untreated nerves of rats can effectively enhance the number of Schwann cell nuclei. LIPUS has been used in conjunction with tissue engineered nerves in repairing peripheral nerve defect, Chang (Chang, Hsu et al. 2005) et al demonstrated that applying low-intensity ultrasound on seeded Schwann cells within poly (D, L-lactic acid-co-glycolic acid) conduits have a significantly greater number and area of regenerated axons compared to the sham groups. Although this secondary response by Schwann cells has been well characterized, there is still limited information as to how Schwann cells would directly respond to LIPUS stimulation. Therefore, we cultured Schwann cells in plate as an in vitro model, and applied LIPUS in the model to demonstrate the direct effects of physical stimulation on Schwann cells (Zhang, Lin et al. 2009).

LIPUS stimulation of cultured Schwann cells induced an alteration in cell function as demonstrated by promoted cell proliferation and NT-3 gene expression, which is consistent with that LIPUS enhances peripheral nerve regeneration that was observed from in vivo models (Lowdon, Seaber et al. 1988; Raso, Barbieri et al. 2005). It has been documented that NT-3 has a strong effect on neurite outgrowth (Markus, Patel et al. 2002; Sahenk, Nagaraja et al. 2005). Additionally, some studies using genetically modified Schwann cells to overexpress the NT-3 gene have examined the role of NT-3 in the neuron survival and axonal regeneration/remyelination (Zhang, Zeng et al. 2007; Pettingill, Minter et al. 2008). It has been reported that Schwann cells transduced ex vivo with adenoviral (AdV) or lentiviral

(LV) vectors encoding a functional NT-3 molecule led to the presence of a significantly increased number of axons in the contusion site (Golden, Pearse et al. 2007). The results of Yamauchi(Yamauchi, Miyamoto et al. 2005) et al showed that NT-3 activation of TrkC stimulates Schwann cell migration through two parallel signaling units, Ras/Tiam1/Rac1 and Dbs/Cdc. Poduslo(Poduslo and Curran 1996) et al observed that NT-3 has a higher permeability coefficient across the blood–nerve barrier, and would contact sensory axons soon after reaching the circulation of adult rats. The increase in NT-3 expression might lead to an increase in the number of nerve regeneration in the axons. LIPUS-induced increase in NT-3 expression, as demonstrated in this model, may produce a microenvironment that is permissive for axonal sprouting and Schwann cells migration after peripheral nerve injury.

Neurotrophin–neurotrophin interactions are regulated by neurotrophin levels, NT-3 and BDNF in particular can be co-expressed and each can regulate the levels of the other. The relative expression level of the neurotrophins is thought to be mediated through receptor tyrosine kinase (Trk) activity (Mallei, Rabin et al. 2004). NT-3 infusion caused a significant decrease in the level of BDNF proteins in both kindled and non-kindled hippocampus, likely via down-regulation of TrkA (Yamamoto and Hanamura 2005) Furthermore, a study of Karchewski (Karchewski, Gratto et al. 2002) et al showed that NT-3 can act in an antagonistic fashion to NGF in the regulation of BDNF expression in intact neurons, and mitigate BDNF's expression in injured neurons. It is also consistent with a study in which, in contrast, deletion of the NT-3 gene in transgenic mice increased BDNF and TrkB mRNA synthesis, suggesting that decreased NT-3 may disinhibit BDNF expression (Elmer, Kokaia et al. 1997). Similarly, our model has demonstrated an up-regulation of NT-3 mRNA and down-regulation of BDNF mRNA expression after the LIPUS stimulation. Hence, it is possible that NT-3 acts in an opposite fashion result in a down-regulation in BDNF expression in intact Schwann cells. Further investigation is necessary to determine the molecular mechanisms of NT-3 and BDNF signaling pathway by the data presented in our study (Zhang, Lin et al. 2009).

5.2 LIPUS enhances neurite elongation in rat cortical neurons

LIPUS enhances neurite elongation in rat cortical neurons, indicating that LIPUS could be a potential application for clinical treatment of nerve regeneration in both the central and peripheral nervous systems. The intracellular mechanism also indicates that LIPUS has the same action as the neurotrophic factors, laminin and LiCl. Compared to other pharmacological inhibitors of GSK-3beta, LIPUS has some advantages: (1) LIPUS is considered to be nontoxic, thus it has a wide margin of biologic safety; (2) It directly irradiates target neurons and does not affect other tissues; and (3) The decreased expression comes from a response of neurons and is not affected by the metabolism or blood brain barrier. However, further investigation is required to identify an accurate and continuous application of LIPUS treatment to achieve constant and reproducible results prior to clinical use.

The results suggest that LIPUS may have several different clinical applications in the improvement of peripheral nerve regeneration. First, given its nontoxicity and a wide margin of biologic safety, it may be used as an effective physical stimulant when engineering peripheral nerve tissue. Schwann cell-based therapies that use transplantation techniques for the treatment of nerve tissue repairing are being widely investigated for their potential as clinical applications (Rochkind, Astachov et al. 2004; Li, Ping et al. 2006; Gravvanis, Lavdas et al. 2007). LIPUS applied in conjunction with other forms of biologic stimulation, is worth considering when optimizing an innovative "multi-level" form of treatment. Second, application of LIPUS in vivo is likely to be considered to stimulate repair of damaged peripheral nerve tissues. Some experimental studies supported the result that both end-to-side and tubulization repair of peripheral nerves led to successful axonal regeneration along the severed nerve trunk as well as to a partial recovery of the lost function (Geuna, Nicolino et al. 2007; Lloyd, Luginbuhl et al. 2007). With the availability of the LIPUS as an activator of Schwann cells, it can be an effective alternative in nerve reconstruction and be of great value in various kinds of peripheral nerve microsurgery. Further investigation is required to identify an accurate and continuous application of LIPUS treatment, in order to achieve constant and reproducible results prior to clinical use.

As demonstrated in the current study, NT-3 and BDNF mRNA expression in Schwann cell response to LIPUS may be independent of the reciprocal regulation between the glial cells and neurons. Normally, during development and axonal injury, this reciprocal relationship between the glial cells and neurons causes a response in the glial cells, which occurs secondary to the neuron. However, data from the in vitro model indicate otherwise. The Schwann cells responded robustly in the absence of neurons, suggesting that Schwann cell responses may be directly elicited through LIPUS stimuli in the model.

6. Acknowledgment

This research was funded by The Natural Science Foundation of Beijing (Grant number: 5072020), China.

7. References

Azuma, Y., M. Ito, et al. (2001). "Low-intensity pulsed ultrasound accelerates rat femoral fracture healing by acting on the various cellular reactions in the fracture callus." *Journal of bone and mineral research : the official journal of the American Society for Bone and Mineral Research* 16(4): 671-680.

Cai, F., W. M. Campana, et al. (1999). "Transforming growth factor-beta1 and glial growth factor 2 reduce neurotrophin-3 mRNA expression in cultured Schwann cells via a cAMP-dependent pathway." *Brain research. Molecular brain research* 71(2): 256-264.

Chang, C. J., S. H. Hsu, et al. (2005). "Low-intensity-ultrasound-accelerated nerve regeneration using cell-seeded poly(D,L-lactic acid-co-glycolic acid) conduits: an in vivo and in vitro study." *Journal of biomedical materials research. Part B, Applied biomaterials* 75(1): 99-107.

Chen, Z. L., W. M. Yu, et al. (2007). "Peripheral regeneration." *Annual review of neuroscience* 30: 209-233.

Choi, J. J., M. Pernot, et al. (2007). "Spatio-temporal analysis of molecular delivery through the blood-brain barrier using focused ultrasound." *Physics in medicine and biology* 52(18): 5509-5530.

Conde, C. and A. Caceres (2009). "Microtubule assembly, organization and dynamics in axons and dendrites." *Nature reviews. Neuroscience* 10(5): 319-332.

Crisci, A. R. and A. L. Ferreira (2002). "Low-intensity pulsed ultrasound accelerates the regeneration of the sciatic nerve after neurotomy in rats." *Ultrasound in medicine & biology* 28(10): 1335-1341.

Danialou, G., A. S. Comtois, et al. (2002). "Ultrasound increases plasmid-mediated gene transfer to dystrophic muscles without collateral damage." *Molecular therapy : the journal of the American Society of Gene Therapy* 6(5): 687-693.

Dent, E. W. and F. B. Gertler (2003). "Cytoskeletal dynamics and transport in growth cone motility and axon guidance." *Neuron* 40(2): 209-227.

Dent, E. W. and K. Kalil (2001). "Axon branching requires interactions between dynamic microtubules and actin filaments." *The Journal of neuroscience : the official journal of the Society for Neuroscience* 21(24): 9757-9769.

Doble, B. W. and J. R. Woodgett (2003). "GSK-3: tricks of the trade for a multi-tasking kinase." *Journal of cell science* 116(Pt 7): 1175-1186.

Elmer, E., M. Kokaia, et al. (1997). "Suppressed kindling epileptogenesis and perturbed BDNF and TrkB gene regulation in NT-3 mutant mice." *Experimental neurology* 145(1): 93-103.

Etienne-Manneville, S. and A. Hall (2003). "Cdc42 regulates GSK-3beta and adenomatous polyposis coli to control cell polarity." *Nature* 421(6924): 753-756.

Frame, S. and P. Cohen (2001). "GSK3 takes centre stage more than 20 years after its discovery." *The Biochemical journal* 359(Pt 1): 1-16.

Funk, D., C. Fricke, et al. (2007). "Aging Schwann cells in vitro." *European journal of cell biology* 86(4): 207-219.

Geuna, S., S. Nicolino, et al. (2007). "Nerve regeneration along bioengineered scaffolds." *Microsurgery* 27(5): 429-438.

Golden, K. L., D. D. Pearse, et al. (2007). "Transduced Schwann cells promote axon growth and myelination after spinal cord injury." *Experimental neurology* 207(2): 203-217.

Goold, R. G., R. Owen, et al. (1999). "Glycogen synthase kinase 3beta phosphorylation of microtubule-associated protein 1B regulates the stability of microtubules in growth cones." *Journal of cell science* 112 (Pt 19): 3373-3384.

Gravvanis, A. I., A. A. Lavdas, et al. (2007). "The beneficial effect of genetically engineered Schwann cells with enhanced motility in peripheral nerve regeneration: review." *Acta neurochirurgica. Supplement* 100: 51-56.

Grimes, C. A. and R. S. Jope (2001). "The multifaceted roles of glycogen synthase kinase 3beta in cellular signaling." *Progress in neurobiology* 65(4): 391-426.

Hatami, H., S. Oryan, et al. (2007). "Alterations of BDNF and NT-3 genes expression in the nucleus paragigantocellularis during morphine dependency and withdrawal." *Neuropeptides* 41(5): 321-328.

Heckman, J. D., J. P. Ryaby, et al. (1994). "Acceleration of tibial fracture-healing by non-invasive, low-intensity pulsed ultrasound." *The Journal of bone and joint surgery. American volume* 76(1): 26-34.

Hess, D. M., M. O. Scott, et al. (2007). "Localization of TrkC to Schwann cells and effects of neurotrophin-3 signaling at neuromuscular synapses." *The Journal of comparative neurology* 501(4): 465-482.

Hill, G. E., S. Fenwick, et al. (2005). "The effect of low-intensity pulsed ultrasound on repair of epithelial cell monolayers in vitro." *Ultrasound in medicine & biology* 31(12): 1701-1706.

Hiyama, A., J. Mochida, et al. (2007). "Synergistic effect of low-intensity pulsed ultrasound on growth factor stimulation of nucleus pulposus cells." *Journal of orthopaedic research : official publication of the Orthopaedic Research Society* 25(12): 1574-1581.

Ikeda, K., T. Takayama, et al. (2006). "Effects of low-intensity pulsed ultrasound on the differentiation of C2C12 cells." *Life sciences* 79(20): 1936-1943.

Iwashina, T., J. Mochida, et al. (2006). "Low-intensity pulsed ultrasound stimulates cell proliferation and proteoglycan production in rabbit intervertebral disc cells cultured in alginate." *Biomaterials* 27(3): 354-361.

Jiang, H., W. Guo, et al. (2005). "Both the establishment and the maintenance of neuronal polarity require active mechanisms: critical roles of GSK-3beta and its upstream regulators." *Cell* 120(1): 123-135.

Johns, L. D. (2002). "Nonthermal effects of therapeutic ultrasound: the frequency resonance hypothesis." *Journal of athletic training* 37(3): 293-299.

Karchewski, L. A., K. A. Gratto, et al. (2002). "Dynamic patterns of BDNF expression in injured sensory neurons: differential modulation by NGF and NT-3." *The European journal of neuroscience* 16(8): 1449-1462.

Kim, W. Y., F. Q. Zhou, et al. (2006). "Essential roles for GSK-3s and GSK-3-primed substrates in neurotrophin-induced and hippocampal axon growth." *Neuron* 52(6): 981-996.

Li, Q., P. Ping, et al. (2006). "Nerve conduit filled with GDNF gene-modified Schwann cells enhances regeneration of the peripheral nerve." *Microsurgery* 26(2): 116-121.

Lloyd, B. M., R. D. Luginbuhl, et al. (2007). "Use of motor nerve material in peripheral nerve repair with conduits." *Microsurgery* 27(2): 138-145.

Lowdon, I. M., A. V. Seaber, et al. (1988). "An improved method of recording rat tracks for measurement of the sciatic functional index of de Medinaceli." *Journal of neuroscience methods* 24(3): 279-281.

Lu, H., L. Qin, et al. (2006). "Low-intensity pulsed ultrasound accelerates bone-tendon junction healing: a partial patellectomy model in rabbits." *The American journal of sports medicine* 34(8): 1287-1296.

Lucas, F. R., R. G. Goold, et al. (1998). "Inhibition of GSK-3beta leading to the loss of phosphorylated MAP-1B is an early event in axonal remodelling induced by WNT-7a or lithium." *Journal of cell science* 111 (Pt 10): 1351-1361.

Lykissas, M. G., A. K. Batistatou, et al. (2007). "The role of neurotrophins in axonal growth, guidance, and regeneration." *Current neurovascular research* 4(2): 143-151.

Mallei, A., S. J. Rabin, et al. (2004). "Autocrine regulation of nerve growth factor expression by Trk receptors." *Journal of neurochemistry* 90(5): 1085-1093.

Markus, A., T. D. Patel, et al. (2002). "Neurotrophic factors and axonal growth." *Current opinion in neurobiology* 12(5): 523-531.

McAllister, A. K., L. C. Katz, et al. (1999). "Neurotrophins and synaptic plasticity." *Annual review of neuroscience* 22: 295-318.

McAllister, A. K., D. C. Lo, et al. (1995). "Neurotrophins regulate dendritic growth in developing visual cortex." *Neuron* 15(4): 791-803.

Mukai, S., H. Ito, et al. (2005). "Transforming growth factor-beta1 mediates the effects of low-intensity pulsed ultrasound in chondrocytes." *Ultrasound in medicine & biology* 31(12): 1713-1721.

Munoz-Montano, J. R., F. Lim, et al. (1999). "Glycogen Synthase Kinase-3 Modulates Neurite Outgrowth in Cultured Neurons: Possible Implications for Neurite Pathology in Alzheimer's Disease." *Journal of Alzheimer's disease : JAD* 1(6): 361-378.

Nahirnyak, V., T. D. Mast, et al. (2007). "Ultrasound-induced thermal elevation in clotted blood and cranial bone." *Ultrasound in medicine & biology* 33(8): 1285-1295.

Parvizi, J., C. C. Wu, et al. (1999). "Low-intensity ultrasound stimulates proteoglycan synthesis in rat chondrocytes by increasing aggrecan gene expression." *Journal of orthopaedic research : official publication of the Orthopaedic Research Society* 17(4): 488-494.

Pettingill, L. N., R. L. Minter, et al. (2008). "Schwann cells genetically modified to express neurotrophins promote spiral ganglion neuron survival in vitro." *Neuroscience* 152(3): 821-828.

Poduslo, J. F. and G. L. Curran (1996). "Permeability at the blood-brain and blood-nerve barriers of the neurotrophic factors: NGF, CNTF, NT-3, BDNF." *Brain research. Molecular brain research* 36(2): 280-286.

Raso, V. V., C. H. Barbieri, et al. (2005). "Can therapeutic ultrasound influence the regeneration of peripheral nerves?" *Journal of neuroscience methods* 142(2): 185-192.

Ren, C., J. M. Li, et al. (2010). "LIPUS enhance elongation of neurites in rat cortical neurons through inhibition of GSK-3beta." *Biomedical and environmental sciences : BES* 23(3): 244-249.

Rochkind, S., L. Astachov, et al. (2004). "Further development of reconstructive and cell tissue-engineering technology for treatment of complete peripheral nerve injury in rats." *Neurological research* 26(2): 161-166.

Sahenk, Z., H. N. Nagaraja, et al. (2005). "NT-3 promotes nerve regeneration and sensory improvement in CMT1A mouse models and in patients." *Neurology* 65(5): 681-689.

Schmidhammer, R., T. Hausner, et al. (2007). "In peripheral nerve regeneration environment enriched with activity stimulating factors improves functional recovery." *Acta neurochirurgica. Supplement* 100: 161-167.

Schortinghuis, J., A. L. Bronckers, et al. (2005). "Ultrasound to stimulate early bone formation in a distraction gap: a double blind randomised clinical pilot trial in the edentulous mandible." *Archives of oral biology* 50(4): 411-420.

Schumann, D., R. Kujat, et al. (2006). "Treatment of human mesenchymal stem cells with pulsed low intensity ultrasound enhances the chondrogenic phenotype in vitro." *Biorheology* 43(3-4): 431-443.

Sena, K., R. M. Leven, et al. (2005). "Early gene response to low-intensity pulsed ultrasound in rat osteoblastic cells." *Ultrasound in medicine & biology* 31(5): 703-708.

Takayama, T., N. Suzuki, et al. (2007). "Low-intensity pulsed ultrasound stimulates osteogenic differentiation in ROS 17/2.8 cells." *Life sciences* 80(10): 965-971.

Takeuchi, R., A. Ryo, et al. (2008). "Low-intensity pulsed ultrasound activates the phosphatidylinositol 3 kinase/Akt pathway and stimulates the growth of chondrocytes in three-dimensional cultures: a basic science study." *Arthritis research & therapy* 10(4): R77.

Yamamoto, N. and K. Hanamura (2005). "Formation of the thalamocortical projection regulated differentially by BDNF- and NT-3-mediated signaling." *Reviews in the neurosciences* 16(3): 223-231.

Yamauchi, J., Y. Miyamoto, et al. (2005). "Ras activation of a Rac1 exchange factor, Tiam1, mediates neurotrophin-3-induced Schwann cell migration." *Proceedings of the National Academy of Sciences of the United States of America* 102(41): 14889-14894.

Yoshimura, T., Y. Kawano, et al. (2005). "GSK-3beta regulates phosphorylation of CRMP-2 and neuronal polarity." *Cell* 120(1): 137-149.

Zhang, H., X. Lin, et al. (2009). "Effect of low-intensity pulsed ultrasound on the expression of neurotrophin-3 and brain-derived neurotrophic factor in cultured Schwann cells." *Microsurgery* 29(6): 479-485.

Zhang, J. Y., X. G. Luo, et al. (2000). "Endogenous BDNF is required for myelination and regeneration of injured sciatic nerve in rodents." *The European journal of neuroscience* 12(12): 4171-4180.

Zhang, X., Y. Zeng, et al. (2007). "Co-transplantation of neural stem cells and NT-3-overexpressing Schwann cells in transected spinal cord." *Journal of neurotrauma* 24(12): 1863-1877.

Zhou, S., A. Schmelz, et al. (2004). "Molecular mechanisms of low intensity pulsed ultrasound in human skin fibroblasts." *The Journal of biological chemistry* 279(52): 54463-54469.

3

Disinfection of Human Tissues in Orthopedic Surgical Oncology by High Hydrostatic Pressure

Peter Diehl[1], Johannes Schauwecker[2],
Hans Gollwitzer[2] and Wolfram Mittelmeier[1]
[1]*Department of Orthopedic Surgery, University of Rostock*
[2]*Department of Orthopedic Surgery, Technische Universitaet Muenchen*
Germany

1. Introduction

Liquids show significant temperature-dependent compressibility under high hydrostatic pressure (HHP). For instance, the specific volume of water at atmospheric pressure decreases by 12% when exposed to 400 MPa (1). Self-ionization of water is also promoted by HHP lowering the pH phase transition of water is triggered under excessive HHP; at 1,000 MPa water freezes at room temperature, whereas at 207.5 MPa the freezing point can be lowered to –22°C (1). This allows for pressure shift freezing of foods with instant and small ice crystal formation, storage of food at subzero temperatures without ice formation, or fast thawing of frozen food by pressurization, allowing gentle processing of foods or food constituents with minimal structural damage.

HHP may also cause alterations in biological molecules, associated with a change in their conformation towards of forms which occupy smaller volumes. With increasing pressure, the non-covalent bonds of macromolecules such as proteins are affected leading to changes in their quaternary, tertiary or secondary structure. HHP is presumed to influence the conformational state of lipids as well (2,3) whereas nucleic acids have proved pressure-resistant because their secondary structure is mainly stabilized by H-bonds that are almost pressure insensitive (4;5). HHP-induced changes can be reversible, metastable or irreversible, partly depending on the pressure level itself, but also depending on the duration of the pressure treatment, on the temperature during treatment, on the chemical conditions and on other conditions of the surroundings.

The growth of eukaryotic and prokaryotic cells can be prevented to a large degree by a number of preservation techniques, most of which act by killing the cells or by slowing down cellular growth. Concerning food products, heating, freezing, drying, vacuum packing, acidifying or the addition of preservatives are the predominant method. At present, however, major trends have emerged towards the use of procedures such as HHP to deliver food products that are less 'heavily' preserved but still with high assurance of no microbiological contamination (6,7). HHP as a means of preserving food, without the addition of any kind of preservative, has attracted increasing attention (4,6,8), since it has the advantage of leaving covalent molecular bonds intact without impairing flavors, aromas, vitamins and other pharmacologically active molecules (9).

In the medical field, HHP technology is now in preclinical testing with the aim of inactivating both pathological microorganisms and tumor cells in resected tissue segments, such as bone, cartilage and tendon *ex vivo* (10-15). This is a promising clinically relevant approach, especially with respect to rapid killing of tumor cells in bone and the subsequent possibility of re-implantation of the once tumor-bearing bone segment back into the patient.

2. HHP and orthopaedic surgery

In orthopedic surgery, restoration of bone defects caused by malignant solid tumors is achieved by several methods of treatment such as extracorporeal irradiation or autoclaving the affected bone segment, as an alternative approach to synthetic limb reconstruction (16-19). In contrast, irradiation or autoclaving of osteochondral segments or tendons may lead to severe alteration of their biomechanical and biological properties, a major concern regarding this type of approach (20-22).

A new technology, the administration of short-term HHP to the resected bone segment immediately after surgery, now offers an alternative to the conventional ways of treating tumor-affected bone. At the high pressure value of 600 MPa applied, the biomechanical properties of bones, tendons and cartilage remain unchanged (10-14). Under these conditions, normal eukaryotic cells, and also malignant cells are irreversibly damaged and outgrowth of cells from tumor-afflicted bone and cartilage segments is efficiently blocked (14,15,23).

With regard to the biological properties of treated bone, cartilage or tendon, no obvious changes in the adhesive or growth promoting properties of the extracellular matrix proteins after HHP treatment of the bone were observed (11), and successful revitalization of HHP-treated bone segments *in vitro* was observed. Also, no enhanced activity of proteases, which might be released after HHP treatment of resected human bone tumor and provoke autolytic bone resorption, could be detected (24). This report reviews the basics and technical potential of HHP in orthopaedic surgery and sheds light on the prospects of HHP for the treatment of neoplastic bone and infected bone tissue, cartilage and tendon.

3. HHP-device and treatment of affected bone, cartilage, or tendon

The HHP system (Record Maschinenbau, Koenigsee, Germany) consists of a high pressure autoclave, a pressure generation unit, a temperature and pressure control unit and a material handling unit (Figure 1a). HHP treatment of the tissue samples is accomplished by a pressure-transferring medium, usually water, thus allowing uniform and instantaneous transmission of pressure to the biological sample. To treat infected or tumor-afflicted bone, tendon or cartilage, larger specimens are placed into polyethylene bags and sealed by vacuum-packaging (Komet, Plochingen, Germany) (Figure 2). Sealing in buffer is required to assure uniform and instantaneous pressure transmission throughout the biological sample and to prevent contamination while in contact with the pressure medium. In case of smaller specimens, instead of plastic bags, tissue specimens are placed into 15 ml flexible Falcon tubes (Becton-Dickinson, Heidelberg, Germany) (Figure 3a) or 2 ml Nalgene cryogenic vials (Thermo Fisher Scientific, Wiesbaden, Germany) (Figure 3b). The vials filled with Ringer buffer are carefully capped avoiding air bubbles and then sealed tightly with parafilm (American National Can GmbH, Gelsenkirchen, Germany).

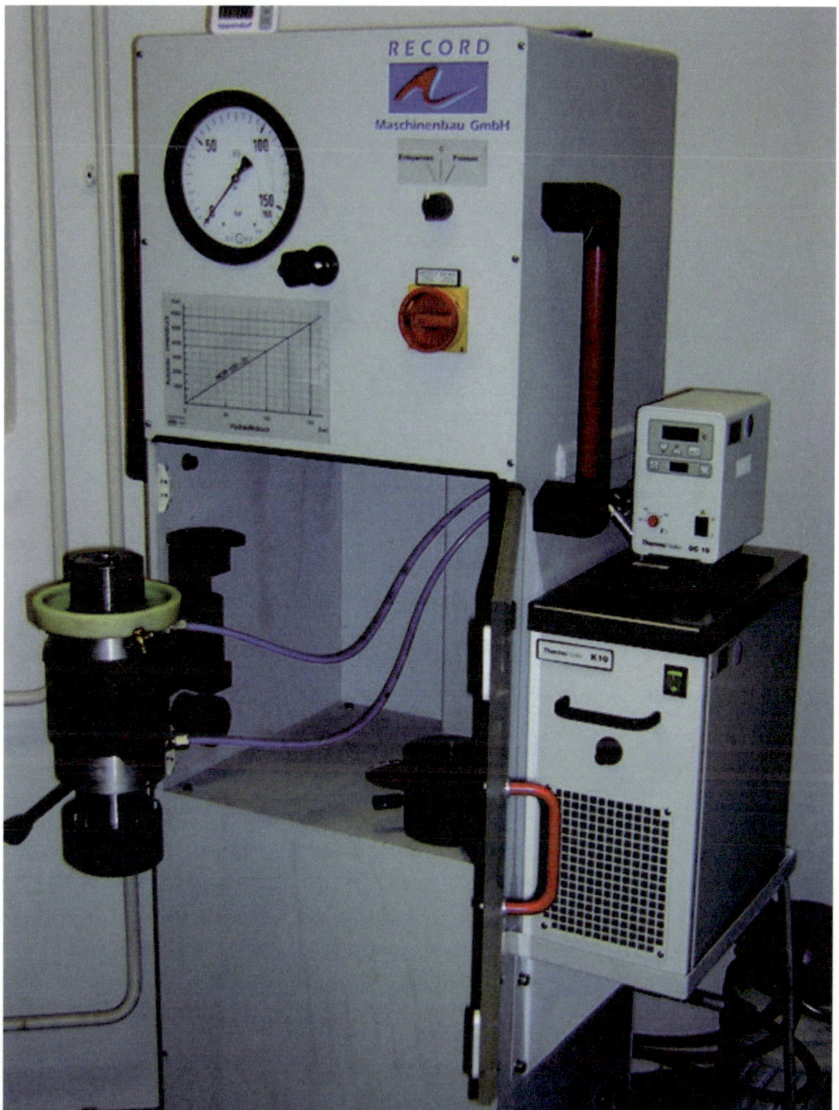

Fig. 1a. High hydrostatic pressure device (Record Maschinenbau, Koenigsee, Germany).

The bags/vials are placed into the central cavity of a water-filled pressure chamber (100 ml) of a custom-made HHP device (Figure 1b). The water is mixed 1:1 with ethylene-glycol to suppress corrosion of the pressure chamber. The temperature of the pressure chamber can

Fig. 1b. The core of the autoclave chamber is made of an amagnetic stainless steel into which a large cylindrical hole has been bored in order to receive the specimen. A metal-on-metal sealing system provides an excellent leak-free closure with minimal mechanical wear. The pressure is transmitted by means of a hydraulic ram and monitored by a pressure gauge. Incubation temperature is monitored by a thermocouple and thermostatic control is ensured by the circulation of water through rubber tubing. On the control panel pressure settings can be adjusted up to 600 MPa).

be adjusted from 0 - 50 °C by a temperature control unit (Thermo Fisher Scientific, Karlsruhe, Germany) (Figure 4). The temperature should be kept constant at any given level since adiabatic compression of water increases the temperature 3 °C per 100 MPa. Pressure levels up to 600 MPa are adjusted manually with a compression / decompression rate of 100-300 MPa/min. The tissue specimens are held under pressure for a defined length of time (plateau phase), then, within a few seconds, pressure is returned to normal.

Fig. 2. Vacuum sealing device (Komet, Plochingen, Germany).
Specimens in polyethylene bags are placed into the vacuum chamber, positioning the open section onto the sealing bar. Once a vacuum is formed a heated wire on the bar seals the plastic bag. On the control board vacuum and sealing settings can be adjusted individually.

Exposure of tendons and ligaments to HHP (300 and 600 MPa; 10 min, 20 °C) did not significantly change their biomechanical features, their Young`s modulus or tensile strength, indicating retention of functional properties after HHP-sterilization (12). Retention of biomechanical properties of tissues after HHP is mainly based on the fact that HHP does not affect covalent molecular bonds, leaving parts of the molecule unchanged whereas exposure to chemicals or high temperature often unfold macromolecules irreversibly (25-27).

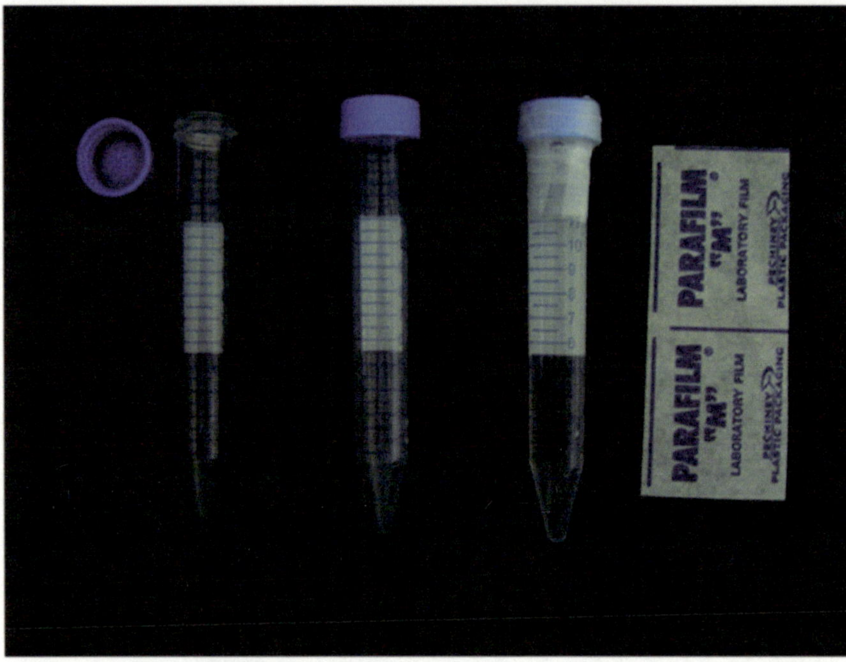

Fig. 3a. Tissue specimens are placed into flexible 15 ml BD Falcon™ conical tubes (Becton-Dickinson, Heidelberg, Germany) or Figure 2

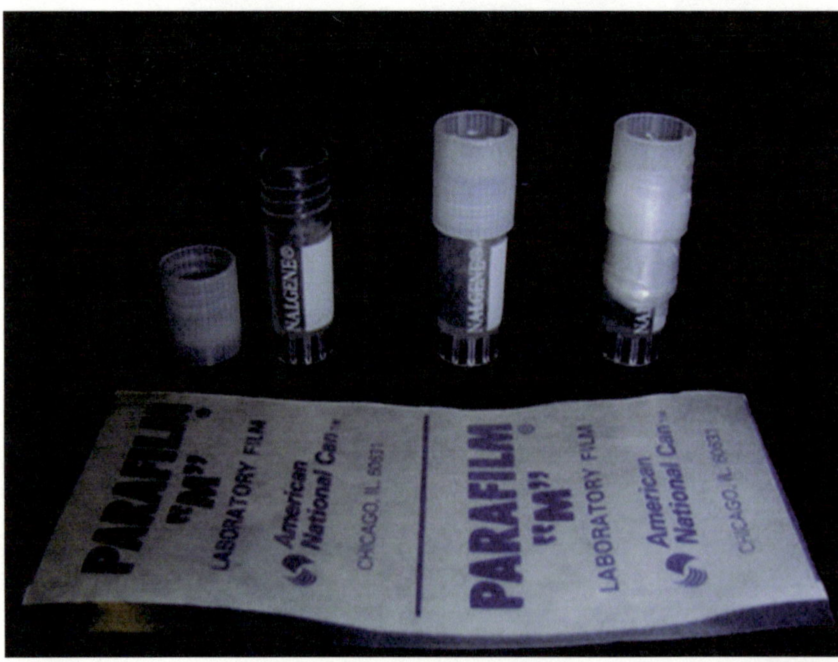

Fig. 3b. 2 ml Nalgene cryogenic vials (Thermo Fisher Scientific, Wiesbaden, Germany) (3b), filled with Ringer buffer and tightly sealed (e. g. with parafilm, American National Can GmbH, Gelsenkirchen, Germany).

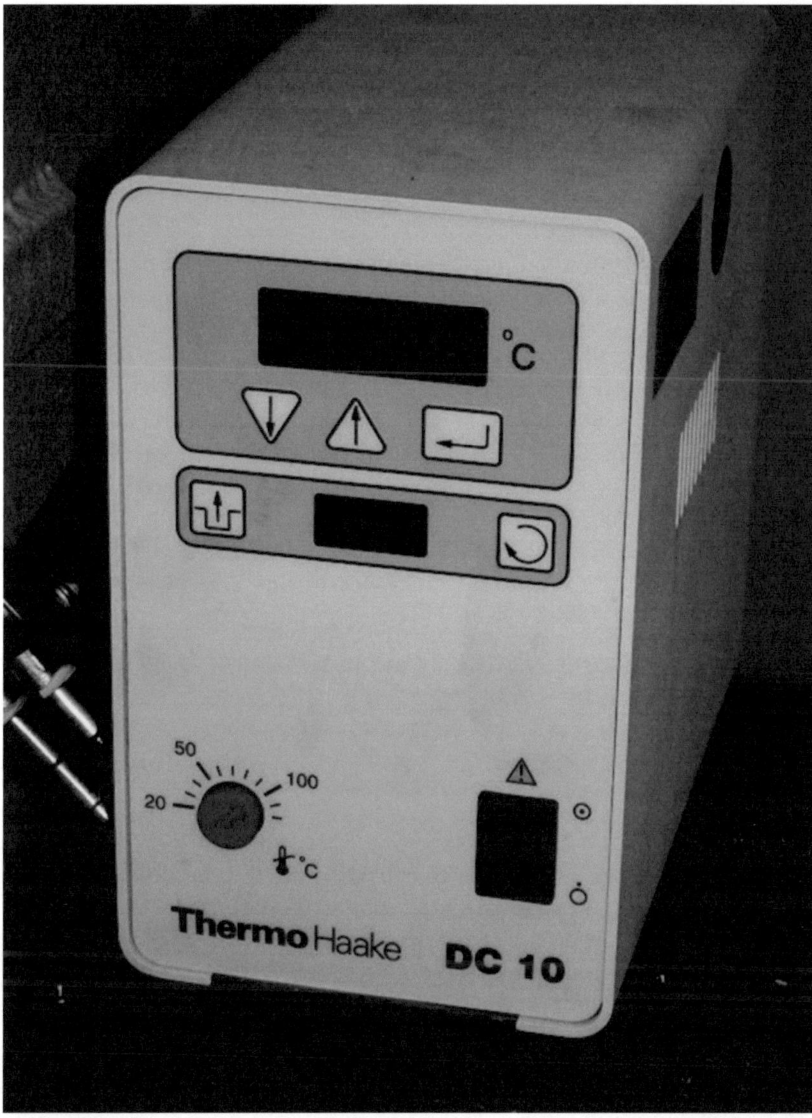

Fig. 4. Temperature control unit (Thermo Fisher Scientific, Karlsruhe, Germany). Open bath circulator with digital settings, bath vessel and bath bridge made of stainless steel and temperature resistant polymer; electronic control using a digital sensor for predetermined reference value, max. temperature 100 °C.

As well as tendons, we have also investigated the biomechanical properties of freshly resected human cortical and trabecular bone specimens or cartilage and menisci exposed to HHP as high as 600 MPa (10 min, 20 °C) (13,28). Under these conditions, no significant alterations relating to the stiffness and relaxation behavior of the osteochondral segments were observed. Unfortunately, inactivation of clinically important bacteria, for instance those present in osteomyelitis was not achieved under these conditions (29,30) although in foods vegetative bacteria, yeasts, and moulds are generally sensitive to pressures of 600 MPa (7).

pre-clinical application of HHP in studies	clinical HHP application	no application in
disinfection of tumor afflicted bone, cartilage, tendons	inactivation of different virus (e.g. HIV) in blood samples	disinfection of inserts of Prosthesis
	food preservation	disinfection of infected tissue

Table 1. The table presents different fields of application of Hygh Hydostatic Pressure. Save inactivation of bacteria in all kind of materials is not possible at the moment.

HHP has also been employed to investigate pressure-related *in vivo*-effects on chondrocytes since hydrostatic pressure is a significant component of the mechanical loading environment within articular cartilage. Chondrocytes within cartilage of diarthrotic joints experience hydrostatic pressure levels of 0.1-20 MPa (31). In *ex vivo* investigations therefore intermittent high pressure of 10 MPa was applied to investigate mechanisms mediating the response of chondrocytes to joint motion and loading (30,31). Under these conditions, a decreased release of matrix metalloproreinases (MMP)-2, tissue inhibitor of matrix metalloproteinase (TIMP)-1 and interleukin-6 by osteoarthritic chondrocytes was observed, suggesting that pressure influences cartilage stability *in vivo* (32).

We have observed that, regarding bone, exposure of normal cells (e.g. osteoblasts) and tumor cells (e.g. osteo-, chondro- and fibrosarcoma cells) to elevated hydrostatic pressure led to irreversibly damaged, non-viable cells, even after short-term exposure to 350 MPa (14,15,23). Under these conditions, eukaryotic cells experience irreversible destruction and permeabilization of cell membranes by HHP causing cell death (33).

Interestingly, suspended tumor cells were more resistant to HHP than adherent tumor cells, yet, normal bone and tissue cells such as fibroblasts and osteoblasts were less resistant to HHP than tumor cells (14,23). We also observed that at 300 MPa *ex vivo* outgrowth of normal or tumor cells from bone ceased concomitant with impairment of the bone-associated cells (15). This finding points to rapid killing of bone-associated tumor cells, potentially allowing re-implantation of the once tumor-bearing bone segment back into the patient.

Looking at other types of cells, Dibb *et al.* investigated the effects of HHP on normal and neoplastic rat cells in culture in the range 0.1 to 150 MPa (34). Morphological changes characterized by cell rounding were observed in secondary fetal brain cells and fibroblasts at about 70 MPa, whereas in the neoplastic neurogenic cell lines tested similar changes occurred at around 100 MPa, again demonstrating that malignant cells may be more resistant to HHP than their normal counterparts. Similar findings were reported by Yamaguchi *et al.* for Ehrlich ascites tumor cells demonstrating that these tumor cells stopped *in vivo* proliferation at HHP above 130 MPa (35).

4. Effect of pressure on extracellular matrix proteins and enzymes

Little is known on the change of biological functions of proteins or other constituents of bone, cartilage, or tendon after exposure to HHP. Our own studies have demonstrated that the extracellular matrix proteins fibronectin, vitronectin and collagen-I present in the bone matrix have not deteriorated after HHP-treatment up to 600 MPa (10 min, room temperature) with respect to cell proliferation, spreading and adherence of human osteoblast-like cells and human osteosarcoma cells (Saos-2) (11). These data encourage further exploration of the potential of HHP to sterilize tumor-affected bone segments prior to re-implantation, since during such treatment eukaryotic bone cells including tumor cells would be irreversibly impaired, while the bone's biomechanical properties and the biological properties of the extracellular matrix proteins fibronectin, vitronectin, and collagen-I would be preserved (11).

HHP causes a stress response in many types of mammalian cells, including chondrocytes and bone tumor cells (36). Further to this, Kopakkala-Tani *et al.* investigated whether some of the well known transduction pathways are activated in human chondrosarcoma cells under stress by exposure to moderate HHP of 15-30 MPa and demonstrated an increased level of active, phosphorylated forms of the extracellular signal-related kinase ERK and phosphoinositide 3-kinase under these pressure conditions (37).

HHP may not only exert an effect on tumor and normal cells present in the bone, but also on the tumor-associated proteases released by these cells, which are conductive to tumor bone turnover. At a pressure level of 600 MPa the latent activity of the inactive zymogens prothrombin, plasminogen, pro-uPA and trypsinogen, in addition to the proteolytically active forms thrombin, plasmin, HMW-uPA, and trypsin was minimally affected by HHP (24). The variation seen between different enzymes is probably due to differences in molecular structures and the resulting modifications after HHP treatment (24). It is worthwhile to note that at this pressure level normal bone cells and tumor cells are irreversibly impaired. Additionally, HHP also influences the activity of other enzymes. With that in mind, Masson *et al.* reviewed HHP technology and its potential applications in medicine and pharmaceutical science (9). The authors explained that HHP may affect both the activity and specificity of enzymes and that HHP is used for the engineering of proteins to allow enzyme-catalyzed synthesis of fine chemicals and pharmaceuticals and the production of modified proteins of medical or pharmaceutical interest. Such reactions can be used for food functionalization and for producing "nutraceuticals" to be used in complementary therapy (38). Pressure processing was found to be efficient in reducing the allergenic activity of food (39).

In general, pressures above 300 MPa cause irreversible protein denaturation at room temperature, whereas lower pressures may result in reversible changes in protein structure. The effects of HHP on enzymes have been divided into two classes: moderate pressure values of 100–200 MPa which activate monomeric enzymes and elevated pressures usually inducing enzyme inactivation (1). Investigations of the impact of moderate HHP up to 200 MPa on alpha-amylase have shown a pressure-dependent stabilization of the enzyme against temperature-induced inactivation (3,40). Interestingly, for some proteases, proteolysis enhancement through HHP (up to 400 MPa) depended on substrate changes and not on changes of the enzyme, as investigated for chymotrypsin in the hydrolysis of beta-lactoglobulin (41).

5. Effect of HHP-treatment on viability of microorganisms in bone

So far, the effect of neoantigens generated during HHP-treatment of bone, cartilage and tendon on the host after re-implantation has not been elucidated and is at present subject to preclinical animal experiments. In spite of that, such physically modified proteins may be new innovative tools in the development of vaccines by making use of the changed immunogenicity of pressure-treated proteins or pressure-killed bacteria, viruses or normal and tumor cells (39,42-44).

Also of importance for HHP-teatment of bone is the fact that viruses are very sensitive to HHP, being inactivated at pressures as low as 100 to 300 MPa. Inactivation of numerous viruses such as herpes viruses, rotaviruses, influenza, picornaviruses as well as immunodeficiency viruses by pressure treatment has been successful in blood (45,46). The use of high pressure in decreasing virus concentration in the blood of patients suffering severe virus infections by *ex vivo* pressure treatment of blood has been proposed (47), but studies on HHP inactivation of viruses present in bone, cartilage or tendon have not been reported yet.

Likewise, different procedures are available to inactivate bacteria and fungi, including their spores, in human bone transplants (48). The most efficient methods of inactivation are gamma irradiation and thermal inactivation as well as chemical sterilization methods such as the peracetic acid-ethanol treatment of bone (49). The direct effect of HHP to achieve killing of vegetative bacterial, yeast and mould cells, has been documented as well (50,51), although much higher pressure values of 500 – 700 MPa are needed than for the inactivation of viruses (52,53). Interestingly, Gram-positive bacteria are more resistant to HHP than Gram-negative bacteria (54). A major advantage of HHP processing over gamma irradiation, thermal inactivation or the use of peracetic acid-ethanol treatment is that it preserves the initial mechanical properties of the bone, cartilage and tendon, a prerequisite for re-implantation of the *ex vivo*-treated tissues.

6. Conclusion

HHP technology has found broad application in the food industry, for instance in activating vegetative microorganisms in meat products, milk, juice, etc.

While viruses and bacteria can be inactivated by moderate to high HHP, outgrowth of tumor cells from tumor-afflicted bone and cartilage segments can be efficiently blocked by extracorporeal HHP, while leaving their biomechanical and key biological properties intact.

These findings raise the hope that HHP can eventually be used in orthopaedic surgery as an alternative technique over other established physical or chemical methods of sterilizing resected bone, cartilage or tendon in order to kill viruses, bacteria and cancer cells to allow autologous re-implantation. Still, before that goal is reached, further pre-clinical studies are required.

7. Acknowledgments

This work was supported in part by the Bavarian Research Foundation (Bayerische Forschungsstiftung, Bayerischer Forschungsverbund for Tissue Engineering and Rapid Prototyping TE3) and by the German Research Association (DFG, SPP 1100).

8. References

[1] Knorr D: Novel approaches in food-processing technology: new technologies for preserving foods and modifying function. Curr Opin Biotechnol *10*: 485-491, 1999.

[2] Abe F, Kato C and Horikoshi K: Pressure-regulated metabolism in microorganisms. Trends Microbiol *7*: 447-453, 1999.
[3] Yamaguchi T, Kawamura H, Kimoto E and Tanaka M: Effects of temperature and pH on hemoglobin release from hydrostatic pressure-treated erythrocytes. J Biochem (Tokyo) *106*: 1080-1085, 1989.
[4] Hoover DG, Metrick C, Papineau AM, Farkas DF, and Knorr D: Biological effects of high hydrostatic-pressure on food microorganisms. Food Tech *43*: 99-107, 1989.
[5] Hoover DG: Pressure effects on biological-systems. Food Tech *47*: 150-155, 1993.
[6] Gould GW: Biodeterioration of foods and an overview of preservation in the food and dairy industries. Int Biodeter Biodeg *36*: 267-277, 1995.
[7] Gould GW: Preservation: past, present and future. Br Med Bull *56*: 84-96, 2000.
[8] Cheftel JC: Review: High-pressure, microbial inactivation and food preservation. Food Sci Tech Int *1*: 75-90, 1995.
[9] Masson P, Tonello C, and Balny C: High-Pressure Biotechnology in Medicine and Pharmaceutical Science. J Biomed Biotechnol *1*: 85-88, 2001.
[10] Naal FD, Schauwecker J, Steinhauser E, Milz S, KnochF, Mittelmeier W and Diehl P: Biomechanical and immunohistochemical properties of meniscal cartilage after high hydrostatic pressure treatment. J Biomed Mater Res B Appl Biomater 2008. in press
[11] Diehl P, Schmitt M, Schauwecker J, Eichelberg K, Gollwitzer H, Gradinger R, Goebel M, Preissner KT, Mittelmeier W and Magdolen U: Effect of high hydrostatic pressure on biological properties of extracellular bone matrix proteins. Int J Mol Med *16*: 285-289, 2005.
[12] Diehl P, Steinhauser E, Gollwitzer H, Heister C, Schauwecker J, Milz S, Mittelmeier W and Schmitt M: Biomechanical and immunohistochemical analysis of high hydrostatic pressure-treated Achilles tendons. J Orthop Sci *11*: 380-385, 2006.
[13] Diehl P, Naal FD, Schauwecker J, Steinhauser E, Milz S, Gollwitzer H and Mittelmeier W: [Biomechanical properties of articular cartilage after high hydrostatic pressure treatment]. Biomed Tech (Berl) *51*: 8-14, 2006.
[14] Naal FD, Mengele K, Schauwecker J, Gollwitzer H, Gerdesmeyer L, Reuning U, Mittelmeier W, Gradinger R, Schmitt M and Diehl P: High hydrostatic pressure-induced cell death in human chondrocytes and chondrosarcoma cells. Anticancer Res *25*: 1977-1982, 2005.
[15] Schauwecker J, Wirthmann L, Schmitt M, Tuebel J, Magdolen U, Gradinger R, Mittelmeier W and Diehl P: Effect of extracorporeal high hydrostatic pressure on cellular outgrowth from tumor-afflicted bone. Anticancer Res *26*: 85-89, 2006.
[16] Araki N, Myoui A, Kuratsu S, Hashimoto N, Inoue T, Kudawara I, Ueda T, Yoshikawa H, Masaki N and Uchida A: Intraoperative extracorporeal autogenous irradiated bone grafts in tumor surgery. Clin Orthop 368: 196-206, 1999.
[17] Bohm P, Springfeld R and Springer H: Re-implantation of autoclaved bone segments in musculoskeletal tumor surgery. Clinical experience in 9 patients followed for 1.1-8.4 years and review of the literature. Arch Orthop Trauma Surg *118*: 57-65, 1998.
[18] Bohm P, Fritz J, Thiede S and Budach W: Reimplantation of extracorporeal irradiated bone segments in musculoskeletal tumor surgery: clinical experience in eight patients and review of the literature. Langenbecks Arch Surg *387*: 355-365, 2003.

[19] Chen WM, Chen TH, Huang CK, Chiang CC and Lo WH: Treatment of malignant bone tumours by extracorporeally irradiated autograft-prosthetic composite arthroplasty. J Bone Joint Surg Br *84*: 1156-1161, 2002.

[20] Knaepler H, Ascherl R, von Garrel T and Gotzen L: [Guidelines for allogeneic bone transplantation]. Chirurg *62*: 763-765, 1991.

[21] Sys G, Uyttendaele D, Poffyn B, Verdonk R and Verstraete L: Extracorporeally irradiated autografts in pelvic reconstruction after malignant tumour resection. Int Orthop *26*: 174-178, 2002.

[22] Urist MR and Hernandez A: Excitation transfer in bone. Deleterious effects of cobalt 60 radiation-sterilization of bank bone. Arch Surg *109*: 586-593, 1974.

[23] Diehl P, Schmitt M, Blumelhuber G, Frey B, van Laak S, Fischer S, Muehlenweg B, Meyer-Pittroff R, Gollwitzer H and Mittelmeier W: Induction of tumor cell death by high hydrostatic pressure as a novel supporting technique in orthopedic surgery. Oncol Rep *10*: 1851-1855, 2003.

[24] Diehl P, Schauwecker J, Eichelberg K, Gollwitzer H, Magdolen U, Gerdesmeyer L, Mittelmeier W and Schmitt M: Quantitative analysis of the impact of short-time high hydrostatic pressure on bone tumor-associated proteases. Int J Mol Med *19*: 667-673, 2007.

[25] Balny C and Masson P: Effects of high-pressure on proteins. Food Rev Int *9*: 611-628, 1993.

[26] Heremans K and Smeller L: Protein structure and dynamics at high pressure. Biochim Biophys Acta *1386*: 353-370, 1998.

[27] Knorr D, Heinz V and Buckow R: High pressure application for food biopolymers. Biochim Biophys Acta *1764*: 619-631, 2006.

[28] Steinhauser E, Diehl P, Hadaller M, Schauwecker J, Busch R, Gradinger R and Mittelmeier W: Biomechanical investigation of the effect of high hydrostatic pressure treatment on the mechanical properties of human bone. J Biomed Mater Res B Appl Biomater *76*: 130-135, 2006.

[29] Weber P, Diehl P, Hofmann GO, Miethke T, Schmitt-Sody M, Mittelmeier W, Gradinger R und Gollwitzer H: Extracorporeal high hydrostatic pressure as a new technology for the disinfection of infected bone specimens. Biomed Tech (Berl) 53: 190-198, 2008.

[30] Ikenoue T, Trindade MC, Lee MS, Lin EY, Schurman DJ, Goodman SB and Smith RL: Mechanoregulation of human articular chondrocyte aggrecan and type II collagen expression by intermittent hydrostatic pressure *in vitro*. J Orthop Res *21*: 110-116, 2003.

[31] Lee MS, Ikenoue T, Trindade MC, Wong N, Goodman SB, Schurman DJ and Smith RL: Protective effects of intermittent hydrostatic pressure on osteoarthritic chondrocytes activated by bacterial endotoxin *in vitro*. J Orthop Res *21*: 117-122, 2003.

[32] Trindade MC, Shida J, Ikenoue T, Lee MS, Lin EY, Yaszay B, Yerby S, Goodman SB, Schurman DJ and Smith RL: Intermittent hydrostatic pressure inhibits matrix metalloproteinase and pro-inflammatory mediator release from human osteoarthritic chondrocytes in vitro. Osteoarthritis Cartilage *12*: 729-735, 2004.

[33] Diehl P, Reuning U, Gollwitzer H, Magdolen U, Gerdesmeyer L, Schauwecker J, Tuebel J, Gradinger R, Mittelmeier W and Schmitt M: Effect of extracorporal high

hydrostatic pressure on tumor cell adherence and viability. Oncol Rep *12*: 369-373, 2004.

[34] Dibb W, Morild E and Laerum OD: Effects of high hydrostatic pressure on normal and neoplastic rat cells in culture. Virchows Arch B Cell Pathol Incl Mol Pathol *38*: 169-176, 1981.

[35] Yamaguchi T, Kuranoshita K, Fujii M, Yokokawa Y, and Kimoto E: Inhibition of the proliferation of *Ehrlich* ascites tumor cells by hydrostatic pressure. Cancer Biochem Biophys *15*: 257-261, 1997.

[36] Kaarniranta K, Elo MA, Sironen RK, Karjalainen HM, Helminen HJ, and Lammi MJ: Stress responses of mammalian cells to high hydrostatic pressure. Biorheology *40*: 87-92, 2003.

[37] Kopakkala-Tani M, Elo MA, Sironen RK, Helminen HJ and Lammi MJ: High hydrostatic pressure induces ERK and PI3 kinase phosphorylation in human HCS-2/8 chondrosarcoma cells. Cell Mol Biol (Noisy -le-grand) *50*: 485-490, 2004.

[38] Butz P and Tauscher B: Recent studies on pressure-induced chemical changes in food constituents. High Pessure Res 11-18, 2000.

[39] Poms RE and Anklam E: Effects of chemical, physical, and technological processes on the nature of food allergens. J AOAC Int *87*: 1466-1474, 2004.

[40] Buckow R, Weiss U, Heinz V and Knorr D: Stability and catalytic activity of alpha-amylase from barley malt at different pressure-temperature conditions. Biotechnol Bioeng *97*:1-11 2007.

[41] Chicon R, Lopez-Fandino R, Quiros A and Belloque J: Changes in chymotrypsin hydrolysis of beta-lactoglobulin A induced by high hydrostatic pressure. J Agric Food Chem *54*: 2333-2341, 2006.

[42] Goldman Y, Peled A and Shinitzky M: Effective elimination of lung metastases induced by tumor cells treated with hydrostatic pressure and N-acetyl-L-cysteine. Cancer Res *60*: 350-358, 2000.

[43] Korn A, Frey B, Sheriff A, Gaipl US, Franz S, Meyer-Pittroff R, Bluemelhuberh G and Herrmann M: High hydrostatic pressure inactivated human tumour cells preserve their immunogenicity. Cell Mol Biol (Noisy -le-grand) *50*: 469-477, 2004.

[44] Perche PY, Cléery C, Bouloy M, Burkhart MF, Masson P and Michel P: Study of inactivation and immunogenicity of Rift valley fever virus type 13 clone treated by high hydrostatic pressure. Am J Trop Med Hyg *57*: 256-257, 1997.

[45] Jurkiewicz E, Villas-Boas M, Silva JL, Weber G, Hunsmann G and Clegg RM: Inactivation of simian immunodeficiency virus by hydrostatic pressure. Proc Natl Acad Sci U S A *92*: 6935-6937, 1995.

[46] Silva JL, Foguel D, Da Poian AT and Prevelige PE: The use of hydrostatic pressure as a tool to study viruses and other macromolecular assemblages. Curr Opin Struct Biol *6*: 166-175, 1996.

[47] Bradley DW, Hess RA, Tao F, Sciaba-Lentz L, Remaley AT, Laugharn JA and Manak M: Pressure cycling technology: a novel approach to virus inactivation in plasma. Transfusion *40*: 193-200, 2000.

[48] Ritz M, Pilet MF, Jugiau F, Rama F, and Federighi M: Inactivation of *Salmonella Typhimurium* and *Listeria* monocytogenes using high-pressure treatments: destruction or sublethal stress? Lett Appl Microbiol *42*: 357-362, 2006.

[49] Pruss A, Baumann B, Seibold M, Kao M, Tintelnot K, von Versen R, Radtke H, Dorner T, Pauli G and Gobel UB: Validation of the sterilization procedure of allogeneic avital bone transplants using peracetic acid-ethanol. Biologicals *29*: 59-66, 2001.

[50] Arroyo G, Sanz PD and Prestamo G: Effect of high pressure on the reduction of microbial populations in vegetables. J Appl Microbiol *82*: 735-742, 1997.

[51] Simpson RK and Gilmour A: The effect of high hydrostatic pressure on *Listeria* monocytogenes in phosphate-buffered saline and model food systems. J Appl Microbiol *83*: 181-188, 1997.

[52] Hayakawa I, Kanno T, Tomita M and Fujio Y: Application of high-pressure for spore inactivation and protein denaturation. Journal of Food Science *59*: 159-163, 1994.

[53] Wuytack EY, Boven S and Michiels CW: Comparative study of pressure-induced germination of *Bacillus subtilis* spores at low and high pressures. Appl Environ Microbiol *64*: 3220-3224, 1998.

[54] Shigehisa T, Ohmori T, Saito A, Taji S and Hayashi R: Effects of high hydrostatic-pressure on characteristics of pork slurries and inactivation of microorganisms associated with meat and meat-products. Int Food Microbiol *12*: 207-216, 1991.

4

Heparan Sulfate Proteoglycan Mimetics Promote Tissue Regeneration: An Overview

Johan van Neck[1], Bastiaan Tuk[1], Denis Barritault[2] and Miao Tong[1]
[1]Department of Plastic and Reconstructive Surgery, Erasmus MC, University Medical Center, Rotterdam,
[2]Laboratoire CRRET, Université Paris Est, Créteil,
[1]The Netherlands
[2]France

1. Introduction

1.1 Normal wound healing

Wound healing is a complex and dynamic process that requires the coordinated completion of a variety of cellular activities, including phagocytosis, chemotaxis, mitogenesis, and synthesis of components of the extracellular matrix. These activities occur in a cascade that correlates with the appearance of multiple cell types and is regulated by soluble mediators such as growth factors and cytokines. In the wound healing process, three phases can be recognized: hemostasis and inflammation, proliferation, and tissue remodeling. These three phases are distinct but overlap in time (Singer and Clark 1999; Diegelmann and Evans 2004; Broughton, Janis et al. 2006).

1.2 Hemostasis and inflammation

Hemostasis and inflammation occur immediately after tissue injury. They prevent ongoing blood and fluid loss and establish an immune barrier against invading micro-organisms. Hemostasis is achieved by vasoconstriction and blood clotting. Platelets initiate the clotting cascade, initially by forming a platelet plug. This platelet plug is followed by a fibrin clot, which provides a provisional matrix scaffold for cell migration. Platelets also secrete a variety of growth factors and cytokines such as fibroblast growth factor (FGF), epidermal growth factor (EGF), platelet derived growth factor (PDGF), transforming growth factor beta (TGF-β), vascular endothelial growth factor (VEGF) and epidermal growth factor (EGF). These growth factors are located in the fibrin clot and act as promoters in the wound healing process by recruiting inflammatory cells to the wound site and initiating angiogenesis (Martin 1997).

Once the bleeding is controlled, inflammatory cells migrate into the wound area. This is the start of the inflammatory phase, which is characterized by the sequential infiltration of neutrophils, macrophages, and lymphocytes (Broughton, Janis et al. 2006).

Neutrophils are recruited to the wound site within 24-36 h after wounding. They are attracted by the growth factors released by degranulating platelets and by the products of complement activation and bacteria degradation (Gurtner, Werner et al. 2008). Infiltrating neutrophils phagocytose contaminating bacteria and release pro-inflammatory cytokines to activate local fibroblasts and keratinocytes (Hubner, Brauchle et al. 1996). Within a few days after injury, neutrophils are extruded as eschar or as a result of apoptosis and finally are replaced by macrophages (Witte and Barbul 1997).

Macrophages migrate into the wound within two or four days after injury and become the predominant cell type. Macrophages are derived from blood monocytes and act as the "orchestra conductor" of wound healing (Lawrence and Diegelmann 1994). In the early stages of wound healing, macrophages phagocytose the remaining debris, bacteria, and apoptotic cells, including neutrophils, thus paving the way for the resolution of inflammation (Guo and Dipietro 2010). Macrophages also secrete a battery of cytokines (e.g., tumor necrosis factor alpha (TNF-α), interleukin-1 (IL-1), and IL-6), growth factors (e.g., VEGF, FGF, PDGF,TGF-β, and EGF), and different types of metalloproteinases that degrade the collagen matrix (Henry and Garner 2003). This way macrophages influence cellular recruitment, cellular activation, angiogenesis, fibroplasia and also regulate the synthesis and formation of the provisional extracellular matrix, which serves as a scaffold for dermal regeneration and epidermal proliferation.

Subsequently, T-lymphocytes enter the wound area and peak during the late-proliferative/early remodeling phage. T-lymphocytes likely are involved in controlling the proliferation phase of wound healing. However, their exact role is not completely understood and is a current area of intensive investigation (Broughton, Janis et al. 2006; Guo and Dipietro 2010).

1.3 Proliferation

The proliferation phase follows and partly overlaps the inflammatory phase. The proliferation phase starts on the third day after injury and lasts for about 2-4 weeks. This phase is characterized by epithelial proliferation and migration over a provisional matrix within the wound (reepithelialization), fibroblast migration, and formation of granulation tissue. With the progression of proliferation, the provisional fibrin/fibronectin based provisional matrix is replaced by newly formed granulation tissue (Broughton, Janis et al. 2006; Velnar, Bailey et al. 2009).

1.3.1 Reepithelialization

Reepithelialization of the wound area starts within hours after injury. Epidermal cells at the wound margin undergo a marked phenotypic alteration and begin to migrate into the wound area (Paladini, Takahashi et al. 1996). Migrating epidermal cells dissect under the fibrin clot across the wound, separating the desiccated eschar from viable tissue. Epidermal cells behind the leading migrating edge proliferate, mature and finally restore the barrier function of the epithelium. The stimulus for the migration and proliferation of epidermal cells during reepithelialization has not been clearly determined, but several possibilities are well documented. The absence of neighbor cells at the margin of wound (i.e., the "free edge" effect) may induce both migration and proliferation of epidermal cells. Local release of growth factors and cytokines also stimulate these processes. The initial stimulus for proliferation and migration of epidermal cells includes the action of EGF, TGF-α, IL-1 and

TNF-α, which are released by activated platelets and/or macrophages (Grotendorst, Soma et al. 1989; Lawrence and Diegelmann 1994). Keratinocyte growth factors (KGFs) and IL-6, which are released by fibroblasts, play a role in attracting neighboring keratinocytes to migrate, proliferate, and differentiate into epithelium (Smola, Thiekotter et al. 1993; Xia, Zhao et al. 1999).

1.3.2 Fibroblast migration

Fibroblast migration occurs two to four days after injury. Fibroblasts are attracted to the wound area by a number of factors, such as PDGF and TGF-β (Goldman 2004), and dominate the wound cell population in the first week. Within the wound area, fibroblasts proliferate and produce multiple structural molecules, including fibrin, fibronectin, glycosaminoglycans (GAGs), later followed by collagen (Witte and Barbul 1997; Robson, Steed et al. 2001; Ramasastry 2005). Together, these components construct the fibrin/fibronectin based provisional matrix (Clark, Lanigan et al. 1982), which contributes to the formation of granulation tissue.

1.3.3 Granulation tissue formation

Granulation tissue formation starts three to five days after injury and is characterized by angiogenesis. The numerous angiogenic factors that are secreted during the hemostatic phase, such as FGF, VEGF, TGF-β, and PDGF, promote angiogenesis (Servold 1991). Four steps can be recognized in this process: (1) proteolytic degradation of the basement membrane of the parent vessels, allowing the formation of "capillary sprouts"; (2) migration of the endothelial cells towards the angiogenic stimulus; (3) proliferation; (4) maturation and remodeling of endothelial cells into capillary tubes (Velnar, Bailey et al. 2009). Capillary sprouts invade the fibrin/fibronectin based provisional matrix within a few days and organize into a dense microvascular network. This is so called vascularized stroma, together with macrophages and proliferating fibroblasts, constitute the acute granulation tissue that replaces the fibrin/fibronectin based provisional matrix (Witte and Barbul 1997; Baum and Arpey 2005). With collagen accumulation, angiogenesis ceases and the density of the microvascular network diminishes. When homeostasis between collagen synthesis and degradation is achieved, tissue remodeling begins.

1.4 Tissue remodeling

Tissue remodeling phase is the final phase of wound healing. It starts one week after injury and lasts over a year or more. The main feature of this phase is the deposition of collagen in an organized network. During this phase, all short term events that were activated after injury cease: most macrophages, endothelial cells, fibroblasts, and myofibroblasts undergo apoptosis or exit from the wound. They leave a mass that consists mostly of collagen and other matrix proteins. Without an increase in collagen content, this largely acellular matrix subsequently is reorganized from a disorganized, mainly type III collagen fibers containing temporary matrix, into a lattice structure which is predominantly composed of type I collagen (Madden and Peacock 1968; Gurtner, Werner et al. 2008). A process that is dependent on collagen synthesis, which in part is the net result of the interaction between matrix metalloproteinases and tissue inhibitors of metalloproteinases (Madlener, Parks et al. 1998). During this phase, the wound progressively continues to increase in tensile strength. Nevertheless, wounds never regain the original strength. At maximal strength, healed wounds are 80% as strong as normal skin (Madden and Peacock 1968).

2. Impaired wound healing

Impaired healing wounds generally failed to progress through the normal stages of wound healing. Impaired healing wounds can be arrested in any of the different healing stages, however, frequently enter a state of pathologic inflammation. As a result, the wounds cannot be repaired in an orderly and timely manner, subsequently resulting in poor anatomical and functional outcome (Lazarus, Cooper et al. 1994). Both acute wounds and chronic wounds can exhibit impaired healing.

2.1 Factors affecting wound healing

Wound healing can be impaired by multiple factors in any of the healing phases. These factors are categorized into local and systemic factors. Local factors are those that directly influence the characteristics of the wound. Systemic factors concern the overall health or disease state of the individual which affects the ability to heal (Table 1). However, these factors often are interrelated so their influences are not mutually exclusive. Single or multiple factors may, therefore, play a role in any one or more individual phases, contributing to the overall outcome of the healing process (Guo and Dipietro 2010).

Local Factors	Systemic Factors
Infection	Age
Tissue maceration	Sex hormones
Foreign bodies	Obesity
Ischemia	Stress
Venous insufficiency	Diseases: diabetes , artery disease, peripheral vascular disease
Desiccation	Immunocompromised conditions: AIDS, radiation therapy, chemotherapy
Necrosis	
Pressure	Congenital healing disorders: epidermolysis bullosa, Ehlers-Danlos syndrome, Marfan's syndrome
Trauma	
Edema	Alcoholism
Local cancer	Smoking
Radiation	Distant cancer
Toxins	Uremia
Iatrogenic factors	Nutritional deficiencies

Table 1. Factors that affect wound healing.

In view of the large variety of factors that can be involved in impaired healing wounds, the types of chronic wounds are numerous. However, the most common chronic wounds are pressure ulcers, diabetic ulcers, and venous ulcers. Together, they constitute approximately 70% of all chronic wounds (Eaglstein and Falanga 1997).

2.2 Pathophysiology of impaired wounds

The differences in physiology and healing dynamics between acute and chronic wounds are numerous (Figure 1) (Schultz and Wysocki 2009). Excessive inflammation and abnormalities in cell-extracellular matrix interaction are considered important mechanisms responsible for the failure of chronic wounds to heal (Eming, Krieg et al. 2007; Menke, Ward et al. 2007; Schultz and Wysocki 2009).

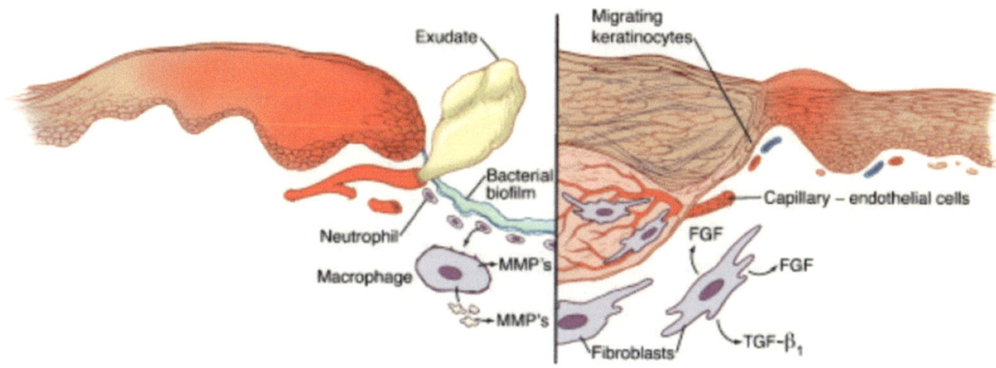

Chronic wound (arrested simultaneous repair)

No clot formation
Increased/excessive inflammation
Leaky capillaries and wound fluid formation
Extracellular matrix / granulation tissue defects
  Impaired/defective/degraded ECM
  Protease-inhibitor imbalance
Impaired cell-ECM interactions
Round fibrotic edges
Impaired keratinocyte migration
Excessive scar formation
Bacterial colonization/infection
Biofilm present

Acute wound (sequential orderly repair)

Clot formation
Minimal inflammation
Migrating capillaries
No/minimal ECM / granulation tissue defects
  Active ECM synthesis
  Protease-inhibitor balance
Active cell-ECM interactions
Flat edges
Active keratinocyte migration
Minimal scar formation
No/minimal bacteria
Biofilm absent

Fig. 1. Comparison of a chronic wound in which repair is arrested and an acute wound in which repair proceeds in an orderly, sequential fashion. Differences between these wounds are seen in clot formation, inflammation, capillary migration, granulation tissue, extracellular matrix, keratinocyte migration, scar formation, bacterial colonization/infection, and biofilm formation (Schultz and Wysocki 2009).

Impaired wounds exhibit an out-of-control prolonged inflammatory response that is self-sustaining (Menke, Ward et al. 2007). An over-abundant neutrophil infiltration is responsible for this chronic inflammation (Diegelmann 2003; Diegelmann and Evans 2004). Neutrophils release significant amounts of enzymes such as metalloproteinases (Yager, Zhang et al. 1996; Nwomeh, Liang et al. 1998; Nwomeh, Liang et al. 1999; Lobmann, Ambrosch et al. 2002), which are not balanced by their respective inhibitors. As a result, the balance between matrix degradation and synthesis shifts towards degradation (Bullen, Longaker et al. 1995). In addition, neutrophils release elastase, an enzyme that is capable to destroy growth factors such as PDGF and TGF-β (Yager, Zhang et al. 1996). This prolonged inflammatory environment also contains excessive reactive oxygen species that further damage cells, growth factors and healing tissues (Wenk, Foitzik et al. 2001). Abnormalities in extracellular matrix - growth factor interactions characterize impaired wound healing (Schultz and Wysocki 2009). Impaired wounds are difficult to heal until the delayed inflammation is reduced and the interactions between extracellular matrix and growth factors are restored. Wound treatment strategies, with a focus on regulating these disrupted interactions, may benefit the treatment of impaired wound healing.

3. 'Standard' management options for wounds

The final goal of any wound management is to achieve wound healing. In the eyes of a cell biologist, these strategies might be seen as attempts to assist the injured tissue in recreating an extracellular matrix and cellular content, to enable tissue regeneration. In a somewhat simplistic view, most of the current routine clinical strategies to improve wound healing, therefore, can be classified into one or more of the categories that are depicted in Table 2.

	TREATMENT / STRATEGY	BENEFIT TO WOUND HEALING	LITERATURE (REVIEWS)
1.	Debridement (e.g. surgical, enzymatic, chemical)	Bring the wound edges into viable tissue in order to allow cells to deposit the right extracellular matrix needed for their migration and differentiation	(Attinger and Bulan 2001; Hess and Kirsner 2003)
2.	Moisture management	Create an environment that assists in propagating cell migration into the wound area	(Okan, Woo et al. 2007; Korting, Schollmann et al. 2011)
3.	Exudate management (e.g. by the application of foams, sponges or vacuum therapy)	Draining the wound fluid to clear extracellular matrix degrading enzymes and bacterial toxins	(Vowden and Vowden 2003)
4.	Local infection management	Create a sterile environment by clearing the wound from bacteria and secreted harmful substances, followed by creating a bacterial balance	(Fung, Chang et al. 2003)
5.	Inflammation management	Reduce the severity and duration of the immune reaction, to limit the production of soluble growth factors by immune cells that attract infiltrating, scar tissue producing, fibroblasts	(Nathan 2002; Diegelmann and Evans 2004)
6.	Growth factor management	Replenish the wound area, extracellular matrix and its cellular components with growth factors	(Krishnamoorthy, Morris et al. 2001; Barrientos, Stojadinovic et al. 2008)

Table 2. Cell biological effects of clinical treatment modalities

4. Advanced treatment options for wounds

Healing-arrested (chronic) wounds seriously lower the patient's quality of life and their treatments are extremely resource consuming. In the USA the costs related to chronic wounds are estimated over $25 billion a year (Fan, Tang et al. 2011). In the European Union the costs related to pressure ulcers and venous ulcers are estimated around €20 billion per year. The severity of the problem, likely accompanied by the substantial financial gain that can be envisioned, triggers the development of a great variety of advanced treatment options. Many are described in literature, although mostly with minimal of proof of efficacy. Also many reviews exist on this topic (recent examples are e.g. (Rizzi, Upton et al. 2010; Fan, Tang et al. 2011).

In the context of this chapter, only a limited discussion will follow that covers the following 'state of art' strategies:

1. Gene therapy
2. Platelet rich plasma therapy
3. Stem cell therapy
4. Biological dressings and skin substitutes

4.1 Gene therapy

Gene therapy clearly holds promise in the repair of soft tissue disorders like wounds (reviews on this topic e.g. Branski, Pereira et al. 2007; Eming, Krieg et al. 2007). Skin is easily accessible for genetic manipulations and has cellular constituents with a high turnover that can be an effective target for the transfer of genetic material. Especially a temporal delivery of growth factors by gene transfer may be helpful in transforming chronic wounds into healing wounds.

July 2011, gene therapy trials are listed at www.clinicaltrials.gov concerning patients with diabetes-related, and lower limb ischemia-related ulcers. However, the application of genetic manipulation to the treatment of chronic wounds is still in its infancy.

4.2 Platelet rich plasma therapy

The alpha granules of platelets are rich in growth factors that are considered important for wound healing, amongst them EGF, FGF, PDGF, TGF-β and VEGF. Therefore, the use of platelet rich plasma is considered superior to any single growth factor application. Topical use of platelet rich plasma is described over 2 decades in several case series. E.g. following a weekly application to patients with cutaneous wounds of variable origin, Crovetti et al. (2004) described a complete response in 9 of 24 patients and a partial response in an additional 9 patients (Crovetti, Martinelli et al. 2004). Schade and Roukis (2008) reported beneficial effects of platelet-rich plasma to the healing of split-thickness skin grafts (Schade and Roukis 2008). Kazakos et al. compared the use of platelet rich plasma versus conventional therapy in 59 patients suffering from acute trauma wounds. And reported a beneficial effect on wound size reduction and pain perception over a three week period (Kazakos, Lyras et al. 2009). Steed et al. (1992) performed a randomised controlled trial in 13 diabetic foot patients and observed, over a period of 15 weeks, increased healing in the platelet rich plasma group (Steed, Goslen et al. 1992).

July 2011, platelet rich plasma trials are listed at www.clinicaltrials.gov concerning patients with burns, skin grafts, and leg ulcers.

4.3 Stem cell therapy

Literature on the use of bone marrow-derived stems cells, adipogenic stem cells and cutaneous mesenchymal stem cells, that are reported around the hair follicles, is manifold (reviews on this topic e.g. (Sellheyer and Krahl 2010; Wu, Zhao et al. 2010). At the preclinical level, evidence in wound areas is accumulating regarding the differentiation of bone marrow-derived stem cells into dermal fibroblasts, fibrocytes and endothelial progenitor cells. However, at the clinical level, reports of sufficient quality are scarce. Dash et al. (2009) reported beneficial effects on ulcer healing and ulcer pain of the addition of bone marrow-derived mesenchymal stem cells to a total of 24 patients with lower limb non-healing ulcers that were randomly allocated to the placebo or stem cell therapy group (Dash, Dash et al. 2009). Walther et al. (2011) reported a pilot on the intraarterial administration of bone marrow derived mononuclear cells to patients with critical ischemia in a phase II randomised-start open label study (Walter, Krankenberg et al. 2011). Forty patients enrolled over a period of 3 years and a significantly improved ulcer healing and reduction of rest pain were found in the bone marrow group.

July 2011, over ten stem cell trials are listed at www.clinicaltrials.gov concerning patients with burn, pressure, diabetes-related and lower limb ischemia-related ulcers.

4.4 Biological dressings and skin substitutes

The complexity of the skin: its cellular constitution, extracellular matrix characteristics and the many growth factors involved in maintaining functional skin layers, let alone their involvement in wound healing (Table 3), largely determined research efforts.

Reports on dermal replacement and dermal addition strategies are manifold (recent reviews on this topic e.g. (Rizzi, Upton et al. 2010; Fan, Tang et al. 2011)). Briefly, research efforts to establish novel treatments can be categorized in tissue engineered skin based therapies, growth-factor based therapies, extracellular matrix based therapies and combinations thereof.

Dermal replacements, e.g. to cover large ulcers or burn injuries, are described using autografts, allografts or tissue engineered skin substitutes. Over a dozen of tissue engineered dermal replacements are on the market (Rizzi, Upton et al. 2010).

Dermal addition strategies are described using collagen, chondroitin-6-sulfate and hyaluronic acid.

Growth factor addition into the wound bed aims to reestablish or accelerate the natural healing process of chronic and acute wounds. Predominantly via preclinical research efforts, with currently PDGF and bFGF on the clinical market (Rizzi, Upton et al. 2010).

Hodde and Johnson (2007) elaborate on the role of the extracellular matrix to stimulate, direct and coordinate healing by storing a variety of growth factors at physiological levels (Hodde and Johnson 2007).

Growth Factors	Cells	Acute Wound	Function	Chronic Wound
EGF	Platelets Macrophages Fibroblasts[44,45]	Increased levels[46,47]	Reepithelialization[48]	Decreased levels[51]
FGF-2	Keratinocytes Mast Cells Fibroblasts Endothelial cells Smooth muscle cells Chondrocytes[58,75,76]	Increased levels[79,81]	Granulation tissue formation Reepithelialization Matrix formation and remodeling[277]	Decreased levels[52]
TGF-β	Platelets Keratinocytes Macrophages Lymphocytes Fibroblasts[92,93,96]	Increased levels[98]	Inflammation Granulation tissue formation Reepithelialization Matrix formation and remodeling[81,101,107]	Decreased levels[52]
PDGF	Platelets Keratinocytes Macrophages Endothelial cells Fibroblasts[58,140,141]	Increased levels[144]	Inflammation Granulation tissue formation Reepithelialization Matrix formation and remodeling[141,142,146,153]	Decreased levels[52]
VEGF	Platelets Neutrophils Macrophages Endothelial cells Smooth muscle cells Fibroblasts[69,160-164]	Increased levels[185]	Granulation tissue formation[177,180]	Decreased levels[52]
IL-1	Neutrophils Monocytes Macrophages Keratinocytes[13,60]	Increased levels[242]	Inflammation Reepithelialization[244]	Increased levels[51]
IL-6	Neutrophils Macrophages[245]	Increased levels[245]	Inflammation Reepithelialization[77,78]	Increased levels[245]
TNF-α	Neutrophils Macrophages[60,242]	Increased levels[51]	Inflammation Reepithelialization[51]	Increased levels[51]

Table 3. Major growth factors and cytokines that participate in wound healing with cell types and their respective roles in both acute and chronic wounds are listed (Barrientos, Stojadinovic et al. 2008). Reference numbers refer to their original publication.

4.5 Concerns about biological products based treatment

Although some of the new techniques mentioned above are promising in small-scale trials, only a minority are evidence based. Important issues that need to be addressed:

- Lack of large double blinded randomised controlled preclinical trials.
  The current evidence for many biological products based treatment is generally based on non-randomised prospective trials, small-scale trials, or single case studies.
- Lack of sufficient evidence of efficacy.
  The efficacy of most biological products is currently not approved in large trials.

- Lack of cost-effectiveness.
  As most biological products are expensive the lack of cost-effectiveness studies may limit their widespread application.
- Confirmation of safety.
  Many sources of the biological products are from human or animal tissues, which hold the, theoretical, risk of transmitting infection and diseases.

Both standard and advanced wound healing strategies frequently aim to recreate a bioactive extracellular matrix. In the remainder of this overview, the focus is on the constituents of the extracellular matrix that secures the stability of growth factors in the matrix. Their role in enabling tissue homeostasis and tissue regeneration is discussed.

5. Tissue homeostasis and the extracellular matrix

Proteoglycans consist of a core protein with one or more covalently linked glycosaminoglycan chains. Glycosaminoglycans are long chain, high molecular weight carbohydrates. Some of these are sulfated (chondroitin sulfate, dermatan sulfate and heparan sulfate), other are non-sulfated (hyaluronic acid) (McGrath and Eady, 1997). When combined with water glycosaminoglycans form a gel and contribute to the viscoeleastic properties of connective tissue. In addition to this mechanical role, proteoglycans may also have regulatory roles of which heparan sulfate is a prominent example.

Heparan sulfates are linear polysaccharides with variable degrees of sulfation, N-sulfation and N-acetylation (Dreyfuss, Regatieri et al. 2009) (Tumova, Woods et al. 2000). Heparan sulfates are widely spread throughout the animal kingdom ranging from invertebrates to mammals. In organs and tissues, they are a ubiquitous part of the extracellular matrix since many of the matrix scaffold proteins, such as collagens, fibronectin and laminin, possess heparan sulfate binding sites (Dreyfuss, Regatieri et al. 2009).

A large variety of proteins can bind heparan sulfates. Amongst these are cell surface proteins, extracellular matrix proteins, growth factors, cytokines, chemokines and morphogens. Heparan sulfates protect these proteins from degradation and secure their presence in the extracellular matrix. Due to the ubiquitous nature of heparan sulfates, the large amount of proteins they sequester and their fine-tuning effect in growth factor bioavailability, heparan sulfates participate in many physiological activities (e.g. cell proliferation, -migration, -differentiation and cell - cell interaction). And, therefore, play a prominent role in enabling tissue homeostasis, the process in which the tissues and organs pursue a constant internal environment and cellular composition (Watt and Fujiwara 2011).

However, the tissues response to 'stress', with stress being any form of integrity disturbance such as injury, inflammation, overuse, auto-immune response, is in releasing cleavage enzymes that degrade the proteins and glycosaminoglycans of the extracellular matrix, including heparan sulfates. Through this degradation, the orchestrating role of heparan sulfate in growth factor and cytokine sequestration is lost which ends tissue homeostasis (Barritault, Garcia-Filipe et al. 2010).

In wound healing, particularly in chronic wounds, dysregulation within the extracellular matrix and between cells and the extracellular matrix has gained importance (Cook, Davies et al. 2000) and do therapeutic initiatives to restore the defective extracellular matrix and to reposition heparan sulfates (Agren and Werthen 2007; Gandhi and Mancera 2010).

6. ReGeneraTing Agents

6.1 Structure of RGTA

ReGeneraTing Agents (RGTAs) are synthetic heparan sulfate mimics, resistant to glycanase digestion (Figure 2) (Barbier-Chassefiere, Garcia-Filipe et al. 2009; Ikeda, Charef et al. 2011).

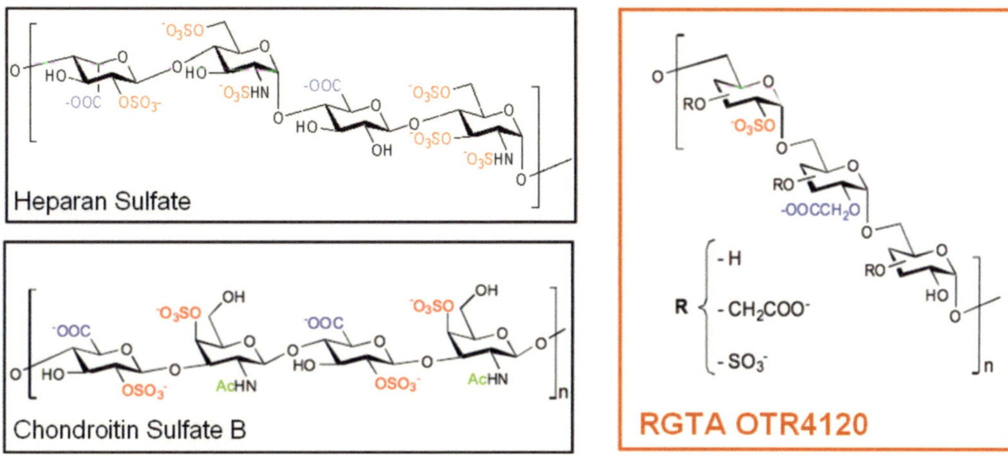

Copyright 2011 by OTR3,

Fig. 2. RGTA OTR4120 is a structural analogue of glycosaminoglycans.

6.2 Working concept of RGTA / RGTA OTR4120

In wound areas, RGTAs can replace heparan sulfates by binding the free heparan binding sites that become available following heparan sulfate degradation. This way, RGTAs can regulate the bioavailability of the large variety of, locally synthesized, heparin binding proteins which allow the cellular tissue components to re-unfold their natural mechanism to achieve wound regeneration.

RGTA OTR4120 is a RGTA member specifically designed to treat chronic wounds and marketed as CACIPLIQ20® (OTR3, Paris, France). The affinity constant of RGTA OTR4120 towards the vacant heparan sulfate binding sites of the extracellular matrix proteins allows a tight binding. This makes a short-term exposure to RGTA OTR4120 sufficient. Once RGTA OTR4120 is in place in the matrix scaffold, the growth factors, cytokines and other heparin binding signaling peptides can be repositioned through RGTA OTR4120 binding in this restored micro-environment. In this way, RGTA OTR4120 is thought to offer a matrix therapy that restores the natural cellular microenvironment. This allows the endogenous signaling of cell communications needed for tissue regeneration to resume their original function thereby halting the self-perpetuating cycles, particularly in impaired healing wounds (Figure 3).

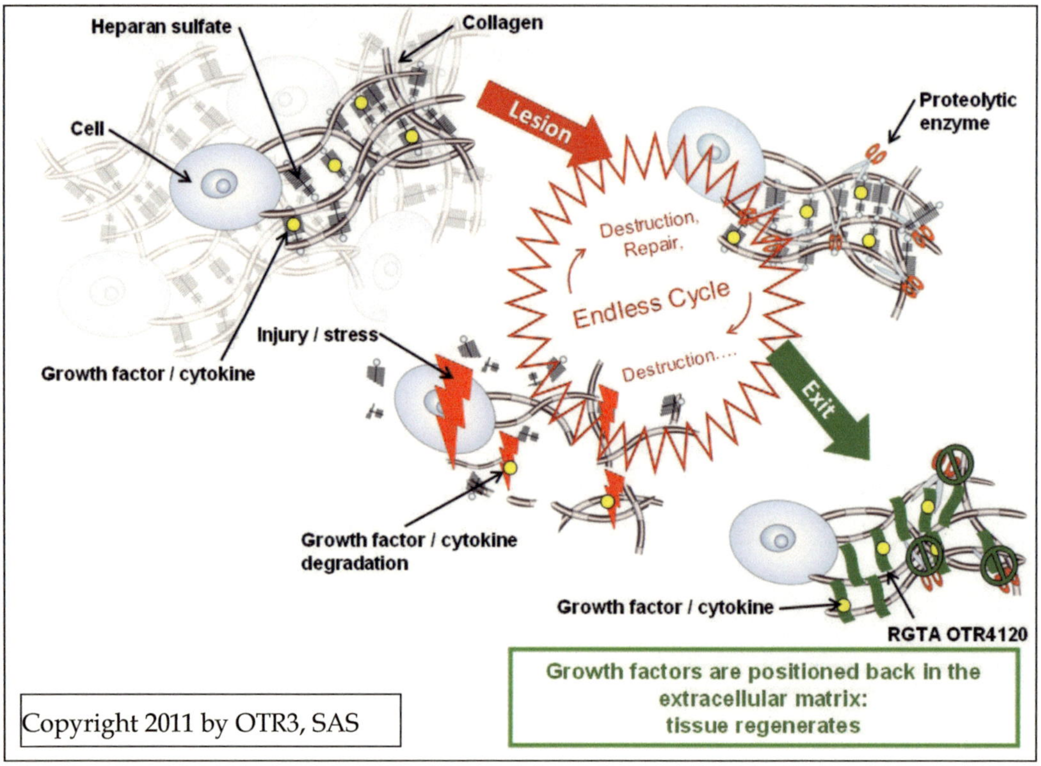

Fig. 3. RGTA OTR4120, mode of action.

6.3 Synthesis, dosing and timing

RGTA OTR4120 is prepared as a 85kD molecular weight fraction from T40 dextran by carboxymethylation and O-sulfonation (Barbier-Chassefiere, 2009; Ikeda, Charef et al. 2011). RGTA OTR4120 was proven completely resistant to digestion with multiple endoglycanases: heparanase, chondroitinase, hyaluronidase, dextranase (Table 4) (Ikeda, Charef et al. 2011).

Following a single i.v. injection in mice at a dose of 5 mg/kg the half life of unbound OTR4120 in plasma was less than 60 minutes. A i.p. bolus injection of 50 mg/kg in this same model created a peak plasma concentration of 88 μg/ml after 90 min (Charef, Papy-Garcia et al. 2010). A study on the acute and subacute toxicity, following i.p. administration, revealed no significant toxicological changes for doses up to 50 mg/kg (Charef, Tulliez et al. 2010). Injected doses used in a preclinical setting routinely are in the range of 1 - 2 mg/kg.

When topically administered on dermal wounds, a bell-shaped dose effect-curve was found for RGTA OTR4120 with an optimal dose of 0.1 mg/ml. A similar dose was found when treating skull defects (Colombier, Lafont et al. 1999) and parodontitis (Lallam-Laroye, Escartin et al. 2006).

RGTA OTR4120 is by itself a non acting molecule that enables the cascade of signals, that propagate wound regeneration, to resume with proper timing. However, the frequency of applications should be timed as the number of free heparan binding sites are limited in wound tissue. RGTA OTR4120 administration every 3 days is proven sufficient to maintain the healing effect in the early phases. Excess RGTA OTR4120 may compete with sites on the matrix-bound RGTA for heparan binding growth factors. In following phases of wound regeneration a weekly administration also proofed to be effective (Tong, Zbinden et al. 2008; Barbier-Chassefiere, Garcia-Filipe et al. 2009). No specific studies are reported to further optimize the timing of RGTA OTR4120 application.

Polysaccharide	Heparanases	Chondroitinase ABC	Hyaluronidase	Dextranase
HM4120	–	–	–	–
Heparin	+++	–	–	–
Heparan sulfate	+++	–	–	–
Chondroitin sulfate A	–	+++	+++	–
Hyaluronic acid	–	+++	+++	–
Dextran	–	–	–	+++

–: Undigested product, starting polysaccharide was recovered. +++: Completely digested polysaccharide.

Table 4. Polysaccharide digestion by endoglycanases. HM4120 = RGTA OTR4120 (Ikeda, Charef et al. 2011)

6.4 Preclinical evidence for dermal RGTA OTR4120 actions

Effects of RGTA OTR4120 administration on tissue regeneration at the preclinical level are numerous and reported in over 70 scientific reports in close to 10 animal species. Dermal effects of RGTA OTR4120 administration are described in necrotic skin ulcers in mice (Barbier-Chassefiere, Garcia-Filipe et al. 2009), incisional dermal wounds in rats (Barritault, Garcia-Filipe et al. 2010), second degree burns in rats (Garcia-Filipe, Barbier-Chassefiere et al. 2007) (Zakine, Barbier et al. 2011), a rat surgical excision model (Tong, Zbinden et al. 2008; Tong, Tuk et al. 2009; Tong, Tuk et al. 2011) and in rat dermal ischemia ulcers (Tong, Tuk et al. 2011).

6.4.1 Necrotic skin ulcers

Barbier-Chassefiere et al. (2009) determined the effects of RGTA OTR4120 in a necrotic skin ulcer model in mice following doxorubicin administration. In the short term, the necrotic surface area was found to be decreased by 40% with an almost 50% reduction in leukocyte count, as a measure for the strength of the inflammatory reaction. RGTA OTR4120 administration increased type I and decreased type III collagen which restoring these values to those found in normal skin (Barbier-Chassefiere, Garcia-Filipe et al. 2009). The results obtained from this ulcer model indicate that RGTA OTR4120 matrix therapy can initiate tissue regeneration. This finding illustrates that matrix contains the proper information to regenerate and confirms the central role for a good-quality extracellular matrix in tissue regeneration (Barbier-Chassefiere, Garcia-Filipe et al. 2009).

6.4.2 Anti scar effects of RGTA

Barritault et al. (2010) described the effects of RGTA OTR4120 administration to a dermal incisional wound model in hairless rats (Barritault, Garcia-Filipe et al. 2010). Topical administration of RGTA OTR4120 to the incision at days 0, 3 and 6 revealed a scar free healing at day 10 whereas in the placebo control scar formation was clearly present (Figure 4).

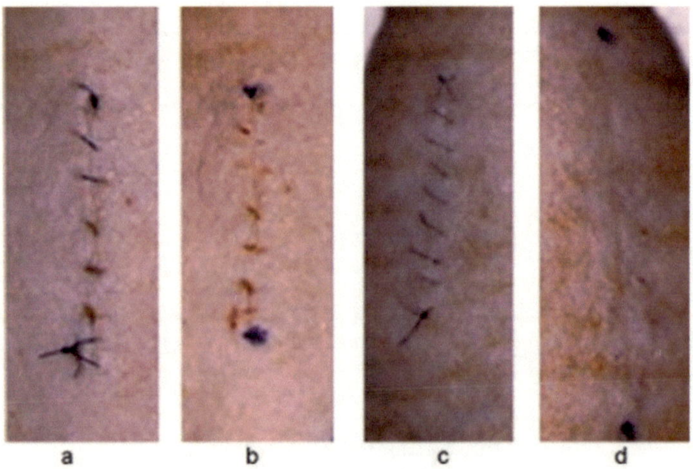

Fig. 4. Macroscopic view of the scar, treated or not treated with RGTA OTR4120 after skin incision. The incision is induced at day 0. RGTA OTR4120 treatment is by topical application (0.1 mg/ml), using a cotton swab, at day 0 before closure by suture and at days 3 and 6. a: untreated skin incision (saline) at day 0; b: untreated skin incision (saline) at day 9; c: RGTA OTR4120 treated skin incision at day 0; d: RGTA OTR4120 treated skin incision at day 9. (Barritault, Garcia-Filipe et al. 2010)

Also severe dermal burns are characterized fibrosis and excessive scarring. On a cellular level aberrant proliferation, inflammation and a changed extracellular matrix architecture are important characteristics for dermal fibrosis. Especially the increased presence of type III collagen is thought to link to the extend of fibrosis (Ulrich, Noah et al. 2002). Ulrich et al. (2003) demonstrated a long term increased type III collagen in fibrous tissue (Ulrich, Noah et al. 2003). In this view, treatments that normalize type III collagen expression without compromising wound healing are of utmost importance. Garcia-Filipe et al. (2007) determined the effects of RGTA OTR4120 administration to a second degree experimental burn on the skin of a hairless rat (Garcia-Filipe, Barbier-Chassefiere et al. 2007). They observed a profoundly improved fibrotic index: this is the ratio of type III collagen over type I collagen. Normalization of the type III collagen / type I collagen ratio also lasted at their final experimental endpoint at 10 months and was caused by a decreased type III collagen production and created a collagen balance that resembled normal skin (Figure 5).

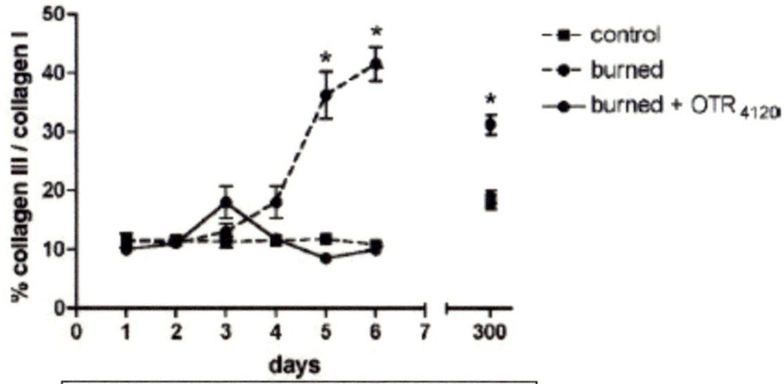

Fig. 5. Effect of OTR4120 on the relative proportions of collagen I and collagen III in control skin and burn sites. Three experimental groups were studied: control (healthy skin from untreated rats), burn sites from untreated rats, and burn sites from rats treated with OTR4120. Figure shows the ratio of collagen III over collagen I (fibrotic index) computed using the data in Figures 1 (A,B). Differences compared with control were evaluated using Student's paired t test; *p values < 0.05 were considered significant (Garcia-Filipe, Barbier-Chassefiere et al. 2007).

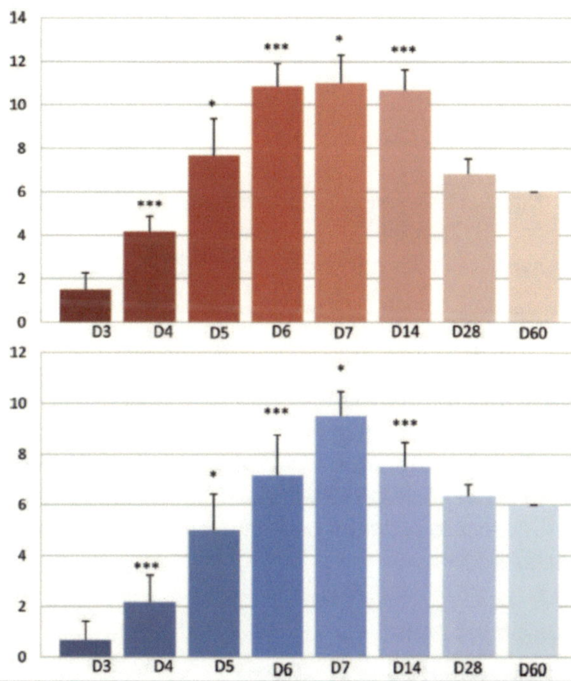

Fig. 6. Evolution of the mean number of keratinocyte layers. RGTA administration increases the mean number of keratinocyte layers between day 3 and day 7 (*above*) in comparison with the control group (*below*). At day 1 and day 2, this number was 0; on and after day 60, this number was 6. (Zakine, Barbier et al. 2011)

A study by Zakine et al. (2011), using this same model, revealed an increased reepithelialisation together with an increased number of keratinocyte layers and blood vessels in the RGTA OTR4120 administered group at the early stages of tissue regeneration that returned to control levels after 1 month post wounding (Figure 6) (Zakine, Barbier et al. 2011).

6.4.3 Full-thickness excisional wounds

Tong et al. studied the effects of RGTA OTR4120 administration to rat full-thickness excisional wounds (Tong, Zbinden et al. 2008; Tong, Tuk et al. 2009). RGTA OTR4120 administration to surgical wounds in rapid healing normal rats significantly improved wound regeneration. RGTA OTR4120 administration promotes epidermal proliferation, increased neodermal granulation tissue deposition, inflammation resolution, improved the vascular response to local heating (Figure 7a), improved collagen maturation and improved the wound breaking strength at all measurement times up to 3 months post wounding (Figure 7b) (Tong, Zbinden et al. 2008; Tong, Tuk et al. 2009).

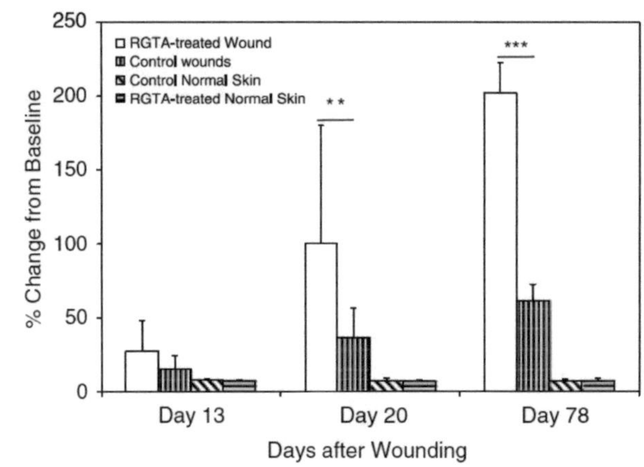

Fig. 7a. The wounded skin and normal skin vascular responses to local heating (44°C for 10 minutes), expressed as the percentages of the increase blood flow over baseline flow, measured by laser Doppler flow with local heat provocation on days 13, 20, and 78 after wounding. Error bars represent the standard deviation. $^{**}p<0.01$ and $^{***}p < 0.001$ indicate the significant differences between treated groups and control groups. (Tong, Zbinden et al. 2008)

6.4.4 Ischemia-reperfusion wounds

Tong et al. (2011) also studied the effects of RGTA OTR4120 administration to ischemia-reperfusion wounds (Tong, Tuk et al. 2011). Similar findings as for the excisional wounds were observed in a cutaneous ischemia-reperfusion model obtained by magnet clamping of a skin fold in the neck area of the animal although with a delayed timing due to the clearance of the necrotic tissue (Tong, Tuk et al. 2011). In addition, monocyte/macrophage staining and CD68 detection on Western blots revealed that RGTA OTR4120 administration facilitated an inflammation controlled environment that progressed to the normal stages of wound healing.

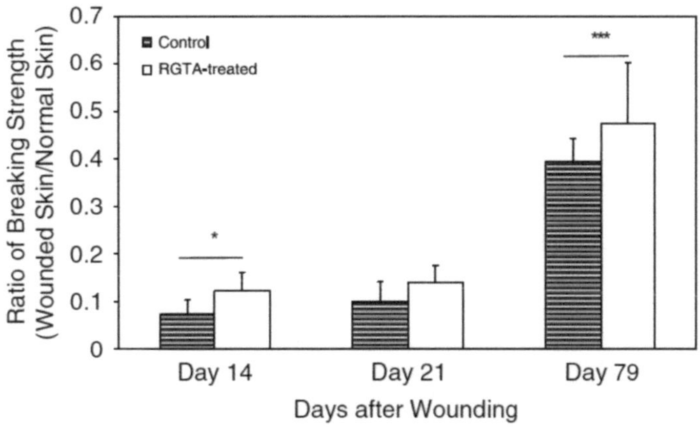

Fig. 7b. Wound strength measured as the ratio of the breaking strength of the wounded skin versus that of normal skin on days 14, 21, and 79 after wounding. Error bars represent the standard deviation. $*p < 0.05$ and $***p < 0.001$ indicate the significant differences between treated groups and control groups (Tong, Zbinden et al. 2008; Tong, Tuk et al. 2009).

In summary: the ability of RGTA OTR4120 administration to reduce inflammation; increase angiogenesis; improved healing quality, amongst others reflected by the recurrence of normal dermal collagen balance and an increased tissue breaking strength have now been demonstrated in multiple models (Alexakis, Guettoufi et al. 2001; Alexakis, Caruelle et al. 2004; Alexakis, Mestries et al. 2004; Barbier-Chassefiere, Garcia-Filipe et al. 2009; Tong, Tuk et al. 2011). All are signs of matrix reconstruction to a formulation that more closely resembles normal matrix. Specifically, RGTA OTR4120 decreasing collagen III accumulation, will dramatically reduce the fibrosis that frequently accompanies wound healing. These promising preclinical findings warrant studies on human subjects.

6.5 Clinical evidence

The first scientific report on the clinical use of RGTA OTR4120 was a study by Chebbi et al. (2008) who described the effects of local RGTA OTR4120 administration for one month to patients with treatment-resistant corneal-ulcers and to patients with treatment-resistant corneal dystrophy (Chebbi, Kichenin et al. 2008). RGTA OTR4120 administration resulted in the majority of cases in the complete healing of the ulcer. The effect on the keratitis was moderate, however, a significant pain reduction was observed that highly improved the patient's quality of life (Chebbi, Kichenin et al. 2008).

In a within-subject study, Groah et al. (2011) demonstrated the effect of RGTA OTR4120 administration in a patient population of largely persistent pressure ulcers and vascular/venous ulcers (Groah, Libin et al. 2011). The mean duration of the ulcers was 2.5

years. RGTA OTR4120 was administered on the debrided wound, twice weekly for 5 minutes each time. After 4 weeks both a significant reduction in the wound size as well as in the pain perception was found (Groah, Libin et al. 2011).

Van Neck et al. (2011) described the complete healing following of recurrent scalp ulcers RGTA OTR4120 administration (Van Neck, Zuidema et al. 2011).

RGTA OTR4120 also was administered to a 60-year old male patient with a complex medical history. He suffered from obesity (BMI 35), developed type II diabetic over a decade ago and was on insulin treatment. Furthermore, he was known with alcohol and nicotine abuse, heart failure, pacemaker, cardiac and vascular disease and kidney and liver failure. In 1992 and 2002 he has had several toe amputations on both his feet likely caused by his poor cardiac and vascular condition. The patient had developed the diabetic pressure ulcer under investigation on the palmar side of his right foot over 8 months ago. So far, he had been unsuccessfully treated with a wide variety of wound dressings (foams, alginate, hydro colloids, silver dressings, foils, impregnated gauzes and collagen) all applied up to three times weekly if needed. At the start of the OTR4120 treatment the wound measured 2.5 cm2 (Figure 8A). The wound displayed an almost immediate response to OTR4120 treatment

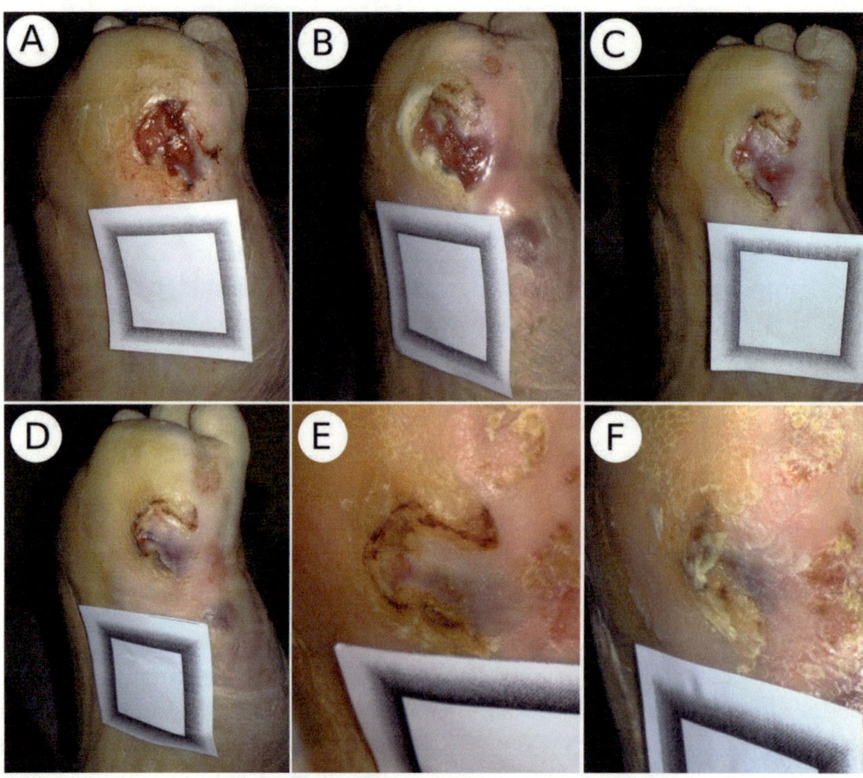

Fig. 8. Diabetic foot of a 60 year old male patient in a state of non-healing for over 8 months. A) day 0, the start of the twice weekly RGTA OTR4120 topical application. At this stage the wound measured 2.5 cm2; B) day 7, healthy granulation tissue immediately formed following RGTA OTR4120 treatment; C) day 16; D) day 31, near to complete wound closure; E) day 46; F) day 56, long term closure control. The inner dimensions of the grey shaded square are 30x30mm.

(Figure 8B). Healthy granulation tissue formed and the wound closed in the 5th week of treatment (Figure 8C-E). Further documentation of the wound area proved it remained completely healed (Figure 8F).

6.6 Future perspectives

The findings presented support the use of RGTA OTR4120 (CACIPLIQ20®) in treating (chronic) wounds by means of restoring the damaged extracellular matrix.

Research to reduce the complexity of the molecule, to facilitate its synthesis and to increase its effectiveness, is ongoing (Ikeda, Charef et al. 2011). Also oral administration, as a patient friendly means of administration with a RGTA OTR4120 delivery to the side of injury via (micro)circulation, are tested (Charef, Papy-Garcia et al. 2010)

7. Concluding remarks

Matrix therapy restores the natural extracellular microenvironment which allows the local cascade of signals to resume in the proper time and order to trigger tissue regeneration. (Barbier-Chassefiere, Garcia-Filipe et al. 2009). Therefore, matrix therapy with engineered biopolymers such as RGTA OTR4120 is simpler and easier to use than cell or gene therapy and is a new alternative in regenerative medicine.

8. Acknowledgement

Mr Ronald Lau, Amersfoort, the Netherlands, is thanked for his dedication while applying CACIPLIQ20® to a diabetic foot patient.

The majority of the results presented here were published.

9. References

Agren, M. S. and M. Werthen (2007). "The extracellular matrix in wound healing: a closer look at therapeutics for chronic wounds." *Int J Low Extrem Wounds* 6(2): 82-97.

Alexakis, C., J. P. Caruelle, et al. (2004). "Reversal of abnormal collagen production in Crohn's disease intestinal biopsies treated with regenerating agents." *Gut* 53(1): 85-90.

Alexakis, C., A. Guettoufi, et al. (2001). "Heparan mimetic regulates collagen expression and TGF-beta1 distribution in gamma-irradiated human intestinal smooth muscle cells." *FASEB J* 15(9): 1546-1554.

Alexakis, C., P. Mestries, et al. (2004). "Structurally different RGTAs modulate collagen-type expression by cultured aortic smooth muscle cells via different pathways involving fibroblast growth factor-2 or transforming growth factor-beta1." *FASEB J* 18(10): 1147-1149.

Attinger, C. E. and E. J. Bulan (2001). "Debridement. The key initial first step in wound healing." *Foot Ankle Clin* 6(4): 627-660.

Barbier-Chassefiere, V., S. Garcia-Filipe, et al. (2009). "Matrix therapy in regenerative medicine, a new approach to chronic wound healing." *J Biomed Mater Res A* 90(3): 641-647.

Barrientos, S., O. Stojadinovic, et al. (2008). "Growth factors and cytokines in wound healing." *Wound Repair Regen* 16(5): 585-601.

Barritault, D., S. Garcia-Filipe, et al. (2010). "[Basement of matrix therapy in regenerative medicine by RGTA((R)): From fundamental to plastic surgery]

Les bases de la therapie matricielle en medecine regenerative par les RGTA((R)) : du fondamental a la chirurgie plastique." *Ann Chir Plast Esthet* 55(5): 413-420.

Baum, C. L. and C. J. Arpey (2005). "Normal cutaneous wound healing: clinical correlation with cellular and molecular events." *Dermatol Surg* 31(6): 674-686; discussion 686.

Branski, L. K., C. T. Pereira, et al. (2007). "Gene therapy in wound healing: present status and future directions." *Gene Ther* 14(1): 1-10.

Broughton, G., 2nd, J. E. Janis, et al. (2006). "The basic science of wound healing." *Plast Reconstr Surg* 117(7 Suppl): 12S-34S.

Bullen, E. C., M. T. Longaker, et al. (1995). "Tissue inhibitor of metalloproteinases-1 is decreased and activated gelatinases are increased in chronic wounds." *J Invest Dermatol* 104(2): 236-240.

Charef, S., D. Papy-Garcia, et al. (2010). "Absorption and tissue distribution of a novel carboxymethyldextran after oral administration." *Biomed Pharmacother* 64(9): 627-632.

Charef, S., M. Tulliez, et al. (2010). "Toxicological evaluation of RGTA OTR4120, a heparan sulfate mimetic." *Food Chem Toxicol* 48(7): 1965-1968.

Chebbi, C. K., K. Kichenin, et al. (2008). "[Pilot study of a new matrix therapy agent (RGTA OTR4120) in treatment-resistant corneal ulcers and corneal dystrophy]

Etude pilote d'un nouvel agent de therapie matricielle (RGTA OTR4120) dans les ulceres de cornee et les dystrophies corneennes rebelles." *J Fr Ophtalmol* 31(5): 465-471.

Clark, R. A., J. M. Lanigan, et al. (1982). "Fibronectin and fibrin provide a provisional matrix for epidermal cell migration during wound reepithelialization." *J Invest Dermatol* 79(5): 264-269.

Colombier, M. L., J. Lafont, et al. (1999). "A single low dose of RGTA, a new healing agent, hastens wound maturation and enhances bone deposition in rat craniotomy defects." *Cells Tissues Organs* 164(3): 131-140.

Cook, H., K. J. Davies, et al. (2000). "Defective extracellular matrix reorganization by chronic wound fibroblasts is associated with alterations in TIMP-1, TIMP-2, and MMP-2 activity." *J Invest Dermatol* 115(2): 225-233.

Crovetti, G., G. Martinelli, et al. (2004). "Platelet gel for healing cutaneous chronic wounds." *Transfus Apher Sci* 30(2): 145-151.

Dash, N. R., S. N. Dash, et al. (2009). "Targeting nonhealing ulcers of lower extremity in human through autologous bone marrow-derived mesenchymal stem cells." *Rejuvenation Res* 12(5): 359-366.

Diegelmann, R. F. (2003). "Excessive neutrophils characterize chronic pressure ulcers." *Wound Repair Regen* 11(6): 490-495.

Diegelmann, R. F. and M. C. Evans (2004). "Wound healing: an overview of acute, fibrotic and delayed healing." *Front Biosci* 9: 283-289.

Dreyfuss, J. L., C. V. Regatieri, et al. (2009). "Heparan sulfate proteoglycans: structure, protein interactions and cell signaling." *An Acad Bras Cienc* 81(3): 409-429.

Eaglstein, W. H. and V. Falanga (1997). "Chronic wounds." *Surg Clin North Am* 77(3): 689-700.

Eming, S. A., T. Krieg, et al. (2007). "Gene therapy and wound healing." *Clin Dermatol* 25(1): 79-92.

Fan, K., J. Tang, et al. (2011). "State of the art in topical wound-healing products." *Plast Reconstr Surg* 127 Suppl 1: 44S-59S.

Fung, H. B., J. Y. Chang, et al. (2003). "A practical guide to the treatment of complicated skin and soft tissue infections." *Drugs* 63(14): 1459-1480.

Gandhi, N. S. and R. L. Mancera (2010). "Heparin/heparan sulphate-based drugs." *Drug Discov Today* 15(23-24): 1058-1069.

Garcia-Filipe, S., V. Barbier-Chassefiere, et al. (2007). "RGTA OTR4120, a heparan sulfate mimetic, is a possible long-term active agent to heal burned skin." *J Biomed Mater Res A* 80(1): 75-84.

Goldman, R. (2004). "Growth factors and chronic wound healing: past, present, and future." *Adv Skin Wound Care* 17(1): 24-35.

Groah, S. L., A. Libin, et al. (2011). "Regenerating matrix-based therapy for chronic wound healing: a prospective within-subject pilot study." *Int Wound J* 8(1): 85-95.

Grotendorst, G. R., Y. Soma, et al. (1989). "EGF and TGF-alpha are potent chemoattractants for endothelial cells and EGF-like peptides are present at sites of tissue regeneration." *J Cell Physiol* 139(3): 617-623.

Guo, S. and L. A. Dipietro (2010). "Factors affecting wound healing." *J Dent Res* 89(3): 219-229.

Gurtner, G. C., S. Werner, et al. (2008). "Wound repair and regeneration." *Nature* 453(7193): 314-321.

Henry, G. and W. L. Garner (2003). "Inflammatory mediators in wound healing." *Surg Clin North Am* 83(3): 483-507.

Hess, C. T. and R. S. Kirsner (2003). "Orchestrating wound healing: assessing and preparing the wound bed." *Adv Skin Wound Care* 16(5): 246-257; quiz 258-249.

Hodde, J. P. and C. E. Johnson (2007). "Extracellular matrix as a strategy for treating chronic wounds." *Am J Clin Dermatol* 8(2): 61-66.

Hubner, G., M. Brauchle, et al. (1996). "Differential regulation of pro-inflammatory cytokines during wound healing in normal and glucocorticoid-treated mice." *Cytokine* 8(7): 548-556.

Ikeda, Y., S. Charef, et al. (2011). "Synthesis and biological activities of a library of glycosaminoglycans mimetic oligosaccharides." *Biomaterials* 32(3): 769-776.

Kazakos, K., D. N. Lyras, et al. (2009). "The use of autologous PRP gel as an aid in the management of acute trauma wounds." *Injury* 40(8): 801-805.

Korting, H. C., C. Schollmann, et al. (2011). "Management of minor acute cutaneous wounds: importance of wound healing in a moist environment." *J Eur Acad Dermatol Venereol* 25(2): 130-137.

Krishnamoorthy, L., H. L. Morris, et al. (2001). "Specific growth factors and the healing of chronic wounds." *J Wound Care* 10(5): 173-178.

Lallam-Laroye, C., Q. Escartin, et al. (2006). "Periodontitis destructions are restored by synthetic glycosaminoglycan mimetic." *J Biomed Mater Res A* 79(3): 675-683.

Lawrence, W. T. and R. F. Diegelmann (1994). "Growth factors in wound healing." *Clin Dermatol* 12(1): 157-169.

Lazarus, G. S., D. M. Cooper, et al. (1994). "Definitions and guidelines for assessment of wounds and evaluation of healing." *Wound Repair Regen* 2(3): 165-170.

Lobmann, R., A. Ambrosch, et al. (2002). "Expression of matrix-metalloproteinases and their inhibitors in the wounds of diabetic and non-diabetic patients." *Diabetologia* 45(7): 1011-1016.

Madden, J. W. and E. E. Peacock, Jr. (1968). "Studies on the biology of collagen during wound healing. I. Rate of collagen synthesis and deposition in cutaneous wounds of the rat." *Surgery* 64(1): 288-294.

Madlener, M., W. C. Parks, et al. (1998). "Matrix metalloproteinases (MMPs) and their physiological inhibitors (TIMPs) are differentially expressed during excisional skin wound repair." *Exp Cell Res* 242(1): 201-210.

Martin, P. (1997). "Wound healing--aiming for perfect skin regeneration." *Science* 276(5309): 75-81.

McGrath, J.A. and Eady, R.A. (1997). "Heparan sulphate proteoglycan and wound healing in skin." *J Pathol* 183(3):251-252.

Menke, N. B., K. R. Ward, et al. (2007). "Impaired wound healing." *Clin Dermatol* 25(1): 19-25.

Nathan, C. (2002). "Points of control in inflammation." *Nature* 420(6917): 846-852.

Nwomeh, B. C., H. X. Liang, et al. (1998). "Dynamics of the matrix metalloproteinases MMP-1 and MMP-8 in acute open human dermal wounds." *Wound Repair Regen* 6(2): 127-134.

Nwomeh, B. C., H. X. Liang, et al. (1999). "MMP-8 is the predominant collagenase in healing wounds and nonhealing ulcers." *J Surg Res* 81(2): 189-195.

Okan, D., K. Woo, et al. (2007). "The role of moisture balance in wound healing." *Adv Skin Wound Care* 20(1): 39-53; quiz 53-35.

Paladini, R. D., K. Takahashi, et al. (1996). "Onset of re-epithelialization after skin injury correlates with a reorganization of keratin filaments in wound edge keratinocytes: defining a potential role for keratin 16." *J Cell Biol* 132(3): 381-397.

Ramasastry, S. S. (2005). "Acute wounds." *Clin Plast Surg* 32(2): 195-208.

Rizzi, S. C., Z. Upton, et al. (2010). "Recent advances in dermal wound healing: biomedical device approaches." *Expert Rev Med Devices* 7(1): 143-154.

Robson, M. C., D. L. Steed, et al. (2001). "Wound healing: biologic features and approaches to maximize healing trajectories." *Curr Probl Surg* 38(2): 72-140.

Schade, V. L. and T. S. Roukis (2008). "Use of platelet-rich plasma with split-thickness skin grafts in the high-risk patient." *Foot Ankle Spec* 1(3): 155-159.

Schultz, G. S. and A. Wysocki (2009). "Interactions between extracellular matrix and growth factors in wound healing." *Wound Repair Regen* 17(2): 153-162.

Sellheyer, K. and D. Krahl (2010). "Skin mesenchymal stem cells: prospects for clinical dermatology." *J Am Acad Dermatol* 63(5): 859-865.

Servold, S. A. (1991). "Growth factor impact on wound healing." *Clin Podiatr Med Surg* 8(4): 937-953.

Singer, A. J. and R. A. Clark (1999). "Cutaneous wound healing." *N Engl J Med* 341(10): 738-746.

Smola, H., G. Thiekotter, et al. (1993). "Mutual induction of growth factor gene expression by epidermal-dermal cell interaction." *J Cell Biol* 122(2): 417-429.

Steed, D. L., J. B. Goslen, et al. (1992). "Randomized prospective double-blind trial in healing chronic diabetic foot ulcers. CT-102 activated platelet supernatant, topical versus placebo." *Diabetes Care* 15(11): 1598-1604.

Tong, M., B. Tuk, et al. (2011). "Heparan sulfate glycosaminoglycan mimetic improves pressure ulcer healing in a rat model of cutaneous ischemia-reperfusion injury." *Wound Repair Regen* 19(4):505-514.

Tong, M., B. Tuk, et al. (2009). "Stimulated neovascularization, inflammation resolution and collagen maturation in healing rat cutaneous wounds by a heparan sulfate glycosaminoglycan mimetic, OTR4120." *Wound Repair Regen* 17(6): 840-852.

Tong, M., M. M. Zbinden, et al. (2008). "RGTA OTR 4120, a heparan sulfate proteoglycan mimetic, increases wound breaking strength and vasodilatory capability in healing rat full-thickness excisional wounds." *Wound Repair Regen* 16(2): 294-299.

Tumova, S., A. Woods, et al. (2000). "Heparan sulfate proteoglycans on the cell surface: versatile coordinators of cellular functions." *Int J Biochem Cell Biol* 32(3): 269-288.

Ulrich, D., E. M. Noah, et al. (2002). "Serum concentration of amino-terminal propeptide of type III procollagen (PIIINP) as a prognostic marker for skin fibrosis after scar correction in burned patients." *Burns* 28(8): 766-771.

Ulrich, D., E. M. Noah, et al. (2003). "TIMP-1, MMP-2, MMP-9, and PIIINP as serum markers for skin fibrosis in patients following severe burn trauma." *Plast Reconstr Surg* 111(4): 1423-1431.

Van Neck, J. W., Y. S. Zuidema, et al. (2011). "Unexpected healing of radiation-induced scalp lesions with OTR4120, a heparan sulfate mimetic." *Eur J Dermatol* 21(3): 451-452.

Velnar, T., T. Bailey, et al. (2009). "The wound healing process: an overview of the cellular and molecular mechanisms." *J Int Med Res* 37(5): 1528-1542.

Vowden, K. and P. Vowden (2003). "Understanding exudate management and the role of exudate in the healing process." *Br J Community Nurs* 8(11 Suppl): 4-13.

Walter, D. H., H. Krankenberg, et al. (2011). "Intraarterial administration of bone marrow mononuclear cells in patients with critical limb ischemia: a randomized-start, placebo-controlled pilot trial (PROVASA)." *Circ Cardiovasc Interv* 4(1): 26-37.

Watt, F. M. and H. Fujiwara (2011). "Cell-extracellular matrix interactions in normal and diseased skin." *Cold Spring Harb Perspect Biol* 3(4).

Wenk, J., A. Foitzik, et al. (2001). "Selective pick-up of increased iron by deferoxamine-coupled cellulose abrogates the iron-driven induction of matrix-degrading metalloproteinase 1 and lipid peroxidation in human dermal fibroblasts in vitro: a new dressing concept." *J Invest Dermatol* 116(6): 833-839.

Witte, M. B. and A. Barbul (1997). "General principles of wound healing." *Surg Clin North Am* 77(3): 509-528.

Wu, Y., R. C. Zhao, et al. (2010). "Concise review: bone marrow-derived stem/progenitor cells in cutaneous repair and regeneration." *Stem Cells* 28(5): 905-915.

Xia, Y. P., Y. Zhao, et al. (1999). "Effects of keratinocyte growth factor-2 (KGF-2) on wound healing in an ischaemia-impaired rabbit ear model and on scar formation." *J Pathol* 188(4): 431-438.

Yager, D. R., L. Y. Zhang, et al. (1996). "Wound fluids from human pressure ulcers contain elevated matrix metalloproteinase levels and activity compared to surgical wound fluids." *J Invest Dermatol* 107(5): 743-748.

Zakine, G., V. Barbier, et al. (2011). "Matrix therapy with RGTA OTR4120 improves healing time and quality in hairless rats with deep second-degree burns." *Plast Reconstr Surg* 127(2): 541-550.

5

Angiogenesis in Wound Healing

Ricardo José de Mendonça
Department of Biological Sciences
Federal University of Triangulo Mineiro
Brazil

1. Introduction

Angiogenesis, the formation of new blood vessels from pre-existing, vessels is a crucial process for tumor growth and metastasis (Folkman 1990; Kaafarani, Fernandez-Sauze et al. 2009). The new vessels supply the tumor cells with nutrients and oxygen and ensure efficient drainage of metabolites. Under normal conditions, a tissue or tumor cannot grow beyond 1 to 2 mm in diameter without neovascularization. This distance is defined by limits in the diffusion of oxygen and metabolites, such as glucose and amino acids (Folkman 1971).

In addition to supplying nutrients for tumor growth, angiogenesis is also a gateway for tumor cells and signals to the bloodstream. This direct communication with the bloodstream is essential for the dissemination and metastasis of cancer. After their arrival and deployment in distant organs, metastatic cells again induce angiogenesis in order to support tumor growth (Eichhorn, Kleespies et al. 2007).

As well as this important role of angiogenesis in tumor growth, the whole process of tissue regeneration depends on a new intake of oxygen and metabolites. Growth of new cells for regeneration involves a large energy demand that occurs for the process of cellular mitosis. Therefore, understanding the biochemical mechanisms involved in angiogenesis is necessary for developing interventions in complex tissue regeneration processes.

Since the hypothesis proposed by the surgeon Judah Folkman in the early 70's, which indicated that the inhibition of angiogenesis as a therapeutic target that could halt or even reduce tumor growth (Folkman 1971), intense and successful research on the molecular mechanisms of angiogenesis tumor began. In recent decades, numerous pro- and anti-angiogenic molecules, as well as their ligands and intracellular signaling pathways, have been identified.

2. The wound healing process

The main aim of wound treatment is achieving a rapid closure of the lesion combined with a functional and aesthetically satisfactory scar. To improve current practice, it is essential to gain a better understanding of the biological processes involved in wound healing and tissue regeneration. Many studies have investigated the complex process of wound repair, and the cell behaviors, chemical signals and extracellular matrices that together lead to scarring.

With the disruption of tissue integrity in vertebrates, so begins the repair process, which comprises a sequence of molecular events that either restores or at least secures the damaged tissue. After birth, the body loses its ability to replace damaged tissue without leaving a scar. Only during the fetal stage of life does repair of damage occurs without scar formation, with a true restoration of tissue by a neoformation process (Martin and Leibovich 2005).

Healing has been conveniently divided into three phases, that overlap temporally: the inflammatory, proliferative and remodeling phases (Mendonca and Coutinho-Netto 2009), as shown in figure 1.

2.1 Inflammatory phase (Latent)

After the occurrence of an injury, tissue begins to leak blood that fills the injured area with plasma and cellular elements, mainly platelets. Platelet aggregation and blood clotting generate a plug rich in fibrin; this, in addition to restoring hemostasis and form a barrier against invading microorganisms, organizes a provisional matrix necessary for cell migration. This matrix will also cache growth factors required during the next stages of the healing process (Werner and Grose 2003; Eming, Krieg et al. 2007).

Platelets, essential to the formation of a hemostatic plug, also secrete multiple mediators into the injured area. Platelets are essential in the coagulation cascade, and undergo degranulation induced by thrombin, releasing growth factors, such as platelet-derived growth factor (PDGF), transforming growth factor-β (TGF-β), epidermal growth factor (EGF), transforming growth factor-α (TGF-α), vascular endothelial growth factor (VEGF) and a adhesive glycoproteins such as fibronectin and thrombospondin, which are important constituents of the provisional extracellular matrix (Streit, Velasco et al. 2000; Nguyen, Hoang et al. 2009; Ribatti 2009). In fact, the coagulation cascade and growth factors released by platelets, together with the activation of the complement cascade and activation of parenchymal cell by injury, produce numerous vasoactive mediators and chemotactic factors, which together assist in the recruitment of inflammatory cells to the wound (Delavary, van der Veer et al. 2011).

In addition to phagocytosing bacteria, cellular debris and foreign bodies, these inflammatory cells produce growth factors that prepare the wound for the proliferative phase, at which time fibroblasts and endothelial cells will continue to be recruited (Singer and Clark 1999; Mendonca and Coutinho-Netto 2009).

Despite the overlap of the healing phases, there is a basic sequence of events: plasma soluble and cellular components exit vessels, followed by platelets, neutrophils and monocytes (Delavary, van der Veer et al. 2011). Subsequently, many neutrophils adhere to the endothelium and migrate to the region of the wound. However, depletion of neutrophils in the blood does not significantly affect the repair process (Simpson and Ross 1972; Werner and Grose 2003; Eming, Werner et al. 2007).

Peripheral blood monocytes, both initially and throughout the course of the healing process, continue to infiltrate the wound in response to chemotactic agents for monocytes, such as PDGF. In the tissue, monocytes are activated and transform into macrophages, which are probably the main cells involved in control of the repair process (Singer and Clark 1999; Delavary, van der Veer et al. 2011).

Macrophage activation has implications for various aspects of wound healing, as in the phagocytosis of cellular debris, synthesis of extracellular matrix and release of cytokines that stimulate increased vascular permeability, angiogenesis and epithelialization. The release of factors from platelets is the main stimulus for the migration and macrophage activation, while the phagocytosis of cellular components such as fibronectin or collagen also contribute (Henderson, Nair et al. 2011).

The activated macrophage is the main cellular effector in the tissue repair process, degrading and removing damaged tissue components such as collagen, elastin and proteoglycans. As well as removing cellular debris, macrophages secrete chemotactic factors that attract other inflammatory cells to the wound site and produce prostaglandins, which act as potent vasodilators and affect the permeability of microvessels (Singer and Clark 1999; Eming, Werner et al. 2007).

Macrophages produce several growth factors such as PDGF, TGF-β, fibroblast growth factor (FGF) and VEGF, which stand out as the key cytokines necessary to stimulate the formation of granulation tissue. Thus, macrophages mediate the initial phase of the inflammatory response during the wound healing process (Singer and Clark 1999; Barrientos, Stojadinovic et al. 2008).

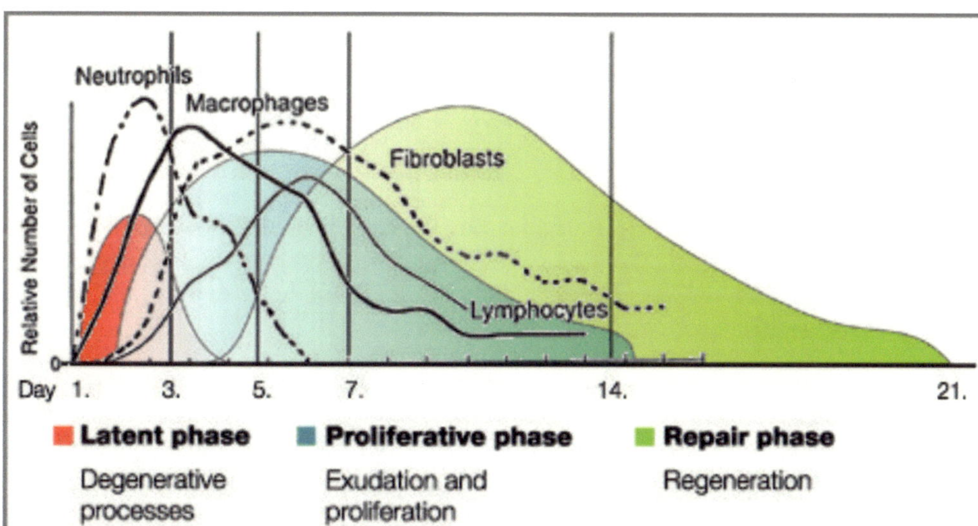

Fig. 1. Migration of immune cell populations correlated with the phases of wound healing

2.2 Proliferative phase

The stage of epithelial proliferation, in the case of the skin, begins with mitogenic and chemotactic stimulation of keratinocytes by EGF and TGF-α. As important as epithelialization, which begins at this stage of the repair process, is the formation of granulation tissue, a name given mainly on account of the characteristic granularity resulting from the presence of new capillaries. Granulation tissue is essential to repair (Mendonca and Coutinho-Netto 2009).

Before describing angiogenesis, however, it is necessary to note that increased microvascular permeability is the first stage of this process and causes, through the leakage of proteins, cytokines and cellular elements, the formation of a provisional extracellular matrix that is necessary for migration and proliferation of endothelial cells (Dvorak 2002; Dvorak 2010).

2.2.1 Vascular permeability

The production of new blood vessels from pre-existing vessels is accompanied by an increase in vascular permeability (Bates and Harper 2002; Dvorak 2010). In pathological angiogenesis, increased vascular permeability to water and macromolecules is important to the sequence of events that follow injury, being directly responsible for edema. This increased capillary permeability seems to have a minor effect during physiological angiogenesis but it causes considerable damage in pathologies such as diabetic retinopathy (Vaquero, Zurita et al. 2000).

VEGF-A, for example, was discovered in ascites tumor and was originally noticed for its ability to increase the permeability of microvessels and extravasation of macromolecules, including fibrinogen and coagulation proteins, to result in extravascular fibrin deposition which favors both wound healing and tumor development (Dvorak 2010).

The mechanisms of vascular permeability regulation, controlled mainly by growth factors, have not yet been fully elucidated. The function of these growth factors, and the mechanism by which exert their effect, are objects of study of great interest and their metabolic pathways are being elucidated (Dvorak 2005).

2.2.2 Angiogenesis

Angiogenesis is a fundamental step in the healing process by which new blood vessels are formed from preexisting vessels (Folkman and Shing 1992). The new vessels involved in the formation of granulation tissue supply the growing tissue with oxygen and nutrients (Schafer and Werner 2008).

In an adult organism, under normal conditions, angiogenesis occurs only in the reproductive cycle of females (in utero, with the formation of the endometrium and ovaries, with the formation of corpus luteum). Generally, adult vasculature remains quiescent but it has the ability to initiate angiogenesis, especially during healing (Schafer and Werner 2008).

Under physiological conditions, angiogenesis is finely regulated; activated for short periods (days) and then completely inhibited. However, many pathologies are a consequence of lack of regulation, for example, rheumatoid arthritis, where new blood capillaries invade the joint and destroy cartilage. In diabetes, new capillaries present in the retina invade the vitreous humor, bleed, and cause blindness. Tumor growth and metastasis are angiogenesis-dependent diseases, (Folkman 1991). Most tumors remains a constant stimulus to the growth of new capillaries to allow their own growth. The blood vessels also provide a route of communication that allows tumor cells to invade the bloodstream and cause metastases in locations distant from the primary (Folkman and Shing 1992).

The induction of angiogenesis was initially attributed to the acidic or basic FGF. Subsequently, many other molecules have been identified as angiogenic, including VEGF, TGF-β, angiogenin, angiotropin and angiopoetin-1 (Folkman and D'Amore 1996). Low oxygen tension (Detmar, Brown et al. 1997) and high levels of lactate and bioactive amines (Remensnyder and Majno 1968) can also stimulate angiogenesis. Many of these molecules are proteins that induce angiogenesis indirectly by stimulating the production of acidic or basic FGF and VEGF by macrophages and endothelial cells, direct inducers of angiogenesis.

2.2.3 Growth factors

The identification, characterization and purification of VEGF (Vascular Endothelial Growth Factor) in 1989 contributed significantly to understanding the regulation of blood flow and vascular permeability in angiogenesis (Ferrara and Henzel 1989; Glass, Harper et al. 2006). VEGF has three main mechanisms of action: 1) it can increase the vessel permeability to water, small solutes and macromolecules (Adamson, Lenz et al. 2004; Nagy, Benjamin et al. 2008); 2) it can reduce the distance of the tissue cells from to the nearest blood vessel, by stimulating angiogenesis, and 3) it can increase blood flow to tissue by acting as potent vasodilators (Bates and Harper 2002).

VEGF exerts its biological activity predominantly through transmembrane receptors with tyrosine kinase activity present in endothelial cells and participates as a principal mediator of angiogenesis. The VEGF protein family currently includes VEGF-A, VEGF-B, VEGF-C, VEGF-D, VEGF-E and placental growth factor (PlGF) (Werner and Grose 2003). VEGF-A is a homodimer glycoprotein whose subunits are linked by two disulfide bonds, and is synthesized from internal rearrangements ("alternative splicing") of a mRNA, thus there is the production of seven isoforms with 121 to 206 amino acids (Ferrara 2001; Bates and Harper 2002; Ferrara 2004). Among these, the VEGF121, VEGF165, VEGF189 and VEGF206 are the predominant isoforms (Kessler, Fehrmann et al. 2007). These isoforms show similar biological activities, but differ in their binding properties to heparin and extracellular matrix (Roth, Piekarek et al. 2006). The smaller isoforms (121 to 165 amino acids) are secreted in soluble form, while larger ones have transmembrane domains, being initially associated with cells, where they are released and activated by proteolysis. The VEGF121 is an acid protein, while the others have basic isoelectric point.

VEGF is also known as vascular permeability factor (VPF) due to its potent action in increase of vasopermeability, allowing leakage of proteins such as fibrinogen and fibronectin, that are essential for the formation of the provisional extracellular matrix (Nagy, Benjamin et al. 2008), besides increasing the hydraulic conductivity (Bates and Curry 1997) and fenestration (Esser, Wolburg et al. 1998). VEGF also acts as a potent mitogen for endothelial cells of the microvasculature inducing endothelial cell migration and sprouting of new blood vessels through the regulation of several endothelial integrin receptors (Primo, Seano et al. 2010). Furthermore, VEGF also acts as a survival factor for endothelial cells by inducing the expression of Bcl-2, an anti-apoptotic protein (Rao, Zhong et al. 2011).

This family of VEGF exerts its biological functions by differential interactions with three transmembrane receptor tyrosine kinase: VEGF receptor-1 (VEGFR-1) [similar to fms tyrosine kinase (Flt-1)], VEGFR-2 [fetal liver kinase (Flk-1)] and VEGFR-3 (Flt-4). Expression of these receptors is driven primarily by hypoxia. The receptors VEGFR-1 and VEGFR-2 are restricted to the vascular endothelium, while VEGFR-3, together with its preferred ligand, VEGF-C and VEGF-D seem to be involved in the growth of lymphatic endothelium (Barrientos, Stojadinovic et al. 2008).

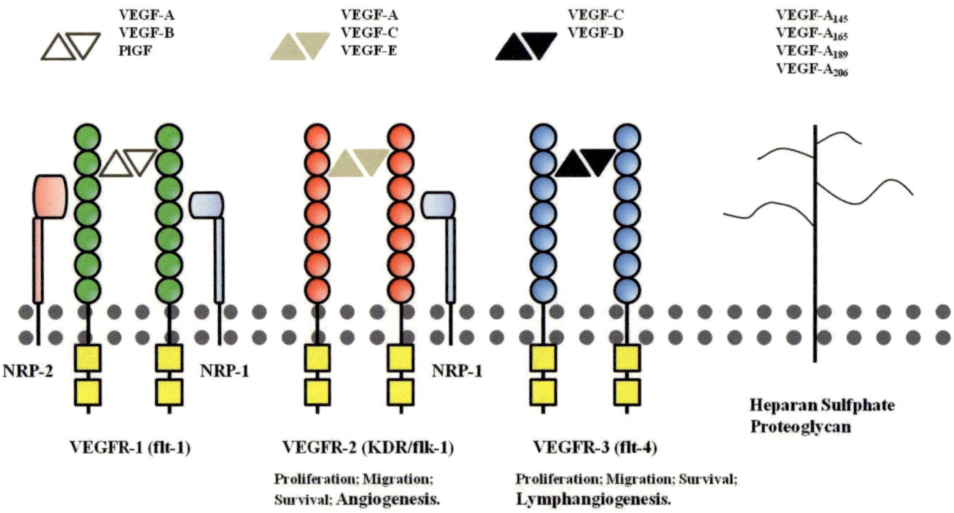

Fig. 2. Vascular endothelial growth factor (VEGF) ligands and receptors. VEGF tyrosine kinase receptors are subfamily of receptor protein tyrosine kinases (RTKs) and possess an extracellular domain containing 7 immunoglobulin-like loops, a single hydrophobic membrane-spanning domain, and a large cytoplasmic domain comprising a single catalytic domain containing all the conserved motifs found in other RTKs. The extracellular domain of VEGFR1 is also independently expressed as a soluble protein (not shown). VEGF-A binds with high affinity to both VEGFR2 (KDR/Flk-1) and VEGFR1 (Flt-1) receptors. Placenta growth factor (PlGF) and VEGF-B exhibit high-affinity binding to VEGFR1 only. VEGF-C and -D are VEGF-related factors that bind to a related receptor, Flt-4 (VEGFR3), and also to VEGFR2. Neuropilin-1 (NRP-1) is a novel non-RTK receptor for VEGF165. Neuropilins and heparan sulfate proteoglycans act as coreceptors that lack enzymatic activity, yet modulate signal output by VEGF receptors.

VEGF is most likely to act through receptors in the endothelium to increase production of nitric oxide (NO) and prostacyclin (PGI2) and augment intracellular endothelial cell survival signaling. NO and PGI2 are predicted to have other biological consequences: decreased platelet aggregation, thrombosis, and, in the case of NO, inhibition of leukocyte adhesion. The combined effect of these biological actions is vascular protection (Zachary 2001).

Many different cell types, fibroblasts, endothelial cells, macrophages and keratinocytes, are able to produce VEGF, and mainly the latter two, are types cells responsible for the production during healing (Barrientos, Stojadinovic et al. 2008). The addition of anti-VEGF inhibits the formation of granulation tissue in the wound (Howdieshell, Callaway et al. 2001) indicating an important function of VEGF in angiogenesis that occurs during the proliferative phase. Low oxygen tension, as occurs during tissue injury, constitutes the greatest inducer of the production of this growth factor (Andrikopoulou, Zhang et al. 2011).

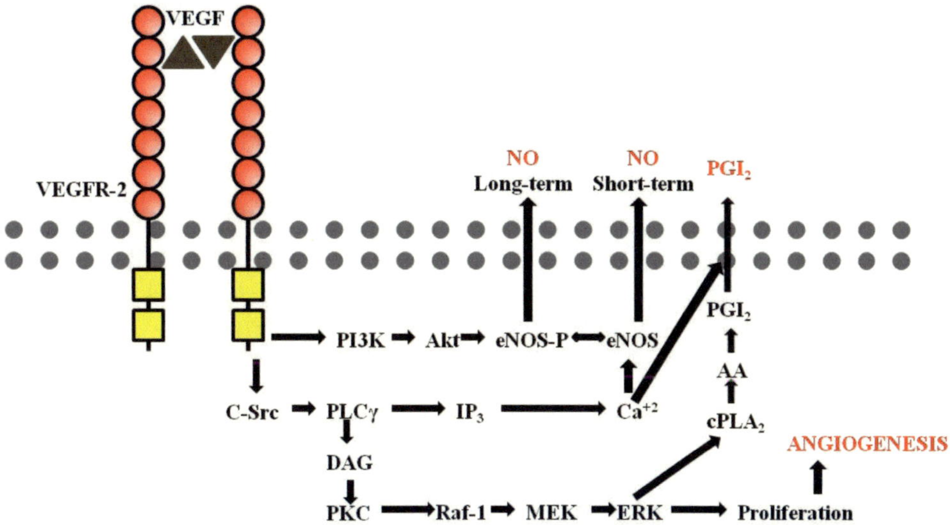

Fig. 3. Mechanisms mediating VEGF-induced NO and PGI2 synthesis. Short-term NO production induced by VEGF is mediated via increased cytosolic Ca^{+2}, resulting from activation of phospholipase C (PLC-gama) and subsequent generation of inositol 1,4,5-trisphosphate (IP_3). c-Src has been implicated in signaling upstream of PLC-γ. Activation of Akt leads to phosphorylation and activation of endothelial NO synthase (eNOS-P), providing a mechanism for sustained Ca^{+2}-independent NO synthesis. PLC-γ -mediated production of diacylglycerol (DAG) leads to activation of PKC, and this pathway plays an important role in mediating VEGF-induced activation of extracellular signal-regulated kinases (ERKs). In turn, ERK activation mediates cytosolic phospholipase A2 ($cPLA_2$)-mediated PGI_2 synthesis. Increased cytosolic Ca^{+2} also stimulates the cellular release of PGI_2 (AA, arachidonic acid).

FGFs are a family of proteins named for their biological activity in promoting the proliferation of fibroblasts in culture. Although the FGF family members follow a numerical designation (Ornitz and Itoh 2001), the designation FGF has only historical value, since FGFs are not only growth factors and their effects are not specifically or universally on fibroblasts. The FGF class of proteins comprises 23 members of homologous structure, all being fairly small polypeptides with a central core containing 140 amino acids. FGF1 (acidic FGF) and FGF2 (basic FGF) are preferentially involved in the process of angiogenesis (Ornitz and Itoh 2001; Barrientos, Stojadinovic et al. 2008). These compounds are polypeptides of about 18 kDa, single chained and non-glycosylated. They transmit their signals through FGF receptor-4 high-affinity, protein family of transmembrane tyrosine kinases (FGFR-1 to FGFR-4), which bind to different FGFs with different affinities. One characteristic of FGF1 and FGF2 is a strong interaction with glycosaminoglycans such as heparan sulfate, present in the extracellular matrix (Folkman, Klagsbrun et al. 1988). This interaction stabilizes FGFs against thermal and proteolytic denaturation, also limit its diffusibility. Thus, the extracellular matrix acts as a reservoir for pro-angiogenic factors. However, neither the use of signal peptide necessary for secretion or release mechanism of these growth factors have been determined to date (Werner and Grose 2003).

Most members of the FGF family act as a broad spectrum mitogen. They stimulate the proliferation of mesenchymal cells of mesodermal origin, as well as ectodermal and endodermal cells. In addition to their mitogenic effects, FGFs regulate the migration and differentiation of their target cells, also showing the cytoprotective function, which increases the survival of cells on adverse conditions (Ornitz and Itoh 2001; Werner and Grose 2003).

The factors FGF1 and FGF2 are synthesized by a variety of cell types involved in angiogenesis and wound healing, including inflammatory cells and dermal fibroblasts. They act on endothelial cells in a paracrine manner, liberated from the extracellular matrix, or in an autocrine way, when released by the endothelial cells themselves, promoting cell proliferation and differentiation. During the formation of granulation tissue, FGF2 promotes cell migration through surface receptors for integrins, which mediate the binding of endothelial cells to extracellular matrix (Barrientos, Stojadinovic et al. 2008).

In addition, many other growth factors and proteins interact during the orchestrated and complex healing process. Proteins such as TGF-beta also act as chemoattractants for neutrophils, macrophages and fibroblasts, stimulate the formation of granulation tissue, demonstrating its importance throughout the healing process. TGF-beta is an important modulator of angiogenesis during wound healing by regulating cell proliferation, migration, capillary tube formation and deposition of extracellular matrix (Brunner and Blakytny 2004; Verrecchia and Mauviel 2007).

2.2.4 Extracellular matrix

For the occurrence of endothelial cell migration and development of new tubular capillaries there is a dependence, not only on cells and cytokines present, but also of the production and organization of extracellular matrix components including fibronectin, collagen, vibronectina, tenascin and laminin, both in the granulation tissue and in the endothelial basement membrane. The extracellular matrix is important for normal growth and maintenance of vessels because, in addition to acting as a scaffold to cell migration, also acts as a reservoir and modulator of the release of growth factors such as FGF2 and TGF-β (Ruoslahti and Yamaguchi 1991; Brunner and Blakytny 2004).

Proliferation of endothelial cells, adjacent to and within the wound, leading to the deposition of the large amounts of fibronectin in the vessel wall (Pankajakshan and Krishnan 2009). Thus, angiogenesis requires the expression of receptors for fibronectin by endothelial cells (Brooks, Clark et al. 1994), organizing fibronectin as a conduit to allow their movement. Expression and activity of proteases are also necessary for angiogenesis, especially during remodeling.

2.3 Remodeling phase (Repair)

At this stage of healing, an attempt is made to recover the normal tissue structure. It is a period marked by maturation of the elements and by changes in the extracellular matrix, resulting in the deposition of collagen and proteoglycans. In a later stage, the fibroblasts of the granulation tissue are transformed into myofibroblasts responsive to contractile agonists that stimulate smooth muscle. As this occurs, a reorganization of the extracellular matrix takes place, making a final matrix, The balance between the processes that shape this determine the balance between regeneration and scarring (Desmouliere, Chaponnier et al. 2005).

In the process of maturation and remodeling, most vessels, fibroblasts and inflammatory cells disappear from the wound site through migration, apoptosis or other mechanisms of cell death. This leads to the formation of scar with a small number of cells. On the other hand, if the cells persist at the site, the formation of hypertrophic scars or keloids will occur (Mendonca and Coutinho-Netto 2009).

The main cytokines involved in this phase are tumor necrosis factor (TNF-α), interleukin (IL-1), PDGF and TGF-β produced by fibroblasts, and those produced by epithelial cells such as EGF and TGF-β (Karukonda, Flynn et al. 2000).

Re-epithelialization, which covers the wound with new epithelium and involves both migration and proliferation of keratinocytes from the periphery of the lesion, also occurs during the proliferative phase. These events are regulated by three main agents: growth factors, integrins and metalloproteases (Santoro and Gaudino 2005).

During the inflammatory phase, the release of growth factors in the plasma, fibroblasts and macrophages/neutrophils activate keratinocytes located at the margins of the wound. Among the growth factors stand out the PDGF that induces the proliferation of fibroblasts with consequent production of the extracellular matrix during wound contraction and reorganization of the matrix, the keratinocyte growth factor (KGF7) which is considered the main regulator of the proliferation of keratinocytes, and TGF-beta, the principal stimulus for the initial migration of epithelial cells. The activation of integrins by keratinocytes allows cellular interaction with a variety of extracellular matrix proteins in the margin and wound bed. On the other hand, the expression and activation of metalloproteinases promotes the degradation and modification of extracellular matrix proteins in the wound site, facilitating cell migration. The proteolytic activity of these enzymes can release growth factors bound to the extracellular matrix in order to maintain a constant stimulus for proliferation and migration of keratinocytes, accelerating the process of reepithelialization (Santoro and Gaudino 2005).

There are many diseases that interfere with the tissue repair process; they include diabetes, systemic sclerosis, anemia, malnutrition, among others. There are also many conditions that make this process difficult to resolve, preventing or delaying a complete tissue restoration. By obstructing tissue repair, such diseases or conditions potentially contributing to increased morbidity and mortality (Mrué, Coutinho-Netto et al. 2004; Mendonca, Mauricio et al. 2010).

3. Drugs

In recent decades, several studies have been carried out to identify substances capable of promoting the repair process. A search for substances with angiogenic activity has been intense, for their great potential for clinical application.

Among the substances that have direct action in the repair process there are some growth factors that, when applied topically to the wound, demonstrat a good ability to accelerate tissue repair in animal experiments (Mustoe, Pierce et al. 1991; Pierce, Tarpley et al. 1994). In this group, products based on recombinant human PDGF interfere directly in order to favor the repair process, showing good results in healing of ulcers in diabetic patients (Steed 1998). Other substances containing agents such as enzyme-based ointments DNAse and collagenase act to promote wound debridement (Hebda, Klingbeil et al. 1990) and in this way assist the course of restoration of tissue. The latter are widely used in clinical practice, but have low efficacy in healing chronic wounds. Some angiogenic growth factors and inhibitors are listed on table 1 (Ribatti 2009).

Angiogenic growth factors	**Angiogenesis inhibitors**
Angiogenin	Anastellin
Angiopoietin-1	Angioarrestin
Del-1	Angiostatin
Fibroblast growth factors	Antiangiogenic antithrombin III
Granulocyte colony-stimulating factor	CD59
Hepatocyte growth factor	Chondromodulin
Interleukin-8	Endostatin
Leptin	Heparinases I and III
Midkine	Human chorionic gonadotropin
Placental growth factor	Interferon alfa/beta/gamma
Plasminogen activator inhibitor-1 (low concentrations)	Interferon-inducible protein-10
Platelet-derived endothelial cell growth factor	Interleukin-12
Platelet-derived growth factor	2-methoxyestradiol
Pleiotrophin	Placental ribonuclease inhibitor
Progranulin	Plasminogen activator inhibitor-1 (high concentrations)
Proliferin	Proliferin-related protein
Transforming growth factor-alfa and beta	Retinoids
Tumor necrosis factor-alfa	Tetrahydrocortisol-S
Vascular endothelial growth factor	Thrombospondin-1
	Tissue inhibitors of matrix metalloproteinases
	Troponin
	Vasculostatin and Vasostatin

Table 1. Angiogenic growth factors and inhibitors*

Drug name	Company	Effect on angiogenesis	Indications
Bevacizumab (Avastin)	Genentech	Monoclonal antibody against VEGF-A.	Metastatic coloretal
Ranibizumab (Lucentis)	Genentech	Monoclonal antibody that binds active forms of VEGF-A.	Age-related macular degeneration
Pegaptanib (Macugen)	Pfizer	Selective VEGF inhibitor that binds extracellular VEGF(165)	Age-related macular degeneration
Cetuximab (Erbitux)	Imclone/ Bristol-Myers Squibb	Human-murine monoclonal antibody to EGFR	Metastatic colorectal cancer and squamous cell carcinoma of the head and neck
Panitumumab (Vectibix)	Amgen	Human monoclonal antibody to EGFR	Metastatic colorectal cancer
Trastuzumab (Herceptin)	Genentech	Human monoclonal antibody to HER-2	Adjuvant treatment of HER-2 overexpressing breast cancer and metastasis.
Sunitinib (Sutent)	Pfizer	Inhibitor of multiples RTKs (VEGFRs).	Advanced renal cell and gastrointestinal stromal tumors.
Sorafenib (Nexavar)	Bayer/Onyx	Inhibitor of multiples RTKs (VEGFRs and PDGFR).	Advanced renal cell and inoperable hepatocellular cancers
Erlotinib (Tarceva)	Genentech/OSI	Tyrosine kinase inhibitor of EGFR	Non-small cell lung and pancreatic cancers
Batimastat (British)	Biotech	MMP inhibitor	Vascular stents
Sirolimus/Rapamycin (Rapamune)	Wyeth-Ayerst	mTOR inhibitor, immunosuppressant	Prophylaxis of organ rejection
Temsirolimus (Torisel)	Wyeth	mTOR inhibitor	Advanced renal cell cancer
Everolimus (Xience V)	Abbot Afinitor, Novartis	mTOR inhibitor	Advanced renal cell cancer and vascular stents
Bortezomib (Velcade)	Millenium	Proteassome inhibitor, down regulation VEGF expression	Multiple myeloma and mantle cell lymphoma
Imiquimob (Aldara)	Graceway Pharmaceuticals	Immune modulator, induces production of angiogenic inhibitors	Actinic keratosis, superficial BCC and external genital warts
Thalidomide (Thalomid)	Celgene	Immune modulator, down-regulates expression of bFGF and VEGF	Multiple myeloma anda erythema nodosum leprosum

EGFR –endothelial growth factor receptor; HER-2 – human estrogen receptor 2; RTK – tyrosine kinase receptor; VEGFR – vascular endothelial growth factor receptor; PDGFR – platelet-derived growth factor receptor; MMP – matrix metalloproteinase; mTOR – mammalian targed of rapamycin; BCC – basal cell carcinoma.

Table 2. Antiangiogenics agents approved by FDA (Nguyen, Hoang et al. 2009).

On the other hand, there is much research on substances that could act to inhibit angiogenesis due mainly to their potential for the treatment of cancer. The first angiogenesis inhibitors were reported in the 1980s from the Folkman laboratory (interferon-gamma, administered at low doses). Subsequently, platelet-factor 4, tetrahydrocortisol, and by 1990, a fumagillin analogue were found to have potent antiangiogenic activity.

Angiostatin, an internal fragment of plasminogen, first revealed that an antiangiogenic peptide could be enzymatically released from a parent protein that lacked this inhibitory activity. Endostatin, an internal fragment of collagen XVIII, provided the first evidence that a basement-membrane collagen contained an angiogenesis inhibitory peptide. Thus new drugs with anti-angiogenic activity entered clinical trials. These drugs began to receive U.S. Food and Drug Administration (FDA) approval in the United States by 2003. Bevacizumab was the first angiogenesis inhibitor approved by the FDA (for colon cancer), and the first to demonstrate prolongation of survival in patients with advanced cancer (Folkman 2007). It is an anti-VEGF antibody, and the story of its discovery and manufacture describes a monumental achievement. However, certain non-endothelial cells (haematopoietic-derived cells that colonize tumour stroma and some cancer cells, such as those in pancreatic cancer) can also express receptors for vascular endothelial growth factor (VEGF; also known as VEGFA), raising the possibility that this drug might also have direct anti-tumor effects.

A new target in therapeutic treatment is the hypoxia-inducible factor 1 (HIF-1) an important regulator of cellular response to oxygen deprivation. HIF-1 is a heterodimeric protein that consists of alpha (HIF-1 alpha) and beta (HIF-1 beta) subunits. Under regular oxygen conditions, HIF-1alpha is continuously expressed but is rapidly destroyed by the proteasome pathway. Low oxygen tension results in a decrease in the rate of HIF-1 alpha polyubiquination and proteolysis, and consequent accumulation of the protein. Thus, HIF-1-alpha-HIF-1-beta heterodimers promote angiogenesis, tumor growth, and metastasis regulating the expression of many angiogenic factors. Some studies position mTOR as an upstream activator of HIF-1 function in cancer cells and suggest that antitumor activity of sirolimus (see table 2) is mediated through the inhibition of cellular responses to hypoxic stress (Nguyen, Hoang et al. 2009).

As the treatment range of angiogenesis inhibitors covers not only many types of cancer, but also unrelated diseases such as age-related macular degeneration and possibly others, angiogenesis inhibitors, or drugs that have varying degrees of antiangiogenic activity, might be defined as a class of drugs that specifically target an organizing principle in biomedicine.

4. References

Adamson, R. H., J. F. Lenz, et al. (2004). "Oncotic pressures opposing filtration across non-fenestrated rat microvessels." *J Physiol* 557(Pt 3): 889-907.

Andrikopoulou, E., X. Zhang, et al. (2011). "Current Insights into the role of HIF-1 in cutaneous wound healing." *Curr Mol Med* 11(3): 218-35.

Barrientos, S., O. Stojadinovic, et al. (2008). "Growth factors and cytokines in wound healing." *Wound Repair Regen* 16(5): 585-601.

Bates, D. O. and F. E. Curry (1997). "Vascular endothelial growth factor increases microvascular permeability via a Ca(2+)-dependent pathway." *Am J Physiol* 273(2 Pt 2): H687-94.

Bates, D. O. and S. J. Harper (2002). "Regulation of vascular permeability by vascular endothelial growth factors." *Vascul Pharmacol* 39(4-5): 225-37.

Brooks, P. C., R. A. Clark, et al. (1994). "Requirement of vascular integrin alpha v beta 3 for angiogenesis." *Science* 264(5158): 569-71.

Brunner, G. and R. Blakytny (2004). "Extracellular regulation of TGF-beta activity in wound repair: growth factor latency as a sensor mechanism for injury." *Thromb Haemost* 92(2): 253-61.

Delavary, B. M., W. M. van der Veer, et al. (2011). "Macrophages in skin injury and repair." *Immunobiology* 216(7): 753-62.

Desmouliere, A., C. Chaponnier, et al. (2005). "Tissue repair, contraction, and the myofibroblast." *Wound Repair Regen* 13(1): 7-12.

Detmar, M., L. F. Brown, et al. (1997). "Hypoxia regulates the expression of vascular permeability factor/vascular endothelial growth factor (VPF/VEGF) and its receptors in human skin." *J Invest Dermatol* 108(3): 263-8.

Dvorak, H. F. (2002). "Vascular permeability factor/vascular endothelial growth factor: a critical cytokine in tumor angiogenesis and a potential target for diagnosis and therapy." *J Clin Oncol* 20(21): 4368-80.

Dvorak, H. F. (2005). "Angiogenesis: update 2005." *J Thromb Haemost* 3(8): 1835-42.

Dvorak, H. F. (2010). "Vascular permeability to plasma, plasma proteins, and cells: an update." *Curr Opin Hematol* 17(3): 225-9.

Eichhorn, M. E., A. Kleespies, et al. (2007). "Angiogenesis in cancer: molecular mechanisms, clinical impact." *Langenbecks Arch Surg* 392(3): 371-9.

Eming, S. A., T. Krieg, et al. (2007). "Gene therapy and wound healing." *Clin Dermatol* 25(1): 79-92.

Eming, S. A., S. Werner, et al. (2007). "Accelerated wound closure in mice deficient for interleukin-10." *Am J Pathol* 170(1): 188-202.

Esser, S., K. Wolburg, et al. (1998). "Vascular endothelial growth factor induces endothelial fenestrations in vitro." *J Cell Biol* 140(4): 947-59.

Ferrara, N. (2001). "Role of vascular endothelial growth factor in regulation of physiological angiogenesis." *Am J Physiol Cell Physiol* 280(6): C1358-66.

Ferrara, N. (2004). "Vascular endothelial growth factor: basic science and clinical progress." *Endocr Rev* 25(4): 581-611.

Ferrara, N. and W. J. Henzel (1989). "Pituitary follicular cells secrete a novel heparin-binding growth factor specific for vascular endothelial cells." *Biochem Biophys Res Commun* 161(2): 851-8.

Folkman, J. (1971). "Tumor angiogenesis: therapeutic implications." *N Engl J Med* 285(21): 1182-6.

Folkman, J. (1990). "What is the evidence that tumors are angiogenesis dependent?" *J Natl Cancer Inst* 82(1): 4-6.

Folkman, J. (1991). Angiogenesis. *Biologic therapy of cancer*. V. De Vita, S. Hellman and S. A. Rosemberg. Philadelphia, J. B. Lippincott Co.: 743-753.

Folkman, J. (2007). "Angiogenesis: an organizing principle for drug discovery?" *Nat Rev Drug Discov* 6(4): 273-86.

Folkman, J. and P. A. D'Amore (1996). "Blood vessel formation: what is its molecular basis?" *Cell* 87(7): 1153-5.

Folkman, J., M. Klagsbrun, et al. (1988). "A heparin-binding angiogenic protein--basic fibroblast growth factor--is stored within basement membrane." *Am J Pathol* 130(2): 393-400.

Folkman, J. and Y. Shing (1992). "Angiogenesis." *J Biol Chem* 267(16): 10931-4.

Glass, C. A., S. J. Harper, et al. (2006). "The anti-angiogenic VEGF isoform VEGF165b transiently increases hydraulic conductivity, probably through VEGF receptor 1 in vivo." *J Physiol* 572(Pt 1): 243-57.

Hebda, P. A., C. K. Klingbeil, et al. (1990). "Basic fibroblast growth factor stimulation of epidermal wound healing in pigs." *J Invest Dermatol* 95(6): 626-31.

Henderson, B., S. Nair, et al. (2011). "Fibronectin: a multidomain host adhesin targeted by bacterial fibronectin-binding proteins." *FEMS Microbiol Rev* 35(1): 147-200.

Howdieshell, T. R., D. Callaway, et al. (2001). "Antibody neutralization of vascular endothelial growth factor inhibits wound granulation tissue formation." *J Surg Res* 96(2): 173-82.

Kaafarani, I., S. Fernandez-Sauze, et al. (2009). "Targeting adrenomedullin receptors with systemic delivery of neutralizing antibodies inhibits tumor angiogenesis and suppresses growth of human tumor xenografts in mice." *Faseb J* 23(10): 3424-35.

Karukonda, S. R., T. C. Flynn, et al. (2000). "The effects of drugs on wound healing--part II. Specific classes of drugs and their effect on healing wounds." *Int J Dermatol* 39(5): 321-33.

Kessler, T., F. Fehrmann, et al. (2007). "Vascular endothelial growth factor and its receptor as drug targets in hematological malignancies." *Curr Drug Targets* 8(2): 257-68.

Martin, P. and S. J. Leibovich (2005). "Inflammatory cells during wound repair: the good, the bad and the ugly." *Trends Cell Biol* 15(11): 599-607.

Mendonca, R. J. and J. Coutinho-Netto (2009). "Cellular aspects of wound healing." *An Bras Dermatol* 84(3): 257-62.

Mendonca, R. J., V. B. Mauricio, et al. (2010). "Increased vascular permeability, angiogenesis and wound healing induced by the serum of natural latex of the rubber tree Hevea brasiliensis." *Phytother Res* 24(5): 764-8.

Mrué, F., J. Coutinho-Netto, et al. (2004). "Evaluation of the biocompatibility of a new biomembrane." *Materials Research* 7(2): 277-283.

Mustoe, T. A., G. F. Pierce, et al. (1991). "Growth factor-induced acceleration of tissue repair through direct and inductive activities in a rabbit dermal ulcer model." *J Clin Invest* 87(2): 694-703.

Nagy, J. A., L. Benjamin, et al. (2008). "Vascular permeability, vascular hyperpermeability and angiogenesis." *Angiogenesis*.

Nguyen, A., V. Hoang, et al. (2009). "Angiogenesis in cutaneous disease: part I." *J Am Acad Dermatol* 61(6): 921-42; quiz 943-4.

Ornitz, D. M. and N. Itoh (2001). "Fibroblast growth factors." *Genome Biol* 2(3): REVIEWS3005.

Pankajakshan, D. and L. K. Krishnan (2009). "Design of fibrin matrix composition to enhance endothelial cell growth and extracellular matrix deposition for in vitro tissue engineering." *Artif Organs* 33(1): 16-25.

Pierce, G. F., J. E. Tarpley, et al. (1994). "Tissue repair processes in healing chronic pressure ulcers treated with recombinant platelet-derived growth factor BB." *Am J Pathol* 145(6): 1399-410.

Primo, L., G. Seano, et al. (2010). "Increased expression of alpha6 integrin in endothelial cells unveils a proangiogenic role for basement membrane." *Cancer Res* 70(14): 5759-69.

Rao, X., J. Zhong, et al. (2011). "Loss of methyl-CpG-binding domain protein 2 enhances endothelial angiogenesis and protects mice against hind-limb ischemic injury." *Circulation* 123(25): 2964-74.

Remensnyder, J. P. and G. Majno (1968). "Oxygen gradients in healing wounds." *Am J Pathol* 52(2): 301-23.

Ribatti, D. (2009). "Endogenous inhibitors of angiogenesis: a historical review." *Leuk Res* 33(5): 638-44.

Risau, W. (1997). "Mechanisms of angiogenesis." *Nature* 386(6626): 671-4.

Roth, D., M. Piekarek, et al. (2006). "Plasmin modulates vascular endothelial growth factor-A-mediated angiogenesis during wound repair." *Am J Pathol* 168(2): 670-84.

Ruoslahti, E. and Y. Yamaguchi (1991). "Proteoglycans as modulators of growth factor activities." *Cell* 64(5): 867-9.

Santoro, M. M. and G. Gaudino (2005). "Cellular and molecular facets of keratinocyte reepithelization during wound healing." *Exp Cell Res* 304(1): 274-86.

Schafer, M. and S. Werner (2008). "Cancer as an overhealing wound: an old hypothesis revisited." *Nat Rev Mol Cell Biol* 9(8): 628-38.

Simpson, D. M. and R. Ross (1972). "The neutrophilic leukocyte in wound repair a study with antineutrophil serum." *J Clin Invest* 51(8): 2009-23.

Singer, A. J. and R. A. Clark (1999). "Cutaneous wound healing." *N Engl J Med* 341(10): 738-46.

Steed, D. L. (1998). "Modifying the wound healing response with exogenous growth factors." *Clin Plast Surg* 25(3): 397-405.

Streit, M., P. Velasco, et al. (2000). "Thrombospondin-1 suppresses wound healing and granulation tissue formation in the skin of transgenic mice." *Embo J* 19(13): 3272-82.

Vaquero, J., M. Zurita, et al. (2000). "Expression of vascular permeability factor in glioblastoma specimens: correlation with tumor vascular endothelial surface and peritumoral edema." *J Neurooncol* 49(1): 49-55.

Verrecchia, F. and A. Mauviel (2007). "Transforming growth factor-beta and fibrosis." *World J Gastroenterol* 13(22): 3056-62.

Werner, S. and R. Grose (2003). "Regulation of wound healing by growth factors and cytokines." *Physiol Rev* 83(3): 835-70.

Zachary, I. (2001). "Signaling mechanisms mediating vascular protective actions of vascular endothelial growth factor." *Am J Physiol Cell Physiol* 280(6): C1375-86.

6

Platelet and Liver Regeneration

Nobuhiro Ohkohchi, Soichiro Murata and Kazuhiro Takahashi
Department of Surgery, University of Tsukuba
Japan

1. Introduction

Platelets are the smallest structures in circulating blood and have a convex disc construction with an equatorial diameter of 2-3 microns and have no nucleus. They are derived from megakaryocytes in the bone marrow. Following their normal life span of 8-10 days, they are removed from the circulation when passing through the spleen. Platelets have three types of secretory granules, i.e., alpha-granules, dense-granules, and lysosomal granules in the cytosol (Fig. 1). Each granule contains a specific mix of soluble factors, such as platelet-derived growth factor (PDGF), hepatocyte growth factor (HGF), insulin-like growth factor-1 (IGF-1), vascular endothelial growth factor (VEGF), serotonin, adenosine diphosphate (ADP), adenosine tri-phosphate (ATP), epidermal growth factor (EGF), and transforming growth factor beta (TGF-beta) (Blair et al., 2009; McNicol & Israels., 1999; Polasek et al., 2005). After being activated by physiological substances such as thrombin, collagen, thromboxane A2 (TXA2), epinephrine, and platelet-activating factor (PAF), or by non-physiological substances such as divalent cationophores and phorobol esters, platelets release these biologically active substances that exert various effects depending on the specific context (Holmsen, 1989; Suzuki H et al., 1992; Broos et al., 2011)

The main physiological role for circulating platelets is hemostasis when a vessel is injured (Holmsen, 1989). This process involves rapid adhesion of the platelets to the exposed subendothelium followed by platelet aggregation which culminates in the formation of a platelet plug that temporarily seals off the injured vessel walls. As they undergo this process, platelet activation leads to exocytosis of granular substances, release of newly synthesized mediators, and discharge of membrane-bond trans-cellular signaling molecules (Holmsen, 1989; Broos et al., 2011). Numbers of the various kinds of mechanisms facilitate platelet participation in other physiological or pathological process including inflammation (McNicol et al., 2008), malignancy (Mehta, 1984; Nash et al., 2002), immune response (Elzey et al., 2005; Sowa et al., 2009; Klinger & Jelkmann 2002; Sprague et al., 2008), wound healing (Mazzucco et al., 2010; Ranzato et al., 2009; Rozman & Bolta., 2007; Yamaguchi et al., 2010), and tissue regeneration (Radice et al., 2010; Dugrillon et al., 2002; Hartmann et al., 2010; de Vos et al., 2010; Rodeo et al., 2010).

Platelets have been reported to accumulate in the liver under some kinds of pathologic conditions, such as ischemia/reperfusion injury (Khandoga et al., 2003, 2006; Pak et al., 2010), liver cirrhosis (Zaldivar et al., 2010), cholestatic liver (Laschke et al., 2008) and viral hepatitis (Lang et al., 2008). Furthermore, platelets flow out slow, with rolling and adhesion in the liver sinusoids, under stressed situations such as ischemia/reperfusion injury (Nakano et al., 2008). Previous works on such conditions have focused on platelets as producers of inflammatory cytokines and therefore being pro-inflammatory (Pereboom et al., 2008). However, recent clinical and basic studies have revealed other ways in which they affect liver biology and pathology.

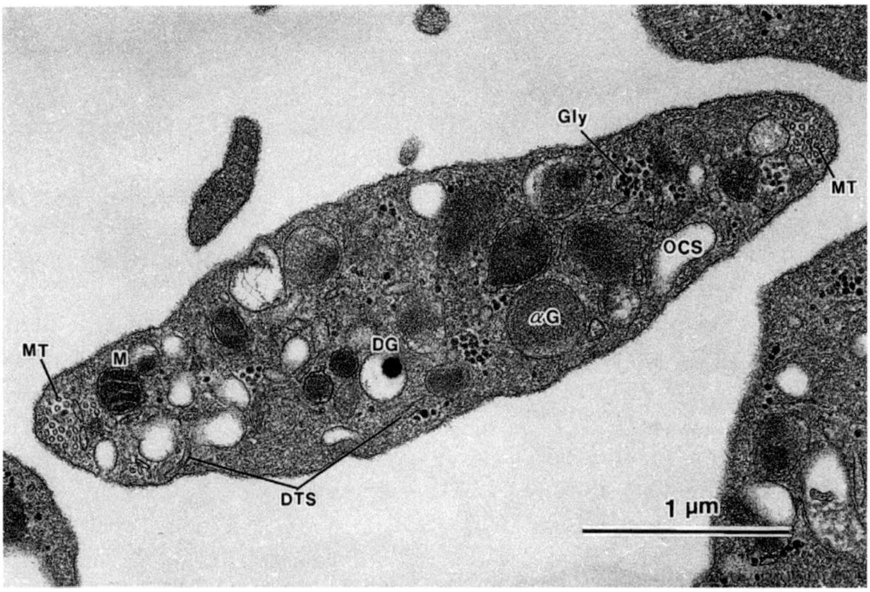

Fig. 1. Platelet ultrastructure. Transmission electron microscopic representation of a human platelets; the microtubules (MT), open cannalicular system (OCS), dense tubular system (DTS), mitochondria (M), alpha-granules (αG), dense granules (DG), and glycogen particles (Gly) are indicated. This electromicrograph is produced by kind permission of Dr. Hidenori Suzuki, Division of Morphological and Biomolecular Research, Graduate School of Medicine, Nippon Medical School.

In clinical studies, Marubashi et al. reported that there was a positive correlation between graft size and post-transplant thrombocytosis (Marubashi et al., 2006). Alkozai et al. described that a peri-operative low platelet count after partial hepatectomy was a predictor of delayed postoperative recovery of liver function and was associated with an increased risk of post-operative mortality (Alkozai et al., 2010). Kim et al. stated that total amount of platelet transfusion was positively associated with graft regeneration (Kim et al., 2010). In basic research, Lesurtel et al. showed that platelet-derived serotonin mediated liver regeneration in mice (Lesurtel et al., 2006). Nocito et al. demonstrated that platelets and platelet-derived serotonin promoted tissue repair after normothermic hepatic ischemia in mice (Nocito et al., 2007). In addition, we have obtained several types of evidence for platelets promoting liver regeneration using different experimental models of liver dysfunction in small and large animals.

In this chapter, we describe our evidence for platelets in promoting liver regeneration. Furthermore, we explain three different mechanisms i.e., 1) a direct effect on hepatocytes, 2) a cooperative effect with liver sinusoidal endothelial cells (LSEC), and 3) a collaborative effect with Kupffer cells.

2. Growth factors, cytokines, and signal transduction related to platelets' effect on liver regeneration

Liver regeneration occurs by proliferation of all of the existing mature cellular populations including hepatocytes, biliary epithelial cells, LSEC, Kupffer cells, and hepatic stellate cells. Of these, hepatocytes are the first cells to proliferate (Malik et al., 2002); they usually replicate once or twice following partial hepatectomy and return to the quiescent state. The kinetics of cell proliferation differ between species, the peak of DNA synthesis in hepatocytes usually being at 24 hours in rats but at 36 hours in mice (Michalopoulos & DeFrances., 1997; Michalopoulos, 2010; Fausto et al, 1995, 2000). Intercellular interactions mediated by many growth factors and cytokines, including HGF, tumor necrosis factor-alpha (TNF-apha), interleukin-6 (IL-6), transforming growth factor-beta (TGF-beta), EGF etc. appear to play important role in this process. Each growth factor leads subsequent activation of downstream transcription cascades that drive transition of the quiescent hepatocytes into the cell cycle and ensure progression beyond the restriction point in the G1 phase of the cycle. Several transcription factors are involved, and nuclear factor-kappa B (NF-KB) (Tewari et al., 1992; Cressman et al., 1994; FitzGerld et al., 1995), activator protein 1 (Ap-1) (Stepniak et al., 2006), CCAAT/enhancer binding protein-beta (C/EBPbeta) (Wang et al., 2008), extracellular signal-regulated kinase 1/2 (ERK 1/2) (Borowiak et al., 2004; Bard-Chapeau et al., 2006; Factor et al., 2010), signal transducer and activator of transcription 3 (STAT3) (Cressman et al., 1995; Li et al., 2002; Moh et al., 2007), and phosphatidylinositol-3-kinase (PI3K)/Akt (Jackson et al., 2008; Haga et al., 2005; Nechemia-Arbely et al., 2011) are representatives. Among these transcription factors and corresponding signaling transductions, the TNF-alpha/NF-KB, IL-6/STAT3, and PI3K/Akt pathways are considered the three major cascades through which platelets exert their effects on liver regeneration (Fig. 2).

The TNF-alpha/NF-KB pathway is activated within 30 minutes of partial hepatectomy and the activation usually lasts no longer than 4-5 hours (Michalopoulos & DeFrances, 1997). NF-KB is found in almost every cell including hepatocytes and non-paranchymal cells. It is a heterodimer composed of two subunits, p65 and p50, which are assembled in the cytosol (Solt & May, 2008). It is inactivated by Inhibitor of NF-KB (IKB) which binds to the p65 subunit. After being stimulated by TNF-alpha, NF-kB is activated by the removal of IkB from the p65 subunit; it migrates to the cell nucleus, where it binds to the promoter of cyclin-D1, which regulates G0/G1-to-S-phase transition (Hinz et al., 1999).

STAT3 is activated more slowly; it becomes detectable 1 to 2 hours after partial hepatectomy and lasts about 4-6 hours (Michalopoulos & DeFrances, 1997). IL-6 binding causes dimerization of the corresponding receptor and the activation of intracellular tyrosine kinase which phosphorylates gp130 and creates a docking site fof STAT3 (Heinrich et al., 1998). STAT3 is phosphorylated and translocates to the nucleus where it promotes the expression of cyclin-D1 and p21 to control the progression of the cell cycle (Turkson & Jove, 2000; Terui et al., 2005). It has been reported that hepatocytic mitosis of STAT3-knockout mice was significantly suppressed after partial hepatectomy in liver regeneration (Haga et al., 2009). The absence of STAT3 in hepatocytes exacerbates liver fibrosis during cholestasis (Shigekawa et al., 2011).

The PI3K/Akt pathway is activated immediately after partial hepatectomy (Murata et al., 2007). The pathway is initiated by the activation of the receptor tyrosine kinases or receptors coupled with G proteins by HGF, IL-6, TNF-alpha, TGF-beta and many other signaling molecules (Osawa et al., 2002; Okano et al., 2003; Tulasne & Foveau, 2008; Kato et al., 2009; Nechemia-Arbely et al., 2011). Met is a tyrosine kinase receptors on the surface of hepatocytes tha binds HGF (Bottaro et al., 1991; Tulasne & Foveau, 2008). HGF/c-met signaling activates PI3K which recruits Akt to the site of membranes, and subsequently phosphorylates Akt (Fresno et al., 2004). Glycogen synthase kinase 3-beta (GSK3-beta) acts downstream of Akt and plays a critical role in liver regeneration by regulating cell growth along with other downstream Akt factors, such as mTOR and 70^{S6K} (Faridi et al., 2003; Latronico et al., 2004; Haga et al., 2005). Phosphorylation of Akt results in activation of GSK3-beta by phosphorylation at serine-9, resulting in accumulation of beta-catenin and cyclin-D1 in the nucleus, which induce DNA synthesis and cellular mitosis of hepatocytes (Gotoh et al., 2003; Chen et al., 2005).

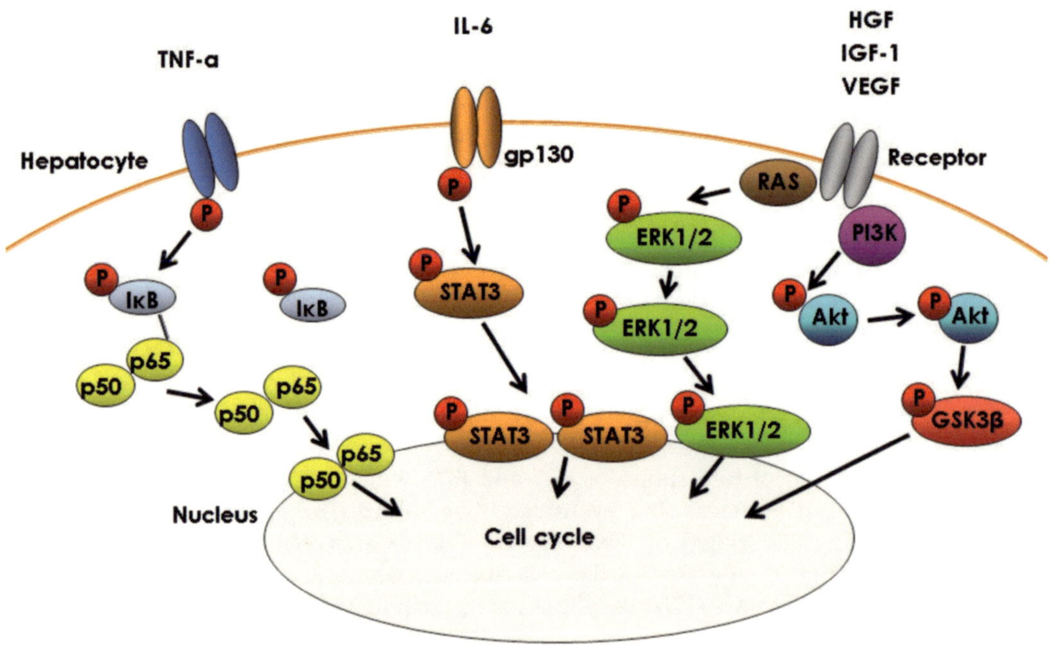

Fig. 2. Cytokines and growth factors for liver regeneration.

3. Effect of platelets on liver regeneration

Our first study was focused on liver regeneration under thrombocytotic conditions induced by thrombopoietin (TPO) (Murata et al., 2007). A 70% partial hepatectomy was carried out and mice were subsequently divided into three groups as follows; untreated mice as normal group, a thrombocytotic group, and a thrombocytopenic group. To induce thrombocytosis, mice were injected TPO. Anti-mouse platelet monoclonal antibody (Pm-1) was administrated to induce thrombocytopenia. Liver regeneration, cytokine and signaling pathways in the three groups were compared. Differences of platelet accumulation in the liver by using immunohistochemical staining technique were also observed.

The liver/body weight ratios in the thrombocytotic group and normal group were significantly higher than in the thrombocytopenic group, 48 hours after partial hepatectomy. In the thrombocytotic group, the liver/body weight ratio 48 hours after partial hepatectomy was significantly higher than that in normal group (Fig. 3A). The hepatocyte Ki-67 labeling index and hepatocyte mitotic index 48 hours after partial hepatectomy in the thrombocytotic group was obviously higher than that of normal and thrombocytopenic groups (Fig. 3B). Furthermore, the hepatocyte proliferating cell nuclear antigen (PCNA) labeling index 48 hours after partial hepatectomy in the thrombocytopenic group was remarkably lower than that in normal and thrombocytotic groups.

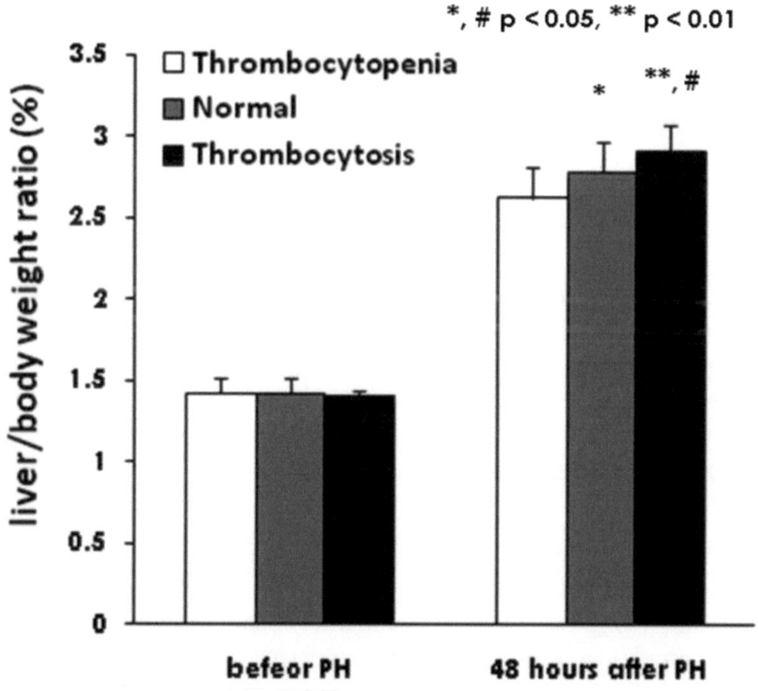

(A)

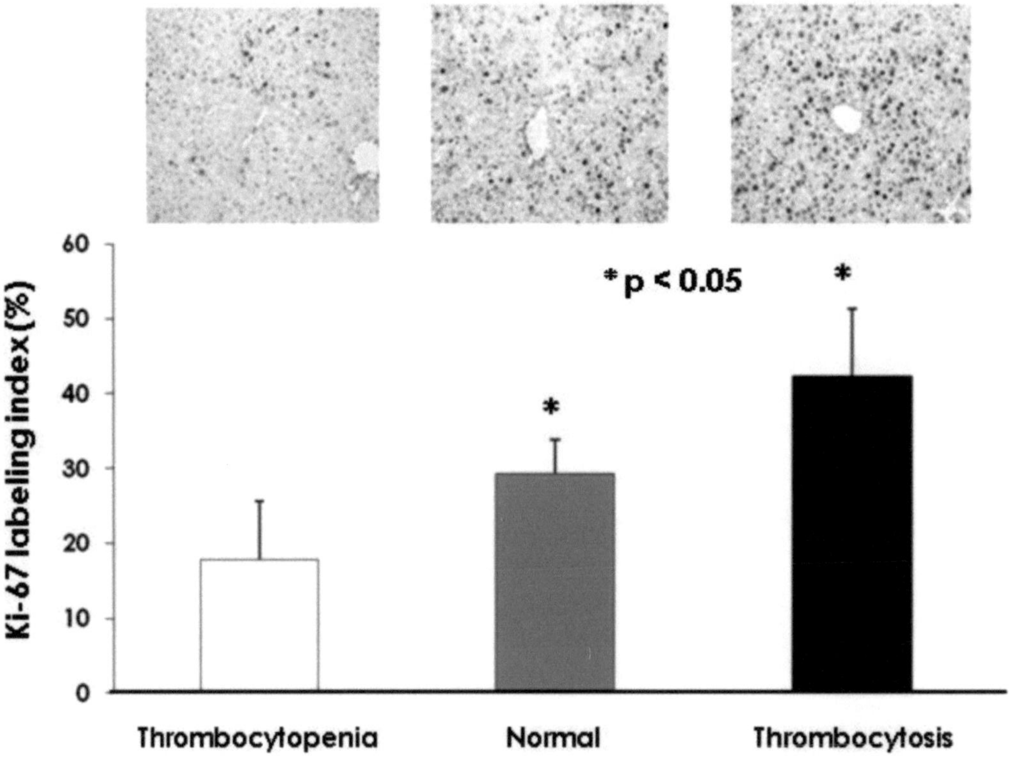

(B)

Fig. 3. Effect of thrombocytosis on liver regeneration after 70% of partial hepatectomy. (A) Liver/body weight ratio before and 48 hours after partial hepatectomy (PH) in thrombocytotic , normal and thrombocytopenic groups. Data were expressed as mean ± SD. *p < 0.05, **p < 0.01 versus thrombocytopenic group. #p < 0.05 normal group versus thrombocytotic group. (B) Ki-67 labeling index 48 hours after partial hepatectomy in thrombocytotic, normal and thrombocytopenic groups. Representative immunohistochemical images are shown. Data were expressed as mean ± SD. *p < 0.05 versus thrombocytopenic group. (Reproduced from Murata et al., 2007, World J Surg with permission.)

HGF and PDGF expression in the liver tissue in thrombocytotic group were significantly higher than in the normal and thrombocytopenic groups. Akt was strongly phosphorylated in the thrombocytotic group compared with the thrombocytopenic group. Activation of Akt in the thrombocytotic group started 5 minutes after partial hepatectomy and persisted for 2 hours. On the other hand, although activation of Akt was seen from 5 minutes after partial hepatectomy in the normal group, activation reduced in 2 hours after partial hepatectomy. In the thrombocytopenic group, Akt was not activated during the first 6 hours after partial hepatectomy. There was no difference in activation in ERK 1/2 and STAT3 among the three groups after partial hepatectomy (Fig. 4).

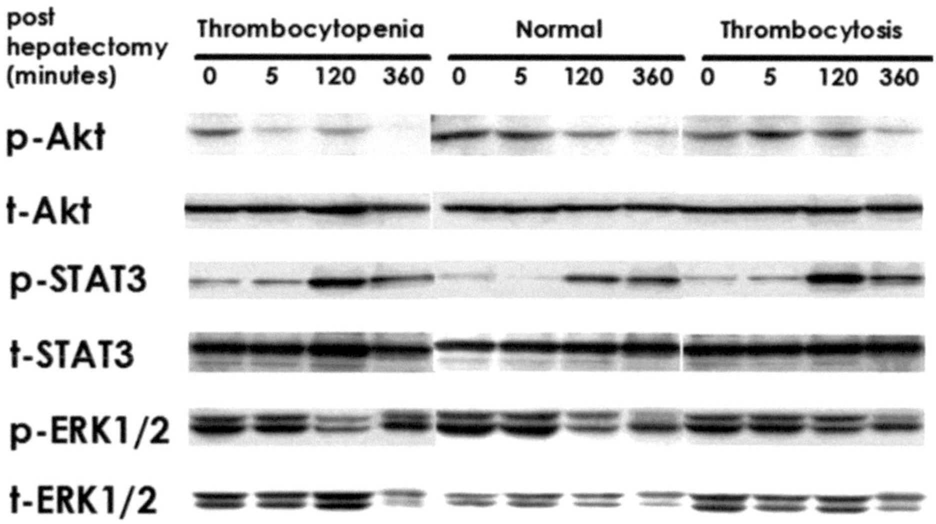

Fig. 4. Effect of platelet increment or reduction on Akt, STAT3, and ERK 1/2 after partial hepatectomy. (Reproduced from Murata et al., 2007, World J Surg. with permission).

Platelet accumulation in the liver was investigated in all groups before and 5 minutes after partial hepatectomy. Platelets accumulated in the residual liver within 5 minutes after partial hepatectomy and a two-fold increase in platelet levels was observed in the normal and thrombocytotic groups (Fig. 5A). On the other hand, in thrombocytotic group, platelets in the residual liver increased remarkably compared with normal and thrombocytopenic groups 5 minutes after partial hepatectomy. However, no increment was observed in thrombocytopenic group. In addition, under transmission electron microscopy, platelets translocated from the liver sinusoidal space into the space of Disse, and they had direct contact with hepatocytes in the thrombocytotic group (Fig. 5B).

These results suggest that platelets accumulate in the liver within a few minutes of partial hepatectomy and cause rapid hepatocyte proliferation through direct contact with hepatocytes, by translocating into the space of Disse.

Taken together, the results described above demonstrate that platelets affect liver regeneration in the acute phase after partial hepatectomy and suggested that the PI3K/Akt pathway is the main signaling pathway involved in platelet mediated liver regeneration.

The following study was done to investigate the role of platelets in liver regeneration using a thrombocytosis model in mice after 90% partial hepatectomy (Myronovych et al. 2008). All mice in the normal group died within 30 hours, predominantly between 20 and 30 hours. In contrast, the survival rate at 30 hours and at one week after partial hepatectomy was 54.5% and 27.3%, respectively in thrombocytotic group (Fig. 6A). Phosphorylation of Akt and STAT3 started earlier and stronger in thrombocytotic group than normal group. Serum albumin levels decreased in both groups after partial hepatectomy, but more rapidly in normal group, and there was a significant difference with higher levels being detected at 24 hours post-hepatectomy in the livers of the thrombocytotic group (Fig 6B). Serum

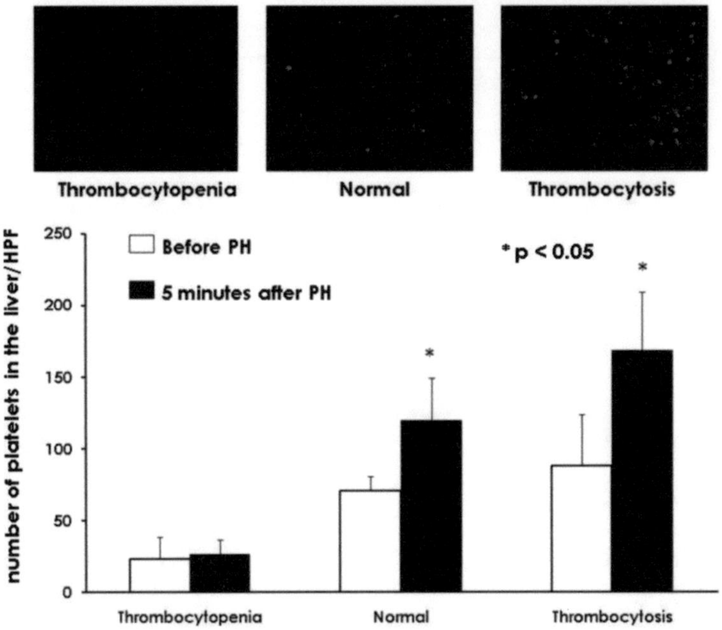

(A)

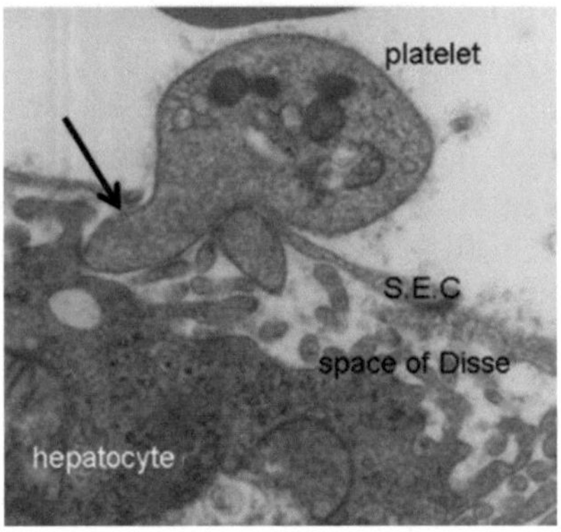

(B)

Fig. 5. Immunohistochemistry and Transmission electron microscopic photograph of the residual liver. (A) Immunohistochemistry of liver frozen section. Red; platelets. Platelets are stained by Pm-1 antibody. Representative images 5 minutes after partial hepatectomy (PH) are shown. Platelets were counted before and 5 minutes after partial hepatectomy in thrombocytopenic, normal, and thrombocytotic groups. Data were expressed as mean ± SD. *$p < 0.05$ versus before partial hepatectomy. Original magnification X 400. (B) Transmission electron microscopic image of partial hepatectomy in the residual liver 5 minutes after partial hepatectomy in thrombocytotic group. Arrow indicates platelet translocations into the space of Disse through the porosity of a flattened process in a sinusoidal endothelial cell (SEC). Original magnification X 7500. (Reproduced from Murata et al., 2007, World J Surg with permission.)

cholesterol levels were higher in thrombocytotic group at all time points with a significant difference at 24 hours after partial hepatectomy (Fig. 6C). In our measurement of insulin-like growth factor binding protein (IGFBP-1) by real-time PCR, the peak value of IGFBP-1 expression was reached sooner in the thrombocytotic group than in the normal group after partial hepatectomy and decreased moderately afterwards.

The findings described above indicated that liver regeneration occurs even in 90% hepatectomized mice under conditions of thrombocytosis. Platelets contribute to cell cycle progression and metabolic pathways, and maintain liver function after the extended hepatectomy.

We also evaluated the effect of TPO on liver regeneration after partial hepatectomy and on fibrosis under conditions of liver cirrhosis in rats (Murata et al., 2008). Rats were divided into three groups as follows; a normal group without any treatment, a liver cirrhosis (LC) group, and an LC group with a single administration of TPO (LC+TPO). 70% of partial hepatectomy was performed and liver regeneration and anti-fibrotic effects were compared.

In the LC group, the platelet count in the blood was significantly lower than that in the normal group. In the LC+TPO group, platelet count increased 2-fold higher than that in the normal group (Fig. 7A). The hepatocyte PCNA labeling index 24 hours after partial hepatectomy in the LC group was significantly lower than that in the normal group; the PCNA labeling index in the LC+TPO group was significantly higher than that in the LC group and the same level as that in normal group (Fig. 7B). HGF concentration in liver tissue in the LC+TPO group at the time of partial hepatectomy was clearly higher than that in the normal group. IGF-1 concentration in the liver tissue in LC+TPO group was significantly higher than that in normal group. Fibrotic change around the portal regions in the LC group was more prominent than that in the normal group. In contrast, fibrotic change decreased remarkably in the LC+TPO group (Fig. 7C)

These results described above indicated that a single administration of TPO in cirrhotic liver induces the remarkable increment of the platelets and then improves liver regeneration and fibrosis of cirrhotic liver after 70% of partial hepatectomy.

We examined whether the TPO itself or increased platelets have a hepatocyte-proliferative effect and anti-fibrotic effect on the fibrotic liver. We injected anti-platelet serum (APS) into LC+TPO group (LC+TPO+APS). The platelet count of LC+TPO+APS group decreased remarkably compared with LC and LC+TPO groups (Fig. 8A). PCNA labeling index 24 hours after partial hepatectomy was markedly lower in LC+TPO+APS group than that in LC and LC+TPO groups (Fig. 8B). Furthermore, liver fibrotic area before partial hepatectomy increased significantly in LC+TPO+APS group compared with LC+TPO group (Fig. 8C).

These results clearly indicate that acceleration of liver regeneration and anti-fibrotic effects of TPO administration are induced by increment of platelets, not by TPO itself.

We investigated whether exogenous platelets have the similar encouraging effect on liver regeneration. Platelet-rich plasma (PRP) was infused via the portal vein after 70% partial

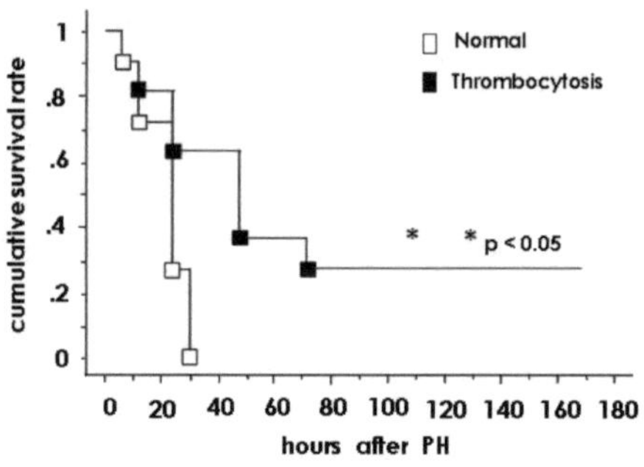

(A)

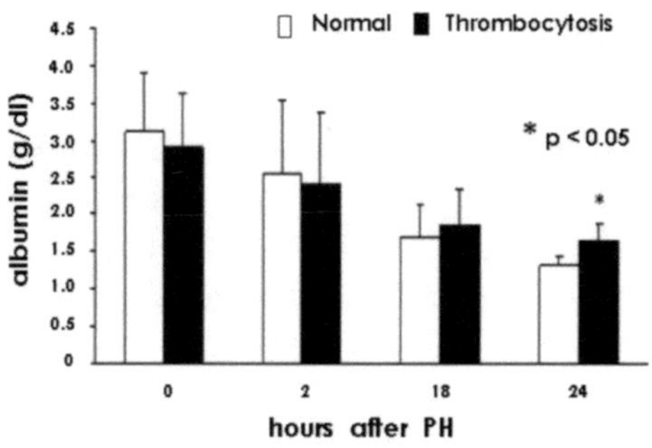

(B)

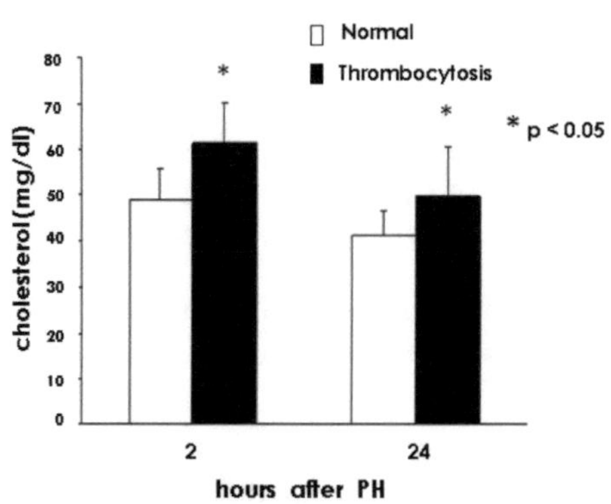

(C)

Fig. 6. Effect of thrombocytosis on survival and metabolic pathways after 90% of partial hepatectomy (PH). (A) Survival rate with Kaplan-Meier method. *$p < 0.05$ versus normal group. (B) Change in serum albumin concentration. Data were expressed as mean ± SD. *$p < 0.05$ versus normal group. (C) Change in serum total cholesterol concentration. Data were expressed as mean ± SD. *$p < 0.05$ versus normal group.
(Reproduced from Myronovych et al., 2008, J Hepatol with permission.)

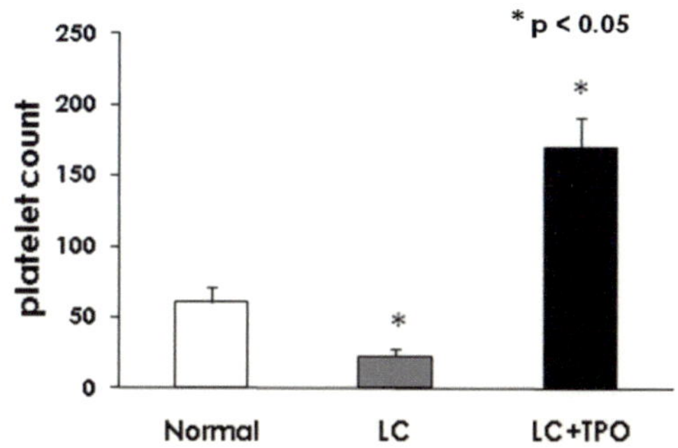

(A)

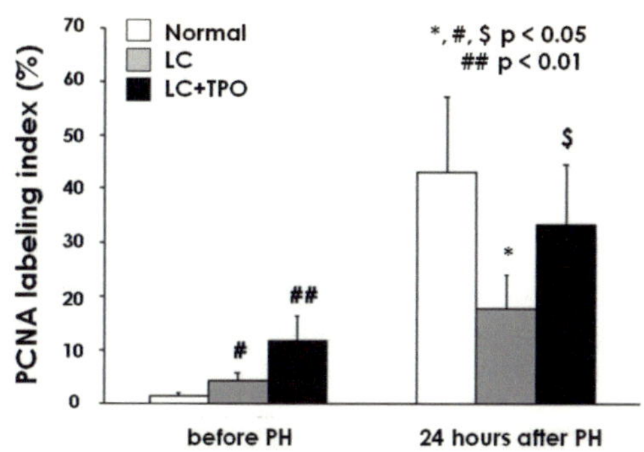

(B)

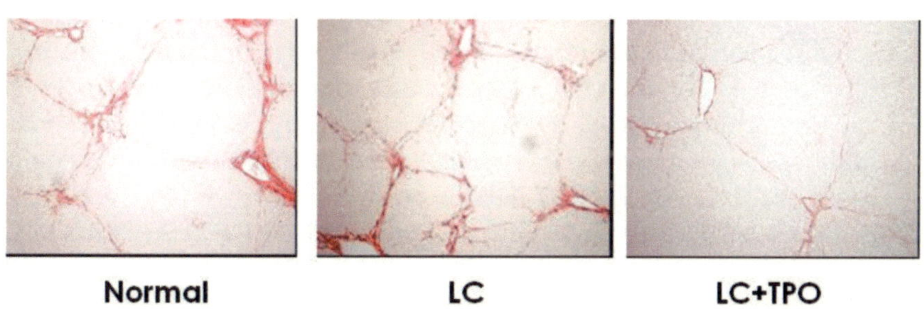

(C)

Fig. 7. Effect of TPO on platelet count, liver regeneration and fibrosis. (A) Platelet count before partial hepatectomy in normal, LC, and LC+TPO groups. Data were expressed as mean ± SD. *$p < 0.05$ versus normal group. (B) The hepatocyte PCNA labeling index before and 24 hours after partial hepatectomy (PH) in normal, LC, and LC+TPO groups. Data were expressed as mean ± SD. #$p < 0.05$, ##$p < 0.01$ versus normal group before partial hepatectomy. *$p < 0.05$ versus normal group 24 hours after partial hepatectomy. $\$p < 0.05$ versus LC group 24 hours after partial hepatectomy.
(C) Fibrotic change in the liver in normal, LC, and LC+TPO groups. Representative image in each group. Sirius red staining of liver sections. Original magnification × 200. (Reproduced from Murata et al., 2008, Ann Surg with permission.)

hepatectomy and residual liver regeneration was evaluated in rats (Matsuo et al., 2011). To clarify the mechanisms by which platelet promote liver regeneration, we also analyzed the dynamics of platelets infused in the liver before and after partial hepatectomy using an intravital microscope (IVM).

The liver/body weight ratio 24 hours after partial hepatectomy was significantly higher in PRP transfused group (PRP+) than in the normal saline administered group (PRP-). The hepatocyte Ki-67 labeling index was significantly higher in the PRP+ group than that in the PRP- group. Akt and ERK 1/2 became phosphorylated earlier in the PRP+ group than in the PRP- group, whereas phosphorylation of STAT3 did not apparently differ between the two groups. Under IVM, although platelets flowed fast and few of them rolled and adhered in the liver sinusoids before partial hepatectomy, a significant proportion of platelets accumulated in the liver sinusoids and flowed slowly with adhesion and rolling after partial hepatectomy. The findings in this experiment indicate that exogenous platelets also promote liver regeneration.

Since the anatomy of porcine liver is similar to that of the human, and porcine liver is useful for mimicking human liver surgery, we evaluated the effect of platelets in anti-liver damage and liver regeneration using pigs (Hisakura et al., 2010). To induce thrombocytosis, pigs received TPO administration (TPO+) or were performed splenectomy (Sp+). Pigs underwent 80% partial hepatectomy and were assigned to either TPO-, TPO+, Sp-, or Sp+ groups; liver damage, histological findings including necrotic changes, ballooning, cholestasis, and liver regeneration were compared among these groups. Serum aspartate aminotransferase levels in the TPO+ group were significantly lower than that in the TPO- group on day 2 after partial hepatectomy. Serum alanine aminotransferase levels in the Sp+ group were significantly lower than that in the Sp- group on day 2 after partial hepatectomy. Serum alkaline phosphatase levels in the TPO+ and the Sp+ groups were significantly lower than those in the TPO- and the Sp- group at 6 hours and on day 2 after partial hepatectomy. Histological analysis for cholestasis, ballooning, and hepatocyte necrosis was carried out by using a scoring system (Table. 1). Although cholestasis, ballooning, and hepatocyte necrosis were observed in zone 2 in TPO- and Sp- groups, structure was mostly preserved in TPO+ and Sp+ group (Fig. 9). On the other hand, the liver/body weight ratio and the hepatocyte PCNA labeling index showed no significant difference among the groups on day 2 and 7 after partial hepatectomy.

Under transmission microscopy, the sinusoidal endothelial linings were destroyed and detached into sinusoidal space with the enlargement of the spaces of Disse and the cytoplasm of sinusoidal endothelial cells was swollen with secondary lysosomes 2 hours after partial hepatectomy in TPO- or Sp- group (Fig. 10). In contrast, the structure of the endothelial lining was well preserved in TPO+ and Sp+ group.

Although there was no direct evidence of platelets in promoting liver regeneration in this experiment, the results indicated that increase in the number of platelets protect sinusoidal linings from disturbance and prevent acute liver damage after extended hepatectomy.

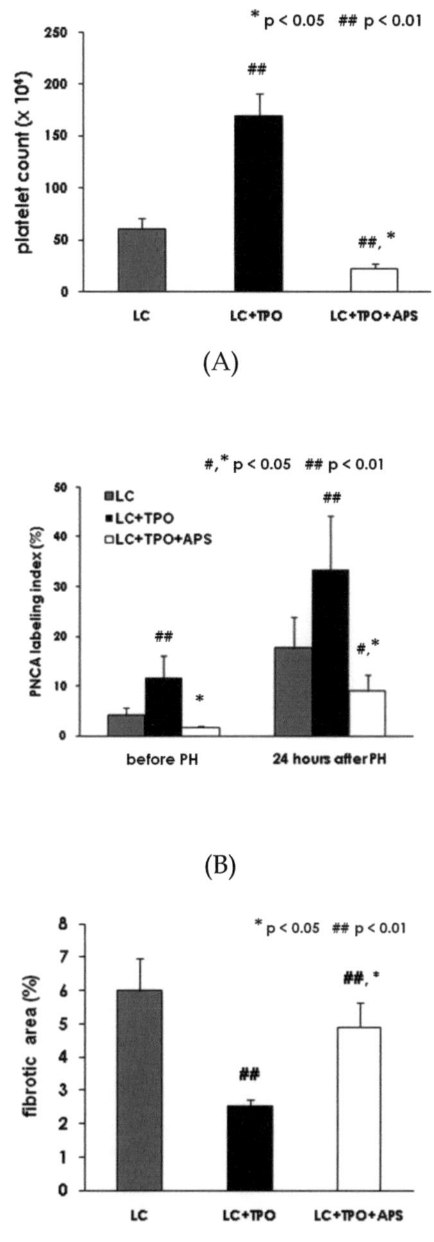

Fig. 8. Effect of platelets on liver regeneration and fibrosis of the liver. (A) Platelet count before partial hepatectomy (PH). (B) PCNA labeling index before and 24 hours after partial hepatectomy. (C) Fibrotic area of the liver in LC, LC+TPO, and LC+TPO+APS groups. Data were expressed as mean ± SD. #p < 0.05, ##p < 0.01 versus LC group. *p < 0.05 versus LC+TPO group. (Reproduced from Murata et al., 2008, Ann Surg with permission.)

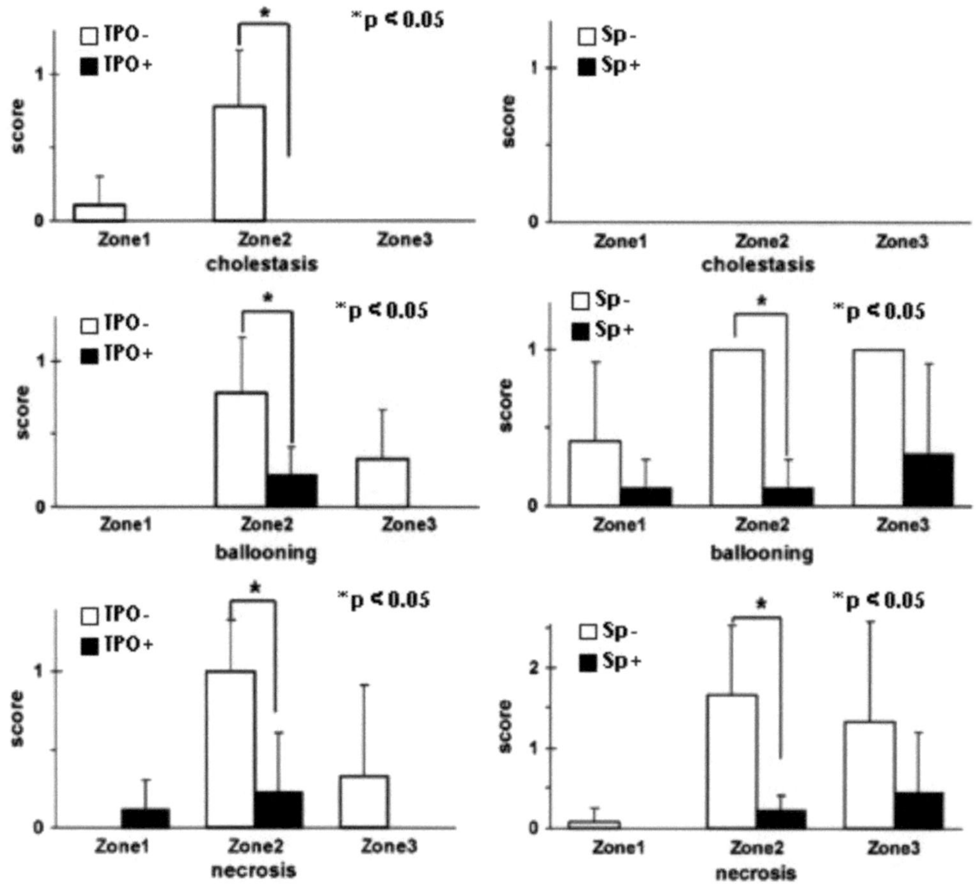

Fig. 9. Semiquantitative scoring for cholestasis, ballooning, and hepatocyte necrosis in TPO-, TPO+, Sp-, and Sp+ groups. Data are expressed as means ± SD. *$p < 0.05$ versus TPO- group or Sp- group. (Reproduced from Hisakura et al, 2010, J Hepatobiliary Pancreat Sci with permission.)

Feature		Scoring system
cholestasis	0	No
	1	Yes
Ballooning	0	No
	1	Yes
Necrosis	0	None
	1	Small foci
	2	Confluent areas
	3	Bridging necrosis

Table 1. Scoring system of cholestasis, ballooning, and hepatocyte necrosis. (Reproduced from Hisakura et al., 2010, *J Hepatobiliary Pancreat Sci* with permission.)

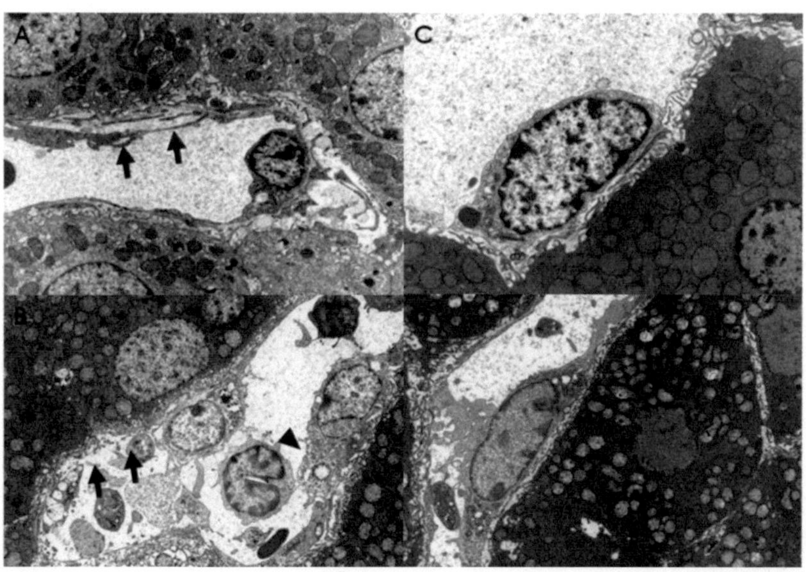

Fig. 10. Transmission electron microscopic findings 2 hours after partial hepatectomy. Magnification × 6000. In TPO- (A), and Sp- groups (B), the sinusoidal endothelial lining was destroyed and detached into the sinusoidal space with enlargement of the space of Disse (arrows), and the cytoplasm of sinusoidal endothelial cells were swollen with secondary lysosomes (arrow head). In contrast, in TPO+ (C), and Sp+ groups (D), the sinusoidal endothelial cells were well preserved. (Reproduced from Hisakura et al., 2010, *J Hepatobiliary Pancreat Sci* with permission.)

4. Mechanisms of direct effect of platelets on liver regeneration

Up to the beginning of the 21st century, there was no report regarding the effect of platelets on liver regeneration. Two studies were reported in which platelets promoted liver regeneration (Murata et al., 2004; Lesurtel et al., 2006). We reported that platelets accumulate in the liver and translocate actively to the space of Disse through fenestrae of LSECs after partial hepatectomy, which enables platelets to contact directly with hepatocytes (Murata et al., 2007). To clarify the role of the direct contact of platelets with hepatocytes, we investigated by using co-culturing chamber systems where platelets and hepatocytes were separated by a permeable membrane (Matsuo et al., 2008). To elucidate characteristics of the direct contact, four groups of separated co-culture were prepared as follows: a without platelet group (platelet-), a mixed co-culture group (co-mix), a separated co-culture group (co-sep), a group with mixed cells (the upper mix group: upper-mix), and the thrombin-stimulated group (thrombin stimulated) were prepared (Fig. 11). TLR2 cells, the murine immortalized primary hepatocytes, in the lower chamber were counted 72 hours after incubation. In the upper-mix group, platelets induced significant proliferation of hepatocytes in the lower chamber, whereas the proliferation in co-sep group was almost the

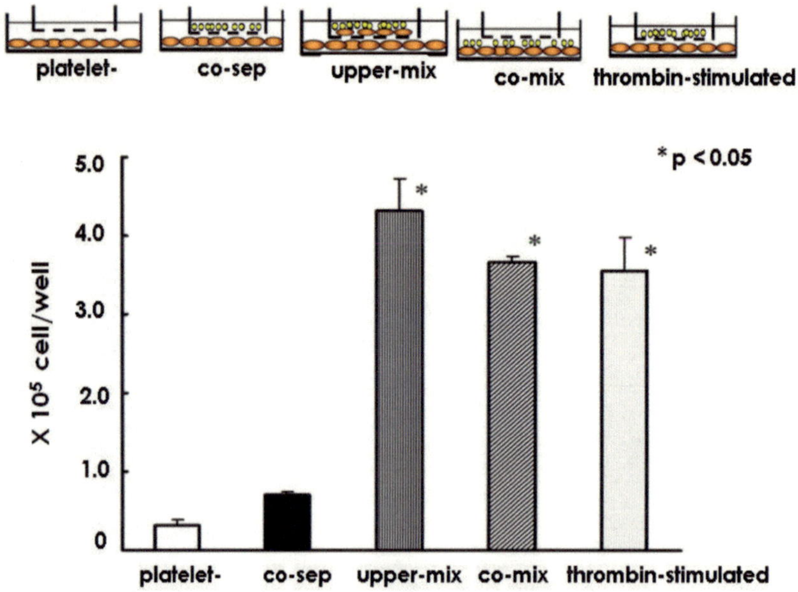

Fig. 11. Co-culture system to elucidate the characteristics of direct contact.
Without platelet group (platelet-): neither hepatocytes nor platelets were seeded in the upper chamber. Separated co-cultured group (co-sep): platelets were seeded in the upper chamber. Upper mix group (upper-mix): hepatocytes were seeded in the upper chamber and overlayered with platelets. Co-mixed group (co-mix): platelets and hepatocytes were seeded in the lower chamber. Thrombin stimulated group (thrombin-stimulated): platelets in the upper chamber were stimulated with thrombin to release soluble factors such as cytokines and growth factors. Hepatocytes in the lower chamber are counted after 72 hours of incubation. Data are expressed as means SD. *$p < 0.05$ versus platelet-.
(Reproduced from Matsuo et al., *J Surg Res* with permission.)

same level as that of platelet- group. Moreover, this effect in the upper-mix group was of the same level as in the co-mix group. These results indicate that, upon direct contact with hepatocytes, platelets released soluble factors that induce hepatocyte proliferation. A proliferative effect was also observed in the thrombin-stimulated group, despite there being no direct contact between platelets and hepatocytes (Fig. 11).

To clarify which component of platelets had an effect on hepatocyte proliferation, the mitogenic activity of the whole disrupted platelets, the soluble fraction, and the membrane fraction were evaluated. The whole disrupted platelets and the soluble fraction had significant proliferative effects, whereas the membrane fraction did not have the effect. To determine which element of the platelet soluble factor exerted the proliferative effect, platelet extracts were gel-excluded into 18 fractions, and mitogenic activity of each fraction was evaluated on BrdU assay. Mitogenic activity was strongly induced in the fraction of HGF, VEGF, and IGF-1 (Fig. 12A). In addition, when hepatocyte signals were analyzed in response to growth factors, HGF, IGF-1, and VEGF strongly activated the Akt and the ERK1/2 pathways, whereas PDGF and serotonin did not induced activation (Fig. 12B). For further confirmation of the platelets soluble factors, IGF-1 and HGF inhabitation using anti-IGF-1 and anti-HGF antibodies significantly inhibited hepatocyte proliferation.

The results of this examination indicated that the direct contact between platelets and hepatocytes triggered the release of soluble factors from the platelets such as IGF-1 and HGF, which caused a proliferative effect on the hepatocytes.

We assessed the direct effect of platelets using Kupffer cell depletion model (Murata et al., 2008). Liposome-encapsulated dichloromethylene diphosphonate (Cl2-MDP) was used for the depletion of Kupffer cells. Mice were divided into four groups as follows: mice without any treatment (normal), mice with Kupffer cell depletion (KD), mice with thrombocytosis caused by injection of thrombopoietin (thrombocytosis), and mice undergoing Kupffer cell depletion and thrombocytosis by injection of thrombopoietin (TKD). Each group of mice underwent 70% partial hepatectomy, and liver regeneration, cytokine and growth factors expression, and phosphorylation of Akt were assayed in the groups.

The liver/body weight ratio in KD group was significantly lower than that in normal group 48 hours after partial hepatectomy. The liver/body weight ratio in TKD group was almost the same as that in normal group. In thrombocytotic group, the liver/body weight ratio was significantly higher than that in normal group (Fig. 13 A). The hepatocyte mitotic index of the KD group 48 hours after partial hepatectomy was significantly lower than that in normal group. And, the hepatocyte mitotic index in the TKD group was almost the same as that in normal group. Furthermore, the hepatocyte mitotic index in thrombocytotic group was significantly higher than those in other groups. Moreover, the hepatocyte PCNA labeling index 48 hours after partial hepatectomy in the KD group was significantly lower compared with normal group. And, in the TKD group, it was significantly higher than that in KD group and the same as that in thrombocytotic and normal groups (Fig. 13 B).

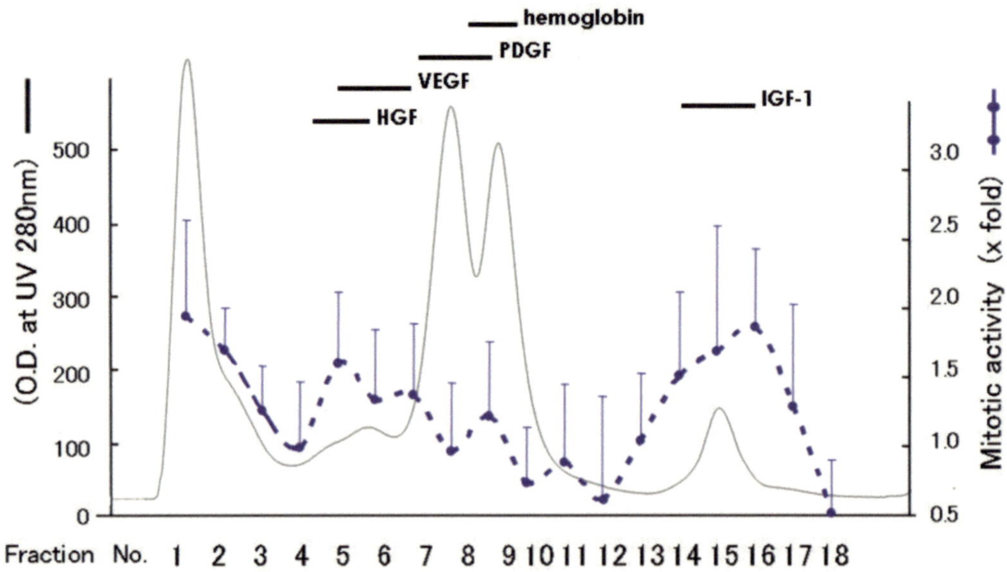

(A)

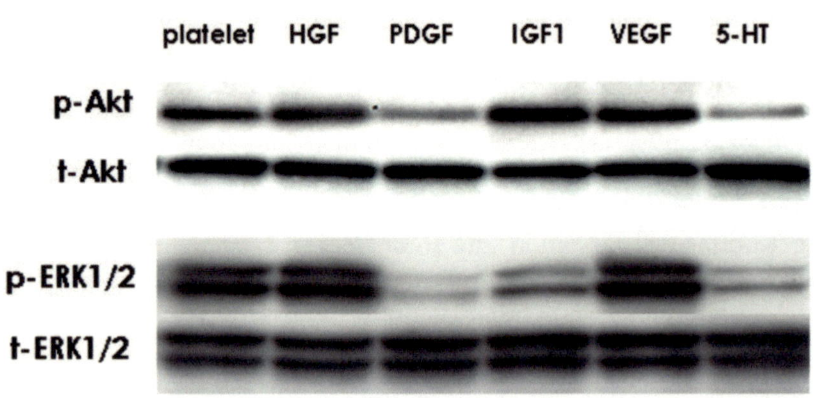

(B)

Fig. 12. Gel exclusion chromatography of platelet extracts and mitotic activities.
(A) The platelet extracts were gel filtrated on Superdex G200 gel. The solid line shows the resulting absorbance profile at 280 nm. The broken line shows the mitogenic activity of each fraction. Fraction 1 and 2 were nonspecifically macroaggregated proteins. Significant mitogenic activity was observed in fraction 1 ,2, 5-7, and 14-17. On western blotting, fractions 4-6 were rich in HGF, fraction 5-7 were rich in VEGF, fractions 7-9 were rich in PDGF, and fraction 14-17 were rich in IGF-1. Data were expressed as means ± SD of each experiments. (B) Cellular signals of hepatocytes stimulated by platelets and PDGF, HGF, PDGF, IGF-1, VEGF, and Serotonin (5-HT). (Reproduced from Matsuo et al., 2008, J Surg Res with permission.)

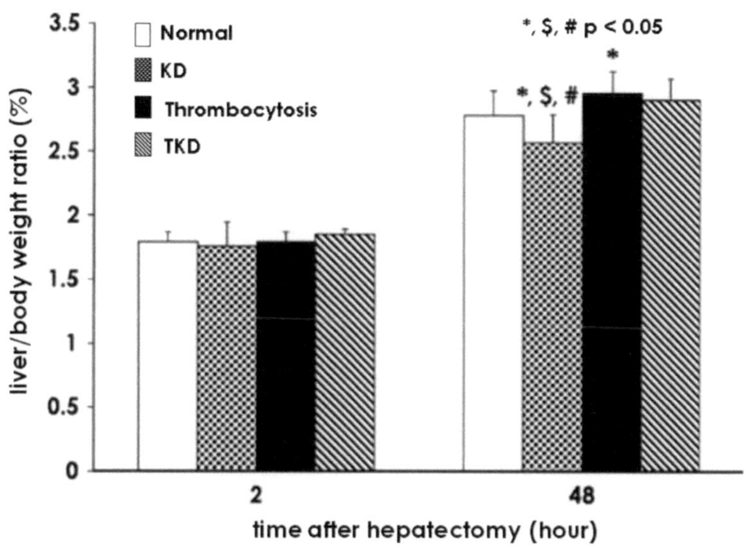

(A)

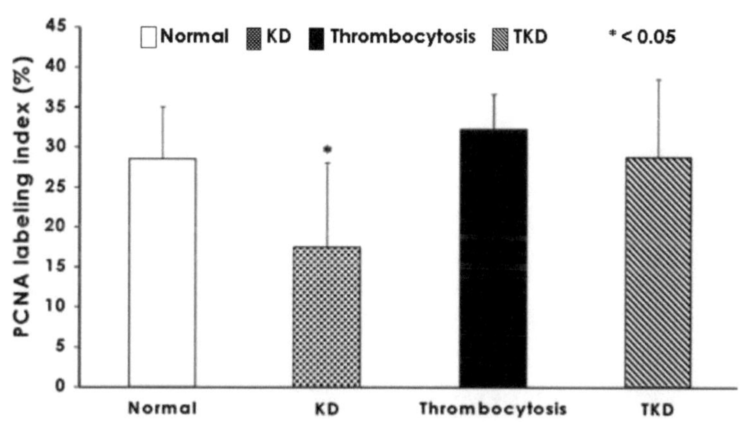

(B)

Fig. 13. Liver regeneration indexes under Kupffer cell depletion and thrombocytosis. (A) Liver/body weight ratio 2 and 48 hours after partial hepatectomy. Data are expressed as means ± SD. *p < 0.05 versus normal group, $p < 0.05 versus thrombocytosis; #p < 0.05 versus TKD group. (B) The hepatocyte PCNA labeling index 48 hours after partial hepatectomy. Data are expressed as means ± SD. *p < 0.05 versus normal group. (Reproduced from Murata et al., 2008, World J Surg with permission.)

The liver content of TNF-alpha, HGF, and IGF-1 was assessed in normal, KD, and TKD groups. TNF-alpha expression increased and reached the peak 2 hours after partial hepatectomy in normal group, whereas it remained low in KD and TKD groups. HGF concentration in the liver tissue in TKD group at the time of partial hepatectomy was significantly higher than that in normal group, and it persisted 6 hours after partial hepatectomy. At the same time, IGF-1 concentration in the liver tissue in KD and TKD groups at the time of partial hepatectomy was significantly lower than that in normal group, and IGF-1 concentration in the TKD groups was higher than that in the KD group. Furthermore, Akt was strongly phosphorylated in normal group compared with the KD group. In the TKD group, phosphorylation of Akt was started at the time of PH and lasted until 120 minutes after PH, and it was almost the same level as it was in normal group.

Platelet accumulation 2 hours after partial hepatectomy was investigated in each group. Platelet accumulation in the KD group demonstrated a significant decrease compared with the normal group. Moreover, platelet accumulation in the TKD group showed a significantly higher level than that in the KD group, and it was almost the same level as that in the normal group. In the thrombocytotic group, platelet accumulation increased significantly compared with the normal group. Transmission electron microscopy demonstrated that in the thrombocytotic group, platelets translocated from the liver sinusoidal space to the space of Disse and were in direct contact with hepatocytes at 5 minutes after hepatectomy.

These results clearly demonstrate that platelets promote liver regeneration under conditions of Kupffer cell depletion. Increase of platelets recruited platelets in the liver tissue and elevated and HGF concentrations in the liver, which activated downstream signaling transduction and hepatocyte mitosis.

In conclusion, our previous studies clarified the direct effect of platelets in promoting liver regeneration. The mechanism is explained as follows; after partial hepatectomy, platelets accumulate in the liver, they translocate to the space of Disse and release growth factors such as IGF-1 and HGF through direct contact with hepatocytes. The growth factors stimulate initiation of hepatocyte mitosis, which eventually promote liver regeneration. Especially in human, since it was reported that human platelets do not contain a significant amount of HGF (Nakamura et al., 1989), IGF-1 is the most important mediator for liver regeneration, which is contained in human platelets (Fig. 14).

5. The effect with liver sinusoidal endothelial cells

LSECs comprise 70% of the sinusoidal cells (Knook & Sleyster, 1976; Smedsrod et al., 1990). By construction of a thin and continuous layer, the sinusoidal endothelium forms the structural barrier, separating the hepatic parenchyma from blood constituents passing through the liver. Unlike other vascular endothelial cells, LSECs have large cytoplasmic gaps without basal membranes. These enable maximal contact between circulating blood and hepatocytes to help exchange various soluble macromolecules and nano-particles such as lipoproteins and endocytosis (Braet & Wisse, 2002). LSECs are involved in liver regeneration as well as Kupffer cells and hepatic stellate cells, and they are known to produce immunoregulatory and pro-inflammatory cytokines including HGF, interleukin-1 (IL-1), IL-6, and interferon. In addition, they synthesize eicosanoids, particularly TXA_2, prostaglandin E_2, as well as synthesizing important regulators of vascular tone, such

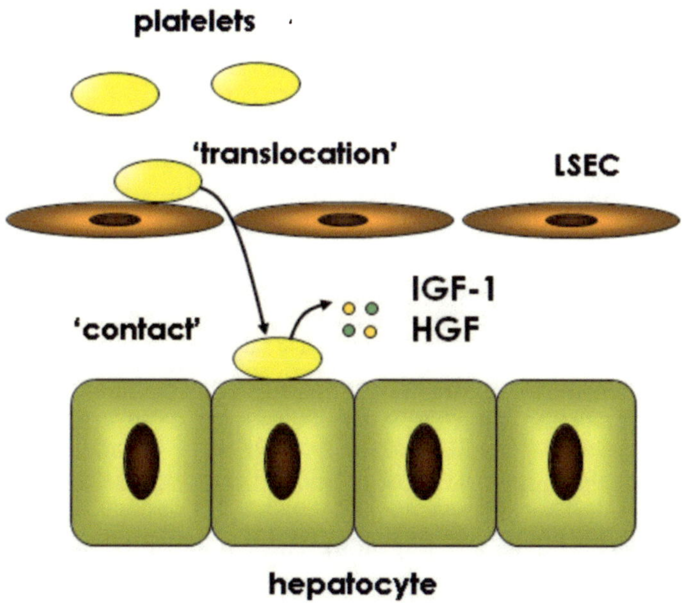

Fig. 14. Scheme for liver regeneration promoted directly by platelets.
Platelets translocate to the space of Disse and release growth factors such as IGF-1 and HGF through direct contact with hepatocytes. The growth factors subsequently induce initiation of mitosis.

as nitric oxide and endothelin (Wisse et al., 1996; Vollmar & Menger, 2009; Ping et al., 2006). IL-6 produced by LSECs and Kupffer cells is one of the important components of early signaling pathways in liver regeneration, and it activates the acute phase of protein synthesis by hepatocytes as part of the overall inflammatory response (Gauldie et al., 1992; Michalopoulos & DeFrances, 1997). After hepatectomy, plasma IL-6 concentration is reported to increase from 6 hours to a peak by 24 hours (Rai et al., 1996; Badia et al., 1998). IL-6 binds to its receptor on hepatocytes, which subsequently leads to phosphorylation of STAT3 monomers (Fausto et al., 2006). STAT3 homodimerizes and translocates to the nucleus, where it stimulates transcription of a number target genes such as cyclin-D1 and p21.

The relationship between platelets and LSECs has been well-documented in ischemia/reperfusion injury models. Rolling and adhering of leukocytes on LSECs with subsequent interaction with platelets is the important pathogenesis of ischemia/reperfusion injury (Montalvo-Jave et al., 2008; Croner et al., 2006, Pak et al., 2010). It was also reported that transient interaction, i.e., rolling, and permanent adhesion of platelets to the post-ischemic hepatic endothelium stimulate platelet activation and expression of endothelial adhesion molecules (Massberg et al., 1998; Khandoga et al., 2003). There have not, however, been any prior studies focused on the relationship between human platelets and LSECs with regards to liver regeneration.

To clarify the role of platelets in liver regeneration in relation to LSECs, we used co-culturing chamber systems where platelets and LSECs could be separated by a permeable membrane (Kawasaki et al., 2010). We used TMNK-1 cells (immortalized human LSECs), instead of primary LSECs, since their utility and efficiency was confirmed in the previous basic research (Matsumura et al., 2004).

Proliferation of LSECs and concentrations of IL-6 and VEGF in the supernatant were significantly higher in the group in which LSECs were co-cultured with human platelets (platelet+ or platelet+ mixed) than they were in the group in which LSECs were cultured without human platelets (platelet-) (Fig. 15A,B). However, when the platelets and LSECs were cultured separately (platelet+separated), no significant increase of IL-6 was observed (Fig. 15B). These results indicated that human platelets increase proliferation of LSECs and induced IL-6 release from LSECs and that the direct contact between platelets and LSECs is required for the production of IL-6. BrdU uptake of the primary hepatocytes in the group administered with the supernatant co-cultured with platelets and LSECs was significantly higher than that in the group administered with the supernatant cultured without platelets. When a specific antagonist for sphingosine 1-phosphate (S1P) 2 receptors were added to LSECs and co-cultured with platelets, the concentration of IL-6 showed significant decrease (Fig. 16A). On the contrary, the concentration of IL-6 was clearly increased in the group administered with S1P compared with those without S1P (Fig. 16B). These results revealed that S1P in platelets played important roles in liver regeneration by release of IL-6 from LSECs.

S1P is generally expressed in human plasma. It belongs to the class of lipid mediators and has been shown to regulate diverse biological processes, including proliferation, survival migration, or cytoskeletal reorganization (Yatomi et al., 2000; Xia & Wadham., 2011). S1P is produced from platelets and interacts with endothelial cells under the conditions of critical platelet-endothelial interactions, i.e., thrombosis, angiogenesis, and atherosclerosis (Yatomi et al., 2000). It was reported that the biological effect of S1P is partially mediated by endothelial nitric oxide synthetic activation and subsequent nitric oxide formation; extracellular S1P could contribute to sinusoidal protection and remodeling in alcoholic liver injury (Zheng et al., 2006). However, it was also described that S1P in human hepatic myofibroblast has an anti-mitogenic effect by increasing expression of TGF-β (Ikeda et al., 2003). As described above, S1P has various kinds of biophysical effects.

From the results of our experiment, the promotive effect of platelets on liver regeneration could be explained by follows; the direct contact between platelets and LSECs induce S1P release from platelets, which subsequently induce excretion of IL-6 from LSECs. LSEC-derived IL-6 promotes DNA synthesis of hepatocytes through STAT3 pathway (Fig. 17).

6. The role of Kupffer cells on liver regeneration

Kupffer cells are the principal constituents of the non-paranchymal cells of the liver (Malik et al., 2002). They locate within the lumen of the liver sinusoids, and are adherent to the LSECs. Kupffer cells play a role as macrophages against bacteria, bacterial endotoxins and microbial debris derived from gastrointestinal tract (Bilzer at al., 2006). Kupffer cells have been postulated to play a key role in liver regeneration after partial hepatectomy, and they could produce important biologically-active mediators that have both stimulatory and inhibitory influence on hepatocyte proliferation after hepatectomy (Boulton et al., 1998). Except for a report stating augmentation of the early phase of liver regeneration with Kupffer cell depletion (Meijer et al., 2000), depletion of Kupffer cells is basically well-known

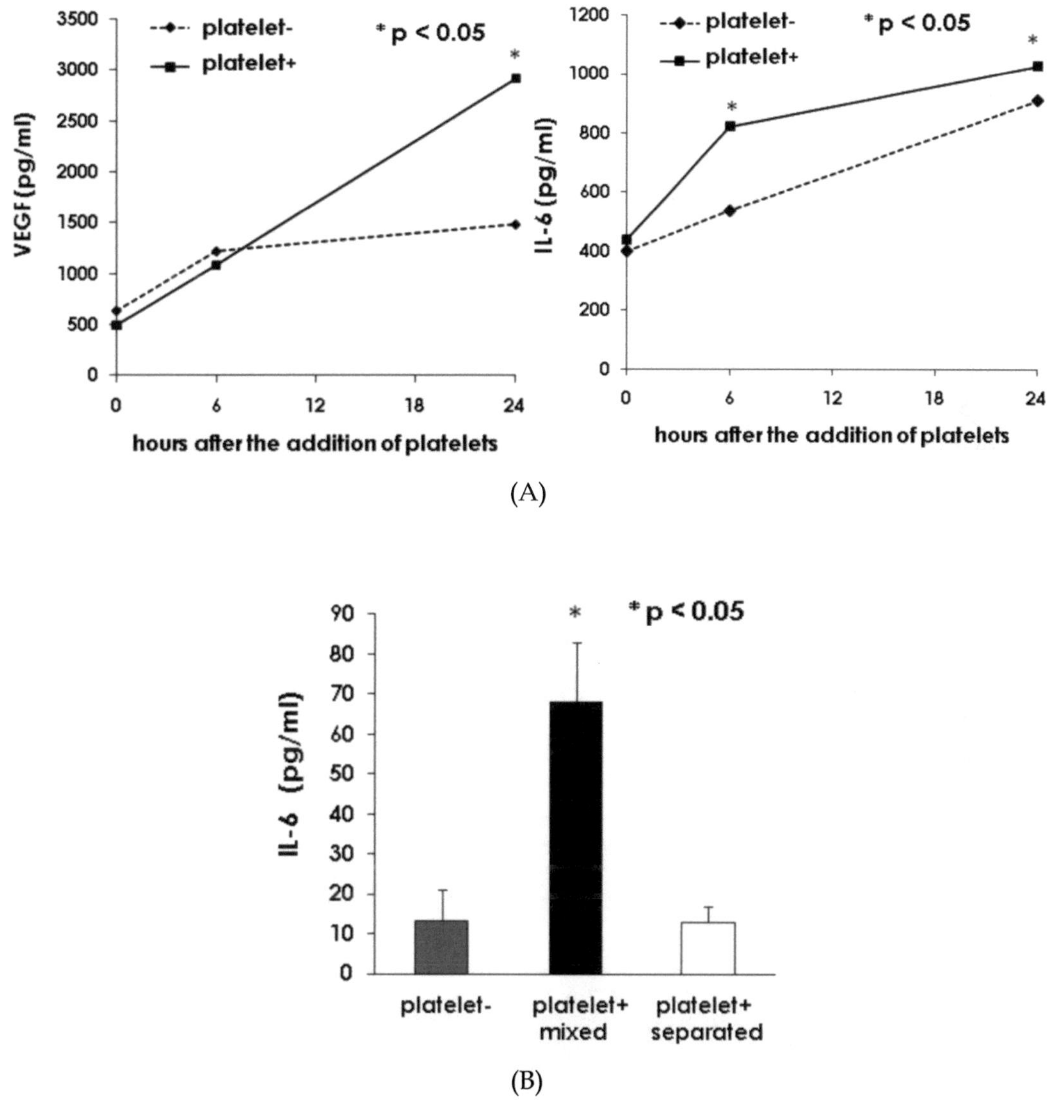

Fig. 15. Assay of IL-6 and VEGF in the supernatant of cultured LSECs after the addition of platelets, and the necessity of contact with platelets for excretion of IL-6 from LSECs. (A) The amounts of IL-6 and VEGF in the supernatant of LSECs were measured 0, 6, and 24 hours after the addition of platelets. Data are expressed as means ± SD. *p < 0.05 versus platelet– group. (B) To investigate the necessity of direct contact between platelets and LSECs, LSECs were cultured for 6 hours with platelets mixed (platelet+mixed) or platelets separated (platelet + separated), and the excretion of IL-6 from LSECs was measured. Data are expressed as means ± SD. *p < 0.05 versus platelet+separated group. (Reproduced from Kawasaki et al., 2010, *J Hepatol* with permission.)

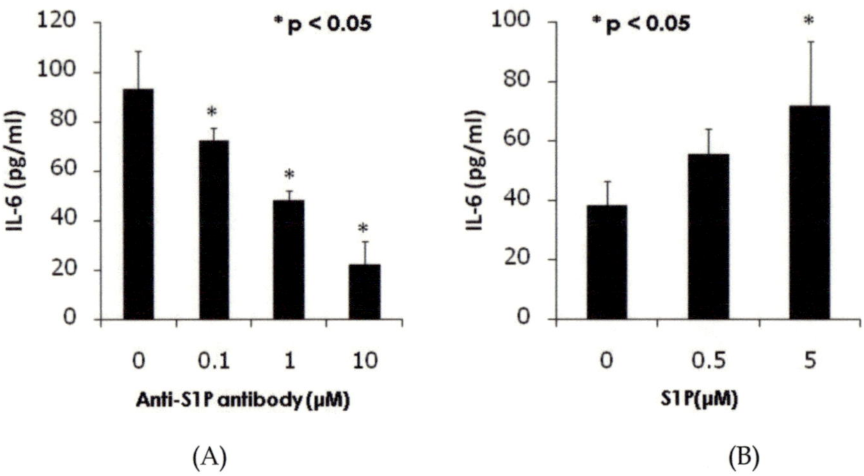

Fig. 16. Effects of inhibitor of S1P and S1P on excretion of IL-6 from LSECs
(A) Excretion of IL-6 from LSECs was evaluated using a specific antagonist for S1P2 receptors. LSECs were cultured with platelets for 6 hours, and the amount of IL-6 in the supernatant of LSECs was measured. Data are expressed as means ± SD. *p < 0.05 versus inhibitor- group. (B) To determine whether S1P had an effect on excretion of IL-6 from LSECs, LSECs were cultured with S1P for 6 hours, and the amount of IL-6 in the supernatant of TMNK-1 cells was measured. Data are expressed as means ± SD. *p < 0.05 versus S1P- group. (Reproduced from Kawasaki et al., 2010, *J Hepatol* with permission.)

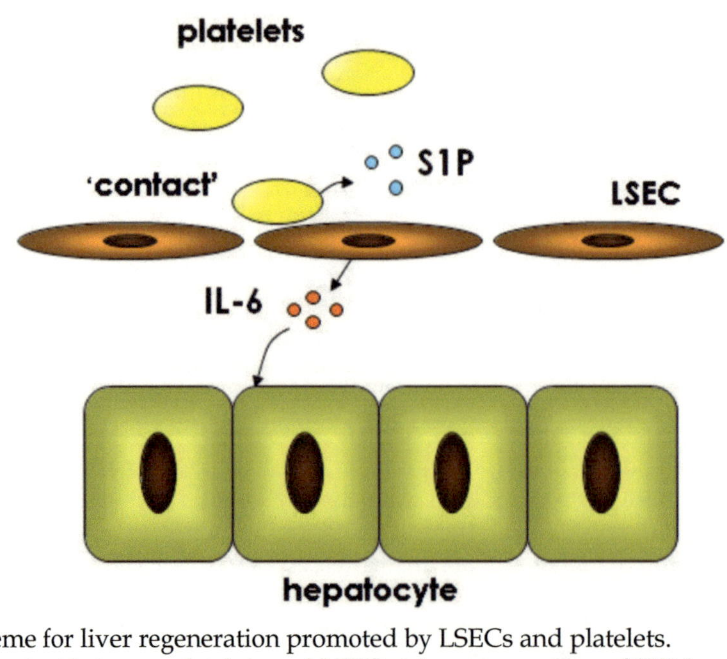

Fig. 17. Scheme for liver regeneration promoted by LSECs and platelets.
The direct contact between platelets and LSECs triggers excretion of S1P from platelets, which subsequently causes excretion of IL-6 from LSECs. IL-6 from LSECs promotes DNA synthesis of hepatocytes.

to exert an inhibitory influence on liver regeneration by alteration of hepatic cytokine expression (Takeishi et al, 1999). A critical early event following partial hepatectomy is the increase in plasma levels of TNF-alpha. In support of this view, an experiment using antibody against TNF-alpha has demonstrated significant reduction of hepatocyte proliferation (Akerman et al., 1992). Mice lacking TNF receptor-1 were shown to demonstrate severe impairment in liver regeneration (Yamada et al., 1998). Activation of the TNF receptor increases hepatic level of the NF-KB in both hepatocytes and non-paranchymal cells, and it is followed by production and release of IL-6 from Kupffer cells. Kupffer cells are assumed to be one of the most important sources of both TNF-alpha and IL-6 (Kahn et al., 1994; Decker., 1998). This is supported by the report that Kupffer cell-depleted mice failed to increase TNF-alpha, and IL-6 levels were equivalent to the level of Kupffer cell-competent mice after partial hepatectomy (Abshagen et al., 2007).

The relationship between platelet and Kupffer cells has been also well-documented in ischemia/reperfusion injury models. Platelets act in concert with the activated Kupffer cells and leukocytes, and a triangular interaction between these cells has been demonstrated as the main mechanism of the injury (Vollmar & Menger, 2009; Sindram et al., 2001). It was reported that when rats with depletion of Kupffer cells were subjected to ischemia and reperfusion, platelet adhesion in sinusoids was suppressed and, as consequence, attenuation of sinusoidal perfusion failure and endothelial damage were seen (Nakano et al., 2008). It is also reported that Kupffer cells produce PAF, which is a potent phospholipid mediator of platelet aggregation (Karidis et al., 2006). PAF is also believed to play important roles in the acute liver injury with ischemia/reperfusion (Karidis et al., 2006; Toledo-Pereyra & Suzuki, 1994), liver graft dysfunction (Hashikura et al., 1994), and post-operative liver failure after extended hepatectomy (Mizuno et al., 2001). As shown above, the role of platelets in relation to Kupffer cells have been described mainly with inflammatory injuries of the liver.

Nakamura et al. described a different character of Kupffer cell function associated with platelets. They reported that in response to LPS, IL-1, and TNF-alpha, platelets accumulated in the liver and a large number of platelets were found in the space of Disse (Endo et al., 1992, 1993; Nakamura et al., 1998). They also observed that platelets in the liver sinusoids were mostly surrounded by well-developed cell processes of Kupffer cells without being phagocytosed (Nakamura et al., 1998). However, depletion of Kupffer cells resulted in abolition of hepatic accumulation and migration of platelets (Nakamura et al., 1998). Although the precise mechanism was not clear, these reports indicated that cellular interaction between platelets and Kupffer cells plays an important role in platelet behavior in the liver.

Previously, we reported that even under condition of Kupffer cell depletion, platelets accumulated in the liver in the thrombocytotic state and promoted liver regeneration by direct contact with hepatocytes through their migration from the liver sinusoidal space to the space of Disse (Murata et al., 2008). In our recent study using SCID mice with human platelet transfusion, we demonstrated that concentrations of mouse-derived TNF-alpha and IL-6 in the liver tissue after 70% of partial hepatectomy were significantly higher in the mice with platelet transfusion than in the mice without transfusion. These results may indicate that platelet transfusion enhances secretory activity of Kupffer cells after hepatectomy. Furthermore, in the mice with platelet transfusion, significant accumulation and activation of platelets transfused were observed in the liver after hepatectomy. Although a few platelets transfused were adhering to the Kupffer cells in the mice without hepatectomy, the majority of platelets adhered to the surface of Kupffer cells in the mice with hepatectomy. It is insufficient to conclude only from these findings, however, it was assumed that platelets promote liver regeneration by interactions with Kupffer cells after hepatectomy. In other words, after hepatectomy, Kupffer cells induce accumulation and activation of platelets in the liver by direct adhering, and function of Kupffer cells are enhanced by the accumulated platelets. As described above, liver regeneration is promoted by the direct effect of growth factors released from platelets and by the paracrine effect of Kupffer cells enhanced by platelets (Fig. 18).

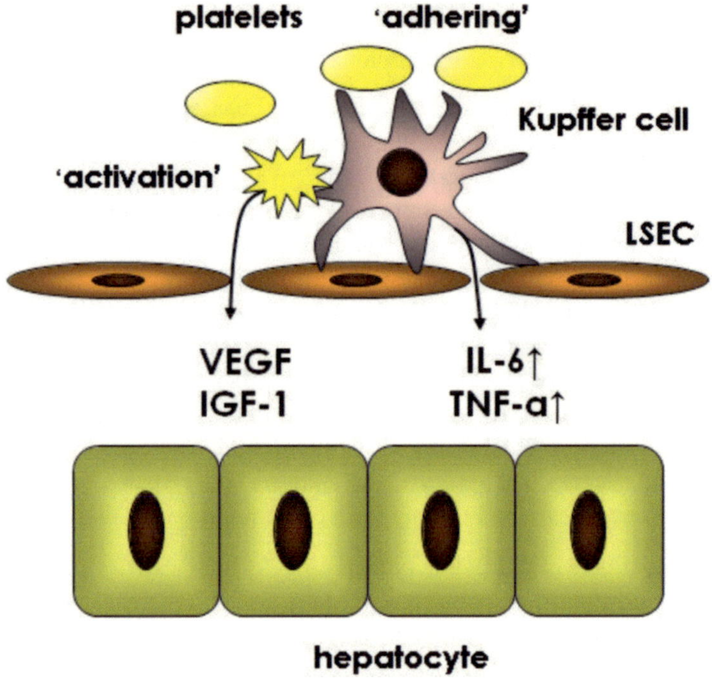

Fig. 18. Scheme for liver regeneration promoted by Kupffer cells

After partial hepatectomy, the activated Kupffer cells induce accumulation and activation of platelets through direct adhering. By the direct effect of growth factors released from platelets, and by the paracrine effect of Kupffer cells enhanced by platelets, liver regeneration is promoted.

7. Conclusion

In this chapter we have described our previous reports of platelets in promoting liver regeneration and the three different mechanisms by which platelete promote liver regeneration, i.e., 1) the direct effect on hepatocytes, 2) the cooperative effect with LSECs, and 3) the collaborative effect with Kupffer cells. Platelets are blood components that contain various kinds of biologically-active growth factors and cytokines. Nowadays artificial platelets (Bode & Fischer, 2007; Okamura et al., 2009), TPO formulae (Rhodes & Stasi, 2010), and freeze-dried platelets (Hoshi et al., 2007) are being developed and are beginning to be utilized in clinical settings; the importance and effects of platelets will become more apparent in the near future. With several lines of evidence showing platelets to be effective in anti-fibrosis (Watanabe et al., 2008; Kodama et al., 2010), anti-apoptosis (Hisakura et al., 2011), and liver regeneration, platelet therapy would open a new avenue to develop novel strategies for the treatments of liver diseases. Through these researches, we believe that platelet therapy could offer a therapeutic strategy for liver regeneration after extended hepatectomy, liver injuries or small grafts in liver transplantation.

8. Acknowledgement

The authors thank Dr. N. Kobayashi, Department of Surgery, Okayama University Graduate School of Medicine and Dentistry, for providing the sinusoidal endothelial cell line-TMNK-1. The authors also thank Dr. N. Yanai, University of Tohoku, for providing the hepatocyte cell line-TLR2.

9. References

Abshagen, K.; Eipel, C.; Kalff, JC.; Menger, MD. & Vollmar, B. (2007). Loss of NF-kappaB activation in Kupffer cell-depleted mice impairs liver regeneration after partial hepatectomy. *Am J Physiol Gastrointest Liver Physiol,* Vol.292, No.6 (June 2007), pp.1570-1577.

Akerman, P.; Cote, P.; Yang, SQ.; McClain, C.; Nelson, S.; Bagby, GJ. & Diehl, AM. (1992). Antibodies to tumor necrosis factor-alpha inhibit liver regeneration after partial hepatectomy. *Am J Physiol,* Vol.263, No.4Pt1, (October 1992), pp. 579-585.

Alkozai, EM.; Nijsten, MW.; de Jong, KP.; de Boer, MT.; Peeters, PM.; Slooff, MJ.; Porte, RJ. & Lisman, T. (2010). Immediate postoperative low platelet count is associated with delayed liver function recovery after partial liver resection. *Ann Surg,* Vol.251, No.2, (February 2010), pp. 300-306.

Badia, JM.; Ayton, LC.; Evans, TJ.; Carpenter, AJ.; Nawfal, G.; Kinderman, H.; Zografos, G.; Uemoto, S.; Cohen, J. & Habib, NA. (1998). Systemic cytokine response to hepatic resections under total vascular exclusion. *Eur J Surg,* Vol.164, No.3, (March 1998), pp. 185-190.

Bard-Chapeau, EA.; Yuan, J.; Droin, N.; Long, S.; Zhang, EE.; Nguyen, TV. & Feng, GS. (2006). Concerted functions of Gab1 and Shp2 in liver regeneration and hepatoprotection. *Mol Cell Biol,* Vol.26, No.12, (June 2006), pp. 4664-4674.

Bilzer, M.; Roggel, F. & Gerbes, AL. (2006). Role of Kupffer cells in host defense and liver disease. *Liver Int,* Vol.26, No.10, (December 2006), pp. 1175-1186.

Blair, P. & Flaumenhaft, R. (2009). Platelet alpha-granules: basic biology and clinical correlates. *Blood Rev,* Vol.23, No.4, (July 2009), pp. 177-189.

Bode, AP. & Fischer, TH. (2007). Lyophilized platelets: fifty years in the making. *Artif Cells Blood Substit Immobil Biotechnol,* vol.35, No.1, (2007), pp. 125-133.

Borowiak, M.; Garratt, AN.; Wüstefeld, T.; Strehle, M.; Trautwein, C. & Birchmeier, C. (2004). Met provides essential signals for liver regeneration. *Proc Natl Acad Sci U S A,* Vol.101, No.29, (July 2004), pp. 10608-10613.

Bottaro, DP.; Rubin, JS.; Faletto, DL.; Chan, AM.; Kmiecik, TE.; Vande Woude, GF. & Aaronson, SA. (1991). Identification of the hepatocyte growth factor receptor as the c-met proto-oncogene product. *Science,* Vol.251, No.4995, (February 1991), pp. 802-804.

Boulton, RA.; Alison, MR.; Golding, M.; Selden, C. & Hodgson, HJ. (1998). Augmentation of the early phase of liver regeneration after 70% partial hepatectomy in rats following selective Kupffer cell depletion. *J Hepatol,* Vol.29, No.2, (August 1998), pp. 271-280.

Braet, F. & Wisse, E. (2002). Structural and functional aspects of liver sinusoidal endothelial cell fenestrae: a review. *Comp Hepatol,* Vol.1, No.1, (August 2002), pp. 1.

Broos, K.; Feys, HB.; De Meyer, SF.; Vanhoorelbeke, K. & Deckmyn, H. (2011). Platelets at work in primary hemostasis. *Blood Rev,* Vol.25, No.4, (July 2011), pp.155-167.

Chen, B.; Pan, H.; Zhu, L. Deng, Y. & Pollard, JW. (2005). Progesterone inhibits the estrogen-induced phosphoinositide 3-kinase-->AKT-->GSK-3beta-->cyclin D1-->pRB pathway to block uterine epithelial cell proliferation. *Mol Endocrinol,* Vol.19, No.8, (August 2005), pp. 1978-1990.

Cressman, DE.; Greenbaum, LE.; Haber, BA. & Taub, R. (1994). Rapid activation of post-hepatectomy factor/nuclear factor kappa B in hepatocytes, a primary response in the regenerating liver. *J Biol Chem,* Vol.269, No.48, (December 1994), pp. 30429-30435.

Cressman, DE.; Diamond, RH. & Taub, R. (1995). Rapid activation of the Stat3 transcription complex in liver regeneration. *Hepatology,* Vol. 21, No.5, (May 1995), pp. 1443-1449.

Croner, RS.; Hoerer, E.; Kulu, Y.; Hackert, T.; Gebhard, MM.; Herfarth, C. & Klar, E. (2006). Hepatic platelet and leukocyte adherence during endotoxemia. *Crit Care,* Vol.10, No.1, (February 2006), pp. R15.

Decker, K. (1998). The response of liver macrophages to inflammatory stimulation. *Keio J Med,* Vol.47, No.1, (March 1998), pp. 1-9.

de Vos, RJ.; Weir, A.; van Schie, HT.; Bierma-Zeinstra, SM.; Verhaar, JA.; Weinans, H. & Tol, JL. (2010). Platelet-rich plasma injection for chronic Achilles tendinopathy: a randomized controlled trial. *JAMA,* Vol.303, No.2, (January 2010), pp. 144-149.

Dugrillon, A.; Eichler, H.; Kern, S. & Klüter, H. (2002). Autologous concentrated platelet-rich plasma (cPRP) for local application in bone regeneration. *Int J Oral Maxillofac Surg,* Vol.31, No.6, (December 2002), pp. 615-619.

Elzey, BD.; Sprague, DL. & Ratliff, TL. (2005). The emerging role of platelets in adaptive immunity. *Cell Immunol,* Vol.238, No.1, (November 2005), pp. 1-9.

Endo, Y. & Nakamura, M. (1992). The effect of lipopolysaccharide, interleukin-1 and tumour necrosis factor on the hepatic accumulation of 5-hydroxytryptamine and platelets in the mouse. *Br J Pharmacol,* Vol.105, No.3, (March 1992), pp. 613-619.

Endo, Y. & Nakamura, M. (1993). Active translocation of platelets into sinusoidal and Disse spaces in the liver in response to lipopolysaccharides, interleukin-1 and tumor necrosis factor. *Gen Pharmacol,* Vol.24, No.5, (September 1993), pp. 1039-1053.

Factor, VM.; Seo, D.; Ishikawa, T.; Kaposi-Novak, P.; Marquardt, JU.; Andersen, JB.; Conner, EA. & Thorgeirsson, SS. (2010). Loss of c-Met disrupts gene expression program required for G2/M progression during liver regeneration in mice. *PLoS One,* Vol.5, No.9, (September 2010), pp. e12739.

Faridi, J.; Fawcett, J.; Wang, L. & Roth, RA. (2003). Akt promotes increased mammalian cell size by stimulating protein synthesis and inhibiting protein degradation. *Am J Physiol Endocrinol Metab,* Vol.285 No.5, (November 2003), pp. E964-972.

Fausto, N.; Laird, AD. & Webber, EM. (1995). Liver regeneration. 2. Role of growth factors and cytokines in hepatic regeneration. *FASEB J*, Vol.9, No.15, (December 1995), pp. 1527-1536.

Fausto, N. (2000). Liver regeneration. J Hepatol, Vol.32, No. 1 suppl, (2000), pp. 19-31.

Fausto, N.; Campbell, JS. & Riehle, KJ. (2006). Liver regeneration. Hepatology, Vol.43, No.2 suppl 1, (February 2006), pp. S45-53.

FitzGerald, MJ.; Webber, EM.; Donovan, JR. & Fausto,N. (1995). Rapid DNA binding by nuclear factor kappa B in hepatocytes at the start of liver regeneration. Cell Growth Differ, Vol.6, No.4, (April 1995), pp.417-427.

Fresno Vara, JA.; Casado, E.; de Castro, J.; Cejas, P.; Belda-Iniesta, C. & González-Barón, M. (2004). PI3K/Akt signalling pathway and cancer. Cancer Treat Rev, Vol.30, No.2, (April 2004), pp. 193-204.

Fujiyoshi, M. & Ozaki, M. (2011). Molecular mechanisms of liver regeneration and protection for treatment of liver dysfunction and diseases. J Hepatobiliary Pancreat Sci, Vol.18, No.1, (January 2011) pp. 13-22.

Gauldie, J.; Richards, C. & Baumann, H. (1992). IL6 and the acute phase reaction. *Res Immunol,* Vol.143, No.7, (September 1992), pp. 755-759.

Gotoh, J.; Obata, M.; Yoshie, M.; Kasai, S. & Ogawa, K. (2003). Cyclin D1 over-expression correlates with beta-catenin activation, but not with H-ras mutations, and phosphorylation of Akt, GSK3 beta and ERK1/2 in mouse hepatic carcinogenesis. *Carcinogenesis,* Vol.24, No.3, (March 2003), pp. 435-442.

Haga, S.; Ogawa, W.; Inoue, H.; Terui, K.; Ogino, T.; Igarashi, R.; Takeda, K.; Akira, S.; Enosawa, S.; Furukawa, H.; Todo, S. & Ozaki, M. (2005). Compensatory recovery of liver mass by Akt-mediated hepatocellular hypertrophy in liver-specific STAT3-deficient mice. *J Hepatol,* Vol.43, No.5, (November 2005), pp. 799-807.

Haga, S.; Ozaki, M.; Inoue, H.; Okamoto, Y.; Ogawa, W.; Takeda, K.; Akira, S. & Todo, S. (2009). The survival pathways phosphatidylinositol-3 kinase (PI3-K)/phosphoinositide-dependent protein kinase 1 (PDK1)/Akt modulate liver regeneration through hepatocyte size rather than proliferation. *Hepatology,* Vol.49, No.1, (January 2009), pp. 204-214.

Hartmann, EK.; Heintel, T.; Morrison, RH. & Weckbach, A. (2010). Influence of platelet-rich plasma on the anterior fusion in spinal injuries: a qualitative and quantitative analysis using computer tomography. *Arch Orthop Trauma Surg,* Vol.130, No.7, (July 2010), pp. 909-914.

Hashikura, Y.; Kawasaki, S.; Matsunami, H.; Ikegami, T. & Makuuchi, M. (1994). Intraoperative increment of platelet-activating factor in clinical liver transplantation. *Clin Transplant,* Vol.8, No.1, (February 1994), pp. 27-29.

Heinrich, PC.; Behrmann, I.; Müller-Newen, G.; Schaper, F. & Graeve, L. (1998). Interleukin-6-type cytokine signalling through the gp130/Jak/STAT pathway. *Biochem J,* Vol.334, Pt 2, (September 1998), pp. 297-314.

Hinz, M.; Krappmann, D.; Eichten, A.; Heder, A.; Scheidereit, C.& Strauss, M. (1999). NF-kappaB function in growth control: regulation of cyclin D1 expression and G0/G1-to-S-phase transition. *Mol Cell Biol,* Vol.19, No.4, (April 1999) pp. 2690-2698.

Hisakura, K.; Murata, S.; Fukunaga, K.; Myronovych, A.; Tadano, S.; Kawasaki, T.; Kohno, K.; Ikeda, O.; Pak, S.; Ikeda, N.; Nakano, Y.; Matsuo, R.; Konno, K.; Kobayashi, E.; Saito, T.; Yasue, H. & Ohkohchi, N. (2010). Platelets prevent acute liver damage after extended hepatectomy in pigs. *J Hepatobiliary Pancreat Sci,* Vol.17, No.6, (November 2010), pp. 855-864.

Hisakura, K.; Murata, S.; Takahashi, K.; Matsuo, R.; Pak, S.; Ikeda, N.; Kawasaki, T.; Kohno, K.; Myronovych, A.; Nakano, Y.; Ikeda, O.; Watanabe, M. & Ohkohchi, N. (2011). Platelets prevent acute hepatitis induced by anti-fas antibody. *J Gastroenterol Hepatol,* Vol.26, No.2, (February 2011), pp. 348-355.

Holmsen, H. (1989). Physiological functions of platelets. *Ann Med,* Vol.21, No.1, (February 1989), pp. 23-30.

Hoshi, R.; Murata, S.; Matsuo, R.; Myronovych, A.; Hashimoto, I.; Ikeda, H. & Ohkohchi, N. (2007). Freeze-dried platelets promote hepatocyte proliferation in mice. *Cryobiology,* Vol.55, No.3, (December 2007), pp. 255-260.

Ikeda, H.; Satoh, H.; Yanase, M.; Inoue, Y.; Tomiya, T.; Arai, M.; Tejima, K.; Nagashima, K.; Maekawa, H.; Yahagi, N.; Yatomi, Y.; Sakurada, S.; Takuwa, Y.; Ogata, I.; Kimura, S. & Fujiwara, K. (2003). Antiproliferative property of sphingosine 1-phosphate in rat hepatocytes involves activation of Rho via Edg-5. *Gastroenterology,* Vol.124, No.2, (February 2003), pp. 459-469.

Jackson, LN.; Larson, SD.; Silva, SR.; Rychahou, PG.; Chen, LA.; Qiu, S.; Rajaraman, S. & Evers, BM. (2008). PI3K/Akt activation is critical for early hepatic regeneration after partial hepatectomy. *Am J Physiol Gastrointest Liver Physiol,* Vol.294, No.6, (June 2008), pp. 1401-1410.

Kahn, D.; Robson, SC. & Hickman, R. (1994). The effect of a tumour necrosis factor antibody on the regenerative response after partial hepatectomy in rats. *Transpl Int,* Vol.7 Suppl 1, (1994), pp. 181-182.

Karidis, NP.; Kouraklis, G. & Theocharis, SE. (2006). Platelet-activating factor in liver injury: a relational scope. *World J Gastroenterol,* Vol.12, No.23, (June 2006), pp. 3695-3706.

Kato, M.; Putta, S; Wang, M.; Yuan, H.; Lanting, L.; Nair, I.; Gunn, A.; Nakagawa, Y.; Shimano, H.; Todorov, I.; Rossi, JJ. & Natarajan, R. (2009). TGF-beta activates Akt kinase through a microRNA-dependent amplifying circuit targeting PTEN. *Nat Cell Biol,* Vol.11, No.7, (July 2009), pp. 881-889.

Kawasaki, T.; Murata, S.; Takahashi, K.; Nozaki, R.; Ohshiro, Y.; Ikeda, N.; Pak, S.; Myronovych, A.; Hisakura, K.; Fukunaga, K.; Oda, T.; Sasaki, R. & Ohkohchi, N. (2010). Activation of human liver sinusoidal endothelial cell by human platelets induces hepatocyte proliferation. *J Hepatol,* Vol.53, No.4, (October 2010), pp. 648-654.

Khandoga, A.; Biberthaler, P.; Messmer, K. & Krombach, F. (2003). Platelet-endothelial cell interactions during hepatic ischemia-reperfusion in vivo: a systematic analysis. *Microvasc Res,* Vol.65, No.2, (March 2003), pp. 71-77.

Khandoga, A.; Hanschen, M.; Kessler, JS. & Krombach, F. (2006). CD4+ T cells contribute to postischemic liver injury in mice by interacting with sinusoidal endothelium and platelets. *Hepatology,* Vol.43, No.2, (February 2006), pp. 306-315.

Kim, J.; Yi, NJ.; Shin, WY.; Kim, T.; Lee, KU. & Suh, KS. (2010). Platelet transfusion can be related to liver regeneration after living donor liver transplantation. *World J Surg,* Vol.34, No.5, (May 2010), pp. 1052-1058.

Klinger, MH. & Jelkmann, W. (2002). Role of blood platelets in infection and inflammation. *J Interferon Cytokine Res,* Vol.22, No.9, (September 2002), pp. 913-922.

Kmieć, Z. (2001). Cooperation of liver cells in health and disease. *Adv Anat Embryol Cell Biol,* Vol.161, No.III-XIII, (2001), pp. 1-151.

Knook, DL. & Sleyster, EC. (1976). Separation of Kupffer and endothelial cells of the rat liver by centrifugal elutriation. *Exp Cell Res,* Vol.99, No.2, (May 1976), pp. 444-449.

Kodama, T.; Takehara, T.; Hikita, H.; Shimizu, S.; Li, W.; Miyagi, T.; Hosui, A.; Tatsumi, T.; Ishida, H.; Tadokoro, S.; Ido, A.; Tsubouchi, H. & Hayashi, N. (2010). Thrombocytopenia exacerbates cholestasis-induced liver fibrosis in mice. *Gastroenterology,* Vol.138, No.7, (June 2010), pp. 2487-2498, 2498. e1-e7.

Lang, PA.; Contaldo, C.; Georgiev, P.; El-Badry, AM.; Recher, M.; Kurrer, M.; Cervantes-Barragan, L.; Ludewig, B.; Calzascia, T.; Bolinger, B.; Merkler, D.; Odermatt, B.; Bader, M.; Graf, R.; Clavien, PA.; Hegazy, AN.; Löhning, M.; Harris, NL.; Ohashi, PS.; Hengartner, H.; Zinkernagel, RM. & Lang, KS. (2008). Aggravation of viral hepatitis by platelet-derived serotonin. *Nat Med,* Vol.14, No.7, (July 2008), pp. 756-761.

Laschke, MW.; Dold, S.; Menger, MD.; Jeppsson, B. & Thorlacius, H. (2008). Platelet-dependent accumulation of leukocytes in sinusoids mediates hepatocellular damage in bile duct ligation-induced cholestasis. *Br J Pharmacol,* Vol.153, No.1, (January 2008), pp. 148-156.

Latronico, MV.; Costinean, S.; Lavitrano, ML.; Peschle, C. & Condorelli, G. (2004). Regulation of cell size and contractile function by AKT in cardiomyocytes. *Ann N Y Acad Sci,* Vol.1015, (May 2004), pp. 250-260.

Lesurtel, M.; Graf, R.; Aleil, B.; Walther, DJ.; Tian, Y.; Jochum, W.; Gachet, C.; Bader, M. & Clavien, PA. (2006). Platelet-derived serotonin mediates liver regeneration. *Science,* Vol.312, No.5770, (April 2006), pp. 104-107.

Li, W.; Liang, X.; Kellendonk, C.; Poli, V. & Taub, R. (2002). STAT3 contributes to the mitogenic response of hepatocytes during liver regeneration. *J Biol Chem,* Vol.277, No.32, (August 2002), pp. 28411-28417.

Malik, R.; Selden, C. & Hodgson, H. (2002). The role of non-parenchymal cells in liver growth. *Semin Cell Dev Biol,* Vol.13, No.6, (December 2002), pp. 425-431.

Marubashi, S.; Dono, K.; Miyamoto, A.; Takeda, Y.; Nagano, H.; Umeshita, K. & Monden, M. (2007). Impact of graft size on postoperative thrombocytopenia in living donor liver transplant. *Arch Surg,* Vol.142, No.11 (November 2007), pp. 1054-1058.

Massberg, S.; Enders, G.; Leiderer, R.; Eisenmenger, S.; Vestweber, D.; Krombach, F. & Messmer, K. (1998). Platelet-endothelial cell interactions during ischemia/reperfusion: the role of P-selectin. *Blood,* Vol.92, No.2, (July 1998), pp. 507-515.

Matsumura, T.; Takesue, M.; Westerman, KA.; Okitsu, T.; Sakaguchi, M.; Fukazawa, T.; Totsugawa, T.; Noguchi, H.; Yamamoto, S.; Stolz, DB.; Tanaka, N.; Leboulch, P. & Kobayashi, N. (2004). Establishment of an immortalized human-liver endothelial cell line with SV40T and hTERT. *Transplantation,* Vol.77, No.9, (May 2004), pp. 1357-1365.

Matsuo, R.; Ohkohchi, N.; Murata, S.; Ikeda, O.; Nakano, Y.; Watanabe, M.; Hisakura, K.; Myronovych, A.; Kubota, T.; Narimatsu, H. & Ozaki, M. (2008). Platelets Strongly Induce Hepatocyte Proliferation with IGF-1 and HGF In Vitro. *J Surg Res,* Vol.145, No.2,(April 2008), pp. 279-286.

Matsuo, R.; Nakano, Y. & Ohkohchi, N. (2011). Platelet administration via the portal vein promotes liver regeneration in rats after 70% hepatectomy. *Ann Surg,* Vol.253, No.4 (April 2011), pp. 759-763.

Mazzucco, L.; Borzini, P. & Gope, R. (2010). Platelet-derived factors involved in tissue repair-from signal to function. *Transfus Med Rev,* Vol.24, No.3, (July 2010), pp. 218-234.

McNicol, A. & Israels, SJ. (1999). Platelet dense granules: structure, function and implications for haemostasis. *Thromb Res,* Vol.95, No.1,(July 1999), pp. 1-18.

McNicol, A. & Israels, SJ. (2008). Beyond hemostasis: the role of platelets in inflammation, malignancy and infection. *Cardiovasc Hematol Disord Drug Targets,* Vol.8, No.2, (June 2008), pp. 99-117. Review

Mehta, P. (1984). Potential role of platelets in the pathogenesis of tumor metastasis. *Blood,* Vol.63, No.1, (January 1984) pp. 55-63.

Meijer, C.; Wiezer, MJ.; Diehl, AM.; Schouten, HJ.; Schouten, HJ.; Meijer, S.; van Rooijen, N.; van Lambalgen, AA.; Dijkstra, CD. & van Leeuwen, PA. (2000). Kupffer cell depletion by CI2MDP-liposomes alters hepatic cytokine expression and delays liver regeneration after partial hepatectomy. *Liver,* Vol.20, No.1, (February 2000), pp. 66-77.

Michalopoulos, GK. & DeFrances, MC. (1997). Liver regeneration. *Science,* Vol.276, No.5309, (April 1997), pp. 60-66.

Michalopoulos, GK. (2010). Liver regeneration after partial hepatectomy: critical analysis of mechanistic dilemmas. *Am J Pathol,* Vol.176, No.1, (January 2010), pp. 2-13.

Moh, A.; Iwamoto, Y.; Chai, GX.; Zhang, SS.; Kano, A.; Yang, DD.; Zhang, W.; Wang, J.; Jacoby, JJ.; Gao, B.; Flavell, RA. & Fu, XY. (2007). Role of STAT3 in liver regeneration: survival, DNA synthesis, inflammatory reaction and liver mass recovery. *Lab Invest,* Vol.87, No.10, (October 2007), pp. 1018-1028.

Montalvo-Jave, EE.; Escalante-Tattersfield, T.; Ortega-Salgado, JA.; Piña, E. & Geller, DA. (2008). Factors in the pathophysiology of the liver ischemia-reperfusion injury. *J Surg Res,* Vol.147, No.1, (June 2008), pp. 153-159.

Murata, S.; Ohkohchi, N.; Abe, T.; Enomoto, Y.; Doi, H. & Satomi, S. (2004). Platelets promote G1-S progression of liver regeneration after hepatectomy. *Congress of the European society for surgical research,* (2004), pp. 107-112.

Murata, S.; Ohkohchi, N.; Matsuo, R.; Ikeda, O.; Myronovych, A. & Hoshi, R. (2007). Platelets promote liver regeneration in early period after hepatectomy in mice. *World J Surg,* Vol.31, No.4, (April 2007), pp. 808-816.

Murata, S.; Matsuo, R.; Ikeda, O.; Myronovych, A.; Watanabe, M.; Hisakura, K.; Nakano, Y.; Hashimoto, I. & Ohkohchi, N. (2008). Platelets promote liver regeneration under conditions of Kupffer cell depletion after hepatectomy in mice. *World J Surg,* Vol.32, No.6, (June 2008), pp. 1088-1896.

Murata, S.; Hashimoto, I.; Nakano, Y.; Myronovych, A.; Watanabe, M. & Ohkohchi, N. (2008). Single administration of thrombopoietin prevents progression of liver fibrosis and promotes liver regeneration after partial hepatectomy in cirrhotic rats. *Ann Surg,* Vol.248, No.5 (November 2008), pp. 821-828.

Myronovych, A.; Murata, S.; Chiba, M.; Matsuo, R.; Ikeda, O.; Watanabe, M.; Hisakura, K.; Nakano, Y.; Kohno, K.; Kawasaki, T.; Hashimoto, I.; Shibasaki, Y.; Yasue, H. &

Ohkohchi, N. (2008). Role of platelets on liver regeneration after 90% hepatectomy in mice. *J Hepatol,* Vol.49, No.3, (September 2008), pp. 363-372.

Nakamura, M.; Shibazaki, M.; Nitta, Y. & Endo, Y. (1998). Translocation of platelets into Disse spaces and their entry into hepatocytes in response to lipopolysaccharides, interleukin-1 and tumour necrosis factor: the role of Kupffer cells. *J Hepatol,* Vol.28, No.6, (June 1998), pp. 991-999.

Nakamura, T.; Nishizawa, T.; Hagiya, M.; Seki, T.; Shimonishi, M.; Sugimura, A.; Tashiro, K. & Shimizu, S. (1989). Molecular cloning and expression of human hepatocyte growth factor. *Nature,* Vol.342, No.6248, (November 1989), pp. 440-443.

Nakano, Y.; Kondo, T.; Matsuo, R.; Hashimoto, I.; Kawasaki, T.; Kohno, K.; Myronovych, A.; Tadano, S.; Hisakura, K.; Ikeda, O.; Watanabe, M.; Murata, S.; Fukunaga, K. & Ohkohchi, N. (2008). Platelet dynamics in the early phase of postischemic liver in vivo. *J Surg Res,* Vol.149, No.2, (October 2008), pp. 192-198.

Nash, GF.; Turner, LF.; Scully, MF. & Kakkar, AK. (2002). Platelets and cancer. *Lancet Oncol,* Vol.3, No.7, (July 2002), pp. 425-430.

Nechemia-Arbely, Y.; Shriki, A.; Denz, U.; Drucker, C.; Scheller, J.; Raub, J.; Pappo, O.; Rose-John, S.; Galun, E. & Axelrod, JH. (2011). Early hepatocyte DNA synthetic response posthepatectomy is modulated by IL-6 trans-signaling and PI3K/AKT activation. *J Hepatol,* Vol.54, No.5,(May 2011), pp. 922-999.

Nocito, A.; Georgiev, P.; Dahm, F.; Jochum, W.; Bader, M.; Graf, R. & Clavien, PA. (2007). Platelets and platelet-derived serotonin promote tissue repair after normothermic hepatic ischemia in mice. *Hepatology,* Vol.45, No.2, (February 2007), pp. 369-376.

Okamura, Y.; Takeoka, S.; Eto, K.; Maekawa, I.; Fujie, T.; Maruyama, H.; Ikeda, Y. & Handa, M. (2009). Development of fibrinogen gamma-chain peptide-coated, adenosine diphosphate-encapsulated liposomes as a synthetic platelet substitute. *J Thromb Haemost,* Vol.7, No.3, (March 2009), pp. 470-477.

Okano, J.; Shiota, G.; Matsumoto, K.; Yasui, S.; Kurimasa, A.; Hisatome, I.; Steinberg, P. & Murawaki, Y. (2003). Hepatocyte growth factor exerts a proliferative effect on oval cells through the PI3K/AKT signaling pathway. *Biochem Biophys Res Commun,* Vol.309, No.2, (September 2003), pp. 298-304.

Osawa, Y.; Nagaki, M.; Banno, Y.; Brenner, DA.; Asano, T.; Nozawa, Y.; Moriwaki, H. & Nakashima, S. (2002). Tumor necrosis factor alpha-induced interleukin-8 production via NF-kappaB and phosphatidylinositol 3-kinase/Akt pathways inhibits cell apoptosis in human hepatocytes. *Infect Immun,* Vol.70, No.11, (November 2002), pp. 6294-6301.

Pak, S.; Kondo, T.; Nakano, Y.; Murata, S.; Fukunaga, K.; Oda, T.; Sasaki, R. & Ohkohchi, N. (2010). Platelet adhesion in the sinusoid caused hepatic injury by neutrophils after hepatic ischemia reperfusion. *Platelets,* Vol.21, No.4, (2010), pp. 282-288.

Pereboom, IT.; Lisman, T. & Porte, RJ. (2008). Platelets in liver transplantation: friend or foe? *Liver Transpl,* Vol.14, No.7, (July 2008), pp. 923-931.

Ping, C.; Xiaoling, D.; Jin, Z.; Jiahong, D.; Jiming, D. & Lin, Z. (2006). Hepatic sinusoidal endothelial cells promote hepatocyte proliferation early after partial hepatectomy in rats. *Arch Med Res,* Vol.37, No.5, (July 2006), pp. 576-583.

Polasek, J. (2005). Platelet secretory granules or secretory lysosomes? *Platelets,* Vol.16, No.8, (December 2005), pp. 500-501.

Radice, F.; Yánez, R.; Gutiérrez, V.; Rosales, J.; Pinedo, M. & Coda, S. (2010). Comparison of magnetic resonance imaging findings in anterior cruciate ligament grafts with and without autologous platelet-derived growth factors. *Arthroscopy,* Vol.26, No.1, (January 2010), pp. 50-57.

Rai, RM.; Yang, SQ.; McClain, C.; Karp, CL.; Klein, AS. & Diehl, AM. (1996). Kupffer cell depletion by gadolinium chloride enhances liver regeneration after partial hepatectomy in rats. *Am J Physiol,* Vol.270, No.6Pt1, (June 1996), pp. 909-918.

Ranzato, E.; Balbo, V.; Boccafoschi, F.; Mazzucco, L. & Burlando, B. (2009). Scratch wound closure of C2C12 mouse myoblasts is enhanced by human platelet lysate. *Cell Biol Int,* Vol.33, No.9, (September 2009), pp. 911-917.

Rhodes, E. & Stasi, R. (2010). Current status of thrombopoietic agents. *Expert Rev Hematol,* Vol.3, No.2, (April 2010), pp. 217-225.

Rodeo, SA.; Delos, D.; Weber, A.; Ju, X.; Cunningham, ME.; Fortier, L. & Maher, S. (2010). What's new in orthopaedic research. *J Bone Joint Surg Am,* Vol.92, No.14, (October 2010), pp. 2491-2501.

Rozman, P. & Bolta, Z. (2007). Use of platelet growth factors in treating wounds and soft-tissue injuries. *Acta Dermatovenerol Alp Panonica Adriat,* Vol.16, No.4, (December 2007), pp. 156-165.

Shigekawa, M.; Takehara, T.; Kodama, T.; Hikita, H.; Shimizu, S.; Li, W.; Miyagi, T.; Hosui, A.; Tatsumi, T.; Ishida, H.; Kanto, T.; Hiramatsu, N. & Hayashi, N. (2011). Involvement of STAT3-regulated hepatic soluble factors in attenuation of stellate cell activity and liver fibrogenesis in mice. *Biochem Biophys Res Commun,* Vol.406, No.4, (March 2011), pp. 614-620.

Sindram , D.; Porte, RJ.; Hoffman, MR.; Bentley, RC. & Clavien, PA. (2000). Platelets induce sinusoidal endothelial cell apoptosis upon reperfusion of the cold ischemic rat liver. *Gastroenterology,* Vol.118, No.1, (January 2000), pp.183-191,

Smedsrød, B.; Pertoft, H.; Gustafson, S. & Laurent, TC. (1990). Scavenger functions of the liver endothelial cell. *Biochem J,* Vol.266, No.2, (March 1990), pp. 313-327.

Solt, LA. & May, MJ. (2008). The IkappaB kinase complex: master regulator of NF-kappaB signaling. *Immunol Res,* Vol.42, No.1-3, (2008), pp. 3-18.

Sowa, JM.; Crist, SA.; Ratliff, TL. & Elzey, BD. (2009). Platelet influence on T- and B-cell responses. *Arch Immunol Ther Exp (Warsz),* Vol.57, No.4, (July 2009), pp. 235-241.

Sprague, DL.; Elzey, BD.; Crist, SA.; Waldschmidt, TJ.; Jensen, RJ. & Ratliff, TL. (2008). Platelet-mediated modulation of adaptive immunity: unique delivery of CD154 signal by platelet-derived membrane vesicles. *Blood,* Vol.111, No.10, (May 2008), pp. 5028-5036.

Stepniak, E.; Ricci, R.; Eferl, R.; Sumara, G.; Sumara, I.; Rath, M.; Hui, L. & Wagner, EF. (2006). c-Jun/AP-1 controls liver regeneration by repressing p53/p21 and p38 MAPK activity. *Genes Dev,* Vol.20, No.16, (August 2006), pp. 2306-2314.

Suzuki, H.; Nakamura, S.; Itoh, Y.; Tanaka, T.; Yamazaki, H. & Tanoue, K. (1992). Immunocytochemical evidence for the translocation of alpha-granule membrane glycoprotein IIb/IIIa (integrin alpha IIb beta 3) of human platelets to the surface membrane during the release reaction. *Histochemistry,* Vol.97, No.5, (1992), pp. 381-388.

Takeishi, T.; Hirano, K.; Kobayashi, T.; Hasegawa, G.; Hatakeyama, K. & Naito M. (1999). The role of Kupffer cells in liver regeneration. *Arch Histol Cytol*, Vol.62, No.5, (Dec 1999), pp. 413-422.

Terui, K. & Ozaki, M. (2005). The role of STAT3 in liver regeneration. *Drugs Today (Barc)*, Vol.41, No.7, (July 2005), pp. 461-469.

Tewari, M.; Dobrzanski, P.; Mohn, KL.; Cressman DE.; Hsu, JC.; Bravo, R. & Taub, R. (1992). Rapid induction in regenerating liver of RL/IF-1 (an I kappa B that inhibits NF-kappa B, RelB-p50, and c-Rel-p50) and PHF, a novel kappa B site-binding complex. *Mol Cell Biol*, Vol.12, No.6, (Jun 1992), pp. 2898-2908.

Toledo-Pereyra, LH. & Suzuki, S. (1994). Cellular and biomolecular mechanisms of liver ischemia and reperfusion injury. *Transplant Proc*, Vol.26, No.1, (Feb 1994), pp. 325-327.

Turkson, J. & Jove R. (2000). STAT proteins: novel molecular targets for cancer drug discovery. *Oncogene*, Vol.19, No.56, (Dec 2000), pp. 6613-6626.

Tulasne, D. & Foveau, B. (2008). The shadow of death on the MET tyrosine kinase receptor. *Cell Death Differ*, Vol.15, No.3, (Mar 2008), pp. 427-434.

Vollmar, B. & Menger, MD. (2009). The hepatic microcirculation: mechanistic contributions and therapeutic targets in liver injury and repair. *Physiol Rev*, Vol89, No.4, (Oct 2009), pp. 1269-1339.

Wang, GL.; Salisbury, E.; Shi, X.; Timchenko, L.; Medrano, EE. & Timchenko, NA. (2008). HDAC1 cooperates with C/EBPalpha in the inhibition of liver proliferation in old mice. *J Biol Chem*, Vol.283, No.38, (Sep 2008), pp. 26169-26178.

Wisse, E.; Braet, F.; Luo, D.; De Zanger, R.; Jans, D.; Crabbé, E. & Vermoesen, A. (1996). Structure and function of sinusoidal lining cells in the liver. *Toxicol Pathol*, Vol.24, No.1, (Jan-Feb 1996), PP. 100-111.

Yamada, Y.; Webber, EM.; Kirillova, I.; Peschon, JJ. & Fausto, N. (1998). Analysis of liver regeneration in mice lacking type 1 or type 2 tumor necrosis factor receptor: requirement for type 1 but not type 2 receptor. *Hepatology*, Vol.28, No.4, (Oct 1998), pp. 959-970.

Yamada, Y.; Kirillova, I.; Peschon, JJ. & Fausto, N. (1997). Initiation of liver growth by tumor necrosis factor: deficient liver regeneration in mice lacking type I tumor necrosis factor receptor. *Proc Natl Acad Sci U S A*, Vol.94, No.4, (Feb 1997), pp. 1441-1446.

Yamaguchi, R.; Terashima, H.; Yoneyama, S.; Tadano, S.; & Ohkohchi, N. (2010) Effects of Platelet-Rich Plasma on Intestinal Anastomotic Healing in Rats: PRP Concentration is a Key Factor. *J Surg Res*, (Nov 2010), [Epub ahead of print].

Yatomi, Y.; Ohmori, T.; Rile, G.; Kazama, F.; Okamoto, H.; Sano, T.; Satoh, K.; Kume, S.; Tigyi, G.; Igarashi, Y. & Ozaki, Y. (2000) Sphingosine 1-phosphate as a major bioactive lysophospholipid that is released from platelets and interacts with endothelial cells. *Blood*, Vol.96, No.10, (Nov 2000), pp. 3431-3438.

Watanabe, M.; Murata, S.; Hashimoto, I.; Nakano, Y.; Ikeda, O.; Aoyagi, Y.; Matsuo, R.; Fukunaga, K.; Yasue, H. & Ohkohchi, N. (2008). Platelets contribute to the reduction of liver fibrosis in mice. *J Gastroenterol Hepatol*, Vol.24, No.1, (Jan 2009), pp. 78-89.

Xia, P.; & Wadham, C. (2011). Sphingosine 1-phosphate, a key mediator of the cytokine network: juxtacrine signaling. *Cytokine Growth Factor Rev*, Vol.22, No.1, (Feb 2011), pp. 45-53.

Zaldivar, MM.; Pauels, K.; von Hundelshausen, P.; Berres, ML.; Schmitz, P.; Bornemann, J.; Kowalska, MA.; Gassler, N.; Streetz, KL.; Weiskirchen, R.; Trautwein, C.; Weber, C. & Wasmuth, HE. (2010) CXC chemokine ligand 4 (Cxcl4) is a platelet-derived mediator of experimental liver fibrosis. *Hepatology*, Vol.51, No.4, (Apr 2010), pp. 1345-1353.

Zheng, DM.; Kitamura, T.; Ikejima, K.; Enomoto, N.; Yamashina, S.; Suzuki, S.; Takei, Y. & Sato, N. (2006). Sphingosine 1-phosphate protects rat liver sinusoidal endothelial cells from ethanol-induced apoptosis: Role of intracellular calcium and nitric oxide. *Hepatology*, Vol.44, No.5, (Nov 2006), pp. 1278-1287.

7

Shared Triggering Mechanisms of Retinal Regeneration in Lower Vertebrates and Retinal Rescue in Higher Ones

Eleonora Grigoryan
Kol'tzov Institute of Developmental Biology, Russian Academy of Sciences, Moscow, Russia

1. Introduction

Neural retina (NR) of the eye of vertebrates is underlined by retinal pigmented epithelium (RPE). NR↔RPE interconnection is critical for development, regeneration and function of both compartments of the retina. A disturbance of the normal quantitative correlation of RPE cells and photoreceptors, their structural and functional integrity unavoidably breaks visual cycle and induces retinal pathology. The majority of retinal diseases - inherent, age-related or systemic, links with a disturbance of NR↔RPE relationship.

In the lab NR↔RPE disintegration can be achieved under some experimental conditions, such as NR separation and explantation, elimination of photoreceptors by bright light, chemical or mechanical damage of RPE and photoreceptors, and etc. A usage of experimental models in studies on the retina of lower and higher vertebrates endows a lot for understanding of cellular and molecular mechanisms of retinal pathology, on one part, and natural mechanisms of its rescue, on the other.

In tailed amphibian (Urodela) complete removal of NR or NR artificial detachment leads to RPE cell transdifferentiation that two months later results in regeneration of functioning retina (Chiba & Mitashov, 2007; Grigoryan & Mitashov, 1979; Hasegawa, 1958; Mitashov, 1997; Stroeva & Mitashov, 1983). In mammals RPE-based NR regeneration has not been reported. It is well known that NR detachment causes serious complications and blinding diseases despite of switching on some protective mechanisms for NR rescue (Fisher et al., 2005; Fisher & Lewis, 2010a, b; Pasto, 1998). This review represents an attempt to study early cellular and molecular mechanisms triggering NR regeneration in amphibians and NR rescue/pathology in mammals.

RPE of all vertebrates being localized between choroidal coat and NR has a big range of very important functions. RPE protects NR photoreceptors against overabundant light, participate in visual cycle, releases growth factors, regulates ion balance, transports nutrients, etc. (Strauss, 2005). In development RPE and NR have a common origin and both derived from the neuroepithelium of the optic cup. The latter delaminates into two layers, NR and RPE. Differentiation of these two tissues is a result of the expression of complex molecular network that is recently named the "*oculome*" (Lachke & Maas, 2010). In the development as well as in the adult state both retinal tissues are in precise, well coordinated interconnection. On the outer side of the eye RPE is separated from vascular (choroid) and scleral coat by Bruch "membrane" that is formed by basal membranes of RPE and choroidal endothelium (fig.1).

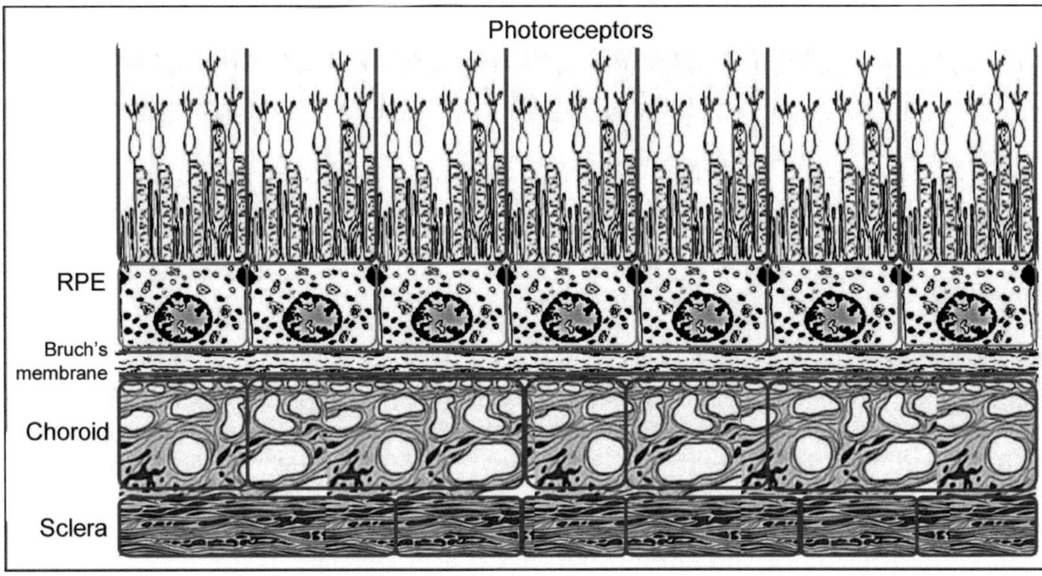

Fig. 1. Retinal pigmented epithelium interplay with other tissues of the eye back wall. Schematic representation, modified from: Strauss, 2005

In this review we concentrated on the early cellular and molecular events induced by disturbance of NR↔RPE interconnections. We should point out that even the very early changes induced by NR↔RPE separation are complex and effect not only on RPE and NR but also other tissues of the eye back wall. Inner part of NR: interneurons, Müller glial cells and ganglion cells, although somewhat later, undergo changes in their function and behavior as a response for alterations in the outer NR. In this review we do not consider these, secondary in time course, processes.

Initiation of retinal regeneration in Urodela and its rescue in mammals as well as progress of following events are represented by comprehensive processes, in which different molecular families and signaling pathways participate. These molecules introduced by NR as well as adjoining tissues, and, in particular by RPE, play important, controlling, and regulative roles. The study of the complexity of intercellular and intertissue communications requires approaches using animals whose RPE↔NR disintegration does not lead to retinal loss but, in contrast, induces epimorphic regeneration. These approaches allow us to understand what kind of molecular changes are introduced by RPE and NR and what subsequent cellular and molecular events they are able to induce. As we'll see the changes initiating retinal regeneration in lower vertebrates and retinal rescue in higher ones have a high degree of universality. This gives us a hope for an existing preservation of some regenerative responses in evolutionary row and not complete block of them on the top, in human.

2. Results and discussion

2.1 A disturbance of retinal cell contacts and behavior

One of the earliest events caused by NR↔RPE separation is the loss of the adhesion and communication between two tissues. Adhesion of NR to RPE is provided by interaction of the RPE apical processes with the outer segments of the photoreceptors. An adhesion force between NR and RPE is ensured by constant elimination of water from the subretinal space. The latter remains closely tight in the normal retina and is essential for retinal functions and visual processing (Ghazi & Green, 2003; Marmor, 1993). The interface between RPE and NR is the interphotoreceptor matrix (IPM) that serves for chemical cross-talk between two tissues for their coordinated function. IPM consists of ECM components but a disruption of them often causes NR detachment. For instance, when IPM chondroitin 6-sulfate proteoglycan is perturbed *in vivo* by intravitreal injections of xyloside (a sugar inhibited chondroitin sulfate proteoglycan synthesis), shallow NR detachment could be observed. This suggests that adhesion between NR and RPE is dependent on continuous presence and synthesis of IPM proteoglycans (Lazarus & Hageman, 1992). Moreover, there are evidences that IPM molecules responsible for adhesion have a neuroprotective effect as well. For instance in the rat galectin-3, participating in RPE↔NR adhesion, inhibits apoptosis through the bcl-2 or cysteine protease pathways and, in contrast, intravitreous injection of anti-galectin-3 antibody accelerates photoreceptor degeneration due to constant light (Uehara et al., 2001).

IPM disruption, subsequent weakening or even loss of RPE lateral contacts and attachment to NR and Bruch membrane, induces a change of RPE cell behavior. In parallel, after NR detachment when part of photoreceptors degenerate a large amount of their debris overwhelm the phagocytic ability of the RPE cells. As a response for altered conditions RPE manifests their multipotential capacity. After retinal removal in newts RPE cells stop to synthesize melanin and increase their proliferative activity (Grigoryan & Mitashov, 1979). In the same animals after NR detachment RPE also can display the unique capacity to transdifferentiate into retinal cells and to form new additional NR (Grigoryan & Mitashov, 1985) (fig. 2).

Another of known differentiation potencies of urodelean RPE is the transformation to mobile cells with many of the characteristics of macrophages. The observation was made long time ago in the course of NR regeneration after retinal removal or optic nerve cutting in the newt (Keefe 1973) and in the case of NR experimental detachment in mammals (Johnson & Faulds, 1977). In those early works it was shown that after NR removal in Urodela some RPE cells withdrawn from the layer, moved along vitreal direction and phagocytosed retinal cell remnants (figure 3a, b). Due to this ability they were named "melanophages" (Keefe, 1973). Something comparable was carrying out in the rodent models of NR detachment (figure 3 c, d). When NR was separated from RPE in the rat and rabbit, epithelium underwent a phenotypic change resulted in the formation of macrophages.

The morphology of RPE cell conversion to macrophages was well described by electron microscopy (Johnson & Faulds, 1977). Nowadays using organotypic 3D culturing of the posterior sector (RPE+choroids+sclera) of the rat eye we found that RPE also gave rise to macrophages: double nuclei cells, morphologically different from typical monocytes, and expressing macrophage-specific antigenes (Grigoryan et al., 2007; Novikova et al., 2010b). Alternatively or additionally, the RPE of mammals can proliferate and then participate in the formation of multilaminar layer of cells with characteristics of mesenchimal ones in connective tissue. The process of RPE transition to myofibroblasts is an attribute of well known ocular fibrotic disease, proliferative vitreoretinopathy (Saika et al., 2008).

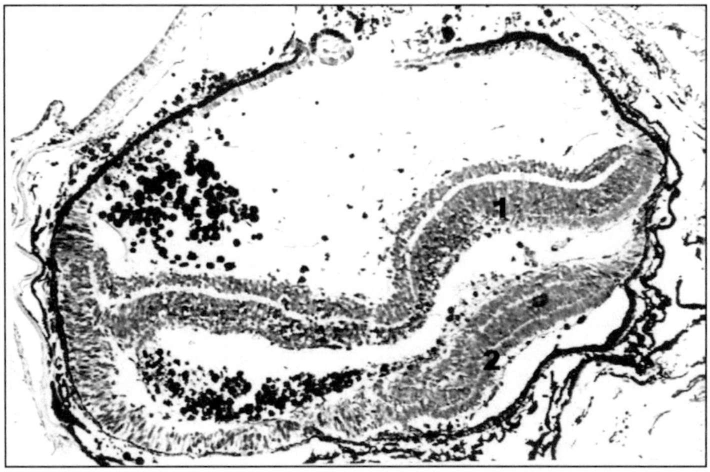

Fig. 2. Growth of additional neural retina (2) derived from RPE after detachment of the initial one (1) in the adult newt.

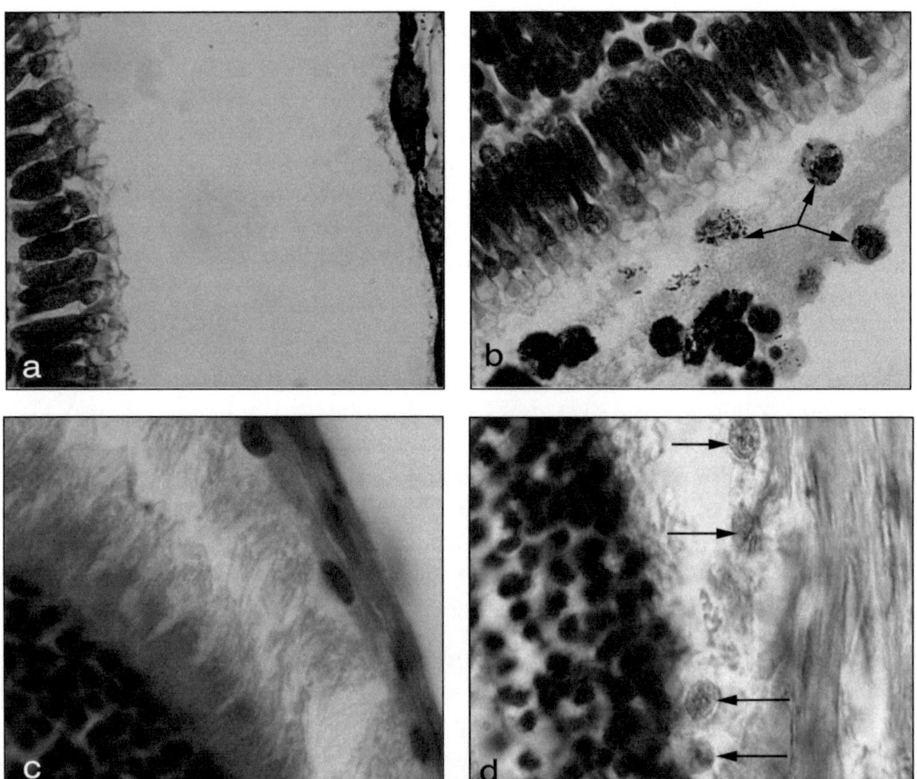

Fig. 3. Detachment of the neural retina in the newt (a,b) and albino rat (c,d) eye. In both cases RPE cells (arrows) withdraw from the layer and migrate in the vitreous direction.

Pigmented macrophage-like cells of RPE origin may be seen in a wide variety of pathological processes affecting the interface between NR and RPE. As we mentioned above a stimulus to cellular activity in the RPE is a loss of the milieu stabilizing RPE differentiation. Such an influence produced by adjacent tissues now finds explanations in terms of molecular biology. Various molecules, as like as: Na-K-ATPase, soluble component of the IPM interphotoreceptor retinoid binding protein (IRBP) (Duffy et al., 1993; Gonzalez-Fernandez, 2003), mucin-type components of IPM for the retinal adhesion, and the serum-type ones for the transport of metabolites (Uehara et al., 1991), are suggested to play a role in the maintenance of intact photoreceptor-RPE complex. In other words, in outer retina cell-to-cell and tissue-to-tissue contacts are participants of the maintenance of retinal integrity and destruction of which leads to NR cell death and RPE cell type transformation both in lower and higher vertebrates.

2.2 Visual cycle disturbance and apoptosis

Another event taking place soon after RPE↔NR separation is a disturbance of visual cycle. In the norm the latter represents a complex of biochemical reactions for regeneration of visual chromophore 11-*cis*-retinal from all-*trans*-retinol. It is well known that visual cycle is based on the renewal of photoreceptor outer segment disks. In this process disks are newly built from the base of the segments, at the connecting cilium and then, at the tips of the outer segments they are shed from the photoreceptors. Shed disks are phagocytosed by the RPE cells where they are digested. *Retinal* undergoes the RPE-specific part of visual cycle and then is redelivered as *11-cis-retinal* to photoreceptors (Bok, 1993). This interaction is essential for maintaining not only the visual function, but also structural integrity of photoreceptors. The processes of disk shedding, phagocytosis, and chromophore renewal must be tightly coordinated between RPE and NR photoreceptors. After disturbance of RPE↔NR interconnection caused by different reasons - from inherent to acquired pathology - the visual cycle gets unavoidably disturbed and brings a part into initiation and following progress of eye diseases.

Recent molecular biology studies of NR detachment suggest an essential inhibition of genes coding visual cycle proteins. "Genomic response" (Rattner et al., 2008) after NR detachment was studied in comprehensive analysis of changes in transcript abundance in the murine RPE. In that work all RPE transcripts coding visual cycle components (Rpe65, Lrat, Cralbp, Rdh5, Rdh10, and Rbp1) showed down-regulation. In parallel, an increase of a small set of transcripts for secreted proteins and cell surface receptors were registered. In accordance with Rattner and co-workers (2008) the decrease in RPE transcripts coding for the visual cycle proteins could be a protective strategy of retinal cells in conditions of detachment when slowing the light-dependent cycling of retinoids takes place. In this case the damage response of the RPE showed similarity to that described for the NR. In particular, a fraction of certain transcripts (e.g., *Cebpd, Osmr, Serpin a3n*) is induced in both tissues (Rattner & Nathans, 2005).

Meanwhile, there are numerous data suggested a decrease of the expression of particular components of visual cycle. Thus, in human IRBP (interphotoreceptor retinoid- binding protein) principal for transport of visual cycle proteins is down-regulated after NR detachment associated with diabetes retinopathy (Garcia-Ramirez et al., 2009). When we compared RPE in the intact eye and RPE soon after NR removal in Urodela we found by

PCR a considerable decrease of RPE65 transcript abundance (Avdonin et al., 2008). Other results strongly suggest that RPE65 in the RPE-derived cells of early retinal regenerate in Urodela is the only reminder after protein degradation or discharge (Chiba et al., 2006).

When RPE and NR are disintegrated, there is not only a decrease of visual cycle protein synthesis but also protein translocation into the inner compartments of photoreceptors. We observed the phenomenon not at once in experiments on the model of NR detachment in the newt *in vivo* or in isolated NR under conditions *in vitro*. In those experiments we used recoverin (the protein involved in calcium-dependent regulation of rhodopsin phosphorylation) as a marker protein of photoreceptors (Grigoryan, 2007; Grigoryan et al., 2009; Krasnov et al., 2003; Novikova et al., 2010a). In both cases: *in vivo* after NR detachment and *in vitro* newt retina retained the ability to express recoverin but its immunoreactivity was displaced from the segments to perikarya and even axons of photoreceptors. As suggested by Liljekvist-Larsson et al. (2003) who observed the same phenomenon in cultured retina of newborn rats, the synthesis of recoverin in the cytoplasm of retinal cells continues, but the transport of its newly synthesized molecules within the cell is impaired. It seems in vertebrates a decrease of expression of visual protein genes and a change of synthesized protein location in RPE and photoreceptors is the characteristic closely related with RPE↔NR separation.

One more event in cohort of those induced by experimental NR separation is an activation of mitochondrial DNA synthesis in retinal photoreceptors. Recently we showed that in the rat NR isolated from RPE and cultivated "whole amount" in 3D conditions *in vitro* the intensive incorporation of DNA synthesis precursor (BrdU) is localized in photoreceptor inner segments - cell compartments extremely rich with mitochondria (Novikova et al., 2010b). We believe that increased synthetic activity of mitochondrial DNA in photoreceptors is an attempt of photoreceptor cells to rescue and avoid apoptosis. It should be also noted that there are some other mechanisms, for instance protein degradation, taking a part in the decrease of the expression or content of visual cycle proteins in RPE and NR after loss of these tissue integrity.

A suppression of biochemical machinery of visual cycle in RPE and NR undoubtedly affects their vitality and differentiation stability. Disturbance of visual cycle and/or other RPE↔NR co-reactions lead to photoreceptor cell apoptosis. This can be a result of accumulation of photochemically active molecules and ROS which in the absence of regular visual cycle trigger an apoptotic cascade. Apoptosis includes a formation of AP-1 complex (transcription factor) and up-regulation of genes coding apoptotic enzymes - caspases (Reme et al., 2003). An evaluation of the possible induction of RPE cell apoptosis by transforming growth factor-beta (TGF-beta) was undertaken by Esser and colleagues (1997). Proapoptotic effect of TGF-beta was well demonstrated in cultured human RPE cells by electron microscopy, in situ DNA end labeling, comet assay, and a photometric enzyme immunoassay for histone associated DNA fragments. In Urodela the occurrence of apoptosis following ablation of the retina was examined by an in situ technique for detecting DNA fragmentation. It was shown that apoptosis occurs not only just after retinal removal but at the initial phase of NR regeneration as well. Authors of the work (Kaneko et al., 1999) came to a conclusion that the apoptosis is closely related to the phenomena of retinal regeneration in Urodela. Therefore, at the early time subsequent to NR↔RPE separation, both in Urodela and in mammals, the programming cell death is one more of cellular responses to destroyed RPE↔NR communication in the retina. However, it's necessary to note that NR detachment from the RPE does not lead to immediate death of the cells and retinal apoptosis is a secondary event conditioned by a number of processes initiated just after NR↔RPE separation (Moriya et al., 1986).

2.3 Changes in vascular and immune systems

In the eye of vertebrates, RPE and NR represent a structural unit that acts only in case when two tissues are interactive. Other tissues of the eye back wall, namely Bruch membrane and choroidal coat underlying RPE, also endow retinal integrity and function (figure 1). It is known that vascular occlusion, thrombus formation, accumulation of fibrinopeptides and inflammatory associated cells and proteins all are participants of many eye diseases accompanying RPE↔NR separation. In our experiments on Urodela when eye surgery was applied for retinal removal the occlusion of choroidal small vessels, an inflammation, and a decrease of a tension of the eye back wall tissues represented the very first events that brought later to RPE cell-type conversion, cell proliferation, and withdrawal from the layer (Grigoryan & Mitashov, 1979). Recently it was shown that in newts the thrombin (a participant of hemostasis and other immediate responses to any damage) pretends to be a regulator of iris cell transdifferentiation (Imokawa & Brockes, 2003; Imokawa et al., 2004). It is known that thrombin derives from prothrombin when activated by coagulation factors and, in particular, by transmembrane protein TF (tissue factor). In the work of Godwin and co-authors (2010) it was found that TF expression correlates topologically and in the time course with lens regeneration. TF and other molecules responsible for clot formation are pretending now to be initiators of tissue regeneration in lower vertebrates. In urodelean amphibians the role of complement system was proposed also in limb and lens regeneration (Kimura et al., 2003) and recently in the chick in retinal regeneration (Haynes et al., 2010). It is not inconceivable that comparable mechanism participates in triggering of retinal rescue in mammals. On the other hand, activated leucocytes associated with TF/thrombin/fibrin system can be also important participants in the initiation of NR epimorphic regeneration in amphibians and NR rescue in mammals. It is proposed (Song et al., 2010) that soon after damage they can be a source of FGF, the key role of which in NR regeneration and rescue is discussed below.

In mammals similar changes in vascular and immune systems can be induced experimentally or come out from eye diseases. For instance, there are data obtained by proteomic analysis of subretinal fluid and vitreous body of patients suffered with retinopathy of different kind and NR detachment, in particular. Authors found an increase of the content of fibrinogenic and inflammatory associated proteins for all types of pathology (Shimata et al., 2008). There are data suggested tPA (tissue plasminogen activator) may be involved in remodeling of the extracellular milieu during eye development (Collinge et al., 2005). tPA was found at the apical interface between the developing RPE and NR and then began to down-regulate once the photoreceptors have differentiated. Therefore, tPA as well as other components of the fibrinolytic system can be involved in regulation of the processes subsequent to retinal tissue disintegration, and specifically ECM changes (see below).

2.4 Expression of growth factors and major signaling pathways

Anatomic and functional relationship between NR, RPE and RPE underlying tissues (Bruch membrane, choroidal coat) is consistent with the idea that signals pass between tissues for coordinated processes in the eye back wall. The RPE secretes a variety of growth factors that support photoreceptor survival and ensure a structural basis for optimal circulation and nutrients' supply (Campochiaro, 1993). One of the signal molecules released by RPE is PEDF (pigment epithelium-derived factor) that plays a broad spectrum of developmental and neuroprotective roles (Tombran-Tink et al., 1995). In particular, it was shown that PEDF can act as an antiangiogenic factor that inhibits endothelial cell proliferation in the choriocapillaris. VEGF is another vasoactive factor of RPE known as preventing endothelial cell apoptosis (Saint-Geniez et al., 2009). In a healthy eye, PEDF and VEGF are secreted at opposite sides of the RPE cell. At the apical side PEDF acts on neurons and photoreceptors but the majority of VEGF is secreted to the basal side where it acts on the choroidal endothelium. It was found that the balance between PEDF and VEGF is disturbed in the early course of retinopathy. Thus, in subretinal fluid of patients suffered with early PVR the concentrations of both factors essentially increase changing a normal balance where PEDF counteracts to angiogenic potential of VEGF (Dieudonné et al., 2007). It is important to note that, as a rule, NR↔RPE disintegration is accompanying by oxidative stress that, in turn, induces a decrease of PEDF correct level (Ohno-Matsui et al., 2001).

In response to retinal damage or injury RPE also secretes the row of neuroprotective factors including those of FGF, CNTF, IGF, and TGF families, all of which are included in the regulative network of the eye and retina (for review: Strauss, 2005). For instance, many extracellular stimuli have been proposed to induce an increasing of VEGF secretion. This signaling exploits growth factors such as IGF-I that can contribute to a pathway in which photoreceptors can stimulate VEGF secretion by RPE cells. Fibroblast growth factor basic (FGF2) is one of several agents that elicits most profound effects in RPE and NR cells. Since 90s the role of FGF2 in RPE transdifferentiation and NR regeneration after RPE↔NR separation in adult amphibians and bird embryos is received the intensive study (Araki, 2007; Mitsuda et al., 2005; Park & Hollenberg, 1993). In accordance with the data including our own, FGF2 and FGF2R coupled with that of transcription factor Pax6 control urodelean RPE cell dedifferentiation and proliferation after NR removal (Avdonin, 2010; Chiba & Mitashov, 2007). FGF-FGFR-MEK cascade and Pax6 up-regulation depended on changes of the cell-ECM and/or cell-cell interaction are supposed important for realization of the first steps of NR regeneration (Avdonin, 2010; Susaki & Chiba, 2007). In the *in vitro – in vivo* like systems it was shown that cells of isolated RPE could be induced to faster dedifferentiation by additing of FGF2 to culture medium (Ikegami et al., 2001; Novikova et al. 2010b).

In mammals, soon after NR detachment FGF2 gene up-regulation also takes place in parallel with high expression of FGF receptors (FGFR) (Hackett et al., 1997; Ozaki et al., 2000). When the retina is perturbed, significant changes occur in the expression of FGFR1 by photoreceptors: FGFR1 immunoreactivity increases rapidly (in 24 hours after injury) and steadily (Ozaki et al., 2000). That appears to be accompanied by similar increase of FGF2 in the IPM. Ozaki and co-workers suggest that this describes a paracrine mechanism: FGF2 is released or activated after retinal injury and then binds to FGFR1 on photoreceptor target cells. The latter, in turn, initiates an intracellular cascade that "protects" the cells from further damage.

The study of the effect of light, various types of stress, neurotrophic factors, and cytokines on FGF2 levels in human RPE cultured *in vitro* showed that many agents of photoreceptor protection (for instance, BDNF, CNTF, IL-1β) can up regulate FGF2 mRNA in RPE cells. An increase in FGF2 protein level was demonstrated by ELISA in RPE cell supernatants after incubation with BDNF or exposure to intense light or oxidizing agents. These data indicate that in RPE cells FGF2 is modulated by stress and by agents that provide protection from stress (Hackett et al., 1997). In addition, it was found that FGF2 immunoreactivity in the interphotoreceptor matrix tends to increase during first 24 hours after retinal detachment in the rat. It is proposed that the interphotoreceptor matrix has its own endogenous local source(s) of FGF2 (Ozaki et al., 2000). Therefore, it is possible to consider that in both cases, at the initiation of NR regeneration in amphibia and NR rescue in mammals, FGF2 signaling pathway participate in neuroprotection and regulation of RPE cell differentiation and proliferation.

Other signaling pathways, as like as IGF-1, CNTF, and TGFβ represent also a part of the molecular network, regulating RPE and NR cell behavior after separation of these tissues in mammals. However, for today there are only few data on their activity in NR regeneration in Urodela. There is the evidence that IGF-1 (as like as FGF2) can accelerate proliferation and proneuronal differentiation of amphibian RPE under *in vitro* conditions (Yoshii et al., 2007). Meanwhile, proapoptotic growth factor TGFβ more likely plays prohibitive role in RPE cell type conversion. Activin, a TGF-β family signaling protein has been shown to contribute to the loss in competence of the RPE to regenerate retina. Sakami and co-authors (2008) have found that additing of activin blocked regeneration from the RPE, even when the cells were competent. Conversely, a small molecule inhibitor of the activin/TGF-β/nodal receptors could delay and reverse the developmental restriction in FGF-stimulated NR regeneration in embryonic chicken (Sakami et al., 2008).

Earlier it was shown that TGFβ inhibits proliferation at the vitreoretinal interface after NR detachment in human (Esser et al., 1997). Nowadays the study of the role of TGFβ is carried out on the model of retinal detachment in experiments using mice null for Smad3, TGFβ functional cooperator, a key signaling intermediate downstream of TGFβ and activin receptors. Obtained results showed that Smad3 is essential for the epithelial-mesenchymal transition of RPE cells induced by NR detachment. *De novo* accumulation of fibrous tissue derived from multilayered RPE cells was seen in experimental NR detachment in eyes of wild type, but not in Smad3-null mice (Saika et al., 2004). Activation of several signaling pathways, particularly TGFβ /Smad, was also fixed by Zacks and coworkers (2006). Soon after NR detachment the interleukin-6/STAT, TGFβ-Smad, and stress response pathway (aryl hydrocarbon receptor) - all were transcriptionally and translationally upregulated, suggesting that retina produces survival factors after detachment and that there is a possible cross-talk between up-regulated pathways (Zacks et al., 2006). In sum, knowing of signaling pathways with proliferative and anti-proliferative as well as pro-apoptotic and anti-apoptotic effects is very important, because in both, retinal epimorphic regeneration in amphibian and proliferative retinopathy after detachment in mammals, changes of RPE cell phenotype, cell proliferation and apoptosis take place.

2.5 Up-regulation of heat shock proteins and immediate-early response genes

RPE↔NR disintegration results in the early activation of stress-response genes and specific signaling pathways which may enable retinal cells to survive at the most acute period of time. During NR detachment/regeneration in Urodela and detachment in mammals, heat shock proteins (HSPs) are involved in fast regenerative responses. Our preliminary (unpublished) results show an accumulation and co-distribution of HSP70, 90 and FGF2 in the NR soon after its detachment in the newt. In the experiments we observed well correlated changes in the intensity of HSPs and FGF2 expression and in the localization of these proteins in the retina. These data preliminary show that besides well known role of HSPs in the protection of newly synthesizing proteins from degradation a regulative link between HSPs and FGF may play a role in triggering of early retinal cell death/survival events. It is interesting also when infected with MC29, a myc expressing virus, the RPE cells in developing eye can be induced to transdifferentiate to neuroretinal epithelium. Beside genes whose work is involved in regulating neuronal differentiation myc also induced a transient expression of Mitf, well-known regulator of the pigmented differentiation (Beche-Belsot et al., 2001). HSPs, growth factors, and mitogen-activated protein kinase (MAPK) signaling are capable of immediate-early response gene up-regulation in different systems. Retinal detachment in the rat results in early up-regulation of genes, coding HSPs, FGF, early emergency genes (c-Fos and c-Jun), and transcription factor AP-1 complex (Faktorovich et al., 1992; Geller et al., 2001). Authors hypothesize that NR detachment causes the rapid release of FGF2 from intra- and/or extra-cellular stores, leading to the activation of FGFR1 and ERK, and proximate induction of c-Fos and c-Jun protein expression in RPE. Up-regulation of these intracellular components linked with FGF expression [HSPs → FGF2→FGFR → ERK&MAPK (MEK) pathway → c-fos&c-jun (AP-I) →] pretends to be an important early step on the way to RPE cell type transformation, migration and proliferation in amphibian and mammals. It is likely that increased AP-1 expression besides entering to apoptosis can regulate a variety of genetic and cellular responses induced by NR↔RPE separation.

2.6 Remodeling of extracellular matrix (ECM) and RPE cytoskeleton

It is quite possible that RPE↔NR separation associated changes of cytoskeleton are involved in regulation of HSPs and AP-1 complex in RPE and NR cells. Our early studies showed fast down-regulation of epithelium-specific intermediate filament expression and up-regulation of pan-neuronal one in RPE soon after NR removal in the newt (Grigoryan & Anton, 1993, 1995; Grigoryan, 1995). Keratins of the cytoskeletal intermediate filaments have been identified immunohistochemically in RPE of the adult newt retina. In conditions of NR surgical removal or complete detachment the expression of keratins markedly decreased. Similar observation has been made immediately after dissociation of the RPE cells isolated from nonoperated newt eyes. The results obtained provide an evidence for the inhibition of cytokeratin expression just after destabilization of RPE cell phenotype. *In vivo* in RPE disappeared cytokeratins were replaced by NF-200 neurofilament proteins that testified an existence of the mechanism responsible for gradual change of cytoskeleton in modified RPE in amphibians.

Changes suggested cytoskeleton rearrangement were also registered in mammalian animal models simulating RPE epithelial-mesenchymal transition specific for NR detachment. It was found that RPE cells lost their initial phenotype, dedifferentiated and acquired mesenchymal migratory morphology and cytoskeleton proteins (Casaroli–Marano et al., 1999). Recently thrombin (see above) pretends to play a promoting role in actin stress fiber formation, an important determinant in eye diseases involving transformation and migration of RPE cells (Ruiz-Loredo et al., 2011). On the other hand cytoskeleton changes in

Kind of events	Prior NR epimorphic regeneration	Prior NR rescue
Disturbance of retinal cell contacts and change of RPE cell behavior	IPM disruption, weakening of RPE cell lateral contacts, RRE cell withdrawal from the layer, high proliferative activity of RPE cells, RPE cell conversion to macrophagal and proneuronal phenotypes	IPM disruption Weakening of RPE cell lateral contacts, RRE cell withdrawal from the layer, low proliferative activity of RPE cells, RPE cell conversion to macrophagal and fibroblast-like phenotypes
Visual cycle disturbance and cell apoptosis	Blockage of melanin synthesis in RPE cells, down regulation of visual cycle proteins, apoptosis of small set of RPE and photoreceptor cells	Up regulation of secreted proteins in RPE cells, down regulation of visual cycle proteins, apoptosis of small set of RPE and photoreceptor cells
Changes in vascular and immune systems in the eye back wall	Possible involvement of fibrinolytic (TF, thrombin) and complement systems in triggering of RPE cell transdifferentiation	Possible involvement of fibrinogenic (tPA), inflammatory associated proteins, and activated lymphocytes at the first stage of NR rescue
Participation of growth factors and major signaling pathways	FGF2, IGF1, TGFβ (activin)	PEDF, VEGF, FGF1,2, CNTF, BDNF, IGF1, IL-1β, and TGFβ
Up-regulation of heat shock proteins and immediate-early response genes	HSP70,90 proteins; *c-Myc* gene	HSPs, *c-Fos* and *c-Jun* genes, AP1-complex
Remodeling of extracellular matrix and RPE cytoskeleton.	Redistribution of fibronectin, laminin stimulating effect on RPE conversion, epithelial–neuronal transition of RPE cells: shift of specific intermediate filaments (cytokeratins → neurofilaments).	Role of laminin and integrins in modulation of RPE cells, epithelial-mesenchymal transition of RPE cells: change in composition of specific intermediate filaments (cytokeratins, vimentin, GFAP).

Table 1. A comparison of known NR detachment-induced cellular and molecular events preceding retinal regeneration in Urodelean amphibians and retinal rescue in mammals

RPE reflect an alteration of cell micro-surrounding. The latter, in turn, is a response for mechanical and chemical changes which are produced inevitably by RPE↔NR separation. Earlier we showed a decrease of fibronectin in Bruch membrane and its redistribution in RPE after NR detachment in the newt (Grigoryan et al., 1990). Similar results were obtained by Ortiz and co-authors (1992): at the beginning of RPE cell transdifferentiation in the eye of the adult newt, fibronectin was the first to appear in the cell border of the newforming neuroepithelium. A dependence of RPE phenotype on changes of ECM was also observed by Reh and co-authors in *in vitro* experiments (1987). They reported that RPE transdifferentiation is profoundly influenced by the substrate on which the cells are cultured. RPE cells plated on laminin-containing substrates frequently were conversed into neurons. Recently some data suggest that interaction of laminins and integrins in Bruch membrane leads to differential behavior of RPE cells in mammals (Aisenbrey et al., 2006).

Degradation of ECM is one more important stimulus for the initiation of RPE cell migration and phenotype transformation. Metalloproteases are known molecules for ECM changes and, vice versa, metalloprotease inhibitors (TIMPs) are factors that stabilize ECM. Mechanical trauma induced by NR↔RPE separation is associated with an increased activity of proteolytic enzymes. To ascertain whether RPE cells release proteases due to mechanical stress special tests *in vitro* were performed by Kain and Reuter (1995). In traction conditions created *in vitro* RPE might release proteases to cut intercellular adhesions in order to escape mechanical strain. Authors suggest that release of proteases from RPE may be involved in the pathology of traction detachment, facilitating the disconnection between RPE and photoreceptor outer segments. In human RPE cultured *in vitro* stromelysin which degrades important constituents of the ECM was found (Schönfeld, 1997). Therefore, the action of lysosomal proteases may change the surrounding that, in turn, can induce further detachment of RPE cells from the basement membrane and initiate RPE proliferation and dedifferentiation under conditions of RPE↔NR separation.

3. Conclusion

In the review we summarized our own and literature data on the early cellular and molecular events taking place after separation of neural retina (NR) from the retinal pigmented epithelium (RPE) in the eyes of vertebrates (Table 1). In amphibians RPE↔NR disintegration leads to the formation of the new NR by means of RPE cell transdifferentiation into retinal cells, while in mammals NR detachment triggers a retinal pathology. A comparison of these two opposite phenomena unexpectedly reveals a similarity of early cell and molecular processes induced by RPE↔NR separation (figure 4). In both cases alterations of RPE cell contacts, changes in cytoskeleton and ECM composition as well as perturbations in blood circulation and immune system can be found. These alterations lead to RPE cell type destabilization, phenotypic transformation, cell withdrawal from the layer, and migration. In parallel, down-regulation of the expression of visual cycle molecules takes place. In contrast heat shock proteins, FGF signaling, immediate-early response genes, and AP-1 complex demonstrate up-regulation. In all animals and in human these, NR detachment associated events represent a limited, universal range of retinal cell responses to the stress. Among them RPE dedifferentiation, proliferation and migration seems most important for both, subsequent NR epimorphic regeneration in amphibians and a progress of detachment induced eye diseases in mammals.

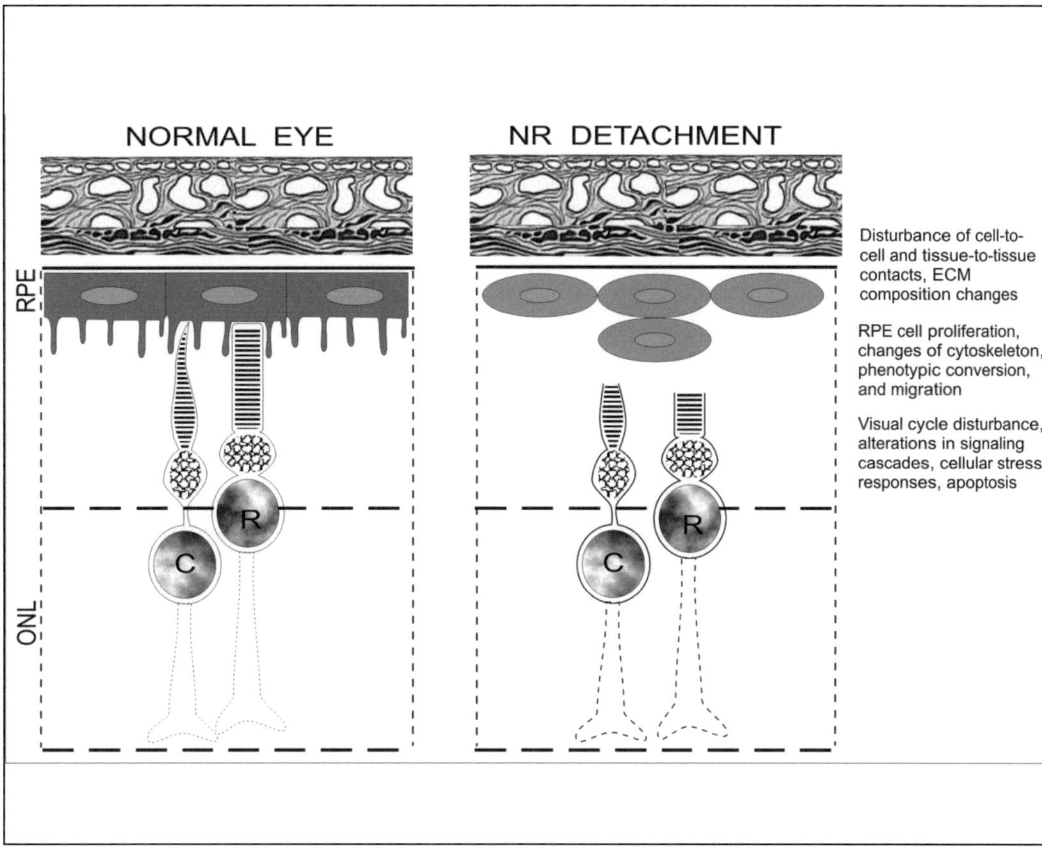

Fig. 4. Main shared processes taking place in the eye back wall soon after neural retina detachment in amphibians (NR epimorphic regeneration) and mammals (NR rescue/pathology). ONL - outer nuclear layer, R - rod, C - cone photoreceptors.

A comparison of initial mechanisms triggering retinal regeneration and retinal rescue/pathology seems efficient. It could be very helpful for understanding why quite similar intrinsic protective responses, occurring at early stages of NR regeneration in amphibians and NR detachment in mammals, give such contrast final results. In regards of this, being based on the accumulated data we can make several suppositions. The first one is the difference in the level of RPE cell differentiation. In Urodela it has some developmental traits whose expression in permissive conditions *in vivo* leads to acquiring of neuronal phenotype. Contrast to amphibians, in adult mammals changes of RPE differentiation *in vivo* imply the epithelial-mesenchymal transition and cell transformation into migrated macrophages, the processes resembling an inflammation and scarring. The second assumption is a difference in the external molecular network, its signals, and cross-talks which regulate RPE cell differentiation in amphibians and mammals. The search of key factors which distinguish detachment induced signaling for amphibian RPE from that for mammalian and human RPE is rather difficult though also necessary step for future work. Finally and more likely, both: RPE cell intrinsic competence (including epigenetic features) and molecular regulation by microsurrounding are different in lower and higher vertebrates. This difference can be proved in only some characteristics, even epigenetic ones, meanwhile crucial for RPE conversion into neuroepithelium. We can also speculate that cellular and molecular mechanisms which adult amphibians use for initiating of NR regeneration, in evolution have been recruited by higher animals for NR rescue attempts. Answers for these fundamental questions should improve our ability to elucidate the maintaining and pathogenic mechanisms triggered by NR detachment and to facilitate a development of therapies for different types of detachment associated diseases of the eye.

4. Acknowledgment

This study was supported by the Russian Foundation for Basic Research, project no. 11-04-00125-a (for E.N. Grigoryan).

5. References

Aisenbrey, S.; Zhang, M.; Bacher D.; Yee J.; Brunken W. & Hunter D. (2006). Retinal pigment epithelial cells synthesize laminins, including laminin 5, and adhere to them through alpha3- and alpha6-containing integrins. *Investigative Ophthalmology and Visual Science*, Vol. 47, N0.12, pp.5537-5544, ISSN 0146-0404

Araki, M. (2007). Regeneration of the amphibian retina: role of tissue interaction and related signaling molecules on RPE transdifferentiation. *Development, Growth & Differentiation*, Vol. 49, No.2, pp.109-120, ISSN 1440-169X

Avdonin, P.; Markitantova, Iu.; Zinov'eva, R. & Mitashov, V. (2008). Expression of regulatory genes: Pax6, Otx2, Six3, and FGF2 during newt retina regeneration. *Biology Bulletin*, No.4, pp. 414-421, ISSN 0002-3329

Avdonin, P. (2010). Expression of regulatory genes Fgf2, Pax6, Six3, Otx2, Pitx1, and Pitx2 in epimorphic regeneration of the eye retina in the newt Pleurodeles waltl, Ph.D. thesis, IDB RAS, Moscow, (28 October, 2010), pp. 1-174.

Beche-Belsot, J.; Planque, N.; Martin, P. & Saule, S. (2001). The *myc* oncogene and transdifferentiation of the retinal pigment epithelium. *Journal de la Société de Biologie*, V.195, No.2, pp.107-113, ISSN 1295-0661

Bok, D. The retinal pigment epithelium: a versatile partner in vision. (1993). *Journal of Cell Science Suppl.* Vol.17, pp.189-195, ISSN 0269-3518

Campochiaro, P. (1993). Cytokine production by retinal pigmented epithelial cells. *International Review of Cytology*, Vol. 146, pp. 75- 83, ISBN 9780123743329

Casaroli-Marano, R.; Pagan, R. & Vilaró S. (1999). Epithelial-mesenchymal transition in proliferative vitreoretinopathy: intermediate filament protein expression in retinal pigment epithelial cells. *Investigative Ophthalmology and Visual Science*, Vol. 40, No.9, pp. 2062-2072, ISSN 0146-0404

Chiba C. & Mitashov V. (2007). Cellular and molecular events in the adult newt retinal regeneration. In: *Strategies for retinal tissue repair and regeneration in vertebrates: from fish to human,* Ch. Chiba, (Ed.), 15-33, Research Singnpost. ISBN 978-81-308-0200-8. Trivandrum. India

Chiba, C.; Hoshino, A.; Nakamura, K.; Susaki, K.; Yamano, Y.; Kaneko, Y.; Kuwata, O.; Maruo, F., Saito T. (2006). Visual cycle protein RPE65 persists in new retinal cells during retinal regeneration of adult newt. *Journal of Comparative Neurology*, V.495, pp. 391-407, ISSN: 1096-9861

Collinge, J.; Simirskii, V. & Duncan, M. Expression of tissue plasminogen activator during eye development. (2005). *Experimantal Eye Research*, V. 81, No.1, pp. 90-96. ISSN 0014-4835

Dieudonné, S.C.; La Heij, E.C.; Diederen, R.M.; Kessels, A.G.; Liem, A.T.; Kijlstra, A. & Hendrikse, F. (2007). Balance of vascular endothelial growth factor and pigment epithelial growth factor prior to development of proliferative vitreoretinopathy. *Ophthalmic Research*, Vol. 39, No.3, pp. 148-154, ISSN 0030-3747

Duffy, M.; Sun Y.; Wiggert, B.; Duncan, T.; Chader G., Ripps, H. (1993). Interphotoreceptor retinoid binding protein (IRBP) enhances rhodopsin regeneration in the experimentally detached retina, *Experimental Eye Research*, Vol. 57, No. 6, pp. 771-782, ISSN 0014-4835

Esser, P.; Heimann, K.; Bartz-schmidt, K.; Fontana, A.; Schraermeyer, U.; Thumann G. & Weller, M. (1997). Apoptosis in proliferative vitreoretinal disorders: possible involvement of TGF-beta-induced RPE cell apoptosis. *Experimantal Eye Research*, Vol.65, No.3, pp.365-378, ISSN 0014-4835

Faktorovich, E.; Steinberg, R.; Yasumura, D.; Matthes, M. & LaVail M. (1992). Basic fibroblast growth factor and local injury protect photoreceptors from light damage in the rat. *Journal of Neuroscience*, Vol.12, No.9, pp. 3554-3567

Fisher, S. & Lewis, G. (2010a). Retinal detachment. *Ocular Disease*, pp. 554-561

Fisher, S. & Lewis, G. (2010b). Injury and Repair Responses: Retinal Detachment. *Encyclopedia of the Eye*, pp. 428-438, ISBN: 978-0-12-374203-2

Fisher, S.; Lewis, G.; Linberg, K. & Verardo, M. (2005). Cellular remodeling in mammalian retina: results from studies of experimental retinal detachment, *Progress in Retinal and Eye Research*, Vol. 24, No.3, pp. 395-431, ISSN 1350-9462

Garcia-Ramírez, M.; Hernández, C.; Villarroel, M.; Alonso M.; Fortuny, R.; Masmiquel, L.; Navarro, A.; García-Arumí J. & Simó R. (2009). Interphotoreceptor retinoid-binding protein (IRBP) is downregulated at early stages of diabetic retinopathy, *Diabetologia*, Vol. 52, No.12, pp. 2633-2641.

Ghazi, N. & Green, W. (2002). Pathology and pathogenesis of retinal detachment, *Eye*, V.16, pp.411–421, ISSN 0950-222X

Geller, S.; Lewis, G. & Fisher S. (2001). FGFR1, signaling, and AP-1 expression after retinal detachment: reactive Müller and RPE cells, *Investigative Ophthalmolology and Visual Science*, Vol. 42, No.6, pp.1363-1369, ISSN 0146-0404

Godwin, J.; Liem, K. & Brockes, J. (2010). Tissue factor expression in newt iris coincides with thrombin activation and lens regeneration, *Mechanisms of Development*, Vol. 127, pp. 321-328, ISSN 0925-4773

Gonzalez-Fernandez, F. (2003). Interphotoreceptor retinoid-binding protein - an old gene for new eyes, *Vision Research*, Vol. 43, No. 28, pp. 3021-3036, ISSN 0042-6989

Grigoryan, E. (1995). Complete retinal detachment induces changes of cytokeratin expression in retinal pigment epithelial cells in newts. *Biology Bulletin,* No.4, pp. 412-421, ISSN 0002-3329

Grigoryan, E. (2007). Alternative intrinsic cell sources for neural retina regeneration in adult urodelean amphibians. In: *Strategies for retinal tissue repair and regeneration in vertebrates: from fish to human,* Ch. Chiba, (Ed.), 35-62, ISBN 978-81-308-0200-8. Trivandrum. India

Grigorian, E. & Anton, H. (1993). The appearance and distribution of the NF-200 neurofilament protein in transdifferentiating cells of the pigment epithelium and in other eye cells during retinal regeneration in tritons, *Ontogenez,* Vol. 24, No.4, pp. 39-51, ISSN 0475-1450

Grigorian, E. & Anton, H. (1995). An analysis of keratin expression in the cells of the retinal pigment epithelium during transdifferentiation in newts. *Ontogenez.* Vol. 26, No.4, pp. 310-323, ISSN 0475-1450

Grigorian, E. & Mitashov, V. (1979). Radioautographic investigation of proliferation and melanin synthesis in retinal pigment epithelium cells during eye regeneration in the newt, *Ontogenez,* Vol. 10, pp. 137-144, ISSN 0475-1450

Grigoryan, E. & Mitashov, V. (1985). Cultivation of retinal pigment epithelium in the cavity of the lensectomized newt eye, Vol. 16, No. 1, pp. 34-43, ISSN 0475-1450

Grigoryan, E.; Bazhin, A.; Krasnov, M. & Philippov, P. (2009). Expression of Ca-binding protein recoverin in normal, surviving, and regenerating retina of Pleurodeles waltl adult triton, *Bulletin of Experimental Biology and Medicine,* Vol. 148, No. 1, pp. 155-162, ISSN 0007-4888

Grigorian, E.; Dol'nikova, A. & Belkin, V. (1990). Fibronectin distribution during the transdifferentiation and proliferation of eye cells after retinal detachment and removal of the crystalline lens in tritons. *Ontogenez,* Vol. 21, No.4, P.403-408, ISSN 0475-1450

Grigoryan, E.; Novikova, Yu.; Kilina, O. & Philippov P. (2007). New method of in vitro culturing of pigment retinal epithelium in the structure of the posterior eye sector of adult rat, *Bulletin of Experimental Biology and Medicine,* Vol.144, No. 4, pp. 618-625. ISSN 0007-4888

Hackett, S.; Schoenfeld, C.; Freund, J; Gottsch, J.; Bhargave, S. & Campochiaro, P. (1997). Neurotrophic factors, cytokines and stress increase expression of basic fibroblast growth factor in retinal pigmented epithelial cells. *Experimental Eye Research,* Vol. 64, No.6, pp. 865-873, ISSN 0014-4835

Hasegawa, M. (1958). Restitution of the eye after removal of the retina and lens in the newt Triturus pyrrhogaster, *Embryologia,* V.4, No.1, pp.1-32

Ikegami, Y.; Mitsuda, S. & Araki, M. (2001). Neural cell differentiation from retinal pigment epithelial cells of the newt: an organ culture model for the urodele retinal regeneration, *Journal of Neurobiology,* Vol.50, pp. 209-220, ISSN 1932-846X

Imokawa, Y. & Brockes, J. (2003). Selective activation of thrombin is a critical determinant for vertebrate lens regeneration, *Current Biology,* Vol.13, No.10, pp. 877-881, ISSN 0960-9822

Imokawa, Y.; Simon, A. & Brockes, J. (2004). A critical role for thrombin in vertebrate lens regeneration, *Philosophical Transactions of the Royal Society (London) Biological Sciences,* Vol. 359, pp. 765–776. ISSN 1471-2970

Jablonski, M.M.; Tombran-Tink, J.; Mrazek, D.A. & Iannaccone, A. (2001). Pigment epithelium-derived factor supports normal Müller cell development and glutamine synthetase expression after removal of the retinal pigment epithelium. *Glia,* Vol. 35, No.1, pp. 14-25, ISSN 1098-1136

Johnson, N. & Foulds, W. (1977). Observations on the retinal pigment epithelium and retinal macrophages in experimental retinal detachment. *British Journal of Ophthalmology,* Vol. 61, pp. 564-572, ISSN 1468-2079

Kain, H. & Reuter, U. (1995). Release of lysosomal protease from retinal pigment epithelium and fibroblasts during mechanical stresses, *Graefe's Archive for Clinical and Experimental Ophthalmology,* Vol. 233, No.4, pp. 236-243, ISSN 0721-832X

Kaneko, Y.; Matsumoto, G. & Hanyu, Y. (1999). The occurrence of apoptosis during retinal regeneration in adult newts, *Brain Research. Developmental Brain Research,* Vol. 117, No.2, pp.225-228, ISSN 0165-3806

Keefe J.R. An analysis of urodelian retinal regeneration. (1973). *Journal of Experimental Zoology,* Vol. 184, pp. 185-257, ISSN 1932-5231

Kimura, Y.; Madhavan, M.; Call, M.K.; Santiago, W.; Tsonis P.; Lambris, J. & Del Rio-Tsonis, K. (2003). Expression of complement 3 and complement 5 in newt limb and lens regeneration, *Journal of Immunology,* Vol. 170, No.5, pp. 2331-2339, ISSN 0022-1767

Krasnov, M.; Grigoryan, E. & Yamskova V. (2003). Model of organotypic cultivation of retina together with the newt eye posterior sector tissues for studying the effects of adhesive glycoproteins, *Biology Bulletin,* No.1, pp. 22-36, ISSN 0002-3329

Lachke, S. & Maas, R. (2010). Building the developmental oculome: systems biology in vertebrate eye development and disease, Wiley interdisciplinary reviews. Systems biology and medicine. Vol. 2, No.3, pp. 305-323, ISSN 1939-5094

Lazarus, H. & Hageman G. (1992). Xyloside-induced disruption of inferphororeceptor matrix proteoglycans results in retinal detachment, *Investigative Ophthalmology and Visual Science,* Vol. 33, No.2, pp. 364-376, ISSN 0146-0404

Liljekvist-Larsson, I.; Torngren, M.; Abrahamson, M. & Johanson, K. (2003). Growth of the postnatal rat retina in vitro: quantitative RT-PCR analyses of mRNA expression for photoreceptor proteins, *Molecular Vision,* Vol.9, pp.657-664, ISSN 1090-0535

Marmor, M. (1993). Mechanisms of retinal adhesion, *Progress in Retinal and Eye Research,* Vol.12, pp. 179–204, ISSN 1350-9462

Mitashov, V. (1997). Retinal regeneration in amphibians, *International Journal of Developmental Biology,* Vol. 41. pp. 893 - 905. ISSN 0214-6282

Mitsuda, S.; Yoshii, Ch.; Ikegami, Y. & Araki, M. (2005). Tissue interaction between the retinal pigment epithelium and the choroids triggers retinal regeneration of the newt Cynops pyrrhogaster, *Developmental Biology,* Vol. 280, pp. 122-132, ISSN 0012-1606

Moriya, M.; Baker B. & Williams, T. (1986). Progression and reversibility of early light-induced alterations in rat retinal rods. *Cell and Tissue Research,* Vol. 246, pp. 607-614, ISSN 0302-766X

Novikova, Yu.; Aleĭnikova, K.; Krasnov, M.; Poplinskaia V. & Grigorian E. (2010a). In vitro organotypic cultivation of adult newt and rat retinas, *Biology Bulletin,* No.4, pp. 389-402, ISSN 0002-3329

Novikova, Yu.; Aleinikova, K.; Poplinskaya, V. & Grigoryan E. (2010b). The retinal pigment epithelial cells of the adult newt and rat under conditions of *in vitro* organotypic culture, *Biology Bulletin,* No.3, pp. 221–230, ISSN 0002-3329

Ohno-Matsui, K.; Morita, I.; Tombran-Tink, J.; Mrazek D.; Onodera M.; Uetama T.; Hayano M.; Murota S.I.& Mochizuki, M. (2001). Novel mechanism for age-related macular degeneration: an equilibrium shift between the angiogenesis factors VEGF and PEDF. *Journal of Cell Physiology,* Vol. 189, No.3, pp. 323-333, ISSN 1097-4652

Ortiz, J.; Vigny, M.; Courtois, Y. & Jeanny J. (1992). Immunocytochemical study of extracellular matrix components during lens and neural retina regeneration in the adult newt, *Experimental Eye Research,* Vol.54, No.6, pp. 861-870, ISSN 0014-4835

Ozaki, S.; Radeke, M. & Anderson, D. (2000). Rapid upregulation of fibroblast growth factor receptor 1 (flg) by rat photoreceptor cells after injury. *Investigative Ophthalmology and Visual Science,* Vol. 41, No.2, pp.568-579, ISSN 1552-5783

Park, C. & Hollenberg, M. (1993). Growth factor-induced retinal regeneration in vivo. *International Review of Cytology,* Vol. 146, pp.49-74, ISSN 9780123743329

Pasto, J. Proliferative Vitreoretinopathy: An Overview. (1998). *Survey of Ophthalmology,* Vol. 43, pp. 3-18, ISSN 0039-6257

Rattner, A. & Nathans, J. (2005). The genomic response to retinal disease and injury: evidence for endothelin signaling from photoreceptors to glia, *Journal of Neuroscience,* Vol. 25, No.18, pp. 4540-4549, ISSN 1529-2401

Rattner, A.; Toulabi, L.; Williams, J.; Yu H. & Nathans J. (2008). The genomic response of the retinal pigment epithelium to light damage and retinal detachment, *Journal of Neuroscience,* Vol. 28, pp. 9880-9889, ISSN 1529-2401

Remé, C.; Grimm, C.; Hafezi, F.; Iseli, H. & Wenzel, A. (2003). Why study rod cell death in retinal degenerations and how? *Documenta Ophthalmologica,* Vol. 106, No.1, pp. 25-29, ISSN 0012-4486

Reh, T.; Nagy, T. & Gretton, H. (1987). Retinal pigmented epithelial cells induced to transdifferentiate to neurons by laminin, *Nature,* Vol. 330, pp. 68-71, ISSN 0028-0836

Ruiz-Loredo, A.; López, E. & López-Colomé, A. (2011). Thrombin promotes actin stress fiber formation in RPE through Rho/ROCK-mediated MLC phosphorylation, *Journal of Cellular Physiology,* Vol. 226, No.2, pp. 414-423, ISSN 1097-4652

Saika, S.; Kono-Saika, S.; Tanaka, T.; Yamanaka, O.; Ohnishi, Y.; Sato, M.; Muragaki, Y.; Ooshima, A.; Yoo J.; Flanders, K. & Roberts, A. (2004). Smad3 is required for dedifferentiation of retinal pigment epithelium following retinal detachment in mice, *Laboratory Investigation,* Vol. 84, No.10, pp.1245-1258, ISSN 0023-6837

Saika, S.; Yamanaka, O.; Flanders, K.; Okada, Y.; Miyamoto, T.; Sumioka, T.; Shirai, K.; Kitano, A.; Miyazaki, K.; Tanaka, S. & Ikeda, K. (2008). Epithelial-mesenchymal transition as a therapeutic target for prevention of ocular tissue fibrosis, *Endocrine, Metabolic & Immune Disorders - Drug Targets,* Vol. 8, No.1, pp. 69-76. ISSN 1871-5303

Saint-Geniez, M.; Kurihara, T.; Sekiyama, E., Maldonado, A. & D'Amore, P. (2009). An essential role for RPE-derived soluble VEGF in the maintenance of the choriocapillaris, *Proceedings of National Academy of Sciences USA,* Vol. 106, No.44, pp.18751–18756, ISSN: 1091-6490

Sakami, S.; Etter, P. & Reh, T. (2008). Activin signaling limits the competence for retinal regeneration from the pigmented epithelium, *Mechanisms of Development,* Vol.125, No.1-2, pp.106-16, ISSN 0925-4773

Schönfeld, C. (1997). Metalloproteinase stromelysin. Expression in human retinal pigment epithelium (RPE), *Ophthalmology,* Vol. 94, No.9, pp. 629-633, ISSN 0161-6420

Shitama, T.; Hayashi, H.; Noge, S.; Uchio, E.; Oshima, K.; Haniu, H.; Takemori, N.; Komori, N. & Matsumoto, H. (2008). Proteome profiling of vitreoretinal diseases by cluster analysis, *Proteomics. Clinical applications,* Vol.2, No.9, pp. 1265–1280, ISSN 1862-8346

Song, J-S.; Lee, J. & Kay, E. (2010). Induction of FGF-2 synthesis by IL-1β in aqueous humor through P13-Kinase and p38 in rabbit corneal endothelium, *Investigative Ophthalmology and Visual Science,* Vol. 51, No.2, pp. 822–829, ISSN 1552-5783

Strauss, O. (2005). The retinal pigment epithelium in visual function, *Physiological Reviews,* Vol. 85, pp. 845–881, ISSN 0031-9333

Stroeva, O. & Mitashov, V. (1983). Retinal pigment epithelium: proliferation and differentiation during development and regeneration, *International Review of Cytology,* Vol. 83, pp. 221-293, ISBN 9780123743329

Susaki, K. & Chiba, C. (2007). MEK mediates in vitro neural transdifferentiation of the adult newt retinal pigment epithelium cells: Is FGF2 an induction factor? *Pigment Cell Research,* Vol. 20, pp. 364–379, ISSN 17551471

Tombran-Tink, J.; Shivaram, S.; Chader, G.; Johnson, L. & Bok, D. (1995). Expression, secretion, and age-related down-regulation of pigment epithelium-derived factor, a serpin with neurotrophic activity, *Journal of Neuroscience,* Vol. 15, pp. 4992-5003, ISSN 0270-6474

Uehara, F.; Ohba, N.; Yasumura, D. & La Vail M. (1991). Lectin binding in the interphotoreceptor matrix in neuraminidase-induced retinal detachment, *Nippon Ganka Gakkai Zasshi,* Vol. 95, No.4, pp. 332-335, ISSN 0029-0203

Uehara, F.; Ohba, N. & Ozawa, M. (2001). Isolation and characterization of galectins in the mammalian retina, *Investigative Ophthalmology and Visual Science,* Vol. 42, pp. 2164-2172, ISSN 0146-0404

Yoshii, C.; Ueda, Y.; Okamoto, M. & Araki, M. (2007). Neural retinal regeneration in the anuran amphibian *Xenopus laevis* post-metamorphosis: transdifferentiation of retinal pigmented epithelium regenerates the neural retina, *Develomental Biology,* Vol. 303, No.1, pp.45-56, ISSN 0012-1606

Zacks, D.; Han, Y.; Zeng, Y. & Swaroop, A. (2006). Activation of signaling pathways and stress-response genes in an experimental model of retinal detachment, *Investigative Ophthalmology and Visual Science*, Vol. 47, No.4, pp.1691-1695, ISSN 1552-5783

Part 2

Application of Stem Cells

8

The Therapeutic Potential of Stimulating Endogenous Stem Cell Mobilization

Christian Drapeau[1], George Eufemio[2], Paola Mazzoni[1], Gerhard D. Roth[3] and Susan Strandberg[4]
[1]*Stemtech International, San Clemente, CA,*
[2]*Cardinal Santos Medical Center, Medical Arts Building, Manila,*
[3]*Neurology and Psychiatry Center, Ostfildern,*
[4]*Stem Cell Center, Spokane, WA*
[1,4]*USA*
[2]*Philippines*
[3]*Germany*

1. Introduction

The past decade has seen a fast and extensive development of various therapies and treatment protocols based on Adult Stem Cells (ASC) and their application to various diseases. While some of these treatment protocols have been well documented in the scientific literature and used in well controlled clinical set ups, others have been developed and are being used by a growing numbers of clinics throughout the world, without thorough documentation though nevertheless with good clinical care and with the reports of very compelling results.

Despite the wide variety of methods, the general procedure guiding these various protocols follows a series of common steps. The first step is the isolation of stem cells from a source. For the purpose of banking or clinical application, stem cells can be isolated from a variety of sources including umbilical cord (Can and Balci, 2011; Zhang et al., 2011), adipose tissue-derived stem cells (Insausti et al., 2011; Zachar et al., 2011), peripheral blood stem cells (Kolbe et al., 2010; Hofmann et al., 2009), amniotic and placental stem cells (Klein and Fauza, 2011; Tsagias et al., 2011), dental pulp stem cells (Gronthos et al., 2011; Tirino et al., 2011), olfactory stem cells (Chen et al., 2006; Viktorov et al., 2008), and even human limbal epithelial stem cells (Vasania et al., 2011).

The second step is proliferation. This is not a necessary step with regard to stem cell function, however the small number of stem cells present in one umbilical cord, one placenta, one blood sample, one liposuction or one dental pulp makes clinical application difficult without the ability to expand the harvested stem cells. Methods to expand embryonic stem cells have been developed more than a decade ago, however it is only a few years ago that methods to significantly expand ASC have been developed, leading to an expansion of the stem cell banking market and greater clinical application (Ivanovic et al., 2011; Dos Santos et al., 2011; Pineault et al., 2011).

The third step is pre-conditioning or treatment to trigger commitment of the stem cells into a specific cellular lineage. For example, stem cells can be led to differentiate into dopamine-producing neuron by an exposure to a cocktail containing sonic hedgehog (SHH), fibroblast growth factor 8 (FGF8), and basic fibroblast growth factor (bFGF) (Trzaska and Rameshwar, 2011; Wang et al., 2011) or into neurons responding to multiple neurotransmitters by a simple exposure to retinoic acid and other growth factors (Greco et al., 2008). Likewise, stem cells can be guided to differentiate in cardiomyocytes by exposure to a cocktail containing transforming growth factor-beta(1), bone morphogenetic protein-4, activin A, vascular endothelial growth factor (VEGF), insulin-like growth factor-1, fibroblast growth factor-2, Epidermal growth factor (EGF), and interleukin-6 (Behfar et al., 2010; Behfar et al., 2008). Pre-conditioning with these cytokines can also enhance the formation of gap junction and improve therapeutic efficacy (Hahn et al., 2008). Exposure of stem cells to a cocktail containing insulin, transferrin, selenium and the GLP-1 (glucagon-like peptide-1) analogue exendin-4 leads to the formation of insulin-producing pancreatic cells (Docherty, 2009; Chandra et al., 2009). Various cocktails have been shown to trigger the differentiation of mesenchymal stem cells in into a wide variety of cell types (Snykers et al., 2011; Arufe et al., 2009; Keilhoff et al., 2006). Nevertheless, this pre-conditioning step is not essential since ASC will naturally differentiate into the cell type with which they find themselves, upon contact with cellular debris or cell marker specific to that cell type. For example, as they migrate into the heart, stem cells can be triggered to differentiate into cardiomyocytes (Orlic et al., 2001), or into keratinocytes and skin appendages, insulin-producing pancreatic cells or hepatocytes as they respectively migrate in a skin wound (Zhang and Fu, 2008), the pancreas (Hasegawa et al., 2007) or the liver (Theise et al., 2000).

The third and final step is the injection of stem cells into the target organ, in the main artery leading to the target organ or in the bloodstream from where a number of them will migrate on their own to the affected organ. In the case of a heart attack for example, stem cells can be injected in coronary artery (Wollert et al., 2004) or directly in the border zone of the infarct (Stamm et al., 2003; Tse et al., 2003). Treatment efficacy can vary significantly with the various methods of injection. For the treatment of acute myocardial infarction, injection in the border zone of the infarct seems to yield the best results, followed by intracoronary and intravenous injection, respectively (Karra and Wu, 2008). For the treatment of spinal cord injury, injection directly in the lesion or in the cerebrospinal fluid seems far superior to intravenous injection (Lima et al., 2010), yet very compelling cases have been documented following intravenous stem cell injection or simple bone marrow stem cell mobilization (see Section 4). For the treatment of diabetes however, intravenous injection seems to yield better results than stem cells transplantation directly into the pancreas (Hasegawa et al., 2007).

In this multistep procedure in which each step can be accomplished according to a wide variety of protocols and methods, it remains that peripheral blood stem cells (PBSC) can, without expansion, pre-conditioning or injection reach various target organs and participate to the process of tissue repair. This observation has led a number of researchers to look at the therapeutic potential of simply stimulating Endogenous Bone Marrow Stem Cell Mobilization (ESCM). This chapter will look in detail into the clinical and therapeutic potential of ESCM by describing its physiological basis, by reviewing the existing literature on the clinical application of ESCM and by presenting a few clinical cases.

2. The repair system of the body

For ESCM to have any clinical relevance, the demonstration must be made that one of the natural roles of stem cells in the body is to participate to tissue repair in cases of injury or degenerative diseases. Therefore, the clinical relevance of mobilizing endogenous bone marrow-derived stem cells (BMSC) would be to increase the number of circulating stem cells available to migrate into affected tissues and contribute to tissue repair. For this phenomenon to be natural: 1) the body must have a mechanism that triggers BMSC mobilization after an injury; 2) BMSC must traffic in the blood and be recruited by the injured tissue; 3) in the injured tissue BMSC must proliferate and 4) a mechanism must exist to trigger the differentiation of BMSC into cells of that tissue.

2.1 Signaling for mobilization

The most common compound known to naturally stimulate BMSC mobilization is Granulocyte-Colony Stimulating Factor (G-CSF). Discovered in 1985 by Welte et al., G-CSF is a cytokine secreted by various tissues that stimulates the proliferation, differentiation and function of neutrophil precursors and mature neutrophils. But G-CSF was also shown to stimulate BMSC mobilization (Petit et al., 2003; Cottler-Fox et al., 2003), making it a common tool in protocols of stem cell apheresis for the purpose of cryopreservation and stem cell transplant (Gordon et al., 2008; Croop et al., 2001).

Given the vital importance of the heart and the fact that cardiovascular diseases are a leading cause of death in the world, much of the clinical stem cell research has focused its efforts on the role of stem cells in cardiac repair taking place after acute myocardial infarction (AMI). A number of studies have revealed the sequence of events taking place after AMI. A few hours after AMI, the cardiac tissue releases or causes to release G-CSF (Leone et al., 2006). As its concentration slowly increases in the bloodstream, G-CSF triggers the release of stem cells from the bone marrow, increasing the number of PBSC which peaks at around 4-7 days after AMI (Shintani et al., 2001; Leone et al., 2006). It is worth mentioning that the serum level of G-CSF and the number of PBSC are also increased in cases of chronic angina (Leone et al., 2006). Similar stem cell mobilization and increase in PBSC have been documented following skeletal muscle injury (Stout et al., 2007).

Other chemokines such as interleukine-8 (IL-8), Stromal-Derived Factor-1 (SDF-1), Stem Cell Factor (SCF), Groß, and vascular endothelial factor (VEGF) have been shown to trigger BMSC mobilization (King et al., 2001; Lapidot & Petit, 2002; Fukuda et al., 2007; Lapid et al., 2009). Contrary to G-CSF and SCF, which lead to a slow increase in the number of PBSC over a period of a few days, other cytokines such as IL-8 lead to a significant increase in the number of PBSC within hours (Fibbe et al., 1999).

As has been described with the heart following AMI, a stroke also triggers the release of cytokines that induce the mobilization of BMSC and their migration into the brain. For example, the number of PBSC in stroke patients nearly tripled within 7 days after the stroke (Hennemann et al., 2008; Paczkowska et al., 2005). In one study, the magnitude of stem cell release was actually correlated with the functional recovery of the patients (Dunac et al, 2007). Interestingly, the number of circulating stem cells did not increase in patients who received thrombolysis therapy immediately after their stroke. Therefore, it appears that it is the lingering injury that slowly leads to the mobilization of stem cells from the bone marrow.

Finally, injuries to the skin and bones were also shown to trigger mobilization of BMSC and their migration into the injured tissue. For example, within 24 hours of a severe burn, a rapid increase of up to 9-fold in the number of PBSC has been observed in the blood of burn patients (Fox et al., 2008). Furthermore, the size of the area of the body affected by the burns strongly correlated with the magnitude of the mobilization. The affected skin also released cytokines such as SDF-1 and VEGF, which are involved in the migration of PBSC to the skin and their differentiation into blood vessels, respectively (Mansilla et al., 2006). In one study (Lee et al., 2008), the number of PBSC peaked around 3 days after a bone fracture and rapidly returned to basal level within a few days. These results were confirmed in another study in which stem cells were shown to migrate to the fracture site and to promote neovascularization. The formation of new vessels was shown to peak at the fracture site 7 days after the fracture, which corresponds to the early phase of ossification of the fracture line (Matsumoto et al., 2008). Therefore BMSC mobilization was documented to naturally follow an injury or even be associated with chronic conditions.

The natural process by which stem cells are mobilized from the bone marrow is still not fully understood. Contrary to most tissues in which SDF-1 is secreted consequent to an injury or a degenerative condition, in the bone marrow SDF-1 is constitutively produced and released, and binding of SDF-1 to its exclusive receptor CXCR4 leads to the externalization of adhesion molecules, namely integrins, which allow for the adherence of stem cells to the bone marrow matrix. The binding of SDF-1 to CXCR4 is referred to as the SDF-1/CXCR4 axis. The general understanding is that disruption of the SDF-1/CXCR4 axis reduces the expression of adhesion molecules, leading to a reduction in the adherence of stem cells to the bone marrow matrix and the consequent mobilization of stem cells. Various compounds known to trigger stem cell mobilization all affect the SDF-1/CXCR4 axis in various ways.

For example, G-CSF disrupts the SDF-1/CXCR4 axis by activating a series of proteolytic enzymes including elastase, cathepsin G, and various matrix metalloproteinases (MMP2 and MMP9) that inactivate SDF-1 (Mannello et al., 2006; Jin et al., 2006; Carion et al., 2003). AMD3100 is a newly developed BMSC mobilizer and it acts by blocking CXCR4, disrupting the SDF-1/CXCR4 axis (Broxmeyer et al., 2005). A blocker of L-selectin was recently isolated from the cyanophyta *Aphanizomenon flos-aquae* and shown to trigger BMSC mobilization (Jensen et al., 2007). Inhibition of L-selectin leads to a down-regulation of CXCR4 expression, partially disrupting the SDF-1/CXCR4 axis. The mobilization mechanism of IL-8, SCF, VEGF and Groß are not well understood.

Therefore, the human body has a mechanism to naturally mobilize BMSC which can then traffic to various areas of the body and contribute to tissue regeneration and repair.

2.2 Extravasation & recruitment

Recruitment is the process by which PBSC are recruited by a specific tissue signaling for repair. The process of PBSC recruitment in a tissue takes place predominantly at the level the postcapillary venule, in a manner similar to neutrophils (Henschler et al., 2008).

In brief, the sudden drop in blood pressure taking place at the postcapillary venule triggers turbulence whose shear force mechanically activates L-selectin which in turn triggers the externalization of CXCR4, making PBSC more responsive to signals coming from tissues. If the tissue is in need of repair, it is secreting SDF-1 as well as other compounds such as stem cell factor (SCF) and hepatocyte growth factor (HGH) (Kucia et al., 2004; , Neuss et al., 2004) that diffuse locally to the capillaries. Binding of SDF-1 and SCF to their specific receptors (e.g. CXCR4 and c-Kit) leads to the expression of adhesion molecules at the surface of stem cells (Voermans et al., 2000; Peled et al., 1999; Peled et al., 2000). Through a complex interaction of microvilli at the surface of both the capillary and PBSC, the stem cells initiate the process of tethering and then arrest on the capillary endothelium (Middleton et al., 1997; 2002).

Following firm attachment, SDF-1 and HGH continue to bind to their respective receptors CXCR4 and c-met at the surface of SC, which triggers the release and activation of matrix metalloproteinases (MMPs) which digest the endothelial lining, allowing for the extravasation of PBSC (Mannello et al., 2006; Janowska-Wieczorek et al., 2000; Neuss et al., 2004; Ries et al., 2007).

When studying the migration behavior of SCs to a wide array of chemokines, SCs were found to migrate only toward SDF-1 (Wright et al., 2002; Jo et al., 2000). The migration to SDF-1 is a polarized phenomenon toward the chemokine source (chemotactic) and not a simple random motion (chemokinetic), as SCs only migrated in a gradient of SDF-1 and not when SDF-1 is uniformly distributed in the media.

SDF-1 is normally secreted, to some extent, by cardiomyocytes (Askari et al., 2003), skeletal muscles (Ratajczak et al., 2003), liver (Hatch et al., 2002; Kollet et al., 2003), brain (Bagri et al., 2002; Lazarini et al., 2003; Zou et al., 1998), and kidney (Schrader et al., 2002). However, its secretion increases during tissue damage such as AMI (Wojakowski et al., 2004; Abbott et al., 2004), ischemia (Takahashi et al., 1999, Iwaguro et al., 2002), toxic liver damage (Kollet et al., 2003; Swenson et al., 2008; Hatch et al., 2002), and excessive bleeding (Ratajczak et al., 2004).

In the tissue, the process of migration toward the site of injury relies on the interaction between CD44 and its ligand hyaluronic acid (HYA). HYA is one of a family of polysaccharides known as glycosaminoglycans (GAGS), typically found in the connective tissues of vertebrates. Alongside other proteins such as collagen, elastin, fibrillin, fibronectin and laminin, GAGS and HYA constitute the extracellular matrix (ECM) of most tissues. Studies in rats have shown that half of the HYA found in the body is in the skin, while 25% is found in joints and bones together (Reed et al., 1988). The rest is distributed somewhat equally in muscles and viscera (Clarris and Fraser,1968; Comper and Laurent, 1978) where the highest concentrations are found in connective tissue that form the ECM of most tissues.

After extravasation, CXCR4 on the surface of SC and SDF-1 secreted by the effected tissue continue their interaction which, in the tissue, leads to the formation of pseudopodia (see Figure 1) toward the source of SDF-1 secretion. As SDF-1 binds to CXCR4 in the tissue, CD44 adhesion molecules are externalized at the tip of the pseudopodia, leading to adhesion to HYA pathways within the tissue. In the tissue, the binding of CD44 to HYA is transient, as CD44 molecules shed soon after binding to HYA (Friedl et al., 1995), thus enabling pseudopodia detachment from the ECM. CD44 can also be cleaved by specific enzymes whose secretion is enhanced by SDF-1 (Heissig et al., 2002; Okamoto et al., 1999;

Janowska-Wieczorek et al., 1999). Through this process, stem cells can migrate within the tissue toward the site of injury.

The fact that PBSC primarily migrate to organs affected by an injury or a degenerative process has been documented in several studies. For example, in the cases of male recipients of liver transplants from female donors, biopsies performed 4-13 months after transplantation contained a significant number of Y-chromosome positive hepatocytes (16% to 43%). In one patient who suffered from hepatitis C after liver transplant and died 4.5 months after transplantation, up to 43% of the transplanted liver was made of Y-chromosome positive hepatocytes in comparison with 5% after 4.5 months in a woman with a sound liver who received a bone marrow transplant from a man (Theise et al., 2000). Investigations carried out on archival autopsy and biopsy liver specimens obtained from two women who received a bone marrow transplant from male donors for the treatment of leukemia revealed that 5 to 10% of the liver had been replaced by donor derived BMSCs after 4.5 and 13 months, respectively (Theise et al., 2000).

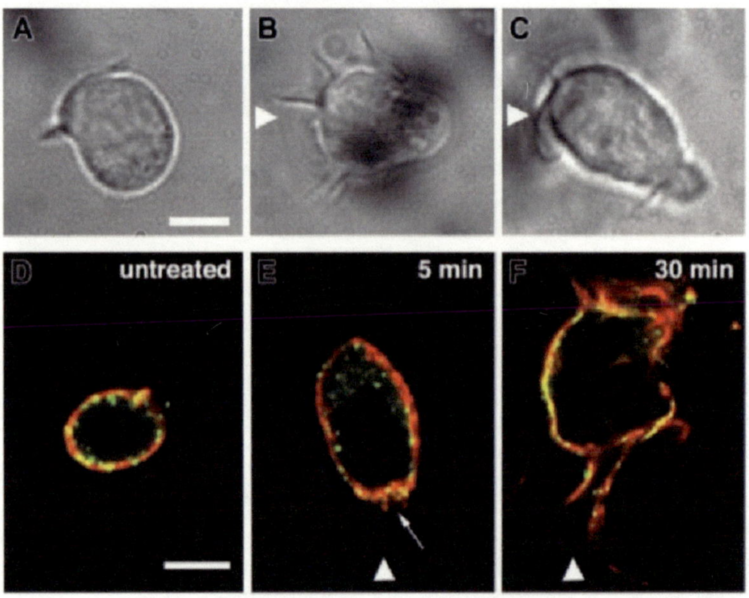

Fig. 1. **CD44 is localized to the leading edge of polarized human stem cells migrating toward SDF-1.** Cord blood-derived CD34+ cells were plated on HA coverslips and allowed to adhere for 30 minutes before recording cell movement. The position of SDF-1 source is indicated by arrowheads. (A-C) Phase contrast microscopy of untreated cells (A), cells stimulated with polarized source of SDF-1 (B), and cells treated with anti-CD44 mAb F10-44-2 and stimulated with polarized source of SDF-1 (C). (D-F) Cells treated as above were fixed 5 and 30 minutes after exposure to polarized source of SDF-1 and indirectly immunolabeled with antihuman CD44 mAb (red) and anti-CXCR4 mAb (green). An arrow is pointing to the fine CD44-positive protrusions at the direction of SDF-1. Bars = 5μm.
(Taken from Avigdor et al., 2004)

Similar observations were made on men who received cardiac transplants from female donors. Analyses of tissue samples from biopsies revealed that an average of 0.1% (Muller et al., 2002) to 15% of cardiomyocytes were Y-chromosome positive (Quaini et al., 2002; Laflamme et al., 2002). In one patient who died of cardiac rejection, 29% of the cardiomyocytes contained the Y chromosome in "hot spots" of high cardiac repair. In male patients who received lung transplants from female donors, recipient-derived BMSCs cells were detected as bronchial epithelial cells, type II pneumocytes, and seromucous glands in the transplanted lungs of all tested patients. In patients suffering from chronic damage, up to 24% of bronchial epithelial cells carried the Y-chromosome, indicating ongoing repair of damaged tissue by the recipient's own stem cells (Kleeberger et al., 2003).

In one study investigating this process of tissue repair by BMSC, irradiated mice were transplanted with GFP-positive SC before being injected in the right tibialis muscle with a large dose of cardiotoxin, which led to a loss of mobility within a few days. Yet, after eight weeks the injured muscle showed massive regeneration, with the right tibialis muscle significantly reconstructed with GFP-positive myocytes. By contrast, the contralateral non-injured leg showed very small incorporation of GFP-positive myocytes (Drapeau et al., 2010). Similar observations were made by Abedi et al. (2004), using smaller injections of cardiotoxin. Four weeks after the injection of cardiotoxin in the right leg, the area of the injection contained 1-2% of GFP-positive muscle cells while the left leg showed no fluorescence at all.

Similar observations were made in models of skin injury (Abedi et al., 2004). In a similar protocol, irradiated mice were transplanted with GFP-positive stem cells. BMSCs were then mobilized during five days using G-CSF. On the fourth day, mice were subjected to punch biopsies on their flank. The area of the injury was rebiopsied and sutured 48 hours and 1 month after the injury, in order to assess the incorporation of GFP-positive cells in the healing skin. While analysis at 48 hours showed significant infiltration by GFP-positive undifferentiated stem cells in the deep layer of the skin (hypodermis), after 4 weeks there was a large number of GFP-positive tissue cells in the dermis composing the structure of the healed skin, such as keratinocytes, sebaceous glands, blood vessels, and some rare muscle fibers and hair follicles. In the control animals that received G-CSF but no punch biopsy, none of these skin structures were positive for GFP, indicating that the few SC that had migrated in the skin had done so randomly.

In other studies looking at the incorporation of BMSCs in injured tissues, directed migration was demonstrated in the gut after section of an intestinal segment, (Hayakawa et al., 2003) in the heart after AMI (Orlic et al., 2001a; Fukuhara et al., 2004) or induced cardiomyopathy, (Hisashi et al., 2004) in the brain after stroke, (Sanchez-Ramos et al., 2002; Hoehn et al., 2002) and in the liver after drug-induced liver damage (Abedi et al., 2004). Taken altogether these studies clearly establish that BMSCs primarily migrate to areas subjected to injury, damage or simple degeneration.

2.3 Proliferation

When stem cells reach the site of an injury they must proliferate and expand, as there are not enough PBSC to accomplish full repair of any significant injury or degenerative process.

Several chemokines such as SDF-1 have been reported to enhance stem cell proliferation (Bonavia et al., 2003). The direct effect of SDF-1 on cell proliferation and survival is not well understood, but SDF-1 has been found to stimulate the proliferation and survival of stem cells under certain experimental conditions (Broxmeyer et al., 2003). In tissues, SDF-1 appears to act as a "cellular survival factor" (Hwang et al., 2006). Insulin-like growth factor (IGF-1), when coupled with epidermal growth factor (EGF) or fibroblast growth factor-2 (FGF-2), was shown to support the proliferation and survival of neural (Arsenijevic et al., 2001) and muscle stem cells (Deasy et al., 2002). Extracellular nucleotides were also shown to support the proliferation of brain stem cell (Mishra et al., 2006).

2.4 Differentiation

The ability of adult stem cells to differentiate into various cell types has been well documented, though the mechanism behind such transformation is still not well understood.

As reported by Krause et al. (2001), 11 months after injection of male stem cells in female mice, Y-chromosome bearing cells were found in various tissues including the liver, muscle, skin, lung, and intestine. It has been well demonstrated that BMSC can differentiate into a wide variety of cell types including myocytes (Ferrari et al, 1998), hepatocytes (Lee et al., 2004), epithelial cells (Krause et al., 2001), neurons (Mezey et al., 2003; Sanchez-Rarnos et al., 2000; Woodbury et al., 2000), retinal cells (Tomita et al., 2002), endothelial cells and cardiomyocytes (Jackson et al., 2001; Orlic et al, 2001), gastrointestinal epithelium (Krause et al., 2001; Okamoto et al., 2002), pancreatic endocrine cells (Ianus et al., 2003), bone and cartilage (Pereira et al., 1995; 1998).

Although little work has been done in this field and many questions remain to be answered, two possible mechanisms have been proposed for SC differentiation.

One proposed mechanism is cellular fusion, which takes place when two cells fuse together to become one cell. A few studies have suggested that SC have the ability to fuse with somatic cells, rescuing the target cell (Tarada et al., 2002; Vassilopoulos et al., 2003; Spees et al., 2003). Although there is clear evidence that this phenomenon did take place in a few experiments and that it may take place naturally in certain tissue such as the heart (Nygren et al., 2004), it is unlikely to be a significant physiological phenomenon (Wurmser and Gage, 2002). For example, while the process of fusion involves the interaction of one single SC with one somatic cell, therefore a ratio of 1:1, the extent of SC-mediated tissue repair that has been documented in numerous studies, involving various tissues, far exceeds the actual number of SCs migrating in the tissue. Furthermore, the process of cellular fusion results in a cell that is tetraploid. Although this has been observed in a few in vitro studies, it has been a very rare observation in vivo. In fact, even in vitro, relatively harsh conditions had to be used in order to obtain cellular fusion. Finally, cellular fusion would imply the merging of two different cellular membranes, a process that in itself is rigged with challenges, as cells are designed not to fuse.

So, although cellular fusion could possibly naturally take place in the body, it is unlikely to contribute significantly to the process of repair that has been documented with ASCs. The other most likely mechanism is differentiation through contact with cellular components when the affected tissue is locally subjected to the action of various matrix metalloproteinases (MMPs) or differentiation induced by cytokines released by neighboring cells.

This process was beautifully put in evidence by the study of Jang et al. (2004) where hematopoietic stem cells (HSC) were co-cultured with either normal or damaged liver tissue separated by a semi-permeable membrane with 0.4 μm pores. Using immunofluorescence to detect markers specific to either HSC (CD45) or hepatocytes (albumin), the authors could follow the transformation of the HSC population. When HSCs were cultured alone for 8 hours, they only expressed CD45 and no albumin. However, when HSCs were exposed to injured liver tissue, they rapidly became positive for albumin. Over time, the population of cells positive for CD45 began to decrease as the population positive for albumin began to increase. Albumin-positive cells were seen as early as 8 hours and constituted 3.0% of the cell population at 48 hours. The conversion was minimal and delayed when HSC were exposed to undamaged liver (control for injury). Therefore the presence of an injury appears to be an important factor in the process of SC differentiation into a specific type of somatic cells.

The authors further investigated the phenomenon of differentiation by tracking the presence of various markers found in developing fetal liver cells, such as αFP, and in mature hepatocytes, such as CK18, albumin, fibrinogen, and transferrin. They found that αFP was expressed only at 8 hours and was lost thereafter. On the other hand, the expression of CK18, albumin, fibrinogen and transferrin each increased over time. While at time 0 the HSCs expressed only CD45, after as little as 8 hours all liver-specific markers were positive. So during the differentiation process, the SC seemed to take a route similar to the development of hepatocytes in the developing fetus, leading to mature hepatocytes within less than 48 hours. This retracing of fetal development has also being documented in cardiopoiesis (Behfar et al., 2008).

In this study, differentiation did not involve cellular fusion, as the new liver cells only contained one set of chromosomes. The differentiation was necessarily triggered by signaling molecules produced by the damaged tissue. It has been suggested that MMPs produced by damaged tissue could be playing an important role in SC differentiation by digesting specific ECM components that would then diffuse and get into contact with SCs. Binding of such compounds to specific receptors would activate internal messengers that would trigger the process of differentiation by activating specific genes, as suggested by the high level of mRNA found in differentiating cells. (Mannello et al., 2006; Chavey et al., 2003)

Gap junction intercellular communication (GJIC) and tunneling nanotubules (TNTs) could constitute other mechanisms playing a significant role in SC differentiation, by direct cell-to-cell contact (Behfar et al., 2010). The diameter of one gap junction is around 2 nm and the molecular size cut-off level is around 1-2 kDa, which is sufficient for the intercellular exchange of ions, nucleotides and even small proteins (Neijssen et al., 2005). GJIC is known to play a crucial role in modulating several cellular functions, to the point that impaired or lack of GJIC has been associated with severe diseases (Dasgupta et al., 1999). It has been suggested that GJIC could play an important role in SC differentiation (Loewenstein and Rose, 1992). Likewise, recent studies indicated that two cells can also exchange information via TNTs (Rustom et al., 2004). TNTs would form a cytoplasma bridge between two cells that could be large enough to allow the transport of large molecules or even whole cell organelles. Such information could play a role in the finalizing process of differentiation, though much work needs to be done in this field before we obtain a better understanding of the role of GJIC and TNTs in stem cell differentiation.

2.5 Paracrine effect

Finally, a number of studies investigating the effect of BMSC on various diagnostic entities have revealed that oftentimes the extent of the benefits observed cannot be accounted for by the sheer number of BMSC that have differentiated into somatic cells, even when there is clear evidence of tissue regeneration mediated by newly formed cells. For example, irradiated female mice transplanted with transgenic GFP+ BMSC showed much better recovery from experimental stroke after G-CSF-induced BMSC mobilization when compared to non-mobilized control mice (Kawada et al., 2006). Both motor and cognitive functions were improved by BMSC mobilization, and the treatment also reduced infarct size. However, using bromodeoxyuridine (BrdU), it was observed that a significant number of the newly formed brain cells did not derive from BMSC but rather from local neural stem cells. Similar results were seen with spinal cord lesion and muscle injury (Kinnaird et al., 2004; Osada et al., 2010). The regenerative effect is believed to be triggered by the secretion of growth factors and paracrine signaling by BMSC (Uccelli et al., 2011).

The paracrine effect has been best put in evidence in the heart after AMI. Delivery of BMSC to ischemic cardiac tissue has led to a significant increase in the concentration of IL-10, interleukin-1β (IL-1β), tumor necrosis factor-α (TNF-α) and other cytokines in the cardiac tissue, which contributed to neovascularization and reduction of infarct size (Kamihata et al., 2001; Burchfield et al., 2008; Gnecchi et al., 2005; Mirotsou et al., 2011). Condition media in which BMSC were exposed to hypoxia proved to be cytoprotective to cardiomyocytes and was able to reduce infarct size (Gnecchi et al., 2006).

In all, a growing body of evidence supports the hypothesis that as BMSC migrate into tissues, aside from differentiating into cells of the target tissue, they also exert their regenerative effect at least in part -and maybe even to a large extent- through the secretion of paracrine signaling compounds.

3. ESCM as a treatment approach

Given the fact that SDF-1 is secreted by various organs and tissues upon injury or degeneration, it would follow that the number of PBSCs should be an important parameter in the overall effectiveness of SC-mediated tissue repair and regeneration. A higher number of PBSC would mean more SC available to respond to SDF-1 signaling and migrate into tissues. In this regard, Tomoda et al. (2003) reported that after a heart attack individuals with a higher number of PBSC showed greater recovery of cardiac functions after 6 months when compared to people having fewer PBSC at the time of the cardiac event.

In a prospective study using more than 500 individuals, Werner et al. (2005, 2007) put in evidence that the number of PBSC is a critical parameter in the role of SCs in tissue repair. The authors quantified the baseline number of PBSC in 519 individuals (average 66.6±10.8 years old) at risk for cardiovascular problems, and monitored their condition for one year. A first major cardiovascular event occurred in 214 patients. After adjustment for age, sex, vascular risk factors, and other relevant variables, increased levels of PBSC were associated with reduced risk of death from cardiovascular causes, lesser risk of a first major cardiovascular event, greater revascularization and lesser frequency of hospitalization.

Similar data has been obtained by Vasa et al. (2001) who documented that a higher number of endothelial progenitor cells (EPC) was associated with greater cardiovascular health. In fact, hypoxia has been associated with the secretion of SDF-1 and VEGF by the ischemic tissue (Schioppa et al., 2003; Bachelder et al., 2002). Circulating EPCs therefore are attracted to the ischemic heart by the action of SDF-1 and the migrating EPCs contribute to the formation of new blood vessels upon the action of VEGF (Lee et al., 2007). Therefore, a higher number of EPCs or PBSC helps maintain optimal blood flow and a strong cardiac tissue.

The link between disease formation and the number of circulating stem cells is not limited to the heart. Similar observations have been made with muscular dystrophy where the rate of progression of the disease has been linked to the number of circulating stem cells and for which the number of PBSC is now considered one of the most important predictors of the disease progression (Marchesi et al., 2008). Likewise, the progression of pulmonary arterial hypertension (Diller et al., 2008; Junhui et al., 2008), arthritis (Herbrig et al., 2006; Grisar et al., 2005), atherosclerosis (Zhu et al., 2006), lupus erythematosus (Westerweel et al., 2007; Moonen et al., 2007), kidney failure (Choi et al., 2004; Herbrig et al., 2004; Eizawa et al., 2003), migraine (Lee et al., 2008), erectile dysfunction (Baumhakel et al., 2007) and other diseases have all been linked to a reduction in the number of PBSC. Recently, a direct relationship has been established between the number of PBSC and the development of diabetes, linking impaired fasting glucose, impaired glucose tolerance, and insulin-dependent diabetes mellitus to progressively lower levels of PBSCs (Fadini et al., 2010).

If the number of PBSC constitutes such a key parameter in the process of SC-mediated tissue repair, therefore increasing the number of PBSC could constitute a therapeutic approach. Following is the description of a number of clinical trials investigating the clinical potential of ESCM in various diseases or injuries.

3.1 Acute myocardial infarction

The heart has been traditionally seen as having little regenerative capabilities after birth, although many recent studies have challenged this view. Evidence clearly suggests that there is a low level of constant regeneration of cardiac cells (Soonpaa and Field, 1998; Bergmann et al., 2009; Quaini et al., 2002), and the number of dividing cells can increase by up to 10-fold in chronic heart disease or after AMI (Kajstura et al., 1998; Beltrami et al., 2001). Yet, this level of proliferation seems insufficient to rescue the cardiac muscle after AMI (Schwartz and Kornowski, 2003) and survivors of heart attacks are left with reduced quality of life and little prospect for improvement.

A number of studies in animals using G-CSF and SCF have shown that ESCM can lead to significant cardiac repair after AMI. Injection of SCF and G-CSF for 8 days after inducing AMI significantly increased the number of PBSCs, which led to the migration of PBSCs into the myocardium (Orlic et al., 2001b). Twenty-seven days after AMI, a band of newly formed cardiac tissue occupied more than 75% of the infarcted region of the ventricule and newly formed blood vessels were supplying the infarcted tissue. The blood vessels were surrounded by smooth muscles and microscopic observations revealed the presence of red blood cells, indicating that the newly formed arterioles integrated structurally with the remaining functional vasculature. By comparison, in control animals the ventricular wall was filled with scar tissue covering the entire area of the infarct and no new blood vessels could be seen.

In summary, while only 17% (9 of 52) of the untreated animals survived AMI, showing severe signs of cardiomyopathy and compromised blood circulation, up to 73% (11 of 15) of the animals treated with G-CSF survived with significantly improved cardiac function and restored blood circulation. After 27 days, ejection fraction was 114% greater in the treated group and other parameters such as end-diastolic pressure, systolic pressure, and other parameters of cardiovascular function were all improved in treated versus non-treated mice.

Injection of G-CSF, however, can have significant negative effect in humans if done at large dose for more than 5-6 days (Bensinger et al., 1996; Shimoda et al., 1993). At lower dose and with shorter treatment duration, human trials have so far delivered mitigated results, though the approach remains promising. While some groups did report very promising results (Ince et al., 2005; Sesti et al., 2005), others reported no effect at all (Ellis et al., 2006; Zohlnhofer et al., 2007; Ripa and Kastrup, 2008) . A comprehensive review of the various studies however reveals that each study used slightly different protocols with regard to the time of treatment after AMI (from hours to 3 months), as well as the intensity and duration of the treatment, suggesting that ESCM could indeed hold great promise once the most effective treatment protocol has been developed (Abdel-Latif et al., 2008).

For example, Wojakowski et al. (2006) reported in 43 cardiac patients that if the patients were treated early after AMI (<12 hours) with G-CSF, the number of PBSCs following BMSC mobilization correlated with the extent of cardiac repair. In another study, G-CSF was injected within 5 days post-AMI in 41 patients at high risk for unfavorable left ventricular remodeling. Five months after G-CSF treatment, ejection fraction had improved 12.5% compared to no improvement in the control group (Leone et al., 2007). The improvements in cardiac function appeared to be linked to the prevention of left ventricular remodeling.

A meta-analysis reviewing the effectiveness of BMSC mobilization for the treatment of AMI included 7 studies and a total of 364 patients. The analysis concluded that treatment with G-CSF can improve LV ejection fraction if the treatment is administered early after the heart attack (Kang et al., 2007). However, in spite of the improvements in ejection fraction, other general parameters of cardiovascular health such as ventricular arrhythmia, rehospitalization for heart failure, and the composite of other cardiovascular events (i.e., death from heart attack, recurrent heart attack, and stroke), were not significantly different in the G-CSF treatment groups compared with the control groups. Similar results were reported by another meta-analysis that included eight studies and 385 patients (Abdel-Latif et al., 2008).

So, it remains unclear whether the simple mobilization of BMSCs can constitute an effective treatment for AMI. While some studies have yielded promising results, others suggest no benefits at all. However, positive results obtained in some studies should not be denied on the basis of the negative results obtained in others. Reconciliation of all this data and the development of an effective treatment protocol will most probably come through the determination of optimal treatment parameters: 1) intensity of ESCM, 2) duration of the treatment, 3) time after AMI, 4) number of treatments received over time, and 5) other yet unidentified parameters. Compounds other than G-CSF might also be discovered that could provide more consistent results (Broxmeyer et al., 2005; DeClercq, 2005).

3.2 Stroke

Many studies have shown that extensive neuronal death in the brain after a stroke triggers the migration of neural stem cells to the site of injury, followed by their proliferation and differentiation into neurons and glial cells (Peterson, 2002; Fallon et al., 2000; Arvidsson et al., 2002; Nakatomi et al., 2002; Schmidt and Reymann 2002). However, this natural process does not appear to be sufficient to produce significant functional recovery (Yamamoto et al., 2001; Magavi and Macklis, 2002).

As with the heart after AMI, stroke has been associated with BMSC mobilization. Studies have shown that the number of PBSC in stroke patients can increase up to 3-fold within 7 days after the stroke (Hennemann et al., 2008; Paczkowska et al., 2005). In one study, the magnitude of BMSC mobilization was correlated with the patients' functional recovery (Dunac et al, 2007).

When rats were injected with rat (Chen et al., 2001; Pavlichenko et al., 2008; Willing et al., 2003) or human SC (Li et al. 2002) after an induced stroke, significant motor and cognitive improvements were observed. Although a significant number of BM-derived cells could be identified as newly formed neurons and glial cells in the stroke foci, they accounted for only a small percentage of the total number of newly formed brain cells. Most of the newly formed brain cells are believed to be derived from neural SC upon the action of paracrines secreted by the migrating SC. Similar results were obtained using human umbilical cord stem cells (HUCSCs) where intravenous injection of HUCSCs 24 hours after a stroke greatly improved functional recovery (Chen et al., 2001). Injection of HUCSCs 7 days after the stroke still led to significant functional recovery, though the extent of the recovery was less than with treatment at 24 hours.

Mobilization of BMSCs induced by G-CSF was shown in a number of studies to improve the outcome of a stroke. For example, when tested 14 and 28 days after a stroke, animals treated with G-CSF showed much greater body coordination than control animals (Shyu et al., 2004). When the brains were analyzed using imaging, the infarcted area was much smaller in the treated group (61 mm^3) when compared to the control group (176 mm^3). All these benefits were greatly reduced when the animals were pre-treated with a blocker of CXCR4, indicating that the observed effects were dependent upon the migration of BMSCs into the brain.

Similar results have been reported by other scientific teams (Six et al., 2003; Kawada et al., 2006). For example, after repopulating the bone marrow with GFP-positive stem cells, G-CSF-induced BMSCs mobilization immediately and roughly 2 weeks after inducing a stroke dramatically improved motor and cognitive performances four weeks after the stroke, as measured by using the Morris water maze (Kawada et al., 2006). In this study, while all the mice in the treated group reached the platform within 40 seconds, none of the control mice reached the submerged platform within the allotted 120 seconds. As in the study by Shyu et al. (2004), the infarct size was much smaller in the treated animals when compared to control. Using BrdU it was observed that the number of new brain cells found in the infarcted area was much higher in the treated group than in the control group. Yet very few of the new brain cells were GFP-positive, supporting the view that as they migrate in the brain BMSCs secrete growth factors that support the proliferation and differentiation of neural stem cells (Yoo et al., 2008). BMSCs also support neovascularization, which further contributes to the regeneration of the brain tissue (Lee et al., 2005; Hess et al., 2002; Kan et al., 2005).

Although much of this work needs to be reproduced in humans, ESCM for the treatment of stroke appears promising and would constitute a safe approach to the treatment of stroke.

3.3 Diabetes

The ability of BMSCs to leave the bone marrow, migrate to the pancreas and become insulin-producing cells was beautifully shown by Ianus et al. (2003). In brief, female mice were lethally irradiated and then transplanted with male BMSCs that express, using a CRE-LoxP system, GFP if the insulin gene is actively transcribed. When analyzed 4-6 weeks after the transplantation, GFP-positive cells were found in the pancreas (Figure 2). The GFP-positive

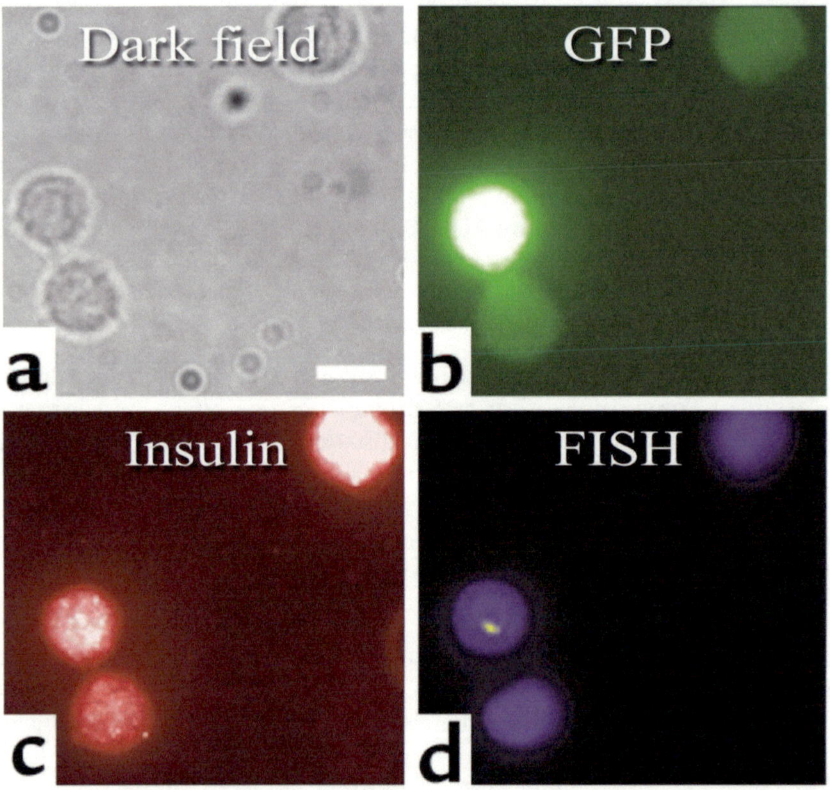

Fig. 2. **FISH and immunofluorescence marking of BM-derived insulin-producing cells.** Immunofluorescence and FISH of isolated, dispersed pancreatic islet cells after transplantation of lethally irradiated female mice with male BMSCs that express, using a CRE-LoxP system, GFP if the insulin gene is actively transcribed. a) Bright-field phase, b) GFP imaging note slight autofluorescence of control isolated islet cells; c) Immunostaining with rhodamine X-labeled secondary antibody for insulin; d) FISH for Y chromosome (in yellow) and nucleus stain with DAPI (blue). Y chromosome is present only in GFP-positive cells. Scale bar, 5 μm; X630. (Taken from Ianus et al., 2003)

cells were also positive for insulin, for insulin RNA, and for Y-chromosome, demonstrating that they originated from the transplanted BMSCs. These cells showed functional characteristics typical of normal pancreatic ß-cells, such as fluctuations of intracellular calcium upon exposure to various concentrations of glucose. Within the time frame of that study (4-6 weeks), 1.7-3% of BM-derived GFP-positive cells were detected in the pancreatic islets. In a similar study, BMSCs were also shown to participate into the development of new blood vessels, further supporting the regeneration of the pancreatic tissue (Mathews et al., 2004; Gao et al., 2008).

Then, using a protocol similar to that used by Ianus et al., Hasegawa et al. (2007) further demonstrated that mobilization of BMSCs was not only effective but essential for pancreatic regeneration. Hasegawa et al. induced diabetes by injection of streptozotocin (STZ) in lethally irradiated female mice followed by infusion of BMSC from GFP transgenic mice. Infusion of BMSCs led to the incorporation of GFP-positive BMSCs into islets of Langerhans in the pancreatic tissue, partially restoring pancreatic islet number and size, and improving STZ-induced hyperglycemia. However, when the same experiment was done while simply infusing the pancreas with BMSCs, without preirradiation, no improvement was obtained. Furthermore, when the experiment was repeated with full BMSC transplant in a model of mice with impaired ability to mobilize stem cells, no benefits were obtained. Therefore, natural mobilization of BMSCs from the bone marrow appears essential for the regeneration of pancreatic function after inducing diabetes with STZ.

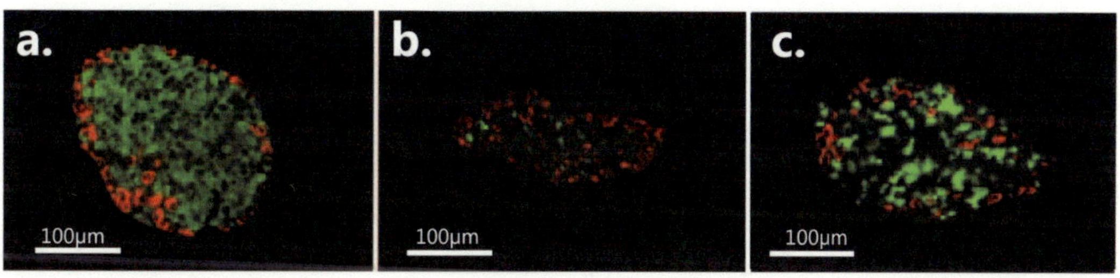

Fig. 3. **Pancreatic islets of STZ-treated mice receiving subsequent bone marrow transplant (BMT)**. Double immunostaining of pancreases with anti-insulin and anti-glucagon antibodies. *Green* indicates insulin-positive and *red* glucagon-positive cells. Pancreases from normoglycemic control mouse (a), hyperglycemic control mouse (b), and STZ-treated mouse receiving BMT (c). BMT improved STZ-induced hyperglycemia. (Taken from Hasegawa et al., 2007)

In one recent study in humans, ESCM showed great promise in the treatment of diabetes. The study selected individuals recently diagnosed for diabetes and the treatment consisted of both stem cell mobilization and autologous stem cell transplant. The patients first received injections of G-CSF in order to harvest PBSC, followed later by autologous stem cell thansplant and, 5 days post-transplant, a second round of G-CSF treatment. The endpoints monitored in the study were overall morbidity along with temporal changes in exogenous insulin requirements. Before the treatment, all patient required daily insulin injection. By the end of the study, 14 of the 15 patients had experienced insulin-free episodes ranging between 1 and 35 months (mean 16.2 months) (Voltarelli et al., 2007).

In this study the patients benefited from two instances of ESCM and one instance of autologous SC transplant. It is not possible to determine what were the respective contributions of the ESCM and SC injection, however it is likely that the mobilizations by themselves significantly contributed to the benefits experienced. While the first mobilization lasted several days and the second mobilization lasted about one week, there was only one instance of stem cell injection.

Diabetes is an interesting disease to study the link between disease progression and the number of circulating PBSCs, as it follows a series of relatively well defined stages with regard to carbohydrate metabolism status, namely normal glucose tolerance (NGT), impaired fasting glucose (IFG), impaired glucose tolerance (IGT), and newly diagnosed diabetes mellitus (DM). Fadini et al. (2010) quantified the number of circulating CD34+ cells by flow cytometry in 425 individuals divided among these four stages of disease progression. The data showed a clear trend of decreased number of PBSCs with disease progression through IFG, IGT and DM (Figure 4). The number of circulating PBSCs was significantly lower in the IGT and DM groups when compared to the NGT group. The reduction in the number of PBSCs can either be a consequence of higher blood glucose levels that might affect the ability of stem cells to mobilize from the bone marrow or a causal factor in the development of DM whereby a reduced number of circulating PBSCs reduces the ability of the pancreas to renew itself over the years, or both. This supports the view previously suggested that diabetes could be a stem cell disease (Fadini et al., 2009).

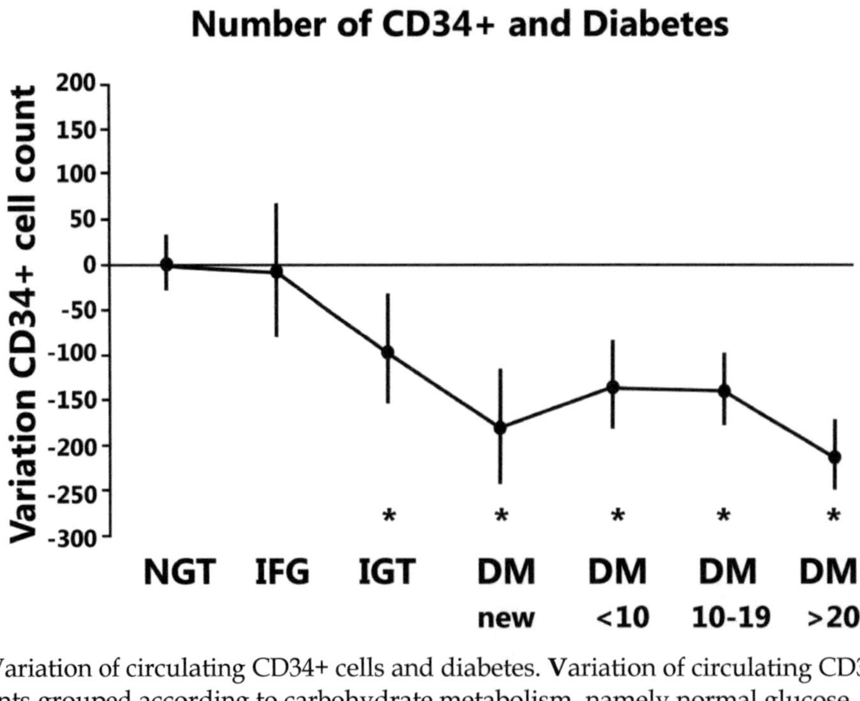

Fig. 4. Variation of circulating CD34+ cells and diabetes. Variation of circulating CD34+ cells in patients grouped according to carbohydrate metabolism, namely normal glucose tolerance (NGT), impaired fasting glucose (IFG), impaired glucose tolerance (IGT) or diabetes (DM) duration, as appropriate. The mean value of patients with NGT was taken to represent the zero point. Bars indicate 95% CIs of means. * Values significantly different when compared to NGT. (Taken from Fadini et al., 2009)

4. Clinical application (SE)

Investigation of the clinical potential of ESCM has been limited, largely due to the significant risk associated with the use of G-CSF, the main stem cell mobilizer used in clinical trial, for extended periods of time (Bensinger et al., 1996; Shimoda et al., 1993). Recently, a new stem cell mobilizer (StemEnhance®; SE) has been developed that triggers a much milder increase in the number of PBSC, but its safety allows for a sustained oral daily consumption over long periods of time, allowing for safe daily ESCM (Jensen et al., 2007).

In brief, SE is an extract from the cyanophyta *Aphanizomenon flos-aquae* that concentrates a protein with an estimated molecular weight of 160–180 kDa, which was shown to be a selective L-selectin blocker. Oral consumption of 1 gram of SE was been shown to trigger an average 25% increase in the number of PBSC within 60 minutes after consumption. The magnitude of the mobilization induced by SE is much smaller than that triggered by G-CSF, however its safety allows for continuous use and therefore offers a novel approach in the study of ESCM. To test its therapeutic potential, SE was used in a number of preliminary clinical trials involving a number of diagnostic entities.

4.1 Parkinson

MC is a 62 year old male with a 17 year history of Parkinson's Disease. MC was initially diagnosed with Parkinson's type resting pill-rolling tremor affecting his left hand and then progressing to the limbs on left side of his body. Over the course of five years the tremors increased gradually affecting the right side as well. At ten years into the disease process, MC was no longer able to practice as an attorney, mobility was affected and limitations increased compromising his personal and professional abilities.

In 2009, before the treatment of SE, the left side tremors had significantly worsened, symptoms of stiffness and bradykinesia which introduced a shuffling gait and poor balance made it unsafe for HM to ambulate without the use of a cane. At the time MC began consuming SE he was unable to dress himself, put on his watch, write, feed himself or drive his car. After 45 days on the product, consuming 1 gram of SE, three times per day, MC showed significant improvement with decreased tremors, less stiffening and bradykinesia; at this time MC was able to get around without the use of his cane. After 60 days on the product, judging that the benefits had plateaued, MC discontinued the use of SE. Forty - five days later, signs of tremor and bradykinesia returned and MC was seeking medical attention, once again. MC resumed SE with a dose of 2 grams, three times per day and within six weeks showed much improvement with less tremors, stiffness and bradykinesia. MC was able to participate in activities of daily living, such as dressing and feeding himself, he was also able to tie his own tie, put on his watch, and once again, he could ambulate without using a cane. To date MC has been on SE for two years, he has returned to driving his car without limitations, he is independent with all activities of daily living and he also participates in some level of professional activity.

Another patient, MT, is a 52 year old woman with an early onset of Parkinson at age 36, with tremor as the primary symptom. Early treatment consisted of Pergolid, then Budipin up to 30 mg three times a day. However due to QTc increase, Budipin was later reduced to 10 mg three times a day. The patient was also treated with Methydopamin 62.5 mg four times a day. With this treatment, MT's main problem was the experience of fluctuations and the beginning of ON-OFF syndrome. One year ago MT began consuming StemEnhance 1 gram three times a day. Today MT's experiences only minor tremor accompanied with some dyskinetic syndrome and sometimes propulsion. Her quality of life has increased and she is much more socially active.

4.2 Traumatic spinal cord injury

A preliminary trial with 8 individuals with spinal cord injuries was performed in a community center in Hawaii. The cases all involved paraplegia and various degrees of quadriplegia. Of the 8 cases, 4 dropped due to various circumstances unrelated to the consumption of SE. Of the 4 remaining participants, 2 had repetitive periods of hospitalization that made their consumption of SE irregular. Of the two remaining participants, one experienced mild though significant improvement in mobility while the other participant experienced significant improvements in mobility.

The latter participant, VS, had a serious car accident 17 years prior to SE consumption and was left with traumatic brain injury that affected her speech and a significant spinal cord lesion. At the beginning of the trial, VS was able to lift her right leg approximately 15 centimeter from her chair with no lateral movement, and showed a total absence of movement in her left leg. She could move her arms, hands and fingers though the movements were very slow with little dexterity and precision. She had some ability to move in her bed but was unable to turn herself without assistance. No data was available to document peripheral sensory perception of the lower limbs or nerve conductivity. After 6 months of daily consumption of 3 grams of SE three times a day, VS could lift her right leg more than 30 centimeters from her chair, with lateral movement outside of her chair. She could also lift her left leg off the chair and laterally outside of the chair. VS had less control over the movement of her left leg, though the magnitude of the movements was comparable to the movements seen with the right leg. After 10 months, VS was able to rotate herself in her bed unassisted and from a supine position she could lift both legs to a 90 degree angle and sit in her bed unassisted. Over the period of the trial her upper limbs also improved in dexterity and her speech showed mild though significant improvement. VS comes from a disadvantage socio-economic environment and did not have access to physical therapy beyond the first few years of her injury, she therefore developed leg and feet deformities that prevented her from bearing weight and possibly resuming physical therapy.

4.3 Coronary Artery Disease

JP is a 60 year old South African male who experienced a heart attack at 51 years of age. After diagnosis of Coronary Artery Disease was made a stent placement was performed. Dietary and lifestyle changes were implemented immediately by JP following the hospitalization. Unfortunately, 3 years later, JP suffered four additional heart attacks. During the hospitalization, the angiogram revealed the right and left coronary arteries were obstructed 100% and 40%, respectively, thus determining that JP was not a good candidate for bypass surgery. At the time of the last hospitalization, JP had decreased energy and experienced "stable" angina with any exertion, and his overall quality of life was greatly compromised. JP was put on a medication regimen which consisted of Atenolol 50mg daily,

Perindopril 4 mg daily, Elantan 20 mg twice per day, Simvostatin 20 mg daily, Adelat xl 30 mg daily, and Aspirin 1 per day. JP was then put on a list for possible heart transplant.

Three weeks after discharge JP began consuming SE, 1 gram three times per day. After 3 months of taking the product, he received a call for a possible heart transplant and returned to see his cardiologist for evaluation and comprehensive testing. The cardiologist reported that JP was making a remarkable recovery, the heart transplant surgery was postponed. Four months later JP was re-evaluated and found to have made a complete recovery. JP has since been returning at 6 month intervals for follow up evaluation, and to date he has been stable with no further coronary incidents. The evaluations and most recent ECG conclude that JP has normal heart function. Presently, JP shares that he is not on any medications, blood pressure is 126/65 mm Hg, he continues to take SE and experiences good quality of life.

4.4 Diabetes and rheumatoid arthritis

NA is a 47 year old Colombian woman who was diagnosed at 18 year of age with deforming rheumatoid arthritis. Over the years she managed her condition with the use of Prednisone, Methotrexate, Sulfasalazina, and Diclofenac. Four years ago NA was diagnosed with diabetes mellitus with glycemia reaching 308 mg/dL and assumed treatment with Euglucon 5 mg twice a day along with NPH insulin. About one year ago, due to her arthritic condition she became wheelchair bound and required assistance for bathing and dressing. At that time her medical records show levels of C-Reactive Protein (CRP) of 96.2 mg/L, Erythrocytes sedimentation rate (ESR) of 81 mm/H, Platelet count of 535 K/ul, fasting glucose of 147 mg/dL and HgbA1c (or Glycosylated Hemoglobin) of 8.07. A few months after becoming wheelchair bound NA began consuming 1 gram of SE once a day. After one month she subjectively reported a reduction in pain and inflammation, while 3 months later her mobility had improved to the point that she could bathe and dress alone. After 6 months she began using a walker, then switched to a cane and one year later she was walking unaided. Her last medical records indicate a level of CRP of 2.1 mg/L, ESR of 34 mm/H, Platelet count of 485 K/ul. NA has also discontinued the use of any diabetes medication; her records indicate a fasting glucose of 106 mg/dL with a glycemia not exceeding 120 mg/dL and HgbA1c of 5.62.

At the time of writing this report, NA does not take any anti-inflammatory drug, undergoes a remarkable improvement in her quality of life, keeping daily consumption for 15 consecutive months of 1 gram of SE, bone support supplementation of 500 mg Calcitriol (1,25-dihydroxycholecalciferol) every month, milk of magnesia and an annual dose of Zoledronic Acid.

4.5 Cerebrovascular accident (Stroke)

In September 2008, GE, a 78 year old male surgical oncologist who was otherwise in good health, had a stroke. The MRI/MRA revealed an acute infarct involving the right lentiform nucleus, moderately extensive chronic small vessel ischemic changes, chronic lacunar infarct involving the right ventromedial thalamus, and intracranial atherosclerotic vascular disease. The stroke left GE with aphasia and a reduced ability to perform any physical activity. October 2009, 13 months after the stroke, GE began taking SE; GE consumed 1 gram twice per day and after 8 weeks on the product GE noticed improvement with his speech and experienced more energy with improved balance. In April 2010 a repeat MRI/MRA showed no evidence of an acute infarct. GE's aphasia was completely resolved at this point and his overall mobility was improved allowing him to perform all activities of daily living. In January of 2011 a repeat MRI/MRA of the anterior and posterior cerebral arteries demonstrated no evidence of hemodynamically significant stenosis, and revealed normal vertebrobasilar arteries with no evidence of intracranial aneurysm or vascular malformation.

To date, GE has resumed playing tennis at age 81, walks around the mall in the neighborhood with other senior citizens and has returned to his professional activity as a surgeon.

4.6 Kidney failure

The whole paragraph should read: "GW is 36 months old; he was born with a malfunctioning valve in the urethra, compromising the flow of urine at birth. At ten days old he had surgery to repair the urethra, however, the damage to the kidneys and the bladder were already evident, caused from the retention of urine at birth. Following the surgical procedure, scar tissue was observed to the wall of the bladder and one of the kidneys did not respond, eventually that kidney quit working and began to atrophy. A scar was also left on his hand caused from an IV being ripped out after surgery. Before SE was introduced, GW had stopped growing at 15 months of age and multiple health issues were leading him toward a kidney transplant. Furthermore, GW had been on antibiotics for one year due to multiple infections. His body was not eliminating fluid as it should, as evidenced by the worsening of edema resulting in the use of diuretics. GW had purple feet caused from compromised circulation, dark spots were observed on his back, his eyes were blood shot and his overall presentation was pale in appearance. GW's activity was nothing normal for a 1 ½ year old child; his energy level was very depleted and he was unable to play like a normal toddler.

GW was 21 months old when SE was introduced at 250 mg per day. After 3 days on SE his mother reported that his eyes were crystal clear and he was observed running through the house full of energy. The dark spots on his back began to fade after one week of SE consumption. Red color was seen on GW cheeks for the first time since he was born. After 75 days on SE, GW had grown 6cm taller and the shrinking kidney at birth had measured 0.8cm larger. GW is off all pharmaceutical medications to date and his bladder and kidney are back to normal functioning to their fullest capacity, the scar to his hand is completely gone.

5. Conclusion

The benefits of ESCM on various degenerative conditions have been documented in several animal models and in humans. In some cases, BMSC clearly migrate into tissues and directly contribute to the formation of new functional somatic cells of the target tissue. However in other cases, especially diseases affecting the heart and central nervous system, a primary mechanism of action appears to be the secretion of paracrines that stimulate the proliferation and differentiation of tissue stem cells. Much work remains to be done to clearly elucidate the mechanisms of action behind the benefits of ECSM in various diagnostic entities, however from a clinical standpoint it remains that regardless of the mechanism of action, ESCM appears to be a valuable approach to increase the quality of life of patients affected by various degenerative diseases.

Various stem cell mobilizers have been documented in the scientific literature and many of them have been associated with side effects that prevent the application in humans of what has been documented as effective in various animal models. Such mobilizers include G-CSF, Stem Cell Factor, interleukin-8 and plerixafor (Lemery et al., 2011), which have all been associated with side effects going from diarrhea, nausea, pain and numbness to pericarditis and thrombosis. In spite of the potential benefits, such side effects have largely prevented the use of such compounds for ESCM in humans, and the lack of safe stem cell mobilizers largely explains the limited interest so far in this therapeutic approach. The main challenge in further investigating the therapeutic potential of ESCM remains therefore the development of safe stem cell mobilizers. In the meanwhile, SE appears to be a valuable tool to study the clinical benefits of ESCM.

6. References

Abbott JD, Huang Y, Liu D, Hickey R, Krause DS, Giordano FJ. (2004) Stromal cell-derived factor-1alpha plays a critical role in stem cell recruitment to the heart after myocardial infarction but is not sufficient to induce homing in the absence of injury. Circulation. 110(21):3300-5.

Abdel-Latif A, Bolli R, Zuba-Surma EK, Tleyjeh IM, Hornung CA, Dawn B. (2008) Granulocyte colony-stimulating factor therapy for cardiac repair after acute myocardial infarction: a systematic review and meta-analysis of randomized controlled trials. Am Heart J. 2008 Aug;156(2):216-226.e9.

Abedi M, Greer DA, Colvin GA. Demers DA, Dooner MS, Harpel JA, Pimentel J, Menon MK and Quesenberry PJ (2004) Tissue injury in marrow transdifferentiation Blood Cells Mol Dis 32(1):42 - 46

Arufe MC, De la Fuente A, Fuentes-Boquete I, De Toro FJ, Blanco FJ. (2009) Differentiation of synovial CD-105(+) human mesenchymal stem cells into chondrocyte-like cells through spheroid formation. *J Cell Biochem*. 108(1):145-55.

Arsenijevic Y, Weiss S, Schneider B, Aebischer P. (2001) Insulin-like growth factor-I is necessary for neural stem cell proliferation and demonstrates distinct actions of epidermal growth factor and fibroblast growth factor-2. J Neurosci. 21(18):7194-202.

Arvidsson A, T Collin, D Kirik, Z Kokaia and O Lindvall. (2002). Neuronal replacement from endogenous precursors in the adult brain after stroke. Nature Med 8:963–970.

Askari AT, Unzek S, Popovic ZB, Goldman CK, Forudi F, Kiedrowski M, Rovner A, Ellis SG, Thomas JD, DiCorleto PE, Topol EJ, Penn MS. (2003) Effect of stromal-cell-derived factor 1 on stem-cell homing and tissue regeneration in ischaemic cardiomyopathy. Lancet. 362(9385):697-703.

Avigdor A, Goichberg P, Shivtiel S, Dar A, Peled A, Samira S, Kollet O, Hershkoviz R, Alon R, Hardan I, Ben-Hur H, Naor D, Nagler A, Lapidot T. (2004) CD44 and hyaluronic acid cooperate with SDF-1 in the trafficking of human CD34+ stem/progenitor cells to bone marrow. Blood 103(8):2981-9.

Bachelder RE, Wendt MA, Mercurio AM. (2002) Vascular endothelial growth factor promotes breast carcinoma invasion in an autocrine manner by regulating the chemokine receptor CXCR4. Cancer Res 62:7203–7206.

Bagri A, Gurney T, He X, Zou YR, Littman DR, Tessier-Lavigne M, Pleasure SJ. (2002) The chemokine SDF-1 regulates migration of dentate granule cells. Development 129:4249–4260.

Baumhäkel M, Werner N, Böhm M, Nickenig G. (2006) Circulating endothelial progenitor cells correlate with erectile function in patients with coronary heart disease. Eur Heart J. 27(18):2184-8.

Behfar A, Faustino RS, Arrell DK, Dzeja PP, Perez-Terzic C, Terzic A. (2008) Guided stem cell cardiopoiesis: discovery and translation. J Mol Cell Cardiol. 45(4):523-9. Epub 2008 Sep 26.

Behfar A, Yamada S, Crespo-Diaz R, Nesbitt JJ, Rowe LA, Perez-Terzic C, Gaussin V, Homsy C, Bartunek J, Terzic A. (2010) Guided cardiopoiesis enhances therapeutic benefit of bone marrow human mesenchymal stem cells in chronic myocardial infarction. J Am Coll Cardiol. 56(9):721-34.

Beltrami AP, Urbanek K, Kajstura J, Yan SM, Finato N, Bussani R, Nadal-Ginard B, Silvestri F, Leri A, Beltrami CA, Anversa P. (2001) Evidence that human cardiac myocytes divide after myocardial infarction. *N Engl J Med* 344:1750–1757.

Bensinger W, Buckner W, Rowley S, Storb R, Appelbaum F. (1996) Treatment of normal donors with recombinant growth factors for transplantation of allogeneic blood stem cells. Bone Marrow Transplant. 17:S19-21.

Bergmann O, Bhardwaj RD, Bernard S, Zdunek S, Barnabé-Heider F, Walsh S, Zupicich J, Alkass K, Buchholz BA, Druid H, Jovinge S, and Frisén J. (2009) Evidence for cardiomyocyte renewal in humans. Science 324: 98-102.

Bonavia R, Bajetto A, Barbero S, Porcile C, Pirani P, Ravetti JL, Zona GL, Spaziante R, Florio T, Schettini G. (2003) Chemokines and their receptors in the CNS: expression of CXCL12/SDF-1 and CXCR4 and their role in astrocyte proliferation. Toxicol Lett 139:181–189.

Broxmeyer HE, Kohli L, Kim CH, Lee Y, Mantel C, Cooper S, Hangoc G, Shaheen M, Li X, Clapp DW. (2003) Stromal cell-derived factor-1/CXCL12 directly enhances survival/antiapoptosis of myeloid progenitor cell through CXCR4 and G proteins and enhances engraftment of competitive, repopulating stem cells. J Leukoc Biol 73:630–638.

Broxmeyer HE, Orschell CM, Clapp DW, Hangoc G, Cooper S, Plett PA, Liles WC, Li X, Graham-Evans B, Campbell TB, Calandra G, Bridger G, Dale DC, Srour EF. (2005) Rapid mobilization of murine and human hematopoietic stem and progenitor cells with AMD3100, a CXCR4 antagonist. J Exp Med. 201(8):1307-18.

Burchfield JS, Iwasaki M, Koyanagi M, Urbich C, Rosenthal N, Zeiher AM, Dimmeler S. (2008) Interleukin-10 from transplanted bone marrow mononuclear cells contributes to cardiac protection after myocardial infarction. Circ Res 103(2):203-11.

Can A, Balci D. (2011) Isolation, Culture, and Characterization of Human Umbilical Cord Stroma-derived Mesenchymal Stem Cells. *Methods Mol Biol.* 698:51-62.

Carion A, Benboubker L, Hérault O, Roingeard F, Degenne M, Senecal D, Desbois I, Colombat P, Charbord P, Binet C, Domenech J. (2003) Stromal-derived factor 1 and matrix metalloproteinase 9 levels in bone marrow and peripheral blood of patients mobilized by granulocyte colony-stimulating factor and chemotherapy. Relationship with mobilizing capacity of haematopoietic progenitor cells. Br J Haematol. 122(6):918-26.

Chandra V, G S, Phadnis S, Nair PD, Bhonde RR. (2009) Generation of pancreatic hormone-expressing islet-like cell aggregates from murine adipose tissue-derived stem cells. *Stem Cells.* 27(8):1941-53.

Chavey C, Mari B, Monthouel MN, Bonnafous S, Anglard P, Van Obberghen E, Tartare-Deckert S. (2003) Matrix metalloproteinases are differently expressed in adipose tissue during obesity and modulate adipocyte differentiation. J Biol Chem 278:11888 -11896.

Chen J, Zhang ZG, Li Y, Wang L, Xu YX, Gautam SC, Lu M, Zhu Z, Chopp M. (2003) Intravenous administration of human bone marrow stromal cells induces angiogenesis in the ischemic boundary zone after stroke in rats. Circ Res. 92(6):692-9.

Chen X, Fang H, Schwob JE. (2004) Multipotency of purified, transplanted globose basal cells in olfactory epithelium. *J Comp Neurol.* 469(4):457-74.

Choi JH, Kim KL, Huh W, Kim B, Byun J, Suh W, Sung J, Jeon ES, Oh HY, Kim DK. (2004) Decreased number and impaired angiogenic function of endothelial progenitor cells in patients with chronic renal failure. Arterioscler Thromb Vasc Biol. 24(7):1246-52.

Clarris BJ, Fraser JR. (1968) On the pericellular zone of some mammalian cells in vitro. Exp Cell Res. 49(1):181-93.

Comper WD, Laurent TC. (1978) Physiological function of connective tissue polysaccharides. Physiol Rev. 58(1):255-315.

Contreras JL, Smyth CA, Eckstein C, Bilbao G, Thompson JA, Young CJ, Eckhoff DE. (2003) Peripheral mobilization of recipient bone marrow-derived endothelial progenitor cells enhances pancreatic islet revascularization and engraftment after intraportal transplantation. Surgery 134(2):390-8.

Cottler-Fox MH, Lapidot T, Petit I, Kollet O, DiPersio JF, Link D, Devine S. (2003) Stem cell mobilization. *Hematology Am Soc Hematol Educ Program.* 419-37.

Croop JM, Cooper R, Fernandez C, Graves V, Kreissman S, Hanenberg H, Smith FO, Williams DA. (2001) Mobilization and collection of peripheral blood CD34+ cells from patients with Fanconi anemia. *Blood.* 98(10):2917-21.

Dasgupta C, Escobar-Poni B, Shah M, Duncan J, Fletcher WH. (1999) Misregulation of connexin43 gap junction channels and congenital heart defects. Novartis Found Symp. 219:212-21.

Deasy BM, Qu-Peterson Z, Greenberger JS, Huard J. (2002) Mechanisms of muscle stem cell expansion with cytokines. Stem Cells. 20(1):50-60.

De Clercq E. (2005) Potential clinical applications of the CXCR4 antagonist bicyclam AMD3100. Mini Rev Med Chem. 5(9):805-24.

Diller GP, van Eijl S, Okonko DO, Howard LS, Ali O, Thum T, Wort SJ, Bédard E, Gibbs JS, Bauersachs J, Hobbs AJ, Wilkins MR, Gatzoulis MA, Wharton J. (2008) Circulating endothelial progenitor cells in patients with Eisenmenger syndrome and idiopathic pulmonary arterial hypertension. Circulation 117(23):3020-30.

Docherty K. (2009) Pancreatic stellate cells can form new beta-like cells. Biochem J. 421(2): e1-4.

Dos Santos F, Andrade PZ, Eibes G, da Silva CL, Cabral JM. (2011) Ex Vivo Expansion of Human Mesenchymal Stem Cells on Microcarriers. *Methods Mol Biol.* 698:189-198.

Drapeau C, Antarr D, Ma H, Yang Z, Tang L, Hoffman RM, Schaeffer DJ. (2010) Mobilization of bone marrow stem cells with StemEnhance improves muscle regeneration in cardiotoxin-induced muscle injury. *Cell Cycle* 9(9):1819-23.

Dunac A, Frelin C, Popolo-Blondeau M, Chatel M, Mahagne MH, Philip PJ. (2007) Neurological and functional recovery in human stroke are associated with peripheral blood CD34+ cell mobilization. J Neurol. 2007 Mar;254(3):327-32.

Eizawa T, Murakami Y, Matsui K, Takahashi M, Muroi K, Amemiya M, Takano R, Kusano E, Shimada K, Ikeda U. (2003) Circulating endothelial progenitor cells are reduced in hemodialysis patients. Curr Med Res Opin. 19(7):627-33.

Ellis SG, Penn MS, Bolwell B, Garcia M, Chacko M, Wang T, Brezina KJ, McConnell G, Topol EJ. (2006) Granulocyte colony stimulating factor in patients with large acute myocardial infarction: results of a pilot dose-escalation randomized trial. Am Heart J. 152(6):1051.e9-14.

Fadini GP, de Kreutzenberg S, Agostini C, Boscaro E, Tiengo A, Dimmeler S, Avogaro A (2009) *Low CD34+ cell count and metabolic syndrome synergistically increase the risk of adverse outcomes. Atherosclerosis 2009; 207: 213– 219*

Fadini GP, Boscaro E, de Kreutzenberg S, Agostini C, Seeger F, Dimmeler S, Zeiher A, Tiengo A, Avogaro A. (2010) Time course and mechanisms of circulating progenitor cell reduction in the natural history of type 2 diabetes. Diabetes Care. 2010 May;33(5):1097-102.

Fallon J, S Reid, R Kinyamu, I Opole, R Opole, J Baratta, M Korc, TL Endo, A Duong, G Nguyen, M Karkehabadhi, D Twardzik, S Patel and S Loughlin. (2000). In vivo induction of massive proliferation, directed migration, and differentiation of neural cells in the adult mammalian brain. PNAS 96:14686–14691.

Ferrari G, Cusella-De Angelis G, Coletta M, Paolucci E, Stornaiuolo A, Cossu G, Mavilio F. (1998) Muscle regeneration by bone marrow-derived myogenic progenitors. Science. 279(5356):1528-30.

Fibbe WE, Pruijt JF, Velders GA, et al. (1999) Biology of IL-8-induced stem cell mobilization. Ann N Y Acad Sci. 872:71-82.

Fox A, Smythe J, Fisher N, Tyler MP, McGrouther DA, Watt SM, Harris AL. (2008) Mobilization of endothelial progenitor cells into the circulation in burned patients. Br J Surg. 95(2):244-51

Fraser JR, Laurent TC, Laurent UB. (1997) Hyaluronan: its nature, distribution, functions and turnover. J Intern Med. 242(1):27-33.

Friedl P, Maaser K, Klein CE, Niggemann B, Krohne G, Zanker KS. Migration of highly aggressive MV3 melanoma cells in 3-dimensional collagen lattices results in local matrix reorganization and shedding of alpha2 and beta1 integrins and CD44. Cancer Res.57:2061-2070.

Fukuda S, Bian H, King AG, Pelus LM. (2007) The chemokine GRObeta mobilizes early hematopoietic stem cells characterized by enhanced homing and engraftment. *Blood* 110(3):860-9.

Fukuhara S, Tomita S, Nakatani T, Ohtsu Y, Ishida M, Yutani C, Kitamura S. (2004) G-CSF promotes bone marrow cells to migrate into infarcted mice heart, and differentiate into cardiomyocytes. Cell Transplant. 13(7-8):741-8

Gao X, Song L, Shen K, Wang H, Niu W, Qin X. (2008) Transplantation of bone marrow derived cells promotes pancreatic islet repair in diabetic mice. Biochem Biophys Res Commun. 371(1):132-7.

Gnecchi M, He H, Liang OD, Melo LG, Morello F, Mu H, Noiseux N, Zhang L, Pratt RE, Ingwall JS, Dzau VJ. (2005) Paracrine action accounts for marked protection of ischemic heart by Akt-modified mesenchymal stem cells. Nat Med. 11(4):367-8.

Gnecchi M, He H, Noiseux N, Liang OD, Zhang L, Morello F, Mu H, Melo LG, Pratt RE, Ingwall JS, Dzau VJ. (2006) Evidence supporting paracrine hypothesis for Akt-modified mesenchymal stem cell-mediated cardiac protection and functional improvement. FASEB J. 20(6):661-9.

Gordon MK, Sher D, Karrison T, Kebriaei P, Chuang K, Zhang Y, McDonnell D, Artz A, Godley L, Odenike O, Rich E, Michaelis L, Thirman MJ, Wickrema A, van Besien K, Larson RA, Stock W. (2008) Successful autologous stem cell collection in patients with chronic myeloid leukemia in complete cytogenetic response, with quantitative measurement of BCR-ABL expression in blood, marrow, and apheresis products. *Leuk Lymphoma*. 49(3):531-7.

Greco SJ, Zhou C, Ye JH, Rameshwar P. (2008) A method to generate human mesenchymal stem cell-derived neurons which express and are excited by multiple neurotransmitters. Biol Proced Online. 10:90-101.

Grisar J, Aletaha D, Steiner CW, Kapral T, Steiner S, Seidinger D, Weigel G, Schwarzinger I, Wolozcszuk W, Steiner G, Smolen JS. (2005) Depletion of endothelial progenitor cells in the peripheral blood of patients with rheumatoid arthritis. Circulation. 111(2):204-11.

Gronthos S, Arthur A, Bartold PM, Shi S. (2011) A method to isolate and culture expand human dental pulp stem cells. *Methods Mol Biol*. 698:107-21.

Hahn JY, Cho HJ, Kang HJ, Kim TS, Kim MH, Chung JH, Bae JW, Oh BH, Park YB, Kim HS. (2008) Pre-treatment of mesenchymal stem cells with a combination of growth factors enhances gap junction formation, *J Am Coll Cardiol*. 51(9):933-43.

Hasegawa Y, Ogihara T, Yamada T, Ishigaki Y, Imai J, Uno K, Gao J, Kaneko K, Ishihara H, Sasano H, Nakauchi H, Oka Y, Katagiri H. (2007) Bone marrow (BM) transplantation promotes beta-cell regeneration after acute injury through BM cell mobilization. Endocrinology. 148(5):2006-15.

Hatch HM, Zheng D, Jorgensen ML, Petersen BE. (2002) SDF-1alpha/CXCR4: a mechanism for hepatic oval cell activation and bone marrow stem cell recruitment to the injured liver of rats. Cloning Stem Cells. 4(4):339-51.

Hayakawa J, Migita M, Ueda T, Shimada T, Fukunaga Y. (2003) Generation of a chimeric mouse reconstituted with green fluorescent protein-positive bone marrow cells: a useful model for studying the behavior of bone marrow cells in regeneration in vivo. Int J Hematol. 77(5):456-62.

Heissig B, Hattori K, Dias S, Friedrich M, Ferris B, Hackett NR, Crystal RG, Besmer P, Lyden D, Moore MA, Werb Z, Rafii S. (2002) Recruitment of stem and progenitor cells from the bone marrow niche requires MMP-9 mediated release of kit-ligand. Cell. 109:625-637.

Hennemann B, Ickenstein G, Sauerbruch S, Luecke K, Haas S, Horn M, Andreesen R, Bogdahn U, Winkler J. (2008) Mobilization of CD34+ hematopoietic cells, colony-forming cells and long-term culture-initiating cells into the peripheral blood of patients with an acute cerebral ischemic insult. Cytotherapy. 10(3):303-11.

Henschler R, Deak E, Seifried E.(2008) Homing of Mesenchymal Stem Cells. Transfus Med Hemother. 35(4):306-312.

Herbrig K, Haensel S, Oelschlaegel U, Pistrosch F, Foerster S, Passauer J. (2006) Endothelial dysfunction in patients with rheumatoid arthritis is associated with a reduced number and impaired function of endothelial progenitor cells. Ann Rheum Dis. 65(2):157-63.

Herbrig K, Pistrosch F, Oelschlaegel U, Wichmann G, Wagner A, Foerster S, Richter S, Gross P, Passauer J. (2004) Increased total number but impaired migratory activity and adhesion of endothelial progenitor cells in patients on long-term hemodialysis. Am J Kidney Dis. 44(5):840-9.

Hess DC, Hill WD, Martin-Studdard A, Carroll J, Brailer J, Carothers J. (2002) Bone marrow as a source of endothelial cells and NeuN-expressing cells after stroke. Stroke 33(5):1362-8.

Hisashi Y, Tomita S, Nakatani T, Fukuhara S, Yutani C, Kitamura S. (2004) Granulocyte-colony stimulating factor enhanced the recruitment of bone marrow cells into the heart: time course evaluation of phenotypic differentiation in the doxorubicin-induced cardiomyopathic model. Jpn J Thorac Cardiovasc Surg. 52(10):451-5.

Hoehn M, Küstermann E, Blunk J, Wiedermann D, Trapp T, Wecker S, Föcking M, Arnold H, Hescheler J, Fleischmann BK, Schwindt W, Bührle C. (2002) Monitoring of implanted stem cell migration in vivo: a highly resolved in vivo magnetic resonance imaging investigation of experimental stroke in rat. Proc Natl Acad Sci U S A. 99(25):16267-72.

Hofmann NA, Reinisch A, Strunk D. (2009) Isolation and large scale expansion of adult human endothelial colony forming progenitor cells. J Vis Exp. 28;(32).

Hwang JH, Kim SW, Park SE, Yun HJ, Lee Y, Kim S, Jo DY. (2006) Overexpression of stromal cell-derived factor-1 enhances endothelium-supported transmigration, maintenance, and proliferation of hematopoietic progenitor cells. Stem Cells Dev. 15(2):260-8.

Ince H, Petzsch M, Kleine HD, Eckard H, Rehders T, Burska D, Kische S, Freund M, Nienaber CA. (2005) Prevention of left ventricular remodeling with granulocyte colony-stimulating factor after acute myocardial infarction: final 1-year results of the Front-Integrated Revascularization and Stem Cell Liberation in Evolving Acute Myocardial Infarction by Granulocyte Colony-Stimulating Factor (FIRSTLINE-AMI) Trial. Circulation 112(9 Suppl):I73-80.

Insausti CL, Blanquer M, Meseguer L, López-Martínez MC, Ferez X, Rodriguez FJ, Cabañas V, Funes C, Nicolás FJ, Majado MJ, Moraleda JM. Isolation and characterization of mesenchymal stem cells from the fat layer on the density gradient separated bone marrow. Stem Cells Dev. 2011 Apr 19. [Epub ahead of print]

Ivanovic Z, Duchez P, Chevaleyre J, Vlaski M, Lafarge X, Dazey B, Robert-Richard E, Mazurier F, Boiron JM. (2011) Clinical-scale cultures of cord blood CD34+ cells to amplify committed progenitors and maintain stem cell activity. Cell Transplant. 2011 Feb 3. [Epub ahead of print]

Iwaguro H, Yamaguchi J, Kalka C, Murasawa S, Masuda H, Hayashi S, Silver M, Li T, Isner JM, Asahara T. (2002) Endothelial progenitor cell vascular endothelial growth factor gene transfer for vascular regeneration. Circulation 105:732–738.

Jackson KA, Majka SM, Wang H, Pocius J, Hartley CJ, Majesky MW, Entman ML, Michael LH, Hirschi KK, Goodell MA. (2001) Regeneration of ischemic cardiac muscle and vascular endothelium by adult stem cells. J Clin Invest. 107(11):1395-402.

Jang YY, Collector MI, Baylin SB, Diehl AM, Sharkis SJ. (2004) Hematopoietic stem cells convert into liver cells within days without fusion. Nat Cell Biol. 6(6):532-9.

Janowska-Wieczorek A, Marquez LA, Nabholtz JM, Cabuhat ML, Montaño J, Chang H, Rozmus J, Russell JA, Edwards DR, Turner AR. (1999) Growth factors and cytokines upregulate gelatinase expression in bone marrow CD34(+) cells and their transmigration through reconstituted basement membrane. Blood. 93:3379-3390.

Janowska-Wieczorek A, Marquez LA, Dobrowsky A, Ratajczak MZ, Cabuhat ML. (2000) Differential MMP and TIMP production by human marrow and peripheral blood CD34- cells in response to chemokines. Exp Hematol 28:1274–1285.

Jensen GS, Hart AN, Zaske LA, Drapeau C, Gupta N, Schaeffer DJ, Cruickshank JA. (2007) Mobilization of human CD34+ CD133+ and CD34+ CD133(-) stem cells in vivo by consumption of an extract from Aphanizomenon flos-aquae--related to modulation of CXCR4 expression by an L-selectin ligand? Cardiovasc Revasc Med. 8(3):189-202.

Jin FY, Qiu LG, Li QC, Meng HX, Wang YF, Yu Z, Li Q, Han JL. (2006) The effect of matrix metalloproteinase-9 in granulocyte colony stimulation factor-induced stem cell mobilization. Zhonghua Yi Xue Za Zhi. 86(42):2966-70.

Jo DY, Rafii S, Hamada T, Moore MA. (2000) Chemotaxis of primitive hematopoietic cells in response to stromal cell-derived factor-1. J Clin Invest. 105(1):101-11.

Junhui Z, Xingxiang W, Guosheng F, Yunpeng S, Furong Z, Junzhu C. (2008) Reduced number and activity of circulating endothelial progenitor cells in patients with idiopathic pulmonary arterial hypertension. Respir Med. 102(7):1073-9.

Kajstura J, Leri A, Finato N, Di Loreto C, Beltrami CA, Anversa P. (1998) Myocyte proliferation in end-stage cardiac failure in humans. *Proc Natl Acad Sci USA* 95:8801–8805.

Kamihata H, Matsubara H, Nishiue T, Fujiyama S, Tsutsumi Y, Ozono R, Masaki H, Mori Y, Iba O, Tateishi E, Kosaki A, Shintani S, Murohara T, Imaizumi T, Iwasaka T. (2001) Implantation of bone marrow mononuclear cells into ischemic myocardium enhances collateral perfusion and regional function via side supply of angioblasts, angiogenic ligands, and cytokines. Circulation. 104(9):1046-52.

Kan I, Melamed E, Offen D. (2005) Integral therapeutic potential of bone marrow mesenchymal stem cells. Curr Drug Targets. 6(1):31-41.

Kang S, Yang Y, Li CJ, Gao R. (2007) Effectiveness and tolerability of administration of granulocyte colony-stimulating factor on left ventricular function in patients with myocardial infarction: a meta-analysis of randomized controlled trials. Clin Ther. 2007 Nov;29(11):2406-18.

Karra R, Wu SM. (2008) Multipotent stem cells in cardiac regenerative therapy. Regen Med. 3(2):189-98.

Kawada H, Takizawa S, Takanashi T, Morita Y, Fujita J, Fukuda K, Takagi S, Okano H, Ando K, Hotta T. (2006) Administration of hematopoietic cytokines in the subacute phase after cerebral infarction is effective for functional recovery facilitating proliferation of intrinsic neural stem/progenitor cells and transition of bone marrow-derived neuronal cells. Circulation. 113(5):701-10.

Keilhoff G, Stang F, Goihl A, Wolf G, Fansa H. (2006) Transdifferentiated mesenchymal stem cells as alternative therapy in supporting nerve regeneration and myelination. Cell Mol Neurobiol. 26(7-8):1235-52.

King AG, Horowitz D, Dillon SB, Levin R, Farese AM, MacVittie TJ, Pelus LM. (2001) Rapid mobilization of murine hematopoietic stem cells with enhanced engraftment properties and evaluation of hematopoietic progenitor cell mobilization in rhesus monkeys by a single injection of SB-251353, a specific truncated form of the human CXC chemokine GRObeta. Blood. 97(6):1534-42.

Kinnaird T, Stabile E, Burnett MS, Shou M, Lee CW, Barr S, Fuchs S, Epstein SE. (2004) Local delivery of marrow-derived stromal cells augments collateral perfusion through paracrine mechanisms. Circulation. 109(12):1543-9.

Kleeberger W, Versmold A, Rothämel T, Glöckner S, Bredt M, Haverich A, Lehmann U, Kreipe H. (2003) Increased chimerism of bronchial and alveolar epithelium in human lung allografts undergoing chronic injury. Am J Pathol. 162(5):1487-94.

Klein JD, Fauza DO. (2011) Amniotic and placental mesenchymal stem cell isolation and culture. *Methods Mol Biol.* 698:75-88.

Kolbe M, Dohle E, Katerla D, Kirkpatrick CJ, Fuchs S. (2010) Enrichment of outgrowth endothelial cells in high and low colony-forming cultures from peripheral blood progenitors. *Tissue Eng Part C Methods.* 16(5):877-86.

Kollet O, Shivtiel S, Chen YQ, Suriawinata J, Thung SN, Dabeva MD, Kahn J, Spiegel A, Dar A, Samira S, Goichberg P, Kalinkovich A, Arenzana-Seisdedos F, Nagler A, Hardan I, Revel M, Shafritz DA, Lapidot T. (2003) HGF, SDF-1, and MMP-9 are involved in stress-induced human CD34+ stem cell recruitment to the liver. J Clin Invest. 112(2):160-9.

Krause DS, Theise ND, Collector MI, Henegariu O, Hwang S, Gardner R, Neutzel S, and Sharkis SJ (2001) Multi-organ, multi-lineage engraftment by a single bone marrow-derived stem cell. Cell 105:369-377

Kucia M, Ratajczak J, Reca R, Janowska-Wieczorek A, Ratajczak MZ.. (2004) Tissue-specific muscle, neural and liver stem/progenitor cells reside in the bone marrow, respond to an SDF-1 gradient and are mobilized into peripheral blood during stress and tissue injury. Blood Cells Mol Dis. 32(1):52–57.

Laflamme MA, Myerson D, Saffitz JE, Murry CE. (2002) Evidence for cardiomyocyte repopulation by extracardiac progenitors in transplanted human hearts. Circ Res. 90(6):634-40.

Lapid K, Vagima Y, Kollet O, Lapidot T. (2009) Egress and mobilization of hematopoietic stem and progenitor cells. StemBook. Cambridge (MA): Harvard Stem Cell Institute; 2008-.2009 Mar 15.

Lapidot T and Petit I. (2002) Current understanding of stem cell mobilization: the roles of chemokines, proteolytic enzymes, adhesion molecules, cytokines, and stromal cells. Exp Hematol. 30(9):973-81.

Lazarini F, Tham TN, Casanova P, Arenzana-Seisdedos F, Dubois-Dalcq M. (2003) Role of the alpha-chemokine stromal cell-derived factor (SDF-1) in the developing and mature central nervous system. Glia 2003;42:139–148.

Lee DY, Cho TJ, Kim JA, Lee HR, Yoo WJ, Chung CY, Choi IH. (2008) Mobilization of endothelial progenitor cells in fracture healing and distraction osteogenesis Bone. 42(5):932-41.

Lee HJ, Kim KS, Park IH, Kim SU (2007) Human neural stem cells over-expressing VEGF provide neuroprotection, angiogenesis and functional recovery in mouse stroke model. PLoS ONE 2(1): e156.

Lee KD, Kuo TK, Whang-Peng J, Chung YF, Lin CT, Chou SH, Chen JR, Chen YP and Lee OK. (2004) In vitro hepatic differentiation of human mesenchymal stem cells. Hepatology 40:1275-1284.

Lee ST, Chu K, Jung KH, Ko SY, Kim EH, Sinn DI, Lee YS, Lo EH, Kim M, Roh JK. (2005) Granulocyte colony-stimulating factor enhances angiogenesis after focal cerebral ischemia. Brain Res. 1058(1-2):120-128.

Lee ST, Chu K, Jung KH, Kim DH, Kim EH, Choe VN, Kim JH, Im WS, Kang L, Park JE, Park HJ, Park HK, Song EC, Lee SK, Kim M, Roh JK. (2008) Decreased number and function of endothelial progenitor cells in patients with migraine. Neurology. 70(17):1510-7.

Lemery SJ, Hsieh MM, Smith A, Rao S, Khuu HM, Theresa D, Viano JM, Cook L, Goodwin R, Boss C, Calandra G, Geller N, Tisdale J, Childs R. (2011) A pilot study evaluating the safety and CD34+ cell mobilizing activity of escalating doses of plerixafor in healthy volunteers. Br J Haematol. 153(1):66-75.

Leone AM, Rutella S, Bonanno G, Contemi AM, de Ritis DG, Giannico MB, Rebuzzi AG, Leone G, Crea F. (2006) Endogenous G-CSF and CD34+ cell mobilization after acute myocardial infarction. Int J Cardiol. 111(2):202-8.

Leone AM, Galiuto L, Garramone B, Rutella S, Giannico MB, Brugaletta S, Perfetti M, Liuzzo G, Porto I, Burzotta F, Niccoli G, Biasucci LM, Leone G, Rebuzzi AG, Crea F. (2007)

Usefulness of granulocyte colony-stimulating factor in patients with a large anterior wall acute myocardial infarction to prevent left ventricular remodeling (the rigenera study). Am J Cardiol. 100(3):397-403.

Li Y, Chen J, Chen XG, Wang L, Gautam SC, Xu YX, Katakowski M, Zhang LJ, Lu M, Janakiraman N, Chopp M. (2002) Human marrow stromal cell therapy for stroke in rat: neurotrophins and functional recovery. Neurology. 59(4):514-23.

Lima C, Escada P, Pratas-Vital J, Branco C, Arcangeli CA, Lazzeri G, Maia CA, Capucho C, Hasse-Ferreira A, Peduzzi JD. (2010) Olfactory mucosal autografts and rehabilitation for chronic traumatic spinal cord injury. Neurorehabil Neural Repair. 24(1):10-22.

Loewenstein WR, Rose B. (1992) The cell-cell channel in the control of growth. Semin Cell Biol. 3(1):59-79.

Mannello F, Tonti GA, Bagnara GP, Papa S. (2006) Role and function of matrix metalloproteinases in the differentiation and biological characterization of mesenchymal stem cells. Stem Cells. 24(3):475-81.

Mansilla E, Marín GH, Drago H, Sturla F, Salas E, Gardiner C, Bossi S, Lamonega R, Guzmán A, Nuñez A, Gil MA, Piccinelli G, Ibar R, Soratti C. (2006) Bloodstream cells phenotypically identical to human mesenchymal bone marrow stem cells circulate in large amounts under the influence of acute large skin damage: new evidence for their use in regenerative medicine. Transplant Proc. 38(3):967-9.

Marchesi C, Belicchi M, Meregalli M, Farini A, Cattaneo A, Parolini D, Gavina M, Porretti L, D'Angelo MG, Bresolin N, Cossu G, Torrente Y. (2008) Correlation of circulating CD133+ progenitor subclasses with a mild phenotype in Duchenne muscular dystrophy patients. PLoS ONE. 3(5):e2218.

Magavi SS and JD Macklis. (2002). Immunocytochemical analysis of neuronal differentiation. Meth Mol Biol 198:291–297.

Mathews V, Hanson PT, Ford E, Fujita J, Polonsky KS, Graubert TA. (2004) Recruitment of bone marrow-derived endothelial cells to sites of pancreatic beta-cell injury. Diabetes 53(1):91-8.

Matsumoto T, Mifune Y, Kawamoto A, Kuroda R, Shoji T, Iwasaki H, Suzuki T, Oyamada A, Horii M, Yokoyama A, Nishimura H, Lee SY, Miwa M, Doita M, Kurosaka M, Asahara T. (2008) Fracture induced mobilization and incorporation of bone marrow-derived endothelial progenitor cells for bone healing. J Cell Physiol. 215(1):234-42

Mezey E, Key S, Vogelsang G, Szalayova I, Lange GD and Crain G. (2003). Transplanted bone marrow generates new neurons in human brains. Proc. Natl. Acad. Sci. USA 100, 1364-1369.

Middleton J, Neil S, Wintle J, Clark-Lewis I, Moore H, Lam C, Auer M, Hub E, Rot A. (1997) Transcytosis and surface presentation of IL-8 by venular endothelial cells. Cell. 91(3):385-95.

Mirotsou M, Jayawardena TM, Schmeckpeper J, Gnecchi M, Dzau VJ. (2011) Paracrine mechanisms of stem cell reparative and regenerative actions in the heart. J Mol Cell Cardiol. 50(2):280-9.

Mishra SK, Braun N, Shukla V, Füllgrabe M, Schomerus C, Korf HW, Gachet C, Ikehara Y, Sévigny J, Robson SC, Zimmermann H. (2006) Extracellular nucleotide signaling in adult neural stem cells: synergism with growth factor-mediated cellular proliferation. Development. 133(4):675-84.

Moonen JR, de Leeuw K, van Seijen XJ, Kallenberg CG, van Luyn MJ, Bijl M, Harmsen MC. (2007) Reduced number and impaired function of circulating progenitor cells in patients with systemic lupus erythematosus. Arthritis Res Ther. 2007;9(4):R84.

Müller P, Pfeiffer P, Koglin J, Schäfers HJ, Seeland U, Janzen I, Urbschat S, Böhm M. (2002) Cardiomyocytes of noncardiac origin in myocardial biopsies of human transplanted hearts. Circulation. 106(1):31-5.

Nakatomi H, T Kuriu, S Okabe, S Yamamoto, O Hatano, N Kawahara, A Tamura, T Kirino and M Nakafuku. (2002) Regeneration of hippocampal pyramidal neurons after ischemic brain injury by recruitment of endogenous neural progenitors. Cell 110:429-441.

Neijssen J, Herberts C, Drijfhout JW, Reits E, Janssen L, Neefjes J. (2005) Cross-presentation by intercellular peptide transfer through gap junctions. Nature. 434(7029):83-8.

Neuss S, Becher E, Wöltje M, Tietze L, Jahnen-Dechent W. (2004) Functional expression of HGF and HGF receptor/c-met in adult human mesenchymal stem cells suggests a role in cell mobilization, tissue repair, and wound healing. Stem Cells 22:405- 414.

Nygren JM, Jovinge S, Breitbach M, Säwén P, Röll W, Hescheler J, Taneera J, Fleischmann BK, Jacobsen SE. (2004) Bone marrow-derived hematopoietic cells generate cardiomyocytes at a low frequency through cell fusion, but not transdifferentiation. Nat Med.10(5):494-501.

Okamoto I, Kawano Y, Tsuiki H, Sasaki J, Nakao M, Matsumoto M, Suga M, Ando M, Nakajima M, Saya H. (1999) CD44 cleavage induced by a membrane-associated metalloprotease plays a critical role in tumor cell migration. Oncogene. 18:1435-1446.

Orlic, D., Kajstura, J., Chimenti, S., Jakoniuk, I., Pickel, J., McKay, R., Nadal-Ginard, B., Bodine, D. M., Leri, A. & Anversa, P. (2001a) Bone marrow cells regenerate infracted myocardium. *Nature (London)* 410, 701–705.

Orlic D, Kajstura J, Chimenti S, Limana F, Jakoniuk I, Quaini F, Nadal-Ginard B, Bodine DM, Leri A, Anversa P. (2001b) Mobilized bone marrow cells repair the infarcted heart, improving function and survival. Proc Natl Acad Sci U S A. 98(18):10344-9.

Osada T, Watanabe M, Hasuo A, Imai M, Suyama K, Sakai D, Kawada H, Matsumae M, Mochida J. (2010) Efficacy of the coadministration of granulocyte colony-stimulating factor and stem cell factor in the activation of intrinsic cells after spinal cord injury in mice. J Neurosurg Spine. 13(4):516-23.

Paczkowska E, Larysz B, Rzeuski R, Karbicka A, Jałowiński R, Kornacewicz-Jach Z, Ratajczak MZ, Machaliński B. (2005) Human hematopoietic stem/progenitor-enriched CD34(+) cells are mobilized into peripheral blood during stress related to ischemic stroke or acute myocardial infarction. Eur J Haematol. 75(6):461-7.

Pavlichenko N, Sokolova I, Vijde S, Shvedova E, Alexandrov G, Krouglyakov P, Fedotova O, Gilerovich EG, Polyntsev DG, Otellin VA Mesenchymal stem cells transplantation could be beneficial for treatment of experimental ischemic stroke in rats. Brain Res. 2008 Jul 14.

Peled A, Grabovsky V, Habler L, Sandbank J, Arenzana-Seisdedos F, Petit I, Ben-Hur H, Lapidot T, Alon R. (1999) The chemokine SDF-1 stimulates integrin-mediated arrest of CD34(+) cells on vascular endothelium under shear flow. J Clin Invest. 104(9):1199-211.

Peled A, Kollet O, Ponomaryov T, Petit I, Franitza S, Grabovsky V, Slav MM, Nagler A, Lider O, Alon R, Zipori D, Lapidot T. (2000) The chemokine SDF-1 activates the integrins LFA-1, VLA-4, and VLA-5 on immature human CD34(+) cells: role in transendothelial/stromal migration and engraftment of NOD/SCID mice. Blood. 95(11):3289-96.

Pereira RF, O'Hara MD, Laptev AV, Halford KW, Pollard MD, Class R, Simon D, Livezey K and Prockop DJ. (1998) Marrow stromal cells as a source of progenitor cells for nonhematopoietic tissues in transgenic mice with a phenotype of osteogenesis imperfecta. Proc. Natl. Acad. Sci. USA 95, 1142-1147.

Peterson DA. (2002). Stem cells in brain plasticity and repair. Curr Opin Pharmacol 2:34–42.

Petit I, Szyper-Kravitz M, Nagler A, Lahav M, Peled A, Habler L, Ponomaryov T, Taichman RS, Arenzana-Seisdedos F, Fujii N, Sandbank J, Zipori D, Lapidot T. (2003) G-CSF induces stem cell mobilization by decreasing bone marrow SDF-1 and up-regulating CXCR4. *Nat Immunol.* 3(7):687-94.

Pineault N, Cortin V, Boyer L, Garnier A, Robert A, Thérien C, Roy DC. (2011) Individual and synergistic cytokine effects controlling the expansion of cord blood CD34(+) cells and megakaryocyte progenitors in culture. Cytotherapy. 13(4):467-80.

Quaini F, Urbanek K, Beltrami AP, Finato N, Beltrami CA, Nadal-Ginard B, Kajstura J, Leri A, Anversa P. (2002) Chimerism of the transplanted heart. N Engl J Med. 346(1):5-15.

Ratajczak MZ, Majka M, Kucia M, Drukala J, Pietrzkowski Z, Peiper S, Janowska-Wieczorek A. (2003) Expression of functional CXCR4 by muscle satellite cells and secretion of SDF-1 by muscle-derived fibroblasts is associated with the presence of both muscle progenitors in bone marrow and hematopoietic stem/progenitor cells in muscles. Stem Cells. 21(3):363-71.

Reed RK, Lilja K, Laurent TC. (1988) Hyaluronan in the rat with special reference to the skin. Acta Physiol Scand. 134(3):405-11.

Ries C, Egea V, Karow M, Kolb H, Jochum M, Neth P. (2007) MMP-2, MT1-MMP, and TIMP-2 are essential for the invasive capacity of human mesenchymal stem cells: differential regulation by inflammatory cytokines. Blood. 109(9):4055-63.

Ripa RS, Kastrup J. (2008) G-CSF therapy with mobilization of bone marrow stem cells for myocardial recovery after acute myocardial infarction--a relevant treatment? Exp Hematol. 36(6):681-6.

Rustom A, Saffrich R, Markovic I, Walther P, Gerdes HH. (2004) Nanotubular highways for intercellular organelle transport. Science. 303(5660):1007-10.

Sanchez-Ramos JR. (2002) Neural cells derived from adult bone marrow and umbilical cord blood. J Neurosci Res. 69(6):880-93

Schioppa T, Uranchimeg B, Saccani A, Biswas SK, Doni A, Rapisarda A, Bernasconi S, Saccani S, Nebuloni M, Vago L, Mantovani A, Melillo G, Sica A. (2003) Regulation of the chemokine receptor CXCR4 by hypoxia. J Exp Med 2003;198:1391–1402.

Schmidt W and KG Reymann. (2002). Proliferating cells differentiate into neurons in the hippocampal CAI region of gerbils after global cerebral ischemia. Neurosci Lett 334:153–156.

Schrader AJ, Lechner O, Templin M, Dittmar KE, Machtens S, Mengel M, Probst-Kepper M, Franzke A, Wollensak T, Gatzlaff P, Atzpodien J, Buer J, Lauber J. (2002) CXCR4/CXCL12 expression and signalling in kidney cancer. Br J Cancer. 86(8):1250-6.

Schwartz Y and Kornowski R. (2003) Autologous stem cells for functional myocardial repair. Heart Fail Rev. 8(3):237-45.

Sesti C, Hale SL, Lutzko C, Kloner RA. (2005) Granulocyte colony-stimulating factor and stem cell factor improve contractile reserve of the infarcted left ventricle independent of restoring muscle mass. J Am Coll Cardiol. 46(9):1662-9.

Shimoda K, Okamura S, Haarada N, Kondo S, Niho Y. (1993) Identification of a functional receptor for granulocyte colony-stimulating factor on platelets. J Clin Invest. 91:1310-1313.

Shintani S, Murohara T, Ikeda H, Ueno T, Honma T, Katoh A, Sasaki K, Shimada T, Oike Y, Imaizumi T. (2001) Mobilization of endothelial progenitor cells in patients with acute myocardial infarction. Circulation. 103(23):2776-9.

Shyu WC, Lin SZ, Yang HI, Tzeng YS, Pang CY, Yen PS, Li H. (2004) Functional recovery of stroke rats induced by granulocyte colony-stimulating factor-stimulated stem cells. Circulation 110(13):1847-54.

Six I, Gasan G, Mura E, Bordet R. (2003) Beneficial effect of pharmacological mobilization of bone marrow in experimental cerebral ischemia. Eur J Pharmacol. 458(3):327-8.

Snykers S, De Kock J, Tamara V, Rogiers V. (2011) Hepatic differentiation of mesenchymal stem cells: in vitro strategies. Methods Mol Biol. 698:305-14.

Soonpaa MH and Field LJ. (1998) Survey of studies examining mammalian cardiomyocyte DNA synthesis. *Am Heart Assoc* 1998;83:15–26.

Spees JL, Olson SD, Ylostalo J, Lynch PJ, Smith J, Perry A, Peister A, Wang MY, Prockop DJ. (2003) Differentiation, cell fusion, and nuclear fusion during ex vivo repair of epithelium by human adult stem cells from bone marrow stroma. Proc Natl Acad Sci U S A. 100(5):2397-402.

Stamm C, Westphal B, Kleine HD, Petzsch M, Kittner C, Klinge H, Schümichen C, Nienaber CA, Freund M, Steinhoff G. (2003) Autologous bone-marrow stem-cell transplantation for myocardial regeneration. Lancet. 361(9351):45-6.

Stout CL, Ashley DW, Morgan JH 3rd, Long GF, Collins JA, Limnios JI, Lochner F, McCommon G, Hixson D, Black AC Jr, Young HE. (2007) Primitive stem cells residing in the skeletal muscle of adult pigs are mobilized into the peripheral blood after trauma. *Am Surg*. 73(11):1106-10.

Suratt BT, Cool CD, Serls AE, Chen L, Varella-Garcia M, Shpall EJ, Brown KK, Worthen GS. (2003) Human pulmonary chimerism after hematopoietic stem cell transplantation. Am J Respir Crit Care Med. 168(3):318-22.

Swenson ES, Kuwahara R, Krause DS, Theise ND. (2008) Physiological variations of stem cell factor and stromal-derived factor-1 in murine models of liver injury and regeneration. Liver Int. 28(3):308-18

Takahashi T, Kalka C, Masuda H, Chen D, Silver M, Kearney M, Magner M, Isner JM, Asahara T. (1999) Ischemia- and cytokine-induced mobilization of bone marrow-derived endothelial progenitor cells for neovascularization. Nat Med. 5(4):434-8.

Terada N, Hamazaki T, Oka M, Hoki M, Mastalerz DM, Nakano Y, Meyer EM, Morel L, Petersen BE, Scott EW. (2002) Bone marrow cells adopt the phenotype of other cells by spontaneous cell fusion. Nature. 416(6880):542-5.

Theise ND, Nimmakayalu M, Gardner R, Illei PB, Morgan G, Teperman L, Henegariu O, Krause DS. (2000) Liver from bone marrow in humans. Hepatology. 32(1):11-6.

Tirino V, Paino F, d'Aquino R, Desiderio V, De Rosa A, Papaccio G. (2011) Methods for the Identification, Characterization and Banking of Human DPSCs: Current Strategies and Perspectives. Stem Cell Rev. Feb 12. [Epub ahead of print]

Tomita M, Adachi Y, Yamada H, Takahashi K, Kiuchi K, Oyaizu H, Ikebukuro K, Kaneda H, Matsumura M, Ikehara S. (2002) Bone marrow-derived stem cells can differentiate into retinal cells in injured rat retina. Stem Cells. 20(4):279-83.

Tomoda H and Aoki N. (2003) Bone marrow stimulation and left ventricular function in acute myocardial infarction. Clin. Cardiol. 26:455-457

Trzaska KA, Rameshwar P. (2011) Dopaminergic Neuronal Differentiation Protocol for Human Mesenchymal Stem Cells. *Methods Mol Biol*. 698:295-303.

Tsagias N, Koliakos I, Lappa M, Karagiannis V, Koliakos GG. (2011) Placenta perfusion has hematopoietic and mesenchymal progenitor stem cell potential. *Transfusion*. 51(5):976-85.

Tse HF, Kwong YL, Chan JK, Lo G, Ho CL, Lau CP. (2003) Angiogenesis in ischaemic myocardium by intramyocardial autologous bone marrow mononuclear cell implantation. Lancet. 361(9351):47-9.

Vasa M, Fichtlscherer S, Aicher A, Adler K, Urbich C, Martin H, Zeiher AM, and Dimmeler S. (2001) Number and migratory activity of circulating endothelial progenitor cells inversely correlate with risk factors for coronary artery disease. *Circ Res*. 89:e1-e7.

Vasania VS, Prasad P, Gill RK, Mehta A, Viswanathan C, Sarang S, Majumdar AS. (2010) Molecular and cellular characterization of expanded and cryopreserved human limbal epithelial stem cells reveal unique immunological properties. *Exp Eye Res*. 92(1):47-56.

Vassilopoulos G, Wang PR, Russell DW. (2003) Transplanted bone marrow regenerates liver by cell fusion. Nature. 422(6934):901-4.

Viktorov IV, Savchenko EA, Ukhova OV, Alekseyeva NY, Chekhonin VP. (2006) Multipotent stem and progenitor cells of the olfactory epithelium. *Bull Exp Biol Med*. 142(4):495-502.

Voermans C, Rood PM, Hordijk PL, Gerritsen WR, van der Schoot CE. (2000) Adhesion molecules involved in transendothelial migration of human hematopoietic progenitor cells. Stem Cells. 18(6):435-43.

Voltarelli JC, Couri CEB, Stracieri ABPL, Oliveira MC, Moraes DA, Pieroni F, Coutinho M, Malmegrim KCR, Foss-Freitas MC, Simoes BP, Foss MC, Squiers E, and Burt RK.

(2007) Autologous nonmyeloablative hematopoietic stem cell transplantation in newly diagnosed Type 1 diabetes mellitus. *JAMA*. 297(14):1568-1576.

Wang J, Wang X, Sun Z, Wang X, Yang H, Shi S, Wang S. (2011) Stem cells from human-exfoliated deciduous teeth can differentiate into dopaminergic neuron-like cells. *Stem Cells Dev*. 19(9):1375-83.

Welte K, Platzer E, Lu L, Gabrilove JL, Levi E, Mertelsmann R, and Moore MA. (1985) Purification and biochemical characterization of human pluripotent hematopoietic colony-stimulating factor. Proc Natl Acad Sci U S A. 82(5):1526-30.

Werner N, Kosiol S, Schiegl T, Ahlers P, Walenta K, Link A, Böhm M, Nickenig G. (2005) Circulating endothelial progenitor cells and cardiovascular outcomes. N Engl J Med. 353(10):999-1007.

Westerweel PE, Luijten RK, Hoefer IE, Koomans HA, Derksen RH, Verhaar MC. (2007) Haematopoietic and endothelial progenitor cells are deficient in quiescent systemic lupus erythematosus. Ann Rheum Dis. 66(7):865-70.

Willing AE, Vendrame M, Mallery J, Cassady CJ, Davis CD, Sanchez-Ramos J, Sanberg PR. (2003) Mobilized peripheral blood cells administered intravenously produce functional recovery in stroke. Cell Transplant.12(4):449-54.

Wojakowski W, Tendera M, Michałowska A, Majka M, Kucia M, Maślankiewicz K, Wyderka R, Ochała A, Ratajczak MZ. (2004) Mobilization of CD34/CXCR4+, CD34/CD117+, c-met+ stem cells, and mononuclear cells expressing early cardiac, muscle, and endothelial markers into peripheral blood in patients with acute myocardial infarction. Circulation. 110(20):3213-20.

Wojakowski W, Tendera M, Zebzda A, Michalowska A, Majka M, Kucia M, Maslankiewicz K, Wyderka R, Król M, Ochala A, Kozakiewicz K, Ratajczak MZ. (2006) Mobilization of CD34(+), CD117(+), CXCR4(+), c-met(+) stem cells is correlated with left ventricular ejection fraction and plasma NT-proBNP levels in patients with acute myocardial infarction. Eur Heart J. 27(3):283-9.

Wollert KC, Meyer GP, Lotz J, Ringes-Lichtenberg S, Lippolt P, Breidenbach C, Fichtner S, Korte T, Hornig B, Messinger D, Arseniev L, Hertenstein B, Ganser A, Drexler H. (2004) Intracoronary autologous bone-marrow cell transfer after myocardial infarction: the BOOST randomised controlled clinical trial. Lancet. 364(9429):141-8.

Woodbury D, Schwarz EJ, Prockop DJ and Black IB. (2000) Adult rat and human bone marrow stromal cells differentiate into neurons. J. Neurosci. Res. 61, 364-370.

Wright DE, Bowman EP, Wagers AJ, Butcher EC, Weissman IL. (2002) Hematopoietic stem cells are uniquely selective in their migratory response to chemokines. J Exp Med. 195(9):1145-54.

Wurmser AE, Gage FH. (2002) Stem cells: cell fusion causes confusion. Nature. 416(6880):485-7.

Yamamoto S, N Yamamoto, T Kitamura and M Nakafuku. (2001). Proliferation of parenchymal neural progenitors in response to injury in the adult rat spinal cord. Exp Neurol 172:1115–1127.

Yoo SW, Kim SS, Lee SY, Lee HS, Kim HS, Lee YD, Suh-Kim H. (2008) Mesenchymal stem cells promote proliferation of endogenous neural stem cells and survival of newborn cells in a rat stroke model. Exp Mol Med. 40(4):387-97.

Zachar V, Rasmussen JG, Fink T. (2011) Isolation and growth of adipose tissue-derived stem cells. Methods Mol Biol. 698:37-49.

Zhang CP, Fu XB. (2008) Therapeutic potential of stem cells in skin repair and regeneration. Chin J Traumatol. 11(4):209-21.

Zhang X, Hirai M, Cantero S, Ciubotariu R, Dobrila L, Hirsh A, Igura K, Satoh H, Yokomi I, Nishimura T, Yamaguchi S, Yoshimura K, Rubinstein P, Takahashi TA. (2011) Isolation and characterization of mesenchymal stem cells from human umbilical cord blood: reevaluation of critical factors for successful isolation and high ability to proliferate and differentiate to chondrocytes as compared to mesenchymal stem cells from bone marrow and adipose tissue. *J Cell Biochem.* 112(4):1206-18.

Zhu J, Wang X, Chen J, Sun J, Zhang F. (2006) Reduced number and activity of circulating endothelial progenitor cells from patients with hyperhomocysteinemia. Arch Med Res. 37(4):484-9.

Zohlnhöfer D, Dibra A, Koppara T, de Waha A, Ripa RS, Kastrup J, Valgimigli M, Schömig A, Kastrati A. (2008) Stem cell mobilization by granulocyte colony-stimulating factor for myocardial recovery after acute myocardial infarction: a meta-analysis. J Am Coll Cardiol. 51(15):1429-37.

Zou Y, Kottmann AH, Kuroda M, Taniuchi I, Littman DR. (1998) Function of the chemokine receptor CXCR4 in haematopoiesis and in cerebellar development. Nature 393:595–599

9

Spermatogonial Stem Cells: An Alternate Source of Pluripotent Stem Cells for Regenerative Medicine

Liz Simon[1], Marie-Claude Hofmann[2] and Paul S. Cooke[3]
[1]*Department of Biomedical Sciences, College of Veterinary Medicine, Nursing and Allied Health, Tuskegee University, Tuskegee, AL*
[2]*Department of Comparative Biosciences, College of Veterinary Medicine, University of Illinois at Urbana-Champaign, Urbana, IL*
[3]*Department of Physiological Sciences, College of Veterinary Medicine, University of Florida, Gainesville, FL*
USA

1. Introduction

There is immense scientific and medical interest in stem cells as a potential source material for regenerative medicine to replace or restore lost, damaged, or aging cells, tissues or organs (Mason and Dunnill 2008). The ideal stem cell candidate for regenerative medicine is a pluripotent stem cell that is easily obtainable, has a stable developmental potential even after prolonged culture, forms derivatives of all three embryonic germ layers from the progeny of a single cell and generates teratomas after injection into immunosuppressed mice (Mason and Dunnill 2008). Stem cell-based therapy has the potential to offer important new treatment options for insulin-dependent diabetes, Parkinson's disease, cardiovascular, renal, musculoskeletal and retinal diseases and spinal cord diseases and trauma, among others. A critical question in this field is to establish which stem cells could be efficiently used clinically.

In 1981, Martin and Evans achieved a milestone in stem cell biology with the derivation of mouse embryonic stem cells (ESCs; Evans and Kaufman 1981). The subsequent derivation of pluripotent human ESCs in 1998 by James Thomson (Thomson et al. 1998) and pluripotent stem cells from human primordial germ cells (Shamblott et al. 1998) ushered in a revolution in the field of regenerative medicine and tissue engineering. However, the destruction of human embryos to obtain ESCs and the need for therapeutic cloning to use them optimally made their clinical application highly controversial. In addition to ethical, legal and moral issues, ESCs have inherent limitations that must be overcome before their clinical use. One prime concern is the potential tumorigenicity of these cells in vivo. Although efforts to eliminate this possibility are underway, it remains a serious issue (Blum and Benvenisty 2008; Fujikawa et al. 2005; Strulovici et al. 2007; Wu, Boyd, and Wood 2007). Other concerns are immune rejection (Drukker et al. 2002; Wobus and Boheler 2005), genetic instability and incomplete epigenetic reprogramming (Wobus and Boheler 2005). These concerns have sparked interest within the scientific community in finding alternate sources of pluripotent/multipotent cells that have similar can as ESCs, but potentially circumvent the problems associated with these cells.

The derivation of induced pluripotent stem cells (iPSCs) through the introduction of a small combination of transcription factors into terminally differentiated cells has raised the possibility of producing an alternate pluripotent stem cell source for use in regenerative medicine. Researchers at Kyoto University, Japan, were first to identify conditions that allowed mouse skin fibroblasts to be induced into pluripotency (Takahashi and Yamanaka 2006) and a year later human skin fibroblasts were also induced to pluripotency using this same approach (Takahashi, Okita et al. 2007; Takahashi, Tanabe et al. 2007; Yu et al. 2007). The iPSC technology has immense potential for clinical therapy as these cells avoid some ethical and moral issues associated with ESCs. The technology involves producing critical levels of proteins needed for pluripotency through introducing the genes into the target cells exogenously using viral vectors (Takahashi, Tanabe et al. 2007; Takahashi and Yamanaka 2006; Yu et al. 2007), adenoviral vectors (Stadtfeld et al. 2008; Okita et al. 2008) or non-viral plasmids (Okita et al. 2008) or introducing the actual recombinant proteins themselves into cells (Zhou et al. 2009). Although initially there was concern over use of viral vectors integrating into the human genome and potentially inducing neoplastic changes, the use of alternate methods to deliver the pluripotency genes appears to circumvent that problem. Nonetheless, because there is genetic manipulation involved (Stadtfeld et al. 2010; Stadtfeld and Hochedlinger 2010) and an increased risk of tumorigenicity (Nakagawa et al. 2008; Ben-David and Benvenisty 2011; Kooreman and Wu 2010) extensive research has to be conducted before iPSCs are used clinically. Since iPSCs can be derived from patient-specific cells, one potential benefit of these cells was a low or no risk of immune rejection. However, a recent finding suggests that this might not be true: In contrast to derivatives of ESCs, abnormal gene expression in iPSCs introduced into syngeneic mice in vivo induced T-cell-dependent immune responses, leading to immune infiltration and ultimately rejection of these cells (Zhao et al. 2011). This raises concerns for use of these cells in regenerative medicine. As with any new technology, there are a number of other hurdles to overcome before using iPSCs clinically, such as refractoriness of many adult cells to reprogramming, transient epigenetic memory of donor cells (Kim et al. 2010; Polo et al. 2010) and possible non-recurrent mutations. Despite these concerns, the tremendous progress and obvious potential of iPSCs over the past few years has overshadowed work on other potentially pluripotent stem cells, although some of these cells may ultimately offer therapeutic potential equal to that of iPSCs.

2. Progress in regenerative medicine

In 2010, two ESC-based human clinical trials were approved by the U. S. Food and Drug Administration, one initiated by Geron Corporation and the other by Advanced Cell Technology. Though the clinical trials are underway, there is concern that introducing stem cells that have not transformed into specialized cells into patients may pose the risk of teratoma formation. However, remarkable progress has been made in cell-based regenerative medicine over the past decade. The cell-therapy based industry now has an annual revenue of over a billion dollars and it is projected to rise to $5.1 billion by 2014

(Mason and Manzotti 2010). The initial research and monetary investment to develop stem-cell based therapy is high. However, since the treatment involves transplantation of cells or tissues that can function normally for extended periods of time or even potentially for the remainder of a patient's life, there is negligible need for ongoing treatment, rendering this modality of treatment potentially cost-effective (Mason and Dunnill 2008).

The limitations and problems associated with iPSCs suggest that they may not offer the optimal solution for all aspects of regenerative medicine, raising the question of whether other stem cells may be preferable in some or potentially all regenerative medicine applications. Other stem cells that have received widespread attention in this regard include hematopoietic (HSC), mesenchymal and fetal cord blood stem cells. An adult allogenic stem cell source with developmental potential similar to ESCs and iPSCs would potentially be the most promising clinical approach if an optimal starting cell was found.

Another stem cell that has been studied extensively for years and has recently been shown to have potential in regenerative medicine is the spermatogonial stem cell (SSC). This chapter focuses on recent developments in this field and the advantages and limitations of SSC use in regenerative medicine. We also briefly discuss other multipotent stem cells that may have significant clinical use.

3. Spermatogonial Stem Cells

In mammalian testes, SSCs are found along the basement membrane of seminiferous tubules. They produce the spermatogenic lineage, ensuring lifelong fertility of the individual. Like other stem cells, SSCs are undifferentiated and capable of self-renewal. While in the testicular microenvironment, they differentiate only into one specialized cell lineage, spermatozoa. However, when SSCs are isolated from the testis and placed into a different environment, they acquire or manifest pluripotency and differentiate into tissues belonging to all three embryonic germ layers (Kanatsu-Shinohara et al. 2008; Golestaneh et al. 2009; Simon et al. 2009; Ning et al. 2010). This property makes SSCs a powerful potential source of cells for regenerative therapy.

3.1 Development and differentiation potential of SSCs

In the embryo, the germ line arises from primordial germ cells (PGCs). PGCs are initially identifiable as a small cluster of cells in the proximal epiblast near the extra-embryonic ectoderm. This lineage differentiates under the influence of bone morphogenetic proteins (BMPs) and diverges from the somatic lineage in the late embryonic and early fetal stage (Lawson et al. 1999; Ying et al. 2000). Due to their extra-embryonal origin, they are not classified as belonging to a particular embryonic germ layer. Therefore, they are not subjected to many of the differentiation signals other stem cells receive (Lawson and Hage 1994; Simon, Hess, and Cooke 2010); this may allow them to remain more undifferentiated than other stem cells. Following their initial development, PGCs subsequently migrate into the mesoderm, the endoderm (hindgut) and across the dorsal mesentery to reach the developing gonads at about 4-5 weeks of gestation in humans and 11-13 day post-coitum in rodents (Culty 2009). In the testis, PGCs then become mitotically quiescent until birth, and are called gonocytes. Shortly after birth, gonocytes resume mitosis and migrate to the basement membrane of seminiferous tubules, where they form SSCs and remain throughout life. Thus, the embryological origin of SSCs is unique. This may facilitate their potential for differentiation into cell types of different germ cell layers and underlines their clinical potential for regenerative medicine. Furthermore, teratomas occur exclusively in gonads (Stevens 1964) and are of germ cell origin, and gene expression in early germ cells is very similar to ESCs (Zwaka and Thomson 2005; Simon, Hess, and Cooke 2010) , which emphasizes the broad developmental potential of SSCs. SSCs express genes such as POU domain class 5, transcription factor 1 (Pou5f1; Huang et al. 2009; Bhartiya et al.), Lin 28 (Zheng et al. 2009), undifferentiated embryonic cell transcription factor 1 (UTF-1) and Zinc finger protein 42 (Rex-1; Kristensen et al. 2008), which impart pluripotency. However, expression of another pluripotency gene, Nanog, is repressed in the testis by transformation related protein 53 (TRP53) and phosphatase and tensin homolog (PTEN). Both proteins belong to a critical signaling pathway preventing SSCs from being pluripotent while in the testis (Kuijk et al. 2009). Overall, this gene expression pattern suggests that SSCs have a gene profile similar to ESCs and thus are more undifferentiated than other adult stem cells.

3.2 Current methods of isolation and propagation of SSCs with emphasis on human SSC isolation

Spermatogonial stem cell constitute only 0.03% of the total germ cell population in rodent and human testis (Tegelenbosch and de Rooij 1993). Thus, their small numbers and the lack of specific markers are the main hurdles to their characterization. Nonetheless, significant progress has recently been made in the isolation and propagation of cells with SSCs properties from rodent (Dym et al. 2009; Guan et al. 2009; Kanatsu-Shinohara, Takehashi, and Shinohara 2008; Kubota, Avarbock, and Brinster 2004; Oatley and Brinster 2006, 2008; Ogawa et al. 2003; Hofmann et al. 2005; Kanatsu-Shinohara et al. 2010) and human testes (Conrad et al. 2008; Glaser et al. 2008; Golestaneh et al. 2009; Kossack et al. 2009; Zovoilis et al. 2008; Izadyar et al. 2011; He et al. 2009). Human testicular tissue is currently obtained from testicular biopsies (Izadyar et al. 2011; Kossack et al. 2009), orchiectomies (Izadyar et al. 2011) and organ donors (Golestaneh et al. 2009). Testicular biopsies of approximately one gram can yield sufficient number of human SSCs for most clinical applications. Although our knowledge of mouse and human SSCs phenotype is still limited, studies suggest that human SSCs express proteins such as cluster of differentiation antigens 49f, 90 and 133 (CD49f, CD90, and CD133, respectively), glial cell line-derived neurotrophic factor family receptor alpha 1 (GFRA1), G protein-coupled receptor 125 (GPR125), melanoma antigen family A 4 (MAGE4), promyelocytic leukemia zinc finger (PLZF) and stage-specific embryonic antigen-4 (SSEA-4; Costoya et al. 2004; Conrad et al. 2008; He et al. 2009; Izadyar et al. 2011). Using these markers in magnetic- or fluorescent-activated cell sorting, SSCs can be isolated with a high degree of purity (Gassei et al. 2009; Izadyar et al. 2011; Kokkinaki et al. 2009; Simon et al. 2010). Although there have been no definitive culture conditions for propagation of either mouse or human SSCs, culture systems established by different groups seem to be conducive for their propagation. At least in rodents, glial cell line-derived neurotrophic factor (GDNF) was found to be essential to maintain SSCs in an undifferentiated state in vivo and in vitro (Tyagi et al. 2009; Hofmann 2008; Sariola and Immonen 2008; Oatley, Avarbock, and Brinster 2007; Oatley et al. 2006; Naughton et al. 2006; Kubota, Avarbock, and Brinster 2004; Meng et al. 2000).

3.3 Spermatogonial stem cells are pluripotent

In the testicular microenvironment, SSCs produce only the spermatogenic lineage, and the assumption was that this was their sole potential developmental fate. However, the iconoclastic finding of Kanatsu-Shinohara et al. (Kanatsu-Shinohara et al. 2004) demonstrated that neonatal murine SSCs produced ESC-like cells when isolated from the testis and grown for extended periods in ESC culture conditions. Although initial work was done with neonatal SSCs (Kanatsu-Shinohara et al. 2004; Kanatsu-Shinohara et al. 2008; Simon et al. 2009), subsequent work showed that adult murine SSCs (Glaser et al. 2008; Izadyar et al. 2008; Seandel et al. 2007; Guan et al. 2006) grown for 4-7 weeks in vitro also produced a low frequency of ESC-like colonies. These ESC-like cells can give rise to cell types derived from all three embryonic germ layers, and also produce teratomas when injected subcutaneously into nude mice. In addition, these ESC-like cells contribute to embryonic development when injected into blastocysts. ESC-like cells derived from SSCs have been termed multipotent germline stem cells (mGSCs) and these cells have been differentiated into many cell types. For example, two separate groups demonstrated that mGSCs could differentiate into mature cardiac and endothelial cells and that these cardiac cells were contractile and had electric potentials and ion channels (Baba et al. 2007; Guan et al. 2007). mGSCs derived from adult mouse SSCs could be differentiated into functional neurons and glia (Glaser et al. 2008; Streckfuss-Bomeke et al. 2009). Pluripotent stem cells were derived from adult mGSCs that not only could differentiate into a variety of cell types both in vivo and in vitro, but also showed germline transmission to the next generation when injected into blastocysts (Ko et al. 2009). Moreover, mGSCs have pluripotency characteristics similar to ESCs such as telomerase activity, telomere length and hypomethylation of pluripotency marker genes (Zechner et al. 2009)

In recent years, several research groups reported methodologies for isolation and culture of human SSCs and also demonstrated that these cells were pluripotent/multipotent (Conrad et al. 2008; Golestaneh et al. 2009; Kossack et al. 2009; Dym et al. 2009; Izadyar et al. 2011). SSCs isolated from human testicular tissues and cultured for a week or more spontaneously produced ESC-like small colonies, which were then transferred into ESC media and cultured for about 4 weeks to get sufficient numbers of ESC-like colonies and these ESC-like cells could then be differentiated into specific cell types. These results indicate that human SSCs have the potential to be used as an alternate source of pluripotent stem cells in regenerative cell therapy, without the ethical concerns of ESC and the concerns involving exogenous gene introduction in iPSCs.

4. Direct differentiation of spermatogonial stem cells

4.1 Epithelial-mesenchymal interactions

Epithelial-mesenchymal interactions are critical for organogenesis in many organs such as lung, prostate, mammary gland, liver, pancreas and salivary glands (Grobstein 1953). Mesenchyme, which is undifferentiated connective tissue, signals to the epithelium to induce epithelial morphogenesis and cytodifferentiation in a wide variety of organs. The central role of epithelial-mesenchymal interactions in organ development was postulated initially by Pander and later experimentally demonstrated by Spemann and Saunders (in Simon, Hess, and Cooke 2010).

Classical tissue recombination experiments conducted by Cunha and coworkers with reproductive tissue demonstrated that the mesenchyme dictates the fate of the epithelium in various reproductive tissues. The urogenital sinus is an ambisexual fetal organ that gives rise to the prostate in males and a portion of the vagina in females. Under the influence of androgen, urogenital sinus mesenchyme (UGM) induces urogenital sinus epithelium (UGE) to differentiate into prostatic epithelium and UGM also regulates epithelial ductal morphogenesis and cytodifferentiation in prostate. UGM also instructively induces prostatic morphogenesis in other epithelia (Cunha et al. 1983; Cunha, Lung, and Reese 1980; Cunha, Sekkingstad, and Meloy 1983). For example, UGM instructively induces adult bladder epithelium to form prostatic epithelium in tissue recombinants in vivo (Cunha et al. 1983). Similarly, neonatal uterine mesenchyme instructively induces neonatal vaginal epithelium to form uterine epithelium (Cunha 1976).

4.1.1 Potential signaling pathways involved in epithelial-mesenchymal interactions

Although mesenchymal paracrine signaling is essential for determining epithelial fate, the specific signaling pathway(s) involved in this phenomenon remain unclear. Some possible signaling molecules are Wnt7a (Miller and Sassoon 1998), hepatocyte growth factor (HGF) and fibroblast growth factor (FGF) family molecule(s) in the uterus (Chen, Spencer, and Bazer 2000, 2000), HGF and FGF family molecule(s) for lung organogenesis (Ohmichi et al. 1998) (for review see Kumar et al. 2005,) and FGF and Wnt and Hedgehog signaling in the prostate (for review see Thomson, Cunha, and Marker 2008; Cunha, Cooke, and Kurita 2004; Lin and Wang, 2010; Taylor et al. 2009).

4.1.2 Mesenchyme dictates the fate of stem cell differentiation

The demonstrations that epithelial-mesenchymal interactions are essential for organogenesis also triggered interest in using this approach in regenerative therapy. ESCs were differentiated into prostatic epithelium by recombining mouse UGM and human ESCs and growing these tissue recombinations in vivo (Taylor et al. 2006). Using a similar methodology (Oottamasathien et al. 2007; Anumanthan et al. 2008), human ESCs or bone marrow-derived mesenchymal stem cells have been differentiated into bladder epithelium by exposing these cells to the inductive influence of bladder mesenchyme in a tissue recombinant. More recently, Taylor et al. demonstrated that stroma could induce adult stem cells to express dual phenotypes (Taylor et al. 2009). Prostatic stroma induced putative mammary epithelial stem cells to generate glandular epithelia expressing both prostatic and mammary markers. These results demonstrate that the mesenchyme can instructively direct the differentiation of ESCs or other stem cells into a specific cell fate.

4.2 Spermatogonial stem cells differentiate into tissues of all three embryonic germ layers in response to instructive inducers

Based on demonstrations of the importance of the stem cell niche (Tyagi et al. 2009; de Rooij 2009; Hess et al. 2006; Simon et al. 2010; Oatley, Racicot, and Oatley 2010), the pluripotential nature of SSCs and the instructive potential of various mesenchymes, we postulated and subsequently demonstrated that neonatal mouse SSCs could directly differentiate into prostatic, uterine and skin epithelium (Simon et al. 2009) when recombined with the appropriate mesenchyme and grafted in vivo (Fig. 1). To track cell lineages derived from SSCs in tissue recombinations and verify that these cells were undergoing differentiation, wt transgenic C57BL/6 mice expressing enhanced green fluorescent protein (GFP) ubiquitously were used. For example, UGM derived from wild-type mice (wt-UGM) were recombined with SSCs derived from mice expressing GFP (G-SSC) and grafted under the renal capsule of syngeneic male hosts. After 4 weeks of growth, the epithelium in these, wt-UGM + G-SSC, tissue recombinants expressed NKX3.1, a prostatic epithelial marker and androgen receptor but not germ cell nuclear antigen 1 (GCNA1), a germ cell marker. The tissue recombinants had an epithelium that stained intensely for GFP (Fig. 1A), indicating that it was of SSC origin, while stromal cells lacked GFP staining. Similarly uterine mesenchyme (UtM) from mice expressing GFP (G-UtM) recombined with SSCs derived from wt-mice (wt-SSC) differentiated into uterine epithelium that expressed cytokeratin 8 (CK8), estrogen receptor-alpha and progesterone receptor. In these G-UtM + wt-SSC tissue recombinants, stromal cells strongly expressed GFP, while epithelium did not, indicating that the epithelium was of SSC origin (Fig. 1B). This approach provides a method to directly differentiate SSCs into specific cell types from all three embryonic germ layers without the extended culture period needed in vitro to produce ESC-like cells that can subsequently differentiate into various derivatives.

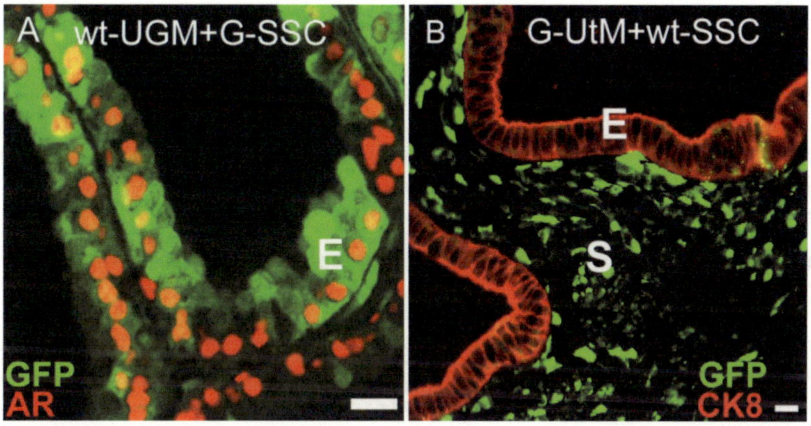

Fig. 1. Differentiation of SSCs into prostatic and uterine epithelium. A) Urogenital sinus mesenchyme (UGM) from wild-type (wt) mice was recombined with SSCs derived from mice that expressed enhanced green fluorescent protein (G) ubiquitously and was grafted under the renal capsule of syngeneic male hosts. After 4 weeks of in vivo growth, the wt-UGM + G-SSC tissue recombinants formed prostatic epithelium (E) that was of SSC origin expressing GFP and androgen receptor (AR, red nuclei). B) Uterine mesenchyme (UtM) from GFP mice was recombined with SSCs derived from wt mice and grafted under the renal capsule of syngeneic female hosts. After 4 weeks of growth, the G-UtM + wt-SSC tissue recombinants formed uterine epithelium (E) that was of SSC origin expressing cytokeratin 8 (CK8). Stromal cells (S) express GFP but E does not. SSCs do not express AR or CK8.

When SSCs were mixed with mammary epithelial cells and grafted into the mammary fat pad in vivo, SSCs differentiated into mammary epithelial cells, but the stem cells alone could not be differentiated into mammary cells (Boulanger et al. 2007). This suggests that the inductive potential of the adult mammary fat pad might not be sufficient for directing the differentiation of SSCs into mammary epithelial cells. This is a potential hurdle for the use of this methodology, as the inductive mesenchyme for most of the organs is present during fetal life, and as they transition to form adult stromal cells they might lose this ability to instructively induce other epithelia. However, a recent study demonstrated direct differentiation of SSCs into hematopoietic cells (Ning et al., 2010). When SSCs were injected into the bone marrow of adult female mice, these mice had Y chromosome positive-cells that had phenotypical and functional characteristics of hematopoietic cells both in vivo and in vitro, emphasizing both the pluripotential nature of SSCs and that the microenvironment even in adult organs plays a decisive role in directing the differentiation of SSCs.

As discussed earlier, our understanding of the mechanistic basis of epithelial-mesenchymal interactions is fragmentary. Fetal mesenchymes are potent instructive inducers, and one of the challenges for using this methodology will be to determine whether or not adult stroma is capable of similar instructive inductions. Once we are able to dissect out the molecular mechanisms used by mesenchyme to instructively induce epithelial morphogenesis, it will be feasible to expose SSCs to the inductive signals produced by these tissues in vitro, on artificial scaffolds or some other arrangement. Another major limitation of this methodology is that it is inapplicable to tissues that do not involve epithelial-mesenchymal interactions during development. However, the direct differentiation of SSCs into hematopoietic cells (Ning et al., 2010) discussed above suggests that SSCs will differentiate into a specific tissue type when exposed to an appropriate microenvironment, even in the absence of epithelial-mesenchymal interactions. Thus further research is necessary to determine both the full developmental potential of SSCs and the most appropriate methodology for inducing specific cell types, but SSCs appear to have great potential in this regard.

4.3 Mechanism of differentiation of spermatogonial stem cells into other cell types

The mechanism of differentiation of SSCs into other cell types under the influence of instructive inducers is poorly understood. One possibility is that in response to an inductive mesenchyme, the SSCs de-differentiate into ESC-like cells (as has been reported in vitro), and then subsequently differentiate into a new epithelia. However, our preliminary studies indicate that SSCs may not undergo a de-differentiation step in the presence of an inductive mesenchyme, but instead may differentiate directly from SSCs to another epithelial type without going through an intermediate ESC-like cell stage. Shinohara and coworkers (Kanatsu-Shinohara et al. 2008) demonstrated that a single spermatogonial stem cell could produce an embryonic stem-like line that was multipotent and germline stem cells that were committed to spermatogenesis, indicating that all SSCs may be capable of becoming pluripotent. This is supported by other studies (Ko et al. 2009). Conversely, Izadyar et al. suggested that there are two distinct populations of SSCs, one that is $OCT4^{+}$ and $c\text{-}KIT^{-}$ that gives rise to multipotent cells and another that is $OCT4^{+}$ and $c\text{-}KIT^{+}$ that gives rise to the spermatogenic lineage (Izadyar et al. 2008). A definitive elucidation of how SSCs differentiate into other tissue types, as well as definitively establishing whether all SSCs or a specific subpopulation can be converted into other tissues, will be an essential prerequisite for successful use of this approach in a clinical setting. Nonetheless, SSCs can be differentiated into specific cell types using different approaches and be a potential source for pluripotent cells for stem cell-based therapy (Fig. 2).

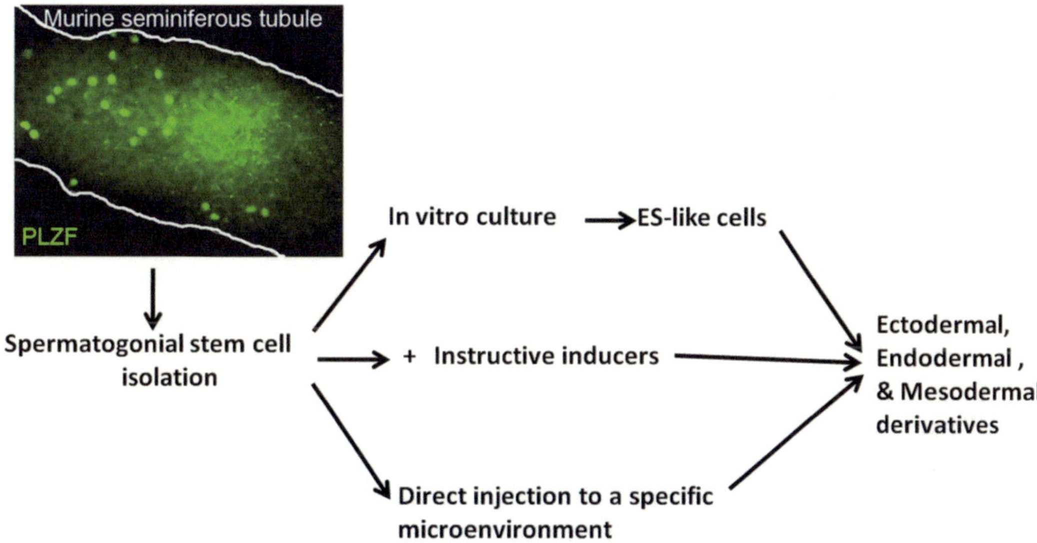

Fig. 2. Potential use of spermatogonial stem cells (SSCs) in regenerative medicine. Spermatogonia in the murine seminiferous tubule expressing PLZF, a SSC marker. SSCs are isolated from the testis using magnetic- or fluorescent-activated cell sorting. Some of the possible methods of differentiating SSCs into specific cell types are 1) Long-term culture of SSCs, selection for cells that form embryonic stem cell-like colonies and subsequent culture and differentiation of these pluripotent cells into specific cell types. 2) Recombination of SSCs with instructive inducers to directly differentiate SSCs into specific epithelial cell types; 3) Direct injection to a specific site of injury or specific microenvironment. Since SSCs are unspecialized, they could home and respond to signals in the new microenvironment and differentiate into cells of all three embryonic germ layers.

5. Advantages and limitations of SSCs over other pluripotent stem cell sources

SSCs, as well as other adult derived stem cells, may be safer to use therapeutically than ESCs or iPSCs. Since SSCs are more differentiated than ESCs, they are less likely to induce teratomas (Kossack et al. 2009). However, the risk of malignant transformation cannot be totally rejected since SSCs are relatively more undifferentiated than other adult stem cells. This type of problem is illustrated by the formation of brain tumors from donor-derived cells in patients who received fetal stem cells for treatment of ataxia-telangiectasia (Amariglio et al. 2009), emphasizing that tumorigenicity is the biggest impediment for the use of pluripotent stem cells in cell therapy. Another obstacle is the immunogenicity of SSCs and potential risk of rejection of the cells (Dressel et al. 2009). But immune rejection of autologous or allogenic stem cell transplants can be minimized by routine immunosuppression treatments as is used for organ transplantation. Another possible alternative is genetic manipulation of stem cells and elimination of genes responsible for immune rejection (Wobus and Boheler 2005).

6. Alternate sources of stem cells for use in regenerative medicine

The use of stem cells in regenerative medicine began when bone marrow cells were transplanted to treat acute leukemia (Thomas et al. 1959), and presently HSC therapy is the only stem cell therapy widely used clinically (Helmy et al. 2010). There is growing evidence that HSCs are plastic—and can give rise to tissues other than those of the blood system, e.g., liver cells (Lagasse et al. 2000). Another readily available source is mesenchymal stem cells that can be isolated from adult tissue, fetal tissue and umbilical cord blood. These cells can differentiate into osteoblasts, chondrocytes and adipocytes (Friedenstein et al. 1974; Pittenger et al. 1999) neurons (Cho et al. 2005), astrocytes (Kopen, Prockop, and Phinney 1999) and hepatocyte-like cells (Petersen et al. 1999). The use of adipose stem cells as a source for cell therapy is increasing rapidly as methods of isolation and culture are standardized; subcutaneous depots are easily accessible, replenishable and are often abundant. Adult stem cells derived from adipose tissues can differentiate in vitro into many cell types including adipocyte, chondrocyte, endothelial, epithelial, hematopoietic support, hepatocyte, neuronal, myogenic, and osteoblast lineages (Gimble and Guilak 2003; Halvorsen et al. 2001; Safford et al. 2002; Zuk et al. 2001). Fetal stem cells are self-renewing cells located in various types of fetal tissue, including umbilical cord blood, umbilical cord matrix, fetal blood and the amniotic membrane (Reinisch and Strunk 2009; Jager et al. 2009; Zeddou et al., 2010). Umbilical cord blood contain multiple populations of stem cells that can be effective in treating many diseases such as hematological malignancies, hemoglobinopathies, metabolic disorders and the greatest advantage of these cells is decreased immune rejection (Liao et al., 2011).

Stem Cells	Advantages	Major limitations
Embryonic stem cells	Pluripotent Indefinite self-renewal potential	Ethical concerns Tumorigenicity Therapeutic cloning involved
Induced pluripotent cells	Pluripotent Initial source of cells are easy to obtain No ethical or moral concerns Indefinite self-renewal potential	Introduction of exogenous genes Tumorigenicity Genetic instability Use of viral vectors to introduce genes
Spermatogonial stem cells	Pluripotent No ethical or moral concerns Relatively easy to obtain Less tumorigenic potential	Relatively small numbers Difficult to maintain in cultures
Fetal stem cells (Fetal cord blood, umbilical cord tissue)	Relatively easy to obtain Minimal immunorejection during autologous transplantation	Restricted differentiation and self-renewal potential
Adult stem cells (hematopoietic, mesenchymal, adipose)	Easy to obtain Reduced risk of tumorigenicity	Restricted differentiation and self-renewal potential

Table 1. Potential advantages and limitations of stem cells in regenerative medicine

7. Conclusions

Among pluripotent/multipotent stem cells (Table 1), spermatogonial stem cells have great potential and some unique advantages. Despite their promise, numerous hurdles must be overcome before clinical use of SSCs. The small population of SSCs in the testis and the difficulty in propagating and maintaining them in culture is one major hurdle. The methodology proposed here is promising but extensive work is needed before its application in regenerative medicine.

8. References

Amariglio, N., A. Hirshberg, B. W. Scheithauer, Y. Cohen, R. Loewenthal, L. Trakhtenbrot, N. Paz, M. Koren-Michowitz, D. Waldman, L. Leider-Trejo, A. Toren, S. Constantini, and G. Rechavi. 2009. Donor-derived brain tumor following neural stem cell transplantation in an ataxia telangiectasia patient. *PLoS Med* 6 (2):e1000029.

Anumanthan, G., J. H. Makari, L. Honea, J. C. Thomas, M. L. Wills, N. A. Bhowmick, M. C. Adams, S. W. Hayward, R. J. Matusik, J. W. Brock, and J. C. Pope. 2008. Directed differentiation of bone marrow derived mesenchymal stem cells into bladder urothelium. *J Urol* 180 (4 Suppl):1778-83.

Baba, S., T. Heike, K. Umeda, T. Iwasa, S. Kaichi, Y. Hiraumi, H. Doi, M. Yoshimoto, M. Kanatsu-Shinohara, T. Shinohara, and T. Nakahata. 2007. Generation of cardiac and endothelial cells from neonatal mouse testis-derived multipotent germline stem cells. *Stem Cells* 25 (6):1375-83.

Ben-David, U., and N. Benvenisty. 2011. The tumorigenicity of human embryonic and induced pluripotent stem cells. *Nat Rev Cancer* 11 (4):268-77.

Bhartiya, D., S. Kasiviswanathan, S. K. Unni, P. Pethe, J. V. Dhabalia, S. Patwardhan, and H. B. Tongaonkar 2010. Newer insights into premeiotic development of germ cells in adult human testis using Oct-4 as a stem cell marker. *J Histochem Cytochem* 58 (12):1093-106.

Blum, B., and N. Benvenisty. 2008. The tumorigenicity of human embryonic stem cells. *Adv Cancer Res* 100:133-58.

Boulanger, C. A., D. L. Mack, B. W. Booth, and G. H. Smith. 2007. Interaction with the mammary microenvironment redirects spermatogenic cell fate in vivo. *Proc Natl Acad Sci U S A* 104 (10):3871-6.

Chen, C., T. E. Spencer, and F. W. Bazer. 2000. Expression of hepatocyte growth factor and its receptor c-met in the ovine uterus. *Biol Reprod* 62 (6):1844-50.

Chen, C., T. E. Spencer, and F. W. Bazer. 2000. Fibroblast growth factor-10: a stromal mediator of epithelial function in the ovine uterus. *Biol Reprod* 63 (3):959-66.

Cho, K. J., K. A. Trzaska, S. J. Greco, J. McArdle, F. S. Wang, J. H. Ye, and P. Rameshwar. 2005. Neurons derived from human mesenchymal stem cells show synaptic transmission and can be induced to produce the neurotransmitter substance P by interleukin-1 alpha. *Stem Cells* 23 (3):383-91.

Conrad, S., M. Renninger, J. Hennenlotter, T. Wiesner, L. Just, M. Bonin, W. Aicher, H. J. Buhring, U. Mattheus, A. Mack, H. J. Wagner, S. Minger, M. Matzkies, M. Reppel, J. Hescheler, K. D. Sievert, A. Stenzl, and T. Skutella. 2008. Generation of pluripotent stem cells from adult human testis. *Nature* 456 (7220):344-9.

Costoya, J. A., R. M. Hobbs, M. Barna, G. Cattoretti, K. Manova, M. Sukhwani, K. E. Orwig, D. J. Wolgemuth, and P. P. Pandolfi. 2004. Essential role of Plzf in maintenance of spermatogonial stem cells. *Nat Genet* 36 (6):653-9.

Culty, M. 2009. Gonocytes, the forgotten cells of the germ cell lineage. *Birth Defects Res C Embryo Today* 87 (1):1-26.

Cunha, G. R. 1976. Stromal induction and specification of morphogenesis and cytodifferentiation of the epithelia of the Mullerian ducts and urogenital sinus during development of the uterus and vagina in mice. *J Exp Zool* 196 (3):361-70.

Cunha, G. R., P. S. Cooke, and T. Kurita. 2004. Role of stromal-epithelial interactions in hormonal responses. *Arch Histol Cytol* 67 (5):417-34.

Cunha, G. R., H. Fujii, B. L. Neubauer, J. M. Shannon, L. Sawyer, and B. A. Reese. 1983. Epithelial-mesenchymal interactions in prostatic development. I. morphological observations of prostatic induction by urogenital sinus mesenchyme in epithelium of the adult rodent urinary bladder. *J Cell Biol* 96 (6):1662-70.

Cunha, G. R., B. Lung, and B. Reese. 1980. Glandular epithelial induction by embryonic mesenchyme in adult bladder epithelium of BALB/c mice. *Invest Urol* 17 (4):302-4.

Cunha, G. R., M. Sekkingstad, and B. A. Meloy. 1983. Heterospecific induction of prostatic development in tissue recombinants prepared with mouse, rat, rabbit and human tissues. *Differentiation* 24 (2):174-80.

de Rooij, D. G. 2009. The spermatogonial stem cell niche. *Microsc Res Tech* 72 (8):580-5.

Dressel, R., K. Guan, J. Nolte, L. Elsner, S. Monecke, K. Nayernia, G. Hasenfuss, and W. Engel. 2009. Multipotent adult germ-line stem cells, like other pluripotent stem cells, can be killed by cytotoxic T lymphocytes despite low expression of major histocompatibility complex class I molecules. *Biol Direct* 4 (1):31.

Drukker, M., G. Katz, A. Urbach, M. Schuldiner, G. Markel, J. Itskovitz-Eldor, B. Reubinoff, O. Mandelboim, and N. Benvenisty. 2002. Characterization of the expression of MHC proteins in human embryonic stem cells. *Proc Natl Acad Sci U S A* 99 (15):9864-9.

Dym, M., Z. He, J. Jiang, D. Pant, and M. Kokkinaki. 2009. Spermatogonial stem cells: unlimited potential. *Reprod Fertil Dev* 21 (1):15-21.

Evans, M. J., and M. H. Kaufman. 1981. Establishment in culture of pluripotential cells from mouse embryos. *Nature* 292 (5819):154-6.

Friedenstein, A. J., R. K. Chailakhyan, N. V. Latsinik, A. F. Panasyuk, and I. V. Keiliss-Borok. 1974. Stromal cells responsible for transferring the microenvironment of the hemopoietic tissues. Cloning in vitro and retransplantation in vivo. *Transplantation* 17 (4):331-40.

Fujikawa, T., S. H. Oh, L. Pi, H. M. Hatch, T. Shupe, and B. E. Petersen. 2005. Teratoma formation leads to failure of treatment for type I diabetes using embryonic stem cell-derived insulin-producing cells. *Am J Pathol* 166 (6):1781-91.

Gassei, K., J. Ehmcke, R. Dhir, and S. Schlatt. 2009. Magnetic activated cell sorting allows isolation of spermatogonia from adult primate testes and reveals distinct GFRa1-positive subpopulations in men. *J Med Primatol* 39 (2):83-91.

Gimble, J. M., and F. Guilak. 2003. Differentiation potential of adipose derived adult stem (ADAS) cells. *Curr Top Dev Biol* 58:137-60.

Glaser, T., T. Opitz, T. Kischlat, R. Konang, P. Sasse, B. K. Fleischmann, W. Engel, K. Nayernia, and O. Brustle. 2008. Adult germ line stem cells as a source of functional neurons and glia. *Stem Cells* 26 (9):2434-43.

Golestaneh, N., M. Kokkinaki, D. Pant, J. Jiang, D. DeStefano, C. Fernandez-Bueno, J. D. Rone, B. R. Haddad, G. I. Gallicano, and M. Dym. 2009. Pluripotent stem cells derived from adult human testes. *Stem Cells Dev* 18 (8):1115-26.

Grobstein, C. 1953. Inductive epitheliomesenchymal interaction in cultured organ rudiments of the mouse. *Science* 118 (3054):52-5.

Guan, K., K. Nayernia, L. S. Maier, S. Wagner, R. Dressel, J. H. Lee, J. Nolte, F. Wolf, M. Li, W. Engel, and G. Hasenfuss. 2006. Pluripotency of spermatogonial stem cells from adult mouse testis. *Nature* 440 (7088):1199-203.

Guan, K., S. Wagner, B. Unsold, L. S. Maier, D. Kaiser, B. Hemmerlein, K. Nayernia, W. Engel, and G. Hasenfuss. 2007. Generation of functional cardiomyocytes from adult mouse spermatogonial stem cells. *Circ Res* 100 (11):1615-25.

Guan, K., F. Wolf, A. Becker, W. Engel, K. Nayernia, and G. Hasenfuss. 2009. Isolation and cultivation of stem cells from adult mouse testes. *Nat Protoc* 4 (2):143-54.

Halvorsen, Y. D., D. Franklin, A. L. Bond, D. C. Hitt, C. Auchter, A. L. Boskey, E. P. Paschalis, W. O. Wilkison, and J. M. Gimble. 2001. Extracellular matrix mineralization and osteoblast gene expression by human adipose tissue-derived stromal cells. *Tissue Eng* 7 (6):729-41.

He, Z., M. Kokkinaki, J. Jiang, I. Dobrinski, and M. Dym. 2009. Isolation, characterization, and culture of human spermatogonia. *Biol Reprod* 82 (2):363-72.

Helmy, K. Y., S. A. Patel, K. Silverio, L. Pliner, and P. Rameshwar. 2010. Stem cells and regenerative medicine: accomplishments to date and future promise. *Ther Deliv* 1 (5):693-705.

Hess, R. A., P. S. Cooke, M. C. Hofmann, and K. M. Murphy. 2006. Mechanistic insights into the regulation of the spermatogonial stem cell niche. *Cell Cycle* 5 (11):1164-70.

Hofmann, M. C. 2008. Gdnf signaling pathways within the mammalian spermatogonial stem cell niche. *Mol Cell Endocrinol* 288 (1-2):95-103.

Hofmann, M.C., L. Braydich-Stolle, and M. Dym. 2005. Isolation of male germ-line stem cells; influence of GDNF. *Dev Biol* 279 (1):114-124.

Huang, Y. H., C. C. Chin, H. N. Ho, C. K. Chou, C. N. Shen, H. C. Kuo, T. J. Wu, Y. C. Wu, Y. C. Hung, C. C. Chang, and T. Y. Ling. 2009. Pluripotency of mouse spermatogonial stem cells maintained by IGF-1- dependent pathway. *FASEB J* 23 (7):2076-87.

Izadyar, F., F. Pau, J. Marh, N. Slepko, T. Wang, R. Gonzalez, T. Ramos, K. Howerton, C. Sayre, and F. Silva. 2008. Generation of multipotent cell lines from a distinct population of male germ line stem cells. *Reproduction* 135 (6):771-84.

Izadyar, F., J. Wong, C. Maki, J. Pacchiarotti, T. Ramos, K. Howerton, C. Yuen, S. Greilach, H. H. Zhao, M. Chow, Y. C. Chow, J. Rao, J. Barritt, N. Bar-Chama, and A. Copperman. 2011. Identification and characterization of repopulating spermatogonial stem cells from the adult human testis. *Hum Reprod* 26 (6):1296-306.

Jager, M., C. Zilkens, B. Bittersohl, and R. Krauspe. 2009. Cord blood--an alternative source for bone regeneration. *Stem Cell Rev* 5 (3):266-77.

Kanatsu-Shinohara, M., K. Inoue, J. Lee, M. Yoshimoto, N. Ogonuki, H. Miki, S. Baba, T. Kato, Y. Kazuki, S. Toyokuni, M. Toyoshima, O. Niwa, M. Oshimura, T. Heike, T.

Nakahata, F. Ishino, A. Ogura, and T. Shinohara. 2004. Generation of pluripotent stem cells from neonatal mouse testis. *Cell* 119 (7):1001-12.

Kanatsu-Shinohara, M., K. Inoue, N. Ogonuki, H. Morimoto, A. Ogura, and T. Shinohara. 2010. Serum- and feeder-free culture of mouse germline stem cells. *Biol Reprod* 84 (1):97-105.

Kanatsu-Shinohara, M., J. Lee, K. Inoue, N. Ogonuki, H. Miki, S. Toyokuni, M. Ikawa, T. Nakamura, A. Ogura, and T. Shinohara. 2008. Pluripotency of a single spermatogonial stem cell in mice. *Biol Reprod* 78 (4):681-7.

Kanatsu-Shinohara, M., M. Takehashi, and T. Shinohara. 2008. Brief history, pitfalls, and prospects of mammalian spermatogonial stem cell research. *Cold Spring Harb Symp Quant Biol* 73:17-23.

Kim, K., A. Doi, B. Wen, K. Ng, R. Zhao, P. Cahan, J. Kim, M. J. Aryee, H. Ji, L. I. Ehrlich, A. Yabuuchi, A. Takeuchi, K. C. Cunniff, H. Hongguang, S. McKinney-Freeman, O. Naveiras, T. J. Yoon, R. A. Irizarry, N. Jung, J. Seita, J. Hanna, P. Murakami, R. Jaenisch, R. Weissleder, S. H. Orkin, I. L. Weissman, A. P. Feinberg, and G. Q. Daley. 2010. Epigenetic memory in induced pluripotent stem cells. *Nature* 467 (7313):285-90.

Ko, K., N. Tapia, G. Wu, J. B. Kim, M. J. Bravo, P. Sasse, T. Glaser, D. Ruau, D. W. Han, B. Greber, K. Hausdorfer, V. Sebastiano, M. Stehling, B. K. Fleischmann, O. Brustle, M. Zenke, and H. R. Scholer. 2009. Induction of pluripotency in adult unipotent germline stem cells. *Cell Stem Cell* 5 (1):87-96.

Kokkinaki, M., T. L. Lee, Z. He, J. Jiang, N. Golestaneh, M. C. Hofmann, W. Y. Chan, and M. Dym. 2009. The molecular signature of spermatogonial stem/progenitor cells in the 6-day-old mouse testis. *Biol Reprod* 80 (4):707-17.

Kooreman, N. G., and J. C. Wu. 2010. Tumorigenicity of pluripotent stem cells: biological insights from molecular imaging. *J R Soc Interface* 7 Suppl 6:S753-63.

Kopen, G. C., D. J. Prockop, and D. G. Phinney. 1999. Marrow stromal cells migrate throughout forebrain and cerebellum, and they differentiate into astrocytes after injection into neonatal mouse brains. *Proc Natl Acad Sci U S A* 96 (19):10711-6.

Kossack, N., J. Meneses, S. Shefi, H. N. Nguyen, S. Chavez, C. Nicholas, J. Gromoll, P. J. Turek, and R. A. Reijo-Pera. 2009. Isolation and characterization of pluripotent human spermatogonial stem cell-derived cells. *Stem Cells* 27 (1):138-49.

Kristensen, D. M., J. E. Nielsen, N. E. Skakkebaek, N. Graem, G. K. Jacobsen, E. Rajpert-De Meyts, and H. Leffers. 2008. Presumed pluripotency markers UTF-1 and REX-1 are expressed in human adult testes and germ cell neoplasms. *Hum Reprod* 23 (4):775-82.

Kubota, H., M. R. Avarbock, and R. L. Brinster. 2004. Growth factors essential for self-renewal and expansion of mouse spermatogonial stem cells. *Proc Natl Acad Sci U S A* 101 (47):16489-94.

Kuijk, E. W., A. van Mil, B. Brinkhof, L. C. Penning, B. Colenbrander, and B. A. Roelen. 2009. PTEN and TRP53 independently suppress Nanog expression in spermatogonial stem cells. *Stem Cells Dev* 19 (7):979-88.

Kumar, V. H., S. Lakshminrusimha, M. T. El Abiad, P. R. Chess, and R. M. Ryan. 2005. Growth factors in lung development. *Adv Clin Chem* 40:261-316.

Lagasse, E., H. Connors, M. Al-Dhalimy, M. Reitsma, M. Dohse, L. Osborne, X. Wang, M. Finegold, I. L. Weissman, and M. Grompe. 2000. Purified hematopoietic stem cells can differentiate into hepatocytes in vivo. *Nat Med* 6 (11):1229-34.

Lawson, K. A., N. R. Dunn, B. A. Roelen, L. M. Zeinstra, A. M. Davis, C. V. Wright, J. P. Korving, and B. L. Hogan. 1999. Bmp4 is required for the generation of primordial germ cells in the mouse embryo. *Genes Dev* 13 (4):424-36.

Lawson, K. A., and W. J. Hage. 1994. Clonal analysis of the origin of primordial germ cells in the mouse. *Ciba Found Symp* 182:68-84; discussion 84-91.

Liao, Y., M. B. Geyer, A. J. Yang, and M. S. Cairo. 2011. Cord blood transplantation and stem cell regenerative potential. *Exp Hematol* 39 (4):393-412.

Lin, Y., and F. Wang 2010. FGF signalling in prostate development, tissue homoeostasis and tumorigenesis. *Biosci Rep* 30 (5):285-91.

Mason, C., and P. Dunnill. 2008. A brief definition of regenerative medicine. *Regen Med* 3 (1):1-5.

Mason, C., and P. Dunnill. 2008. The strong financial case for regenerative medicine and the regen industry. *Regen Med* 3 (3):351-63.

Mason, C., and E. Manzotti. 2010. Regenerative medicine cell therapies: numbers of units manufactured and patients treated between 1988 and 2010. *Regen Med* 5 (3):307-13.

Meng, X., M. Lindahl, M. E. Hyvonen, M. Parvinen, D. G. de Rooij, M. W. Hess, A. Raatikainen-Ahokas, K. Sainio, H. Rauvala, M. Lakso, J. G. Pichel, H. Westphal, M. Saarma, and H. Sariola. 2000. Regulation of cell fate decision of undifferentiated spermatogonia by GDNF. *Science* 287 (5457):1489-93.

Miller, C., and D. A. Sassoon. 1998. Wnt-7a maintains appropriate uterine patterning during the development of the mouse female reproductive tract. *Development* 125 (16):3201-11.

Nakagawa, M., M. Koyanagi, K. Tanabe, K. Takahashi, T. Ichisaka, T. Aoi, K. Okita, Y. Mochiduki, N. Takizawa, and S. Yamanaka. 2008. Generation of induced pluripotent stem cells without Myc from mouse and human fibroblasts. *Nat Biotechnol* 26 (1):101-6.

Naughton, C. K., S. Jain, A. M. Strickland, A. Gupta, and J. Milbrandt. 2006. Glial cell-line derived neurotrophic factor-mediated RET signaling regulates spermatogonial stem cell fate. *Biol Reprod* 74 (2):314-21.

Ning, L., E. Goossens, M. Geens, D. Van Saen, I. Van Riet, D. He, and H. Tournaye. 2010. Mouse spermatogonial stem cells obtain morphologic and functional characteristics of hematopoietic cells in vivo. *Hum Reprod* 25 (12):3101-9.

Oatley, J. M., M. R. Avarbock, and R. L. Brinster. 2007. Glial cell line-derived neurotrophic factor regulation of genes essential for self-renewal of mouse spermatogonial stem cells is dependent on Src family kinase signaling. *J Biol Chem* 282 (35):25842-51.

Oatley, J. M., M. R. Avarbock, A. I. Telaranta, D. T. Fearon, and R. L. Brinster. 2006. Identifying genes important for spermatogonial stem cell self-renewal and survival. *Proc Natl Acad Sci U S A* 103 (25):9524-9.

Oatley, J. M., and R. L. Brinster. 2006. Spermatogonial stem cells. *Methods Enzymol* 419:259-82.

Oatley, J. M., and R. L. Brinster. 2008. Regulation of spermatogonial stem cell self-renewal in mammals. *Annu Rev Cell Dev Biol* 24:263-86.

Oatley, M. J., K. E. Racicot, and J. M. Oatley. 2010. Sertoli cells dictate spermatogonial stem cell niches in the mouse testis. *Biol Reprod* 84 (4):639-45.

Ogawa, T., M. Ohmura, Y. Yumura, H. Sawada, and Y. Kubota. 2003. Expansion of murine spermatogonial stem cells through serial transplantation. *Biol Reprod* 68 (1):316-22.

Ohmichi, H., U. Koshimizu, K. Matsumoto, and T. Nakamura. 1998. Hepatocyte growth factor (HGF) acts as a mesenchyme-derived morphogenic factor during fetal lung development. *Development* 125 (7):1315-24.

Okita, K., M. Nakagawa, H. Hyenjong, T. Ichisaka, and S. Yamanaka. 2008. Generation of mouse induced pluripotent stem cells without viral vectors. *Science* 322 (5903):949-53.

Oottamasathien, S., Y. Wang, K. Williams, O. E. Franco, M. L. Wills, J. C. Thomas, K. Saba, A. R. Sharif-Afshar, J. H. Makari, N. A. Bhowmick, R. T. DeMarco, S. Hipkens, M. Magnuson, J. W. Brock, 3rd, S. W. Hayward, J. C. th Pope, and R. J. Matusik. 2007. Directed differentiation of embryonic stem cells into bladder tissue. *Dev Biol* 304 (2):556-66.

Petersen, B. E., W. C. Bowen, K. D. Patrene, W. M. Mars, A. K. Sullivan, N. Murase, S. S. Boggs, J. S. Greenberger, and J. P. Goff. 1999. Bone marrow as a potential source of hepatic oval cells. *Science* 284 (5417):1168-70.

Pittenger, M. F., A. M. Mackay, S. C. Beck, R. K. Jaiswal, R. Douglas, J. D. Mosca, M. A. Moorman, D. W. Simonetti, S. Craig, and D. R. Marshak. 1999. Multilineage potential of adult human mesenchymal stem cells. *Science* 284 (5411):143-7.

Polo, J. M., S. Liu, M. E. Figueroa, W. Kulalert, S. Eminli, K. Y. Tan, E. Apostolou, M. Stadtfeld, Y. Li, T. Shioda, S. Natesan, A. J. Wagers, A. Melnick, T. Evans, and K. Hochedlinger. 2010. Cell type of origin influences the molecular and functional properties of mouse induced pluripotent stem cells. *Nat Biotechnol* 28 (8):848-55.

Reinisch, A., and D. Strunk. 2009. Isolation and animal serum free expansion of human umbilical cord derived mesenchymal stromal cells (MSCs) and endothelial colony forming progenitor cells (ECFCs). *J Vis Exp* (32).

Safford, K. M., K. C. Hicok, S. D. Safford, Y. D. Halvorsen, W. O. Wilkison, J. M. Gimble, and H. E. Rice. 2002. Neurogenic differentiation of murine and human adipose-derived stromal cells. *Biochem Biophys Res Commun* 294 (2):371-9.

Sariola, H., and T. Immonen. 2008. GDNF maintains mouse spermatogonial stem cells in vivo and in vitro. *Methods Mol Biol* 450:127-35.

Seandel, M., D. James, S. V. Shmelkov, I. Falciatori, J. Kim, S. Chavala, D. S. Scherr, F. Zhang, R. Torres, N. W. Gale, G. D. Yancopoulos, A. Murphy, D. M. Valenzuela, R. M. Hobbs, P. P. Pandolfi, and S. Rafii. 2007. Generation of functional multipotent adult stem cells from GPR125+ germline progenitors. *Nature* 449 (7160):346-50.

Shamblott, M. J., J. Axelman, S. Wang, E. M. Bugg, J. W. Littlefield, P. J. Donovan, P. D. Blumenthal, G. R. Huggins, and J. D. Gearhart. 1998. Derivation of pluripotent stem cells from cultured human primordial germ cells. *Proc Natl Acad Sci U S A* 95 (23):13726-31.

Simon, L., G. C. Ekman, T. Garcia, K. Carnes, Z. Zhang, T. Murphy, K. M. Murphy, R. A. Hess, P. S. Cooke, and M. C. Hofmann. 2010. ETV5 regulates sertoli cell chemokines involved in mouse stem/progenitor spermatogonia maintenance. *Stem Cells* 28 (10):1882-92.

Simon, L., G. C. Ekman, N. Kostereva, Z. Zhang, R. A. Hess, M. C. Hofmann, and P. S. Cooke. 2009. Direct transdifferentiation of stem/progenitor spermatogonia into reproductive and nonreproductive tissues of all germ layers. *Stem Cells* 27 (7):1666-75.

Simon, L., R. A. Hess, and P. S. Cooke. 2010. Spermatogonial stem cells, in vivo transdifferentiation and human regenerative medicine. *Expert Opin Biol Ther* 10 (4):519-30.

Stadtfeld, M., E. Apostolou, H. Akutsu, A. Fukuda, P. Follett, S. Natesan, T. Kono, T. Shioda, and K. Hochedlinger. 2010. Aberrant silencing of imprinted genes on chromosome 12qF1 in mouse induced pluripotent stem cells. *Nature* 465 (7295):175-81.

Stadtfeld, M., and K. Hochedlinger. 2010. Induced pluripotency: history, mechanisms, and applications. *Genes Dev* 24 (20):2239-63.

Stadtfeld, M., M. Nagaya, J. Utikal, G. Weir, and K. Hochedlinger. 2008. Induced pluripotent stem cells generated without viral integration. *Science* 322 (5903):945-9.

Stevens, L. C. 1964. Experimental Production of Testicular Teratomas in Mice. *Proc Natl Acad Sci U S A* 52:654-61.

Streckfuss-Bomeke, K., A. Vlasov, S. Hulsmann, D. Yin, K. Nayernia, W. Engel, G. Hasenfuss, and K. Guan. 2009. Generation of functional neurons and glia from multipotent adult mouse germ-line stem cells. *Stem Cell Res* 2 (2):139-54.

Strulovici, Y., P. L. Leopold, T. P. O'Connor, R. G. Pergolizzi, and R. G. Crystal. 2007. Human embryonic stem cells and gene therapy. *Mol Ther* 15 (5):850-66.

Takahashi, K., K. Okita, M. Nakagawa, and S. Yamanaka. 2007. Induction of pluripotent stem cells from fibroblast cultures. *Nat Protoc* 2 (12):3081-9.

Takahashi, K., K. Tanabe, M. Ohnuki, M. Narita, T. Ichisaka, K. Tomoda, and S. Yamanaka. 2007. Induction of pluripotent stem cells from adult human fibroblasts by defined factors. *Cell* 131 (5):861-72.

Takahashi, K., and S. Yamanaka. 2006. Induction of pluripotent stem cells from mouse embryonic and adult fibroblast cultures by defined factors. *Cell* 126 (4):663-76.

Taylor, R. A., P. A. Cowin, G. R. Cunha, M. Pera, A. O. Trounson, J. Pedersen, and G. P. Risbridger. 2006. Formation of human prostate tissue from embryonic stem cells. *Nat Methods* 3 (3):179-81.

Taylor, R. A., H. Wang, S. E. Wilkinson, M. G. Richards, K. L. Britt, F. Vaillant, G. J. Lindeman, J. E. Visvader, G. R. Cunha, J. St John, and G. P. Risbridger. 2009. Lineage enforcement by inductive mesenchyme on adult epithelial stem cells across developmental germ layers. *Stem Cells* 27 (12):3032-42.

Tegelenbosch, R. A., and D. G. de Rooij. 1993. A quantitative study of spermatogonial multiplication and stem cell renewal in the C3H/101 F1 hybrid mouse. *Mutat Res* 290 (2):193-200.

Thomas, E. D., H. L. Lochte, Jr., J. H. Cannon, O. D. Sahler, and J. W. Ferrebee. 1959. Supralethal whole body irradiation and isologous marrow transplantation in man. *J Clin Invest* 38:1709-16.

Thomson, A. A., G. R. Cunha, and P. C. Marker. 2008. Prostate development and pathogenesis. *Differentiation* 76 (6):559-64.

Thomson, J. A., J. Itskovitz-Eldor, S. S. Shapiro, M. A. Waknitz, J. J. Swiergiel, V. S. Marshall, and J. M. Jones. 1998. Embryonic stem cell lines derived from human blastocysts. *Science* 282 (5391):1145-7.

Tyagi, G., K. Carnes, C. Morrow, N. V. Kostereva, G. C. Ekman, D. D. Meling, C. Hostetler, M. Griswold, K. M. Murphy, R. A. Hess, M. C. Hofmann, and P. S. Cooke. 2009. Loss of Etv5 decreases proliferation and RET levels in neonatal mouse testicular germ cells and causes an abnormal first wave of spermatogenesis. *Biol Reprod* 81 (2):258-66.

Wobus, A. M., and K. R. Boheler. 2005. Embryonic stem cells: prospects for developmental biology and cell therapy. *Physiol Rev* 85 (2):635-78.

Wu, D. C., A. S. Boyd, and K. J. Wood. 2007. Embryonic stem cell transplantation: potential applicability in cell replacement therapy and regenerative medicine. *Front Biosci* 12:4525-35.

Ying, Y., X. M. Liu, A. Marble, K. A. Lawson, and G. Q. Zhao. 2000. Requirement of Bmp8b for the generation of primordial germ cells in the mouse. *Mol Endocrinol* 14 (7):1053-63.

Yu, J., M. A. Vodyanik, K. Smuga-Otto, J. Antosiewicz-Bourget, J. L. Frane, S. Tian, J. Nie, G. A. Jonsdottir, V. Ruotti, R. Stewart, Slukvin, II, and J. A. Thomson. 2007. Induced pluripotent stem cell lines derived from human somatic cells. *Science* 318 (5858):1917-20.

Zechner, U., J. Nolte, M. Wolf, K. Shirneshan, N. E. Hajj, D. Weise, B. Kaltwasser, A. Zovoilis, T. Haaf, and W. Engel. 2009. Comparative methylation profiles and telomerase biology of mouse multipotent adult germline stem cells and embryonic stem cells. *Mol Hum Reprod* 15 (6):345-53.

Zeddou, M., A. Briquet, B. Relic, C. Josse, M. G. Malaise, A. Gothot, C. Lechanteur, and Y. Beguin. 2010. The umbilical cord matrix is a better source of mesenchymal stem cells (MSC) than the umbilical cord blood. *Cell Biol Int* 34 (7):693-701.

Zhao, T., Z. N. Zhang, Z. Rong, and Y. Xu. 2011. Immunogenicity of induced pluripotent stem cells. *Nature*. 474:212-5

Zheng, K., X. Wu, K. H. Kaestner, and P. J. Wang. 2009. The pluripotency factor LIN28 marks undifferentiated spermatogonia in mouse. *BMC Dev Biol* 9:38.

Zhou H, Wu S, Joo JY, Zhu S, Han DW, Lin T, Trauger S, Bien G, Yao S, Zhu Y, Siuzdak G, Schöler HR, Duan L, Ding S. 2009. Generation of induced pluripotent stem cells using recombinant proteins. *Cell Stem Cell* 4 (5):381-384.

Zovoilis, A., J. Nolte, N. Drusenheimer, U. Zechner, H. Hada, K. Guan, G. Hasenfuss, K. Nayernia, and W. Engel. 2008. Multipotent adult germline stem cells and embryonic stem cells have similar microRNA profiles. *Mol Hum Reprod* 14 (9):521-9.

Zuk, P. A., M. Zhu, H. Mizuno, J. Huang, J. W. Futrell, A. J. Katz, P. Benhaim, H. P. Lorenz, and M. H. Hedrick. 2001. Multilineage cells from human adipose tissue: implications for cell-based therapies. *Tissue Eng* 7 (2):211-28.

Zwaka, T. P., and J. A. Thomson. 2005. A germ cell origin of embryonic stem cells? *Development* 132 (2):227-33.

10

Therapeutic Application of Allogeneic Fetal Membrane-Derived Mesenchymal Stem Cell Transplantation in Regenerative Medicine

Shin Ishikane[1], Hiroshi Hosoda[1] and Tomoaki Ikeda[1,2]
[1]*Department of Regenerative Medicine and Tissue Engineering, National Cerebral and Cardiovascular Center Research Institute,*
[2]*Department of Perinatology and Gynecology, National Cerebral and Cardiovascular Center, Japan*

1. Introduction

In 1968, Friedenstein et al. isolated clonogeneic spindle-shaped cells from bone marrow (BM) in monolayer cultures, which they called colony-forming-unit fibroblasts (Friedenstein et al., 1974). These cells showed the ability to self-renew and to differentiate toward a mesodermal lineage as adipocytes, chondrocytes, osteocytes and connective stromal cells. Several studies reported that BM-derived multipotential stromal precursor cells can also differentiate into lineages such as ectodermal cells and endodermal cells (Kopen et al., 1999; Pittenger et al., 1999). For this reason, BM-derived stromal cells were first considered to be stem cells by Caplan and were named mesenchymal stem cells (MSCs) (Caplan, 1991). The Mesenchymal and Tissue Stem Cell Committee of the International Society for Cellular Therapy proposed the following minimal criteria for defining human MSCs: (1) MSCs must be plastic-adherent when maintained under standard culture conditions, (2) MSCs must express CD105, CD73 and CD90, and lack expression of CD45, CD34, CD14 or CD11b, CD79a or CD19 and HLA-DR surface molecules, (3) MSCs must differentiate into osteoblasts, adipocytes and chondroblasts *in vitro* (Dominici et al., 2006; Sensebe et al., 2010).

MSCs have been obtained from adipose tissue, cord blood and many other tissues, and can differentiate into a variety of cells, including adipocytes, osteocytes, chondrocytes, endothelial cells and myocytes (Campagnoli et al., 2001; Kim et al., 2006; Zuk et al., 2001). MSCs secrete a variety of angiogenic, antiapoptotic and mitogenic factors, such as vascular endothelial growth factor (VEGF), hepatocyte growth factor (HGF) and insulin-like growth factor-1 (IGF-1) (Kinnaird et al., 2004; Nagaya et al., 2005). Among MSCs derived from various tissues, BM-derived MSCs (BM-MSCs) are widely used in the field of stem cell transplantation. We previously reported that autologous BM-MSC transplantation induced therapeutic angiogenesis in a rat model of hind-limb ischemia and improved cardiac function in rat models of dilated cardiomyopathy and acute autoimmune myocarditis (Iwase et al., 2005; Nagaya et al., 2005; Ohnishi et al., 2007). However, there are several limitations to using an autologous cell source for cell transplantation, such as the invasiveness of the cell collection procedure, inadequate numbers of cells and donor-site morbidity, and the functionality of precursor cells in patients with cardiovascular risk factors has been questioned. The frequency and differentiation capacity of BM-MSCs decrease with age (D'Ippolito et al., 1999; Mareschi et al., 2006). An alternative source of MSCs that could provide large quantities of cells would be advantageous. One way to circumvent these limitations could be to use allogeneic MSCs. If allogeneic MSCs could be isolated from healthy young donors, and if they had a therapeutic effect similar to that of autologous MSCs, they would be considered a superior new cell source because it would be possible to overcome the problems noted above, and wider clinical applications of cell therapy would become available. Therefore, we focused on fetal membranes (FMs), which are generally discarded as medical waste after delivery, as an alternative source of autologous MSCs. Several studies reported that human FMs contain multipotent cells similar to BM-MSCs and are easy to expand (Alviano et al., 2007; In't Anker et al., 2004; Portmann-Lanz et al., 2006). If FM-MSCs could be used in allogeneic transplantation, FMs would be a useful source of cells for transplantation and regenerative medicine.

In this review, we compare the cellular characteristics and utilization of FM-MSCs with those of BM-MSCs and discuss the potential of allogeneic FM-MSC transplantation therapy in the tissue regeneration (Ishikane et al. 2008, 2010).

2. Fetal membrane-derived mesenchymal stem cells

The two FMs, the amnion and the chorion, marginate outward from the basal surface of the placenta and encase the amniotic fluid in which the fetus is suspended during pregnancy. The FMs facilitate gas and waste exchange and play a critical role as defense barriers, in maintenance of pregnancy and in parturition (Bourne, 1962). Human FMs, which are generally discarded as medical waste after delivery, were recently shown to be rich sources of MSCs. Because fetal tissues are routinely discarded postpartum, FMs are inexpensive and easy to obtain and their availability is virtually limitless, avoiding the need for mass tissue banking. Human amnion membrane-derived MSCs (hAM-MSCs) were isolated for the first time from second and third trimester AMs by In't Anker et al., who demonstrated their potential for differentiation into osteogenic and adipogenic cells (In't Anker et al., 2004). Later, Portmann-Lanz et al. demonstrated their capacity for differentiation into chondrogenic, myogenic and neurogenic lines (Portmann-Lanz et al., 2006). In 2007, Alviano et al. reported that hAM-MSCs are superior in proliferation and differentiation potential to adult hBM-MSCs, providing the first evidence of the angiogenic potential of hAM-MSCs (Alviano et al., 2007). A large quantity of MSCs was isolated from hFMs by serial passaging them prior to senescence at about 15 passages (Kim et al., 2007; Soncini et al., 2007). The availability of a fetal tissue that is usually discarded without any ethical conflict and the high yield in stem cell recovery make FMs a truly exciting alternative source that offers new prospects for expanding the range of clinical applications for stem cells.

In our study, FM-MSCs derived from Lewis rats did not express the hematopoietic or endothelial surface markers CD11b/c, CD31, CD34 and CD45, but stained positive for CD29, CD73 and CD90 (Ishikane et al., 2008). These rat FM-MSCs differentiated into adipocytes, osteocytes and chondrocytes (Figure 1). In culture medium, FM-MSCs secreted the angiogenic factors, VEGF and HGF. In an angiogenic gene polymerase chain reaction array analysis, FM-

MSCs expressed compounds characteristic of several angiogenesis-related genes, including VEGF-C, platelet-derived growth factor-B, angiopoietins, chemokines and interleukins. These results show that FM-MSCs have properties similar to those of BM-MSCs and suggest that transplantation of FM-MSCs may induce therapeutic angiogenesis in cases of ischemic disease.

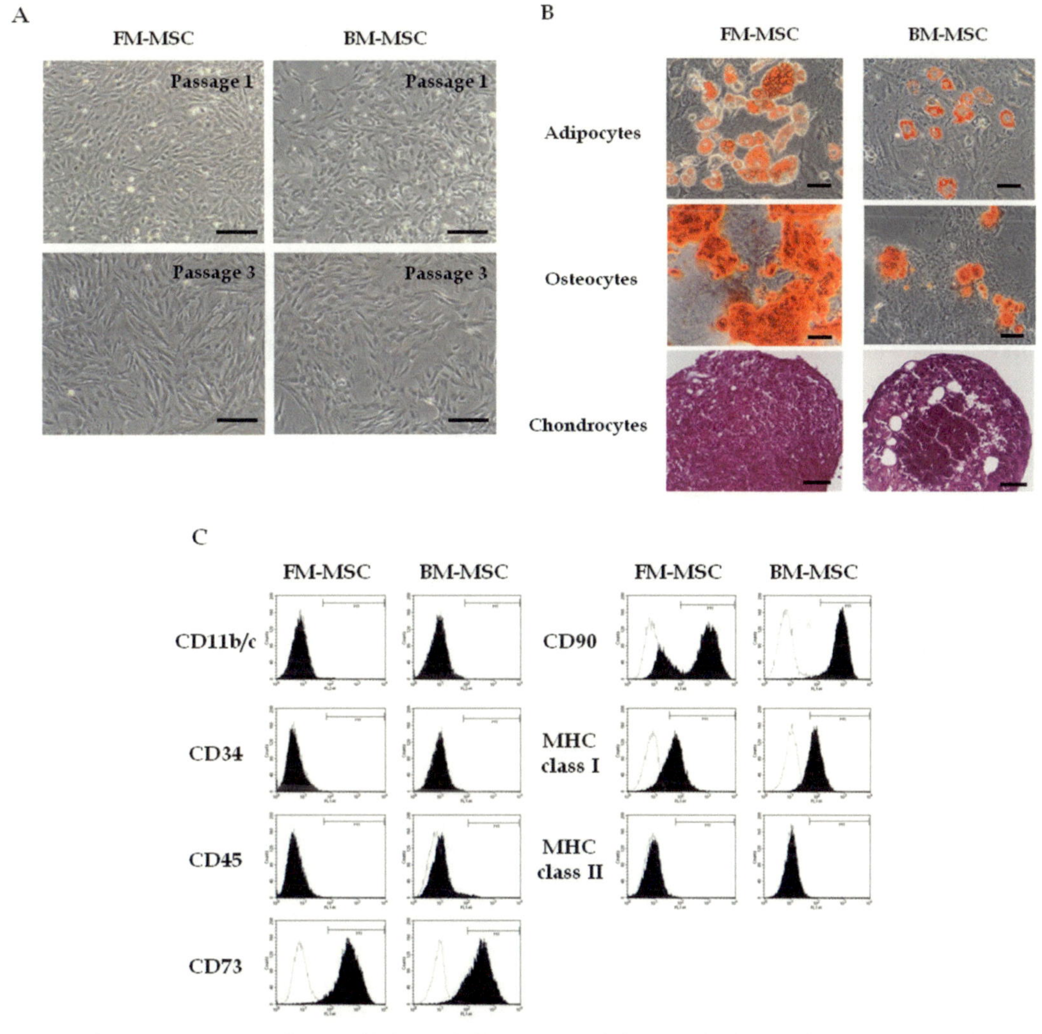

Fig. 1. Characterization of FM-MSCs and BM-MSCs: (A) Morphology of FM-MSCs and BM-MSCs derived from Lewis rats. In the early passages, FM- and BM-MSC derived cells appeared microscopically heterogeneous. After several passages, these cells formed a morphologically homogenous population of fibroblast-like cells, which was similar to BM-MSCs. Scale bars: 100 µm. (B) Multipotency of FM-MSCs and BM-MSCs. Differentiation into adipocytes was observed by oil red O. Differentiation into osteocytes was observed by alizarin red S. Differentiation into chondrocytes was observed by safranin O. Scale bars: 50 µm. (C) Flow cytometric analysis of FM-MSCs and BM-MSCs at passage 3. Closed areas indicate staining with a specific antibody, whereas open areas represent staining with isotype control antibodies.

FM-MSC	Characteristic	BM-MSC
Noninvasive	MSC harvest procedure	Invasive
Placenta	Donor tissue	Adult bone marrow
High	Number of obtained cells	Low
CD11–, CD29+, CD31–, CD34–, CD45–, CD73+, CD90+, MHC class I+, MHC class II–	Immunophenotype	CD11–, CD29+, CD31–, CD34–, CD45–, CD73+, CD90+, MHC class I+, MHC class II–
Adipogenic Osteogenic Chondrogenic	In vitro multipotency	Adipogenic Osteogenic Chondrogenic
VEGF, HGF	Growth factor secretion	VEGF, HGF, IGF-1, adrenomedullin
In hind limb ischemia: induced In acute myocarditis: not	Angiogenesis	In hind limb ischemia: induced In acute myocarditis: induced
Low	Engraftment of transplanted cells	Low
Vascular endothelial cells: none Myocardium: none	In vivo differentiation	Vascular endothelial cells: very low or none Smooth muscle cells: very low Myocardium: very low
Evade	Alloreactive T cell activation (rejection)	Evade
Suppress	CD4+T cell activation (immunomodulatory effect)	Suppress
Suppress	Fibrosis	Suppress
Suppress	Inflammatory cell infiltration	Suppress

Table 1. Comparison of the characteristics of FM-MSCs and BM-MSCs observed in our studies. Abbreviations: BM-MSC, bone marrow-derived mesenchymal stem cell; FM-MSC, fetal membrane-derived mesenchymal stem cell; HGF, hepatocyte growth factor; IGF-1, insulin-like growth factor-1; MSC, mesenchymal stem cell; MHC, major histocompatibility complex; VEGF, vascular endothelial growth factor.

2.1 Immunomodulatory effect of fetal membrane-derived mesenchymal stem cells

MSCs have received renewed interest, particularly for their use in transplantation medicine. Although the main driving force responsible for interest in the regenerative capacity of MSCs in the past was their presumptive plasticity, their ability to modulate the immune response is now attracting greater interest. MSCs are positive for major histocompatibility complex (MHC) class I but negative for MHC class II and for costimulatory factors such as CD40, CD80 and CD86, and are therefore considered nonimmunogenic (Chamberlain et al.

2007). Allogeneic BM-MSC transplantation has been used in several preclinical and clinical studies, in which allogeneic MSCs were not rejected in the absence of immunosuppression (Amado et al., 2005; Hare et al., 2009; Le Blanc et al., 2008).

The use of BM-MSCs not only avoids allogeneic rejection but also may confer immunosuppressive effects. Several studies demonstrated that MSCs modulate the function of T cells, major executors of the adaptive immune response (Krampera et al., 2003; Le Blanc et al., 2003). Di Nicola et al. showed that BM-MSCs strongly suppressed T cell proliferation in a mixed lymphocyte culture (MLC) test (Di Nicola et al., 2002).

In our study of rats, FM-MSCs had immunological properties similar to those of BM-MSCs. In an MLC test with haplotype-mismatched allogeneic cells, FM-MSCs did not provoke alloreactive lymphocyte proliferation. Interleukin (IL)-2 plays a role in the activation and proliferation of T cells. IL-2 concentrations in supernatants of FM-MSC and allogeneic lymphocyte co-cultures and in the MLC were lower than those in lymphocyte and allogeneic lymphocyte co-cultures.

To investigate T cell alloreactivity to transplanted allogeneic FM-MSCs, FM-MSCs, BM-MSCs or splenic lymphocytes obtained from GFP-transgenic Lewis rats were injected into the hind-limb tissue of MHC-mismatched August–Copenhagen Irish (ACI) rats. One week after cell injection, slight T cell infiltration was observed at the injection site of allogeneic FM-MSC-injected hind-limb muscles, but the degree of infiltration was less marked than that after allogeneic splenic lymphocyte transplantation and was equivalent to that induced by allogeneic BM-MSCs. Use of non-autologous cells for transplant also requires that one consider the possibility of graft rejection. Although most clinical applications of FM-MSC transplantation apply to allogeneic transplantation, our results suggest that FM-MSCs evade T cell alloreactivity and may be successfully transplanted across MHC barriers.

2.2 Therapeutic angiogenesis in allogeneic fetal membrane-derived mesenchymal stem cell transplantation in a hind-limb ischemia model

Therapeutic angiogenesis, a strategy to treat tissue ischemia by promoting the proliferation of collateral vessels, has emerged as one of the most promising therapies developed to date (Carmeliet, 2003). In a rat model of hind-limb ischemia, autologous BM-MSC transplantation enhanced angiogenesis and peripheral blood flow in the ischemic limb, and these cells were incorporated into sites of angiogenesis after tissue ischemia (Iwase et al., 2005). MSC transplantation was shown to be a promising approach for restoring tissue vascularization after ischemic events (Moon et al., 2006; Nakagami et al., 2005).

In a previous study, we demonstrated that allogeneic transplantation of FM-MSCs induced angiogenesis in a rat model of hind-limb ischemia (Ishikane et al., 2008). One day after left common iliac artery resection, FM-MSCs obtained from Lewis rats were transplanted into the ischemic thigh muscle of MHC-mismatched ACI rats with hind-limb ischemia (5×10^5 cells/animal). The blood perfusion of the ischemic limb and the capillary density of the ischemic muscle were increased 2 and 3 weeks, respectively, after allogeneic FM-MSC transplantation (Figure 2). It is noteworthy that the therapeutic gain was similar to that of allogeneic BM-MSC transplantation.

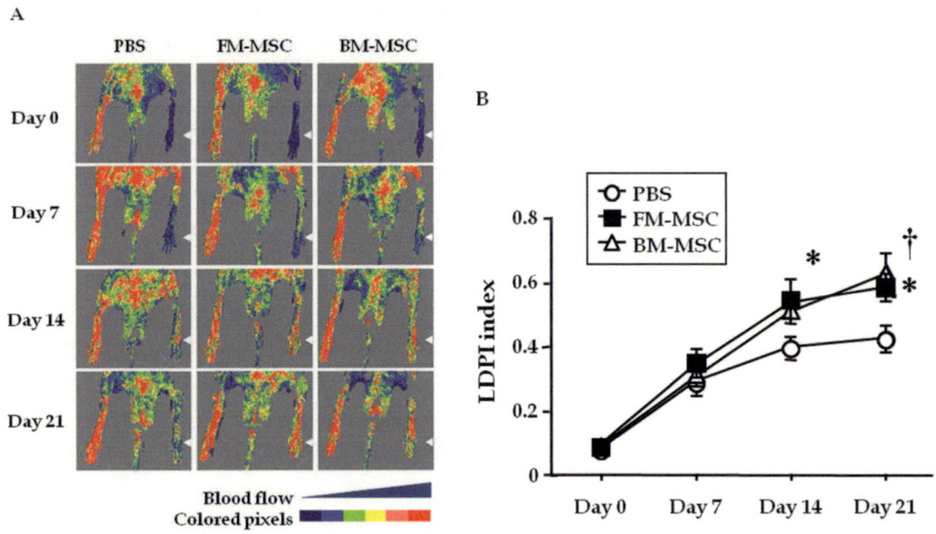

Fig. 2. Comparison of angiogenesis after allogeneic FM-MSC and BM-MSC transplantation in rats with hindlimb ischemia
(A) Representative examples of serial A laser doppler perfusion image (LDPI). Blood perfusion of the ischemic hindlimb was markedly increased in the allogeneic FM-MSCs and BM-MSCs transplanted group 3 weeks after cell injection (red to orange). (B) Quantitative analysis of hindlimb blood perfusion. LDPI index was significantly higher in the allogeneic FM-MSCs and BM-MSCs transplanted groups than in the phosphate-buffer saline (PBS) treated control group 3 weeks after cell injection. The LDPI index was determined as the ratio of ischemic to nonischemic hindlimb blood perfusion. Data are mean ± S.E.M. *$P < 0.05$ FM-MSC vs. PBS; †$P < 0.05$ BM-MSC vs. PBS.

The allogeneic FM- and BM-MSCs in the ischemic hind-limb tissue survived for 3 weeks after transplantation, but the number of engrafted cells decreased significantly in both cases (Figure 3). In a previous trial, intramuscularly transplanted allogeneic BM-MSCs were observed 6 months after transplantation (Dai et al., 2005). In other studies, the number of engrafted autologous and allogeneic MSCs gradually decreased, and MSCs were absent after several weeks (Fouillard et al., 2007; Kraitchman et al., 2005; Shake et al., 2002). Muller-Ehmsen et al. reported the observed transplanted MSC loss was predominantly caused by cell death rather than migration of cells to other organs (Muller-Ehmsen et al., 2006).

To investigate differentiation of transplanted FM-MSCs into blood vessel endothelial cells, we performed immunofluorescent staining of MSC-transplanted ischemic hind-limb sections. GFP-positive transplanted FM-MSCs and BM-MSCs and lectin-positive endothelial cells were observed in hind-limb tissue, but GFP/lectin double-positive cells were not observed. Some studies reported that transplanted BM-MSCs directly differentiated into the vascular endothelial cells and vascular smooth muscles in ischemic models (Al-Khaldi et al., 2003; Moon et al., 2006). However, recent studies demonstrated that the direct contribution of grafted MSCs is minimal or even absent, and that paracrine actions are of major importance in mediating their regenerative effects (Aranguren et al., 2008; Au et al., 2008; Muller-Ehmsen et al., 2006). MSCs were considered to induce neovascularization by secreting large amounts of humoral factors involved in angiogenesis, such as VEGF and HGF (Kinnaird et al., 2004; Nagaya et al., 2005). VEGF is one of the more powerful angiogenic cytokines and can also mobilize endothelial progenitor cells (EPCs) from BM and inhibit EPC apoptosis (Asahara et al., 1999). HGF plays important roles in tissue regeneration, morphogenesis and angiogenesis (Zarnegar and Michalopoulos, 1995). HGF is thought to stimulate endothelial cell proliferation and to induce angiogenesis, and is a key signaling factor that promotes infiltration of circulating stem cells from the peripheral circulation to an ischemic area (Morishita et al., 1999; Weimar et al., 1998). Further studies are needed to improve the availability of transplanted MSCs for engraftment, but allogeneic FM-MSC transplantation could provide a new therapeutic strategy for the treatment of severe peripheral vascular disease.

2.3 Immunomodulatory effect of allogeneic fetal membrane-derived mesenchymal stem cell transplantation in an autoimmune myocarditis model

Several studies reported that MSCs have immunomodulatory effects mediated by secretion of soluble factors such as prostaglandin E_2, indoleamine 2,3-dioxygenase, IL-6, IL-10, heme oxygenase-1 and galectin (Aggarwal and Pittenger, 2005; Chabannes et al., 2007; Meisel et al., 2004; Sioud et al., 2011). Based on the immunomodulatory property of MSCs, allogeneic FM-MSC transplantation may be an attractive treatment for autoimmune myocarditis.

Experimental autoimmune myocarditis (EAM) is induced by injecting porcine cardiac myosin in Lewis rats. Allogeneic FM-MSCs obtained from MHC-mismatched ACI rats (5×10^5 cells/animal) were transplanted intravenously into EAM rats 1 week after myosin injection. Two weeks after transplantation, the intravenous allogeneic transplantation of FM-MSCs reduced fibrosis, edema, necrosis, granulation and eosinophil infiltration in hearts exhibiting EAM and significantly attenuated infiltration of inflammatory cells (CD68-positive monocytes and macrophages) and MCP-1 expression in the myocardium (Figure 4A and B). Hemodynamic and echocardiographic tests showed a significant improvement in cardiac function as a result of allogeneic FM-MSC transplantation (Ishikane et al., 2010). The extent of the improvement ranged from 30% to 60% according to various indices of the level of dysfunction, which is equivalent to that observed in our previous study on autologous BM-MSC transplantation in EAM (Ohnishi et al., 2007). Allogeneic transplantation of FM-MSCs significantly reduced infiltration of T cells (CD3-positive cells) into EAM hearts (Figure 4C). In a T lymphocyte proliferation assay, splenic T lymphocytes collected from allogeneic FM-MSC-transplanted EAM rats had a reduced proliferative response to myosin compared with the response of splenic T lymphocytes from untransplanted EAM rats. In addition, proliferation of activated T lymphocytes was suppressed by co-culture with allogeneic FM-MSCs *in vitro*.

Okada et al. reported that Th2-type cytokine expression in EAM was increased by HGF, whereas Th1-type cytokine expression was suppressed by intramyocardial transplantation of autologous BM-MSCs (Okada et al., 2007). An increase in HGF expression may reduce the severity of EAM by suppressing the Th1 response. Van Linthout et al. reported that MSCs improved murine acute coxsackievirus B3-induced myocarditis via their immunomodulatory properties in a nitric oxide-dependent manner (Van Linthout et al., 2010).

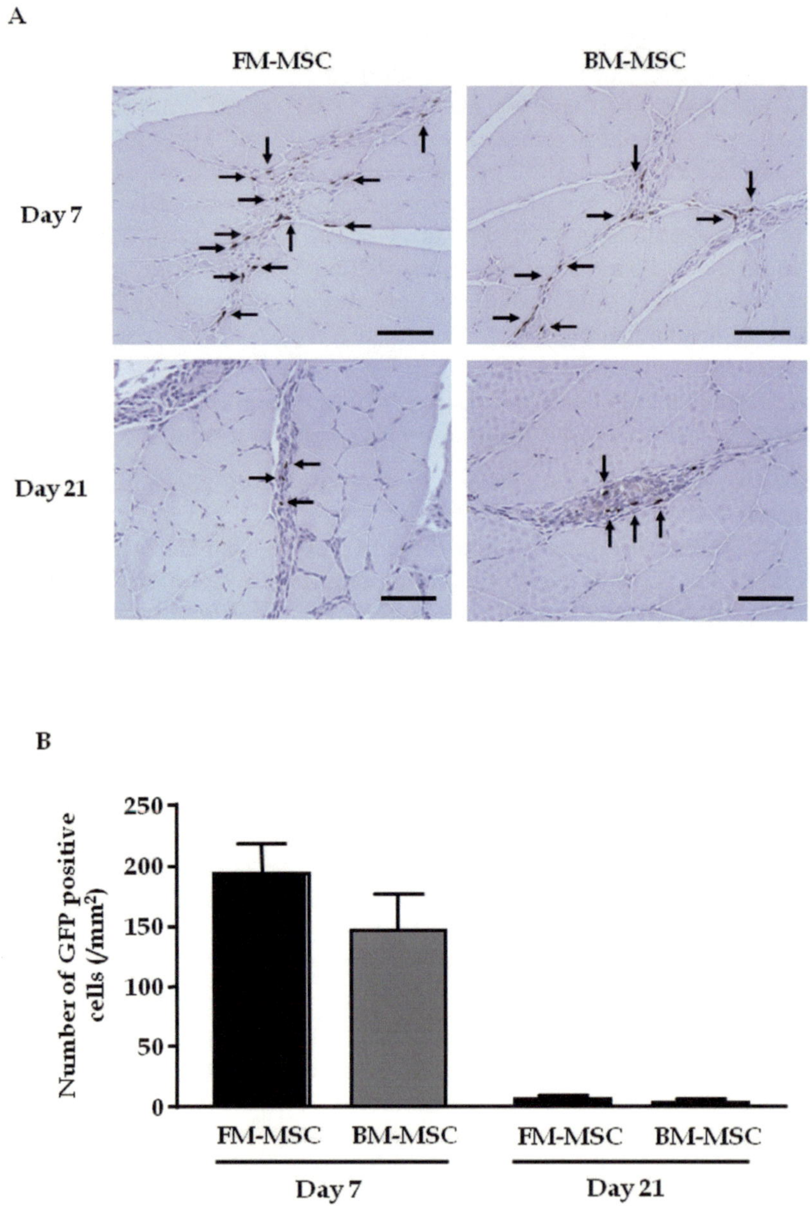

Fig. 3. Engraftment of allogeneic FM-MSCs and BM-MSCs injected into ischemic hindlimb muscles. (A) Representative sections show that GFP-positive allogeneic FM-MSCs and BM-MSCs were present in the hindlimb muscles of rats with hindlimb ischemia 1 and 3 weeks after cell injection (brown stain; black arrows). Scale bars: 50 μm. (B) Quantitative analysis demonstrated that comparable numbers of GFP-positive allogeneic FM-MSCs and allogeneic BM-MSCs were observed in ischemic hindlimbs 1 week after cell injection. Three weeks after cell injection, a few GFP-positive allogeneic FM-MSCs and BM-MSCs were observed. Data are mean ± S.E.M.

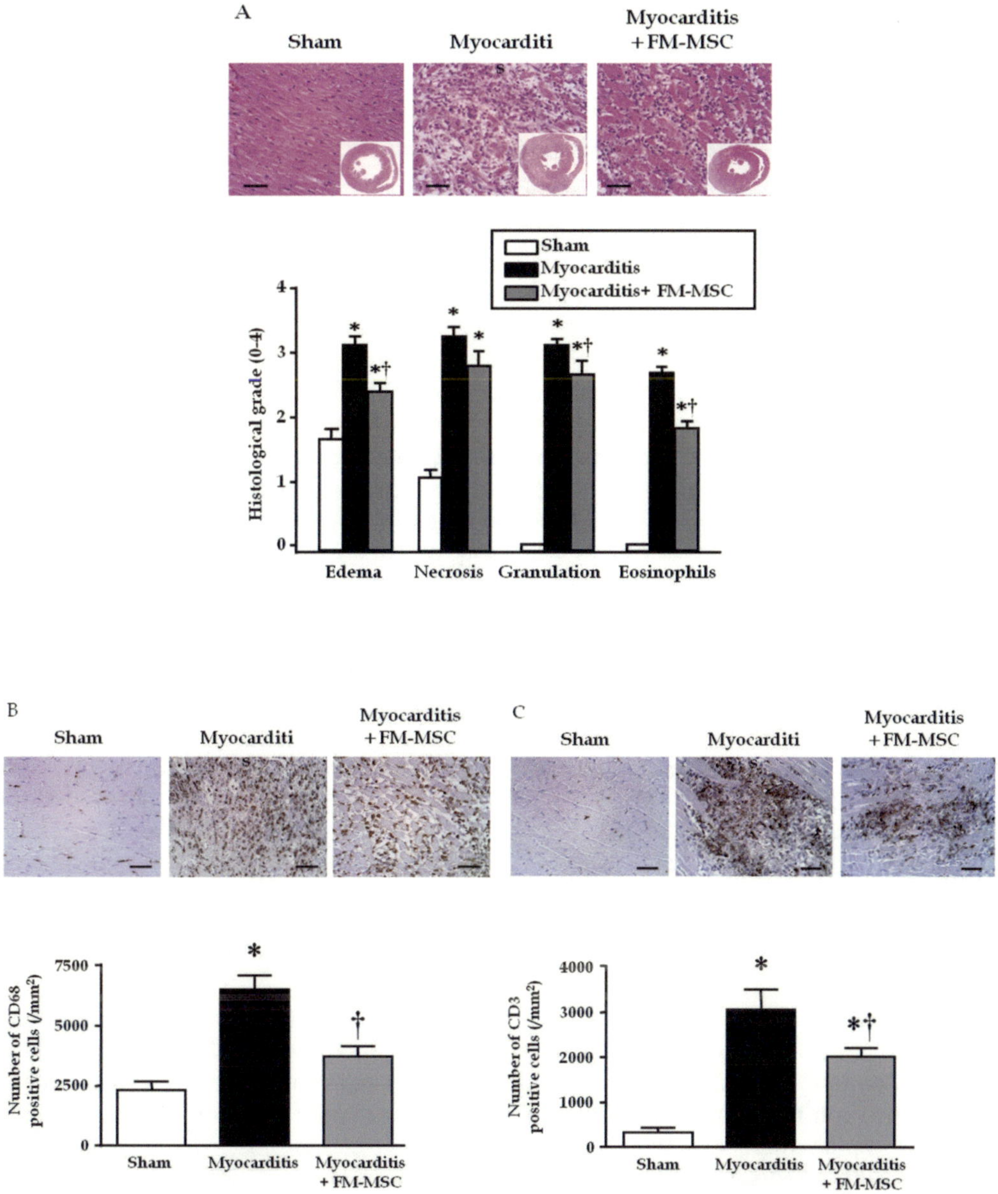

Fig. 4. Histopathological changes in autoimmune myocarditis at 2 weeks after transplantation induced by transplantation of allogeneic FM-MSCs. (A) Myocardial sections showed markedly less inflammation in the allogeneic FM-MSCs transplanted group than in the untransplanted myocarditis group. Insets are transverse sections of the myocardium. The semiquantitative histological grade of edema and eosinophil infiltration were markedly decreased in the allogeneic FM-MSCs transplanted group. (B) CD68-positive macrophage/monocyte infiltration, and (C) CD3-positive T cell infiltration were markedly reduced by allogeneic FM-MSC transplantation. Scale bars = 50 μm. Data are expressed as mean ± SEM. *$P < 0.05$ vs. the sham group; †$P < 0.05$ vs. the untreated myocarditis group.

Allogeneic transplantation of FM-MSCs may be an attractive therapy for the treatment of autoimmune myocarditis. Further studies are needed to elucidate the therapeutic mechanisms.

3. Potential of mesenchymal stem cell sheet transplantation therapy

As discussed above, MSC transplantation has attractive possibilities as a tool for cell transplantation therapy. However, further experiments are needed to develop data obtained with MSCs for application to humans because evidence of an ameliorating effect on angiogenesis and cardiac function is not necessarily sufficient to warrant clinical use. To date, intramuscular and intravenous injections have been used for cell transplantation therapy, but the engraftment rate of MSCs transplanted via these routes was very low (Ishikane et al. 2008, 2010). Although intramuscularly transplanted allogeneic FM-MSCs survived in ischemic hind-limb tissue for 3 weeks after transplantation, the number of engrafted cells decreased significantly. In EAM, some of the intravenously transplanted MSCs were found in the lung, heart, spleen and liver 1 week after transplantation, but these engrafted cells could not be detected 4 weeks after transplantation. Most homing and engraftment studies demonstrated little, if any, long-term (>1 week) engraftment of MSCs after systemic administration (Parekkadan and Milwid, 2010). Studies showed that the majority of administered MSCs (>80%) immediately accumulate in the lung and are cleared with a half-life of 24 h. Although intravenous cell transplantation is very convenient, it is not suitable for transplantation of large numbers of cells. Thus, a more effective transplantation route is needed to enhance angiogenesis and cardiac functional improvement in MSC transplantation.

Recently, cell sheet engineering received attention as a method for heart tissue repair. Okano et al. developed engineered cell sheets containing scaffoldless tissue using temperature-responsive culture dishes (Yamada et al., 1990). These cell sheets enable cell-to-cell connections and maintain the presence of adhesion proteins. The cell sheets preserve extracellular matrix proteins deposited on the basal surface of the cultured cells. These adhesive proteins play an important role in enhancing attachment between stacked cell sheets and between cell sheets and the myocardial surface, thereby enabling stable fixation of the cell sheet constructs to the target tissues. The cell sheets can readily be transferred and grafted to scarred myocardium without additives or suturing. Memon et al. demonstrated that layered skeletal myoblast sheets transplanted to infarcted rat hearts enhanced left ventricular contraction, reduced fibrosis and prevented left ventricular dilation (Memon et al., 2005). Kondoh et al. showed that in hamsters with dilated cardiomyopathy, myoblast sheet graft implantation improved cardiac performance and prolonged life expectancy in association with a reduction in myocardial fibrosis (Kondoh et al., 2006). In our study on rats, adipose tissue-derived MSC sheets improved cardiac function in damaged hearts, with reversal of cardiac wall thinning and prolonged survival after myocardial infarction (Miyahara et al., 2006). These cell sheets enable transplantation of many more cells than with intramuscular or intravenous needle injection. MSC sheet transplantation is expected to increase the number of engrafted cells and to enhance paracrine signaling.

4. Conclusion

This review shows the potential of allogeneic transplantation of FM-MSCs for the treatment of peripheral vascular disease and autoimmune myocarditis. FM-MSCs did not elicit alloreactive T lymphocyte proliferation, and allogeneic FM-MSC transplantation induced therapeutic angiogenesis in a rat model of hind-limb ischemia. The angiogenic effects may be induced in a paracrine manner rather than via vascular differentiation of the transplanted MSCs. It is expected that allogeneic FM-MSC transplantation will be an effective therapy for autoimmune myocarditis with rapidly progressive heart failure. The beneficial effects of allogeneic FM-MSC transplantation are mainly attributable to suppression of T lymphocyte activation and anti-inflammatory effects. FM are potentially promising cell source for clinical use; they are medical waste material, are abundantly available from maternity wards. The unlimited availability of term gestational tissue, large number of cell that can be isolated from FM without invasive procedures, minimal ethical and legal barriers associated with their usage and immune tolerance make these cells highly attractive for stem cell based regenerative and reparative medicine and tissue engineering. Meanwhile, the risk of tumor formation from transplanting allogeneic FM-MSC into patients remains undetermined, and long-term follow-up studies are needed to clarify safety. Although further experiments are needed to adapt the current results for clinical application, we predict that allogeneic FM-MSC transplantation therapy will become a treatment for severe peripheral vascular disease and autoimmune myocarditis.

5. Acknowledgments

Our studies were supported by a Research Grant for Cardiovascular Disease (18C-1) and Human Genome Tissue Engineering 009 from the Ministry of Health, Labor and Welfare. We are grateful to Dr Kenichi Yamahara, Dr Makoto Kodama, Dr Hatsue Ishibashi-Ueda, Dr Shunsuke Ohnishi and Dr Noritoshi Nagaya for their support of our studies. We are thankful to the National BioResource Project for the Rat in Japan (http://www.anim.med.kyoto-u.ac.jp/NBR/) for providing rat strain LEW-TgN(CAG-EGFP)1Ys.

6. References

Aggarwal, S., & Pittenger, M. F. (2005). Human mesenchymal stem cells modulate allogeneic immune cell responses. *Blood*, Vol.105, No.4, (October 2004), pp.1815-1822, ISSN 0006-4971

Al-Khaldi, A.; Al-Sabti, H.; Galipeau, J., & Lachapelle, K. (2003). Therapeutic angiogenesis using autologous bone marrow stromal cells: improved blood flow in a chronic limb ischemia model. *Annals of Thoracic Surgery*, Vol.75, No.1, (January 2003), pp.204-209, ISSN 0003-4975

Alviano, F.; Fossati, V.; Marchionni, C.; Arpinati, M.; Bonsi, L.; Franchina, M.; Lanzoni, G.; Cantoni, S.; Cavallini, C.; Bianchi, F.; Tazzari, P. L.; Pasquinelli, G.; Foroni, L.; Ventura, C.; Grossi, A., & Bagnara, G. P. (2007). Term Amniotic membrane is a high throughput source for multipotent Mesenchymal Stem Cells with the ability to differentiate into endothelial cells in vitro. *BMC Developmental Biology*, Vol.7, (February 2007), pp.11, ISSN 1471-213X

Amado, L. C.; Saliaris, A. P.; Schuleri, K. H.; St John, M.; Xie, J. S.; Cattaneo, S.; Durand, D. J.; Fitton, T.; Kuang, J. Q.; Stewart, G.; Lehrke, S.; Baumgartner, W. W.; Martin, B. J.; Heldman, A. W., & Hare, J. M. (2005). Cardiac repair with intramyocardial injection of allogeneic mesenchymal stem cells after myocardial infarction. *Proceedings of the*

National Academy of Sciences of the United States of America, Vol.102, No.32, (August 2005), pp.11474-11479, ISSN 0027-8424

Aranguren, X. L.; McCue, J. D.; Hendrickx, B.; Zhu, X. H.; Du, F.; Chen, E.; Pelacho, B.; Penuelas, I.; Abizanda, G.; Uriz, M.; Frommer, S. A.; Ross, J. J.; Schroeder, B. A.; Seaborn, M. S.; Adney, J. R.; Hagenbrock, J.; Harris, N. H.; Zhang, Y.; Zhang, X.; Nelson-Holte, M. H.; Jiang, Y.; Billiau, A. D.; Chen, W.; Prosper, F.; Verfaillie, C. M., & Luttun, A. (2008). Multipotent adult progenitor cells sustain function of ischemic limbs in mice. *Journal of Clinical Investigation*, Vol.118, No.2, (January 2008), pp.505-514, ISSN 0021-9738

Asahara, T.; Takahashi, T.; Masuda, H.; Kalka, C.; Chen, D.; Iwaguro, H.; Inai, Y.; Silver, M., & Isner, J. M. (1999). VEGF contributes to postnatal neovascularization by mobilizing bone marrow-derived endothelial progenitor cells. *EMBO Journal*, Vol.18, No.14, (July 1999), pp.3964-3972, ISSN 0261-4189

Au, P.; Tam, J.; Fukumura, D., & Jain, R. K. (2008). Bone marrow-derived mesenchymal stem cells facilitate engineering of long-lasting functional vasculature. *Blood*, Vol.111, No.9, (February 2008), pp.4551-4558, ISSN 1528-0020

Bourne, G. (1962). The foetal membranes. A review of the anatomy of normal amnion and chorion and some aspects of their function. *Postgraduate Medical Journal*, Vol.38, (April 1962), pp.193-201, ISSN 0032-5473

Campagnoli, C.; Roberts, I. A.; Kumar, S.; Bennett, P. R.; Bellantuono, I., & Fisk, N. M. (2001). Identification of mesenchymal stem/progenitor cells in human first-trimester fetal blood, liver, and bone marrow. *Blood*, Vol.98, No.8, (October 2001), pp.2396-2402, ISSN 0006-4971

Caplan, A. I. (1991). Mesenchymal stem cells. *Journal of Orthopaedic Research*, Vol.9, No.5, (September 1991), pp.641-650, ISSN 0736-0266

Carmeliet, P. (2003). Angiogenesis in health and disease. *Nature Medicine*, Vol.9, No.6, (June 2003), pp.653-660, ISSN 1078-8956

Chabannes, D.; Hill, M.; Merieau, E.; Rossignol, J.; Brion, R.; Soulillou, J. P.; Anegon, I., & Cuturi, M. C. (2007). A role for heme oxygenase-1 in the immunosuppressive effect of adult rat and human mesenchymal stem cells. *Blood*, Vol.110, No.10, (August 2007), pp.3691-3694, ISSN 0006-4971

Chamberlain, G.; Fox, J.; Ashton, B., & Middleton, J. (2007). Concise review: mesenchymal stem cells: their phenotype, differentiation capacity, immunological features, and potential for homing. *Stem Cells*, Vol.25, No.11, (July 2007), pp.2739-2749, ISSN 1549-4918

D'Ippolito, G.; Schiller, P. C.; Ricordi, C.; Roos, B. A., & Howard, G. A. (1999). Age-related osteogenic potential of mesenchymal stromal stem cells from human vertebral bone marrow. *Journal of Bone and Mineral Research*, Vol.14, No.7, (July 1999), pp.1115-1122, ISSN 0884-0431

Dai, W.; Hale, S. L.; Martin, B. J.; Kuang, J. Q.; Dow, J. S.; Wold, L. E., & Kloner, R. A. (2005). Allogeneic mesenchymal stem cell transplantation in postinfarcted rat myocardium: short- and long-term effects. *Circulation*, Vol.112, No.2, (July 2005), pp.214-223, ISSN 1524-4539

Di Nicola, M.; Carlo-Stella, C.; Magni, M.; Milanesi, M.; Longoni, P. D.; Matteucci, P.; Grisanti, S., & Gianni, A. M. (2002). Human bone marrow stromal cells suppress T-lymphocyte proliferation induced by cellular or nonspecific mitogenic stimuli. *Blood*, Vol.99, No.10, (May 2002), pp.3838-3843, ISSN 0006-4971

Dominici, M.; Le Blanc, K.; Mueller, I.; Slaper-Cortenbach, I.; Marini, F.; Krause, D.; Deans, R.; Keating, A.; Prockop, D., & Horwitz, E. (2006). Minimal criteria for defining multipotent mesenchymal stromal cells. The International Society for Cellular Therapy position statement. *Cytotherapy*, Vol.8, No.4, (August 2006), pp.315-317, ISSN 1465-3249

Fouillard, L.; Chapel, A.; Bories, D.; Bouchet, S.; Costa, J. M.; Rouard, H.; Herve, P.; Gourmelon, P.; Thierry, D.; Lopez, M., & Gorin, N. C. (2007). Infusion of allogeneic-related HLA mismatched mesenchymal stem cells for the treatment of incomplete engraftment following autologous haematopoietic stem cell transplantation. *Leukemia*, Vol.21, No.3, (January 2007), pp.568-570, ISSN 0887-6924

Friedenstein, A. J.; Chailakhyan, R. K.; Latsinik, N. V.; Panasyuk, A. F., & Keiliss-Borok, I. V. (1974). Stromal cells responsible for transferring the microenvironment of the hemopoietic tissues. Cloning in vitro and retransplantation in vivo. *Transplantation*, Vol.17, No.4, (April 1974), pp.331-340, ISSN 0041-1337

Hare, J. M.; Traverse, J. H.; Henry, T. D.; Dib, N.; Strumpf, R. K.; Schulman, S. P.; Gerstenblith, G.; DeMaria, A. N.; Denktas, A. E.; Gammon, R. S.; Hermiller, J. B., Jr.; Reisman, M. A.; Schaer, G. L., & Sherman, W. (2009). A randomized, double-blind, placebo-controlled, dose-escalation study of intravenous adult human mesenchymal stem cells (prochymal) after acute myocardial infarction. *Journal of the American College of Cardiology*, Vol.54, No.24, (December 2009), pp.2277-2286, ISSN 1558-3597

In 't Anker, P. S.; Scherjon, S. A.; Kleijburg-van der Keur, C.; de Groot-Swings, G. M.; Claas, F. H.; Fibbe, W. E., & Kanhai, H. H. (2004). Isolation of mesenchymal stem cells of fetal or maternal origin from human placenta. *Stem Cells*, Vol.22, No.7, (December 2004), pp.1338-1345, ISSN 1066-5099

Ishikane, S.; Ohnishi, S.; Yamahara, K.; Sada, M.; Harada, K.; Mishima, K.; Iwasaki, K.; Fujiwara, M.; Kitamura, S.; Nagaya, N., & Ikeda, T. (2008). Allogeneic injection of fetal membrane-derived mesenchymal stem cells induces therapeutic angiogenesis in a rat model of hind limb ischemia. *Stem Cells*, Vol.26, No.10, (August 2008), pp.2625-2633, ISSN 1549-4918

Ishikane, S.; Yamahara, K.; Sada, M.; Harada, K.; Kodama, M.; Ishibashi-Ueda, H.; Hayakawa, K.; Mishima, K.; Iwasaki, K.; Fujiwara, M.; Kangawa, K., & Ikeda, T. (2010). Allogeneic administration of fetal membrane-derived mesenchymal stem cells attenuates acute myocarditis in rats. *Journal of Molecular and Cellular Cardiology*, Vol.49, No.5, (August 2010), pp.753-761, ISSN 1095-8584

Iwase, T.; Nagaya, N.; Fujii, T.; Itoh, T.; Murakami, S.; Matsumoto, T.; Kangawa, K., & Kitamura, S. (2005). Comparison of angiogenic potency between mesenchymal stem cells and mononuclear cells in a rat model of hindlimb ischemia. *Cardiovascular Research*, Vol.66, No.3, (May 2005), pp.543-551, ISSN 0008-6363

Kim, J.; Kang, H. M.; Kim, H.; Kim, M. R.; Kwon, H. C.; Gye, M. C.; Kang, S. G.; Yang, H. S., & You, J. (2007). Ex vivo characteristics of human amniotic membrane-derived stem cells. *Cloning Stem Cells*, Vol.9, No.4, (December 2007), pp.581-594, ISSN 1536-2302

Kim, S. W.; Han, H.; Chae, G. T.; Lee, S. H.; Bo, S.; Yoon, J. H.; Lee, Y. S.; Lee, K. S.; Park, H. K., & Kang, K. S. (2006). Successful stem cell therapy using umbilical cord blood-derived multipotent stem cells for Buerger's disease and ischemic limb disease animal model. *Stem Cells*, Vol.24, No.6, (February 2006), pp.1620-1626, ISSN 1066-5099

Kinnaird, T.; Stabile, E.; Burnett, M. S.; Lee, C. W.; Barr, S.; Fuchs, S., & Epstein, S. E. (2004). Marrow-derived stromal cells express genes encoding a broad spectrum of arteriogenic cytokines and promote in vitro and in vivo arteriogenesis through paracrine mechanisms. *Circulation Research*, Vol.94, No.5, (January 2004), pp.678-685, ISSN 1524-4571

Kondoh, H.; Sawa, Y.; Miyagawa, S.; Sakakida-Kitagawa, S.; Memon, I. A.; Kawaguchi, N.; Matsuura, N.; Shimizu, T.; Okano, T., & Matsuda, H. (2006). Longer preservation of cardiac performance by sheet-shaped myoblast implantation in dilated cardiomyopathic hamsters. *Cardiovascular Research*, Vol.69, No.2, (January 2006), pp.466-475, ISSN 0008-6363

Kopen, G. C.; Prockop, D. J., & Phinney, D. G. (1999). Marrow stromal cells migrate throughout forebrain and cerebellum, and they differentiate into astrocytes after injection into neonatal mouse brains. *Proceedings of the National Academy of Sciences of the United States of America*, Vol.96, No.19, (September 1999), pp.10711-10716, ISSN 0027-8424

Kraitchman, D. L.; Tatsumi, M.; Gilson, W. D.; Ishimori, T.; Kedziorek, D.; Walczak, P.; Segars, W. P.; Chen, H. H.; Fritzges, D.; Izbudak, I.; Young, R. G.; Marcelino, M.; Pittenger, M. F.; Solaiyappan, M.; Boston, R. C.; Tsui, B. M.; Wahl, R. L., & Bulte, J. W. (2005). Dynamic imaging of allogeneic mesenchymal stem cells trafficking to myocardial infarction. *Circulation*, Vol.112, No.10, (September 2005), pp.1451-1461, ISSN 1524-4539

Krampera, M.; Glennie, S.; Dyson, J.; Scott, D.; Laylor, R.; Simpson, E., & Dazzi, F. (2003). Bone marrow mesenchymal stem cells inhibit the response of naive and memory antigen-specific T cells to their cognate peptide. *Blood*, Vol.101, No.9, (December 2002), pp.3722-3729, ISSN 0006-4971

Le Blanc, K.; Frassoni, F.; Ball, L.; Locatelli, F.; Roelofs, H.; Lewis, I.; Lanino, E.; Sundberg, B.; Bernardo, M. E.; Remberger, M.; Dini, G.; Egeler, R. M.; Bacigalupo, A.; Fibbe, W., & Ringden, O. (2008). Mesenchymal stem cells for treatment of steroid-resistant, severe, acute graft-versus-host disease: a phase II study. *Lancet*, Vol.371, No.9624, (May 2008), pp.1579-1586, ISSN 1474-547X

Le Blanc, K.; Tammik, C.; Rosendahl, K.; Zetterberg, E., & Ringden, O. (2003). HLA expression and immunologic properties of differentiated and undifferentiated mesenchymal stem cells. *Experimental Hematology*, Vol.31, No.10, (October 2003), pp.890-896, ISSN 0301-472X

Mareschi, K.; Ferrero, I.; Rustichelli, D.; Aschero, S.; Gammaitoni, L.; Aglietta, M.; Madon, E., & Fagioli, F. (2006). Expansion of mesenchymal stem cells isolated from pediatric and adult donor bone marrow. *Journal of Cellular Biochemistry*, Vol.97, No.4, (October 2005), pp.744-754, ISSN 0730-2312

Meisel, R.; Zibert, A.; Laryea, M.; Gobel, U.; Daubener, W., & Dilloo, D. (2004). Human bone marrow stromal cells inhibit allogeneic T-cell responses by indoleamine 2,3-dioxygenase-mediated tryptophan degradation. *Blood*, Vol.103, No.12, (March 2004), pp.4619-4621, ISSN 0006-4971

Memon, I. A.; Sawa, Y.; Fukushima, N.; Matsumiya, G.; Miyagawa, S.; Taketani, S.; Sakakida, S. K.; Kondoh, H.; Aleshin, A. N.; Shimizu, T.; Okano, T., & Matsuda, H. (2005). Repair of impaired myocardium by means of implantation of engineered autologous myoblast sheets. *Journal of Thoracic and Cardiovascular Surgery*, Vol.130, No.5, (November 2005), pp.1333-1341, ISSN 1097-685X

Miyahara, Y.; Nagaya, N.; Kataoka, M.; Yanagawa, B.; Tanaka, K.; Hao, H.; Ishino, K.; Ishida, H.; Shimizu, T.; Kangawa, K.; Sano, S.; Okano, T.; Kitamura, S., & Mori, H. (2006). Monolayered mesenchymal stem cells repair scarred myocardium after myocardial infarction. *Nature Medicine*, Vol.12, No.4, (April 2006), pp.459-465, ISSN 1078-8956

Moon, M. H.; Kim, S. Y.; Kim, Y. J.; Kim, S. J.; Lee, J. B.; Bae, Y. C.; Sung, S. M., & Jung, J. S. (2006). Human adipose tissue-derived mesenchymal stem cells improve postnatal neovascularization in a mouse model of hindlimb ischemia. *Cellular Physiology and Biochemistry*, Vol.17, No.5-6, (June 2006), pp.279-290, ISSN 1015-8987

Morishita, R.; Nakamura, S.; Hayashi, S.; Taniyama, Y.; Moriguchi, A.; Nagano, T.; Taiji, M.; Noguchi, H.; Takeshita, S.; Matsumoto, K.; Nakamura, T.; Higaki, J., & Ogihara, T. (1999). Therapeutic angiogenesis induced by human recombinant hepatocyte growth factor in rabbit hind limb ischemia model as cytokine supplement therapy. *Hypertension*, Vol.33, No.6, (June 1999), pp.1379-1384, ISSN 0194-911X

Muller-Ehmsen, J.; Krausgrill, B.; Burst, V.; Schenk, K.; Neisen, U. C.; Fries, J. W.; Fleischmann, B. K.; Hescheler, J., & Schwinger, R. H. (2006). Effective engraftment but poor mid-term persistence of mononuclear and mesenchymal bone marrow cells in acute and chronic rat myocardial infarction. *Journal of Molecular and Cellular Cardiology*, Vol.41, No.5, (September 2006), pp.876-884, ISSN 0022-2828

Nagaya, N.; Kangawa, K.; Itoh, T.; Iwase, T.; Murakami, S.; Miyahara, Y.; Fujii, T.; Uematsu, M.; Ohgushi, H.; Yamagishi, M.; Tokudome, T.; Mori, H.; Miyatake, K., & Kitamura, S. (2005). Transplantation of mesenchymal stem cells improves cardiac function in a rat model of dilated cardiomyopathy. *Circulation*, Vol.112, No.8, (August 2005), pp.1128-1135, ISSN 1524-4539

Nakagami, H.; Maeda, K.; Morishita, R.; Iguchi, S.; Nishikawa, T.; Takami, Y.; Kikuchi, Y.; Saito, Y.; Tamai, K.; Ogihara, T., & Kaneda, Y. (2005). Novel autologous cell therapy in ischemic limb disease through growth factor secretion by cultured adipose tissue-derived stromal cells. *Arteriosclerosis, Thrombosis, and Vascular Biology*, Vol.25, No.12, (October 2005), pp.2542-2547, ISSN 1524-4636

Ohnishi, S.; Yanagawa, B.; Tanaka, K.; Miyahara, Y.; Obata, H.; Kataoka, M.; Kodama, M.; Ishibashi-Ueda, H.; Kangawa, K.; Kitamura, S., & Nagaya, N. (2007). Transplantation of mesenchymal stem cells attenuates myocardial injury and dysfunction in a rat model of acute myocarditis. *Journal of Molecular and Cellular Cardiology*, Vol.42, No.1, (November 2006), pp.88-97, ISSN 0022-2828

Okada, H.; Suzuki, J.; Futamatsu, H.; Maejima, Y.; Hirao, K., & Isobe, M. (2007). Attenuation of autoimmune myocarditis in rats by mesenchymal stem cell transplantation through enhanced expression of hepatocyte growth factor. *International Heart Journal*, Vol.48, No.5, (November 2007), pp.649-661, ISSN 1349-2365

Parekkadan, B., & Milwid, J. M. (2010). Mesenchymal stem cells as therapeutics. *Annual Reviews Biomedical Engineering*, Vol.12, (April 2010), pp.87-117, ISSN 1545-4274

Pittenger, M. F.; Mackay, A. M.; Beck, S. C.; Jaiswal, R. K.; Douglas, R.; Mosca, J. D.; Moorman, M. A.; Simonetti, D. W.; Craig, S., & Marshak, D. R. (1999). Multilineage potential of adult human mesenchymal stem cells. *Science*, Vol.284, No.5411, (April 1999), pp.143-147, ISSN 0036-8075

Portmann-Lanz, C. B.; Schoeberlein, A.; Huber, A.; Sager, R.; Malek, A.; Holzgreve, W., & Surbek, D. V. (2006). Placental mesenchymal stem cells as potential autologous graft for pre- and perinatal neuroregeneration. *American Journal of Obstetrics and Gynecology*, Vol.194, No.3, (March 2006), pp.664-673, ISSN 1097-6868

Sensebe, L.; Krampera, M.; Schrezenmeier, H.; Bourin, P., & Giordano, R. (2010). Mesenchymal stem cells for clinical application. *Vox Sanguinis*, Vol.98, No.2, (August 2009), pp.93-107, ISSN 1423-0410

Shake, J. G.; Gruber, P. J.; Baumgartner, W. A.; Senechal, G.; Meyers, J.; Redmond, J. M.; Pittenger, M. F., & Martin, B. J. (2002). Mesenchymal stem cell implantation in a swine myocardial infarct model: engraftment and functional effects. *Annals of Thoracic Surgery*, Vol.73, No.6, (June 2002), pp.1919-1925; discussion 1926, ISSN 0003-4975

Sioud, M.; Mobergslien, A.; Boudabous, A., & Floisand, Y. (2011). Mesenchymal stem cell-mediated T cell suppression occurs through secreted galectins. *International Journal of Oncology*, Vol.38, No.2, (December 2010), pp.385-390, ISSN 1791-2423

Soncini, M.; Vertua, E.; Gibelli, L.; Zorzi, F.; Denegri, M.; Albertini, A.; Wengler, G. S., & Parolini, O. (2007). Isolation and characterization of mesenchymal cells from human fetal membranes. *Journal of Tissue Engineering and Regenerative Medicine*, Vol.1, No.4, (November 2007), pp.296-305, ISSN 1932-6254

Van Linthout, S.; Savvatis, K.; Miteva, K.; Peng, J.; Ringe, J.; Warstat, K.; Schmidt-Lucke, C.; Sittinger, M.; Schultheiss, H. P., & Tschope, C. (2010). Mesenchymal stem cells improve murine acute coxsackievirus B3-induced myocarditis. *European Heart Journal*, in press, (December 2010), ISSN 1522-9645

Weimar, I. S.; Miranda, N.; Muller, E. J.; Hekman, A.; Kerst, J. M.; de Gast, G. C., & Gerritsen, W. R. (1998). Hepatocyte growth factor/scatter factor (HGF/SF) is produced by human bone marrow stromal cells and promotes proliferation, adhesion and survival of human hematopoietic progenitor cells (CD34+). *Experimental Hematology*, Vol.26, No.9, (August 1998), pp.885-894, ISSN 0301-472X

Yamada, M.; Koeda, T.; Kikuchi, H.; Nasu, M.; Isagozawa, S.; Mukaida, H.; Yosida, H.; Ahsan, R.; Otokida, K., & Kato, M. (1990). Evaluation of increasing digital blood flow during early period of air-cooled cold test]. *Kokyu to Junkan. Respiration and Circulation*, Vol.38, No.6, (June 1990), pp.571-576, ISSN 0452-3458

Zarnegar, R., & Michalopoulos, G. K. (1995). The many faces of hepatocyte growth factor: from hepatopoiesis to hematopoiesis. *Journal of Cell Biology*, Vol.129, No.5, (June 1995), pp.1177-1180, ISSN 0021-9525

Zuk, P. A.; Zhu, M.; Mizuno, H.; Huang, J.; Futrell, J. W.; Katz, A. J.; Benhaim, P.; Lorenz, H. P., & Hedrick, M. H. (2001). Multilineage cells from human adipose tissue: implications for cell-based therapies. *Tissue Engineering*, Vol.7, No.2, (April 2001), pp.211-228, ISSN 1076-3279

11

Mesenchymal Stem Cells in CNS Regeneration

Arshak R. Alexanian
Medical College of Wisconsin, Neuroscience Research Laboratories, Department of Neurosurgery, VA Medical Center - Research 151, Milwaukee, Wisconsin, USA

1. Introduction

1.1 Mesenchymal stem cells as an ideal source of cells for regenerative medicine

During the last two decades, stem cells have become recognized as a promising tool for various biomedical applications including disease modeling, drug development, and cell replacement therapies. However, identification of the reliable sources of stem cells that can be easily harvested, expanded on a large enough scale, and carry no risk of immune rejection still remains one of the important issues for regenerative medicine.

Mesenchymal stem cells (MSCs) are promising tools for cell therapy (Zeidan-Chulia & Noda, 2009) by autologous and allogeneic transplantation for two significant reasons. Firstly, MSC can easily be isolated and expanded from different adult and postnatal tissues, such as BM (Prockop, 1997), peripheral blood (Kuznetsov et al., 2001), muscle (J. Y. Lee et al., 2000), vasculature (Brighton et al., 1992), skin (Mizuno & Glowacki, 1996), adipose tissue (Zuk et al., 2001) and umbilical cord (O. K. Lee et al., 2004). Secondly, MSCs can differentiate into multiple cell types of mesodermal, endodermal, and epidermal origin such as bone (Pereira et al., 1995), cartilage (Pereira et al., 1998), fat (Umezawa et al., 1991), muscle (Ferrari et al., 1998), cardiomyocytes (Makino et al., 1999), and neurons (Kohyama et al., 2001). Such a surprising high plasticity of MSCs might be explained by the expression of a variety of gene families in undifferentiated MSCs. Several recent studies have shown that MSCs express several embryonic stem cell markers (pluripotent markers) such as Oct4, Nanog, alkaline phosphatase and SSEA-4, and SOX2 (Park & Patel, 2010; Pierantozzi et al., 2011; Riekstina et al., 2009). It also has been demonstrated that the translational and transcriptional machinery in MSCs responsible for the expression of multiple genes typical of several derivatives of three germ layers are not silenced, rather operating at the low level (Blondheim et al., 2006; Tondreau et al., 2008). Most importantly, at appropriate environmental conditions in vitro and in vivo MSCs can upregulate the expression of these genes and exhibit several characteristics of mature cells of different tissues such as heart (Choi, Kurtz, & Stamm, 2011; Hattan et al., 2005; Makino et al., 1999), lever (Stock et al., 2010) and central nervous system (Alexanian, 2010). While it still many controversy concerning transdifferentiation of MSCs these recent data suggest that MSCs could be ideal autologous source of easily reprogrammable cells. Harboring such a high plasticity these cells, in contrast to adult and other tissue specific stem cells and progenitors, could be manipulated more easily (Niibe et al., 2011).

With the promise that MSCs present for the development of new cell therapies, researchers have pursued a broad range of investigations for their therapeutic utilization (Parekkadan & Milwid; Picinich, Mishra, Glod, & Banerjee, 2007; Stappenbeck & Miyoshi, 2009; Wang, Liao, & Tan, 2011). During the last two decades an overwhelming amount of basic and preclinical research has been accumulated that demonstrates the therapeutic usefulness of MSCs in the treatment of several diseases and injuries such as neurodegenerative diseases (Joyce et al.), spinal cord and brain injuries (Y. Jiang et al.), cardiovascular diseases (Trivedi, Tray, Nguyen, Nigam, & Gallicano), diabetes mellitus (Y. H. Zhang et al., 2009) and diseases of the skeleton (Chanda, Kumar, & Ponnazhagan). In most of these studies, treatment with MSCs results in substantial functional benefit and these pre-clinical studies have led to the initiation of a number of clinical trials worldwide.

MSCs have been used in clinical trials since 1995 and, currently, more than 180 trials are registered with ClinicalTrials.gov for the treatment of several diseases including numerous neurological disorders and injuries such amyotrophic lateral sclerosis, stroke, parkinson's disease, Alzheimer's disease, brain and spinal cord injuries.

2. In vitro neural differentiation potential of MSCs

Demonstration of neural differentiation potential of MSCs in several in vitro and in vivo studies suggests the potential usefulness of MSCs in the treatment of various CNS disorders. This potential has led to extensive studies to further explore the neural plasticity of these cells (Azizi, Stokes, Augelli, DiGirolamo, & Prockop, 1998; Kopen, Prockop, & Phinney, 1999; Munoz-Elias, Marcus, Coyne, Woodbury, & Black, 2004).

During the last several years, numerous in vitro neural induction protocols to produce neural cells from MSCs have been reported. In most induction experiments, MSCs were simply exposed to growth factors, neurotrophic factors or factors favoring neural cell differentiation (Bi et al., 2010; M. Chen et al., 2000; Q. Chen et al., 2005; Joannides et al., 2003; B. J. Kim, Seo, Bubien, & Oh, 2002; S. S. Kim et al., 2005; Kondo, Johnson, Yoder, Romand, & Hashino, 2005; Lim et al., 2008; Long, Olszewski, Huang, & Kletzel, 2005; Padovan et al., 2003; Sanchez-Ramos et al., 2000; Zeng et al., 2011). Other studies have used different culture media, supplemented with individual or various combinations of chemical and pharmacological agents, such as DMSO, b-mercaptoethanol, 5-bromo-2-deoxyuridine (BrdU), butylated hydroxyanisole, forskolin, and dibutyryl cyclic AMP (Ankeny, McTigue, & Jakeman, 2004; W. Deng, Obrocka, Fischer, & Prockop, 2001; Episkopou, 2005; Hermann et al., 2006; Jori et al., 2005; S. S. Kim et al., 2005; Lu, Blesch, & Tuszynski, 2004; Munoz-Elias, Woodbury, & Black, 2003; Tio, Tan, Lee, Wang, & Udolph, 2010; Yang, Wu, & Xiao, 2005; L. Zhang, Seitz, Abramczyk, Liu, & Chan, 2011). Other methods to induce MSCs into cells with neural characteristics include: transfection of MSCs with Noggin and Notch transcription factors (Dezawa et al., 2004; Kohyama et al., 2001); manipulation with surface proteins of culture substrate (Qian & Saltzman, 2004); co-culturing MSCs with NSCs or neural cells (Alexanian, 2005; Chu, Yu, Zhang, & Yu, 2008; Krampera et al., 2007; Wislet-Gendebien et al., 2005; Y. Q. Zhang et al., 2010); and growing MSCs as spheres in cultures (Shiota et al., 2007), transfection of MSCs with microRNA-9 (Jing et al., 2011). In several other studies, MSCs were turned into multipotent stage and then induced into neural cell lineages, by exposing them to appropriate neural differentiation conditions (Alexanian, 2007; Kohyama et al., 2001; Qu et al., 2004). Recently, we proposed an original method for efficient generation of neural cells from feline and human BM-derived MSCs (hMSC) (Alexanian, 2010; Z. Zhang, Maiman, Kurpad, Crowe, & Alexanian, 2011). In these studies, neural induction was achieved by exposing cells simultaneously to inhibitors of DNA methylation and histone deacetylation and pharmacological agents that increased cAMP levels. The main idea of this methodological approach was the reactivation of pluripotency-associated genes in MSCs simultaneously exposing them to neural-inducing factors. Neurally modified MSCs by this methodology, in contrast to naïve MSCs, express several neural progenitor and mature neural markers demonstrated by real time RT-PCR, western blot, ELISA and immunocytochemistry Fig.1. and Fig.2.

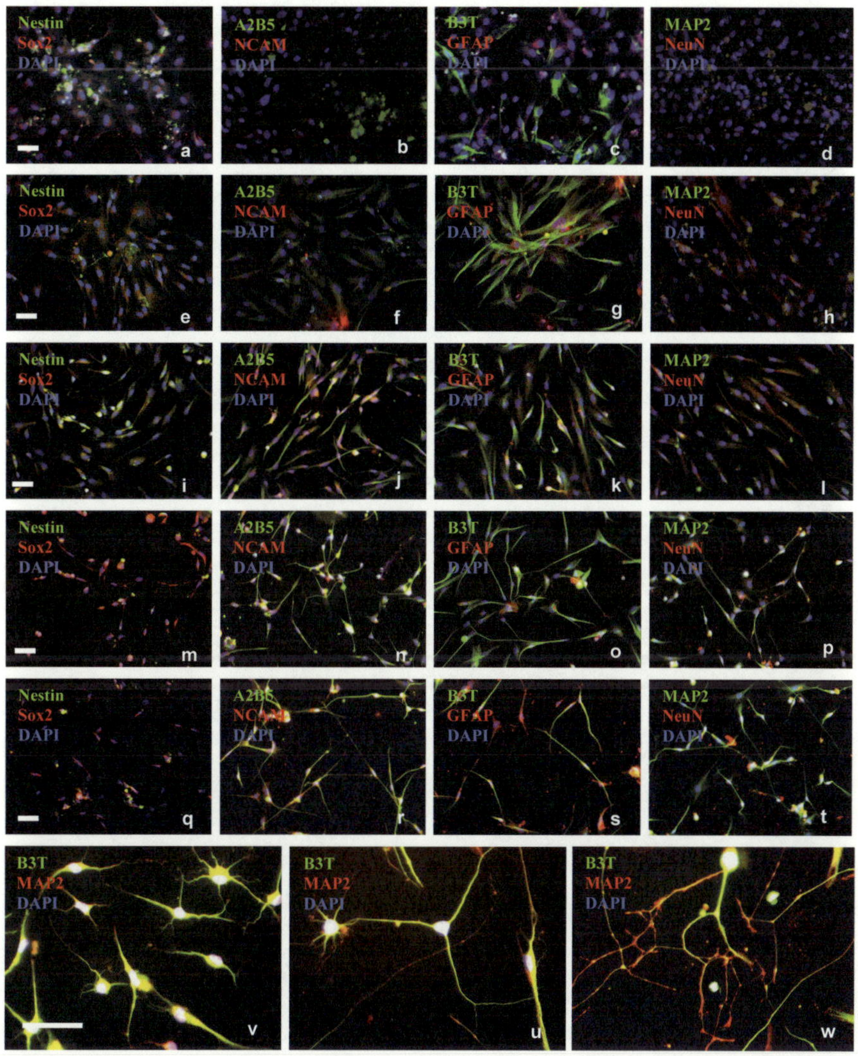

Fig. 1. Expression of neural markers nestin, Sox2, A2B5, NCAM, B3T, GFAP, MAP-2, and NeuN in hMSCs (a-d) and NI-hMSC grown 24h, 1, 2, 3 weeks in neural induction medium (e-t). NI-hMSCs grown an additional week in neuronal induction medium were generated cells with long axon- and dendrite-like extensions (v-w). Bars 40um.

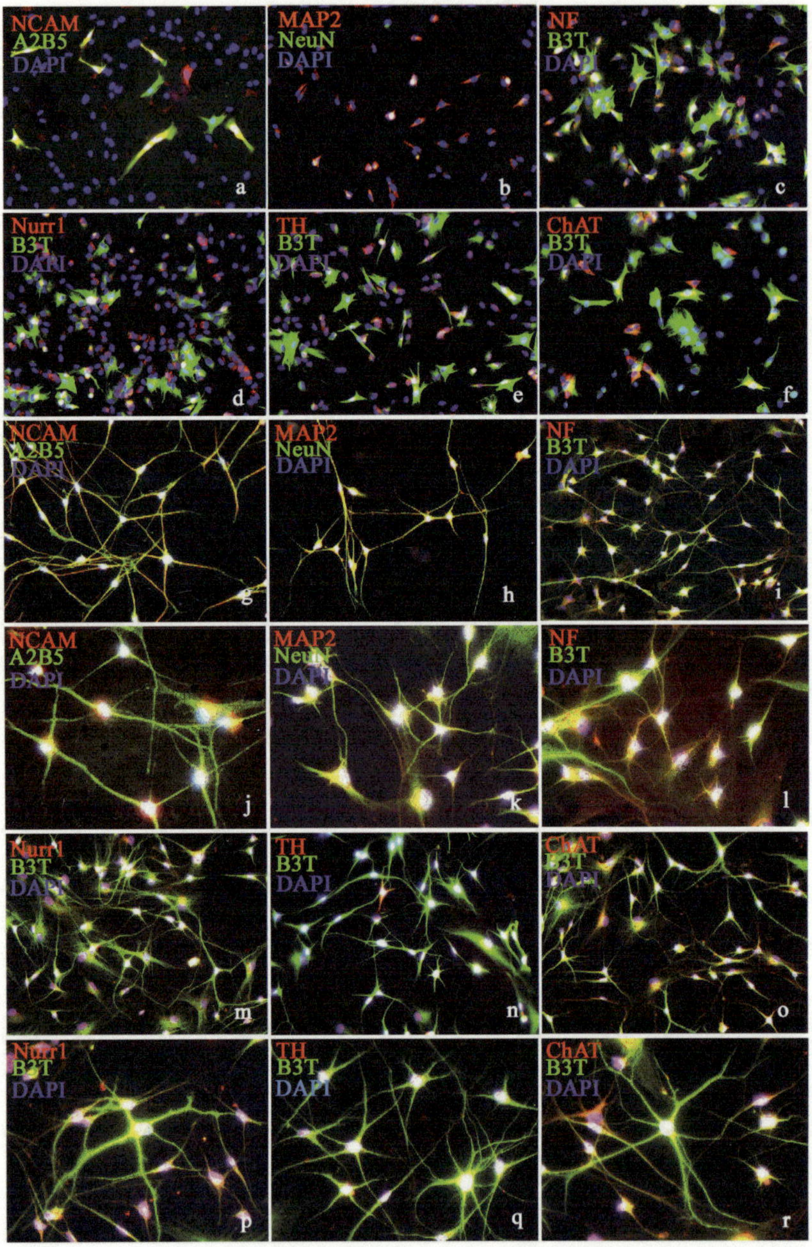

Fig. 2. Morphological and immunocytochemical characterization of unmodified and NI-fMSCs. Expression of neural markers B3T, NCAM, A2B5, MAP2, NeuN, NF, Nurr1, TH and ChAT in unmodified fMSCs (a-f) and in NI-fMSCs grown for 72h in neural induction medium (g-r).

Despite these studies, there is an intense ongoing debate about the nature of these differentiation responses. For example, some recent reports suggested that cell fusion could account for transdifferentiation (Terada et al., 2002). However, spontaneous cell fusion is a very rare event and, therefore, can not be account for massive transdifferentiation demonstrated in numerous recent studies. In addition, MSCs can be induced into neural-like cells with several neural inducing factors, without being grown in co-cultures with NSCs.

A few other reports suggested that some of these investigations suffered from artifacts created by in vitro chemical stress (Lu et al., 2004; Neuhuber et al., 2004). Nevertheless, Tondreau and colleagues have recently found significant upregulation of neural genes and downregulation of chondrogenic, osteogenic, adipogenic and myogenic genes in neurally differentiated MSCs as demonstrated by microarray analysis (Tondreau et al., 2008). In addition, a numerous studies suggest that with appropriate neural induction protocols, MSCs could produce mature neuron-like cells that exhibit multiple neuronal properties and traits, such as action potential, synaptic transmission, secretion of neurotrophic factors and dopamine, and demonstration of spontaneous post-synaptic current (Alexanian, Maiman, Kurpad, & Gennarelli, 2008; Bonilla et al., 2005; Greco, Zhou, Ye, & Rameshwar, 2008; Hermann et al., 2004; Y. Jiang et al., 2003; S. S. Kim et al., 2008; Mareschi et al., 2009; Trzaska et al., 2009; Wislet-Gendebien et al., 2005). Whether these neurally modified MSCs can produce fully functional neural cells in vitro and vivo is still under intensive investigations.

3. In vivo neural differentiation potential of MSCs

One of the first discoveries that demonstrate the pluripotent nature of adult MSCs in vivo, came from Ferrari et al. who clearly showed that adult murine BM contained cells capable of differentiation into skeletal muscle (Ferrari et al., 1998). In the past decade or more, several other studies have documented the ability of adult BM-derived cells to differentiate into liver and epithelium (Petersen et al., 1999; Theise, Badve, et al., 2000; Theise, Nimmakayalu, et al., 2000), endothelium (Kawamoto et al., 2001; Kawamoto et al., 2003; Takahashi et al., 1999), heart (Kucia et al., 2004; Orlic et al., 2001; Tomita et al., 1999), and brain (Brazelton, Rossi, Keshet, & Blau, 2000; Eglitis & Mezey, 1997; Mezey, Chandross, Harta, Maki, & McKercher, 2000). These striking observations indicate that there are BM cells that can migrate to distant sites and participate in repair of tissues across germ layer boundaries. In the most striking examples, BM cells injected in to the blastocyst contributed to most somatic cell lineages, including neural (Y. Jiang et al., 2002). These discoveries have led to extensive studies to further explore the neural differentiation potential of MSCs in intact, injured and diseased CNS.

However, multiple studies conducted during the last decade showed that MSCs transplanted into the intact, injured or diseased CNS environments do not differentiate or only a small portion of cells produce neural phenotypes (Alexanian, Kwok, Pravdic, Maiman, & Fehlings, 2010; Castro et al., 2002; J. Deng, Petersen, Steindler, Jorgensen, & Laywell, 2006). In contrast, MSCs transplanted in developing embryonic brain or in neurogenic areas of the adult brain expressed heterogeneous traits characteristic of radial glia, subventricular zone progenitors, migratory cells, parenchymal neurons, and glia (Azizi et al., 1998; Kopen et al., 1999; J. M. Li et al., 2011; Munoz-Elias et al., 2004). The fate of MSCs consequently appeared to be regulated by multiple influences, presumably including different microenvironments. These are in close analogy with studies in which pluripotent or highly immature NSCs were used. In a similar way, transplanted cells generated different neural phenotypes when transplanted into one of the few neurogenic areas of the brain [35,36] but remained undifferentiated or differentiated predominantly into the glial cells when transplanted into injured or non-neurogenic areas [37]. However, when late-stage precursors and immature neurons were transplanted into non-neurogenic or injured brain and spinal cord, more neural differentiation was observed [38,39]. This indicates that, while microenvironment can play a decisive role in determining the fate of engrafted MSCs or NSCs, the intrinsic state of these transplanted cells is another important factor for the commitment of cells to a particular phenotype. MSCs which presumably committed to mesodermal lineages most probably will not produce neural cells in intact, injured or diseased CNS and therefore, manipulation of cells into neural fate maybe required before transplantation. In fact, several recent studies showed that neurally modified MSCs transplanted into intact or damaged CNS exhibited higher ability to generate cells positive to various neural markers (Alexanian et al., 2008; Alexanian, Michael, Zhang, & Maiman, 2011; Cho et al., 2009).

4. Therapeutic effects of naïve and neurally modified MSCs in CNS disorders and their underlying mechanisms

Experimental treatments of CNS disorders can be broadly grouped into the two distinct but interrelated strategies of neuroprotection and neurorepair/neuroregeneration. Neuroprotection refers to inhibition of the death of CNS parenchymal cells in traumantic and neurodegenerative CNS, neurorepair/neuroregeneration refers to the replacement of lost neural cells, stimulation of endogenous neural progenitors and/or regeneration of severed axons or sprouting of intact axons to innervate denervated targets in injured or diseased CNS. MSCs have been used for all of these strategies and exhibited beneficial therapeutic effect in several animal models of CNS injury and neurodegenerative diseases.

4.1 MSC in CNS injury (traumatic spinal cord and brain injury, ischemia/stroke)

Recent multiple studies demonstrated that naïve or neurally modified MSCs derived from different tissue sources exerted therapeutic effect in several animal models of spinal cord injury (SCI). However, the precise mechanisms by which transplantation of MSCs promote functional recovery after SCI is still unclear. A number of mechanisms have been suggested, including the promotion of axon regeneration, neuroprotection, modulation of the immune responses, and trans-differentiation into neural cell types (Chamberlain, Fox, Ashton, & Middleton, 2007; Dezawa, 2002; Enzmann, Benton, Talbott, Cao, & Whittemore, 2006; Keilhoff, Goihl, Stang, Wolf, & Fansa, 2006). The immunosuppressive properties of MSCs (Bartholomew et al., 2002a; Corcione et al., 2006; Di Nicola et al., 2002; X. X. Jiang et al., 2005) may combine to reduce the acute inflammatory response to SCI and hence reduce cavity formation as well as decrease astrocyte and microglia/macrophage reactivity (Abrams et al., 2009; Himes et al., 2006; Neuhuber, Timothy Himes, Shumsky, Gallo, & Fischer, 2005) in injured spinal cords. The therapeutic effect of MSCs on axonal growth could be exerted by creation of a favorable environments such as cellular bridges, guiding strands and scaffolds, secretion of trophic factors, cytokines and production of extracellular matrix (Fuhrmann et al., 2010; Gu et al., 2010; Hofstetter et al., 2002; Neuhuber et al., 2004). The neuroprotective mechanism of MSCs could be multifactorial, such as modulation of immune response and provision of trophic factors (Uccelli, Benvenuto, Laroni, & Giunti, 2011). Whether MSCs therapeutic effect can be exerted via cell replacement is still one of the most debated issues.

In most reported studies, transplanted MSCs either do not differentiate, or only very small percentage of cells survive and produce neural cells in vivo. This led to studies to elucidate whether neural modification of MSCs will promote cell survival and neural differentiation of transplanted cells in intact and injured CNS.

Several recent studies suggest that neural modification of MSC prior to their transplantation can exhibit even higher beneficial therapeutic effect then naïve MSCs. In one of these studies Sung-Rae Cho et al. showed that transplantation of neurally differentiated MSCs derived from bone marrow promoted functional recovery in spinal cord injured rats and the latency of somatosensory evoked potentials were significantly improved compared with those of naïve MSCs and PBS controls (Cho et al., 2009). Furthermore, transplanted cells prelabeled with BrdU also differentiated into neural lineage cells that expressed specific markers for astrocytes and oligodendrocytes 4 weeks after transplantation, even though the number of integrated cells was not abundant. However, these differentiated cells did not survive longer than 8 weeks post transplantation, which was similar to what was reported in a previous studies (4). Because injured rats showed significant motor recovery at a relatively early stage after transplantation, and only a small number of transplanted cells survived in the injured spinal cord for a limited period, authors concluded that trophic or paracrine support could be the main factors for functional improvement.

Recently, we also demonstrated that transplanted neurally induced hMSCs (NI-hMSCs) promoted tissue preservation and improved locomotor recovery of injured animals (Alexanian et al., 2011). Motor recovery that consisted of hindlimb weight support and consistent hindlimb stepping was significantly different at 2-12 weeks post-recovery in the group that was transplanted with NI-hMSCs when compared with the control groups that received hMSCs and PBS (Fig.3).

Histological studies of spinal cord sections at specified distances rostral and caudal to the epicenter demonstrated that at the epicentre and 1mm caudal and rostral from it the percentage of the eriochrome cyanine-positive spared white matter was significantly larger in NI-hMSCs treated group than that in the PBS group (Fig.4.A,B). While there was no significant difference between naïve hMSCs and PBS groups, there was a modest trend for increased white matter sparing in hMSCs-treated versus PBS-treated spinal cords (Fig.4.B).

Stereological assessments of injured spinal cord tissues demonstrated a modest reduction in the percentage of cystic cavities in the NI-hMSCs and hMSCs treated groups versus PBS group (Fig.4.C) (Fig.5). Although no statistically significant difference had been noticed between groups (Fig.4.C), the difference found between NI-HMSCs and PBS was very close to the significance level adopted in the study ($p<0.05$).

Immunohistochemistry data showed that NI-hMSCs were survived at post transplantation weeks 1-12. Analysis of the spinal cord slices of two weeks treated animals revealed that 85% percent of survived cells were positive to B3T (Fig.6.a,b,c,d). A small percentage of cells (2%) was positive to GFAP (Fig.4.e) and 5% to Sox2 (Fig.6.f). By 12 weeks the number of surviving cells declined to 15-20% of that at week 2 and only 10% of survived cells were positive to B3T (Fig.6.g,h,i).

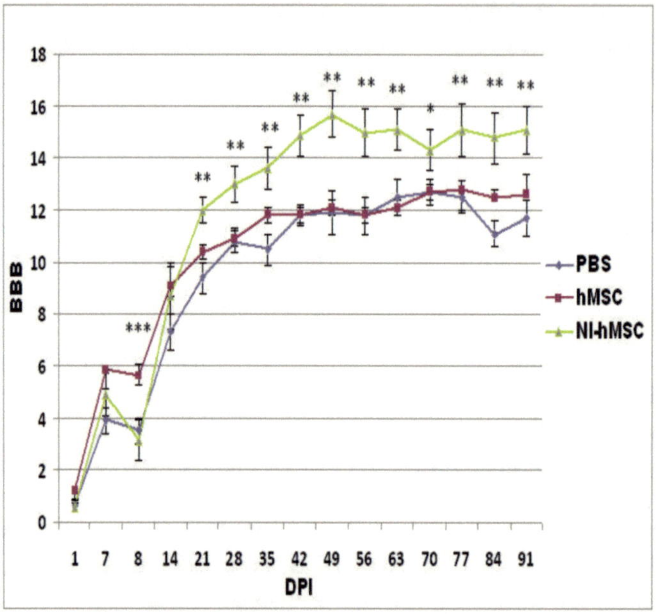

Fig. 3. Locomotor recovery (BBB) scores for the post spinal cord injury (DPI-days post injury) behavioral analysis. The asterisks (*) and (**) indicates a significant differences between the NI-hMSCs transplanted group compaired to the PBS and PBS+HMSCs groups respectively. Asterisk (***) indicates a significant differences between the hMSCs transplanted group compaired to the PBS.

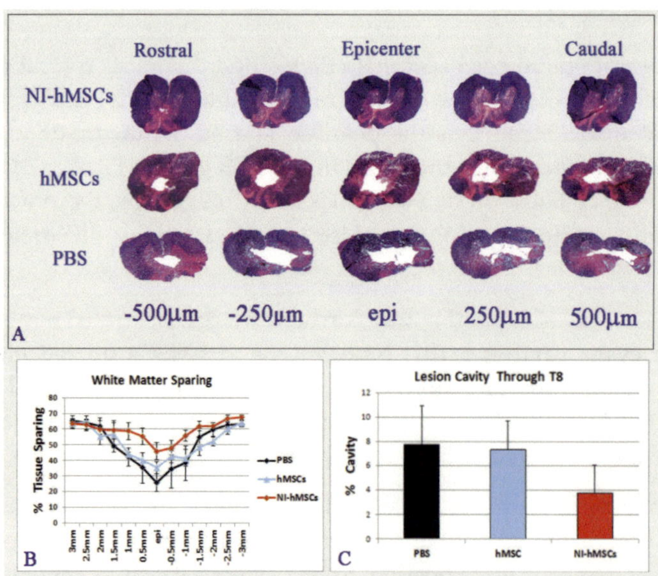

Fig. 4. Analysis of white matter sparing and lesion cavity volumes in NI-hMSCs, hMSCs, and PBS treated groups. (A) Representative spinal cord cross-sections extending 500um rostral and caudal from the lesion epicenter. (B) Graph representing the percentages of

spared white matter through the entire T8 spinal cord segment. (C) Graph representing comparison of the volumes of lesion cavities.

Fig. 5. Representative three-dimensionally reconstructed images of the lesion cavities through T8 injured spinal cord segments of NI-hMSCs and PBS treated animals.

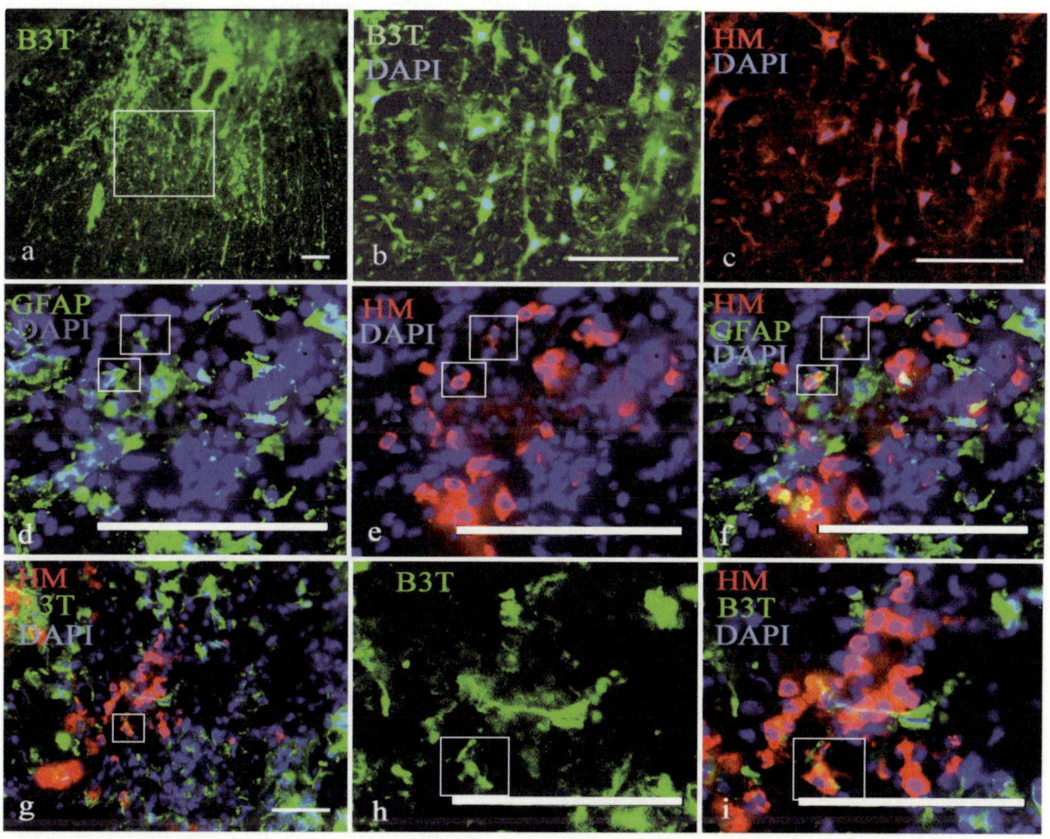

Fig. 6. Transplanted NI-hMSCs survived 2 weeks after transplantation and expressed neural markers such as B3T (a-c, b and c are the higher magnifications of the marked area in the image a), and GFAP (d-f). By 12 weeks the number of surviving cells declined to 15-20% of

that at week 2 and only 10% of survived cells were positive to B3T (g,h,i). The images h and i are higher magnifications of g. HM stands for human anti-mitochondrial antibody.

Thus, MSCs, either neurally modified or not, may provide an alternative source of autologous adult stem cells that could be useful for replacing damaged neural cells in injured spinal cord and/or providing support to spinal cord tissue cells.

Over the last decade or so, MSCs have been also used in experimental repair of the injured brain. Chopp and Li initially demonstrated transplanted MSCs promote functional recovery in rats with traumatic brain injury and attributed the beneficial effects of MSCs to the enhancement of endogenous restorative and regenerative processes (Chopp & Li, 2002). Later, Chopp and his group showed that MSCs treated with neurotrophins NGF and BDNF in vitro led to a higher number of engrafted cells after transplantation into the adult rat brain and improved motor function. A small number of cells stained for either astrocytic or neuronal markers (Mahmood, Lu, Wang, & Chopp, 2002), but were far too few to provide cellular replacement. This group also reported that i.v. administration of MSCs 1 day after brain injury in the rat brain resulted in an increase in BDNF and NGF (Mahmood et al., 2002). Both intracerebral and i.v. MSC administration promoted endogenous progenitor cell proliferation after traumatic brain injury (Mahmood, Lu, & Chopp, 2004), and functional recovery was dose dependent and persisted for at least 3 months (Mahmood, Lu, Qu, Goussev, & Chopp, 2006). Recently another group confirmed the therapeutic effect of human MSCs (hMSCs) in a rat model of TBI and demonstrated that expression of neurotrophic growth factors was induced by MSC treatment (H. J. Kim, Lee, & Kim, 2010). Furthermore, they observed an increase in phosphorylation of the cell survival signaling molecule, Akt, followed by decreased caspase-3 activation. These results suggest that the therapeutic effects of hMSCs transplantation may involve promotion of antiapoptotic activity as a result of secreted growth factors (H. J. Kim et al., 2010).

A single Phase I study using bone marrow-derived MSCs in children after isolated TBI has recently been completed (Cox et al., 2011). In this study, 10 children age 5–14 years with a Glasgow coma scale score of 5–8 were treated with 6 × 106 bone marrow-derived mononuclear cells per kg body weight delivered intravenously within 48 hours of an isolated TBI. To determine the safety of administration, systemic and cerebral hemodynamics, laboratory parameters, chest radiographs, and serial clinical assessments were monitored. Additionally, serial cerebral magnetic resonance imaging neuropsychologic evaluation, and functional outcome measures were obtained as preliminary measures of efficacy. There were no identifiable adverse events with close monitoring of the neurologic, pulmonary, renal, hepatic, and hematologic systems. Functional and neuropsychological testing, including the Glasgow Outcome Scale, the Pediatric Injury Functional Outcome Scale, and the Wechsler Abbreviated Scale of Intelligence, revealed recovery consistent with (or improved from) expected baselines. Magnetic resonance imaging volumetric data revealed no significant change in grey matter, white matter, intracranial volume, or CSF space at 1 and 6 months as measured relative to expected norms. Authors concluded that bone marrow harvest and intravenous mononuclear cell infusion as treatment for severe TBI in children is logistically feasible and safe.

The therapeutic effect of MSCs was also demonstrated in animal models of stroke. Several recent studies showed that transplantation of MSCs, derived from bone marrow, into rodent cerebral ischemia models can reduce infarct size and improve functional outcome (18,27,50,52,83,85,106,110,111,139,144). MSCs derived from adipose tissue (ADSCs) also showed therapeutic effect in rat model of cerebral ischemia (150). Importantly, treatment of ischemic animals with neurally induced ADSCs resulted in better functional recovery and more reduction in hemispheric atrophy in comparison to unmodified ADSCs (150).

To test the clinical relevance of these observations, recently, a phase I clinical trial was conducted. A feasibility and safety of transplantation of autologous human MSCs in stroke patients was the main objective of this trail (51). In this study the autologous MSCs were delivered intravenously 36–133 days post-stroke. All patients had magnetic resonance angiography to identify vascular lesions, and magnetic resonance imaging prior to cell infusion and at intervals up to 1 year after. Neurological status was scored using the National Institutes of Health Stroke Scale and modified Rankin scores. The results of this study showed that the median daily rate of National Institutes of Health Stroke Scale change was 0.36 during the first week post-infusion, compared with a median daily rate of change of 0.04 from the first day of testing to immediately before infusion. No central nervous system tumors, abnormal cell growths or neurological deterioration was observed, and there was no evidence for venous thromboembolism, systemic malignancy or systemic infection in any of the patients following stem cell infusion. Thus the stroke is another potential target for MSCs therapy.

4.2 MSCs in neurodegenerative diseases

There is currently a great deal of interest in the use of MSCs to treat several neurological diseases such as Parkinson's disease, Alzheimer's disease, amyotrophic lateral sclerosis, and multiple sclerosis.

Recently, a number of studies have examined the ability of MSCs to differentiate into dopamine-producing cells, re-innervate the striatum, and ameliorate behavioral deficits in Parkinsonian models. Varying degrees of success have been achieved in vitro, including dopaminergic marker expression, and dopamine secretion in response to depolarization (Dezawa et al., 2004; Fu et al., 2006; Guo et al., 2005; Suon, Yang, & Iacovitti, 2006; Trzaska, Kuzhikandathil, & Rameshwar, 2007; Trzaska & Rameshwar, 2011). In addition, engraftment and functional improvement were demonstrated following transplantation of undifferentiated (Hellmann, Panet, Barhum, Melamed, & Offen, 2006; Y. Li et al., 2001) and neurally differentiated MSCs (Dezawa et al., 2004; Fu et al., 2006) in hemiparkinsonian rodents. However, only relatively low efficiencies of dopaminergic differentiation were achieved, and comparisons between the varying methods have not been performed, resulting in difficulties with identifying the optimal methodology. These studies suggest that complex mechanisms might underline the therapeutic effect of MSCs in these animal parkinsonian models and neuroprotection could be the most important ones (P. H. Lee & Park, 2009). Despite all these promising data several issues remain to be resolved including the optimal method for inducing a dopaminergic phenotype from MSCs, engraftment and survival capabilities of MSCs, optimal sites for transplantation, potential immunological responses to MSC grafts, and whether neural differentiation prior to transplantation provides engraftment advantages.

Unlike Parkinson disease, which is a slower degenerative disease and affects a specific area of the brain, amyotrophic lateral sclerosis (ALS) presents quite a challenge for cellular therapy because of the distributed cell loss throughout the body and the requirement to properly reinnervate muscle tissue. Transplantation of wild-type BM cells into irradiated SOD1 transgenic mouse models of ALS demonstrated a delay in disease onset and an increase in life span (Corti et al., 2004). Minimal neural differentiation was detected, thus the authors concluded that functional improvement was likely due to trophic effects. Another study showed that transplantation of human MSCs into SOD1 ALS mice significantly delayed disease onset and progression, in addition to increasing lifespan (56). The human cells survived more than 20 weeks in the xenogenic model, and were able to migrate into the brain and spinal cord and differentiate into neuroglial cells (Zhao et al., 2007). Initial clinical studies began in 2003, when Mazzini et al took autologous MSCs from seven ALS patients and expanded them in culture (Mazzini et al., 2003). The cells were directly transplanted into the spinal cord, and did not result in toxicity or uncontrolled proliferation. Three months after transplantation, four patients experienced a mild reduction in muscle strength decline in the lower limbs. In a long term follow-up of the patients, the same group reported, after 36 months, that four of the seven patients showed a significant reduction in the linear decline of lung function and ALS functional rating scale (Mazzini et al., 2006). Though these preliminary clinical studies are encouraging, further studies are warranted.

Research on the role of MSCs in Alzheimer's disease (AD) is in its infancy. However, a recent study showed positive results in an AD rat model (Wu, Li, Feng, & Wang, 2007). Transplantation of BM-derived MSCs into the hippocampus of rats injected with ß amyloid protein to mimic AD demonstrated significant improvement based on the Morris Water Maze test (Wu et al., 2007). The authors suggested that the MSCs transdifferentiated into cholinergic cells and improved the cognitive ability of the AD rat models. Another group recently showed that transplanted MSCs exerted anti-apoptotic effect in an acutely-induced AD mice model produced by injecting Abeta intrahippocampally (J. K. Lee, Jin, & Bae, 2010). The same group also showed that intracerebral transplantation of BM-MSCs into APP/PS1 mice significantly reduced amyloid beta-peptide (Abeta) deposition (J. K. Lee, Jin, Endo, et al., 2010). Interestingly, these effects were associated with restoration of defective microglial function, as evidenced by increased Abeta-degrading factors, decreased inflammatory responses, and elevation of alternatively activated microglial markers. Furthermore, APP/PS1 mice treated with BM-MSCs had decreased tau hyperphosphorylation and improved cognitive function. Thus, BM-MSCs can modulate immune/inflammatory responses in AD mice, ameliorate their pathophysiology, and improve the cognitive decline associated with Abeta deposits. These results demonstrate that BM-MSCs are a potential new therapeutic agent for AD. Interestingly, Stroch A. et al and his group recently detected the functional induction of two genes upon neuroectodermal conversion of human adult MSCs, namely F-spondin and neprilysin (CD10), with a 4,992 + or - 697-fold and 692 + or - 226-fold increase of mRNA levels in converted cells compared to MSCs, respectively (Habisch et al., 2010). These genes are known to be involved in the formation and degradation of Abeta peptides, respectively. Consistently, co-incubation of the neuroectodermally converted MSCs with HEK-293 cells stably expressing amyloid precursor protein (APP) lead to a significant cell dose-dependent decrease of Abeta peptides. These in vitro results indicate that neurally modified MSCs might be even more useful vehicles for delivering anti-Abeta activity and thus exhibiting the maxium therapeutic effect on AD (Habisch et al., 2010).

The potentials of MSCs as a therapy for autoimmune neurological diseases arose from some unexpected observations. Therapies with MSCs were originally based on their similarities to most adult stem cells and the possibility that they might regenerate tissues through their ability to differentiate into mesodermal tissues and perhaps other embryonic lineages. The unexpected observation that MSCs inhibited T cell proliferation both in vitro (Di Nicola et al., 2002) and in vivo (Bartholomew et al., 2002b) introduced the possibility that MSCs might be effective in experimental autoimmune encephalomyelitis (EAE), a model for multiple sclerosis (MS) halting the (auto)immune attack to myelin antigens and promoting nervous tissue repair through their integration in the central nervous system (CNS). During the last few years, animal experiments in the EAE mouse model of MS showed that intraventricular, intraperitoneal or intravenous injection of human or murine BM-MSCs significantly improved clinical outcomes (Bai et al., 2009; Gordon et al., 2008; Kassis et al., 2008; Zappia et al., 2005; J. Zhang et al., 2005) (Zappia et al., 2005, Bai et al., 2009, Gordon et al., 2008, Zhang et al., 2005, Kassis et al., 2008). Chopp's group, in addition to observing functional recovery in EAE mice, demonstrated that small percentage of transplanted MSCs was integrated and expressed neural markers (J. Zhang et al., 2006). Their observations therefore suggested that some transdifferentiation had occurred. Overall, these pioneer studies demonstrated the therapeutic efficacy of MSCs in a model of CNS autoimmunity, but they left open the question whether their integration in the nervous system was essential for their therapeutic benefits.

Recently, a phase I trial was initiated to evaluate the safety and feasibility of intrathecal injection of autologous BM-MSCs in MS patients (Karussis et al., 2010). The initial findings of this trial support the possibility of migration of MSCs from their site of injection (lumbar area of the cerebrospinal fluid) to the brain ventricles and spinal cord parenchyma. Preliminary data of this trail also demonstrated the immunomodulatory effect of MSC in human neurological diseases. The authors concluded that the early clinical stabilization and improvement in some of the patients could be related to these immunomodulating effects. The possibility of neuroprotection and neuroregeneration through transdifferentiation of MSCs into cells of the neuronal or glial lineage, although theoretically viable, has yet to be proved by neuroimaging studies.

Promising results from this study will support further clinical trials to evaluate the long term safety and the potential clinical efficacy of MSC transplantation in the treatment of MS.

5. Conclusions

Although the curative effect of MSCs has been demonstrated in several animal models of CNS injury and neurodegeneration as well as in early human clinical trials of neurological disorders, the mechanisms that are responsible for these beneficial therapeutic effects are still poorly understood. Analysis of accumulated literature in this area suggest the following main mechanisms that may underlie the therapeutic effect of naive or neurally modified MSCs: 1) neurorepair (replacement of damaged or diseased neural cells by neurally transdifferentiated MSCs), 2) neuroprotection (modulation of immune response and inflammation, provision of trophic factors that could prevent neural cell apoptosis and demyelination); 3) neuroregeneration (creation of a favorable environment such as cellular bridges, guiding strands and scaffolds, provision of of neurotrophines, growth factors or cytokines that could promote axonal growth and sprouting and endogenous neurogenesis, restoration of blood flow, repair of blood–brain barrier, angiogenesis). However, these mechanisms are not mutually exclusive and it is most likely that combination of several factors accountable for such therapeutic effects.

6. References

Abrams, M. B., Dominguez, C., Pernold, K., Reger, R., Wiesenfeld-Hallin, Z., Olson, L., et al. (2009). Multipotent mesenchymal stromal cells attenuate chronic inflammation and injury-induced sensitivity to mechanical stimuli in experimental spinal cord injury. Restor Neurol Neurosci, 27(4), 307-321.

Alexanian, A. R. (2005). Neural stem cells induce bone-marrow-derived mesenchymal stem cells to generate neural stem-like cells via juxtacrine and paracrine interactions. Exp Cell Res, 310(2), 383-391.

Alexanian, A. R. (2007). Epigenetic modifiers promote efficient generation of neural-like cells from bone marrow-derived mesenchymal cells grown in neural environment. J Cell Biochem, 100(2), 362-371.

Alexanian, A. R. (2010). An efficient method for generation of neural-like cells from adult human bone marrow-derived mesenchymal stem cells. Regen Med, 5(6), 891-900.

Alexanian, A. R., Kwok, W. M., Pravdic, D., Maiman, D. J., & Fehlings, M. G. (2010). Survival of neurally induced mesenchymal cells may determine degree of motor recovery in injured spinal cord rats. Restor Neurol Neurosci, 28(6), 761-767.

Alexanian, A. R., Maiman, D. J., Kurpad, S. N., & Gennarelli, T. A. (2008). In vitro and in vivo characterization of neurally modified mesenchymal stem cells induced by epigenetic modifiers and neural stem cell environment. Stem Cells Dev, 17(6), 1123-1130.

Alexanian, A. R., Michael, F. G., Zhang, Z., & Maiman, D. J. (2011). Transplanted neurally modified bone marrow derived mesenchymal stem cells promote tissue protection and locomotor recovery in spinal cord injured rats. Neurorehabil Neural Repair. 25, 873-880.

Ankeny, D. P., McTigue, D. M., & Jakeman, L. B. (2004). Bone marrow transplants provide tissue protection and directional guidance for axons after contusive spinal cord injury in rats. Exp Neurol, 190(1), 17-31.

Azizi, S. A., Stokes, D., Augelli, B. J., DiGirolamo, C., & Prockop, D. J. (1998). Engraftment and migration of human bone marrow stromal cells implanted in the brains of albino rats--similarities to astrocyte grafts. Proc Natl Acad Sci U S A, 95(7), 3908-3913.

Bai, L., Lennon, D. P., Eaton, V., Maier, K., Caplan, A. I., Miller, S. D., et al. (2009). Human bone marrow-derived mesenchymal stem cells induce Th2-polarized immune response and promote endogenous repair in animal models of multiple sclerosis. Glia, 57(11), 1192-1203.

Bartholomew, A., Sturgeon, C., Siatskas, M., Ferrer, K., McIntosh, K., Patil, S., et al. (2002a). Mesenchymal stem cells suppress lymphocyte proliferation in vitro and prolong skin graft survival in vivo. Experimental hematology, 30(1), 42-48.

Bartholomew, A., Sturgeon, C., Siatskas, M., Ferrer, K., McIntosh, K., Patil, S., et al. (2002b). Mesenchymal stem cells suppress lymphocyte proliferation in vitro and prolong skin graft survival in vivo. Exp Hematol, 30(1), 42-48.

Bi, Y., Gong, M., Zhang, X., Jiang, W., Zhang, Y., Chen, J., et al. (2010). Pre-activation of retinoid signaling facilitates neuronal differentiation of mesenchymal stem cells. Dev Growth Differ, 52(5), 419-431.

Blondheim, N. R., Levy, Y. S., Ben-Zur, T., Burshtein, A., Cherlow, T., Kan, I., et al. (2006). Human mesenchymal stem cells express neural genes, suggesting a neural predisposition. Stem Cells Dev, 15(2), 141-164.

Bonilla, S., Silva, A., Valdes, L., Geijo, E., Garcia-Verdugo, J. M., & Martinez, S. (2005). Functional neural stem cells derived from adult bone marrow. Neuroscience, 133(1), 85-95.

Brazelton, T. R., Rossi, F. M., Keshet, G. I., & Blau, H. M. (2000). From marrow to brain: expression of neuronal phenotypes in adult mice. Science, 290(5497), 1775-1779.

Brighton, C. T., Lorich, D. G., Kupcha, R., Reilly, T. M., Jones, A. R., & Woodbury, R. A., 2nd. (1992). The pericyte as a possible osteoblast progenitor cell. Clin Orthop(275), 287-299.

Castro, R. F., Jackson, K. A., Goodell, M. A., Robertson, C. S., Liu, H., & Shine, H. D. (2002). Failure of bone marrow cells to transdifferentiate into neural cells in vivo. Science, 297(5585), 1299.

Chamberlain, G., Fox, J., Ashton, B., & Middleton, J. (2007). Concise review: mesenchymal stem cells: their phenotype, differentiation capacity, immunological features, and potential for homing. Stem cells, 25(11), 2739-2749.

Chanda, D., Kumar, S., & Ponnazhagan, S. (2010). Therapeutic potential of adult bone marrow-derived mesenchymal stem cells in diseases of the skeleton. J Cell Biochem, 111(2), 249-257.

Chen, M., Ona, V. O., Li, M., Ferrante, R. J., Fink, K. B., Zhu, S., et al. (2000). Minocycline inhibits caspase-1 and caspase-3 expression and delays mortality in a transgenic mouse model of Huntington disease. Nat Med, 6(7), 797-801.

Chen, Q., Long, Y., Yuan, X., Zou, L., Sun, J., Chen, S., et al. (2005). Protective effects of bone marrow stromal cell transplantation in injured rodent brain: synthesis of neurotrophic factors. J Neurosci Res, 80(5), 611-619.

Cho, S. R., Kim, Y. R., Kang, H. S., Yim, S. H., Park, C. I., Min, Y. H., et al. (2009). Functional recovery after the transplantation of neurally differentiated mesenchymal stem cells derived from bone barrow in a rat model of spinal cord injury. Cell Transplant, 18(12), 1359-1368.

Choi, Y. H., Kurtz, A., & Stamm, C. (2011). Mesenchymal stem cells for cardiac cell therapy. Hum Gene Ther, 22(1), 3-17.

Chopp, M., & Li, Y. (2002). Treatment of neural injury with marrow stromal cells. Lancet Neurol, 1(2), 92-100.

Chu, Q., Yu, Z., Zhang, S., & Yu, S. (2008). Astrocytes facilitate the growth and differentiation of co-cultured mesenchymal stem cells. J Huazhong Univ Sci Technolog Med Sci, 28(3), 333-336.

Corcione, A., Benvenuto, F., Ferretti, E., Giunti, D., Cappiello, V., Cazzanti, F., et al. (2006). Human mesenchymal stem cells modulate B-cell functions. Blood, 107(1), 367-372.

Corti, S., Locatelli, F., Donadoni, C., Guglieri, M., Papadimitriou, D., Strazzer, S., et al. (2004). Wild-type bone marrow cells ameliorate the phenotype of SOD1-G93A ALS mice and contribute to CNS, heart and skeletal muscle tissues. Brain, 127(Pt 11), 2518-2532.

Cox, C. S., Jr., Baumgartner, J. E., Harting, M. T., Worth, L. L., Walker, P. A., Shah, S. K., et al. (2011). Autologous bone marrow mononuclear cell therapy for severe traumatic brain injury in children. Neurosurgery, 68(3), 588-600.

Deng, J., Petersen, B. E., Steindler, D. A., Jorgensen, M. L., & Laywell, E. D. (2006). Mesenchymal stem cells spontaneously express neural proteins in culture and are neurogenic after transplantation. Stem Cells, 24(4), 1054-1064.

Deng, W., Obrocka, M., Fischer, I., & Prockop, D. J. (2001). In vitro differentiation of human marrow stromal cells into early progenitors of neural cells by conditions that increase intracellular cyclic AMP. Biochem Biophys Res Commun, 282(1), 148-152.

Dezawa, M. (2002). Central and peripheral nerve regeneration by transplantation of Schwann cells and transdifferentiated bone marrow stromal cells. Anatomical science international, 77(1), 12-25.

Dezawa, M., Kanno, H., Hoshino, M., Cho, H., Matsumoto, N., Itokazu, Y., et al. (2004). Specific induction of neuronal cells from bone marrow stromal cells and application for autologous transplantation. J Clin Invest, 113(12), 1701-1710.

Di Nicola, M., Carlo-Stella, C., Magni, M., Milanesi, M., Longoni, P. D., Matteucci, P., et al. (2002). Human bone marrow stromal cells suppress T-lymphocyte proliferation induced by cellular or nonspecific mitogenic stimuli. Blood, 99(10), 3838-3843.

Eglitis, M. A., & Mezey, E. (1997). Hematopoietic cells differentiate into both microglia and macroglia in the brains of adult mice. Proc Natl Acad Sci U S A, 94(8), 4080-4085.

Enzmann, G. U., Benton, R. L., Talbott, J. F., Cao, Q., & Whittemore, S. R. (2006). Functional considerations of stem cell transplantation therapy for spinal cord repair. Journal of neurotrauma, 23(3-4), 479-495.

Episkopou, V. (2005). SOX2 functions in adult neural stem cells. Trends Neurosci, 28(5), 219-221.

Ferrari, G., Cusella-De Angelis, G., Coletta, M., Paolucci, E., Stornaiuolo, A., Cossu, G., et al. (1998). Muscle regeneration by bone marrow-derived myogenic progenitors. Science, 279(5356), 1528-1530.

Fu, Y. S., Cheng, Y. C., Lin, M. Y., Cheng, H., Chu, P. M., Chou, S. C., et al. (2006). Conversion of human umbilical cord mesenchymal stem cells in Wharton's jelly to dopaminergic neurons in vitro: potential therapeutic application for Parkinsonism. Stem cells, 24(1), 115-124.

Fuhrmann, T., Montzka, K., Hillen, L. M., Hodde, D., Dreier, A., Bozkurt, A., et al. (2010). Axon growth-promoting properties of human bone marrow mesenchymal stromal cells. Neuroscience letters, 474(1), 37-41.

Gordon, D., Pavlovska, G., Glover, C. P., Uney, J. B., Wraith, D., & Scolding, N. J. (2008). Human mesenchymal stem cells abrogate experimental allergic encephalomyelitis after intraperitoneal injection, and with sparse CNS infiltration. Neurosci Lett, 448(1), 71-73.

Greco, S. J., Zhou, C., Ye, J. H., & Rameshwar, P. (2008). A method to generate human mesenchymal stem cell-derived neurons which express and are excited by multiple neurotransmitters. Biol Proced Online, 10, 90-101.

Gu, W., Zhang, F., Xue, Q., Ma, Z., Lu, P., & Yu, B. (2010). Transplantation of bone marrow mesenchymal stem cells reduces lesion volume and induces axonal regrowth of injured spinal cord. Neuropathology, 30(3), 205-217.

Guo, L., Yin, F., Meng, H. Q., Ling, L., Hu-He, T. N., Li, P., et al. (2005). Differentiation of mesenchymal stem cells into dopaminergic neuron-like cells in vitro. Biomedical and environmental sciences : BES, 18(1), 36-42.

Habisch, H. J., Schmid, B., von Arnim, C. A., Ludolph, A. C., Brenner, R., & Storch, A. (2010). Efficient processing of Alzheimer's disease amyloid-Beta peptides by neuroectodermally converted mesenchymal stem cells. Stem Cells Dev, 19(5), 629-633.

Hattan, N., Kawaguchi, H., Ando, K., Kuwabara, E., Fujita, J., Murata, M., et al. (2005). Purified cardiomyocytes from bone marrow mesenchymal stem cells produce stable intracardiac grafts in mice. Cardiovasc Res, 65(2), 334-344.

Hellmann, M. A., Panet, H., Barhum, Y., Melamed, E., & Offen, D. (2006). Increased survival and migration of engrafted mesenchymal bone marrow stem cells in 6-hydroxydopamine-lesioned rodents. Neuroscience letters, 395(2), 124-128.

Hermann, A., Gastl, R., Liebau, S., Popa, M. O., Fiedler, J., Boehm, B. O., et al. (2004). Efficient generation of neural stem cell-like cells from adult human bone marrow stromal cells. J Cell Sci, 117(Pt 19), 4411-4422.

Hermann, A., Liebau, S., Gastl, R., Fickert, S., Habisch, H. J., Fiedler, J., et al. (2006). Comparative analysis of neuroectodermal differentiation capacity of human bone marrow stromal cells using various conversion protocols. J Neurosci Res, 83(8), 1502-1514.

Himes, B. T., Neuhuber, B., Coleman, C., Kushner, R., Swanger, S. A., Kopen, G. C., et al. (2006). Recovery of function following grafting of human bone marrow-derived stromal cells into the injured spinal cord. Neurorehabilitation and neural repair, 20(2), 278-296.

Hofstetter, C. P., Schwarz, E. J., Hess, D., Widenfalk, J., El Manira, A., Prockop, D. J., et al. (2002). Marrow stromal cells form guiding strands in the injured spinal cord and promote recovery. Proc Natl Acad Sci U S A, 99(4), 2199-2204.

Jiang, X. X., Zhang, Y., Liu, B., Zhang, S. X., Wu, Y., Yu, X. D., et al. (2005). Human mesenchymal stem cells inhibit differentiation and function of monocyte-derived dendritic cells. Blood, 105(10), 4120-4126.

Jiang, Y., Henderson, D., Blackstad, M., Chen, A., Miller, R. F., & Verfaillie, C. M. (2003). Neuroectodermal differentiation from mouse multipotent adult progenitor cells. Proc Natl Acad Sci U S A, 100 Suppl 1, 11854-11860.

Jiang, Y., Jahagirdar, B. N., Reinhardt, R. L., Schwartz, R. E., Keene, C. D., Ortiz-Gonzalez, X. R., et al. (2002). Pluripotency of mesenchymal stem cells derived from adult marrow. Nature, 418(6893), 41-49.

Jing, L., Jia, Y., Lu, J., Han, R., Li, J., Wang, S., et al. (2011). MicroRNA-9 promotes differentiation of mouse bone mesenchymal stem cells into neurons by Notch signaling. Neuroreport, 22(5), 206-211.

Joannides, A., Gaughwin, P., Scott, M., Watt, S., Compston, A., & Chandran, S. (2003). Postnatal astrocytes promote neural induction from adult human bone marrow-derived stem cells. J Hematother Stem Cell Res, 12(6), 681-688.

Jori, F. P., Napolitano, M. A., Melone, M. A., Cipollaro, M., Cascino, A., Altucci, L., et al. (2005). Molecular pathways involved in neural in vitro differentiation of marrow stromal stem cells. J Cell Biochem, 94(4), 645-655.

Joyce, N., Annett, G., Wirthlin, L., Olson, S., Bauer, G., & Nolta, J. A. (2010). Mesenchymal stem cells for the treatment of neurodegenerative disease. Regen Med, 5(6), 933-946.

Karussis, D., Karageorgiou, C., Vaknin-Dembinsky, A., Gowda-Kurkalli, B., Gomori, J. M., Kassis, I., et al. (2010). Safety and immunological effects of mesenchymal stem cell transplantation in patients with multiple sclerosis and amyotrophic lateral sclerosis. Arch Neurol, 67(10), 1187-1194.

Kassis, I., Grigoriadis, N., Gowda-Kurkalli, B., Mizrachi-Kol, R., Ben-Hur, T., Slavin, S., et al. (2008). Neuroprotection and immunomodulation with mesenchymal stem cells in chronic experimental autoimmune encephalomyelitis. Arch Neurol, 65(6), 753-761.

Kawamoto, A., Gwon, H. C., Iwaguro, H., Yamaguchi, J. I., Uchida, S., Masuda, H., et al. (2001). Therapeutic potential of ex vivo expanded endothelial progenitor cells for myocardial ischemia. Circulation, 103(5), 634-637.

Kawamoto, A., Tkebuchava, T., Yamaguchi, J., Nishimura, H., Yoon, Y. S., Milliken, C., et al. (2003). Intramyocardial transplantation of autologous endothelial progenitor cells for therapeutic neovascularization of myocardial ischemia. Circulation, 107(3), 461-468.

Keilhoff, G., Goihl, A., Stang, F., Wolf, G., & Fansa, H. (2006). Peripheral nerve tissue engineering: autologous Schwann cells vs. transdifferentiated mesenchymal stem cells. Tissue Eng, 12(6), 1451-1465.

Kim, B. J., Seo, J. H., Bubien, J. K., & Oh, Y. S. (2002). Differentiation of adult bone marrow stem cells into neuroprogenitor cells in vitro. Neuroreport, 13(9), 1185-1188.

Kim, H. J., Lee, J. H., & Kim, S. H. (2010). Therapeutic effects of human mesenchymal stem cells on traumatic brain injury in rats: secretion of neurotrophic factors and inhibition of apoptosis. J Neurotrauma, 27(1), 131-138.

Kim, S. S., Choi, J. M., Kim, J. W., Ham, D. S., Ghil, S. H., Kim, M. K., et al. (2005). cAMP induces neuronal differentiation of mesenchymal stem cells via activation of extracellular signal-regulated kinase/MAPK. Neuroreport, 16(12), 1357-1361.

Kim, S. S., Yoo, S. W., Park, T. S., Ahn, S. C., Jeong, H. S., Kim, J. W., et al. (2008). Neural induction with neurogenin1 increases the therapeutic effects of mesenchymal stem cells in the ischemic brain. Stem Cells, 26(9), 2217-2228.

Kohyama, J., Abe, H., Shimazaki, T., Koizumi, A., Nakashima, K., Gojo, S., et al. (2001). Brain from bone: efficient "meta-differentiation" of marrow stroma-derived mature osteoblasts to neurons with Noggin or a demethylating agent. Differentiation, 68(4-5), 235-244.

Kondo, T., Johnson, S. A., Yoder, M. C., Romand, R., & Hashino, E. (2005). Sonic hedgehog and retinoic acid synergistically promote sensory fate specification from bone marrow-derived pluripotent stem cells. Proc Natl Acad Sci U S A, 102(13), 4789-4794.

Kopen, G. C., Prockop, D. J., & Phinney, D. G. (1999). Marrow stromal cells migrate throughout forebrain and cerebellum, and they differentiate into astrocytes after injection into neonatal mouse brains. Proc Natl Acad Sci U S A, 96(19), 10711-10716.

Krampera, M., Marconi, S., Pasini, A., Galie, M., Rigotti, G., Mosna, F., et al. (2007). Induction of neural-like differentiation in human mesenchymal stem cells derived from bone marrow, fat, spleen and thymus. Bone, 40(2), 382-390.

Kucia, M., Dawn, B., Hunt, G., Guo, Y., Wysoczynski, M., Majka, M., et al. (2004). Cells expressing early cardiac markers reside in the bone marrow and are mobilized into the peripheral blood after myocardial infarction. Circ Res, 95(12), 1191-1199.

Kuznetsov, S. A., Mankani, M. H., Gronthos, S., Satomura, K., Bianco, P., & Robey, P. G. (2001). Circulating skeletal stem cells. J Cell Biol, 153(5), 1133-1140.

Lee, J. K., Jin, H. K., & Bae, J. S. (2010). Bone marrow-derived mesenchymal stem cells attenuate amyloid beta-induced memory impairment and apoptosis by inhibiting neuronal cell death. Curr Alzheimer Res, 7(6), 540-548.

Lee, J. K., Jin, H. K., Endo, S., Schuchman, E. H., Carter, J. E., & Bae, J. S. (2010). Intracerebral transplantation of bone marrow-derived mesenchymal stem cells reduces amyloid-beta deposition and rescues memory deficits in Alzheimer's disease mice by modulation of immune responses. Stem Cells, 28(2), 329-343.

Lee, J. Y., Qu-Petersen, Z., Cao, B., Kimura, S., Jankowski, R., Cummins, J., et al. (2000). Clonal isolation of muscle-derived cells capable of enhancing muscle regeneration and bone healing. J Cell Biol, 150(5), 1085-1100.

Lee, O. K., Kuo, T. K., Chen, W. M., Lee, K. D., Hsieh, S. L., & Chen, T. H. (2004). Isolation of multipotent mesenchymal stem cells from umbilical cord blood. Blood, 103(5), 1669-1675.

Lee, P. H., & Park, H. J. (2009). Bone marrow-derived mesenchymal stem cell therapy as a candidate disease-modifying strategy in Parkinson's disease and multiple system atrophy. J Clin Neurol, 5(1), 1-10.

Li, J. M., Zhu, H., Lu, S., Liu, Y., Li, Q., Ravenscroft, P., et al. (2011). Migration and differentiation of human mesenchymal stem cells in the normal rat brain. Neurol Res, 33(1), 84-92.

Li, Y., Chen, J., Wang, L., Zhang, L., Lu, M., & Chopp, M. (2001). Intracerebral transplantation of bone marrow stromal cells in a 1-methyl-4-phenyl-1,2,3,6-tetrahydropyridine mouse model of Parkinson's disease. Neuroscience letters, 316(2), 67-70.

Lim, J. Y., Park, S. I., Oh, J. H., Kim, S. M., Jeong, C. H., Jun, J. A., et al. (2008). Brain-derived neurotrophic factor stimulates the neural differentiation of human umbilical cord blood-derived mesenchymal stem cells and survival of differentiated cells through

MAPK/ERK and PI3K/Akt-dependent signaling pathways. J Neurosci Res, 86(10), 2168-2178.

Long, X., Olszewski, M., Huang, W., & Kletzel, M. (2005). Neural cell differentiation in vitro from adult human bone marrow mesenchymal stem cells. Stem Cells Dev, 14(1), 65-69.

Lu, P., Blesch, A., & Tuszynski, M. H. (2004). Induction of bone marrow stromal cells to neurons: differentiation, transdifferentiation, or artifact? J Neurosci Res, 77(2), 174-191.

Mahmood, A., Lu, D., & Chopp, M. (2004). Marrow stromal cell transplantation after traumatic brain injury promotes cellular proliferation within the brain. Neurosurgery, 55(5), 1185-1193.

Mahmood, A., Lu, D., Qu, C., Goussev, A., & Chopp, M. (2006). Long-term recovery after bone marrow stromal cell treatment of traumatic brain injury in rats. J Neurosurg, 104(2), 272-277.

Mahmood, A., Lu, D., Wang, L., & Chopp, M. (2002). Intracerebral transplantation of marrow stromal cells cultured with neurotrophic factors promotes functional recovery in adult rats subjected to traumatic brain injury. J Neurotrauma, 19(12), 1609-1617.

Makino, S., Fukuda, K., Miyoshi, S., Konishi, F., Kodama, H., Pan, J., et al. (1999). Cardiomyocytes can be generated from marrow stromal cells in vitro. J Clin Invest, 103(5), 697-705.

Mareschi, K., Rustichelli, D., Comunanza, V., De Fazio, R., Cravero, C., Morterra, G., et al. (2009). Multipotent mesenchymal stem cells from amniotic fluid originate neural precursors with functional voltage-gated sodium channels. Cytotherapy, 11(5), 534-547.

Mazzini, L., Fagioli, F., Boccaletti, R., Mareschi, K., Oliveri, G., Olivieri, C., et al. (2003). Stem cell therapy in amyotrophic lateral sclerosis: a methodological approach in humans. Amyotroph Lateral Scler Other Motor Neuron Disord, 4(3), 158-161.

Mazzini, L., Mareschi, K., Ferrero, I., Vassallo, E., Oliveri, G., Boccaletti, R., et al. (2006). Autologous mesenchymal stem cells: clinical applications in amyotrophic lateral sclerosis. Neurol Res, 28(5), 523-526.

Mezey, E., Chandross, K. J., Harta, G., Maki, R. A., & McKercher, S. R. (2000). Turning blood into brain: cells bearing neuronal antigens generated in vivo from bone marrow. Science, 290(5497), 1779-1782.

Mizuno, S., & Glowacki, J. (1996). Chondroinduction of human dermal fibroblasts by demineralized bone in three-dimensional culture. Exp Cell Res, 227(1), 89-97.

Munoz-Elias, G., Marcus, A. J., Coyne, T. M., Woodbury, D., & Black, I. B. (2004). Adult bone marrow stromal cells in the embryonic brain: engraftment, migration, differentiation, and long-term survival. J Neurosci, 24(19), 4585-4595.

Munoz-Elias, G., Woodbury, D., & Black, I. B. (2003). Marrow stromal cells, mitosis, and neuronal differentiation: stem cell and precursor functions. Stem Cells, 21(4), 437-448.

Neuhuber, B., Gallo, G., Howard, L., Kostura, L., Mackay, A., & Fischer, I. (2004). Reevaluation of in vitro differentiation protocols for bone marrow stromal cells:

disruption of actin cytoskeleton induces rapid morphological changes and mimics neuronal phenotype. J Neurosci Res, 77(2), 192-204.

Neuhuber, B., Timothy Himes, B., Shumsky, J. S., Gallo, G., & Fischer, I. (2005). Axon growth and recovery of function supported by human bone marrow stromal cells in the injured spinal cord exhibit donor variations. Brain Res, 1035(1), 73-85.

Niibe, K., Kawamura, Y., Araki, D., Morikawa, S., Miura, K., Suzuki, S., et al. (2011). Purified mesenchymal stem cells are an efficient source for iPS cell induction. PLoS One, 6(3), e17610.

Orlic, D., Kajstura, J., Chimenti, S., Jakoniuk, I., Anderson, S. M., Li, B., et al. (2001). Bone marrow cells regenerate infarcted myocardium. Nature, 410(6829), 701-705.

Padovan, C. S., Jahn, K., Birnbaum, T., Reich, P., Sostak, P., Strupp, M., et al. (2003). Expression of neuronal markers in differentiated marrow stromal cells and CD133+ stem-like cells. Cell Transplant, 12(8), 839-848.

Parekkadan, B., & Milwid, J. M. (2010). Mesenchymal stem cells as therapeutics. Annu Rev Biomed Eng, 12, 87-117.

Park, E., & Patel, A. N. (2010). Changes in the expression pattern of mesenchymal and pluripotent markers in human adipose-derived stem cells. Cell Biol Int, 34(10), 979-984.

Pereira, R. F., Halford, K. W., O'Hara, M. D., Leeper, D. B., Sokolov, B. P., Pollard, M. D., et al. (1995). Cultured adherent cells from marrow can serve as long-lasting precursor cells for bone, cartilage, and lung in irradiated mice. Proc Natl Acad Sci U S A, 92(11), 4857-4861.

Pereira, R. F., O'Hara, M. D., Laptev, A. V., Halford, K. W., Pollard, M. D., Class, R., et al. (1998). Marrow stromal cells as a source of progenitor cells for nonhematopoietic tissues in transgenic mice with a phenotype of osteogenesis imperfecta. Proc Natl Acad Sci U S A, 95(3), 1142-1147.

Petersen, B. E., Bowen, W. C., Patrene, K. D., Mars, W. M., Sullivan, A. K., Murase, N., et al. (1999). Bone marrow as a potential source of hepatic oval cells. Science, 284(5417), 1168-1170.

Picinich, S. C., Mishra, P. J., Glod, J., & Banerjee, D. (2007). The therapeutic potential of mesenchymal stem cells. Cell- & tissue-based therapy. Expert Opin Biol Ther, 7(7), 965-973.

Pierantozzi, E., Gava, B., Manini, I., Roviello, F., Marotta, G., Chiavarelli, M., et al. (2011). Pluripotency regulators in human mesenchymal stem cells: expression of NANOG but not of OCT-4 and SOX-2. Stem Cells Dev, 20(5), 915-923.

Prockop, D. J. (1997). Marrow stromal cells as stem cells for nonhematopoietic tissues. Science, 276(5309), 71-74.

Qian, L., & Saltzman, W. M. (2004). Improving the expansion and neuronal differentiation of mesenchymal stem cells through culture surface modification. Biomaterials, 25(7-8), 1331-1337.

Qu, T. Y., Dong, X. J., Sugaya, I., Vaghani, A., Pulido, J., & Sugaya, K. (2004). Bromodeoxyuridine increases multipotency of human bone marrow-derived stem cells. Restor Neurol Neurosci, 22(6), 459-468.

Riekstina, U., Cakstina, I., Parfejevs, V., Hoogduijn, M., Jankovskis, G., Muiznieks, I., et al. (2009). Embryonic stem cell marker expression pattern in human mesenchymal stem cells derived from bone marrow, adipose tissue, heart and dermis. Stem Cell Rev, 5(4), 378-386.

Sanchez-Ramos, J., Song, S., Cardozo-Pelaez, F., Hazzi, C., Stedeford, T., Willing, A., et al. (2000). Adult bone marrow stromal cells differentiate into neural cells in vitro. Exp Neurol, 164(2), 247-256.

Shiota, M., Heike, T., Haruyama, M., Baba, S., Tsuchiya, A., Fujino, H., et al. (2007). Isolation and characterization of bone marrow-derived mesenchymal progenitor cells with myogenic and neuronal properties. Exp Cell Res, 313(5), 1008-1023.

Stappenbeck, T. S., & Miyoshi, H. (2009). The role of stromal stem cells in tissue regeneration and wound repair. Science, 324(5935), 1666-1669.

Stock, P., Bruckner, S., Ebensing, S., Hempel, M., Dollinger, M. M., & Christ, B. (2010). The generation of hepatocytes from mesenchymal stem cells and engraftment into murine liver. Nat Protoc, 5(4), 617-627.

Suon, S., Yang, M., & Iacovitti, L. (2006). Adult human bone marrow stromal spheres express neuronal traits in vitro and in a rat model of Parkinson's disease. Brain research, 1106(1), 46-51.

Takahashi, T., Kalka, C., Masuda, H., Chen, D., Silver, M., Kearney, M., et al. (1999). Ischemia- and cytokine-induced mobilization of bone marrow-derived endothelial progenitor cells for neovascularization. Nat Med, 5(4), 434-438.

Terada, N., Hamazaki, T., Oka, M., Hoki, M., Mastalerz, D. M., Nakano, Y., et al. (2002). Bone marrow cells adopt the phenotype of other cells by spontaneous cell fusion. Nature, 416(6880), 542-545.

Theise, N. D., Badve, S., Saxena, R., Henegariu, O., Sell, S., Crawford, J. M., et al. (2000). Derivation of hepatocytes from bone marrow cells in mice after radiation-induced myeloablation. Hepatology, 31(1), 235-240.

Theise, N. D., Nimmakayalu, M., Gardner, R., Illei, P. B., Morgan, G., Teperman, L., et al. (2000). Liver from bone marrow in humans. Hepatology, 32(1), 11-16.

Tio, M., Tan, K. H., Lee, W., Wang, T. T., & Udolph, G. (2010). Roles of db-cAMP, IBMX and RA in aspects of neural differentiation of cord blood derived mesenchymal-like stem cells. PLoS One, 5(2), e9398.

Tomita, S., Li, R. K., Weisel, R. D., Mickle, D. A., Kim, E. J., Sakai, T., et al. (1999). Autologous transplantation of bone marrow cells improves damaged heart function. Circulation, 100(19 Suppl), II247-256.

Tondreau, T., Dejeneffe, M., Meuleman, N., Stamatopoulos, B., Delforge, A., Martiat, P., et al. (2008). Gene expression pattern of functional neuronal cells derived from human bone marrow mesenchymal stromal cells. BMC Genomics, 9, 166.

Trivedi, P., Tray, N., Nguyen, T., Nigam, N., & Gallicano, G. I. (2010). Mesenchymal stem cell therapy for treatment of cardiovascular disease: helping people sooner or later. Stem Cells Dev, 19(7), 1109-1120.

Trzaska, K. A., King, C. C., Li, K. Y., Kuzhikandathil, E. V., Nowycky, M. C., Ye, J. H., et al. (2009). Brain-derived neurotrophic factor facilitates maturation of mesenchymal stem cell-derived dopamine progenitors to functional neurons. J Neurochem, 110(3), 1058-1069.

Trzaska, K. A., Kuzhikandathil, E. V., & Rameshwar, P. (2007). Specification of a dopaminergic phenotype from adult human mesenchymal stem cells. Stem cells, 25(11), 2797-2808.

Trzaska, K. A., & Rameshwar, P. (2011). Dopaminergic neuronal differentiation protocol for human mesenchymal stem cells. Methods Mol Biol, 698, 295-303.

Uccelli, A., Benvenuto, F., Laroni, A., & Giunti, D. (2011). Neuroprotective features of mesenchymal stem cells. Best practice & research. Clinical haematology, 24(1), 59-64.

Umezawa, A., Tachibana, K., Harigaya, K., Kusakari, S., Kato, S., Watanabe, Y., et al. (1991). Colony-stimulating factor 1 expression is down-regulated during the adipocyte differentiation of H-1/A marrow stromal cells and induced by cachectin/tumor necrosis factor. Mol Cell Biol, 11(2), 920-927.

Wang, J., Liao, L., & Tan, J. (2011). Mesenchymal-stem-cell-based experimental and clinical trials: current status and open questions. Expert Opin Biol Ther, 11(7), 893-909.

Wislet-Gendebien, S., Hans, G., Leprince, P., Rigo, J. M., Moonen, G., & Rogister, B. (2005). Plasticity of cultured mesenchymal stem cells: switch from nestin-positive to excitable neuron-like phenotype. Stem Cells, 23(3), 392-402.

Wu, Q. Y., Li, J., Feng, Z. T., & Wang, T. H. (2007). Bone marrow stromal cells of transgenic mice can improve the cognitive ability of an Alzheimer's disease rat model. Neurosci Lett, 417(3), 281-285.

Yang, X. S., Wu, H. X., & Xiao, B. (2005). [Human mesenchymal stem cells differentiate into neuron-like cells and show SMN protein expression]. Zhonghua Yi Xue Za Zhi, 85(16), 1125-1128.

Zappia, E., Casazza, S., Pedemonte, E., Benvenuto, F., Bonanni, I., Gerdoni, E., et al. (2005). Mesenchymal stem cells ameliorate experimental autoimmune encephalomyelitis inducing T-cell anergy. Blood, 106(5), 1755-1761.

Zeidan-Chulia, F., & Noda, M. (2009). "Opening" the mesenchymal stem cell tool box. Eur J Dent, 3(3), 240-249.

Zeng, R., Wang, L. W., Hu, Z. B., Guo, W. T., Wei, J. S., Lin, H., et al. (2011). Differentiation of human bone marrow mesenchymal stem cells into neuron-like cells in vitro. Spine, 36(13), 997-1005.

Zhang, J., Li, Y., Chen, J., Cui, Y., Lu, M., Elias, S. B., et al. (2005). Human bone marrow stromal cell treatment improves neurological functional recovery in EAE mice. Exp Neurol, 195(1), 16-26.

Zhang, J., Li, Y., Lu, M., Cui, Y., Chen, J., Noffsinger, L., et al. (2006). Bone marrow stromal cells reduce axonal loss in experimental autoimmune encephalomyelitis mice. J Neurosci Res, 84(3), 587-595.

Zhang, L., Seitz, L. C., Abramczyk, A. M., Liu, L., & Chan, C. (2011). cAMP initiates early phase neuron-like morphology changes and late phase neural differentiation in mesenchymal stem cells. Cell Mol Life Sci, 68(5), 863-876.

Zhang, Y. H., Wang, H. F., Liu, W., Wei, B., Bing, L. J., & Gao, Y. M. (2009). Insulin-producing cells derived from rat bone marrow and their autologous transplantation in the duodenal wall for treating diabetes. Anat Rec (Hoboken), 292(5), 728-735.

Zhang, Y. Q., Zeng, X., He, L. M., Ding, Y., Li, Y., & Zeng, Y. S. (2010). NT-3 gene modified Schwann cells promote TrkC gene modified mesenchymal stem cells to differentiate into neuron-like cells in vitro. Anat Sci Int, 85(2), 61-67.

Zhang, Z., Maiman, D. J., Kurpad, S. N., Crowe, M. J., & Alexanian, A. R. (2011). Feline Bone Marrow-Derived Mesenchymal Stem Cells Express Several Pluripotent and Neural Markers and Easily Turn into Neural-Like Cells by Manipulation with Chromatin Modifying Agents and Neural Inducing Factors. Cellular reprogramming, 13(5).

Zhao, C. P., Zhang, C., Zhou, S. N., Xie, Y. M., Wang, Y. H., Huang, H., et al. (2007). Human mesenchymal stromal cells ameliorate the phenotype of SOD1-G93A ALS mice. Cytotherapy, 9(5), 414-426.

Zuk, P. A., Zhu, M., Mizuno, H., Huang, J., Futrell, J. W., Katz, A. J., et al. (2001). Multilineage cells from human adipose tissue: implications for cell-based therapies. Tissue Eng, 7(2), 211-228.

12

Therapeutic Potential of MSCs in Musculoskeletal Diseases (Osteoarthritis)

José Ramón Lamas, Pilar Tornero-Esteban
and Benjamín Fernández-Gutiérrez
Hospital Clínico San Carlos, Madrid
Spain

1. Introduction

1.1 Musculoskeletal disorders (MSDs)

The Musculoskeletal or locomotor system is defined by a complex interconnection of different body parts functionally arranged in two main sub-systems: the skeletal system, composed of the bones, and secondly the muscle system. In addition other structures such as tendons, ligaments and other connective tissue joins both sub-systems providing functional and structural support.

Musculoskeletal disorder or more generically Rheumatic disease (RD) is a term used to describe over 200 different disorders involving different connective tissue. Depending on the main target affected, RDs can be grouped into different pathologies. Joints are structures commonly affected in RDs, as it is the location where bone, tendon, ligament and muscle meet. Joint disorders are generically termed as arthropathies, and it is only when inflammation occurs, in one or more joints, that the disorder is called arthritis. However, RDs also include systemic disorders (autoimmune diseases affecting multiple organs), dorsopathies (back disorders), soft tissue disorders (involving muscles, tendons, etc.), and osteopathies/chondropathies (e.g., disorders related to bone density and structure-like osteoporosis).

The prevalence of RDs in the elderly has erroneously been associated only with ageing; however RDs may develop at any time even in childhood (Manners, 2003, Mariller, 2005). Musculoskeletal disorders constitute the most common cause of severe chronic pain and physical disability, thus they are considered a public health problem that affect millions of individuals and constitute a major burden on health care, a situation which is aggravated by an increasingly aging population (Bansback, 2005, Loza, 2008).

1.2 Aetiology of osteoarthritis

Osteoarthritis is the most prevalent type of inflammatory arthritis (Spahn, 2011). Although it has long been considered to be primarily a cartilage disorder, induced by accumulated mechanical stress, as occurs in many other arthropathies the contribution of an inflammatory component is well established; sometimes produced by an autoimmune response, leading to chronic joint inflammation, destruction and cartilage loss. Little is known about underlying molecular mechanisms. Its initiation and progression appear to be independent processes associated with different risk factors (Worthington, 2005). In addition to biomechanical stress on articular cartilage, the involvement of other tissues of the affected joint, such as the synovium, ligaments, periarticular muscles, and nerves, have also been proposed in OA aetiology and progression (Brandt, 2006). Several studies suggest that the subchondral bone is likely to be the most important structural element in both pain generation and disease progression. At least in its generalised form, OA shows features of a systemic musculoskeletal disease with a metabolic component and a genetic predisposition leading to the formation of a defective cartilage matrix (Aspden, 2008, Zhu, 2009). Unfortunately, despite advances in research, little is known about OA's exact etiology and pathogenetic mechanisms.

Currently there is no known cure for OA, and modern treatments only manage to reduce pain and maintain joint movement as much as possible. For many years the only known options for OA treatment were disease-modifying drugs, in mild cases, and several types of surgery depending on the affectation of articular structures. Among the surgery choices, arthroscopy and joint arthroplasty are the most common. The first is used in people with moderate lesions of articular cartilage or bone, in order to alleviate pain for a short time and to allow the joints to move more easily. Although it does not seem to treat the arthritis itself, sometimes the relief can delay the use of other more aggressive surgeries (Laupattarakasem, 2008). Total or partial joint arthroplasty is the ultimate surgical treatment when joint damage can be seen on radiographs. It involves surgery to replace the ends of bones, mostly in the hip, knee and shoulder thereby creating new surfaces. However, surgery is not recommended in those cases where the patient's health is precarious, due to serious risk of infections, and because after surgery, long periods of physical rehabilitation are needed. Moreover, the prostheses have a lifespan of 10 to 20 years, after which they require substitution.

In this therapeutic context, it is easy to understand that there is a real and urgent need for new and alternative treatments to circumvent the relatively low efficiency of existing therapies. This is where the emerging potential of regenerative medicine becomes increasingly important as the most promising method to restore, maintain or improve tissue structure and joint function (Bruder, 1997, Mackay, 1998, Pittenger, 1999, Zavan, 2007).

2. Regenerative medicine in rheumatic diseases

Broadly speaking, the term "regenerative medicine" refers to a new field in biomedical research focused on the development of therapeutic approaches allowing the body to replace and regenerate damaged or diseased cells, and ultimately the function of tissues and organs. This goal is achieved by means of a combination of approaches that include the use of soluble molecules, biomaterials, tissue engineering, gene therapy, stem cell transplantation and the reprogramming of cell and tissue types.

In the context of musculoskeletal disease, and in particular the reconstruction of articular defects caused by trauma or disease, the goal is to deliver cells that become competent in the defect site, initially optimizing biomechanics, and ultimately initiating new tissue production. Sometimes, as occurs in the case of soft tissue repair, an additional implant vehicle(s), is required to transport and constrain the implanted cells in the defect site and to provide mechanical stability to the surgical site. The progressive biodegradation of the vehicle during new tissue formation would be the optimal scenario for the repairing process. However, this seemingly straightforward schema can be complicated depending on the tissue to repair, which in turn determines not only the type of cells to use but also the number and the mode of application. Thus, current challenges in musculoskeletal regenerative medicine cover several topics under study including: (1) the better understanding of cell biology, (2) the synthesis of new biomaterials for extracellular matrices (*scaffolds*), and (3) the definition of the best combination of cells, biologically active molecules and vehicles to promote growth and differentiation.

2.1 Regenerative medicine using mature chondrocytes

Given that cells are the main building blocks of regenerative therapies, their availability and their commitment to a specific lineage are major limitations. The cells can be of autologous (host-derived) or allogeneic origin (non-host derived). Other sources, are cells of xenogeneic (from individuals of another species), syngeneic or isogeneic (isolated from genetically identical organisms or highly inbred individuals, respectively) origin are only constrained to experimental models.

Obviously, the most logical approach for the regeneration of joint degraded cartilage, consists in the direct re-establishment of its main functional component, the chondrocytes. Joint cartilage is a connective tissue with special characteristics. It consists of chondrocytes that secrete a cartilage-specific extracellular matrix (ECM) made of collagens, mainly type II collagen, and different proteoglycans. The chondrocytes do not have direct cell-to-cell contact, thus each cell acts as a functional unit responsible for the production and maintenance of the ECM in its surrounding. These characteristics, in addition to the cartilage avascularity, explain the difficulties involved in repairing this tissue, because chondroprogenitor cell access to the damaged site is very limited.

The first approach, and the gold standard for years in joint orthopaedic surgery, has been the autologous chondrocyte implantation (ACI), after harvesting, from healthy cartilage biopsies and expanded in culture. So far, thousands of ACIs have been clinically applied with encouraging results in the short- and mid-term, but their long-term efficiency needs further confirmation (Alvarez-Dolado, 2007). The effectiveness of the technique is limited by some major drawbacks, including the absence of appropriate sources of suitable hyaline cartilage and the additional damage caused at the site of biopsy. Other important issues arise during chondrocyte expansion *in vitro* and further transplantation. In culture, chondrocytes easily dedifferentiate losing their chondrogenic phenotype and their re-differentiation potential and once it occurs, about half of the ACIs show evidence of chondrocyte hypertrophy, indicating the formation of a bone-like tissue. Finally, the occurrence of poor adhesion between the new and the original tissue is common and in those cases where scaffolds are used, the biomechanical properties obtained do not achieve the expected results. These problems have raised the need for alternative cell sources with chondrogenic potential for cartilage tissue engineering, a requisite accomplished by the stem cells, and in particular by mesenchymal stem cells (MSCs).

2.2 Regenerative medicine using MSCs

Under normal conditions body tissues are subjected to a continuous process of repair and regeneration of damaged and dead cells by means of a pool of progenitor or stem cells, which have the capacity to differentiate into the specialised cell type being replaced. 'Stem cells' is a generic term to describe a variety of cells which share two common characteristics: (1) their self-renewal potential and (2) their capacity to give rise to different tissues. However their "potency", or differentiation potential is variable and therefore there exists a hierarchy according to stem cell types. The most versatile, the totipotent embryonic stem cells (ESCs) give rise to other embryonic or extra embryonic adult stem cells (ASCs) with pluri- multi- or uni- potentiality. Pluripotent stem cells are descendants of totipotent cells and can differentiate into cells derived from the endoderm, mesoderm and ectoderm germ layers. Multipotent stem cells can produce only cells of a closely related family of cells, e.g., hematopoietic stem cells and MSCs. Finally, unipotent cells only produce one cell type, but retain their self-renewal properties, a feature that distinguishes them from other non-stem cells.

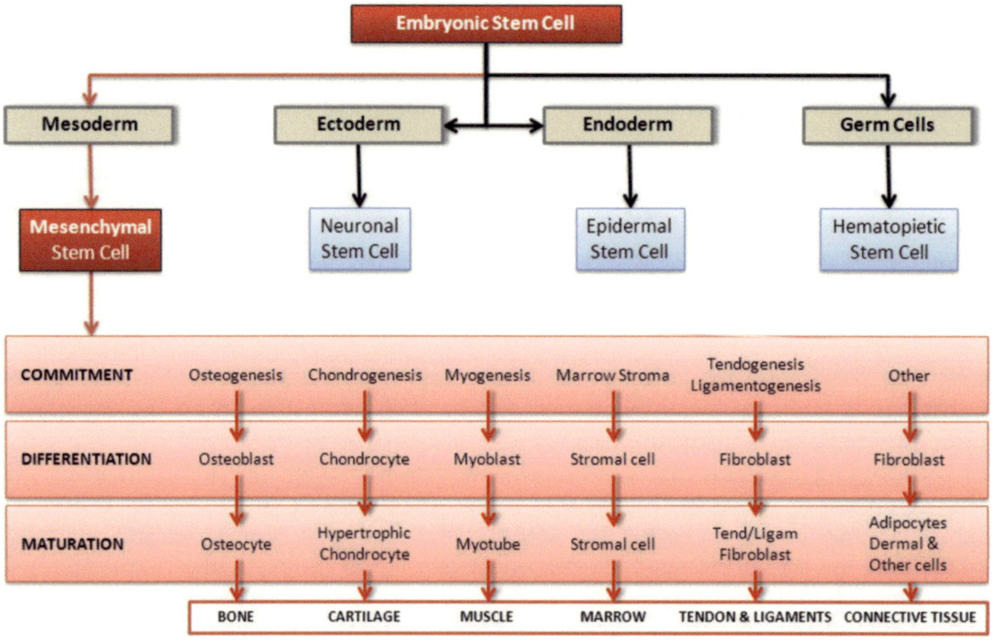

Fig. 1. Diagram of the mesenchymal stem cell lineage and its differentiation potential

MSCs have the potential to differentiate into several cell types of mesodermal origin including bone, cartilage, muscle, bone marrow stroma, tendon/ligament, fat, dermis. Thus these cells are optimal candidates in regenerative medicine strategies intended to restore these connective tissues. Adapted from (Caplan, 2007).

Although for the purposes of regenerative medicine, ESCs could be considered the optimal candidates; their clinical use in human therapies is still controversial due to ethical issues. In the context of musculoskeletal diseases, the least compromised are mesenchymal stem cells (MSCs) and are of particular interest for several reasons. First, they are the progenitors of cells giving rise to a variety of cells which can form connective tissues such as chondrocytes, osteocytes, adipocytes and tenocytes (Pittenger, 1999), (Figure 1) and second: in contrast to most other adult stem cells, they can be isolated from a diversity of accessible tissues, such as bone marrow, fat tissues and umbilical cord blood (Chanda, 2010, Moretti, 2010, Romanov, 2005). Moreover, they can be isolated and identified through their adhesion potential in culture and by the expression of several "positive markers", represented by the transmembrane proteins CD90, CD73, CD105 and CD166. Additionally, MSCs are easily expanded *in vitro* without losing their "stemness" and/or self-renewal capacity (Bianco, 2001, Caplan, 2000, Reiser, 2005).

MSCs have been shown to differentiate *in vitro* into bone, cartilage, muscle, tendon, and fat, and possibly also into cardiomyocytes and hepatocytes (Conget, 1999, Chivu, 2009, Dennis, 1999, Pereira, 1995, Pittenger, 1999, Remy-Martin, 1999). Finally, from an immunological point of view, an important property of MSCs, especially for their use in rheumatic diseases, resides in their potent immunosuppressive and anti-inflammatory functions, the lack of induction of graft rejection and their chemotactic properties, similar to immune cells in response to injury on sites of inflammation (Le Blanc, 2004, Spaeth, 2008). As such, these cells are currently being considered for their potential use in cell and gene therapy, in a large number of human diseases, and particularly in a variety of clinical musculoskeletal conditions, including the repair of cartilage defects, tendon/ligament and bone.

2.3 MSCs and cartilage repair in OA

As occurs during embryogenesis, the generation of new cartilage (chondrogenesis) involves the MSCs progression through different stages in a tightly regulated process coordinated by multiple signalling pathways which include the Wnt, Notch or TGF (Quintana, 2009, Roelen, 2003). In particular, the Wnt/β-catenin signalling pathway plays a crucial role in cartilage reparation, since it participates in the differentiation of MSCs into osteoblasts or chondrocytes during osteogenesis and/or chondrogenesis (Day, 2005, Gaur, 2005, Hill, 2005). Markedly, alteration in any of the aforementioned pathways, as occurs in some diseases, can lead to detrimental effects during the regeneration process.

In vitro chondrogenesis is routinely performed by culturing MSCs in three dimensional scaffolds made of different biomaterials such as collagen, fibrin, agarose, alginate, chitosan or hyaluronic acid of natural or synthetic origin, or a combination of both types (Li, 2005, Lisignoli, 2005, Necas, 2010, Zhou, 2008). In addition, these scaffolds can be supplemented with soluble factors such as TGF-β, growth factors, bone morphogenetic proteins (BMPs), etc. to facilitate the chondrogenic differentiation of MSCs.

Application of engineered MSCs for cartilage regeneration has been addressed by the slight modifications of two main approaches widely tested in different OA animal models with encouraging results (Figure 2).

In the first, MSCs are seeded on 3D scaffolds with the presence or absence of soluble factors (growth factors and/or cytokines) and the resulting structure is used to repair the cartilage defect (Zscharnack, 2010). The second approach, consists of the direct administration of MSCs, (loaded or not in 3D scaffolds) without previous differentiation (Thorpe, 2010).

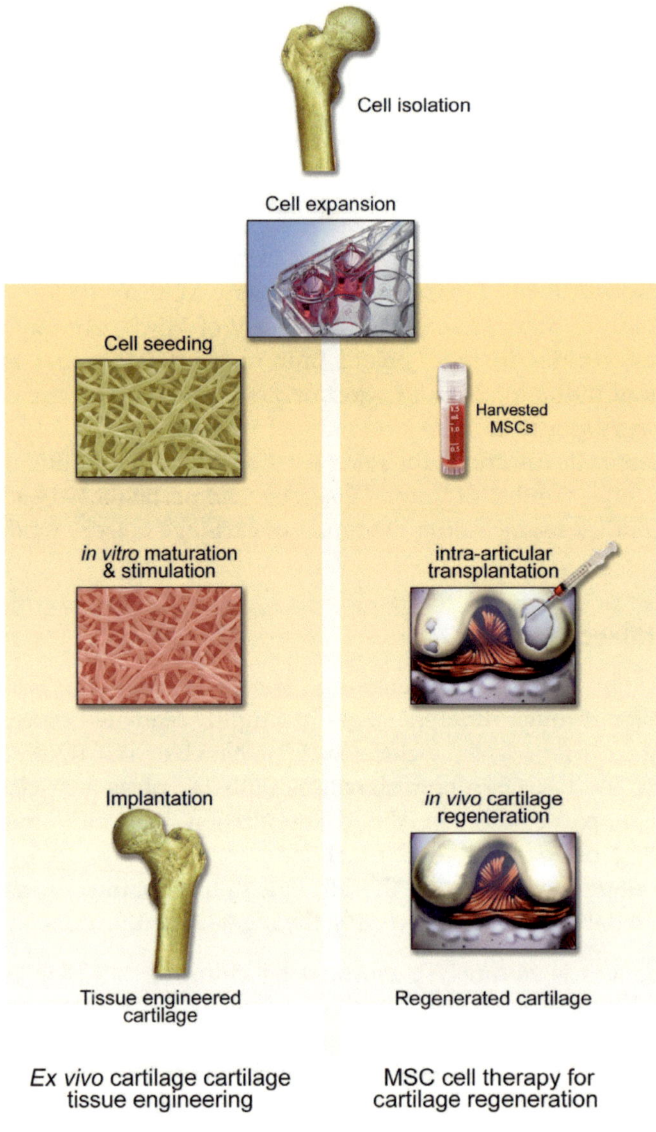

Fig. 2. MSC-based tissue engineering and cell therapy for cartilage repair and regeneration. Once MSCs are isolated and expanded, both tissue engineering and cell therapy approaches are suited for regeneration. In the *ex vivo* approach cells are loaded *in vitro* onto the scaffolds under appropriate stimuli and after a short incubation to insure attachment, the cell-scaffold composites are implanted. Another strategy is based on local injection into the affected joint. Adapted from (Caplan, 2007).

However, there are still some unanswered questions about the mechanism by which MSCs perform the repair; but several possibilities have been outlined, such as the following: (1) the secretion of cytokines to enhance repair (Chen, 2008); (2) the modulation of immune (Aggarwal, 2005, Gerdoni, 2007, Karussis, 2008, Le Blanc, 2007, Ren, 2008) and inflammatory responses (Gupta, 2007, Ortiz, 2007); (3) stimulation of the proliferation of tissue endogenous stem cells (Lee, 2006, Munoz, 2005); and (4) the rescue of damaged cells (Spees, 2006, Spees, 2003). Finally, MSCs are the subject in a controversy where their contradictory effects *in vitro* and *in vivo* on tumour cell growth have been called into question. Recent studies have shown that MSCs can increase the proliferation of tumor cells *in vitro* and promote tumor growth *in vivo* by increasing the neovascularization (Suzuki, 2011, Tian, 2011). This is a major concern that should be carefully considered, particularly in conditions where tumoral malignancies are present.

3. Studies carried out in our group

Our group has been focused for several years on the study of the biology of articular cartilage in the OA pathogenesis and the potential of MSCs in regeneration of damaged cartilage due to this disease. Some of the issues addressed include the basic research and the clinical trials to validate the translational efficacy in the clinic of MSC implantation. Much of our work in this field has been based on the use of modern techniques, that include proteomics and genomics approaches in combination with bioinformatics and genetic validation.

Proteomics is considered and emerging field with widespread potential applications to shape how rheumatic diseases are diagnosed, prognosticated, and clinically managed (Camafeita, 2009, Vanarsa, 2010). A key methodological advance in the classical two-dimensional gel electrophoresis (2-DE) has been the emergence of multiplexing two-dimensional fluorescence difference gel electrophoresis (2D-DIGE) (Unlu, 1997). The 2D-DIGE technique circumvents many of the issues associated with traditional 2-DE, providing more sensitivity, high reproducibility and a wide dynamic range of detection (Alban, 2003, Viswanathan, 2006). It consists in the labelling of lysine groups on protein extracts with fluorescent dyes with different emission spectra before isoelectric focusing (IEF). Protein samples are further labelled with Cy3 and Cy5 fluorescent dyes, while Cy2 dye is used to label the internal standard, which consists of a pooled sample comprising equal amounts of all samples to be compared. Then the three samples are electrophoresed on a single 2D gel, which allows both direct quantitative comparisons within each gel and the normalization of quantitative abundance values for each protein between gels. The combination of 2D-DIGE with matrix-assisted laser desorption ionization time of flight mass spectrometry (MALDI-TOF MS) provides a powerful tool for identifying disease-related proteins (Stults, 2005). (Figure 3). Genomic approaches have been developed through DNA microarray analysis, a proven "state of the art" technology for the simultaneous screening of expression levels in large numbers of genes (Licatalosi, 2010). (Figure 4)

Another strategy, different from the previous two, is the systemic application of MSCs, that have been shown to promote tissue repair by formation of fibrocartilage-like tissues in response to damaged subchondral bone (Chang, 2011), which is likely due to the intrinsic ability of MSCs to migrate into injured or inflamed tissues.

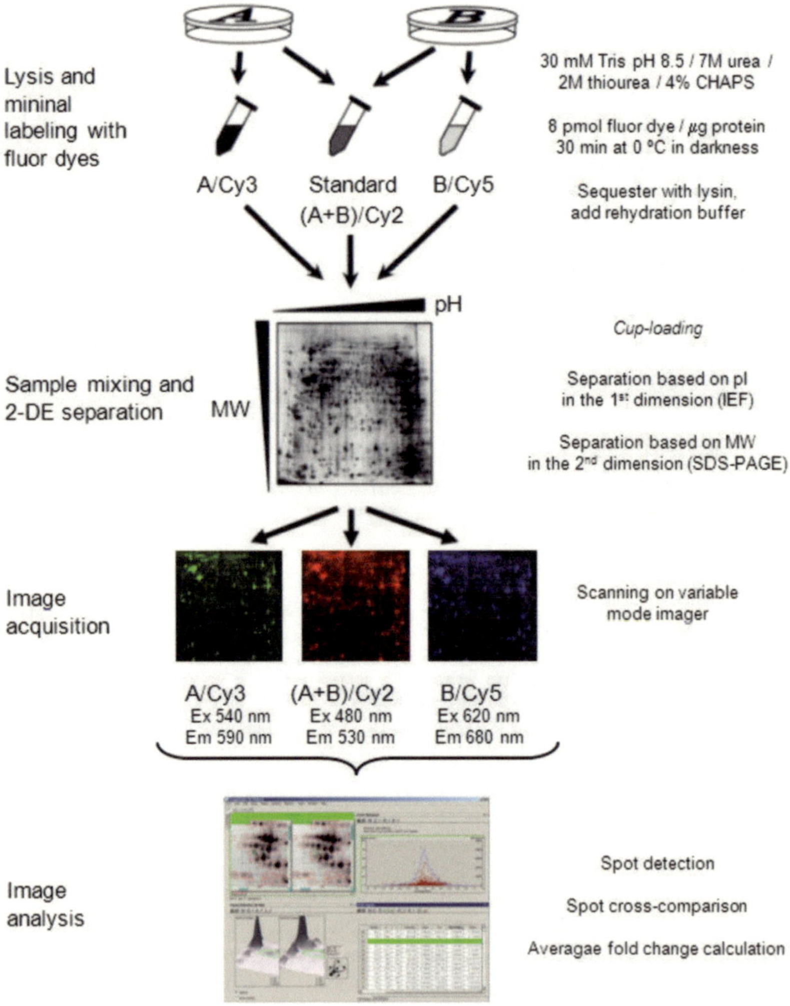

Fig. 3. Schematic representation of the 2D-DIGE methodology
2D-DIGE used for the analysis of protein differential expression in MSCs and chondrocytes of patients with osteoarthritis (sample A) compared to control subjects (sample B). CHAPS, 3-[(3-Cholamidopropyl)dimethylammonio]-1-propanesulfonate; Em, Emission; Ex, Excitation; IEF, isoelectrofocusing. MM, Molecular mass; pI, Isoelectric point; SDS PAGE, polyacrylamide gel electrophoresis in the presence of Sodium Dodecyl Sulphate.

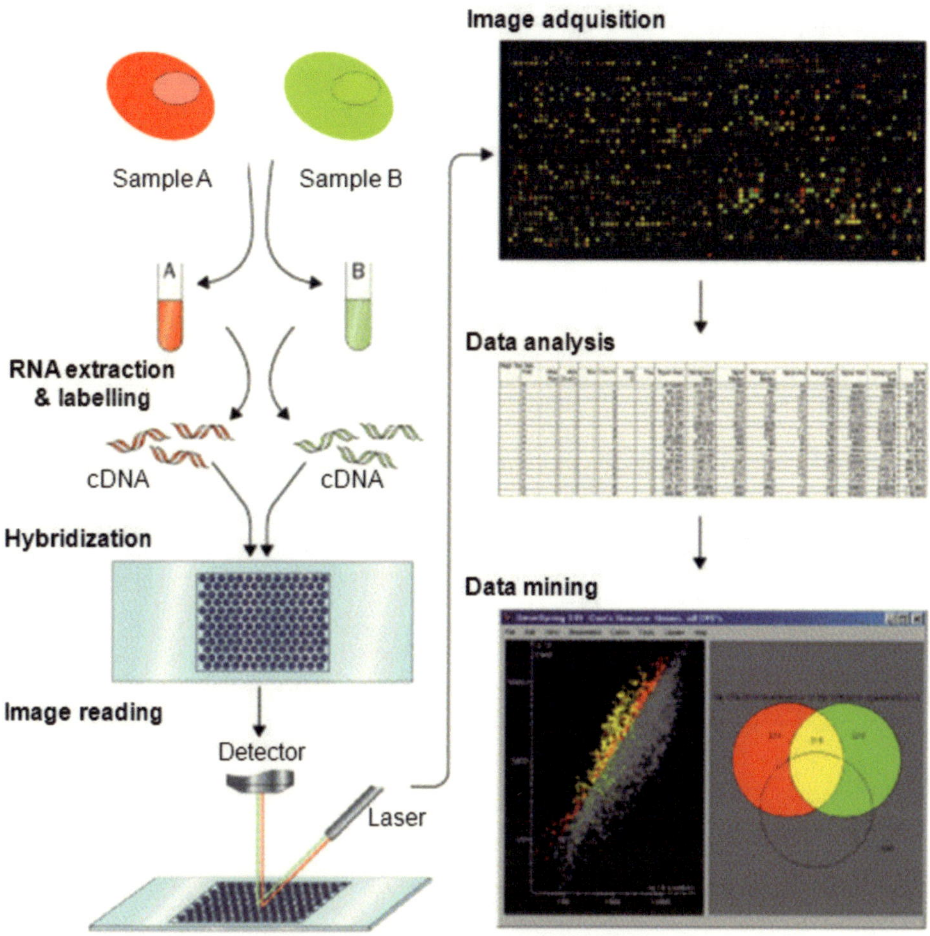

Fig. 4. Schematic representation of the DNA microarray methodology.
DNA microarrays are commonly used to detect messenger RNAs (mRNA), referred to as expression profiling. The method consists in the fluorescently labelling of RNA while the RNA is converted into complementary DNA (cDNA). Amplification of sequences by PCR is sometimes incorporated into this step. Two-color labeling allows two samples or conditions, to be hybridized to the same array and their gene expression profiles compared via the difference in the fluorescence of the two samples. Statistical post-processing of the fluorescence data is usually necessary to eliminate artifacts and false results from the data obtained.

3.1 Proteomic studies in Chondrocytes and MSCs in osteoarthritis

The breakdown of cartilage in OA involves the degradation of the extracellular matrix macromolecules and the altered expression of chondrocyte proteins necessary for normal joint function (Lane Smith, 2000). Thus the screening of proteins with altered expression in chondrocytes from patients with end stage OA compared to control subjects could expand the knowledge of the pathological processes implicated in the damage of articular cartilage in OA. Elucidation of the phenotypical alterations occurring in OA is important for the ascertainment of disease aetiology and for the development of effective treatments for OA.

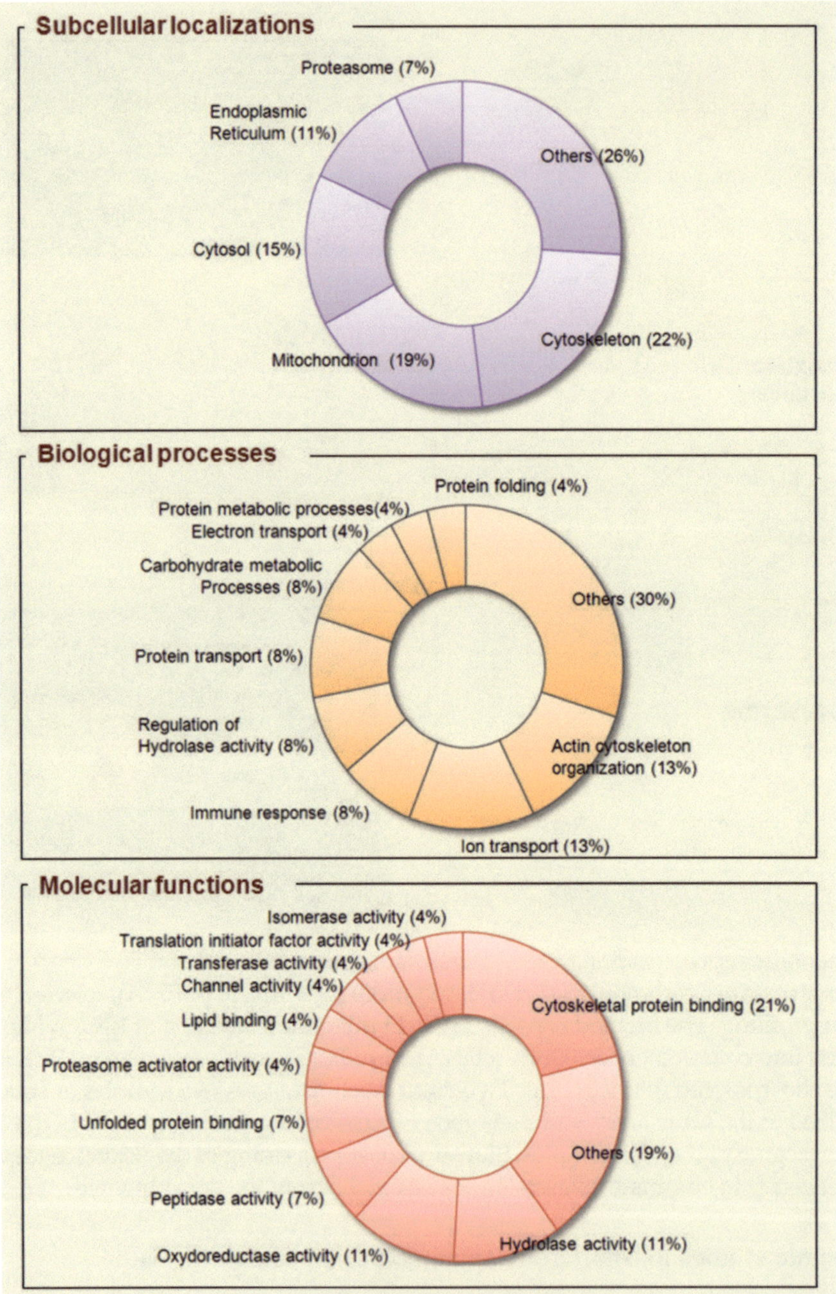

Fig. 5. Gene ontology annotation of the 27 changed proteins identified in OA chondrocytes.

Our first proteomic study dates from 2008 (Rollin, 2008c). Here human knee cartilage was obtained during total joint replacement surgery in six patients with clinical and radiological features of OA and six control samples from adult donors of similar age. Chondrocytes were isolated and cultured during 2-3 weeks at confluence in primary culture, before protein extraction. After 2D-DIGE, differentially expressed proteins were excised from the gel, digested with trypsin and analysed by a MALDI-TOF MS mass spectrometer. Protein identification after peptide mass fingerprinting (PMF) enabled the identification of 27 proteins (14 decreased and 13 increased) in OA chondrocytes. The cellular localization, biological process and molecular functions and of the identified proteins obtained from the online FatiGO ontology database are summarized in (Figure 5).

In this study, a significant differential expression pattern was observed for 27 different chondrocyte proteins. These included an elevated number of cytoskeletal binding proteins cytoskeleton binding, protein disruption, apoptosis and glycolysis proteins displayed a significantly changed expression in OA chondrocytes. Overall, the results suggested the deregulated production in OA cartilage of proteins pertaining to key cellular processes essential for the proper functioning of the chondrocytes, which may have direct effects on OA cartilage biology.

A similar approach was also carried out to study the differential proteome of bone marrow MSCs (BM-MSCs) (Rollin, 2008b) from patients with OA *vs.* MSCs control, obtained from patients with hip fracture without OA signs. In this study we demonstrated the existence of specific alterations in the proteome content of bone marrow MSCs from patients with OA. Once classified into different groups, according to their biological function, the majority of proteins that changed at least 1.5-fold, belonged to the metabolic enzymes, cytoskeleton/motility and transport categories. Markedly, most proteins related to cytoskeleton/motility were down-regulated in MSCs from OA patients. Considering previous evidences supporting that MSCs can home to some tissues, particularly when injured or inflamed, the mechanisms underlying migratory capacity, as a key event for tissue repair by MSCs, were also studied *in vitro* using PDGF as chemoattractant. Our results demonstrated a significant increase in the motility of MSC of OA patients. Together with the differential expression of metabolic and cytoskeleton proteins we concluded that an activation of OA BM-MSCs occurs in response to chemotactic signals sent by the altered subchondral bone in an attempt to heal damaged tissues.

3.2 Gene expression alterations in bone marrow MSCs in osteoarthritis

Our previous experimental data obtained in MSCs proteomic studies indicate an increased migratory capacity of BM-MSCs to the damaged tissues, likely to initiate and/or enhance the wound repair process. In this context, it is known that transforming growth factor-β (TGF-β) plays an important role in directing the cell fate choices in mesenchymal cells (Roelen, 2003). TGF-β induces the chondrogenic differentiation of MSC in the presence of dexamethasone or 3-dimensional cell aggregates (Mackay, 1998) and may act in conjunction with other microenvironmental factors on MSC differentiation. To assess the importance of TGF-β expression in MSCs from OA we comparatively studied by quantitative real-time PCR the expression of genes encoding the total TGF-β and those of the 3 isoforms of TGF-β (1, 2, 3) and TGF-β receptors (TBR-I, TBR-II, TBR-III) in primary cultures of BM-MSCs from patients with end stage OA and healthy control subjects (Rollin, 2008a). Our results showed that only TGF-β1 isoform was significantly increased in MSCs from OA. In addition, we also described an increased expression of TBR-II and TBR-III genes, but not of TBR-I in MSCs from OA. A possible explanation for this upregulated TGF-β upregulation in MSCs could be related to an stimulatory effect on mesenchymal cell proliferation in bone marrow allowing their expansion in response to the bone and cartilage damage characteristic of this disease.

More recently, another experimental approach was carried out by our group based on the comprehensive study of gene expression of MSCs using a DNA microarray expression analysis (Lamas, 2010). Gene expression profiles of MSCs from OA patients were compared to those of MSCs from healthy individuals. After integration of expression profiles into functional categories, by means of a gene ontology (GO)-based statistical analysis using GeneCodis 2.0 (Carmona-Saez, 2007), seventy-five genes from a list of 532 provided for comparison did not show annotations. The remaining 457 genes were grouped into different GO categories based on the subcellular location and functionality (Figure 6). Functional categories showing a major number of genes with downregulated expression in OA-MSCs

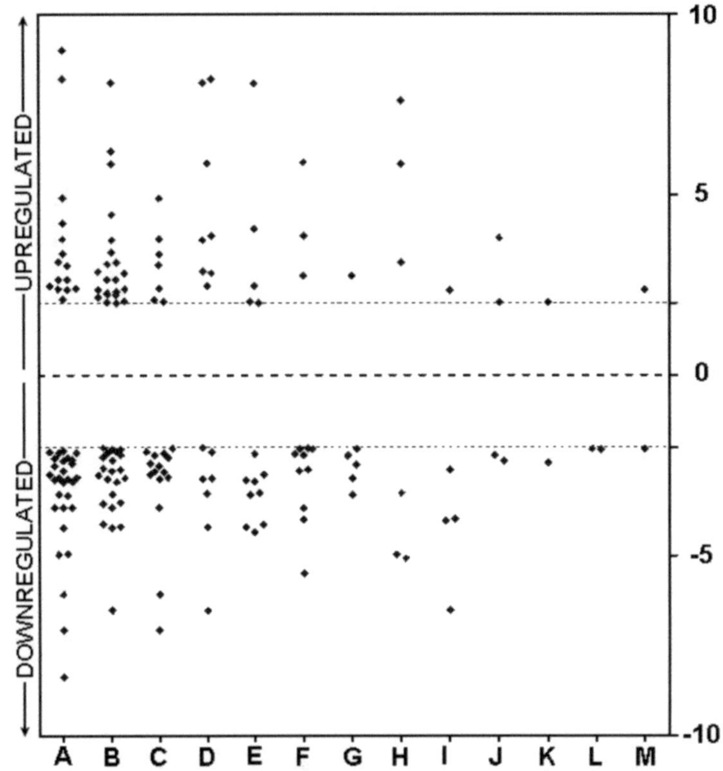

Fig. 6. Differential expression of genes in osteoarthritis mesenchymal stem cells (OA-MSCs) according to gene ontology (GO) categories.
The X-axis represents individual genes classified according to the gene ontology (GO) slim categories provided by the GeneCodis2 (Carmona-Saez, 2007) analysis. The Y-axis represents the fold variation in expression of OA-MSCs compared to control subjects ($p<0.05$). Only genes expressing at least twofold differences in expression were considered. A: Multicellular organismal development; B: Signal transduction; C: Cell differentiation; D: Cell-cell signaling; E: Cell proliferation; F: Metabolic Process; G: Carbohydrate Metabolic Process; H: Anatomical structure morphogenesis; I: Cytoskeleton organization; J: Response to stress; K: Cellular component organization; L: Response to external stimulus; M: Growth

were signal transduction, development and cell differentiation, which in turn are key functions in pluripotential cells. Based on the function of the proteins encoded by these genes, our results suggested that MSCs from patients with OA have a diminished differentiation and regenerative potentials that limit ther ability to generate a functional lineage of cells involved in musculoskeletal tissue homeostasis.

Overall, in this study we provided a reference dataset of genes related to essential functions for the normal biology of MSCs that become altered in OA (Figure 4). We also described for the first time an association between the COL10A1 gene and OA susceptibility suggesting that underlying biological changes which occur during OA disease might be related, at least in part, to defects in the ECM and the formation of subchondral bone, an essential structure for providing joint stability. Moreover, in this work we also demonstrated that the expression of multiple genes related to the wnt pathway were downregulated in OA patients.

3.3 Animal models in regenerative medicine (tendon repair)

Chronic degeneration is the most frequent cause for lesions of the rotator cuff, a set of four tendons connecting the scapula with the humeral head in shoulder. The most affected tendon of the cuff, frequently affected by tears, is the supraspinatus tendon. These lesions are increased in aged patients who frequently need reconstructive surgery. Surgical methods are often unsatisfactory due to inefficient recovery. In this context, MSC-based regenerative medicine offers a hopeful alternative for this type of treatment. In this sense, our group is performing several studies focused on the evaluation of the effectiveness of recovery after treatments consisting in surgical implantations of MSCs alone or in combination with a commercial membrane of type I collagen (Orthoadapt™). This strategy has been previously tested in in a model of acute and chronic injury in rats with promising results allowing us to conduct a clinical trial in humans that is currently under development.

4. MSCs, conventional and other experimental treatments in OA

Treatment of osteoarthritis includes a combination of pharmacological and non-pharmacological measures aimed at relieving pain and the improvement of joint function. Initial treatment is dependent on the extent of the disease, age of patients and the joints affected, in descending order of frequency: Hip, Knee, Foot and Ankle. In mild cases, treatment begins by the use of simple analgesics (eg. paracetamol), (Nonsteroidal anti-inflammatory drugs) NSAIDs (eg. ibuprofen and naproxen) or intermittent intra-articular administration (infiltration) of corticosteroids. However, although the symptoms and pain can be partially alleviated, adverse effects associated with conventional drug therapy is not recommended for long time periods. Moreover, treatment is often accompanied by non-pharmacological treatments, these include patient education and physical exercises to restore joint movement and to increase muscle strength, reduction of weight on painful joints. When joints are severely damaged treatment may require surgery. The most common surgical treatments are arthroscopic surgery, to trim damaged cartilage. Osteotomy, to change the alignment of a bone to relieve stress on the bone or joint. Arthrodesis or surgical fusion of bones, usually in the spine and the total or partial arthroplasty to replace the damaged joint with an artificial one.

Halfway between drug therapy and joint replacement surgery, several arthroscopic strategies combined with cell therapies have been developed for the treatment of cartilage injuries. The techniques used and the results obtained greatly vary depending on the size of the lesion. For smaller and medium sized cartilage defects, autologous osteochondral cylinder transfer or mosaicplasty has been widely used but its efficacy is limited by donor site morbidity and the poor integration of implants.

Techniques of cell therapy in OA were initially based on the stimulation of bone marrow by drilling, microfracture and abrasive chondroplasty to promote better access of pluripotential stem cells from subchondral vascular area, to the site of injury (Steinwachs, 2008; Chen, 2011). Although the microfracture has achieved good results in terms of functionality and reduction of pain, several limitations such as chondral defect size and age of the patient are major constraints. These methods only provide a partial filling of the defect with fibrocartilage without the characteristics of hyaline cartilage. More recently improvements of these cell therapies have been made using the implantation of cultured autologous chondrocytes in the defect site (ACI) and a variation of this technique, using collagen Type III/I scaffolds, MACI (Matrix-induced autologous chondrocyte implantation) (Strauss, 2011; Ventura, 2011). MACI was developed to enable the treatment of larger defects when cell engaged procedures such as ACI cannot be used or it is not indicated. The results of the ongoing studies in chondrocyte implantation show better results in the formation of a hyaline-like cartilage with similar characteristics and durability than normal hyaline cartilage. In any case, the major drawbacks are that the chondrocytes harvesting require additional surgery and only a small number of chondrocytes can be isolated from the explants. In addition these cells lose their phenotypic characteristics in culture, limiting their application in extensive chondral defects, such as those produced in osteoarthritis. Otherwise, allograft transplantation is limited by donor availability.

Table 1 shows a summary of the most common techniques used clinically and experimentally. However the number of combinations of treatment options with each strategy is unlimited and growing every day. A great number of studies involve animal models evaluating different scaffolds, number of cells and ambiental factors used, etc. However, given the complex variety of combinations, there are no well-conducted clinical trials in humans evaluating the efficacy of a particular method.

In summary, regarding OA, advances in research for the development of new technologies in the management of cartilage defects is currently unresolved. Actually any treatment method provides consistent and acceptable long-term clinical results, and in particular for treatment of large chondral defects. With evolving techniques, versatility, availability and differentiation potential of stem cells have become the hope to improve current treatments based on other more committed cells. Alone or in combination with different scaffold materials and environmental factors, including growth factors, signalling molecules and mechanical influence, these cells are exceptional candidates for engineer cartilage constructs *in vitro*. Several studies have shown an improvement in the quality of the new tissue formed, but its long-term efficacy and the mechanism by which it occurs are unknown. In this regard it has been postulated that the low intrinsic immunogenicity of MSCs along with its ability to reduce inflammation, are characteristics that determine the establishment of a less inflammatory environment that facilitates the repair.

Symptom	Current approach	Mode of action	Advantages	Disadvantages
Initial stages (non surgical)				
Pain and Inflammation	- Pharmacological treatment	Analgesic Anti-inflammatory	Surgery not needed	Adverse effects
	- Physical exercice	Joint structures reinforcement		Not recommended in some cases
	- Educational	Prevention		none
Structural damage of joint cartílage (surgical methods)				
Small defects <2.5cm^2	- Subchondral drilling - Microfracture - Abrasive chondroplasty	Bone marrow stimulation	Relatively inexpensive surgery	Age of patients Inferior quality of neoformed tissue
Structural damage of joint cartílage (trasplantation methods)				
Medium defects 2.5 to 4 cm^2	- Osteochondral allografts - Autologous periosteal grafts - Autologous mosaicplasty	Tissue trasplantation from autologous or allogeneic origin	Age of donor eligible Relatively well suited for medium cartilage defects	Availability of implants. Donor morbidity. Poor integration and maintenance of implant.
	- Autologous chondrocyte trasplantation (ACI)	Cellular therapy using committed cell lineages	Relatively well suited for medium cartilage defects Defect filling with hyaline cartilage in a short time	Two sugeries needed Small number of chondrocyte availability Poor integration Do not show improvement over other conventional techniques

Symptom	Current approach	Mode of action	Advantages	Disadvantages
	- Matrix-induced autologous chondrocyte implantation (MACI)		Better early results	Medium and long-term results not available Expensive
	Mesenchymal Stem Cells (MSCs)	Cellular therapy using stem cells	Potential	Expensive
Large defects > 4 cm^2	Total arthroplasty	Joint replacementt	Joint restoration	Infections Expiration

Table 1. Conventional and experimental treatments in OA.

5. Conclusions and future perspectives

The goal in regenerative medicine is based in a conceptually simple scheme: the development of new strategies to replace human cells or induce the regeneration of diseased or injured human tissues. Although during the last twenty years a considerable scientific progress has been done in this field, there are still many unanswered questions about key concepts concerning both tissue engineering and cell therapy.

Stem cells, in its two "flavours": embryonic and adult stem cells are the basis of regenerative medicine; however, biological differences between adult and embryonic stem cells and among adult stem cells found in different tissues is an important aspect which implication for therapeutic uses is not resolved. From the point of view of their clinical application, the source of the cells is of extreme importance. In the case of autologous cells that are not rejected by the patient's immune system their application is potentially safer than in allogeneic cells and more suitable for permanent tissue replacement. However, and for example, in cases where the recipient suffers from a genetic disorder, their application would be inappropriate. Future efforts should be done to minimise rejection, and to favour the banking and use of allogeneic adult cells. Among adult stem cells, the MSCs are of paramount importance for the treatment of several rheumatic diseases. Besides their plasticity and regenerative potential they show immunosuppressive and antiinflammatory characteristics *in vitro* and proven in preclinical and clinical studies.

Future studies will need to focus on the particular cell biology of MSCs including the biochemical signal transduction pathways involved in maintaining and enhancing chondrogenic differentiation, but also in the mechanisms implicated in immunomodulation. Other important aspects that need further research include the evaluation of safety and efficacy of local or systemic modes of admistration of MSCs; the mechanisms of cell to cell communication, such as microvesicles transporting RNAs, cytokines, etc.; the behaviour of MSCs in different niches; the design of specialised engineered scaffolds, to enable the efficient repair of a variety of tissues; and finally, the implementation and use of genetic reprogramming strategies.

6. Acknowledgments

Special thanks to Peter Clarke and his disinterested revision and editing of the manuscript and to Dr. Emilio Camafeita for providing some of the representations included.

This work was supported by a research grant from Roche Farma S.A. and was also partially supported by the RETICS Program, RD08/0075 (RIER) from Instituto de Salud Carlos III.

7. References

Aggarwal S; Pittenger MF. (2005). Human mesenchymal stem cells modulate allogeneic immune cell responses. *Blood*, Vol. 105, No. 4, pp. (1815-1822), ISSN:0006-4971

Alban A; David SO; Bjorkesten L et al. (2003). A novel experimental design for comparative two-dimensional gel analysis: two-dimensional difference gel electrophoresis incorporating a pooled internal standard. *Proteomics*, Vol. 3, No. 1, pp. (36-44), ISSN 1615-9853

Alvarez-Dolado M. (2007). Cell fusion: biological perspectives and potential for regenerative medicine. *Front biosci*, Vol. 12, pp. (1-12) , ISSN 1093-4715

Aspden RM. (2008). Osteoarthritis: a problem of growth not decay? *Rheumatology (Oxford)*, Vol. 47, No. 10, pp. (1452-1460) , ISSN 1462-0332

Bansback NJ; Regier DA; Ara R et al. (2005). An overview of economic evaluations for drugs used in rheumatoid arthritis : focus on tumour necrosis factor-alpha antagonists. *Drugs*, Vol. 65, No. 4, pp. (473-496) , ISSN 0012-6667

Bianco P; Riminucci M; Gronthos S et al. (2001). Bone marrow stromal stem cells: nature, biology, and potential applications. *Stem Cells*, Vol. 19, No. 3, pp. (180-192) , ISSN 1066-5099

Brandt KD; Radin EL; Dieppe PA et al. (2006). Yet more evidence that osteoarthritis is not a cartilage disease. *Ann Rheum Dis*, Vol. 65, No. 10, pp. (1261-1264) , ISSN 0003-4967

Bruder SP; Jaiswal N; Haynesworth SE. (1997). Growth kinetics, self-renewal, and the osteogenic potential of purified human mesenchymal stem cells during extensive subcultivation and following cryopreservation. *J Cell Biochem*, Vol. 64, No. 2, pp. (278-294) , ISSN 0730-2312

Camafeita E; Lamas JR; Calvo E et al. (2009). Proteomics: New insights into rheumatic diseases. *Proteomics Clinical Applications*, Vol. 3, No. 2, pp. (226-241) , ISSN 1862-8346

Caplan AI. (2000). Mesenchymal stem cells and gene therapy. *Clin Orthop Relat Res*, Vol. No. (379 Suppl), pp. (S67-70) , ISSN 0009-921X

Caplan AI. (2007). Adult mesenchymal stem cells for tissue engineering versus regenerative medicine. *J Cell Physiol*, Vol. 213, No. 2, pp. (341-347), ISSN 0021-9541

Carmona-Saez P; Chagoyen M; Tirado F et al. (2007). GENECODIS: a web-based tool for finding significant concurrent annotations in gene lists. *Genome Biol*, Vol. 8, No. 1, pp. (R3), ISSN 1465-6914

Chanda D; Kumar S; Ponnazhagan S. (2010). Therapeutic potential of adult bone marrow-derived mesenchymal stem cells in diseases of the skeleton. *J Cell Biochem*, Vol. 111, No. 2, pp. (249-257) ISSN 1097-4644

Chang CH; Kuo TF; Lin FH et al. (2011). Tissue engineering-based cartilage repair with mesenchymal stem cells in a porcine model. *J Orthop Res*, ISSN 1554-527X

Chen H, Hoemann CD, Sun J, Chevrier A, McKee MD, Shive MS, et al. Depth of subchondral perforation influences the outcome of bone marrow stimulation cartilage repair. J Orthop Res, Vol. 29, No. 8, (Aug), pp. (1178-1184) ISSN 1554-527X

Chen L; Tredget EE; Wu PY et al. (2008). Paracrine factors of mesenchymal stem cells recruit macrophages and endothelial lineage cells and enhance wound healing. *PLoS One*, Vol. 3, No. 4, pp. (e1886), ISSN 1932-6203

Chivu M; Dima SO; Stancu CI et al. (2009). In vitro hepatic differentiation of human bone marrow mesenchymal stem cells under differential exposure to liver-specific factors. *Transl Res*, Vol. 154, No. 3, pp. (122-132), ISSN 1931-5244

Conget PA; Minguell JJ. (1999). Phenotypical and functional properties of human bone marrow mesenchymal progenitor cells. *J Cell Physiol*, Vol. 181, No. 1, pp. (67-73), ISSN 0021-9541

Day TF; Guo X; Garrett-Beal L et al. (2005). Wnt/beta-catenin signaling in mesenchymal progenitors controls osteoblast and chondrocyte differentiation during vertebrate skeletogenesis. *Dev Cell*, Vol. (8), No. (5), pp. (739-750), ISSN 1534-5807

Dennis JE; Merriam A; Awadallah A et al. (1999). A quadripotential mesenchymal progenitor cell isolated from the marrow of an adult mouse. *J Bone Miner Res*, Vol. 14, No. (5, pp. (700-709), ISSN 0884-0431

Gaur T; Lengner CJ; Hovhannisyan H et al. (2005). Canonical WNT signaling promotes osteogenesis by directly stimulating Runx2 gene expression. *J Biol Chem*, Vol. 280, No. 39, pp. (33132-33140), ISSN 0021-9258

Gerdoni E; Gallo B; Casazza S et al. (2007). Mesenchymal stem cells effectively modulate pathogenic immune response in experimental autoimmune encephalomyelitis. *Ann Neurol*, Vol. 61, No. 3, pp. (219-227) , ISSN 0364-5134

Gupta N; Su X; Popov B et al. (2007). Intrapulmonary delivery of bone marrow-derived mesenchymal stem cells improves survival and attenuates endotoxin-induced acute lung injury in mice. *J Immunol*, Vol. 179, No. 3, pp. (1855-1863), ISSN 0022-1767

Hill TP; Spater D; Taketo MM et al. (2005). Canonical Wnt/beta-catenin signaling prevents osteoblasts from differentiating into chondrocytes. *Dev Cell*, Vol. 8, No. 5, pp. (727-738), ISSN 1534-5807

Karussis D; Kassis I; Kurkalli BG et al. (2008). Immunomodulation and neuroprotection with mesenchymal bone marrow stem cells (MSCs): a proposed treatment for multiple sclerosis and other neuroimmunological/neurodegenerative diseases. *J Neurol Sci*, Vol. 265, No. 1-2, pp. (131-135), ISSN 0022-510X

Lamas JR; Rodriguez-Rodriguez L; Vigo AG et al. (2010). Large-scale gene expression in bone marrow mesenchymal stem cells: a putative role for COL10A1 in osteoarthritis. *Ann Rheum Dis*, Vol. 69, No. 10, pp. (1880-1885), ISSN 1468-2060

Lane Smith R; Trindade MC; Ikenoue T et al. (2000). Effects of shear stress on articular chondrocyte metabolism. *Biorheology*, Vol. 37, No. 1-2, pp. (95-107), ISSN 0006-355X

Laupattarakasem W; Laopaiboon M; Laupattarakasem P et al. (2008). Arthroscopic debridement for knee osteoarthritis. *Cochrane Database Syst Rev*, Vol. No. (1), pp. (CD005118), ISSN 1469-493X

Le Blanc K; Rasmusson I; Sundberg B et al. (2004). Treatment of severe acute graft-versus-host disease with third party haploidentical mesenchymal stem cells. *Lancet*, Vol. 363, No. 9419, pp. (1439-1441), ISSN 1474-547X

Le Blanc K; Ringden O. (2007). Immunomodulation by mesenchymal stem cells and clinical experience. *J Intern Med*, Vol. 262, No. 5, pp. (509-525), ISSN 0954-6820

Lee RH; Seo MJ; Reger RL et al. (2006). Multipotent stromal cells from human marrow home to and promote repair of pancreatic islets and renal glomeruli in diabetic NOD/scid mice. *Proc Natl Acad Sci U S A*, Vol. 103, No. 46, pp. (17438-17443), ISSN 0027-8424

Li WJ; Tuli R; Okafor C et al. (2005). A three-dimensional nanofibrous scaffold for cartilage tissue engineering using human mesenchymal stem cells. *Biomaterials*, Vol. 26, No. 6, pp. (599-609), ISSN 0142-9612

Licatalosi DD; Darnell RB. (2010). RNA processing and its regulation: global insights into biological networks. *Nat Rev Genet*, Vol. 11, No. 1, pp. (75-87), ISSN 1471-0064

Lisignoli G; Cristino S; Piacentini A et al. (2005). Cellular and molecular events during chondrogenesis of human mesenchymal stromal cells grown in a three-dimensional hyaluronan based scaffold. *Biomaterials*, Vol. 26, No. 28, pp. (5677-5686), ISSN 0142-9612

Loza E; Abasolo L; Jover JA et al. (2008). Burden of disease across chronic diseases: a health survey that measured prevalence, function, and quality of life. *J Rheumatol*, Vol. 35, No. 1, pp. (159-165), ISSN 0315-162X

Mackay AM; Beck SC; Murphy JM et al. (1998). Chondrogenic differentiation of cultured human mesenchymal stem cells from marrow. *Tissue Eng*, Vol. 4, No. 4, pp. (415-428), ISSN 1076-3279

Manners PJ. (2003). Epidemiology of the rheumatic diseases of childhood. *Curr Rheumatol Rep*, Vol. 5, No. 6, pp. (453-457), ISSN 1523-3774

Mariller MM; Santos-Eggimann B. (2005). The prevalence of rheumatic diseases in the elderly in developed countries and its evolution over time. *Soz Praventivmed*, Vol. 50, No. 1, pp. (45-51), ISSN 0303-8408

Moretti P; Hatlapatka T; Marten D et al. (2010). Mesenchymal stromal cells derived from human umbilical cord tissues: primitive cells with potential for clinical and tissue engineering applications. *Adv Biochem Eng Biotechnol*, Vol. 123, No. pp. (29-54), ISSN 1616-8542

Munoz JR; Stoutenger BR; Robinson AP et al. (2005). Human stem/progenitor cells from bone marrow promote neurogenesis of endogenous neural stem cells in the hippocampus of mice. *Proc Natl Acad Sci U S A*, Vol. 102, No. 50, pp. (18171-18176), ISSN 0027-8424

Necas A; Planka L; Srnec R et al. (2010). Quality of newly formed cartilaginous tissue in defects of articular surface after transplantation of mesenchymal stem cells in a composite scaffold based on collagen I with chitosan micro- and nanofibres. *Physiological research / Academia Scientiarum Bohemoslovaca*, Vol. 59, No. 4, pp. (605-614), ISSN 0862-8408

Ortiz LA; Dutreil M; Fattman C et al. (2007). Interleukin 1 receptor antagonist mediates the antiinflammatory and antifibrotic effect of mesenchymal stem cells during lung injury. *Proc Natl Acad Sci U S A*, Vol. 104, No. 26, pp. (11002-11007), ISSN 1091-6490

Pereira RF; Halford KW; O'Hara MD et al. (1995). Cultured adherent cells from marrow can serve as long-lasting precursor cells for bone, cartilage, and lung in irradiated mice. *Proc Natl Acad Sci U S A*, Vol. 92, No. 11, pp. (4857-4861), ISSN 0027-8424

Pittenger MF; Mackay AM; Beck SC et al. (1999). Multilineage potential of adult human mesenchymal stem cells. *Science*, Vol. 284, No. 5411, pp. (143-147), ISSN 0036-8075

Quintana L; zur Nieden NI; Semino CE. (2009). Morphogenetic and regulatory mechanisms during developmental chondrogenesis: new paradigms for cartilage tissue engineering. *Tissue Eng Part B Rev*, Vol. 15, No. 1, pp. (29-41), ISSN 1937-3368

Reiser J; Zhang XY; Hemenway CS et al. (2005). Potential of mesenchymal stem cells in gene therapy approaches for inherited and acquired diseases. *Expert Opin Biol Ther*, Vol. 5, No. 12, pp. (1571-1584), ISSN 1744-7682

Remy-Martin JP; Marandin A; Challier B et al. (1999). Vascular smooth muscle differentiation of murine stroma: a sequential model. *Exp Hematol*, Vol. 27, No. 12, pp. (1782-1795), ISSN 0301-472X

Ren G; Zhang L; Zhao X et al. (2008). Mesenchymal stem cell-mediated immunosuppression occurs via concerted action of chemokines and nitric oxide. *Cell Stem Cell*, Vol. 2, No. 2, pp. (141-150), ISSN 1875-9777

Roelen BA; Dijke P. (2003). Controlling mesenchymal stem cell differentiation by TGFBeta family members. *J Orthop Sci*, Vol. 8, No. 5, pp. (740-748), ISSN 0949-2658

Rollin R; Alvarez-Lafuente R; Marco F et al. (2008a). Abnormal transforming growth factor-beta expression in mesenchymal stem cells from patients with osteoarthritis. *J Rheumatol*, Vol. 35, No. 5, pp. (904-906), ISSN 0315-162X

Rollin R; Marco F; Camafeita E et al. (2008b). Differential proteome of bone marrow mesenchymal stem cells from osteoarthritis patients. *Osteoarthritis Cartilage*, Vol. 16, No. 8, pp. (929-935), ISSN 1063-4584

Rollin R; Tornero P; Marco F et al. (2008c). Differential Proteome of Articular Chondrocytes From Patients with Osteoarthritis. *J Proteomics Bioinformatics*, Vol. 1, No. pp. (267-280), ISSN 1063-4584

Romanov YA; Darevskaya AN; Merzlikina NV et al. (2005). Mesenchymal stem cells from human bone marrow and adipose tissue: isolation, characterization, and differentiation potentialities. *Bull Exp Biol Med*, Vol. 140, No. 1, pp. (138-143), ISSN 0007-4888

Spaeth E; Klopp A; Dembinski J et al. (2008). Inflammation and tumor microenvironments: defining the migratory itinerary of mesenchymal stem cells. *Gene Ther*, Vol. 15, No. 10, pp. (730-738), ISSN 1476-5462

Spahn G; Schiele R; Hofmann GO et al. (2011). [The Prevalence of Radiological Osteoarthritis in Relation to Age, Gender, Birth-Year Cohort, and Ethnic Origins.]. *Z Orthop Unfall*, Vol. 149, No. 2, pp. (145-152), ISSN 1864-6743

Spees JL; Olson SD; Whitney MJ et al. (2006). Mitochondrial transfer between cells can rescue aerobic respiration. *Proc Natl Acad Sci U S A*, Vol. 103, No. 5, pp. (1283-1288), ISSN 0027-8424

Spees JL; Olson SD; Ylostalo J et al. (2003). Differentiation, cell fusion, and nuclear fusion during ex vivo repair of epithelium by human adult stem cells from bone marrow stroma. *Proc Natl Acad Sci U S A*, Vol. 100, No. 5, pp. (2397-2402), ISSN 0027-8424

Steinwachs M. (2009). New technique for cell-seeded collagen-matrix-supported autologous chondrocyte transplantation. Arthroscopy, Vol. 25 No.2, pp. (208-211).

Strauss, E. J. et al. Fonseca, L. E. et al. Shah, M. R., et al. (2011). Management of focal cartilage defects in the knee - Is ACI the answer? Bull NYU Hosp Jt Dis, Vol. 69, No. 1, pp. (63-72). ISSN 1936-9727

Stults JT; Arnott D. (2005). Proteomics. *Methods Enzymol*, Vol. 402, pp. (245-289), ISSN 0076-6879

Suzuki K; Sun R; Origuchi M et al. (2011). Mesenchymal Stromal Cells Promote Tumor Growth Through the Enhancement of Neovascularization. *Mol Med*, ISSN 1528-3658

Thorpe SD; Buckley CT; Vinardell T et al. (2010). The response of bone marrow-derived mesenchymal stem cells to dynamic compression following TGF-beta3 induced chondrogenic differentiation. *Ann Biomed Eng*, Vol. 38, No. 9, pp. (2896-2909), ISSN 1521-6047

Tian LL; Yue W; Zhu F et al. (2011). Human mesenchymal stem cells play a dual role on tumor cell growth in vitro and in vivo. *J Cell Physiol*, Vol. 226, No. 7, pp. (1860-1867), ISSN 1097-4652

Unlu M; Morgan ME; Minden JS. (1997). Difference gel electrophoresis: a single gel method for detecting changes in protein extracts. *Electrophoresis*, Vol. 18, No. 11, pp. (2071-2077), ISSN 0173-0835

Vanarsa K; Mohan C. (2010). Proteomics in rheumatology: the dawn of a new era. *F1000 Med Rep*, Vol. 2, No. pp. (87), ISSN 1757-5931

Ventura, A. et al. Memeo, A. et al. Borgo, E., et al. (2011). Repair of osteochondral lesions in the knee by chondrocyte implantation using the MACI(®) technique. Knee Surg Sports Traumatol Arthrosc, Vol. No. (Jun 17), pp.

Viswanathan S; Unlu M; Minden JS. (2006). Two-dimensional difference gel electrophoresis. *Nat Protoc*, Vol. 1, No. 3, pp. (1351-1358), ISSN 1750-2799

Worthington J. (2005). Investigating the genetic basis of susceptibility to rheumatoid arthritis. *J Autoimmun*, Vol. (25 Suppl), No. pp. (16-20), ISSN 0896-8411

Zavan B; Giorgi C; Bagnara GP et al. (2007). Osteogenic and chondrogenic differentiation: comparison of human and rat bone marrow mesenchymal stem cells cultured into polymeric scaffolds. *Eur J Histochem*, Vol. (51 Suppl 1), No. pp. (1-8), ISSN 1121-760X

Zhou XZ; Leung VY; Dong QR et al. (2008). Mesenchymal stem cell-based repair of articular cartilage with polyglycolic acid-hydroxyapatite biphasic scaffold. *The International journal of artificial organs*, Vol. 31, No. 6, pp. (480-489), ISSN 0391-3988

Zhu M; Tang D; Wu Q et al. (2009). Activation of beta-catenin signaling in articular chondrocytes leads to osteoarthritis-like phenotype in adult beta-catenin conditional activation mice. *J Bone Miner Res*, Vol. 24, No. 1, pp. (12-21), ISSN 1523-4681

Zscharnack M; Hepp P; Richter R et al. (2010). Repair of chronic osteochondral defects using predifferentiated mesenchymal stem cells in an ovine model. *Am J Sports Med*, Vol. 38, No. 9, pp. (1857-1869), ISSN 1552-3365

13

Stem Cell-Mediated Intervertebral Disc Regeneration

Namath S. Hussain, Vickram Tejwani and Mick Perez-Cruet
Oakland University William Beaumont School of Medicine, Royal Oak, Michigan USA

1. Introduction

Currently, degenerative disk disease (DDD) and the subsequent chronic lower back pain that results from it represent a significant source of morbidity and mortality worldwide. The available treatment modalities such as pain therapy and surgical interventions aim to provide symptomatic relief; however, they do not address the underlying pathophysiology of DDD. The disease also has high societal health care costs (Chan et al., 2006; Cassinelli et al, 2001). Many modalities exist for symptomatic treatment of this condition, including bed rest, massage, stretching, strengthening exercises, physical therapy, epidural injections and other pain management therapies, and spinal surgery. Most conservative therapies are

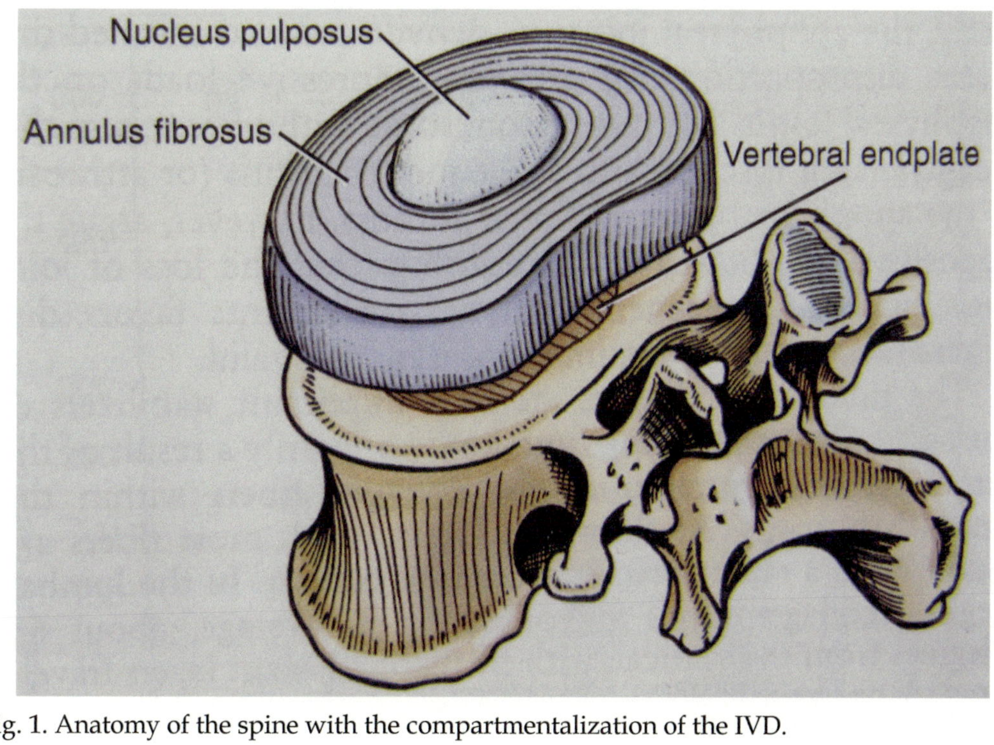

Fig. 1. Anatomy of the spine with the compartmentalization of the IVD.

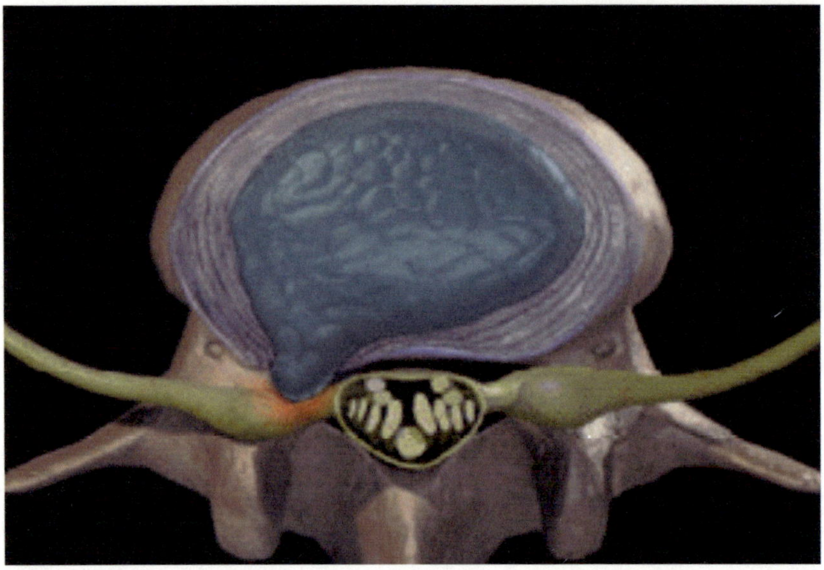

Fig. 2. Axial slice model of the intervertebral disc with an image of a disc herniation

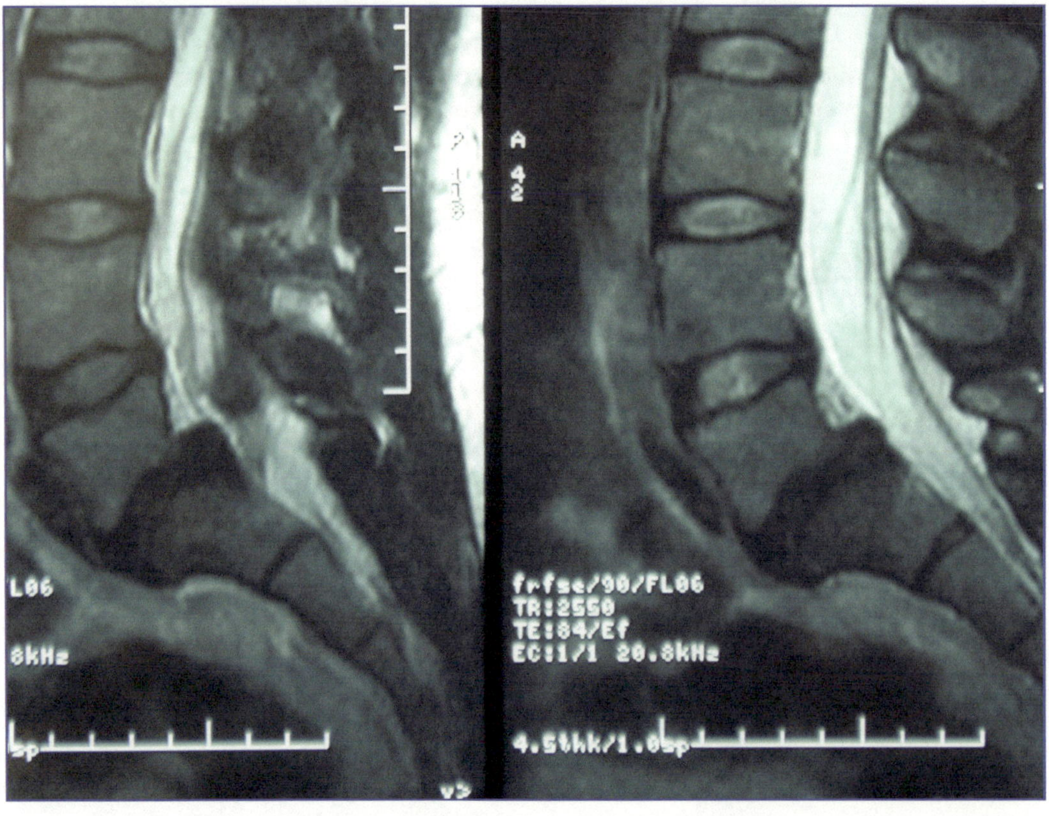

Fig. 3. Sagittal T2-weighted MRI showing degeneration and loss of T2 signal in the L5-S1 IVD

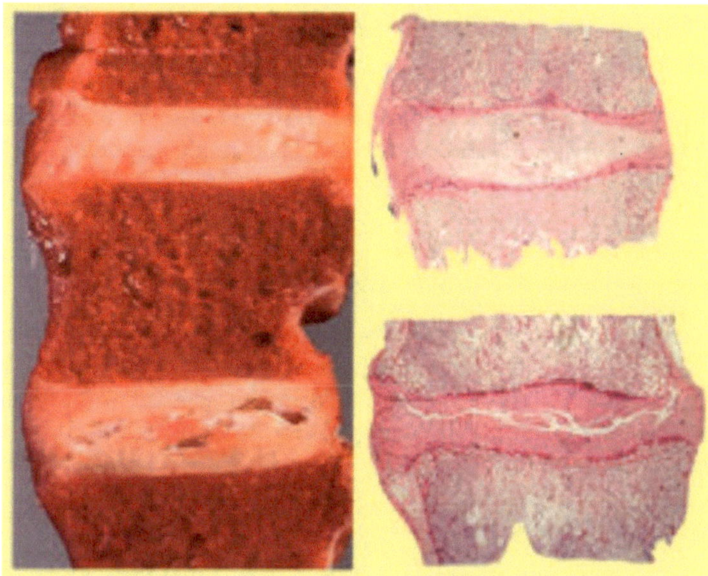

Fig. 4. Postmortem specimen and hematoxylin and eosin staining showing multilevel degeneration of the IVDs with fissuring of the actual disc structure

attempted before surgery with the intent to spare patients the possible complications associated with surgical intervention. However, these conservative measures and even surgery itself with its associated risks only address the symptoms with no impact on the disease process in the disc itself. Recent research has given further insight into the pathogenesis of DDD, which has borne out a renewed interest in biologic therapies centered on the nucleus pulposus (NP) and the annulus fibrosus and the potential of stem cells to reverse the disease process at a histological and cellular level. In this chapter, we will systemically review the current literature and the most salient studies regarding biologic therapies in the regeneration of the intervertebral disc (IVD). We go on to describe the direction this field is heading in and the future potential of the therapies being developed using ESCs.

2. Basic science laboratory studies

Before examining the utility of stem cells in human and animal models, it is important to review several of the basic science benchtop laboratory studies that have provided the rationale for in-vivo testable treatments and hypotheses. These studies examined factors influencing both mesenchymal and embryonic stem cell proliferation and differentiation towards a NP-like phenotype. We will examine how these studies have provided valuable information regarding multiple factors that can stimulate embryonic stem cells (ESCs) and mesenchymal stem cells (MSCs) towards a chondrocytic lineage, as well as factors that can inhibit this differentiation in basic in-vitro models.

2.1 Genetic studies

DDD is a condition that rises from a combination of a genetic predisposition (Chan et al., 2006) along with environmental modifiers (Stokes & Iatridis, 2004). Several causes of age-related degeneration of the IVD include loss of biomechanical support by surrounding muscular and ligamentous structures, uneven force loading as the aging spine deforms while trying to compensate for these changes, cell senescence, loss of viable progenitor cells, accumulation of degraded matrix molecules, and fatigue failure of both the disc matrix and surrounding annulus fibrosus. Correlations have been made between DDD and collagen, aggrecan, and matrix metalloproteinase polymorphisms coding for structural proteins (Ala-Kokko, 2002).

2.2 Factors influencing stem cell proliferation

In order to further study how these cells would interact in various factor environments, it became crucial to more fully characterize these cells. This point is very important with regard to stem cell research because it is essential to characterize and identify what factors provide the best type of environment to stimulate ESCs and MSCs to differentiate toward a chondrocytic-type cell lineage.

2.2.1 Mesenchymal Stem Cells

Transforming growth factor-β3 (TGF-β3) is one factor that has been shown in multiple studies (Steck et al., 2005; Risbud et al., 2004; Shen, 2009) to stimulate cells to differentiate into chondrocytes. Several studies have shown that after TGF-β3 stimulation, MSCs turned positive for collagen type II protein and expressed a large panel of genes characteristic for chondrocytes, such as aggrecan, decorin, fibromodulin, and cartilage oligomeric matrix protein (Steck et al., 2005; Risbud et al., 2004). Shen et al. have shown that bone morphogenic protein-2 (BMP-2) can help to enhance TGF-β3-mediated chondrogenesis in MSCs (Shen, 2009). The combination of BMP-2 and TGF-β3 in alginate culture was found to be superior to the standard differentiation method using TGF-β3 alone as evinced by increased mRNA expression of aggrecan, type II collagen, Sox-9, BMP-2, and BMP-7, all of which are chondrocyte markers. This effect was even more pronounced when TGF-β3 and rhBMP-2 were both added (Kuh et al., 2008). This synergistic effect was consistently found in the study, providing further support as to an as yet unknown pathway towards chondrocytic differentiation.

2.2.2 Embryonic Stem Cells

Hoben et al performed a similar characterization study using human ESCs (Hoben et al., 2009). Growth factors were studied with a coculture method for 3 weeks and evaluated for collagen and glycosaminoglycan (GAG) synthesis. The growth factors studied were TGF-β3, BMP-2, BMP-4, BMP-6, and sonic hedgehog protein. The investigators found that the combination of BMP-4 and TGF-β3 within the fibrochondrocyte coculture led to an increase in cell proliferation and GAG production compared to either treatment alone. Koay et al had similar results with BMP-2 and TGF-β3 leading human ESCs down a differentiation path that produced an end product with high type I collagen content (Koay et al., 2007). However, they also found that human ESCs treated with TGF-β3 followed by TGF-β1 and IGF-1 produced constructs with no collagen I, showing that different growth factor application in different temporal sequences can have a marked impact on end-product composition and biomechanical properties. The importance of temporal sequences cannot be understated with regard to stem cell development and has important implications pertaining to harvesting and large-scale production of these cells for future potential therapeutic uses.

2.3 Stem cell growth in the native IVD microenvironment

Several groups have conducted well-designed in-vitro studies that have gone one step beyond identifying environmental factors that affect differentiation of stem cells into NP-like cells, and have actually studied how these factors may correlate to the current in-vivo microenvironment of the IVD. This was done in order to obtain a clear picture of what would happen if these stem cells were implanted into these native biological conditions. Culturing under IVD-like glucose conditions (1.0 mg/mL glucose) stimulated aggrecan and collagen I expression and deposition. IVD-like osmolarity (485 mOsm) and pH (pH = 6.8) conditions, on the other hand, strongly decreased proliferation and expression of matrix proteins. Combining these conditions resulted in decreased proliferation and gene expression of matrix proteins, demonstrating that, in this case, osmolarity and pH play a larger impact in inhibiting differentiation than glucose does in stimulating it (Wuertz et al., 2008).

Another study by the same group showed that acidity caused an inhibition of aggrecan and collagen I expression, as well as a decrease in proliferation and cell viability. This demonstrates that pH may be the major limitation for stem cell-based IVD repair (Wuertz, 2009). This also illustrates the importance of early intervention and the role of predifferentiation when planning to use stem cells for reparative treatments. However, some studies have shown that implantation of stem cells at a later stage in the DDD process may result in a greater increase in disc height when compared to implantation at an earlier stage (Ho et al., 2008). This finding highlights the importance of studies involving stem cell-based intervertebral disc regeneration being carefully controlled in the context of stage of disc degeneration. Again, this point highlights the importance of temporal sequence when examining therapeutics with stem cells. Additionally, inflammatory processes have been shown to inhibit the chondrogenic differentiation of stem cells, whereas hypoxic conditions exert beneficial effects on chondrogenesis and phenotype stability of transplanted stem cells (Felka et al., 2009).

2.4 Optimizing conditions to promote proliferation

There is currently an avid interest in using our accumulated data and knowledge of the factors influencing stem cell proliferation and the exact conditions in the native IVD microenvironment to optimize the chances for stem cell proliferation.

Multiple studies have investigated culturing MSCs with NP cells in a co-culture system, allowing for cell-to-cell contact (Yang et al., 2009; Le Maitre et al., 2009; Vadalà et al., 2009; Richardson et al., 2006; Richardson et al., 2008). This contact has been shown to stimulate these MSCs to differentiate toward a chondrocytic lineage, therefore removing the need for pre-differentiation in-vitro (Watanabe et al., 2010; Svanvik et al., 2010; Niu et al., 2009; Wei et al., 2009; Tao et al., 2008; Le Visage et al., 2006; Richardson et al., 2006). This was evidenced by mRNA expression levels of Type II collagen and aggrecan being elevated in co-cultured cells and cells undergoing morphological changes to form three-dimensional micromasses expressing collagen-2, aggrecan, and Sox-9 at RNA and protein levels after 14 days of co-culture. These changes were unique and not detected in the samples of stem cells cultured alone (Svanvik et al., 2010; Niu et al., 2009; Wei et al., 2009). Furthermore, MSCs from older individuals differentiate spontaneously into chondrocyte-like NP cells upon insertion into NP tissue in-vitro, and thus may not require additional stimulation to induce differentiation. This is a key finding, as such a strategy would minimize the level of external manipulation required prior to insertion of these cells into the patient, thus simplifying the treatment strategy and reducing costs (Le Maitre et al., 2009).

Adipose-Derived Stem Cells (ADSCs) have also been shown to be able to differentiate into NP cells in multiple in-vitro studies (Xie et al., 2009; Tapp et al., 2008; Lu et al., 2007; Lu et al., 2008; Li et al., 2005). Soluble factors released by NP cells direct chondrogenic differentiation of ADSCs in collagen hydrogels, and combination with a nucleus-mimicking collagen type II microenvironment enhances differentiation towards a more pronounced cartilaginous lineage (Lu et al., 2007; Lu et al., 2008).

Studies using annulus fibrosus cells isolated from nondegenerated intervertebral discs have shown that these cells have the capability of differentiating into adipocytes, osteoblasts, chondrocytes, neurons, and endothelial cells in-vitro. These cells may also be induced to become more plastic, allowing them to differentiate along more mesenchymal lineages (Li et al., 2005; Feng et al., 2010; Saraiya et al., 2010). However, when annulus cells are differentiated into a chondrocyte micromass, it was not as rounded or compact as that which occurs with stem cells induced into chondrocyte differentiation (Saraiya et al., 2010). TGF-β stimulation of fetal cells cultured in high cell density led to the production of aggrecan, type I and II collagens and variable levels of type X collagen, although fetal cells had lower adipogenic and osteogenic differentiation capacity than MSCs and variability in matrix synthesis was observed between specific donors (Quintin et al., 2009; Quintin et al., 2010).

3. Animal studies

Many studies using stem cells for disc regeneration have been performed in a wide array of animal models with promising results. Two recent studies were conducted utilizing ADSCs in a murine (Jeong et al., 2010) and a canine model (Ganey et al., 2009). Staining in both studies demonstrated increased Type II collagen and aggrecan in the transplantation group. Additionally, at 6 weeks after transplantation, discs exhibited a restoration of disc hydration and MRI T2 signal intensity and more closely resembled the healthy controls as evidenced by matrix translucency, compartmentalization of the annulus, and increased cell density within the nucleus pulposus. Discs also showed a significantly smaller reduction in disc height when compared with controls.

Multiple studies have shown that MSCs are able to proliferate and survive inside the IVD, with assessments being made as far out as six months post-transplant (Tan et al., 2009; Jeong et al., 2009; Henriksson et al., 2009; Sobajima et al., 2008; Zhang et al., 2005; Crevensten et al., 2004). Additionally, these cells have been proven to differentiate into cells expressing chondrocytic phenotypes, as evidenced by positive immunostaining of collagen type II, aggrecan, and other markers (Henriksson et al., 2009; Yang et al., 2010; Wei et al., 2009; Sakai et al., 2005). Cells were also shown to exhibit NP phenotypic markers (Sakai et al., 2005). The injected discs had a central NP-like region which had a close similarity to the normal biconvex structure of the IVD and contained viable chondrocytes forming a matrix like that of the normal disc (Sakai et al., 2003; Revell et al., 2007). Omlor et al. studied the practical phenomenon of transplanted stem cell loss through the actual annular puncture which was used to not only simulate disc damage and herniation but also to inject the stem cells themselves. They made a logical conclusion that IVD regeneration strategies should increasingly focus on annulus reconstruction in order to reduce implant loss due to annular failure (Omlor et al., 2010). Most studies focusing on this point are still ongoing.

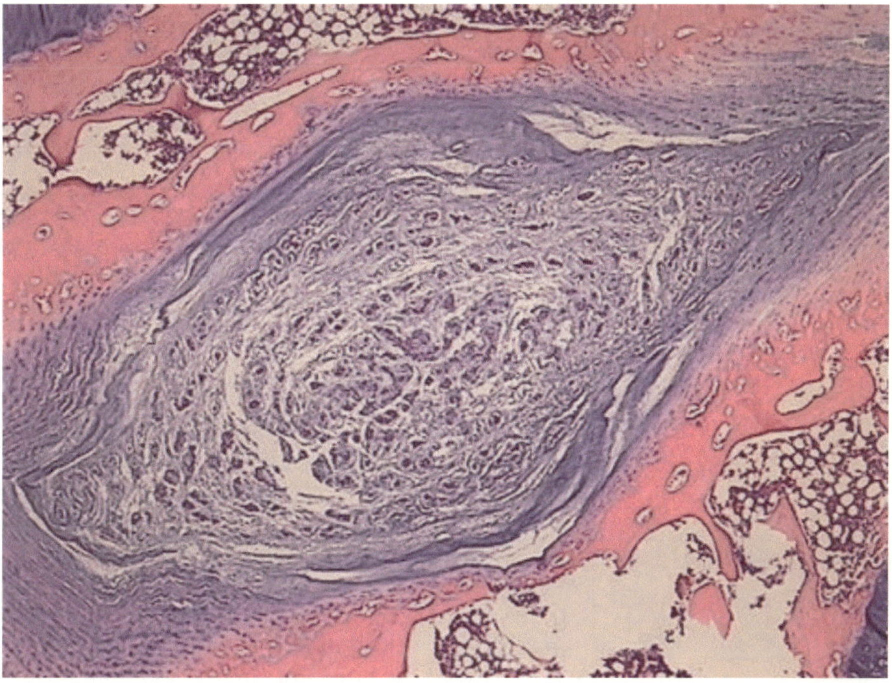

Fig. 5. Hematoxylin and eosin staining of the rabbit IVD, showing healthy notochordal cell rests

Several xenotransplant studies involving ESCs have been conducted with promising results. Jeong et al have shown that rats receiving human ESCs showed relative restoration of the inner annulus structure compared to a control group (Jeong et al., 2010). This finding may help to address the concern of loss of implanted material through the needle puncture.

Many of the stem cells in these studies were xenografted from other species and the recipient animals were not treated with immunosuppressive agents. In spite of this, there was a lack of immune response suggesting an unrecognized immune-privileged site within the intervertebral disc space (Wei et al., 2009; Sheikh et al., 2009). On top of this, there has been some study with MSC showing that transplantation contributes to this immunosuppressive phenomenon by the differentiation of these cells into cells expressing FasL, which has been shown to be an immunosuppressive factor (Hiyama et al., 2008).

Studies	Model	Intervention	Results
Jeong et al.	Rat model	The first coccygeal disc segments of Sprague-Dawley rat were left undamaged as controls, and other two segments were damaged by needle injection. Two weeks later, stem cells or saline were injected into each of the two damaged segments.	At 6 weeks after transplantation, the experimental group showed a significantly smaller reduction in disc height than the saline-injected group and exhibited a restoration of MRI signal intensity. Hematoxylin and eosin staining revealed a greater restoration of the inner annulus structure. There was also increased collagen type II and aggrecan.
Ganey et al.	Canine model	3 discs that had undergone partial nucleotomy were randomized to receive: (1) stem cells in hyaluronic acid carrier (Cells/HA); (2) HA only; or (3) No Intervention.	Disc levels receiving stem cells more closely resembled the healthy controls as evidenced in matrix translucency, compartmentalization of the annulus, and in cell density within the nucleus pulposus. Matrix analysis showed increased Type-II collagen and aggrecan.
Hiyama et al.	Canine model	4 weeks after nucleotomy, MSCs were transplanted into the degeneration-induced discs. The animals were followed for 12 weeks when radiological, histological, biochemical, immunohistochemical, and RT-PCR analyses were performed.	MSC transplantation effectively led to the regeneration of degenerated discs. GFP-positive MSCs detected in the NP region 8 weeks after transplantation expressed FasL protein.
Sobajima et al.	Rabbit model	MSCs were isolated New Zealand White rabbits, retrovirally transduced with the lacZ marker gene, and injected into the nucleus pulposus of the L2-3, L3-4, and L4-5 lumbar discs of 12 other NZW rabbits. Rabbits each were sacrificed at 3, 6, 12, or 24 weeks after cell implantation, and staining	MSCs were detected in histological sections of rabbit discs up to 24 weeks after transplant with engraftment into the inner annulus fibrosus.

Studies	Model	Intervention	Results
		was done to assess cell survival and localization.	
Henriksson et al.	Porcine model	Three lumbar discs in each of 9 were damaged with needle puncture. 2 weeks later human MSCs were injected The animals were sacrificed after 1, 3, or 6 months. Disc appearance was visualized by MRI. Immunohistochemistry was used to detect human MSCs.	All injured discs demonstrated degenerative signs on MRI. Immunostaining for Aggrecan and Collagen type II expression were observed in NP after 3 and 6 months. mRNA expression of Collagen IIA, Collagen IIB, Versican, Collagen 1A, Aggrecan, and SOX9 were detected at 3 and 6 months by real-time PCR.
Sheikh et al.	Rabbit model	16 New Zealand white rabbits underwent needle puncture of the disc with MRIs before and after injection with ESCs expressing green fluorescent protein. At 8 weeks post-ESC implantation, the animals were killed and the intervertebral discs were harvested and analyzed using H & E staining and immunohistochemical analysis.	MRI confirmed intervertebral disc degeneration at needle-punctured segments. Postmortem H & E histological analysis of Group A discs (no intervention) showed mature chondrocytes and no notochordal cells. Group B discs (needle puncture only) displayed an intact annulus fibrosus and generalized disorganization within the NP. Group C discs showed islands of notochordal cell growth (injection of ESCs).
Sakai et al.	Rabbit model	Stem cells labeled with green fluorescent protein, were transplanted into mature rabbits. Consecutive counts of transplanted cells in the nucleus area were performed for 48 weeks with immunohistochemical and proteoglycan content analyses along with PCR detection of mRNA expression of Type I and II collagen, aggrecan and versican.	Cells that were positive for green fluorescent protein were observed in the nucleus pulposus of cell-transplanted rabbit discs 2 weeks after transplantation. GFP-positive cells were positive for Type II collagen, keratan sulfate, chondroitin sulfate, and aggrecan.

Studies	Model	Intervention	Results
Bendtsen et al.	Porcine model	DDD was induced in 15 minipigs. After 12 weeks, the animals underwent percutaneous intradiscal injection of stem cells. MRI was performed before treatment and at 24 weeks.	Stem cell treated animal had increased T2 signal in the disc along with increased relative vertebral blood flow.
Omlor et al.	Porcine model	6 minipigs underwent matrix based cell transfer after partial nucleotomy of lumbar IVDs. Segments were analyzed for retained volume of labeling particles	There was a 90% loss of the implant material under in vivo conditions when the annulus was not reconstructed.

Table 1. Animal Studies

Our group recently reported seminal work with regard to ESC implantation in a rabbit model (Sheikh et al., 2009). This study used a needle puncture model with appropriate controls to simulate disc injury. The effects of implanted murine ESCs were measured at 8 weeks using imaging, histological, and immunohistochemical analyses. In-vivo new notochordal cell populations were seen in ESC-injected discs, providing convincing evidence for stem-cell mediated regeneration of the IVD. Another study established the utility of stem cells implanted at 12 weeks post-injury in regenerating the IVD and maintaining perfusion to the endplate and subchondral bone in a porcine model (Bendtsen et al., 2010). Sobajima et al used a rabbit model to show that IVD cells harvested 48 weeks post-implantation revealed a restoration of both glycoprotein content and matrix characteristics (Sobajima et al., 2008). These analyses all provide further evidence that ESC transplantation does have strong potential for clinical use in regenerating the IVD and reversing the cascade of degeneration that occurs with time.

4. Human studies

To date, there have been only two studies where stem cells were injected into the IVD in humans to stimulate regeneration of the disc. Yoshikawa et al percutaneously grafted MSCs into degenerated IVDs in two women aged 67 and 70 years. After two years, both individuals had alleviation of symptoms and radiographic changes that included improvement of vacuum phenomenon on X-ray and increased signal intensity of IVDs on T2-weighted MRI (Yoshikawa et al., 2010). Another study involved intradiscal injection of hematopoietic stem cells into ten patients that had confirmed disc pain and these patients' pain was assessed at 6-month and 12-month intervals. In contrast to previous study, none of these individuals had any relief of symptoms (Haufe et al., 2006). These trials suggest that stem cells have the potential to relieve symptoms of DDD and restore normal IVD anatomy; however, more human studies are needed to truly establish this. To date, there have been no human ESC implantation studies into the IVD in humans. Further study is needed to verify safety before such work is undertaken.

Studies	Subjects	Intervention	Results	Study Critique
Yoshikawa et al.	2 patients	Percutaneous stem cell grafting	Clinical symptoms improved; increased T2 signal in the disc space on MRI	Few patients
Haufe et al.	10 patients	Percutaneous stem cell grafting	No clinical symptom relief	No imaging conducted

Table 2. Human Studies

5. Future potential of ESCs

Although many laboratory and animal studies have been performed utilizing stem cells for the purposes of cell characterization and inducing chondrocyte formation, much further study is needed before human trials are undertaken on a larger scale. Several studies have already showcased the ability of ESCs to differentiate towards a chondrocytic lineage in-vitro and also to improve DDD in in-vivo animal and human trials, using a combination of imaging and histological analyses. Several benchtop lab studies have been performed to show that ESCs can be successfully stimulated to differentiate into chondrocyte-like cells (Hoben et al., 2009; Fecek et al., 2008; Hegert et al., 2002; Kawaguchi et al., 2005; zur Nieden et al., 2005; Kramer et al., 2000). Similar to the case with MSCs, different factors affect this process in ESCs, such as TGF-β3, BMP-2, and BMP-4 (Hegert et al., 2002; Kawaguchi et al., 2005; zur Nieden et al., 2005; Kramer et al., 2000; Sakai et al., 2005). Biological scaffolds seeded with chondrocytic cells derived from ESCs, when implanted in mice have been shown to generate cartilage tissue in-vivo (Kramer et al., 2000). Injection of ESCs in a DDD-induced rabbit model led to viable notochordal-type cells within the discs (Sheikh et al., 2009). These animal studies demonstrate the ability of ESCs to differentiate into a chondrocytic lineage in-vitro and in-vivo.

Our group is currently developing chondroprogenitor stem cell lines that can restore the functional capability of the IVD (Sheikh et al., 2009). Our rationale stemmed from the idea that currently there is no biologic therapy for repairing a degenerated IVD and that ESCs have a potential to fill this role based on their regenerative potential. Studies have shown that ESCs can be induced to differentiate into specific cell lineages by using selective culture media and growth environments (Kawaguchi et al., 2005).

Relying on the significant strides made by these basic science groups with regard to cell and factor characterization, our lab proceeded for further refine these methods and develop a protocol for both stem cell differentiation along a chondrocytic lineage and also for examining the utility of transplantation of these cells in a rabbit model of DDD. We initially developed a novel percutaneous animal model of disc degeneration using New Zealand white rabbits (Figure 1) and used this model to explore the possibility of ESC implantation for both structural regeneration and for the growth and continued presence of notochordal stem cells in the disc space (Sheikh et al., 2009).

Previous research transplanting MSCs into degenerated rabbit discs has shown consistent biochemical and radiographic (MRI) evidence of IVD restoration (Sakai et al., 2005). Human

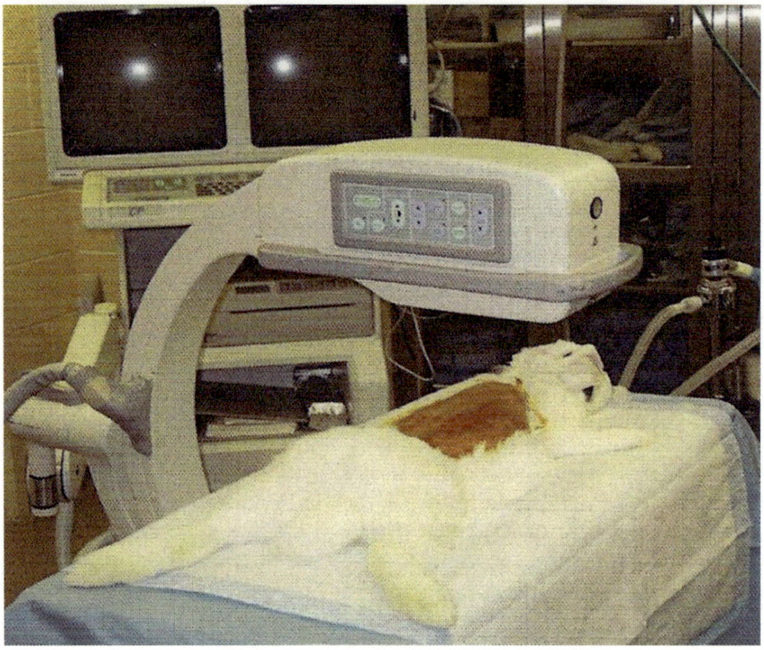

A.

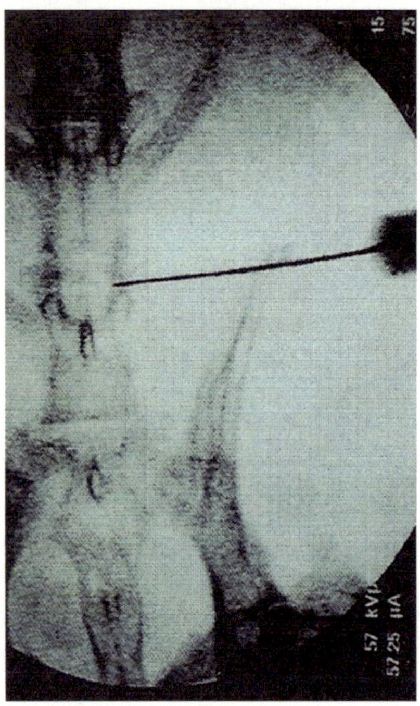

B.

Fig. 6. Photographs of our group's rabbit model for IVD degeneration. The rabbit is positioned prone, its back is shaved and prepared for surgery (A), with a corresponding fluoroscopic view (B).

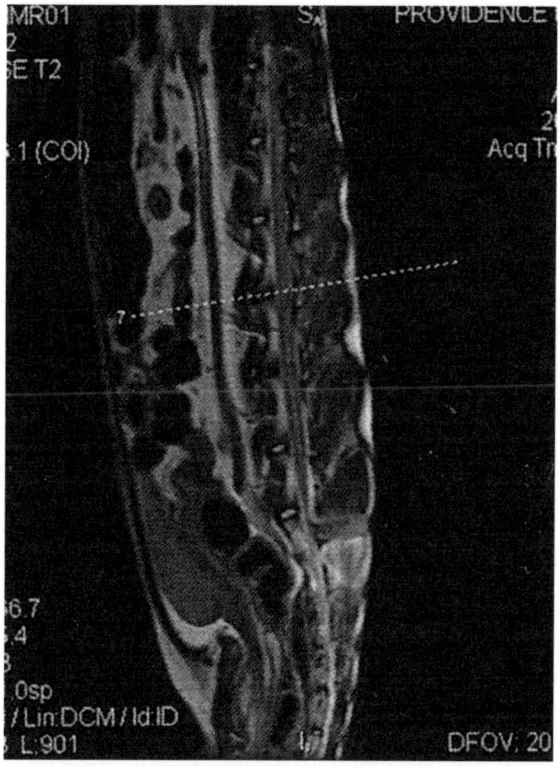
MR01
E T2
1 (COI)
PROVIDENCE
Acq Tr
6.7
.4
.0sp
/ Lin:DCM / Id:ID
L:901
DFOV: 20

A.

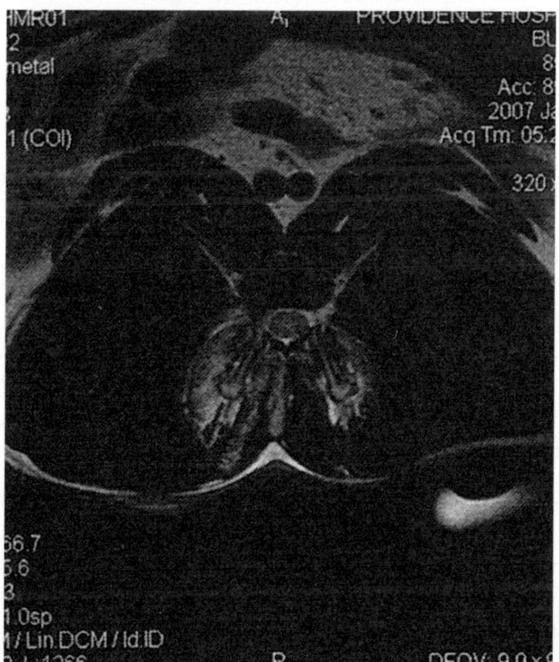
MR01
metal
1 (COI)
PROVIDENCE HOSP
2007 J
Acq Tm: 05:
320
66.7
5.6
1.0sp
/ Lin:DCM / Id:ID
L:1266
DFOV: 9.0 x

B.

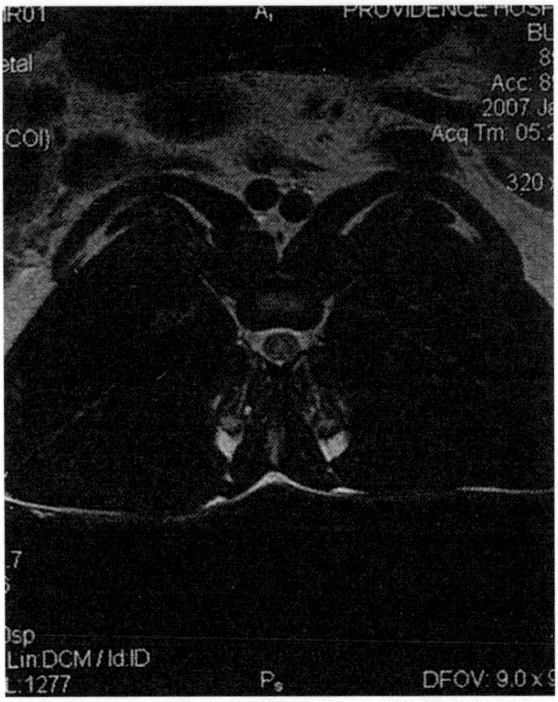

C

Fig. 7. Sagittal T2-weighted MRI of the rabbit spine (A), with a corresponding axial view at the level of the induced disc degeneration (B) and at a separate normal control level (C).

MSCs have also been investigated for their bone-forming capabilities with good results (Jaiswal et al., 1997). Stem cells are already being used in therapeutic applications with placement of cells directly at the site of intended spinal fusion during open surgical procedures.

Our lab has developed chondroprogenitor cells lines that can restore the functional capacity of the IVD, with these cells differentiating into chondrocytes. Using our novel percutaneous model of disc degeneration in a rabbit model, we obtained MRIs preoperatively and at 2, 4, and 8 weeks postoperatively (Figure 2). Before implantation, ESCs were cultured with cis-retinoic acid, TGF-beta, ascorbic acid, and insulin-like growth factor to induce differentiation along a chondrocyte lineage. After MRI confirmation of disc degeneration, the discs were then injected with murine ESCs that were labeled with mutant green fluoroscent protein (GFP). At 8 weeks post-implantation, IVDs were harvested and analyzed with hematoxylin and eosin staining along with immunohistochemical analyses (Figure 3).

Three groups were analyzed: group A consisted of control animals with nonpunctured discs; group B consisted of control animals with experimentally punctured discs; and group C consisted of animals with experimentally punctured discs that were subsequently implanted with ESCs. Gel electrophoresis was used to analyze ESCs for cartilaginous tissue formation. MRI confirmed IVD degeneration after needle puncture starting at 2 weeks postoperatively. Postmortem histological analysis of group A IVDs showed chondrocytes, but no notochordal cells. Group B disc displayed intact annulus fibrosus but disorganized

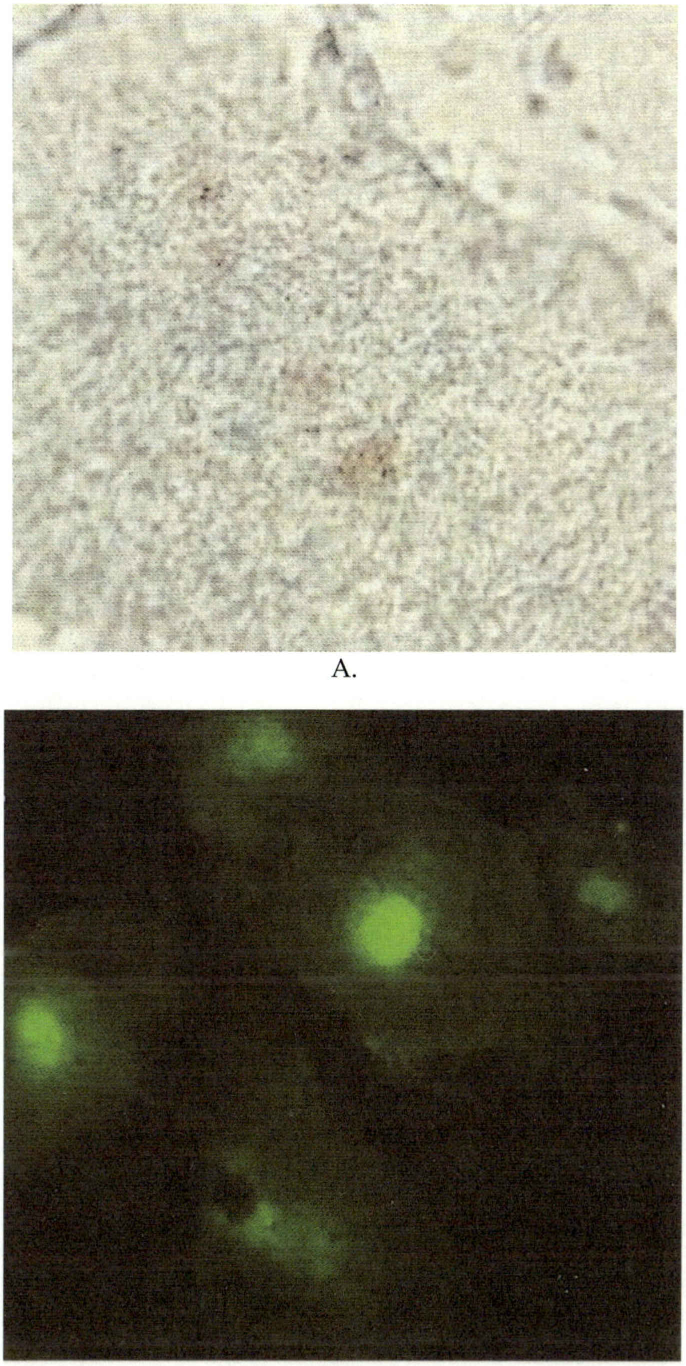

A.

B.

Fig. 8. Photomicrographs of tissue obtained preimplanation for histological analysis of ESCs grown in-vitro with Alcian blue staining showing 86% viability (A) and high power magnification showed adequate GFP cell labeling (B).

fibrous tissue in the NP. Group C discs showed new notochordal cell growth, indicating survival and proper differentiation of the injected ESCs. Fluorescent microscopic analysis was positive in group C tissue, confirming the viability of GFP-labeled ESCs within the injected IVD. In addition, the notochordal cells in group C stained positive for cytokeratin and vimentin, providing further evidence of their chondrocyte origin. There was no inflammatory response in group C discs, indicating no cell-mediated immune response.

Our study provides a novel, reproducible model for the study of disc degeneration. New notochordal cell populations were seen in discs injected with ESCs. The lack of an immune response to xenograft-implanted mouse stem cells in an immune-competent rabbit suggests an immunoprivileged site within the IVD. Although preliminary, this study highlights the possible use of stem cells to promote IVD regeneration. Further ongoing studies are in the process of fully elucidating the processes involved with ESC differentiation along chondrogenic cell lines and how they may be used for new disc formation in the future. These studies will provide a good deal of evidence with regard to the future potential of ESCs for use in restoring the IVD in humans.

6. Summary

DDD is a high-morbidity condition with many modalities of treatment including surgery and more conservative measures such as pain injections, which only provide symptomatic treatment. No therapy has been developed that targets DDD at the cellular level. Recently, many biologic therapies have emerged that may be able to restore the NP and the normal cellular structure of the IVD. This restoration may in turn alleviate the symptoms of DDD through restoration of foraminal height, removing the compression of nerves. In-vitro studies have been performed to identify what cells are capable of differentiating towards a chondrocytic lineage and to best define parameters and factors that influence this differentiation. Multiple laboratory studies have been performed showing that MSCs, ADSCs, fetal cartilaginous cells, and annulus fibrosus cells all have the ability to differentiate towards a chondrocytic pathway. Factors that can induce these cells to differentiate toward a chondrocytic lineage have been identified and include TGF-β3 and BMP-2, which have a synergistic effect when used together. Other factors that may be beneficial include hypoxia, IVD-like glucose conditions (1.0 mg/mL glucose), and cell-to-cell contact with NP cells; the latter negating the need for other soluble factors (i.e. TGF-β3). A major limiting factor may be the acidic pH (6.8) of the IVD, one that may be especially important as acidic pH levels are typical of increasingly degenerated discs. These studies yielded encouraging results with cells in the IVD being positive for markers of chondrocytic differentiation such as collagen type II and aggrecan. Additionally, cells exhibited NP phenotypic markers and had a close similarity to the normal biconvex structure of the NP. In-vitro studies have clearly established that ESCs are capable of differentiating into a chondrocytic lineage and have delineated some of the factors that affect this. The optimal microenvironment needs to be more accurately characterized at this time.

Animal studies of cell implantation have been performed in DDD-induction models. Weeks after injury, stem cells have been implanted and outcomes followed. These outcomes which have included radiographic analyses along with histological and immunohistochemical analyses have provided preliminary data that stem cell therapies are a viable option with regard to IVD regeneration (Sheikh et al., 2009). Human studies have further provided some preliminary evidence that stem cell therapy may be of clinical value (Haufe et al., 2006). The use of ESCs in regenerating IVD shows exciting new possibilities and further studies are needed in humans to establish its efficacy.

ESC-based regeneration of the human IVD is still in its infancy. Much progress has been made regarding laboratory research identifying the correct factors and microenvironment, and initial results from animal studies using stem cells remain promising. ESCs may be useful for repairing DDD as evidenced by their ability to differentiate into a chondrocytic lineage and yield notochordal-type cells in DDD models. ESCs need to be further studied and characterized with respect to safety, and larger human trials with appropriate clinical outcomes such as pain and disability reduction are needed to definitively establish its clinical efficacy.

7. Conclusions

The last half-century has seen an exponential rate of progress with regard to elucidating the mechanisms of degeneration of the IVD and how targeted therapies can help to alleviate this common condition. These studies have provided us with an improved understanding of the IVD and how it behaves under typical biomechanical forces and loads experienced in in-vivo conditions. Novel therapies are being studied, including stem cells with their potential regenerative capabilities in the spine. The development and action of these stem cells can be further modified through gene therapy and microenvironment manipulation. Immunologic markers are being used for more efficient targeting of these cells. With enhanced cell delivery and an improved understanding of the cell differentiation process, true regeneration of the IVD and surrounding supportive structures of the spine will become a reality that can be applied to treat patients with this common, debilitating condition.

8. References

[1] Chan D, Song Y, Sham P. Genetics of disc degeneration. *Eur Spine J.* 15(Suppl 3):S317-S325. 2006.

[2] Cassinelli EH, Hall RA, Kang JD. Biochemistry of intervertebral disc degeneration and the potential for gene therapy applications. *Spine J.* 1:205-214, 2001.

[3] Stokes IA, Iatridis JC. Mechanical conditions that accelerate intervertebral disc degeneration: overload versus immobilization. *Spine* 29:2724-2732, 2004.

[4] Ala-Kokko L. Genetic risk factors for lumbar disc disease. *Ann Med.* 34:42-47, 2002.

[5] Steck E, Bertram H, Abel R, Chen B, Winter A, Richter W. Induction of intervertebral disc-like cells from adult mesenchymal stem cells. *Stem Cells.* 2005 Mar;23(3):403-11.

[6] Risbud MV, Albert TJ, Guttapalli A, Vresilovic EJ, Hillibrand AS, Vaccaro AR,Shapiro IM. Differentiation of mesenchymal stem cells towards a nucleus pulposus-like phenotype in vitro: implications for cell-based transplantation therapy. *Spine (Phila Pa 1976).* 2004 Dec 1;29(23):2627-32.

[7] Shen B, Wei A, Tao H, Diwan AD, Ma DD. BMP-2 enhances TGF-beta3-mediated chondrogenic differentiation of human bone marrow multipotent mesenchymal stromal cells in alginate bead culture. *Tissue Eng Part A.* 2009 Jun;15(6):1311-20.

[8] Kuh SU, Zhu Y, Li J, Tsai KJ, Fei Q, Hutton WC, Yoon ST. Can TGF-beta1 and rhBMP-2 act in synergy to transform bone marrow stem cells to discogenic-type cells? *Acta Neurochir (Wien).* 2008 Oct;150(10):1073-9; discussion 1079.

[9] Hoben GM, Willard VP, Athanasiou KA. Fibrochondrogenesis of hESCs: growth factor combinations and cocultures. *Stem Cells Dev.* 2009 Mar;18(2):283-92.

[10] Koay EJ, Hoben GM, Athanasiou KA. Tissue engineering with chondrogenically differentiated human embryonic stem cells. *Stem Cells.* 2007 Sep;25(9):2183-90.

[11] Wuertz K, Godburn K, Neidlinger-Wilke C, Urban J, Iatridis JC. Behavior of mesenchymal stem cells in the chemical microenvironment of the intervertebral disc. *Spine (Phila Pa 1976).* 2008 Aug 1;33(17):1843-9.

[12] Wuertz K, Godburn K, Iatridis JC. MSC response to pH levels found in degenerating intervertebral discs. *Biochem Biophys Res Commun.* 2009 Feb 20;379(4):824-9.

[13] Ho G, Leung VY, Cheung KM, Chan D. Effect of severity of intervertebral disc injury on mesenchymal stem cell-based regeneration. *Connect Tissue Res.* 2008;49(1):15-21.

[14] Felka T, Schäfer R, Schewe B, Benz K, Aicher WK. Hypoxia reduces the inhibitory effect of IL-1beta on chondrogenic differentiation of FCS-free expanded MSC. *Osteoarthritis Cartilage.* 2009 Oct;17(10):1368-76. Epub 2009 May 8.

[15] Yang F, Leung VY, Luk KD, Chan D, Cheung KM. Mesenchymal stem cells arrest intervertebral disc degeneration through chondrocytic differentiation and stimulation of endogenous cells. *Mol Ther.* 2009 Nov;17(11):1959-66.

[16] Le Maitre CL, Baird P, Freemont AJ, Hoyland JA. An in vitro study investigating the survival and phenotype of mesenchymal stem cells following injection into nucleus pulposus tissue. *Arthritis Res Ther.* 2009;11(1):R20.

[17] Vadalà G, Studer RK, Sowa G, Spiezia F, Iucu C, Denaro V, Gilbertson LG, Kang JD. Coculture of bone marrow mesenchymal stem cells and nucleus pulposus cells modulate gene expression profile without cell fusion. *Spine (Phila Pa 1976).* 2008 Apr 15;33(8):870-6.

[18] Richardson SM, Hughes N, Hunt JA, Freemont AJ, Hoyland JA. Human mesenchymal stem cell differentiation to NP-like cells in chitosan-glycerophosphate hydrogels. *Biomaterials.* 2008 Jan;29(1):85-93.

[19] Richardson SM, Curran JM, Chen R, Vaughan-Thomas A, Hunt JA, Freemont AJ, Hoyland JA. The differentiation of bone marrow mesenchymal stem cells into chondrocyte-like cells on poly-L-lactic acid (PLLA) scaffolds. *Biomaterials.* 2006 Aug;27(22):4069-78.

[20] Watanabe T, Sakai D, Yamamoto Y, Iwashina T, Serigano K, Tamura F, Mochida J. Human nucleus pulposus cells significantly enhanced biological properties in a coculture system with direct cell-to-cell contact with autologous mesenchymal stem cells. *J Orthop Res.* 2010 May;28(5):623-30.

[21] Svanvik T, Henriksson HB, Karlsson C, Hagman M, Lindahl A, Brisby H. Human disk cells from degenerated disks and mesenchymal stem cells in co-culture result in increased matrix production. *Cells Tissues Organs.* 2010;191(1):2-11.

[22] Niu CC, Yuan LJ, Lin SS, Chen LH, Chen WJ. Mesenchymal stem cell and nucleus pulposus cell coculture modulates cell profile. *Clin Orthop Relat Res.* 2009 Dec;467(12):3263-72.

[23] Wei A, Chung SA, Tao H, Brisby H, Lin Z, Shen B, Ma DD, Diwan AD. Differentiation of rodent bone marrow mesenchymal stem cells into intervertebral disc-like cells following coculture with rat disc tissue. *Tissue Eng Part A.* 2009 Sep;15(9):2581-95.

[24] Tao F, Li F, Li G, Pan F. Differentiation of mesenchymal stem cells into nucleus pulposus cells in vitro. *J Huazhong Univ Sci Technolog Med Sci.* 2008 Apr;28(2):156-8.

[25] Le Visage C, Kim SW, Tateno K, Sieber AN, Kostuik JP, Leong KW. Interaction of human mesenchymal stem cells with disc cells: changes in extracellular matrix biosynthesis. *Spine (Phila Pa 1976).* 2006 Aug 15;31(18):2036-42.

[26] Richardson SM, Walker RV, Parker S, Rhodes NP, Hunt JA, Freemont AJ, Hoyland JA. Intervertebral disc cell-mediated mesenchymal stem cell differentiation. *Stem Cells.* 2006 Mar;24(3):707-16.

[27] Xie LW, Fang H, Chen AM, Li F. Differentiation of rat adipose tissue-derived mesenchymal stem cells towards a nucleus pulposus-like phenotype in vitro. *Chin J Traumatol.* 2009 Apr;12(2):98-103.

[28] Tapp H, Deepe R, Ingram JA, Kuremsky M, Hanley EN Jr, Gruber HE. Adipose-derived mesenchymal stem cells from the sand rat: transforming growth factor beta and 3D co-culture with human disc cells stimulate proteoglycan and collagen type I rich extracellular matrix. *Arthritis Res Ther.* 2008;10(4):R89. Epub 2008 Aug 11.

[29] Lu ZF, Doulabi BZ, Wuisman PI, Bank RA, Helder MN. Influence of collagen type II and nucleus pulposus cells on aggregation and differentiation of adipose tissue-derived stem cells. *J Cell Mol Med.* 2008 Dec;12(6B):2812-22.

[30] Lu ZF, Zandieh Doulabi B, Wuisman PI, Bank RA, Helder MN. Differentiation of adipose stem cells by nucleus pulposus cells: configuration effect. *Biochem Biophys Res Commun.* 2007 Aug 10;359(4):991-6. Epub 2007 Jun 8.

[31] Li X, Lee JP, Balian G, Greg Anderson D. Modulation of chondrocytic properties of fat-derived mesenchymal cells in co-cultures with nucleus pulposus. *Connect Tissue Res.* 2005;46(2):75-82.

[32] Feng G, Yang X, Shang H, Marks IW, Shen FH, Katz A, Arlet V, Laurencin CT, Li X. Multipotential differentiation of human anulus fibrosus cells: an in vitro study. *J Bone Joint Surg Am.* 2010 Mar;92(3):675-85.

[33] Saraiya M, Nasser R, Zeng Y, Addya S, Ponnappan RK, Fortina P, Anderson DG, Albert TJ, Shapiro IM, Risbud MV. Reversine enhances generation of progenitor-like cells by dedifferentiation of annulus fibrosus cells. *Tissue Eng Part A.* 2010 Apr;16(4):1443-55.

[34] Gruber HE, Chow Y, Hoelscher GL, Ingram JA, Zinchenko N, Norton HJ, Sun Y, Hanley EN Jr. Micromass culture of human annulus cells: morphology and

extracellular matrix production. *Spine (Phila Pa 1976).* 2010 May 1;35(10): 1033-8.

[35] Quintin A, Schizas C, Scaletta C, Jaccoud S, Applegate LA, Pioletti DP. Plasticity of fetal cartilaginous cells. *Cell Transplant.* 2010;19(10):1349-57.

[36] Quintin A, Schizas C, Scaletta C, Jaccoud S, Gerber S, Osterheld MC, Juillerat L, Applegate LA, Pioletti DP. Isolation and in vitro chondrogenic potential of human foetal spine cells. *J Cell Mol Med.* 2009 Aug;13(8B):2559-69.

[37] Jeong JH, Lee JH, Jin ES, Min JK, Jeon SR, Choi KH. Regeneration of intervertebral discs in a rat disc degeneration model by implanted adipose-tissue-derived stromal cells. *Acta Neurochir (Wien).* 2010 Oct;152(10):1771-7.

[38] Ganey T, Hutton WC, Moseley T, Hedrick M, Meisel HJ. Intervertebral disc repair using adipose tissue-derived stem and regenerative cells: experiments in a canine model. *Spine (Phila Pa 1976).* 2009 Oct 1;34(21):2297-304.

[39] Tan S, Jia C, Liu Z, Liu R, Yang J, Zhang L, Shou F, Ju X. Study on survival time of autogeneic BMSCs labeled with superparamagnetic iron oxide in rabbit intervertebral discs. *Zhongguo Xiu Fu Chong Jian Wai Ke Za Zhi.* 2009 Nov;23(11):1355-9.

[40] Jeong JH, Jin ES, Min JK, Jeon SR, Park CS, Kim HS, Choi KH. Human mesenchymal stem cells implantation into the degenerated coccygeal disc of the rat. *Cytotechnology.* 2009 Jan;59(1):55-64.

[41] Henriksson HB, Svanvik T, Jonsson M, Hagman M, Horn M, Lindahl A, Brisby H. Transplantation of human mesenchymal stems cells into intervertebral discs in a xenogeneic porcine model. *Spine (Phila Pa 1976).* 2009 Jan 15;34(2):141-8.

[42] Sobajima S, Vadala G, Shimer A, Kim JS, Gilbertson LG, Kang JD. Feasibility of a stem cell therapy for intervertebral disc degeneration. *Spine J.* 2008 Nov-Dec;8(6):888-96.

[43] Zhang YG, Guo X, Xu P, Kang LL, Li J. Bone mesenchymal stem cells transplanted into rabbit intervertebral discs can increase proteoglycans. *Clin Orthop Relat Res.* 2005 Jan;(430):219-26.

[44] Crevensten G, Walsh AJ, Ananthakrishnan D, Page P, Wahba GM, Lotz JC, Berven S. Intervertebral disc cell therapy for regeneration: mesenchymal stem cell implantation in rat intervertebral discs. *Ann Biomed Eng.* 2004 Mar;32(3):430-4.

[45] Yang H, Wu J, Liu J, Ebraheim M, Castillo S, Liu X, Tang T, Ebraheim NA. Transplanted mesenchymal stem cells with pure fibrinous gelatin-transforming growth factor-beta1 decrease rabbit intervertebral disc degeneration. *Spine J.* 2010 Sep;10(9):802-10. Epub 2010 Jul 24.

[46] Wei A, Tao H, Chung SA, Brisby H, Ma DD, Diwan AD. The fate of transplanted xenogeneic bone marrow-derived stem cells in rat intervertebral discs. *J Orthop Res.* 2009 Mar;27(3):374-9.

[47] Sakai D, Mochida J, Iwashina T, Watanabe T, Nakai T, Ando K, Hotta T. Differentiation of mesenchymal stem cells transplanted to a rabbit degenerative disc model: potential and limitations for stem cell therapy in disc regeneration. *Spine (Phila Pa 1976).* 2005 Nov 1;30(21):2379-87.

[48] Sakai D, Mochida J, Yamamoto Y, Nomura T, Okuma M, Nishimura K, Nakai T, Ando K, Hotta T. Transplantation of mesenchymal stem cells embedded in Atelocollagen gel to the intervertebral disc: a potential therapeutic model for disc degeneration. *Biomaterials.* 2003 Sep;24(20):3531-41.

[49] Revell PA, Damien E, Di Silvio L, Gurav N, Longinotti C, Ambrosio L. Tissue engineered intervertebral disc repair in the pig using injectable polymers. *J Mater Sci Mater Med.* 2007 Feb;18(2):303-8.

[50] Omlor GW, Bertram H, Kleinschmidt K, Fischer J, Brohm K, Guehring T, Anton M, Richter W. Methods to monitor distribution and metabolic activity of mesenchymal stem cells following in vivo injection into nucleotomized porcine intervertebral discs. *Eur Spine J.* 2010 Apr;19(4):601-12.

[51] Sheikh H, Zakharian K, De La Torre RP, Facek C, Vasquez A, Chaudhry GR, Svinarich D, Perez-Cruet MJ. In vivo intervertebral disc regeneration using stem cell-derived chondroprogenitors. *J Neurosurg Spine.* 2009 Mar;10(3):265-72.

[52] Hiyama A, Mochida J, Iwashina T, Omi H, Watanabe T, Serigano K, Tamura F, Sakai D. Transplantation of mesenchymal stem cells in a canine disc degeneration model. *J Orthop Res.* 2008 May;26(5):589-600.

[53] Bendtsen M, Bunger CE, Zou X, Foldager C, Joregensen HS. Autologous stem cell therapy maintains vertebral blood flow and contrast diffusion through the endplate in experimental IDD. Spine (Phila Pa 1976). 2010 Dec 29.

[54] Yoshikawa T, Ueda Y, Miyazaki K, Koizumi M, Takakura Y. Disc regeneration therapy using marrow mesenchymal cell transplantation: a report of two case studies. *Spine (Phila Pa 1976).* 2010 May 15;35(11):E475-80.

[55] Haufe SM, Mork AR. Intradiscal injection of hematopoietic stem cells in an attempt to rejuvenate the intervertebral discs. *Stem Cells Dev.* 2006 Feb;15(1):136-7.

[56] Fecek C, Yao D, Kaçorri A, Vasquez A, Iqbal S, Sheikh H, Svinarich DM, Perez-Cruet M, Chaudhry GR. Chondrogenic derivatives of embryonic stem cells seeded into 3D polycaprolactone scaffolds generated artilage tissue in vivo. *Tissue Eng Part A.* 2008 Aug;14(8):1403-13.

[57] Hegert C, Kramer J, Hargus G, Müller J, Guan K, Wobus AM, Müller PK, Rohwedel J. Differentiation plasticity of chondrocytes derived from mouse embryonic stem cells. *J Cell Sci.* 2002 Dec 1;115(Pt 23):4617-28.

[58] Kawaguchi J, Mee PJ, Smith AG. Osteogenic and chondrogenic differentiation of embryonic stem cells in response to specific growth factors. *Bone.* 2005 May;36(5):758-69. Epub 2005 Mar 24.

[59] zur Nieden NI, Kempka G, Rancourt DE, Ahr HJ. Induction of chondro-, osteo- and adipogenesis in embryonic stem cells by bone morphogenetic protein-2: effect of cofactors on differentiating lineages. *BMC Dev Biol.* 2005 Jan 26;5:1.

[60] Kramer J, Hegert C, Guan K, Wobus AM, Müller PK, Rohwedel J. Embryonic stem cell-derived chondrogenic differentiation in vitro: activation by BMP-2 and BMP-4. *Mech Dev.* 2000 Apr;92(2):193-205.

[61] Sakai D, Mochida J, Iwashina T. Differentiation of mesenchymal stem cells transplanted to a rabbit degenerative disc model: potential and limitations for stem cell therapy in disc regeneration. *Spine* 30: 2379-2387. 2005.

[62] Jaiswal N, Haynesworht S, Caplan A. Osteogenic differentiation of purified culture expanded human mesenchymal stem cells in vitro. *J Cell Biochem* 64: 295-312, 1997.

14

Towards Clinical Application of Mesenchymal Stromal Cells: Perspectives and Requirements for Orthopaedic Applications

Marianna Karagianni*, Torsten J. Schulze* and Karen Bieback
Institute of Transfusion Medicine and Immunology;
Medical Faculty Mannheim, Heidelberg University;
German Red Cross Blood Donor Service Baden-Württemberg – Hessen
Germany

1. Introduction

Mesenchymal stromal cells (MSC) possess a wide spectrum of interacting properties that contribute to their broad therapeutic potential: In pre- and clinical settings MSC have been demonstrated to reduce tissue damage, to activate the endogenous regenerative potential of tissues and to participate in tissue regeneration (Noort, Feye et al. 2010). Initially, MSC have been described to differentiate into derivates of the mesoderm: bone, adipose and cartilage tissue and were therefore applied to restore damaged tissue (Frohlich, Grayson et al. 2008). Subsequent analyses, however, indicated that the repair process does not only lay in the differentiation potential and plasticity of MSC. As demonstrated in later studies even if only few cells were detectable after MSC transplantation, the therapeutic effect was obvious (Fuchs, Baffour et al. 2001; Shake, Gruber et al. 2002). This could be attributed to paracrine properties with consecutive modification of the tissue microenvironment to decrease inflammatory and immune reactions. MSC are therefore beyond doubt promising candidates for cell therapy in various settings (Horwitz, Prockop et al. 2001; Le Blanc, Rasmusson et al. 2004; Prockop 2009; Pontikoglou, Deschaseaux et al. 2011).

The broad therapeutic efficacy of MSC renders them attractive candidates for cell therapy. However, translating basic research into clinical application is a complex multistep process (Bieback, Karagianni et al. 2011). It necessitates product regulation by the regulatory authorities and accurate management of the expected therapeutic benefits with the potential risks in order to balance the speed of clinical trials with a time-consuming, cautious risk assessment (Sensebe, Bourin et al. 2011). Despite their use in clinical studies, some questions remain open: What are the deviations among the MSC from different tissue sources? How shall MSC be adequately procured, isolated and cultivated? How should their therapeutic propensity, e.g. their homing properties, the secretion of bioactive factors, the differentiation pattern *in vivo* and their plasticity, be defined?

---

* Both authors contributed equally

It is obvious that MSC need to be further characterised in clinical studies with standardized protocols (Bieback, Karagianni et al. 2011; Sensebe, Bourin et al. 2011). Furthermore, despite immense work, still MSC cannot be identified as a distinct cell population by a set of marker proteins as CD34 defines hematopoietic stem cells. The field currently uses "minimal criteria" for MSC to describe them according to their *in vitro* behaviour (osteo-, adipo- and chondrogenic differentiation) and morphology (fibroblastoid, expressing a set of markers) (Dominici, Le Blanc et al. 2006). Nevertheless it has to be taken into account that *in vitro* data do not necessarily predict *in vivo* behaviour: MSC seem to alter their *in vitro* traits after *in vivo* transplantation and this might affect a future therapeutic outcome severely. For example MSC can express HLA-class II antigens and can therefore possibly trigger an immunreaction in the host after transplantation (Vassalli and Moccetti 2011) or may calcify spontaneously in uremic conditions and cause vessel occlusion in case of intravenous application (Kramann, Couson et al. 2011).

Using the example of bone defect regeneration, we will emphasize key parameters relevant for the translation of experimental data to clinical application. The focus on bone defect regeneration exemplifies the possibilities and challenges for MSC in combination with biomaterials in the light of regulatory frameworks in Europe, where MSC may be classified as "Advanced Therapy Medicinal Product - ATMP", or the US, where MSC fall under the term "Human Cells, Tissues, and Cellular and Tissue-Based Products -HCT/Ps". In this context, questions that need to be answered concern an adequate MSC tissue source with superior osteogenic potential compared to other tissues, the degree of cell differentiation prior to implantation and the adequate scaffold for tissue engineering (Seong, Kim et al. 2010).

1.1 MSC definition

Mesenchymal stromal cells (MSC) were initially isolated from bone marrow (BM) as described by Friedenstein and co-workers in 1968 (Friedenstein, Petrakova et al. 1968). They were identified as non hematopoietic, fibroblast-like cells adherent to plastic, with a colony-forming capacity (Friedenstein, Deriglasova et al. 1974), also as feeder cells for hematopoietic precursors (Eaves, Cashman et al. 1991; Wagner, Saffrich et al. 2008). Subsequent characterisation revealed their mesodermal differentiation and immune modulatory capacity, raising the interest in these cells (Le Blanc, Rasmusson et al. 2004; Bieback, Hecker et al. 2009; Mosna, Sensebe et al. 2010). Consequently, numerous terms for these cells were established: mesenchymal stem cells, mesenchymal stromal cells, adult stromal cells, multipotent and non hematopoietic adult precursor cells (Horwitz, Le Blanc et al. 2005; Dominici, Le Blanc et al. 2006). These conflicting nomenclature suggestions in the literature lead to a complex information exchange upon MSC (Prockop 2009). In an attempt to clarify and define the nomenclature, the ISCT (International Society for Cell Therapy) set "minimal criteria" for MSC, such as:

- adherence to plastic when maintained in standard culture conditions,
- expression of CD105, CD73 and CD90, and lack of expression of CD45, CD34, CD14 or CD11b, CD79 alpha or CD19 and HLA-DR surface molecules,
- as well as differentiation ability into osteoblasts, adipocytes and chondroblasts *in vitro* (Dominici, Le Blanc et al. 2006).

In the last decade there has been rapid movement from bench to bedside. Based on their stromal origin, MSC were initially applied in co-transplantation studies with hematopoietic precursor cells (Koc, Day et al. 2002). Later, due to their mesodermal differentiation potential, Horwitz et al. were able to perform seminal studies applying MSC to children with osteogenesis imperfecta (Horwitz, Prockop et al. 2001). MSC were then applied as immunosuppressants in patients with graft versus host disease (Le Blanc, Rasmusson et al. 2004). Further studies introduced them as promising candidates for tissue regeneration in bone and cartilage repair (Frohlich, Grayson et al. 2008), epithelial regeneration (Long, Zuk et al. 2010), cardiovascular regeneration (Noort, Feye et al. 2010; Rangappa, Makkar et al. 2010), immunomodulation in graft versus host disease (GvHD) (Ringden, Uzunel et al. 2006), and inflammatory neurological diseases (Momin, Mohyeldin et al. 2010). MSC are expected to reduce tissue damage, to activate the endogenous regenerative potential of tissues and to participate in the regeneration (Noort, Feye et al. 2010). However, in all these studies it became apparent that MSC function mainly through paracrine effects rather than differentiating into cells or tissues (Caplan and Correa 2011).

1.2 MSC from different tissue sources

Bone marrow (BM) was the first source of MSC identified by Friedenstein and co-workers (Friedenstein, Gorskaja et al. 1976). BM-MSC are already being tested worldwide in clinical studies with currently over 1500 found in the Clinical Trials registry of the NIH (www.clinicaltrials.gov). Due to the long lasting research on BM-MSC they became the gold standard for any MSC research and therapeutic application. Nevertheless, a limitation for BM MSC clinical application is the low cell frequency in source tissue. Thus large volume bone marrow aspiration is necessary even in autologous settings, feasible only in general anaesthesia which is associated with an additional patient morbidity. In consequence, investigators have developed protocols for isolating MSC from a variety of different tissues and sources other than bone marrow. Latest studies led to the conclusion that MSC are not limited to a certain tissue source: the MSC niche is rather localized in the perivascular area of virtually all tissues (Crisan, Yap et al. 2008; da Silva Meirelles, Caplan et al. 2008). Thus numerous tissues containing MSC have been identified, for example adipose tissue (AT), cord blood (CB), fetal membranes and amniotic fluid, pancreatic islet, lung parenchyma, intestinal lamina propria, oral and nasal mucosa, eye limbus, dental tissues and synovial fluid (Jakob, Hemeda et al. ; Karaoz, Ayhan et al. ; Marynka-Kalmani, Treves et al. ; Pinchuk, Mifflin et al. ; Powell, Pinchuk et al. ; Zuk, Zhu et al. 2002; Kern, Eichler et al. 2006; Phinney and Prockop 2007; Jones, Crawford et al. 2008; Polisetty, Fatima et al. 2008; Huang, Gronthos et al. 2009; Ilancheran, Moodley et al. 2009; Karoubi, Cortes-Dericks et al. 2009).

Among all tissue sources, AT shows several important clinical advantages compared to BM: AT procurement can be achieved via tumescent-lipoaspiration in local anaesthesia, a lower risk operating procedure. Adipose tissue is abundant even in older individuals. AT-MSC are shown to have similar functional properties to BM-MSC while their frequency is definitely higher than in BM (Zuk, Zhu et al. 2002; Kern, Eichler et al. 2006). AT-MSC are currently being applied in clinical trials, at least 33 trials can be found in the NIH registry. The high frequency of MSC in AT renders it possible to isolate the mononuclear cell fraction directly at the patients bedside without the need for expansion in a GMP facility (Duckers, Pinkernell et al. 2006). There are divergent outcomes in those studies directly comparing freshly isolated with expanded cells (Garcia-Olmo, Herreros et al. 2009). Despite the advantages of processing at the patient's bedside, direct application of the freshly isolated mononuclear cells in one session procedure gives no opportunity to control the clinical outcome, for an amount of diverse undefined cell populations are effective in these settings. However, this is still being exercised as autologous treatment.

Studies are being performed in order to compare BM-MSC, AT-MSC and MSC of other tissue sources. They show that MSC are not one distinct cell population. Among their tissue sources MSC differ concerning their isolating rate, their expansion potential, their differentiating capacities (Kern, Eichler et al. 2006), their immunosuppressive and migratory properties (Najar, Raicevic et al. ; Constantin, Marconi et al. 2009). These differences have probably an impact on their quality and therapeutic ability, which only can be definitely clarified in "*in vivo*" studies. Summarizing, there is a complex algorithm, which should be followed in order to find the adequate tissue source for MSC cell therapy. Very important are:

- the patient's risk associated with the tissue procurement,
- the MSC frequency in the origin tissue stroma,
- the potential of MSC to be enrolled in its therapeutic function *in vivo*.

All this can rather be answered gradually applying standardized protocols. After procurement and expansion MSC have to be analysed regarding their functional properties through well defined *in vitro* potency assays. Finally functional properties have to be compared *in vivo* through animal studies and phase I clinical trials.

2. MSC protocols for clinical applications

Translating MSC into cell therapy settings requires a manufacturing process and manufacturing authorisation congruent to the local regulatory framework. Regulatory standards in the EU and USA comply with the good manufacturing practice (GMP) regulations and are set in order to control the therapeutics' safety process, e.g. tissue procurement, cell isolation, selection and expansion and have to be validated according to the quality criteria as defined by the manufacturer. Furthermore it is essential to control the quality, purity and potency of the cell product prior to their administration by well defined and validated quality control and potency assays to ensure safety.

2.1 Isolation and expansion of MSC for clinical applications

For clinical applications, MSC shall be isolated under aseptic conditions in GMP facilities. MSC are a subpopulation among the mononuclear cell fraction. They can be isolated after density gradient centrifugation or if MSC are embedded in extracellular matrix after enzymatic digestion. In general, the low frequency of human MSC within their origin tissues necessitates their expansion prior to clinical use. This raises the risk for contaminations (Bieback, Karagianni et al. 2011; Sensebe, Bourin et al. 2011). Furthermore, in long term cell culture the proliferation rate decays, the cell size increases, differentiation potential becomes affected and chromosomal instabilities and neoplastic transformation may arise (Prockop, Brenner et al.; Lepperdinger, Brunauer et al. 2008; Wagner, Horn et al. 2008) raising the risk for adverse reactions.

Similarly, the cultivation media potentially affect MSC, exposing them to pathogens and immunogens (Heiskanen, Satomaa et al. 2007; Sundin, Ringden et al. 2007; Bieback, Hecker et al. 2009). In order to achieve controlled conditions and a safe cell product for clinical use it is necessary to define quality criteria to monitor the cell product (Bieback, Schallmoser et al. 2008; Bieback, Karagianni et al. 2011; Sensebe, Bourin et al. 2011). For expansion aiming at clinical application it is obligatory to use GMP-grade supplements and sera if available. However, these reagents are just under development. Accordingly we, amongst others, tested human blood-derived components, like human serum or platelet derivatives to replace fetal bovine serum commonly used to expand MSC (Kocaoemer, Kern et al. 2007; Mannello and Tonti 2007; Bieback, Schallmoser et al. 2008; Bieback, Hecker et al. 2009). Human blood components offer the advantage that they are both well controlled and already in clinical use for decades. Still, human serum as well as platelet lysate is a very crude protein cocktail. Essential growth factors for optimal MSC culture have not yet been defined. Platelet derived growth factor (PDGF), epidermal growth factor (EGF), transforming growth factor (TGF-ß), and insulin growth factor (IGF) have been subjected to investigation. Basic fibroblast growth factor (bFGF) has demonstrated most promising effects in expanding MSC whilst maintaining stem cell properties and reducing replicative senescence (Tsutsumi, Shimazu et al. 2001). Recently, Pytlik et al described a human serum and growth factor supplemented clinical-grade medium, which allowed high cell expansion mediated by loss of contact inhibition (Pytlik, Stehlik et al. 2009). Anyhow, the ideal solution is a chemically defined clinical-grade medium permitting both adhesion and expansion of MSC and numerous attempts are ongoing to develop this (Mannello and Tonti 2007).

2.2 Quality control

In order to obtain a manufacturing authorization for cell therapeutics the quality criteria ought to meet the regulatory standards. Quality controls are instrumented within the manufacturing process to prove according to the set quality criteria. Essential quality criteria are the traceability of the cell product through donor identification and product labelling, the prevention of introduction and spreading of infection and communicable diseases through donor screening and aseptic cell processing and proof of the therapeutic safety, lot consistency, potency and purity of the cell product (European Parliament 2007; FDA 2010).

2.2.1 Therapeutic safety, purity and potency

Safety is a key issue in cell therapy. In addition to the above mentioned aspects regarding reagents (fetal bovine serum has been elaborated on) and sterility testing (bacterial, fungal, viral, mycoplasma), cellular aspects have to be considered as well. In long term cell culture current testing methods of chromosomal aberrations and neoplastic transformation are fluoerescence in situ hybridization (FISH), karyotype analysis or detection of proto-oncogenes or activators of tumorigenesis like myc-assosiated proteins (Agrawal, Yu et al. 2010). Further lately developed testing methods are BAC-based (Bacterial Artificial Chromosome) Array to detect DNA copy number or oligonucleotide-based Array CGH (Chromosomal Comparative Genomic Hybridization) to detect small genomic regions with amplification or deletion (Wicker, Carles et al. 2007). Additionally, detection of telomerase activation is often performed, as telomerase plays a role in malignant transformation *in vitro* (Yamaoka, Hiyama et al. 2011). All these assays indicate that there is a low risk of transformation of MSC in *in vitro* expansion. However, more safety studies – especially long term follow up *in vivo* - are required to exclude risks and to enable to value risks against therapeutic value.

Further aspects that are critical for the therapeutic safety and need to be analysed are the spontaneous or the induced *in vivo* differentiation potential of MSC. It has to be proven that MSC after *in vivo* application serve their therapeutic function and do not develop into unwanted cell types for example BM-MSC into adipocytes or osteocytes when intended for epithelial or myogenic regeneration. The latter could possibly lead to threatening thrombembolic incidents after intravascular application. In general, intravascular injection is associated with a higher risk than direct application into the site of injury or into the neighbouring parenchyma (Furlani, Ugurlucan et al. 2009).

MSC are not a distinct cell fraction in fresh tissue isolates. Accordingly purity is a key issue to be taken into account. To isolate MSC, mononuclear cells of fresh tissue isolates are seeded on plastic culture dishes, MSC adhere, proliferate and form colonies. Those expanded MSC should have a distinct immune phenotype, defined by the ISCT, they do not express haematopoietic markers and have a characteristic fibroblastoid morphology (Dominici, Le Blanc et al. 2006). Based on these criteria, contaminations of MSC with hematopoietic or endothelial cells can be assessed and consequently purity of the MSC cell product can be proven via flow cytometry. This is further amended by description of expanded MSC morphology and colony assays (CFU-F-assay) to quantify the precursor frequency. Quality controls of MSC expanded in scaffolds or in bioreactors vs. 2D cell culture regarding population purity is probably more complex.

MSC are applied in various clinical settings, as they possess a variety of functional properties. MSC can work as progenitor cells in tissue modelling, due to their adipo-, osteo-, chondrogenic potential, or as immunomodulatory agents in GvHD, autoimmune disease or as anti-inflammatory agents through their paracrine abilities. Due to this extremely broad range it is difficult to establish potency assays. These standardized *in vitro* functional assays have to be performed to predict the consistency of the manufacturing process and the functionality of the cell product. Quality control assays, including potency assays, have to be well established and validated to be capable of addressing the consistent quality of the cellular product. It is certainly difficult to reproduce the *in vivo* setting within *in vitro* conditions,. This is probably why *in vitro* potency assays often fail to predict the *in vivo* outcome (Sensebe, Bourin et al. 2011). Anyhow, it is a demand for the manufacturing facility to implement potency assays capable of predicting therapeutic capacity. These assays have to be quantitative and directly related to the mechanism of action. Where possible surrogate assays can replace time-consuming functional assays (e.g. cell surface marker expression, growth factor release, gene or protein expression analysis). Finally, the manufacturing process in order to conduct clinical trials in Europe and the US has to be validated and approved by the authorities in accordance to the pharmaceutical regulations.

2.3 Pharmaceutical guidelines

2.3.1 Advanced therapy medicinal products as described in the Regulation (EC) No 1394/2007 of the European Parliament

In cases where MSC are to be used in a medicinal product the donation, procurement and testing of the cells are covered in Europe by the Tissues and Cells Directive (2004/23/EC). To make innovative treatments available to patients, and to ensure that these novel treatments are safe, the EU institutions agreed on a "regulation on advanced therapies"

(EC1394/2007). Furthermore, a number of products also combine biological materials, cells and tissues with scaffolds. This regulation defines those products as "advanced therapy medicinal products (ATMP)" that are:

- "a gene therapy medicinal product" (Part IV of Annex I to Directive 2001/83/EC),
- "a somatic cell therapy medicinal product" (Part IV of Annex I to Directive 2001/83/EC) and
- "a tissue engineered product".

Cells or tissues shall be considered 'engineered' if they fulfil at least one of the following conditions:

- "the cells or tissues have been subject to substantial manipulation, in order to unfold their biological characteristics, physiological functions or structural properties" or
- "the cells or tissues are not intended to be used for the same essential function or functions in the recipient as in the donor" (Official Journal of the European Union 10.12.2007).

The scope of this regulation is to set standards for advanced therapy medicinal products which are intended to be placed on the market in European member states. It indicates the setting of manufacturing guidelines specific for ATMP as to properly reflect the particular nature of their manufacturing process. The directive 2004/23/EC amends to this regulation setting standards of quality and safety in tissue procurement and donor testing. Regarding clinical trials on ATMP, they should be conducted in accordance with the Directive 2001/20/EC. Additionally Directive 2005/28/EC laid down principles and detailed guidelines for good clinical practice as well as the requirements for authorisation of the manufacturing and importation of ATMP. Considering tissue engineered cell products, medicinal devices incorporated in the ATMP (combined medicinal products) are regulated by the directive 93/42/and the directive 90/385/ EEC.

2.3.2 Human cells, tissues, and cellular and tissue-based products (HCT/P's) as described by the US Food and Drug Administration (FDA)

The quality system for Food and Drug Administration (FDA) regulated products is known as current good manufacturing practices (cGMP). For globally operating pharmaceutical facilities it is mandatory to fulfil the requirements of both FDA and EU. The Code of Federal Regulation (CFR) Title 21, part 1271 has the purpose to create a unified registration and listing system for human cells, tissues, and cellular and tissue-based products (HCT/P's) and to establish donor-eligibility, current good tissue practice, and other procedures to "prevent the introduction, transmission, and spread of communicable diseases by HCT/P's" (www.FDA.gov).

Whereas cell products, only minimally manipulated or subjected to homologous use without systemic effect, are regulated solely by the Public Health Service (PHS) Act Section 361 and do not require to undergo premarket review (GEN Mar. 15, vol 25, no 6), they still must comply with Good Tissue practice (GTP) (Burger 2003). Clinical trials of higher-risk involving ''more-than-minimally manipulated'' HCT/P's require the Investigational New Drug (IND) mechanism.

3. Example for MSC in regenerative medicine: Attempts for orthopaedic applications in bone defect healing

Orthopaedic surgery provides a fascinating field for the application of MSC (Horwitz, Prockop et al. 2001; Le Blanc, Gotherstrom et al. 2005; Bernhardt, Lode et al. 2009; Chanda, Kumar et al. 2010; Diederichs, Bohm et al. 2010; Mosna, Sensebe et al. 2010; Parekkadan and Milwid 2010; Levi and Longaker 2011). Bone defects appear in increasing numbers in orthopaedic clinics due to aseptic loosening of hip endoprosthesis after 10 to 20 years. These defects are then covered primarily with either bone cement or acellular bone from a bone bank prior to insertion of a new endoprosthesis in order to provide primary stability - that is immediate mechanical support of a new implant (Gruner and Heller 2009).

An ideal scaffold must offer osteoinduction - induction of bone growth - and osteoconduction - providing the guiding structure that paves the way for future bone growth - and eventually osteointegration, becoming part of the bone architecture of a body (Frohlich, Grayson et al. 2008; Ferretti, Ripamonti et al. 2010). The advantages and disadvantages of bone cement have been controversially discussed regarding different rates of implant failure in follow up examinations (Kavanagh, Ilstrup et al. 1985; Izquierdo and Northmore-Ball 1994; Stromberg and Herberts 1996). Recent works suggest to proceed without use of bone cement if possible, and recommend other surgical techniques to implant a total hip endoprosthesis. Bone cement is stiff and strong with a gradual increasing resorption area at its limits. Where bone cement is placed, immediate primary stability is provided, however, at the expense of bone regeneration that does not take place anymore (Izquierdo and Northmore-Ball 1994; Gruner and Heller 2009). Depending on the localization of the bone cement and the mechanical stress, this can gradually lead to a decreased stability. In case another revision operation is needed but great bone defects and osteolysis can impede or even inhibit surgical possibilities (Kavanagh, Ilstrup et al. 1985; Izquierdo and Northmore-Ball 1994; Stromberg and Herberts 1996; Gruner and Heller 2009). Fresh autologous bone or allogenous acellular bone from a bone bank can support bone growth. These preparations are osteoconductive and are, if preserved as a cancellous bone even osteoinductive but fail to provide immediate stability alone. These scaffolds have osteoconductive potential, however regular radiological controls often demonstrate gradually increasing resorption at sites of the implanted acellular bone. In the consequence, stability may be compromised (Gruner and Heller 2009).

Given the potential of MSC to differentiate into bone, MSC became attractive candidates. For hard tissue replacement, cells alone are not adequate. Thus surgical procedures treating bone defects in which a combination of MSC and scaffolds are applied, may provide both immediate stability and permanent integration into the recipient's bone. Different techniques are described for the implantation of MSC. Still it remains unclear if implants shall carry completely osteogenically differentiated MSC, or more likely optimize adaptive possibilities within the host organism. The more differentiated the MSC the more initial stability they provide for implants in areas with high mechanical force exposure (Bernhardt, Lode et al. 2008). Less differentiated MSC on the other prove more plasticity (Niemeyer, Krause et al. 2004; Bieback, Kern et al. 2008). In the worst case, undesired differentiation or even dedifferentiation might occur. Medication, integrated drugs or even genetically engineered cells may prove a possible control *in vivo*.

3.1 *In vitro* 3D culture, choice of scaffold

Tissue engineering aims at regenerating or replacing tissues or even organs. Therefore a complex architecture is needed, which cannot be generated by simple two-dimensional (2D) cultures. Investigation on MSC concentrates on characterization *in vitro* in a 2D culture, as mentioned above, to assess both the differentiation potential and the influence of the biomaterial surface on growth and development. MSC can be driven towards osteogenic differentiation by use of dexamethasone, β-glycerophosphate and ascorbate in addition to osteogenic basal medium (Jaiswal, Haynesworth et al. 1997; Pittenger, Mackay et al. 1999; Augello and De Bari 2010). Cells can be used as undifferentiated, pre- or terminally differentiated cells in combinations with scaffolds to achieve tissue-like conditions. Compared to 2D, 3D cultures better mimic physiological conditions. Static 3D cultures are mainly used to investigate the suitability of a certain biomaterial (Bernhardt, Lode et al. 2009). Increasing attention is recently been paid to dynamic 3D culture, assuring a more homogenous cell distribution within a scaffold, a higher number of cells and all in all less manipulation (Diederichs, Roker et al. 2009; Stiehler, Bunger et al. 2009). Flow perfusion cultures itself, even in absence of dexamethasone, may lead to differentiation into bone tissue (Holtorf, Jansen et al. 2005). Nevertheless, cell expansion of MSC in order to achieve a high cell dose prior to use in animal or humans may not always be advantageous, since uncontrolled growth can also lead to benign or malign tumours.

There is a broad choice of biomaterials for scaffolds for clinical applications. However, only bone cement and bone of bone banks are regularly favoured for bone defect surgeries, when available. Bone itself has become the biomaterial per se as a natural scaffold supply. Bone cement on the other hand can be stored as powder, provides immediate stability and is easily prepared and applied during an operation. Within a few minutes, the cement becomes firm (Gruner and Heller 2009). Although acellular scaffolds prove stability immediately following implantation, a better option would be to seed them with cells. For MSC application a great variety of materials, ranging from sterilised original bone to nanostructures and bioglass-collagen composites are being utilised (Karageorgiou and Kaplan 2005; Tanner 2010). Eventually, in order to approach the therapeutic effect of scaffold-MSC composites, studies are currently being performed on several stages: cell culture either in a dish or in a bioreactor, animal models and individual attempts in human (Bernstein, Bornhauser et al. 2009; Diederichs, Roker et al. 2009; Diederichs, Bohm et al. 2010). Further key parameters for the choice of the suitable biomaterial is the ability to support cell growth, cellular ingrowth, osteogenic differentiation and antimicrobial functions (Costantino, Hiltzik et al. 2002; Bernstein, Bornhauser et al. 2009). For that reason, additional osteogenic cytokines such as bone morphogenetic proteins (BMP) or bioactive peptides that become integrated into scaffolds are of interest (Keibl, Fugl et al. 2011).

An optimum scaffold must allow bone cells to grow into it. Pores of 300 to 500μm are requested (De Long, Einhorn et al. 2007; Stiehler, Bunger et al. 2009). Apart from this an optimum scaffold has to be adapted to bone structures. Defects in facial areas, in the skull, femur or hip require different stabilities and shapes. Only hip re-implantation seems to provide some standardised features (Gruner and Heller 2009).

Building suitable biomaterials to be combined with MSC has led to very different approaches: Collagen as a basis of any bone tissue was modified and calcified at all pore sizes. Integration of MSC is easily achieved but primary stability is comparably low (Bernhardt, Lode et al. 2009; Nienhuijs, Walboomers et al. 2011). Hydroxyapatite is a ubiquitous part of the vertebrate bone. Hydroxyapatite ceramics become easily integrated and also prove enough primary stability (John, Varma et al. 2009; Nair, Bernhardt et al. 2009; Nair, Varma et al. 2009). Beta-tricalciumposphate is a completely resorbable scaffold with high purity. It is available at all sizes, all porous degrees, it can be supplied as granules or as plates and therefore serves as comparison to newly developed biomaterials (Wiedmann-Al-Ahmad, Gutwald et al. 2007). Due to its low tissue reactivity and good stability titanium based structures not only serve well as implants but also as scaffolds. Titanium or TiO_2 does not become degraded or resorbed, instead as a whole it becomes very firmly integrated into any tissue (Gotman 1997; Olmedo, Tasat et al. 2009). Due to the fact that titanium is not resorbable it holds the risks of infection, be it acute or slowly increasing, so that an explantation must be performed. Since titanium becomes very well integrated into the host's body, an explantation is often associated with a great tissue loss. Application of titanium has to be carefully considered. In sum, since tissue reactions to titanium are quite well characterised as an implant it serves well as an example of future challenges and possibilities of other biomaterials. Silver nanoparticles are matter of current discussion due to their antimicrobial and toxic effects that can also be used within polymeric nanocomposites. Titanium nanostructures alone have been proven to act antimicrobially (Dallas, Sharma et al. 2011; Ercan, Taylor et al. 2011).

3.2 Analysis of 3D cultures and biomaterials

Once a 3D scaffold has been seeded, the efficiency of the seeding procedure, cell growth and differentiation must be determined, e.g. by quantifying the DNA content and mineralisation by histochemical stains or RT-PCR (Stiehler, Bunger et al. 2009; Peister, Woodruff et al. 2011). Homogeneic seeding and / or cell growth can be determined by fluorescence microscopy or µCT (Zou, Hunter et al. 2011). Mechanical tests are not standardized. For *in vitro* generated bone tissue from MSC crush tests, i.e. the use of a defined force until a scaffold breaks, are the most simple. For *in vivo* generated bone tissue shear and bending tests give additional data concerning the stability of the MSC composite within the animal's original bone. However, *in vitro* and *in vivo* experiments are only conclusive when scaffolds used are comparable in size and porosity (De Long, Einhorn et al. 2007; Stiehler, Bunger et al. 2009). The same applies to standardisation of surgical procedures and animal models used (Reichert, Saifzadeh et al. 2009).

3.3 Tissue source

As already mentioned, tissue engineering requires a scaffold next to the cells to seed it. Since MSC can be isolated from different tissue sources, the question remains: which cells are best suited? MSC derived from different tissues show different osteogenic differentiation properties: human embryonic stem cells (hESC), CB-MSC, AT-MSC, BM-MSC and even amniotic membrane-derived MSC can undergo osteogenic differentiation. Historically, most work had been performed on BM-MSC, so at least BM-MSC are the source to compare with, when MSC behaviour in a scaffold is analysed (Lindenmair, Wolbank et al. 2010; Guven, Mehrkens et al. 2011; Stockmann, Park et al. 2011; Weinand, Nabili et al. 2011). In recent studies, aspects of differentiation in 2D tissue culture and in 3D tissue culture have been examined. Comparisons between BM-MSC and amniotic fluid derived stem cells (AFS) showed different properties in differentiation in 2D and 3D. In 2D tissue culture, AFS produce more mineralized matrix but delayed peaks in osteogenic markers. Differentiation towards bone tissue occurred faster in BM-MSC, however, after weeks mineralization slowed down. AFS differentiated more slowly but mineralized until the end of the observation period 15 weeks, producing 5 fold higher amounts of mineral matrix. Human term placenta derived MSC seem to be less prone to osteogenic differentiation than BM-MSC (Pilz, Ulrich et al. 2011). These characteristics might be of interest, when fast ingrowth is needed (Peister, Woodruff et al. 2011). As initially mentioned, for some groups AT-MSC are the most promising candidates in bone tissue engineering (Levi and Longaker 2011). Osteogenic capacity does not decrease with age in contrast to BM-MSC (Khan, Adesida et al. 2009). Also due to a relatively high and still increasing rate of obesity in the western hemisphere it can be considered that adipose tissue has a great potential as main source for MSC. So, metabolic disease can be of benefit when it comes to autologous MSC implantation (Diederichs, Bohm et al. 2010). All in all an ideal cell source has yet not been identified. Further research is important to compare the advantages of all tissue sources. Moreover, for each biomaterial the MSC differentiation properties have to be determined. The adequate MSC will depend both on availability and differentiating / functional properties.

3.4 Clinical trials

There is no on-going clinical trial that deals with the use of MSC and a suitable biomaterial in healing of bone defects in humans. Osteogenesis imperfecta has been successfully treated with MSC alone, even with allogenic MSC (Horwitz, Prockop et al. 2001; Le Blanc, Gotherstrom et al. 2005). The Iranian Royan Institute, Teheran, announced a clinical trial in 2008 (http://www.clinicaltrials.gov). The study aimed to establish the influence of MSC in non-union fracture healing. However, in 2011 the state of the study is still unknown and cannot be verified. One case report from 2009 refers to a clinical trial in preparation. The benefit of the use of decellularized bone and MSC was demonstrated in a case of large hip transplant loosening. Follow-up radiological exams could confirm the stable position of a new hip implant (Bernstein, Bornhauser et al. 2009). So far, no clinical trial on the use of MSC for bone fracture healing has been published. Various preclinical studies predict benefits in bone tissue healing and stability by use of MSC (Bernhardt, Lode et al. 2008; Bernhardt, Lode et al. 2009; John, Varma et al. 2009; Nair, Bernhardt et al. 2009; Nienhuijs, Walboomers et al. 2011). However the methods and more importantly the animal models to prove beneficial effects of MSC are not yet standardized. This is of great importance since the forces exerted on a fracture cannot be compared between animal species, nor can it be to humans. Comparisons between different procedures, cells and scaffolds are thus not reliable. A recent article proposes rules for comparable preclinical bone defects model that amongst others affect standardized surgical procedures and measurements. In this work tibia fracture and segmental defect models are preferred (Reichert, Saifzadeh et al. 2009).

3.5 Animal model and interpretation

Unfortunately, the criteria to evaluate the outcome of studies - be it *in vitro* or *in vivo* - differ considerably. Regarding the major requirement of mechanical stability, a variety of mechanical tests exist that determine stability. However, till date none of them has been defined as standard (Hak, Makino et al. 2006; Jones, Atwood et al. 2009; Reichert, Saifzadeh et al. 2009). In animal models success criteria of implanted MSC and scaffold are restricted mainly to analysis of regenerated bone e.g. by histologiacal findings, CT-scan technology, x-ray or simply by measuring the weight of the created bone as well as by mechanical torsion tests (Zou, Hunter et al. 2011). The fate of implanted scaffold and MSC, in terms of material resorption and MSC engraftment into the host body, is rarely studied (Bernstein, Bornhauser et al. 2009). Since there is no standard in animal models, experiments are being carried out on various models. The rat model is broadly used because of availability. Bio-mechanical properties similar to humans are found in sheep, especially in hip arthroplasty (Korda, Blunn et al. 2008). Usually a fracture is induced as described by Matsumoto et al or Mifune et al (Matsumoto, Kawamoto et al. 2006; Mifune, Matsumoto et al. 2008). In a first step a tibia is fractured. Then a collagen scaffold is inserted containing saline and either BM-MSC or hESC. Then Undale et al compared the bone tissue healing properties of BM-MSC and hESC in rats after an induced fracture. BM-MSC resulted to be more efficient than hESC to bridge and heal a critical bone fracture. Moreover, in this setting hESC tended to produce benign bony tumours compromising the use of these cells in clinical settings (Undale, Fraser et al. 2011).

Bone fracture healing or integration into the animal's bone tissue can be demonstrated by follow-up conventional radiology in two weeks intervals. The limbs are both fully extended so that the broken and fractured limb can be compared. In recent studies μCT, a specialized CT for small animal structure, is used. Precise 3D models can be built from the data, allowing a comparison between the original and the newly built bone. Eight weeks after fracturing the animals can be euthanized and the limbs can be analysed histologically or biomechanically. Biomechanical stability of the fracture healing can be assessed by torsional load to evaluate normal and abnormal fracture healing (Undale, Fraser et al. 2011).

In summary, MSC from different sources appear as complementation to biomaterial implants. Depending on the tissue source and culture, different patterns of differentiation into bone, cartilage or fibre can be obtained. Depending on the precise situation different sorts of MSC-biocomposites may facilitate wound healing and functional regeneration of bone defects with high long term stability. However, the handling of biomaterial MSC composites is far more complex than conventional methods and oblige to adhere to regulatory standards: Since living cells are worked with, purity, a lack of bacterial contamination and absence of cell transformation has to be proven before clinical application. Conventional methods, that are acellular implants, may be limited because of rigidity and even lack of stability on the long run, but actually, in contrast to MSC biocomposites, they can be well compared regarding their advantages and disadvantages. MSC may differ much more as a matter of treatment, culture conditions and the cells itself need further investigation, experimental and clinical studies to evaluate their true potential at best in comparative studies. But the prospect of individual medicine with the patients' easily extractable and expandable own cells may support future research and applications in regenerative medicine.

3.6 Future prospects

Future orthopaedic research that may one day provide suitable personalized scaffolds to cover bone defects must integrate vascularisation as well. A balanced attempt to support both bone growth and blood supply must be established to create a stable long lasting graft that becomes completely integrated into bone. Osteoinduction is difficult to obtain. Local application of osteoinductive factors such as FGF, the bone morphogenic proteins BMP-2, BMP-4, BMP-7 and vascular endothelial growth factor VEGF does either not lead to results due to degradation or does lead to too strong responses since it cannot be well regulated. Recent work shows promising results in this regard. However no standard can be proposed in terms of choice of growth factor, dose and modification (Keibl, Fugl et al. 2011). Recent work demonstrated the feasibility of plasmid DNA-integration into a scaffold that lead to a higher bone differentiation ratio (Hosseinkhani, Hosseinkhani et al. 2008). Future research must also deal with possibly breaking the border between autologous and allogenic MSC in treatment, in case patients cannot donate autologous MSC of any source. Allogeneic MSC in treatment of patients with osteogenesis imperfecta defects could be recently demonstrated (Le Blanc, Gotherstrom et al. 2005).

The optimal degree of differentiation in culture prior to implantation in an animal model or a human remains unclear: Should implants carry completely osteogenically differentiated MSC, or more likely quite the opposite to provide an optimum of adaptive possibilities within the host organism? The more differentiated the MSC the more initial stability they provide for implants in areas in which great forces act. Less differentiated MSC on the other hand prove more plasticity. In the worst case undesired differentiation or even dedifferentiation might occur. Medication, integrated drugs or even genetically engineered cells may provide a possible control *in vivo*.

The specifications defined by the regulatory framework focussing on the clinical use of MSC are becoming increasingly detailed (Burger 2003). These are more complex when it comes to MSC and biomaterial composites as there are no standards for quality controls. *In vitro* and *in vivo* interactions between scaffolds and in-growing cells, as well as between scaffolds and host tissues, need to be investigated further.

4. Conclusion

In vitro studies indicate that MSC possess a wide spectrum of properties in tissue regeneration as adult progenitor cells or by secreting immunomodulatory and antiinflammatory factors. Still various manufacturing protocols, cultivating media and methods hinder to correlate and interpret scientific findings. Nevertheless MSC are very promising candidates for cell therapy and have moved extremely quickly in the last ten years from the bench to the bedside. For controlled clinical trials there are several obstacles to overcome in order to define a safe and efficacious therapeutic. There is a need to determine factors that may influence the cell quality and consequently the clinical outcome in terms of the tissue source, the isolating, expansion and cultivating conditions. Above that, protocols and *in vitro* and safety animal studies need to be performed in compliance with GMP requirements. To be able to conduct clinical trials on MSC, the manufacturing process has to fulfil several regulatory standards. Advances in clinical application of MSC can be exemplified in the field of orthopaedic bone regeneration. The osteogenic potential of MSC is seen to be of great benefit in bone defect healing. However, only in rare conditions are MSC alone beneficial. The choice of a suitable biomaterial to both carry MSC and provide good primary stability is crucial for clinical applications in hard tissue regeneration. Different sources of MSC that have different differentiation properties can be used. To assess compatibility of both MSC and biomaterial *in vitro*, MSC can be cultured on 2D or in 3D structures. Stability testing of seeded scaffolds helps determine the biomechanical properties of the biocomposite. Different animal models are being used, but no standard has yet been proposed that allows comparison of biomaterials and biomaterial/MSC. No biomaterial/MSC composite is in regular use in human for bone regeneration at present. Future efforts to establish treatments with these biocomposites must therefore concentrate on standardised procedures both in evaluation of tissue culture experiments and, more importantly, in animal models. The choice of the animal and the precise comparable procedures need to be defined. The prospect is individual autologous healing.

5. References

Agrawal, P., K. Yu, et al. (2010). "Proteomic profiling of Myc-associated proteins." Cell Cycle 9(24): 4908-4921.

Augello, A. and C. De Bari (2010). "The regulation of differentiation in mesenchymal stem cells." Hum Gene Ther 21(10): 1226-1238.

Bernhardt, A., A. Lode, et al. (2008). "Mineralised collagen--an artificial, extracellular bone matrix--improves osteogenic differentiation of bone marrow stromal cells." J Mater Sci Mater Med 19(1): 269-275.

Bernhardt, A., A. Lode, et al. (2009). "In vitro osteogenic potential of human bone marrow stromal cells cultivated in porous scaffolds from mineralized collagen." J Biomed Mater Res A 90(3): 852-862.

Bernstein, P., M. Bornhauser, et al. (2009). "[Bone tissue engineering in clinical application : assessment of the current situation]." Orthopade 38(11): 1029-1037.

Bieback, K., A. Hecker, et al. (2009). "Human alternatives to fetal bovine serum for the expansion of mesenchymal stromal cells from bone marrow." Stem Cells 27(9): 2331-2341.

Bieback, K., M. Karagianni, et al. (2011). "Translating research into clinical scale manufacturing of mesenchymal stromal cells." Stem Cells Int 2010: 193519.

Bieback, K., S. Kern, et al. (2008). "Comparing mesenchymal stromal cells from different human tissues: bone marrow, adipose tissue and umbilical cord blood." Biomed Mater Eng 18(1 Suppl): S71-76.

Bieback, K., K. Schallmoser, et al. (2008). "Clinical Protocols for the Isolation and Expansion of Mesenchymal Stromal Cells." Transfus Med Hemother 35(4): 286-294.

Burger, S. R. (2003). "Current regulatory issues in cell and tissue therapy." Cytotherapy 5(4): 289-298.

Caplan, A. I. and D. Correa (2011). "The MSC: An Injury Drugstore." Cell Stem Cell 9(1): 11-15.

Chanda, D., S. Kumar, et al. (2010). "Therapeutic potential of adult bone marrow-derived mesenchymal stem cells in diseases of the skeleton." J Cell Biochem 111(2): 249-257.

Constantin, G., S. Marconi, et al. (2009). "Adipose-derived mesenchymal stem cells ameliorate chronic experimental autoimmune encephalomyelitis." Stem Cells 27(10): 2624-2635.

Costantino, P. D., D. Hiltzik, et al. (2002). "Bone healing and bone substitutes." Facial Plast Surg 18(1): 13-26.

Crisan, M., S. Yap, et al. (2008). "A perivascular origin for mesenchymal stem cells in multiple human organs." Cell Stem Cell 3(3): 301-313.

da Silva Meirelles, L., A. I. Caplan, et al. (2008). "In search of the in vivo identity of mesenchymal stem cells." Stem Cells 26(9): 2287-2299.

Dallas, P., V. K. Sharma, et al. (2011). "Silver polymeric nanocomposites as advanced antimicrobial agents: Classification, synthetic paths, applications, and perspectives." Adv Colloid Interface Sci.

De Long, W. G., Jr., T. A. Einhorn, et al. (2007). "Bone grafts and bone graft substitutes in orthopaedic trauma surgery. A critical analysis." J Bone Joint Surg Am 89(3): 649-658.

Diederichs, S., S. Bohm, et al. (2010). "Application of different strain regimes in two-dimensional and three-dimensional adipose tissue-derived stem cell cultures induces osteogenesis: implications for bone tissue engineering." J Biomed Mater Res A 94(3): 927-936.

Diederichs, S., S. Roker, et al. (2009). "Dynamic cultivation of human mesenchymal stem cells in a rotating bed bioreactor system based on the Z RP platform." Biotechnol Prog 25(6): 1762-1771.

Dominici, M., K. Le Blanc, et al. (2006). "Minimal criteria for defining multipotent mesenchymal stromal cells. The International Society for Cellular Therapy position statement." Cytotherapy 8(4): 315-317.

Duckers, H. J., K. Pinkernell, et al. (2006). "The Bedside Celution system for isolation of adipose derived regenerative cells." EuroIntervention 2(3): 395-398.

Eaves, C. J., J. D. Cashman, et al. (1991). "Molecular analysis of primitive hematopoietic cell proliferation control mechanisms." Ann N Y Acad Sci 628: 298-306.

Ercan, B., E. Taylor, et al. (2011). "Diameter of titanium nanotubes influences anti-bacterial efficacy." Nanotechnology 22(29): 295102.

European Parliament, E. C. (2007). "REGULATION (EC) No 1394/2007 OF THE EUROPEAN PARLIAMENT AND OF THE COUNCIL." Available from: http://ec.europa.eu/health/files/eudralex/vol-1/reg_2007_1394/reg_2007_1394_en.pdf.

FDA (2010). "FDA : Code of Federal Regulations Title 21." Available from: http://www.accessdata.fda.gov/scripts/cdrh/cfdocs/cfcfr/CFRSearch.cfm?CFRPart=1271&showFR=1

Ferretti, C., U. Ripamonti, et al. (2010). "Osteoinduction: translating preclinical promise into clinical reality." Br J Oral Maxillofac Surg 48(7): 536-539.

Friedenstein, A. J., U. F. Deriglasova, et al. (1974). "Precursors for fibroblasts in different populations of hematopoietic cells as detected by the in vitro colony assay method." Exp Hematol 2(2): 83-92.

Friedenstein, A. J., J. F. Gorskaja, et al. (1976). "Fibroblast precursors in normal and irradiated mouse hematopoietic organs." Exp Hematol 4(5): 267-274.

Friedenstein, A. J., K. V. Petrakova, et al. (1968). "Heterotopic of bone marrow. Analysis of precursor cells for osteogenic and hematopoietic tissues." Transplantation 6(2): 230-247.

Frohlich, M., W. L. Grayson, et al. (2008). "Tissue engineered bone grafts: biological requirements, tissue culture and clinical relevance." Curr Stem Cell Res Ther 3(4): 254-264.

Fuchs, S., R. Baffour, et al. (2001). "Transendocardial delivery of autologous bone marrow enhances collateral perfusion and regional function in pigs with chronic experimental myocardial ischemia." J Am Coll Cardiol 37(6): 1726-1732.

Furlani, D., M. Ugurlucan, et al. (2009). "Is the intravascular administration of mesenchymal stem cells safe? Mesenchymal stem cells and intravital microscopy." Microvasc Res 77(3): 370-376.

Garcia-Olmo, D., D. Herreros, et al. (2009). "Expanded adipose-derived stem cells for the treatment of complex perianal fistula: a phase II clinical trial." Dis Colon Rectum 52(1): 79-86.

Gotman, I. (1997). "Characteristics of metals used in implants." J Endourol 11(6): 383-389.

Gruner, A. and K. D. Heller (2009). "[Revision hip arthroplastiy of the hip joint. Revision of the femur: which implant is indicated when?]." Orthopade 38(8): 667-680.

Guven, S., A. Mehrkens, et al. (2011). "Engineering of large osteogenic grafts with rapid engraftment capacity using mesenchymal and endothelial progenitors from human adipose tissue." Biomaterials 32(25): 5801-5809.

Hak, D. J., T. Makino, et al. (2006). "Recombinant human BMP-7 effectively prevents non-union in both young and old rats." J Orthop Res 24(1): 11-20.

Heiskanen, A., T. Satomaa, et al. (2007). "N-glycolylneuraminic acid xenoantigen contamination of human embryonic and mesenchymal stem cells is substantially reversible." Stem Cells 25(1): 197-202.

Holtorf, H. L., J. A. Jansen, et al. (2005). "Flow perfusion culture induces the osteoblastic differentiation of marrow stroma cell-scaffold constructs in the absence of dexamethasone." J Biomed Mater Res A 72(3): 326-334.

Horwitz, E. M., K. Le Blanc, et al. (2005). "Clarification of the nomenclature for MSC: The International Society for Cellular Therapy position statement." Cytotherapy 7(5): 393-395.

Horwitz, E. M., D. J. Prockop, et al. (2001). "Clinical responses to bone marrow transplantation in children with severe osteogenesis imperfecta." Blood 97(5): 1227-1231.

Hosseinkhani, H., M. Hosseinkhani, et al. (2008). "DNA nanoparticles encapsulated in 3D tissue-engineered scaffolds enhance osteogenic differentiation of mesenchymal stem cells." J Biomed Mater Res A 85(1): 47-60.

Huang, G. T., S. Gronthos, et al. (2009). "Mesenchymal stem cells derived from dental tissues vs. those from other sources: their biology and role in regenerative medicine." J Dent Res 88(9): 792-806.

Ilancheran, S., Y. Moodley, et al. (2009). "Human fetal membranes: a source of stem cells for tissue regeneration and repair?" Placenta 30(1): 2-10.

Izquierdo, R. J. and M. D. Northmore-Ball (1994). "Long-term results of revision hip arthroplasty. Survival analysis with special reference to the femoral component." J Bone Joint Surg Br 76(1): 34-39.

Jaiswal, N., S. E. Haynesworth, et al. (1997). "Osteogenic differentiation of purified, culture-expanded human mesenchymal stem cells in vitro." J Cell Biochem 64(2): 295-312.

Jakob, M., H. Hemeda, et al. "Human nasal mucosa contains tissue-resident immunologically responsive mesenchymal stromal cells." Stem Cells Dev 19(5): 635-644.

John, A., H. K. Varma, et al. (2009). "In vitro investigations of bone remodeling on a transparent hydroxyapatite ceramic." Biomed Mater 4(1): 015007.

Jones, E. A., A. Crawford, et al. (2008). "Synovial fluid mesenchymal stem cells in health and early osteoarthritis: detection and functional evaluation at the single-cell level." Arthritis Rheum 58(6): 1731-1740.

Jones, J. R., R. C. Atwood, et al. (2009). "Quantifying the 3D macrostructure of tissue scaffolds." J Mater Sci Mater Med 20(2): 463-471.

Karageorgiou, V. and D. Kaplan (2005). "Porosity of 3D biomaterial scaffolds and osteogenesis." Biomaterials 26(27): 5474-5491.

Karaoz, E., S. Ayhan, et al. "Isolation and characterization of stem cells from pancreatic islet: pluripotency, differentiation potential and ultrastructural characteristics." Cytotherapy 12(3): 288-302.

Karoubi, G., L. Cortes-Dericks, et al. (2009). "Identification of mesenchymal stromal cells in human lung parenchyma capable of differentiating into aquaporin 5-expressing cells." Lab Invest 89(10): 1100-1114.

Kavanagh, B. F., D. M. Ilstrup, et al. (1985). "Revision total hip arthroplasty." J Bone Joint Surg Am 67(4): 517-526.

Keibl, C., A. Fugl, et al. (2011). "Human adipose derived stem cells reduce callus volume upon BMP-2 administration in bone regeneration." Injury.

Kern, S., H. Eichler, et al. (2006). "Comparative analysis of mesenchymal stem cells from bone marrow, umbilical cord blood, or adipose tissue." Stem Cells 24(5): 1294-1301.

Khan, W. S., A. B. Adesida, et al. (2009). "The epitope characterisation and the osteogenic differentiation potential of human fat pad-derived stem cells is maintained with ageing in later life." Injury 40(2): 150-157.

Koc, O. N., J. Day, et al. (2002). "Allogeneic mesenchymal stem cell infusion for treatment of metachromatic leukodystrophy (MLD) and Hurler syndrome (MPS-IH)." Bone Marrow Transplant 30(4): 215-222.

Kocaoemer, A., S. Kern, et al. (2007). "Human AB serum and thrombin-activated platelet-rich plasma are suitable alternatives to fetal calf serum for the expansion of mesenchymal stem cells from adipose tissue." Stem Cells 25(5): 1270-1278.

Korda, M., G. Blunn, et al. (2008). "Use of mesenchymal stem cells to enhance bone formation around revision hip replacements." J Orthop Res 26(6): 880-885.

Kramann, R., S. K. Couson, et al. (2011). "Exposure to Uremic Serum Induces a Procalcific Phenotype in Human Mesenchymal Stem Cells." Arterioscler Thromb Vasc Biol.

Le Blanc, K., C. Gotherstrom, et al. (2005). "Fetal mesenchymal stem-cell engraftment in bone after in utero transplantation in a patient with severe osteogenesis imperfecta." Transplantation 79(11): 1607-1614.

Le Blanc, K., I. Rasmusson, et al. (2004). "Treatment of severe acute graft-versus-host disease with third party haploidentical mesenchymal stem cells." Lancet 363(9419): 1439-1441.

Lepperdinger, G., R. Brunauer, et al. (2008). "Controversial issue: is it safe to employ mesenchymal stem cells in cell-based therapies?" Exp Gerontol 43(11): 1018-1023.

Levi, B. and M. T. Longaker (2011). "Concise review: adipose-derived stromal cells for skeletal regenerative medicine." Stem Cells 29(4): 576-582.

Lindenmair, A., S. Wolbank, et al. (2010). "Osteogenic differentiation of intact human amniotic membrane." Biomaterials 31(33): 8659-8665.

Long, J. L., P. Zuk, et al. (2010). "Epithelial differentiation of adipose-derived stem cells for laryngeal tissue engineering." Laryngoscope 120(1): 125-131.

Mannello, F. and G. A. Tonti (2007). "Concise review: no breakthroughs for human mesenchymal and embryonic stem cell culture: conditioned medium, feeder layer, or feeder-free; medium with fetal calf serum, human serum, or enriched plasma; serum-free, serum replacement nonconditioned medium, or ad hoc formula? All that glitters is not gold!" Stem Cells 25(7): 1603-1609.

Marynka-Kalmani, K., S. Treves, et al. "The lamina propria of adult human oral mucosa harbors a novel stem cell population." Stem Cells 28(5): 984-995.

Matsumoto, T., A. Kawamoto, et al. (2006). "Therapeutic potential of vasculogenesis and osteogenesis promoted by peripheral blood CD34-positive cells for functional bone healing." Am J Pathol 169(4): 1440-1457.

Mifune, Y., T. Matsumoto, et al. (2008). "Local delivery of granulocyte colony stimulating factor-mobilized CD34-positive progenitor cells using bioscaffold for modality of unhealing bone fracture." Stem Cells 26(6): 1395-1405.

Momin, E. N., A. Mohyeldin, et al. (2010). "Mesenchymal stem cells: new approaches for the treatment of neurological diseases." Curr Stem Cell Res Ther 5(4): 326-344.

Mosna, F., L. Sensebe, et al. (2010). "Human bone marrow and adipose tissue mesenchymal stem cells: a user's guide." Stem Cells Dev 19(10): 1449-1470.

Nair, M. B., A. Bernhardt, et al. (2009). "A bioactive triphasic ceramic-coated hydroxyapatite promotes proliferation and osteogenic differentiation of human bone marrow stromal cells." J Biomed Mater Res A 90(2): 533-542.

Nair, M. B., H. K. Varma, et al. (2009). "Triphasic ceramic coated hydroxyapatite as a niche for goat stem cell-derived osteoblasts for bone regeneration and repair." J Mater Sci Mater Med 20 Suppl 1: S251-258.

Najar, M., G. Raicevic, et al. "Mesenchymal stromal cells use PGE2 to modulate activation and proliferation of lymphocyte subsets: Combined comparison of adipose tissue, Wharton's Jelly and bone marrow sources." Cell Immunol 264(2): 171-179.

Niemeyer, P., U. Krause, et al. (2004). "Evaluation of mineralized collagen and alpha-tricalcium phosphate as scaffolds for tissue engineering of bone using human mesenchymal stem cells." Cells Tissues Organs 177(2): 68-78.

Nienhuijs, M. E., X. F. Walboomers, et al. (2011). "The Evaluation of Mineralized Collagen as a Carrier for the Osteoinductive Material COLLOSS((R))E, In Vivo." Tissue Eng Part A 17(13-14): 1683-1690.

Noort, W. A., D. Feye, et al. (2010). "Mesenchymal stromal cells to treat cardiovascular disease: strategies to improve survival and therapeutic results." Panminerva Med 52(1): 27-40.

Olmedo, D. G., D. R. Tasat, et al. (2009). "The issue of corrosion in dental implants: a review." Acta Odontol Latinoam 22(1): 3-9.

Parekkadan, B. and J. M. Milwid (2010). "Mesenchymal stem cells as therapeutics." Annu Rev Biomed Eng 12: 87-117.

Peister, A., M. A. Woodruff, et al. (2011). "Cell sourcing for bone tissue engineering: Amniotic fluid stem cells have a delayed, robust differentiation compared to mesenchymal stem cells." Stem Cell Res 7(1): 17-27.

Phinney, D. G. and D. J. Prockop (2007). "Concise review: mesenchymal stem/multipotent stromal cells: the state of transdifferentiation and modes of tissue repair--current views." Stem Cells 25(11): 2896-2902.

Pilz, G. A., C. Ulrich, et al. (2011). "Human term placenta-derived mesenchymal stromal cells are less prone to osteogenic differentiation than bone marrow-derived mesenchymal stromal cells." Stem Cells Dev 20(4): 635-646.

Pinchuk, I. V., R. C. Mifflin, et al. "Intestinal mesenchymal cells." Curr Gastroenterol Rep 12(5): 310-318.

Pittenger, M. F., A. M. Mackay, et al. (1999). "Multilineage potential of adult human mesenchymal stem cells." Science 284(5411): 143-147.

Polisetty, N., A. Fatima, et al. (2008). "Mesenchymal cells from limbal stroma of human eye." Mol Vis 14: 431-442.

Pontikoglou, C., F. Deschaseaux, et al. (2011). "Bone Marrow Mesenchymal Stem Cells: Biological Properties and Their Role in Hematopoiesis and Hematopoietic Stem Cell Transplantation." Stem Cell Rev.

Powell, D. W., I. V. Pinchuk, et al. "Mesenchymal Cells of the Intestinal Lamina Propria." Annu Rev Physiol.

Prockop, D. J. (2009). "Repair of tissues by adult stem/progenitor cells (MSCs): controversies, myths, and changing paradigms." Mol Ther 17(6): 939-946.

Prockop, D. J., M. Brenner, et al. "Defining the risks of mesenchymal stromal cell therapy." Cytotherapy 12(5): 576-578.

Pytlik, R., D. Stehlik, et al. (2009). "The cultivation of human multipotent mesenchymal stromal cells in clinical grade medium for bone tissue engineering." Biomaterials 30(20): 3415-3427.

Rangappa, S., R. Makkar, et al. (2010). "Review article: current status of myocardial regeneration: new cell sources and new strategies." J Cardiovasc Pharmacol Ther 15(4): 338-343.

Reichert, J. C., S. Saifzadeh, et al. (2009). "The challenge of establishing preclinical models for segmental bone defect research." Biomaterials 30(12): 2149-2163.

Ringden, O., M. Uzunel, et al. (2006). "Mesenchymal stem cells for treatment of therapy-resistant graft-versus-host disease." Transplantation 81(10): 1390-1397.

Sensebe, L., P. Bourin, et al. (2011). "Good manufacturing practices production of mesenchymal stem/stromal cells." Hum Gene Ther 22(1): 19-26.

Seong, J. M., B. C. Kim, et al. (2010). "Stem cells in bone tissue engineering." Biomed Mater 5(6): 062001.

Shake, J. G., P. J. Gruber, et al. (2002). "Mesenchymal stem cell implantation in a swine myocardial infarct model: engraftment and functional effects." Ann Thorac Surg 73(6): 1919-1925; discussion 1926.

Stiehler, M., C. Bunger, et al. (2009). "Effect of dynamic 3-D culture on proliferation, distribution, and osteogenic differentiation of human mesenchymal stem cells." J Biomed Mater Res A 89(1): 96-107.

Stockmann, P., J. Park, et al. (2011). "Guided bone regeneration in pig calvarial bone defects using autologous mesenchymal stem/progenitor cells - A comparison of different tissue sources." J Craniomaxillofac Surg.

Stromberg, C. N. and P. Herberts (1996). "Cemented revision total hip arthroplasties in patients younger than 55 years old. A multicenter evaluation of second-generation cementing technique." J Arthroplasty 11(5): 489-499.

Sundin, M., O. Ringden, et al. (2007). "No alloantibodies against mesenchymal stromal cells, but presence of anti-fetal calf serum antibodies, after transplantation in allogeneic hematopoietic stem cell recipients." Haematologica 92(9): 1208-1215.

Tanner, K. E. (2010). "Bioactive composites for bone tissue engineering." Proc Inst Mech Eng H 224(12): 1359-1372.

Tsutsumi, S., A. Shimazu, et al. (2001). "Retention of multilineage differentiation potential of mesenchymal cells during proliferation in response to FGF." Biochem Biophys Res Commun 288(2): 413-419.

Undale, A., D. Fraser, et al. (2011). "Induction of fracture repair by mesenchymal cells derived from human embryonic stem cells or bone marrow." J Orthop Res.

Vassalli, G. and T. Moccetti (2011). "Cardiac repair with allogeneic mesenchymal stem cells after myocardial infarction." Swiss Med Wkly 141: w13209.

Wagner, W., P. Horn, et al. (2008). "Replicative senescence of mesenchymal stem cells: a continuous and organized process." PLoS One 3(5): e2213.

Wagner, W., R. Saffrich, et al. (2008). "The Stromal Activity of Mesenchymal Stromal Cells." Transfus Med Hemother 35(3): 185-193.

Weinand, C., A. Nabili, et al. (2011). "Factors of Osteogenesis Influencing Various Human Stem Cells on Third-Generation Gelatin/beta-Tricalcium Phosphate Scaffold Material." Rejuvenation Res 14(2): 185-194.

Wicker, N., A. Carles, et al. (2007). "A new look towards BAC-based array CGH through a comprehensive comparison with oligo-based array CGH." BMC Genomics 8: 84.

Wiedmann-Al-Ahmad, M., R. Gutwald, et al. (2007). "Growth of human osteoblast-like cells on beta-tricalciumphosphate (TCP) membranes with different structures." J Mater Sci Mater Med 18(4): 551-563.

Yamaoka, E., E. Hiyama, et al. (2011). "Neoplastic transformation by TERT in FGF-2-expanded human mesenchymal stem cells." Int J Oncol 39(1): 5-11.

Zou, W., N. Hunter, et al. (2011). "Application of polychromatic microCT for mineral density determination." J Dent Res 90(1): 18-30.

Zuk, P. A., M. Zhu, et al. (2002). "Human adipose tissue is a source of multipotent stem cells." Mol Biol Cell 13(12): 4279-4295.

15

Oral Tissues as Source for Bone Regeneration in Dental Implantology

Dilaware Khan[1], Claudia Kleinfeld[1], Martin Winter[2] and Edda Tobiasch[1]
[1]*University of Applied Sciences Bonn-Rhine-Sieg, Rheinbach*
[2]*Oralchirurgische Praxis, Rheinbach*
Germany

1. Introduction

One of the most common problems in Regenerative Medicine is the regeneration of damaged bone with the aim of repairing or replacing lost or damaged bone tissue by stimulating the natural regenerative process. Particularly in the fields of orthopedic, plastic, reconstructive, maxillofacial and craniofacial surgery there is need for successful methods to restore bone. From a regenerative point of view two different bone replacement problems can be distinguished: large bone defects and small bone defects. Currently, no perfect system exists for the treatment of large bone defects. Autologous bone material from the hip or the split calvarial graft is the gold standard to repair bone defects, as it has osteoinductive and osteoconductive properties (Tessier, 1982; Tessier et al., 2005a; Laurencin et al., 2006). Unfortunately this method is associated with an additional invasive intervention that leads to an increase risk of infection, pain during recovery, morbidity and frequent long periods of convalescence due to surgical trauma. Besides, only a limited amount of tissue can be obtained and harvested (Younger & Chapman, 1989; Tessier et al., 2005b). Also, the outcome is not always satisfactory after surgical treatment using bone splits (Baltzer et al., 2000; Lietman et al., 2000; Sorger et al., 2001). Heterologous transplants on the other hand, bear the risk of infection and rejection of the donor material. If the required amount of implant material cannot be obtained, another source is bovine-derived xenografts. There is, however, a potential risk for prion infection that cannot be totally avoided. Last not least large bone defect replacement needs nutrient and oxygen supply via blood vessels, so angiogenesis must be considered. This is very different in small bone defects: here angiogenesis is not an issue, but most of the other problems addressed above do play a role here too. This chapter will focus on small bone defects, especially those linked to dental implants.

2. Bone structure and regulation

The skeletal system is composed of bones that support the body, protect internal organs, and allow movement. Bone itself can be described as a natural composite material that consists of minerals and collagen that are merged in a complex amalgam. It consists mainly of two structures: an organic component as a matrix that contains collagen and a mineral component that is predominantly hydroxyapatite (Rho et al., 1997). The complex mineral substances give hardness to the bone and the softer organic collagen matrix causes visco-elasticity and toughness (Hutmacher et al., 2007). Together with cartilage, connective tissue, nerves, blood vessels, and marrow, they constitute the bone.

In the mineralized organic bone matrix, living and dead cells are present. Three types are known to play a role in bone homeostasis: osteoblasts, osteocytes and osteoclasts.

Osteoblasts are derived from MSCs and are cuboidal in shape (Fig. 1). They contain prominent Golgi bodies with a well developed rough endoplasmic reticulum, which is a histological sign for prominent protein production. These cells are located on the endosteal and periosteal bone surfaces. They secrete collagen type I and the non-collagenous proteins of the organic bone matrix. These cells also synthesize the enzyme alkaline phosphatase (ALP) that regulates the mineralization of the bone matrix. Their lifetime is about three months, after which they become metabolically inactive, flattened bone lining cells (Fig. 1). Bone lining cells are found covering inactive bone surfaces where they serve as a barrier for certain ions. The osteocytes originate from metabolically inactive osteoblasts and become trapped within the newly formed bone matrix during bone formation. Osteocytes have reduced synthetic activity compared to osteoblasts but maintain their sensitivity to vitamin D while continuing to participate in calcium regulation. On the other hand osteoclasts are derived from the fusion of monocyte and macrophage lineages (Ash, 1980) (Fig. 1). They are multi-nucleated cells that resorb bone. Osteoblasts regulate the differentiation of osteoclasts and osteocytes, which secrete factors in a feedback loop that play a role in regulating the functions of osteoblasts (Hartmann, 2006) and osteoclasts (Seeman & Delmas, 2006). The formation and resorption of bone is a continuous process that is kept in balance by the regulation of these three types of cells, with emphasis on osteoblasts and osteoclasts.

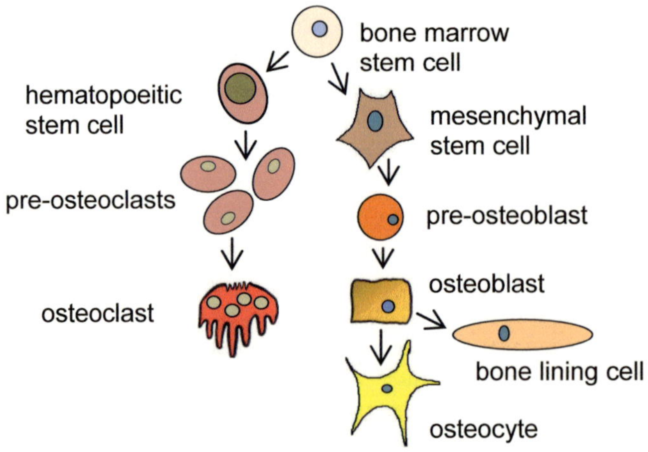

Fig. 1. Development of Bone Cells. Bone marrow stem cells give rise to hematopoietic stem cells and mesenchymal stem cells. Hematopoietic stem cells give rise to osteoclasts and mesenchymal stem cells are differentiated into osteoblasts together with other cell types. Osteoblasts further develop into bone lining cells and osteocytes.

In some diseases this balance is disrupted, as in osteoporosis, where increased osteoclast activity results in more resorption of bone than formation by osteoblasts. Along with osteoporosis, other medical conditions like bone cancer and osteogenesis imperfecta can lead to weakness of bones that can result in fractures. Bone defects can also occur due to trauma after accidents (Schäffler & Büchler, 2007). In addition, changes in recreational behavior especially in young adults lead to more need for bone replacement. Also, improved conditions of public health, nutrition and medicine have increased the life expectancy that resulted in an enhanced need for dental replacement. Taken together there is a growing need for bone regeneration and replacement.

3. Bone regeneration and replacement

3.1 The need for bone regeneration in dental defects

Studies revealed that approximately 70 % of all adults between 35 and 44 years lost at least one permanent tooth and by the age of 74 around 26 % of the adults lost all their permanent teeth (National Institutes of Health, 2001). Additionally, 45 % of the adults between 35 and 44 years and 54 % of the seniors between 65 and 74 years suffered from a middle heavy periodontitis, which is connected with a higher risk of tooth-loss (Holtfreter et al., 2010). To overcome these problems dental implants are one of the most common features to realize oral prosthetic reconstruction.

In order to guarantee a long and successful osseointegration of dental implants, they should be circumferentially covered with bone. Furthermore, it seems advantageous that the intraosseous part of the fixture is longer than the extraosseous prosthetic part. At least, the length of the implant should not be shorter than the abutment. Nowadays correct implant placement is determined by esthetic and prosthetic aspects, which often cannot be realized when only the residually available bone (restoration-driven implant placement) is being used (Garber et al., 1995). (National Institutes of Health, 2001)

There are defects of the alveolar bone which occur as a result of trauma, inflammation, resective surgical intervention such as tumor resection, bone loss after periodontal disease or athrophia after tooth loss or agenesis. In the posterior maxilla the phenomenon of pneumatization of the sinus maxillaris increases after tooth loss, which results in a vertical compromized bone level (Fig. 2A). Thus, bone reconstruction before or simultaneously to implant placement is often necessary (Fig. 2B). To do so guided, bone regeneration with autologous material such as bone graft material or other autologous or artificial grafting

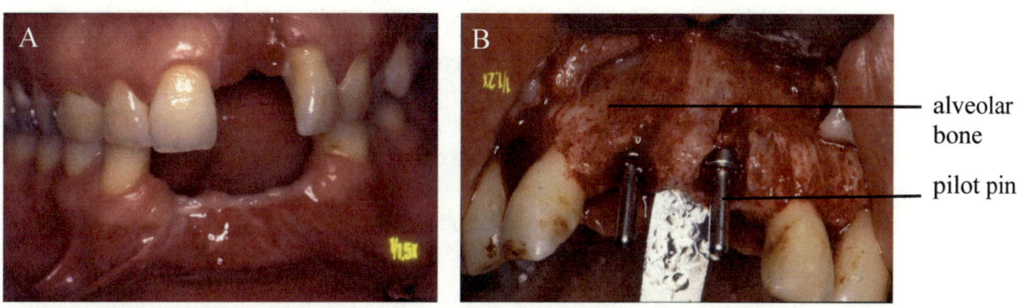

Fig. 2. A) Bone Degeneration. After tooth loss, reduced jawbone is as a result of trauma in a 24 years old male. B) Pilot pins *in situ* demonstrate the compromized bone-situation (male, 43 years old). Stable integration of implants is dependent on a thick jawbone. Stem cells could be used to fill the gaps and increase the thickness and induce osseointegration of implants.

procedures are methods of choice. Nevertheless, there exist many unsolved problems such as a e.g. higher morbidity in conjunction with the second wound of the donor site.

Therefore the use of stem cells (SCs) as source material for bone regeneration could represent an interesting approach for dental implantology.

3.2 Stem cells for bone regeneration

A modern strategy in Regenerative Medicine is the approach to combine living cells and scaffold material to establish a biological alternative for the diseased organ or tissue that can restore the functions. (Sittinger et al., 1996; Vacanti & Langer, 1999; Khademhosseini et al., 2009). Some degradable polymers, ceramics, or a combination of both can provide desirable mechanical and osteoconductive properties as basic scaffold material for bone replacement (Zippel et al., 2010b). Different factors should be considered for the use of such a biomaterial scaffold. It should imitate the three dimensional environment of the extracellular matrix, it should provide stability until replaced by regrown bone tissue and serve as an extended surface area for migration, adhesion, and differentiation of cells to encourage the growth of new tissue (Schultz et al., 2000; Ringe et al., 2002; Moroni et al., 2008).

The proliferating cells cover the scaffold and can grow into three dimensional tissue within. They are also an important factor for forming new tissue through extracellular matrix synthesis (Bonassar & Vacanti, 1998). Due to the development of new blood vessels towards, and to some extent onto, the new tissue, the scaffold begins to degenerate from the outside and is reconstituted by new natural bone tissue. As tissue related cell types cannot always be obtained in an adequate number or quality, SCs are a useful alternative for tissue regeneration.

Stem cells are the precursors of all cells and are involved in the repair system of the body. They are defined by three characteristics: self sustainability, self renewal and the potential of differentiation into different tissue types. For example adipocytes, astrocytes, chondroblasts, or osteoblasts come from mesenchymal stem cells (MSCs) (Pittenger et al., 1999; Pansky et al., 2007). In several publications, it has been suggested that MSCs can differentiate towards lineages that are naturally derived from the endoderm (Zuk et al., 2002; Tobiasch, 2009). Thus, increasing their potential because of these properties, the use of SCs to heal or rebuild damaged organs may provide an approach in future Regenerative Medicine (Zippel et al., 2010a).

SCs have been isolated from embryonic sources and well developed tissues of adult organism such as bone marrow, skin, dental pulp and adipose tissue (Kern et al., 2006). In addition two other sources for SCs have been discovered: cancer stem cells and induced pluripotent stem cells (iPS) (Takahashi et al., 2007; Aoi et al., 2008). Since the higher potency of embryonic stem cells and iPS compared to adult stem cells goes together with a higher risk of tumor formation, and embryonic stem cells are ethically problematic. Therefore, adult stem cells present themselves as an interesting cell source for bone replacement.

Adult stem cells can be divided into two main subpopulations: hematopoietic and mesenchymal stem cells (MSCs). Hematopoietic stem cells derived from bone marrow have been investigated best and could be a source for osteoclasts (Ash, 1980) (see Fig. 1). MSCs have been found in umbilical cord blood, bone marrow, and adipose tissue among others (Zuk et al., 2002). Generally, the isolation of MSCs is accomplished by plastic adherence resulting in colonies that are heterogeneous in size and morphology might contain contaminating non-mesenchymal cells such as macrophages or fibroblasts. The purity of isolated MSCs can be investigated by using the surface markers: CD73, CD90, and CD105 (should be expressed) and CD14, CD34 and CD45 (should not be expressed). These markers serve next to the adherence to plastic as a second feature for the identification and characterization of MSCs as suggested by the 'International Society for Cellular Therapy' (Dominici et al., 2006).

Another group of adult stem cells that has attracted attention are the ectomesenchymal stem cells derived from oral tissues. This stem cell group includes the dental pulp stem cells (DPSCs) and stem cells of human exfoliated deciduous teeth (SHEDs), both deriving from the pulpa, dental periodontal ligament stem cells (DPLSCs), dental follicle cells (DFCs), and stem cells from the apical papilla (SCAPs) (see Fig. 3). These cell types have the potential to differentiate into cells of all dental tissue types and bone as well. They share common phenotypic markers of MSCs (Alipur et al., 2010).

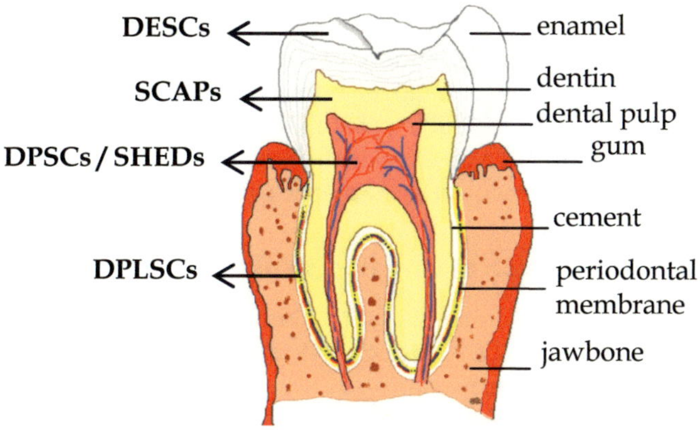

Fig. 3. Stem Cell Types in Tooth. From tooth different stem cell types namely, dental pulp stem cells (DPSCs), stem cells of human exfoliated deciduous teeth (SHEDs), dental periodontal ligament stem cells (DPLSCs), dental enamel derived stem cells (DESCs), stem cells from the apical papilla (SCAPs) and dental follicle cells (DFCs) can be obtained.

In comparison to other dental sources, dental follicle cells (DFCs) can be easily obtained in high amounts from young and healthy donors, since they are isolated from tooth extraction material collected during surgical removal of wisdom teeth. As these cells are derived from young donors, long telomeres extend their lifespan which makes them interesting cells for Regenerative Medicine (Shay & Wright, 2010). The dental follicle develops from ectomesenchyme. It surrounds the developing tooth germ before eruption (Ten Cate, 1997; Wise, 2002). During embryonic development, the ectomesenchyme is partly derived from migrating cells of the cranial neural crest. Therefore, the cells derivative from dental follicle differ from mesenchymal stem cells isolated from other sources (Chung et al., 2004; Slootweg, 2009). Due to having the more ectodermal character, these cells can have a differentiation potential diverse from MSCs. As expected these cells can differentiate into hard tissue such as the periodontal, cementoblastic, chondrocytic, and osteogenic lineages.

ATSCs and DFCs, both show osteogenic differentiation potential and are thus suitable candidates for the use in bone regeneration for stable osseointegration of dental implants. As these cells are obtained from healthy individuals, they might be used as an autograft in the future. The transplantation not only of autologous but also of allogenic sources could provide benefits in comparison to other common procedures in bone regeneration. As MSCs have low immune characteristics, they appear to be suitable for allogenic therapeutic purposes, without activating the immune response in immunocompetent patients (Jung et al., 2009). In different studies the use of MSCs has been investigated to replace lost or damaged bone (Schaefer et al., 2000; Ringe et al., 2002). After tooth loss, jawbone degenerates and stable integration of dental implant needs a thick jawbone. To overcome this problem there are two different alternatives that can be considered for using SCs in dental implants. The reconstruction after bone defects with SCs to achieve a sufficient bone thickness to insert the implants and the loading of an implant or artificial tooth-root with SCs with the aim to realize a sufficient integration in the bone.

SCs have the capability to re-establish cell function, reverse cellular damage, and heal damaged tissue (Conrad and Huss, 2005). SCs could also be a source to regenerate human teeth in the future, as these cells have been successfully used to regenerate living teeth in rabbit extraction sockets (Hung et al., 2011). In some mammals like rodents, rabbits, prairie dogs, and pikas, the teeth can grow throughout life because in these mammals as the pulp cavity remains open permanently. While on the other hand in humans tooth cannot grow continuously as pulp cavity closes when the teeth are fully grown. Therefore this study cannot be adapted easily for the regeneration of teeth or teeth related tissues in humans but it at least provides interesting basic results that can be helpful for use of SCs in dental tissues.

3.3 Bone chips for the stabilization of dental implants

Another approach next to scaffold loaded with stem cells to overcome the problem of unstable dental implants is the use of particulated non-vascularized bone autografts. The particles can be collected during the implant-bed preparation in the process of drilling the hole for the implant into the bone. An advantage of the use of these bone chips is that this material can be expected to facilitate bone regeneration. However, contradictory statements were made about the quality of this material such as if it contains living cells. In addition, it is not clear how to disinfect the bone chips, which are contaminated with bacteria of the oral cavity due to the sampling process. To address these questions bone chips were collected from two different regions of bone: carticular bone and spongy bone (see Fig. 4).

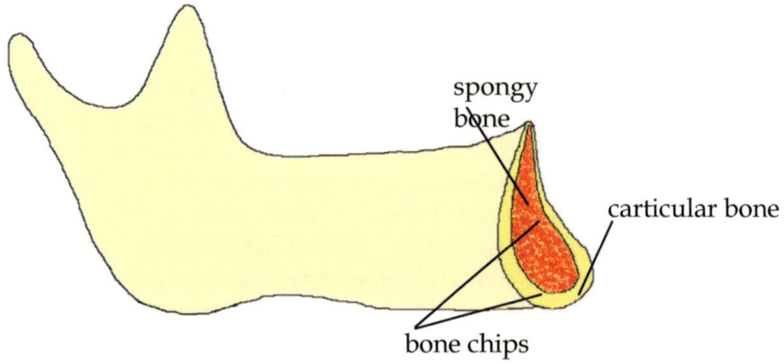

Fig. 4. Schematic Structure of Lower Jaw. Bone is composed of two tissue types mainly: spongious and carticular bone. Bone chips obtained during dental surgery for implant-bed preparation is derived from both bone tissue types.

4. Research methods

4.1 Isolation of primary cells for osteo-differentiation

4.1.1 Isolation of cells from tooth extraction material

For the isolation of ectomesenchymal stem cells, dental follicles were collected from human third molars before tooth eruption after surgical removal. The dental follicles were washed three times with 1 x PBS. Afterwards, the dental follicles were separated from the mineralized tooth and minced with a scalpel under sterile conditions. The tissue was digested in Collagenase (0.1 U / mL) and Dispase (0.8 U / mL) for 2 h at 37 °C in humidified atmosphere with 5 % CO_2. The cells were passed through a 100 µm strainer to obtain single-cell suspensions and seeded in 10 cm dishes in stem cell medium (SCM) that consisted of DMEM supplemented with 10 % FCS, 2 mM L-glutamine, 100 units / mL penicillin, 100 mg / mL streptomycin and 1 % amphotericin and cultured at 37 °C in a humidified atmosphere with 5 % CO_2. After 24 hours, non-adherent cells were removed by washing with 1 x PBS. The medium was changed and the plastic adherent cell fraction was cultured until 80 % confluent for further use.

The bone chip particles were collected with a bone filter integrated into a surgical suction pipe during the implant-bed preparation to isolate primary cells. For the isolation of bone chip derived cells (BCDCs), the same procedure as described above for DFCs was followed.

4.1.2 Isolation of adipose tissue derived stem cells

Human adipose tissue derived stem cells (ATSCs) were isolated from lipoaspirate obtained from plastic surgery. The isolation technique used during surgery was the tumescent liposuction technique. Using this particular technique, diluted epinephrine and lidocaine is infiltrated into the body fat to be removed, which leads to swelling and firmness of the targeted region, providing more accuracy during the liposuction procedure. The protocol was adjusted and modified to the procedure described by Zuk and colleagues (Zuk et al., 2001). The obtained lipoaspirate was augmented with PBS in a 1:2 ratio. After incubation for 30 minutes at room temperature (RT), two phases, a lower aqueous and upper fat phase of the lipoaspirate were obtained.

The lower phase was centrifuged at 200 x g for 10 minutes at RT. The resulting pellets, comprising the cells, were pooled and washed with 1 x PBS. Remaining erythrocytes were removed by applying 10 mL erythrolysis buffer for 10 minutes at RT. After another centrifugation step, under the same conditions as mentioned before, the cells were cultured in 60 cm^2 culture plates in SCM medium.

The upper phase comprising the fat tissue was augmented with 10 mg / mL type I collagenase in 1 x PBS and incubated for 45 minutes at 37 °C with agitation. The following steps for the treatment of the upper phase were according to the treatment of the lower phase. Cells of both phases were incubated at 37 °C with 5 % CO_2 in a humidified atmosphere. ATSCs were isolated due to their adherence to plastic and purified by washing with 1 x PBS after 24 hours, to remove undesired non-adherent cells.

4.2 Fluorescence activated cell sorting

The percentages of ATSCs or DFCs positive for the mesenchymal stem cell markers CD44, CD90 and CD105 and negative for CD14, CD45 and CD34 were measured using FACS analysis. The stem cells were trypsinized, centrifuged at 200 x g for 5 min and counted. 1 x 10^6 cells were resuspended in 1 mL 0.1 % PBSB and passed through a 100 µM cell strainer to obtain a single cell solution. 100 µL of the cell solution (100.000 cells) were incubated for 20 min in the dark with either the isotype control or the antibodies. Cells were washed with 2 mL 0.1 % (w / v) PBSB, centrifuged at 200 x g for 5 min and resuspended in 1 mL 0.1 % (w / v) PBSB. The cytometer settings and cell gates were adjusted to the isotype control, followed by measurement of the stem cell markers using the same conditions.

4.3 Adipogenic differentiation

For adipogenic induction, the isolated cells were seeded in a density of 2.8 x 10^3 cells / cm^2 in SCM. After one day, the medium was changed to adipogenic differentiation medium (AM), containing 1 µM dexamethasone, 1 µM insulin and 200 µM indomethacin. The cells were grown in AM for four weeks at 37 °C with 5 % CO_2 under humidified conditions. The AM was changed once a week. After four weeks, adipogenic differentiation was visualized with Oil Red O after fixing cells for 90 min with formalin (4 %) at 37 °C.

4.4 Osteogenic differentiation

The isolated cells were seeded in a density of 1.3 x 10^3 cells / cm^2 in 6 cm^2 and 12 well plates for osteogenic differentiation. After one day SCM was replaced with osteogenic medium (OM) containing dexamethasone, ascorbic acid and β-glycerophosphate. ATSCs were grown in OM for 4 weeks at 37 °C with 5 % CO_2 under humidified conditions. The OM medium was changed once a week. After four weeks, osteogenic differentiation was visualized by staining with Alizarin Red S after fixing cells for 5 min with formalin (4 %) at 37 °C.

4.5 Microbiological testing

Directly after surgery the obtained dental follicles were transferred into cold sodium chloride (0.9 % w / v) for determining possible microbial contaminations. The samples were kept cold until processing.

The samples were rolled over the surface of Columbia blood agar (CBA) and fastidious anaerobe agar (FAA) plates to isolate microorganisms. In addition the transport solution was put onto CBA and FAA plates. Under aerobic and anaerobic conditions the incubations were conducted over night at 37 °C. The Gas PakTM100-system was used for the incubation under anaerobic conditions. Single colonies were picked and isolated with respect to their morphological differences. Gram stainings, catalase- and oxidase-tests were used for the first characterization. API test strips were used to determine the exact bacteria species.

5. Comparison of stem cell sources

5.1 The characterization of ATSCs, DFCs and BCDCs for bone regeneration

The high plasticity of mesenchymal stem cells has resulted in an increased interest for their use in a variety of cellular therapies. However, different laboratories working with these cells isolate them from various tissue sources by following different protocols and characterizing these cells by different markers. Therefore, to set a standard, the minimal criteria for the definition of human MSCs were suggested by the 'Mesenchymal and Tissue Stem Cell Committee of The International Society for Cellular Therapy' (Dominici et al., 2006). The multipotent character of the isolated adipose tissue derived stem cells, ectomesenchymal dental follicle cells and bone chip derived cells was tested according to these criteria. MSCs were isolated from human adult adipose tissue of different aged female donors. DFCs were isolated from dental follicles and BCDCs from the bone chips collected during implant-bed preparation of male and female donors. The enrichment of specific stem cells was achieved due to their property of plastic adherence that is the first criterion for the testing of aMSCs character (Dominici et al., 2006). Isolated mesenchymal and ectomesenchymal cells of all donors showed a morphology similar to fibroblasts, which is typical for these stem cells (Yoshimura et al., 2006).

According to the above mentioned criteria the isolated cells should express the stem cell specific surface markers CD73, CD90, and CD105, and should not express CD14, CD34, and CD45. All isolated SC types expressed the expected markers (CD73, CD90 and CD105) as assessed by RT-PCR. The mesenchymal character of ATSCs and DFCs was also confirmed using FACS analysis for the presence of CD90, CD105, and in addition CD44. Furthermore, the cell types ATSCs and DFCs did not show the expression of leukocyte marker CD45 and macrophage marker CD14. ATSCs were positive and DFCs were negative for CD34. The presence of the expression of CD34 on ATSCs is controversial discussed. Some studies confirm the absence of CD34 expression on ATSCs (Zuk et al., 2002; Lee et al., 2004; Wagner et al., 2005) while other investigations showed ATSCs expressing CD34 (Mitchell et al., 2006; Yoshimura et al., 2006; De Francesco et al., 2009). These differences could be due to different stem cell isolation protocols, passage number or a different gating strategy during FACS analysis. In this study a subpopulation of ATSCs was stained positive for CD34.

Another typical MSCs character is the multilineage differentiation potential towards various lineages such as adipocytes, chondroblasts and osteoblasts. ATSCs showed a strong adipogenic differentiation potential whereas DFCs and BCDCs could not differentiate towards adipocytes. However, Kémoun and colleagues reported DFCs to differentiate towards the adipogenic lineage (Kémoun et al., 2007). The differences during isolation and precipitation in cell population might be possible reasons for this discrepancy. Also, DFCs can be different in their potency because these cells are derived from ectomesenchyme that is more committed toward hard tissue as tooth enamel.

According to all the findings mentioned above, the isolated ATSCs can be considered to belong to the population of multipotent MSCs, whereas the DFCs and BCDCs have a limited differentiation potential. Haddouti and colleagues showed that DFCs have a strong commitment towards the osteogenic lineage and show a more quantitative osteogenic differentiation (Haddouti et al., 2009). Thus, DFCs and BCDCs seem to be more committed towards osteogenic lineage.

Taken together all these stem cell types are good candidates for bone regeneration. But material from the oral cavity for isolation of primary cells such as DFCs and BCDCs cannot be obtained without microbial contamination. The question arises if this is a draw back on the use of these stem cells.

5.2 Microbial load of the oral tissue material

In order to evaluate the quality of the cells derived from oral tissues, microbiological investigations were performed. Our results revealed that all samples contained microbial species. Pre-treatment of patients with the antibiotics chlorhexidine (0.2 %), which is done anyway to decrease the chances of inflammation after surgery, reduced the number of microorganisms to less than 5 % but did not suffice to eliminate all bacteria. On the other hand pre-surgical, antibiotic treatment seemed to be negative for cell-outgrowth. To reduce contamination of the harvested cell-material, an optimized surgical procedure is more important than pre-surgical irrigation with chlorhexidine (0.2 %), and the use of a stringent dual suction pipe procedure. The predominantly found species were gram-positive cocci being either catalase-positive and oxidase-negative or catalase- and oxidase-negative. Most microorganisms belonged to the families of *Streptococcaceae* and *Staphylococcaceae*. The detected microorganisms did not interfere with cell growth and differentiation. They can be easily suppressed with standard antibiotics, applied routinely in patient treatment during the implantation procedure. Thus, these stem cells can be used for bone regeneration in dental implants.

6. Conclusion

The stability of dental implants is associated with a successful osseointegration into thick jawbone. Due to bone defects, bone regeneration is often needed before an implant can be inserted. For this stem cells can be a suitable candidates.

The stem cells isolated from adipose tissue, dental follicle and bone chips share mainly the multipotent character of mesenchymal stem cells. ATSCs can be successfully differentiated towards adipogenic and osteogenic lineages while DFCs and BCDCs did not show adipogenic differentiation. However, these cell types showed stronger commitment and differentiation towards osteogenic lineage. Therefore all three cell types are promising candidates for the treatment of various bone defects, and therefore also for the incorporation of tooth implants. They can be used to reconstruct jawbone defects to achieve enough bone thickness for the insertion of dental implants. It might be possible to load these cells on a dental implant or an artificial tooth root to increase its integration stability with the bone.

DFCs might be an ideal option if there will be a bank of donor material for these cells in the future, similar to those banks already existing as umbilical cord blood stem cells. If DFCs and BCDCs are not available for a specific patient, ATSCs are a reasonable option as they can differentiate towards the osteogenic lineage and be obtained from the patient itself as well, reducing the risk for rejection. Taken together all these tested stem cell types are suitable to improve the conditions for dental implants. Patients could preserve their dental follicle cells for later use in the future or their stem cells could be isolated from fat tissue directly before use. If a stem cell bank is arranged in the future, stem cells from other stem cell donors for dental follicle and fat tissue derived SCs could be used.

7. Acknowledgements

This work was supported by the BMBF-AIF, AdiPaD; FKZ: 1720X06 for ET and the Higher Education Commission of Pakistan, DAAD Germany for DK.

8. List of abbreviation

ALP	Alkaline phosphatase
AM	Adipogenic medium
ATSCs	Human adipose tissue derived mesenchymal stem cells
BCDCs	Bone chip derived cells
BMP2	Bone morphogenetic protien 2
°C	Degree centigrade
CBA	Columbia blood agar
CD14	Cluster of differentiation 14
CD34	Cluster of differentiation 34
CD45	Cluster of differentiation 45
CD73	Cluster of differentiation 73
CD90	Cluster of differentiation 90
CD105	Cluster of differentiation 105
cm	Centimeter
CO_2	Carbon dioxide
DESCs	Dental enamel derived stem cells
DFCs	Dental follicle cells
DMEM	Dulbecco's modified Eagle medium
DPLSCs	Dental periodontal ligament stem cells
DPSCs	Dental pulp stem cells
ECSs	Embryonic stem cells
FAA	Fastidious anaerobe agar
FACs	Fluorescence activated cell sorting
FCS	Fetal calf serum
iPS	Induced pluripotent stem cells
IGF-1	Insulin-like growth factor 1
LPL	Lipoprotein lipase
mL	Milliliter
mM	Millimolar
μL	Microliter
OM	Osteogenic medium
PBS	Phosphate buffer saline
PPARγ	Peroxisome proliferator-activated receptor gamma
RT	Room temperature
RT-PCR	Reverse transcriptase polymerase chain reaction
Runx2	Runt-related transcription factor 2
SCAPs	Stem cells from the apical papilla
SCM	Stem cell medium
SCs	Stem cells
SHEDs	Stem cells of human exfoliated deciduous teeth
w / v	Weight per volume
x g	Relative centrifugal force

9. References

Alipur, R., Sadeghi, F., Hashemi-Beni, B., Zarkesh-Esfahani, S.H., Heydari, F., Mousavi, S.B., Adib, M., Narimani, M., & Esmaeili, N. (2010). Phenotypic characterizations and comparison of adult dental stem cells with adipose-derived stem cells. Int. J. Prev. Med, 1, 164-171

Aoi, T., Yae, K., Nakagawa, M., Ichisaka, T., Okita, K., Takahashi, K., Chiba, T. & Yamanaka, S. (2008). Generation of pluripotent stem cells from adult mouse liver and stomach cells. Science, 321, 699-702

Ash, P., Loutit, J.F. & Townsend, K.M. (1980). Osteoclasts derived from haematopoietic stem cells. Nature, 283, 669-670

Baltzer, A.W., Lattermann, C., Whalen, J.D., Wooley, P., Weiss, K., Grimm, M., Ghivizzani, S.C., Robbins, P.D., & Evans, C.H. (2000). Genetic enhancement of fracture repair: healing of an experimental segmental defect by adenoviral transfer of the BMP-2 gene. Gene Ther, 7, 734-739

Bonassar, L.J. & Vacanti, C.A. (1998). Tissue engineering: The first decade and beyond. J. Cell. Biochem, 72, 297-303

Chung, U.I., Kawaguchi, H., Takato, T. & Nakamura, K. (2004). Distinct osteogenic mechanisms of bones of distinct origins. J. Orthop. Sci, 9, 410-414

Conrad, C. & Huss, R. (2005). Adult stem cell lines in regenerative medicine and reconstructive surgery. J. Surg. Res, 124, 201-208

De Francesco, F., Tirino, V., Desiderio, V., Ferraro, G., D'Andrea, F., Giuliano, M., Libondi, G., Pirozzi, G., De Rosa, A. & Papaccio, G. (2009). Human CD34+/CD90+ ASCs are capable of growing as sphere clusters, producing high levels of VEGF and forming capillaries. PLoS One, 4, e6537

Diagnosis and management of dental caries throughout Life. Consensus development conference statement, national institutes of health 2001

Dominici, M., Le Blanc, K., Mueller, I., Slaper-Cortenbach, I., Marini, F., Krause, D., Deans, R., Keating, A., Prockop, D. & Horwitz, E. (2006). Minimal criteria for defining multipotent mesenchymal stromal cells. The International Society for Cellular Therapy position statement. Cytotherapy, 8, 315-317

Garber, D.A. & Belser, U. (1995). Restoration-driven implant placement with restoration-generated site development. Compend. Contin. Educ. Dent, 16, 796-804

Haddouti, E.-M., Skroch, M., Zippel, N., Müller, C., Birova, B., Pansky, A., Kleinfeld, C., Winter, M. & Tobiasch, E. (2009). Human Dental Follicle Precursor Cells of Wisdom Teeth: Isolation and Differentiation towards Osteoblasts for Implants with and without Scaffolds. Mater. Sci. Engin. Technol, 40, 732-737

Hartmann, C. (2006). A Wnt canon orchestrating osteoblastogenesis. Trends. Cell. Biol, 16, 151-158

Holtfreter, B., Kocher, T., Hoffmann, T., Desvarieux, M., & Michaelis, W. (2010). Prevalence of periodontal disease and treatment demands based on a German dental survey (DMS IV). J. Clin. Periodontol, 37, 211-219

Hung, C.N., Mar, K., Chang, H.C., Chiang, Y.L., Hu, H.Y., Lai, C.C., Chu, R.M. & Ma, C.M. (2011). A comparison between adipose tissue and dental pulp as sources of MSCs for tooth regeneration. Biomaterial, 32, 6995-7005

Hutmacher, D.W., Schantz, J.T., Lam, C.X., Tan, K.C. & Lim, T.C. (2007). State of the art and future directions of scaffold-based bone engineering from a biomaterials perspective. J. Tissue. Eng. Regen. Med, 1, 245-260

Jung, D.I., Ha, J., Kang, B.T., Kim, J.W., Quan, F.S., Lee, J.H., Woo, E.J. & Park, H.M. (2009). A comparison of autologous and allogenic bone marrowderived mesenchymal stem cell transplantation in canine spinal cord injury. J. Neurol. Sci, 285, 67-77

Kémoun, P., Laurencin-Dalicieux, S., Rue, J., Farges, J.C., Gennero, I., Conte-Auriol, F., Briand-Mesange, F., Gadelorge, M., Arzate, H., Narayanan, A.S., Brunel, G. & Salles, J.P. (2007). Human dental follicle cells acquire cementoblast features under stimulation by BMP-2/-7 and enamel matrix derivatives (EMD) in vitro. Cell. Tissue. Res, 329, 283-294

Kern, S., Eichler, H., Stoeve, J., Kluter, H. & Bieback, K. (2006). Comparative analysis of mesenchymal stem cells from bone marrow, umbilical cord blood, or adipose tissue. Stem Cells, 24, 1294-1301

Khademhosseini, A., Vacanti, J.P. & Langer, R. (2009). Progress in tissue engineering. Sci. Am, 300, 64-71

Laurencin, C., Khan, Y. & El-Amin, S.F. (2006). Bone graft substitutes. Expert. Rev. Med. Devices, 3, 49-57

Lee, R.H., Kim, B., Choi, I., Kim, H., Choi, H.S., Suh, K., Bae, Y.C. & Jung, J.S. (2004). Characterization and expression analysis of mesenchymal stem cells from human bone marrow and adipose tissue. Cell. Physiol. Biochem, 14, 311-324

Lietman, S.A., Tomford, W.W., Gebhardt, M.C., Springfield, D.S. & Mankin, H.J. (2000). Complications of irradiated allografts in orthopaedic tumor surgery. Clin. Orthop. Relat. Res, 375, 214-217

Mitchell, J.B., McIntosh, K., Zvonic, S., Garrett, S., Floyd, Z.E., Kloster, A., Di Halvorsen, Y., Storms, R.W., Goh, B., Kilroy, G., Wu, X. & Gimble, J.M. (2006). Immunophenotype of human adipose-derived cells: temporal changes in stromal-associated and stem cell-associated markers. Stem Cells, 24, 376-385

Moroni, L., Wijn, J. & Van Blitterswijk, A. (2008). Integrating novel technologies to fabricate smart scaffolds. J. Biomater. Sci. Polymer Edn, 19, 543-572

Pansky, A., Roitzheim, B & Tobiasch, E. (2007). Differentiation potential of adult human mesenchymal stem cells. Clin Lab, 53, 81-84

Pittenger, M., Mackay, A., Beck, S., Jaiswal, R., Douglas, R., Mosca, J., Moorman, M., Simonetti, D., Craig, S. & Marshak, D. (1999). Multilineage potential of adult human mesenchymal stem cells. Science, 284, 143-147

Rho, J., Tsui, T. & Pharr, G. (1997). Elastic properties of human cortical and trabecular lamellar bone measured by nanoindentation. Biomaterials, 18, 1325- 1330

Ringe, J., Kaps, C., Burmester, G.-R. & Sittinger, M. (2002). Stem cells for regenerative medicine: advances in the engineering of tissues and organs. Naturwissenschaften, 89, 338-351

Seeman, E & Delmas, P.D. (2006). Bone quality—the material and structural basis of bone strength and fragility. N. Engl. J. Med, 354, 2250-2261

Schaefer, D., Klemt, C., Zhang, X., & Stark, G. (2000). Tissue engineering with mesenchymal stem cells for cartilage and bone regeneration. Chirurg, 71, 1001-1008

Schäffler, A. & Büchler, C. (2007). Concise review: adipose tissue-derived stromal cells--basic and clinical implications for novel cell-based therapies. Stem Cells, 25, 818-827

Schultz, O., Sittinger, M., Haeupl, T. & Burmester, G.R. (2000). Emerging strategies of bone and joint repair. Arthritis. Res, 2, 433 - 436

Sengupta, A. & Cancelas, J.A. (2010). Cancer stem cells: A stride towards cancer cure? J. Cell. Physiol, 225, 7-14

Shay, J.W. & Wright, W.E. (2010). Telomeres and telomerase in normal and cancer stem cells. FEBS Lett, 584, 3819-3825

Sittinger, M., Bujia, J., Rotter, N., Reitzel, D., Minuth, W.W. & Burmester, G.R. (1996). Tissue engineering and autologous transplant formation: practical approaches with resorbable biomaterials and new cell culture techniques. Biomaterials, 17, 237-242

Slootweg, P.J. (2009). Lesions of the jaws. Histopathology, 54, 401-418

Sorger, J.I., Hornicek, F.J., Zavatta, M., Menzner, J.P., Gebhardt, M.C., Tomford, W.W. & Mankin, H.J. (2001). Allograft fractures revisited. Clin. Orthop. Relat. Res, 382, 66-74

Takahashi, K., Okita, K., Nakagawa, M. & Yamanaka, S. (2007). Induction of pluripotent stem cells from fibroblast cultures. Nat. Protoc, 2, 3081-3089

Ten Cate, A.R. (1997). The development of the periodontium - a largely ectomesenchymally derived unit. Periodontology, 13, 9-19

Tessier, P. (1982). Autogenous bone grafts taken from the calvarium for facial and cranial applications. Clin. Plast. Surg, 9, 531-538

Tessier, P., Kawamoto, H., Matthews, D., Posnick, J., Raulo, Y., Tulasne, J.F. & Wolfe, S.A. (2005a). Autogenous bone grafts and bone substitutes--tools and techniques: I. A 20,000-case experience in maxillofacial and craniofacial surgery. Plast. Reconstr. Surg, 116, 6-24

Tessier, P., Kawamoto, H., Posnick, J., Raulo, Y., Tulasne, J.F. & Wolfe, S.A. (2005b). Complications of harvesting autogenous bone grafts: a group experience of 20,000 cases. Plast. Reconstr. Surg, 116, 72-73

Tobiasch, E. (2009) Adult human mesenchymal stem cells as source for future tissue engineering, Forschungsspitzen und Spitzenforschung. Zacharias, C., Horst, KW Ter., Witt, K.-U., Sommer, V., Ant, M. Essmann, U and Mülheim, L. 329-338, Spring Verlag, ISBN: 978-3-7908-2126-0

Vacanti, J. & Langer, R. (1999). Tissue engineering: the design and fabrication of living replacement devices for surgical reconstruction and transplantation. Lancet, 354, I32-34

Wagner, W., Wein, F., Seckinger, A., Frankhauser, M., Wirkner, U., Krause, U., Blake, J., Schwager, C., Eckstein, V., Ansorge, W. & Ho, A.D. (2005). Comparative characteristics of mesenchymal stem cells from human bone marrow, adipose tissue, and umbilical cord blood. Exp. Hematol, 33, 1402-1416

Wise, G.E., Frazier-Bowers, S. & D'Souza, R.N. (2002). Cellular, molecular, and genetic determinants of tooth eruption. Crit. Rev. Oral. Biol. Med, 13, 323-334

Yoshimura, K., Shigeura, T., Matsumoto, D., Sato, T., Takaki, Y., Aiba-Kojima, E., Sato, K., Inoue, K., Nagase, T., Koshima, I. & Gonda, K. (2006). Characterization of freshly isolated and cultured cells derived from the fatty and fluid portions of liposuction aspirates. J. Cell. Physiol, 208, 64-76

Younger, E.M. & Chapman, M.W. (1989). Morbidity at bone graft donor sites. J. Orthop. Trauma, 3, 192-195

Zippel, N., Limbach C.A., Ratajski, N., Urban, C., Luparello, C., Pansky, A., Kassack, M.U., & Tobiasch, E. (2010a) Purinergic Receptors Influence the Differentiation of Human Mesenchymal Stem Cells," Stem cells & Development. Epub ahead of print doi:10.1089/scd.2010.0576.

Zippel, N., Schulze, M. & Tobiasch, E. (2010b). Biomaterials and mesenchymal stem cells for regenerative medicine. Recent. Pat. Biotechnol, 4, 1-22

Zuk, P., Zhu, M., Ashjian, P., De Ugarte, D., Huang, J., Mizuno, H., Alfonso, Z., Fraser, J., Benhaim, P. & Hedrick, M. (2002). Human adipose tissue is a source of multipotent stem cells. Mol. Biol. Cell, 13, 4279-4295

16

Technologies Applied to Stimulate Bone Regeneration

Arnaldo Rodrigues Santos Jr.[1], Christiane Bertachini Lombello[2] and Selma Candelária Genari[3]

[1]*Centro de Ciências Naturais e Humanas (CCNH), Universidade Federal do ABC, Santo André, SP;*
[2]*Centro de Engenharia e Ciências Sociais Aplicadas (CECS), Universidade Federal do ABC, Santo André, SP;*
[3]*Centro Estadual de Educação Tecnológica Paula Souza, Faculdade de Tecnologia de Bauru (FATEC), Bauru, SP;*
Brazil

1. Introduction

Regenerative medicine constantly encounters situations where specific tissues do not have sufficient regenerative capacity to cope with lesions. Routine techniques involve varied surgical procedures, and often involve mechanical, functional and/or aesthetic discomfort. The need to develop alternative techniques for reducing these inconveniences is therefore necessary.

Techniques that stimulate normal tissue repair represent a major advance in biology and regenerative medicine. Frequently applied to the repair of bone lesions and reconstructive surgery, these new biomedical technologies and procedures have afforded technical simplification, elimination of some surgical processes, ease in handling, availability, good levels of predictability and effectiveness, and cost reduction. All this results in an improvement in the quality of life of the patients involved.

All the components of the skeletal system, bones and cartilage as well as the connective tissue present in tendons and ligaments, are capable of repair after an injury. During morphogenesis, bone is formed in a particular sequence of events. First, mesenchymal cells proliferate and differentiate into chondroblasts. This process leads to the production of cartilaginous skeleton. In the following stages, cartilage hypertrophy is observed with mineralization of the cartilaginous matrix. The cartilage cells then replaced by osteoblasts. Vascular invasion is necessary for this stage. The bone is then remodeled. Such events are seen during bone morphogenesis and, in the adult, during fracture consolidation (Reddi, 2001; Tsonis, 2002).

A bone fracture results in the loss of mechanical stability, discontinuity of the bone tissue and partial destruction of its blood supply. Repair is a complex process of tissue regeneration, resulting in stabilization of the fragments, consolidation through bone union, reconstruction of the tips of the avascular and partially necrotic fragments and, finally, internal and external remodeling of the newly formed tissue.

Hemorrhage caused by the blood vessel lesion, destruction of matrix and death of bone cells occurs at the site of a bone fracture. For the repair to begin, the blood clot and the cellular remnants from the matrix must be removed by the macrophages. The *periosteum* and the *endosteum* that are close to the fractured area respond with intense proliferation, forming tissue that is very rich in osteoprogenitor cells which constitute a collar around the fracture that will penetrate between the ruptured bone extremities, making a ring or collar that is located between the fractured bone extremities. This leads to the appearance of immature bone tissue, both through endochondral ossification of small pieces of cartilage that form there and through intramembranous ossification (Kierszenbaum, 2004; Junqueira & Carneiro, 2008). Areas of cartilage, areas of intramembranous ossification and areas of endochondral ossification can be found at the repair site. This process evolves with the appearance after a while of a bone callus covering the extremity of the fractured bones. The callus is formed of immature bone tissue that will temporarily join together the extremities of the fractured bone (Kierszenbaum, 2004; Junqueira & Carneiro, 2008).

Bone repair is also mediated by Mesenchymal stem cells (MSCs) (Bruder et al., 1994). These cells can be stimulated to differentiate into osteoblasts, cultivating them in the presence of serum, dexamethasone, beta-glycerophosphate and ascorbic acid. Moreover, MSCs can differentiate into osteoblasts due to the influence of vitamin D and BMP-2 (Pittenger et al, 1999). Human adipose tissue also contains stroma cells that are able to differentiate into chondrocytes and osteoblasts (Halvorsen et al, 2001).

The traction and pressure applied to the bone during fracture repair, and soon after the patient resumes normal activities, cause the remodeling of the bone callus and its complete substitution with lamellar bone tissue. When these tractions and pressures are identical to those applied to the bone before the fracture, the bone structure returns to its previous state; unlike other connective tissues, bone tissue, despite its rigidity, heals without scar formation (Kessel, 2001; Junqueira & Carneiro, 2008). Bone formation depends on the existence of an extensive vascular network and on the stability of the fracture focus that facilitates local vascularization, giving rise to the differentiation of the osteoprogenitor cells into osteoblasts; moreover, it is well established that the osteoblasts only synthesize the bone matrix in the presence of high oxygen tension.

The mechanical instability of the fracture hinders local vascularization and under these conditions the osteoprogenitor cells differentiate preferentially into chondroblasts. Accordingly, the fracture focus will initially be filled with cartilage (avascular tissue), which will provide a certain degree of stability to this focus, subsequently favoring vascularization of the site (Kierszenbaum, 2004). The quantity of cartilage formed during embryogenesis and at the site of a fracture (bone granulation tissue) is inversely proportional to the quantity of osteoprogenitor cells and of blood capillaries present at the site. If the bone callus tissue and the capillaries develop at the same time, the osteoprogenitor cells will differentiate into a vascularized environment and will consequently form bone. If there are proportionally few vessels or movement of the fracture segments, there will be cartilage formation followed by its substitution with bone tissue (endochondral ossification). If, however, the movement is excessive and the vascular supply limited, the establishment of local fibrous connective tissue (fibrosis) is probable (Kierszenbaum, 2004).

In the *regeneration of a fracture without loss of bone mass*, the repair process occurs in a biologically determined order. The first priority is stabilization and consolidation through callus formation on the edges and between the fragments, followed by its remodeling, besides revascularization and substitution of the necrotic areas. External factors can deeply affect the regeneration process, but the tissues act according to biological rules that control proliferation and cell differentiation as well as the production of matrix, which may occur regardless of, yet influenced by, external interferences.

However, *fractures with loss of bone mass* call for the use of grafts or implants. The latter serve as a support to bone regeneration, interacting with the interface of the receptor fragments and stimulating the tissue restoration process. These devices developed to be implanted are currently known as biomaterials (Hench, 1998), and will be addressed subsequently over the course of this text, constituting the basis of procedures such as guided tissue regeneration and tissue engineering. We will also address other technologies applied to bone regeneration, seeking optimization and acceleration of the process.

2. Regeneration biology without loss of bone mass

The bone repair process can be characterized by 6 physiological stages: impact, induction, inflammation, formation of cartilaginous callus, formation of bone callus and remodeling (Heppenstall, 1980).

Impact consists of the period of energy absorption until the fracture. The quantity of energy absorbed depends on the bone volume and is related to the loading rate. The impact stage of the fracture occurs until energy dissipation.

The *induction* stage involves modulation and differentiation of cellular elements required during the regeneration process. In fractures there is always local hemorrhage caused by injury to the blood vessels of the bone and of the periosteum, besides destruction of the matrix and death of the bone cells adjacent to the fractured site (Fawcett, 1986). This process triggers the inflammatory stage that will persist until the remodeling stage, with phagocytic activity of macrophages that will remove tissue and clot remnants. Cells from the periosteum and from the endosteum, close to the fractured area, will be activated (induction stage) and will respond with intense proliferation of their fibroblasts. Mesenchymal tissues, undifferentiated osteogenic and chondrogenic cells will differentiate into functional osteoblasts and chondrocytes, respectively. The stimulus for this induction can be electrical, low oxygen tension, low pH, release of lysosomal enzymes, release of cytosine and the presence of a series of inductor proteins, including bone morphogenetic proteins (BMP) and cartilage growth factors (Reddi, 1981; Canalis, 1983; Zellin *et al.*, 1996; Lieberman *et al.*, 2002; Oringer, 2002). The induction stage actually occurs through a series of sub-stages in which the inducing phenomena for each subsequent repair stage can be quite distinct, with unique characteristics toward specific target cells (for example: chondrocytes in the cartilaginous callus and osteoblasts in the bone callus).

During the *cartilaginous callus formation* stage, there is a considerable increase of vascularity and cellularity, and of the production of collagen, proteoglycans and lipids.

The callus is electronegative and the osteoclasts continue to remove the necrotic bone. The cartilage formed undergoes modifications, with hypertrophy of the chondrocytes that generate compression on the preexisting cartilaginous matrix, and consequent enlargement of its gaps, being gradually reduced to fenestrated thin septa and spicules of irregular shapes (Probst & Spiegel, 1997). The hyaline matrix from this hypertrophic region becomes calcified, and small granular aggregates and crystals of calcium phosphate are deposited on it.

The *bone callus formation* stage is marked by the substitution of calcified cartilage in primary bone tissue (Heppenstall, 1980; Probst & Spiegel, 1997; Mandracchia *et al.*, 2001). Cells with osteogenic potential, originating from the endosteum and particularly from the periosteum, are activated and a thin layer of bone (a periosteal ring or collar) is deposited around the central portion of the calcified cartilage. At the same time, periosteal blood vessels grow, invading the irregular cavities of the cartilaginous matrix created by the enlargement of the chondrocytes and by the confluence of its gaps. Vessels with thin walls branch out and grow into the cavities of the cartilaginous matrix with blind bottoms. Pluripotent cells are carried to the perivascular tissue of these blood vessels and, some differentiate into hematopoietic elements of the bone marrow. Other cells differentiate into osteoblasts, which deposit an aligned layer similar to an epithelium on the irregular walls of the spicules of the calcified cartilaginous matrix, and start the production of bone matrix. The osteoblasts covered by matrix become osteocytes and start to maintain contact with one another through cytoplasmic processes in a system of canaliculi. The callus remains electronegative, while the osteoclasts finish removing the necrotic bone. The cartilaginous matrix is gradually replaced by primary bone tissue.

During *remodeling*, the conversion of the primary bone tissue into secondary or lamellar bone tissue is completed. The collagen fibers are thicker and present preferential orientation alternating between layers or lamellae. These lamellae can be compacted if deposited on a flattened or concentric surface covering a blood vessel. The collagen fibers extend between the lamellae, thus increasing the bone strength. The blood vessels are contained in central canals (the Haversian canals), which intercommunicate through Volkmann's canals. Moreover, there are several canaliculi that extend to nourish the osteocytes. This assembly is known as the *osteon* or classically as *Haversian system*. Secondary osteons are formed when part of the concentric lamellar bone is converted into Haversian systems (Fawcett, 1986). The medullary canal is reestablished and the diameter and electronegativity of the callus decrease until they disappear.

3. Regeneration biology with loss of bone mass

As mentioned previously, the natural phenomenon of bone regeneration is insufficient, on its own, to reestablish the integrity of fractures with substantial loss of bone mass. For bone regeneration process to take place, it is necessary to have four components: a) a morphogenetic signal, b) host cells that respond to the signal, c) an appropriate carrier of this signal that can deliver it to the specific sites and thus serve as support to the growth of responsive cells of the host and d) a viable and well-vascularized bed (Burg *et al.*, 2000). Consequently, the need to use materials that would serve as support for the regeneration process is present and currently appears as an alternative to the use of grafts, where there are difficulties involved in obtaining bone tissue and in molding for the fracture in question.

These devices produced to serve as implants are known as biomaterials (Hench, 1998) and should exhibit characteristics that stimulate (osteoinduction) and/or guide (osteoconduction) the bone regeneration process.

Osteoinduction consists of a set of chemical, humoral or physical signals that initiate and sustain the various stages of the bone regeneration process and several factors may be involved (Nakagawa & Tagawa, 2000). The concept of osteoinduction was explored in 1965 by Urist, who showed ectopic bone formation when the demineralized bone matrix was implanted in muscles of rabbits, rats, mice and guinea pigs (Urist, 1965). Afterwards, it was concluded that a protein, called *bone morphogenetic protein* (BMP), was involved in the sequence of events involving chemotaxis, mitosis, bone differentiation and formation (Urist *et al.*, 1979). BMP is a glycoprotein of 17,500 daltons, and one of the factors currently identified in bone formation (Adriano *et al.*, 2000; Nakagawa & Tagawa, 2000; Reddi, 2000). Nowadays, several groups have shown that BMPs have the capacity to induce new formation of bone tissue by the endochondral route when implanted in ectopic sites in animals used for experimentation (Habibovic & de Groot, 2007). Besides BMP, other factors such as electromagnetic fields and direct currents also manifest inductive properties. The common result of osteoinduction is the modulation and differentiation of cells for bone production.

Osteoconduction, in turn, is related to the establishment of an appropriate environment model on which osteoprogenitor cells, when adequately stimulated, can produce bone (Heppenstall, 1980). Osteoconduction also facilitates production and bone deposition in the appropriate three-dimensional arrangement and increases the ability of the regeneration process in large segmental defects. Collagen, a natural organic component of bone and of bone surfaces, is the prototype of an osteoconductive substance (Kimura *et al.*, 2000; Lee *et al.*, 2001). A large number of natural and manufactured substances can also stimulate a favorable environment for bone formation (Mandracchia *et al.*, 2001; Pineda *et al.*, 1996), as discussed below. Osteoconduction is the phenomenon in which a single vehicle physically conducts the proliferation of osteogenic cells.

Many studies have investigated the inductive signals for bone morphogenesis, but the greatest emphasis has been placed on BMPs (Reddi, 2000). These proteins were found to have highly specialized patterns of expression during bone repair (Bostrom, 1998; Groeneveld & Burger, 2000). During the initial phases of consolidation, some primordial cells express BMPs in the bone callus. Expression is greater in MSCs and chondrocytes when endochondral ossification occurs. Expression decreases as the cartilaginous component of the callus matures. BMPs are expressed by osteoblasts, but decrease as the primary bone is replaced by lamellar bone. BMP-2, -3, -4, -5, -7, and -8 are responsible for the induction of bone and cartilage formation. BMP-12, -13 and -14, are cartilage derived. BMPs have also been used in clinical trials for the treatment fractures and pseudarthrosis, for example (Reddi, 2001).

A large number of natural substances can be extracted and/or manufactured to stimulate a favorable environment for bone formation (Mandracchi *et al.*, 2001; Pineda *et al.*, 1996). Among all the biomaterials of natural origin that aim to assist in bone regeneration, special emphasis is placed on those of bovine origin, where different protocols of chemical treatment of the bovine bone are evaluated with the purpose of preserving the organic components and inorganic components, such as collagen and hydroxyapatite repectively, which results in a mixed bovine bone (MBB) with an increase of the material's mechanical resistance. The MBB scaffold used in tissue regeneration has the appearance of a porous sponge that will occupy the space of the bone defect, preventing the migration of epithelial and connective cells, so that the osteoblastic cells have access to the regenerating tissue and start to populate the scaffold. Another role of the scaffold is its use as a vehicle for drugs that induce tissue regeneration and inhibit the progression of the disease.

Autogenous bone grafts are considered very advantageous, since they avoid complications of immunological rejection and supply cells that can immediately start the regenerative process (Cunha et al., 2005, 2006). The use of bone grafts is becoming frequent in orthopedics, as a method for resolution of comminuted fractures, (fractures in which the bone is splintered or crushed) thus significantly reducing the need to amputate an affected limb (Cavassini *et al.*, 2001).

One option in the treatment of partial bone defects is to perform the transportation of small bone fragments - called parietal transportation. In this technique (seldom reported in medical literature), the viable bone segment contiguous to the bone cavity is preserved. A bone fragment is created in the healthy region adjacent to the cavity and transported, according to the Llizarov method, filling a cavity of approximately 50% of the bone diameter (Rodrigues & Mercadante, 2005). The option of performing resection of viable bone occupying the complete cortical, transforming the partial defect into segmental for application of the conventional bone transportation technique, appears to us to be absolute nonsense: removing healthy bone when this is what is missing. The main advantage of parietal transportation is bone formation, even during infection.

The exact mechanism of osteoinduction by biomaterials is still largely unknown. Neither is it known whether the mechanisms of osteoinduction by BMPs and biomaterials are the same. In a recent review (Habibovic & de Groot, 2007) striking differences were shown in osteoinduction by BMPs and biomaterials, namely: (1) bone induced by biomaterials is always intramembranous, while bone induced by BMP is formed mainly by the endochondral route; (2) in small animals, just as in rodents, the bone is very rarely induced by biomaterials, but easily by BMPs; (3) bone is never observed on the edge of biomaterials but instead is always formed inside their pores, while bone formation by BMPs is regularly seen on the outside of the carrier and the soft tissue distant from the surface of this carrier.

4. Bone regeneration scaffolds

4.1 Autogenous grafts and allografts

The advantages cited for autogenous grafts in the bone regeneration process allow us to classify it as "gold standard". The presence of osteoprogenitor cells and osteoblasts confer the property of *osteogenesis;* the proteins that are contained in the bone matrix (for example, transforming growth factor β (TGF–β), and the BMPs) confer the aspect of *osteoinduction;* and the actual mineralized bone matrix provides a structural base for the growth of newly formed tissue, favoring the lodging of cells, the growth of blood vessels, and the deposition of bone matrix, characterizing *osteoconduction* (Bauer & Muschler, 2000).

The autogenous graft of *spongy* or *cortical* origin, whether vascularized or not, presents good integration with the adjacent tissue (Khan et al., 2005). In general autogenous grafts are obtained from the iliac crest, due to ease of access and as the obtainment of spongy bone of good quality, which is considered more osteoconductive, osteogenic and osteoinductive than the cortical bone graft, once it favors the diffusion of nutrients and revascularization of the treated area, and presents in its structure osteoprogenitor cells and osteoinductor proteins. Cortical bone graft acts mainly as a support for bone regeneration, changing the direction of the tissue regeneration process (Cypher & Grossman, 1996).

As regards *mechanical resistance*, the spongy graft does not offer immediate resistance at the grafting site. But the osseointegration process favors the acquisition of resistance during bone neoformation, and in an interval of 6 to 12 months it acquires resistance similar to that offered by the cortical graft (Dell et al., 1985; Stevenson, 1999). The opposite occurs with the cortical bone graft, which initially presents good mechanical resistance; however, over the first 6 months of grafting, this resistance is decreased by the presence of a mixture of newly formed bone and necrotic bone at the grafting site (Goldberg & Stevenson, 1992).

The different properties of spongy and cortical autogenous grafts establish the direction of their clinical use. The spongy autogenous graft can be used in cases involving difficulty in the consolidation of long bone fractures, and in the reconstruction of depressed lateral tibial plateau fractures (Marsh, 2006; Marino & Ziran, 2010; Nandi et al., 2010). A case report on autograft use in osteotomy for ulnar lengthening demonstrates the use of the trabecular autograft in functional recovery of the humeroulnar joint, resulting from difficulty in bone union. The patient presented recovery of movements in the two years of follow-up, achieving 105°of humeroulnar movement (Doornberg & Marti, 2010).

The cortical bone graft can be used in cases that require greater initial mechanical resistance at the graft site, even with the need for stabilization of the fracture with implants. Bone defects larger than 5 or 6cm are indications for the use of cortical grafts, as they require immediate mechanical support and a longer period of graft use (Nandi et al., 2010).

In spite of clinical results demonstrating the efficacy and safety of use of autogenous grafts in bone regeneration, some *disadvantages* of their clinical application limit their use (Arrington et al., 1996). We can mention high morbidity of the graft obtainment procedure, the aesthetic discomfort of this route, complications related to the surgical technique that mainly include infection and hemorrhage, and the actual limitations of indication, such as young or elderly patients and cases of recurring surgeries (Seiler & Johnson, 2000; Giannoudis et al., 2005).

As possible alternatives to autograft surgeons have allografts, demineralized bone matrix and natural or synthetic bone graft substitutes at their disposal.

Allografts have a growing clinical use, favored by the upgrading of techniques for obtaining, preparing and storing these materials. Fresh, frozen or freeze-dried grafts can be obtained, but fresh materials are used less often due to their technical difficulty and associated risks (Boyce et al., 1999; Keating & McQueen, 2001). Frozen and freeze-dried allografts are kept in specific tissue banks, properly processed and sterilized prior to storage. Their processing ends up eliminating cells naturally present in the graft, so there is no osteogenic activity here. It is considered that the allograft retains osteoinductive activity, and in spite of its processing some proteins are maintained. But its main activity is conduction of bone formation. However, even the osteoconductive activity can be affected by the material processing. The freezing, drying and sterilization stages, normally using Gamma rays, end up weakening the graft structure, reducing its mechanical properties (Pelker & Friedlaender, 1987; Henman & Finlayson, 2000).

The advantages of allograft use include immediate availability in a sufficient quantity for any treatment and in varied forms, facilitating clinical handling of this graft (Nandi et al., 2010). Li and collaborators described allograft use in the treatment of malignant humeral resection in patient treated between 2005 and 2008, with bone regeneration occurring at 26.3 weeks on average (Li et al., 2011). In another study, Virolainen and collaborators performed a survey of 10 years of allograft use for the treatment of periprosthetic fractures. This type of fracture can entail some surgical complication, and in this case the fractures occur soon after the prosthesis implant surgery, while fractures occurring at a later stage usually result from osteolytic lesions or osteoporosis. In both cases there is bone impairment at the implant site, hindering corrective surgical treatment. There were 71 patients treated between 1999 and 2008 with the use of cortical allograft and stabilization of the site with metal implants, and the patients presented a bone union rate of 91%. Allograft use was considered adequate, allowing biomechanical stability of the site (Virolainen et al., 2010).

4.2 Demineralized bone matrix

An alternative to bone tissue regeneration induction is the use of demineralized bone matrix (DBM) (Pietrzak et al., 2005). This type of biomaterial, obtained by acid hydrolysis of the bone matrix, through the action of hydrochloric acid, basically presents osteoinductive properties (Tuli & Singh, 1978; Katz et al., 2009). Its principle of action is based on preservation of the trabeculated collagen structure of the matrix and of bone formation inductor proteins, even with the processing of the tissue, obtained preferentially from human or bovine bones. The use of DBM to replace grafts should be observed with restrictions, as it does not present osteoconductive properties, due to the absence of the calcified bone matrix, and osteogenic properties, since processing for demineralization ends up killing the cells initially present in the tissue.

Nowadays there is a wide variety of available forms of DBM, either rigid or malleable. One of their main applications is the treatment of unconsolidated fractures (Pietrzak et al., 2005), in addition to the filling of bone cysts and cavities (Docquier & Delloye, 2005) and long bone fractures (Tiedeman et al., 1995; Keating & McQueen, 2001). Pieske and collaborators presented data on 20 patients with unconsolidated diaphyseal long bone fractures, treated between the years 2000 and 2006. The patients received autogenous grafts (n=10) or demineralized bone matrix (n=10), with bone formation having been observed in all the patients treated with DBM, while 20% of the patients treated with autogenous graft did not obtain the expected result (Pieske et al., 2009). The use of demineralized bone matrix has also been indicated for treatments of arthrodesis of the spinal column on account of its bone formation inducing action, and there may be an association with osteoconductive graft substitutes (Morone and Boden, 1998; Park et al., 2009). However, one of the disadvantages of demineralized bone matrix is related to the significant variability of donor sources, and corresponding variability of results obtained (Pietrzak et al., 2005).

Therefore in spite of the *availability* of natural materials as autogenous grafts, allografts and demineralized bone matrix, some limitations of use or clinical disadvantages of these materials drive the development of new technologies for bone tissue regeneration. *Natural and synthetic bone graft substitutes* are available to perform this role. The synthetic bone graft substitutes include ceramic and polymeric biomaterials, while biopolymers represent natural bone graft substitutes, including collagen and chitosan.

4.3 Bioceramics

Bioceramics are biocompatible biomaterials with a long history of clinical applications for bone regeneration. Among the advantages of these biomaterials we can cite their synthetic origin, eliminating the risk of autograft morbidity, or the risk of immunorejection and transmission of diseases of allografts or even biomaterials of human and animal origin. The structural similarity between some bioceramics, such as hydroxyapatite and beta-tricalcium phosphate, and spongy bone, allows us to classify them as biomimetic in relation to physical structure and chemical composition (Giannoudis et al., 2005). This mimicry favors the differentiation of osteoprogenitor cells and the deposition of bone matrix, characterizing bioceramics as essentially *osteoconductive*. The porous structure of bioceramics, or even the crystalline structure of calcium sulfate, also allows neoangiogenesis, which is essential in the osteoconduction process. Bioceramics of interest in the bone tissue regeneration process are those classified as temporary, since they are gradually replaced by newly formed bone (Tormala et al., 1998). Calcium sulfate and beta-tricalcium phosphate do this. Resorption time varies depending on the bioceramic in question, but is generally consistent with the bone callus formation time, sustaining tissue regeneration as osteoconductive agents. Hydroxyapatite is not considered resorbable by many authors, since the resorption process of this bioceramic averages 5 years, which corresponds to the period of natural bone remodeling of the body. Therefore it is considered that this bioceramic is integrated to the newly formed bone tissue and its resorption occurs during the intrinsic remodeling of the tissue.

Calcium Sulfate is one of the synthetic biomaterials with a long history of clinical use as a graft substitute for bone regeneration (Peltier et al., 1957; Tay et al., 1999). The dihydrated form of calcium sulfate, also called "gypsum", presents a crystalline structure that is not very uniform, and is currently used as a raw material in a calcination process that results in hemi-hydrated calcium sulfate ($CaSO4.½ H2O$), also called "Plaster of Paris" (Peltier et al., 1957). Calcium sulfate presents optimal biocompatibility, with reports of sporadic cases of inflammatory reaction after its use, with good evolution and spontaneous resolution in most cases. The length of stay in the organism is 8 weeks on average, a relatively short time, yet sufficient for bone callus formation to begin (Coetzee , 1980; Kelly et al., 2001). Calcium sulfate has ample clinical application potential, including bone defects resulting from trauma or created surgically, such as osteotomies and resection of tumors (Finkemeier, 2002; Kelly et al., 2001), as well as spinal surgery, for filling or bone fusion (Hadjipavlou et al., 2001).

More recently there was a proposal for the expansion of the clinical use of calcium sulfate as an *antibiotic release* agent, since it ensures high local concentration of the drug, avoiding its systemic circulation (Gogia et al., 2009). Reports demonstrate the control of osteomyelitis through the application of calcium sulfate pellets with antibiotics such as tobramycin, vancomycin and gentamicin (Bibbo & Patel, 2006; Chang et al., 2007). A randomized, prospective clinical study, published in 2010, presents data on local control of chronic osteomyelitis of long bones and cases of infection at non-bone consolidation sites. Thirty patients were treated, with half receiving calcium sulfate associated with tobramycin and the other half bone cement (polymethyl methacrylate) impregnated with antibiotic. The results demonstrated the mean follow-up of the patients for 38 months (ranging between 24 and 38 months) with the resolution of 86% of the cases in both experimental groups, concluding on the efficacy of calcium sulfate application in the local control of osteomyelitis (McKee et al., 2010).

In turn, *calcium phosphates* constitute bioceramics with a nanoparticulated physical structure, porous with pores of 100μm to favor the osteoconductive aspect of the biomaterial. The pore density can range between 40% and 60%, with Ca:P stequiometric ratio similar to spongy bone, imitating it (Gautier et al., 1998; Tanaka et al., 2008; Porter et al., 2009). Osteoconduction with calcium phosphate, often used in beta conformation, as *beta-tricalcium phosphate* (β-TCP, CA3(PO4)2), results in resorption of the biomaterial and osseointegration of the treated region in approximately 12 weeks. The bioresorption process occurs through a combination of dissolution and osteoclastic resorption at the implant site (Dong et al., 2002).

The persistence of the biomaterial favors the treatment of cavities resulting from bone resection, filling of osteotomy regions, defects of critical size of the bone (Gaasbeek et al., 2005; Tanaka et al., 2008) or even spinal fusion. Le Huec and collaborators reported the use of β-TCP for spinal fusion in 30 patients in association with bone graft, in comparison to another 24 patients treated with cortical allograft. The authors did not report pseudarthrosis and demonstrated the formation of bone callus 6 months after the β-TCP implant, with full resorption in 2 years (Le Huec et al., 1997).

The physical properties of β-TCP favor its association with liquids such as blood and bone marrow aspirate. In an experimental study with dogs, Bruder and collaborators demonstrated bone formation and the refinement of the bioceramic in association with mesenchymal cells obtained from bone marrow aspirate (Bruder et al., 1998). In 2007 the same author published, together with collaborators, the result of an experimental application of β-TCP grafts in sheep for posterolateral fusion (Gupta et al., 2007). In this experiment the authors compared the results of the fusion process with the use of autograft, biomaterial enriched with mesenchymal cells, biomaterial associated with total bone marrow aspirate and pure biomaterial. The radiological findings, in line with histological data, demonstrated a high rate of bone formation after 6 months in the presence of autograft (25%) and in the presence of the biomaterial enriched with cells (33%), whereas the biomaterial associated with the total bone marrow aspirate presented a low rate of bone formation (8%) and no bone formation was observed with the use of pure biomaterial, reinforcing the need for association of characteristics such as osteogenesis for guided tissue regeneration.

Unlike beta-tricalcium phosphate, *hydroxyapatite* ($Ca_{10}(PO_4)_6(OH)2$) a bioceramic with a low resorption rate and greater mechanical resistance, is commonly used in association with beta-tricalcium phosphate, in the proportion of 60/40 to improve osseointegration of the graft substitute (Balcik et al., 2007). The porosity of the biomaterial is essential for its action, requiring pores of 100-200μm, at a density of 60 to 65% for cellular lodging and vascularization of the treated area, confirming the osteoconductive action of hydroxyapatite (Giannoudis et al., 2005). The regeneration of defects of critical size and defects in long bones, created surgically or resulting from trauma, are general indications for its use, either pure or in association with β-TCP. Hydroxyapatite can also be used in spinal fusion procedures. The report on the use of hydroxyapatite in orthopedic lesions, including the resection of bone tumors, and the treatment of cystic lesions in rheumatoid arthritis, without the occurrence of adverse reactions and with good clinical evolution of the patients, was published by the group of Yoshikawa and collaborators (2009).

Bioactive glass, or *bioglass*, is a biocompatible bioceramic that allows good integration with newly formed tissue (Hench et al., 1971). It is basically composed of silica, sodium oxide, calcium oxide and phosphates. Some factors influence the integration of bioglass with the surrounding environment, such as composition of the biomaterial, pH of the environment, temperature, and porosity, directing its osteoconductive function (Nandi et al., 2010). Bioglass is indicated for filling bone cavities in general, in reconstructive surgery, including craniofacial defects, besides spinal column fusion procedures (Asano et al., 1994; Suominen &, Kinnunen, 1996).

4.4 Polymeric biomaterials

Among the polymeric, biocompatible and bioresorbable biomaterials used for bone tissue regeneration, the poly (L-lactic acid) (PLLA), poly (glycolic acid) (PGA), polycaprolactone (PCL) polyesters, and their copolymers, such as poli (D,L-lactic-co-glycolic acid) (PLGA) (Santos & Wada, 2007; Santos, 2010) deserve special emphasis. These polymers are often associated with bone formation induction proteins, such as BMPs or even with osteoconductive bioceramics, such as hydroxyapatite. One of the advantages of the use of polymers for tissue regeneration resides in the wide variety of possible applications, not just as graft substitutes, but also as fastening elements, including screws and plates. Bone tissue regeneration is guided by the polymer structure used, whereas proliferation induction and cellular differentiation are observed in these specific scaffolds (Ishaug-Riley et al., 1998; Santos et al., 2001; Santos et al., 2004). The PLGA copolymers implanted in bones induce bone tissue neoformation at the implant site, over a variable period of time, depending on the ratio of polyesters present in the copolymers (Reed & Gilding, 1981).

Polymer/bioceramic composites have the advantage of conferring on polymers the intrinsic biomechanical property of calcium phosphates, such as hydroxyapatite, favoring osteoconductive characteristic of the biomaterial (Hutmacher et al., 2007). Osteoblast cell cultures in porous PLLA/hydroxyapatite composites (PLLA-HA) enable cell proliferation, the lodging of cells throughout the scaffold of the biomaterial and the differentiation of these cells with synthesis of mineralized matrix (Ma et al., 2001). These results are corroborated by the study of Rizzi and collaborators with the biomaterial of PLA-HA and PCL-HA (Rizzi et al. 2001). HA induces the activity of the bone cells preferentially adhered to these particles, exposed on the surface of the composite.

The application of HA-PLLA to two cases of mandibular reconstruction after tumor resection was published recently (Matsuo et al., 2010). The plates were designed with the use of computed tomography. In one of the cases there was association of the composite biomaterial with growth factors obtained from platelets harvested from the patient and in the second case there was a dental graft. Both cases presented good clinical evolution, without the observation of bone resorption in two years of follow-up, and with the formation of good quality bone.

4.5 Other biomaterials

Besides the synthetic bioceramics and polymeric biomaterials, some biomaterials obtained in nature present considerable potential for application in bone regeneration: coralline hydroxyapatites and chitosan.

Similar to hydroxyapatite, *coralline hydroxyapatites* have been explored recently for their osteoconductive potential. They derive from marine corals, with a calcium carbonate base, a porous structure, and pore size ranging between 100 and 500μm, suitable for the proposed function. They can be obtained directly in nature (and processed mainly for sterilization), or obtained from hydroxyapatite (Keating & McQueen, 2001). Indications for use include long bone fractures and tibial plateau fractures, presenting a behavior similar to the autogenous graft (Bucholz et al., 1989).

Chitosan is another biocompatible biomaterial with potential for clinical application under analysis, and is considered very promising for the area of tissue regeneration. It is a natural biopolymer, obtained from the polysaccharide chitin, common in the exoskeleton of crustaceans (such as shrimps and lobsters). It presents encouraging results demonstrating its performance as an osteoconductive biomaterial guiding osseointegration. A study published in 2003 uses chitosan glutamate associated with hydroxyapatite for the treatment of defects of critical size in rat calvaria. The results were obtained after 9 and 18 weeks. The association with osteoprogenitor cells obtained from bone marrow proved ideal for tissue regeneration according to the protocol under investigation, including with mineralization of the treated areas (Mukherjee et al., 2003).

A recently published study (Jayasuriya & Kibbe, 2010) demonstrates the preparation of chitosan microparticles on a wide scale, and the incubation of these particles in concentrated physiological fluid for the stimulation of in vitro biomineralization and subsequent incorporation of insulin-like growth factor (IGF-1). The study evidenced the release of IGF-1 over a 30-day period, characterizing the possibility of the biomaterial's use as a drug release agent.

Collagen, in turn, exhibits a series of possible clinical applications, such as a scaffold for the regeneration of various tissues, including skin, cartilage and bone. It is a natural biopolymer, obtained from animal tissue, generally bovine, with low toxicity and immunogenicity. It can be made available in the form of gels, films and sponges, favoring cell adhesion and resorption, driving the regenerative process. In the case of bone regeneration, collagen is often associated with osteoconductive materials such as beta-tricalcium phosphate or hydroxyapatite (Wahl & Czernuszka, 2006). These composites aim to reproduce the natural conditions of bone and thus to drive cell behavior, with the differentiation of osteoblasts and the synthesis of mineralized bone matrix (Zhang et al., 2010). A randomized, prospective clinical study brings data on the clinical application of collagen biomaterial associated with calcium phosphate bioceramic in the treatment of long bone fractures, having the use of autogenous grafts as a form of control. The fractures were stabilized with metal implants suitable for each case. There was a follow-up on 213 patients, and a total of 249 fractures. According to the authors the collagen-based composite had the same performance observed for the autograft as regards fracture union rate and functional measurements, and is a possible treatment alternative (Chapman et al., 1997).

Associations of biomaterials, initially used as *scaffolds* for the conduction of bone formation, with tissue regeneration *inductor proteins,* are not just a promise for regenerative medicine, but are already taking shape as potential and usual clinical applications. At the same time associations with *osteoprogenitor cells* or bone marrow aspirate are also consolidating for the refinement of the functions of these scaffolds.

5. Stem cells and bone regeneration

Cells are the essential elements during repair and regeneration, with *stem cells* playing an important role in this process, as already mentioned previously. Nowadays there are a growing number of studies seeking therapeutic strategies and applications using stem cells to minimize clinical problems caused by injury or diseases in the bone tissue (Meyer, et al., 2006; Charbord, 2010), which present increasing demand, considering the demographic growth of the population and the rise in the number of elderly citizens, where the frequency of diseases in the musculoskeletal system is higher (De Peppo et al., 2010).

6. Stem cells and their application to regeneration and to bioengineering of bone tissue

Stem cells correspond to a group of *undifferentiated cells* with the capacity for unlimited self-renewal, as they are capable of successive divisions throughout the entire lifetime of the organism. Moreover, these cells, once stimulated by specific signals and under ideal conditions, will be able to differentiate into cell types with specialized forms and functions and that will maintain the homeostasis of the body. Therefore the proliferative capacity associated with the potential to differentiate into different specific cell types, confer immense potential for application to different areas of biomedicine including gene therapy and tissue engineering on stem cells (Kirschstein & Skirboll, 2001).

Thus, the success of tissue engineering depends on the use of the appropriate cells, on the ability to predict the cell response and on culture techniques for proliferation and differentiation into specific cell types. Nowadays tissue engineering applications are allowing, among others, the use of cells from the actual patient (autologous cells), from donors (allogenic), from different species (xenogeneic), from immortalized lineages (both allogeneic and xenogeneic) and fetal and adult stem cells (Parenteau, 2002); which can be cultivated on molds of biocompatible materials, and subsequently implanted to the injured tissue or inoculated directly or onto the biomaterials at the implant sites. This methodology opens vast perspectives for application in the medical area, allowing the performance of graft implants in injured tissues leading to a greater benefit to the patient, with the initial use of a small number of cells, which will be expanded *in vitro* by means of culture techniques, and also due to the fact that it will be possible to either minimize or avoid immunological problems such as rejection of non-autogenous transplants (Calvert et al.,

2000; Temenoff & Mikos, 2000a,b). To this effect, several strategies are being applied to improve the efficiency of tissue engineering such as growth factors and recombinant differentiation factors, use of autologous cells, gene therapy through the incorporation of vectors and genetic engineering of cells (Satija et al., 2007).

7. Embryonic stem cells

The self-renewal capacity of *human embryonic stem cells (ESCs)* over prolonged periods and their ability to differentiate into different tissues from the three embryonic layers, were characterized by Thomson and collaborators (Thomson et al., 1998). These oocyte-derived cells fertilized in the morule phase or derived from the inner cell mass of embryos in the blastula phase, are able to divide in an unlimited manner, keeping their original characteristics and genetic information, besides being pluripotent, that is, they can differentiate into practically all cell types, derived from the three embryonic germ layers, mesoderm, ectoderm and endoderm (Figure 1) (Doetschman et al., 1985; Smith, 2001).

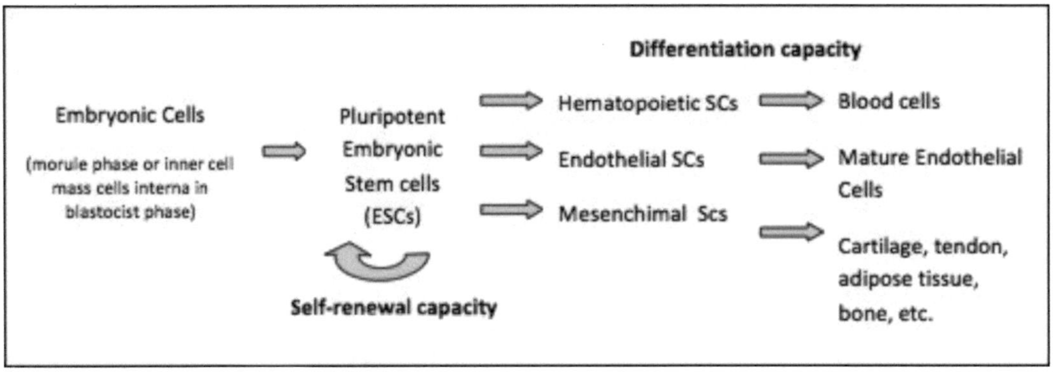

Fig. 1. Diagram showing pluripotent property of embryonic stem cells and their capacity to originate cell types from the three embryonic germ layers.

Although there is immense potential for the use of embryonic stem cells, due to their pluripotency, in practical terms their use is still very limited due to problems including cell regulation, immunological incompatibility, and possible development of neoplasias upon their administration (Passier & Mummery, 2003). These complications are also accentuated by ethical and religious issues and government regulations vis-à-vis the use of human embryonic cells in research (Zuk et al., 2001; Lee et al., 2003, Undale et al. 2009). Such factors led scientists to seek options with greater application potential, such as adult stem cells.

8. Adult stem cells

Although they decrease with age, *adult stem cells* are present in a wide variety of tissues throughout the lifetime of an individual. These cells, like the ESCs, also have the capacity for unlimited self-renewal and the potential to differentiate into cell types with specific morphologic and functional characteristics. This differentiation process generally involves intermediate cell types called precursor or progenitor cells that, although with a reduced self-renewal capacity, can split up to produce specific cell types (Robey, 2000; Gamradt & Lieberman, 2004). Accordingly, adult stem cells are being identified by different methods and there is a growing number of tissues and organs identified as carriers of the so-called mesenchymal stem cells (MSCs), including the bone marrow, peripheral blood, brain, spinal cord, dental pulp, blood vessels, skeletal muscle, epithelium of the skin and of the digestive system, cornea, retina, liver, pancreas and others, whereas the umbilical cord and the placenta are also carriers of cells similar to the mesenchymal stem cells. In spite of the fact that they have similar characteristics, MSCs of different origins present varied differentiation and gene expression potentials. Bone marrow is known to present considerable potential for obtaining stem cells and they have been studied with clinical and therapeutic objectives for fractures with substantial loss of bone mass and metabolic diseases involving the bone tissue.

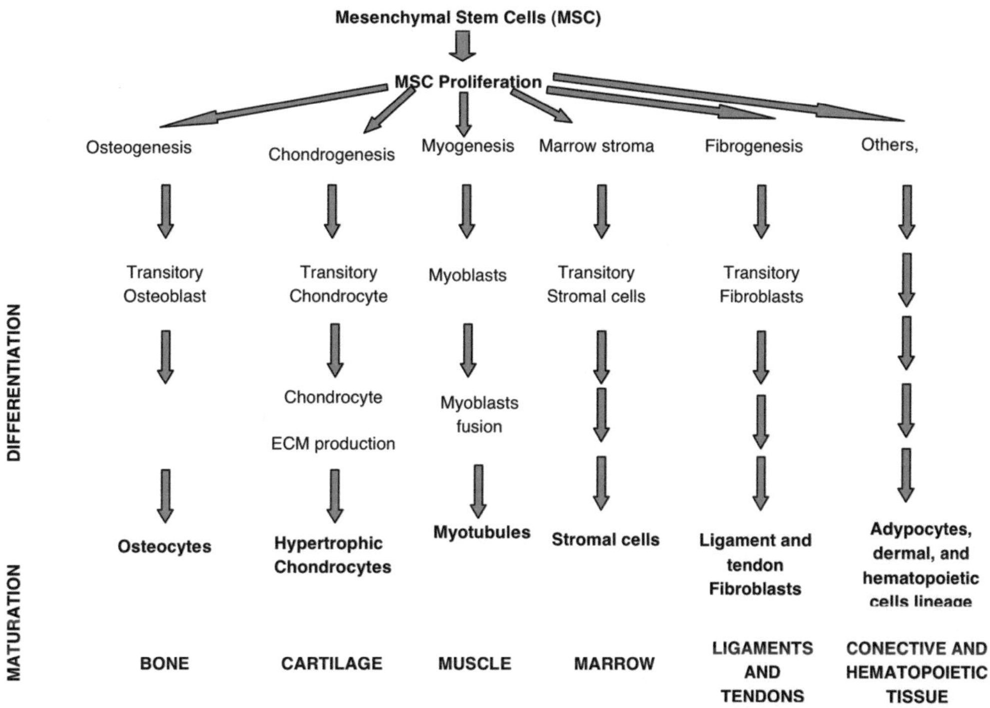

Fig. 2. Summary diagram showing the capacity of mesenchymal stem cells (MSCs) to differentiate into bone, cartilage, muscle, tendons/ligaments and other tissues. Each stage of this differentiation and maturation process involves the control and the interaction of growth factors and cytokines (Caplan, 2010).

9. Bone marrow stem cells

The pioneer studies that evidenced the separation of a population of cells with the capacity to differentiate into a variety of cell types including osteoblasts, chondrocytes, adipocytes and hematopoietic cells, were carried out by Friedenstein *et al*, in the sixties (Friedenstein et. al., 1968). Their studies demonstrated the existence of precursor *mesenchymal cells*, with the potential to differentiate into osteoblasts and fibrous tissue (Figure 2) (Charbord, 2010; Hidalgo-Bastida et al., 2010).

These stem cells in the muridae are frequently obtained from femoral or tibial flushing, obtaining the bone marrow mononuclear cells (BMMNCs) that are isolated by density gradient centrifugation then cultivated *in vitro*. On the other hand, human cells are aspirated from the iliac crest and cultivated directly, since gradient centrifugation techniques have not been seen to increase the separation efficiency and frequently present contamination with hematopoietic cells. This methodology is used to obtain populations that undergo a cloning process, and are characterized by the presence of both positive markers (Stro-1, CD29, CD73, CD90, CD105, CD166 and CD44) and negative markers (CD43, CD45, CD14, CD11b, CD19, CD79a and HLA-DR). The positive expression for Stro-1, identifies cells with osteogenic potential and expression of the three markers for osteoblast differentiation: alkaline phosphatase, 1,25-dihydroxy-vitamin D that has induction dependent on the specific bone protein: osteocalcin and hydroxyapatite production (mineralized matrix). Recent studies have pointed to other markers present in mesenchymal stem cells of the bone marrow that present osteogenic potential (Undale et al., 2009).

The molecular regulatory mechanism involved in the MSC differentiation control process has been extensively studied *in vitro*, whereas *in vivo* control is little known due to the difficulties inherent in the study process. However, the properties of the MSCs *in vivo* and *in vitro* vary according to the method of removal of these cells from their natural environment and the use of chemical and physical factors to keep them in culture, which can lead to alterations in their characteristics. Heterogeneity and diversity of types of MSCs and their ability to undergo phenotypic rearrangements in culture, modifies the expression of markers and hinders the comparison of data or renders it unfeasible in some situations (Augello & De Bari, 2010).

In vitro, the classical methodology to induce osteogenic differentiation in human MSCs consists of incubation with bovine fetal serum, in a medium supplemented with ascorbic acid, β-glycerophosphate and dexamethasone, which leads to the increase of alkaline phosphatase and calcium deposition (Jaiswal et al, 1997; Pittenger et al. 1999).

On the other hand, *in vivo*, the information obtained from studies of embryonic development indicates that different signaling routes and transcription factors may play a critical role in the differentiation of MSCs. Different molecules have been described in the regulation of MSC differentiation, including Wnt and the TGF-β superfamily.

Wnt proteins, coded by a family of 19 genes in humans and in mice, are involved in cell proliferation, differentiation and apoptosis. They act directly on MSCs, and are crucial for embryonic development and regeneration of different tissues in adults, including bone. Etheridge and collaborators (2004) demonstrated that MSCs express a series of ligands including Wnt2, Wnt4, Wnt5a, Wnt11, and Wnt16, and different Wnt receptors, FZD2, 3, 4, 5, and 6, as well as several co-receptors and inhibitors.

Several studies have shown that osteogenic differentiation *in vitro* is upregulated by some molecules related to the Wnt family and downregulated by others. For example, the administration of exogenous Wnt3 leads to osteogenesis repression. The TGF–β superfamily represents a set of growth factors and morphogens that play a role in skeletogenesis and in postnatal skeletal homeostasis. The TGF–β superfamily of ligands includes BMPs, growth and differentiation factors (GDFs), anti-mullerian hormone (AMH), activin, nodal, and TGF–β (Piek et al., 1999; Derynck & Miyazono, 2008).

The expression of growth factors from the TGF–β family is crucial for bone repair in adults and has been described during the embryonic phase as essential for the development of cartilage and bones. TGF–β1 promotes the specific gene expression, initializing chain events with the participation of the SMAD proteins, which lead to the process of chondrogenesis and differentiation of the MSCs (Tuli et al., 2003).

BMPs are important morphogens involved in the regulation of chondrogenesis and osteogenesis during normal embryonic development (Hogan, 1996). The effects of BMPs on MSCs has been investigated, demonstrating that the culture of MSCs in the presence of BMP2 increases alkaline phosphatase activity and osteocalcin expression, both indicators of osteoblast differentiation, whereas this effect is intensified in the presence of dexamethasone. Other factors are also known to influence the differentiation of MSCs, interacting at different levels with the metabolism of the Wnts and/or TGF–β/BMP. One of these factors is FGF-2 (fibroblast growth factor - 2), which promotes cell proliferation and maintains the populations of MSCs undifferentiated for prolonged periods of time (Martin et al., 1997).

10. Applications and clinical potential

Different characteristics of MSCs, including their availability, potential for autologous use and absence of immunological rejection, make them very promising for clinical and therapeutic applications, especially in fractures with significant bone loss and metabolic diseases (Caplan, 2010). In spite of the major advances that have occurred in orthopedic surgery, fractures that involve considerable bone loss and non-union still represent a very important clinical problem. During the normal regeneration of a fracture, as seen previously, undifferentiated MSCs, with the assistance of BMPs and regulatory cytokines, proliferate and differentiate into chondrocytes and osteoblasts, which will form bone tissue reconstituting the lesion. Although related to the site where they occur, around 5 to 20% or more of fractures present failure in regeneration and consolidation (Kimelman et al., 2007; Undale, et al, 2009). Experiments on animal models using autologous MSCs and different scaffolds have resulted in bone regeneration (Arinzeh et al, 2003; Bruder et al. 1998a, b; Kon et al, 2000; Petite et al, 2000). Clinical studies on humans with the use of MSCs aspirated from the iliac crest and subsequently expanded in cultures on different biomaterials (Quarto et al., 2001; Marcacci et al, 2007), or percutaneously injected (Hernigou et al., 2005), have also been conducted, indicating clinical success positively correlated with greater capacity for *in vitro* formation of colonies and concentration of injected MSCs. Clinical applications in humans have also been described in patients with metabolic diseases of the bone tissue such as osteogenesis imperfecta and hypophosphatasia. Cultures of allogeneic MSCs and intravenous administration have mainly been used in these diseases, demonstrating the ability of these stem cells to stimulate bone mineralization and regeneration (Undale et al., 2009)

11. Mesenchymal stem cells of the adipose tissue

Stem cells play a crucial role for the body's homeostasis, as they maintain the functional state of the tissues and also replace cells killed by injury or disease. These cells are very rare in the adult (Kirschstein & Skirboll, 2001). For example, it is estimated that in the bone marrow only one among ten to fifteen thousand cells is a hematopoietic source cell (Weissman, 2000).

Although the bone marrow is the place where the presence and the differentiation process of MSCs is currently best known and characterized, they are also found in other places (Gamradt & Lieberman, 2004).

Studies have indicated that MSCs are also found in animal (Lee et al., 2002) and human (Zuk et al., 2002) *adipose tissue*, and can be obtained by the lipoaspiration process. They are frequently referred to in literature as PLA (processed lipoaspirative) or ADAS (adipose-derived adult stem cells). Different studies have evidenced that mesenchymal stem cells obtained from the adipose tissue, when stimulated by different factors, can also differentiate into adipose cells (Halbleib et al., 2003), osteoblasts (Hicok et al., 2004), chondroblasts, myocytes and neural cells, which means that they draw great interest for applications in regenerative medicine and in tissue engineering (Barry & Murphy, 2004; Ogawa et al., 2004a,b). Since they are easily and abundantly obtained by lipoaspirative process, which is therefore less invasive, using local anesthesia, mesenchymal cells from the adipose tissue offer advantages over the bone marrow (Mizuno & Hyakusoku, 2003; Macleod et al, 2010). In the latter, the obtainment of mesenchymal cells is generally performed using aspiration and flushing of the upper part of the iliac crest, involving a process that is extremely painful for the patient, with the risk of a general or spinal anesthesia, usually implying morbidity of the donor site, resulting in a small number of functional cells. Now the obtainment of mesenchymal cells from the adipose tissue has presented more homogeneous populations with normal karyotype, and can be kept *in vitro* for long periods, with constancy in the cell doubling time and low levels of senescence (Zuk et al., 2002; Aust et al., 2004). Comparative studies between the mesenchymal cells of the bone marrow and of adipose tissue have shown that they both exhibit similarity in their ability to differentiate into adipose cells, from the bone, cartilaginous and muscle tissues; share similarities in the kinetics of growth and senescence, with the capacity for gene transduction and also among the cell surface markers (Mizuno & Hyakusoku, 2003; De Urgate et al., 2003a,b; Mosna et al., 2010)

Thus the autologous mesenchymal stem cells of the adipose tissue are also being used in the construction of three-dimensional scaffolds and applied to patients with severe problems of bone mass loss (Gamradt & Lieberman, 2004).

12. Stem cells and gene therapy: prospects of future applications

Genetic engineering of adult stem cells with genes presenting osteogenic potential has gained considerable emphasis in the repair of fractures and bone tissue formation. Studies have indicated that these genetically modified cells can produce autocrine and paracrine effects on the stem cells present in the actual patient, leading to a greater response in the osteogenic effect. These strategies involve both the use of viral and non-viral vectors, presenting genes that code different BMPs, as well as genetically modified cells containing these implanted transgenes. The advance of these studies may be essential for the future prospects of clinical use of stem cells for bone regeneration (Kimelman et al, 2007), bringing more efficient solutions in the field of orthopedics.

The proliferation capacity of MCSs is a measure of the number of cell divisions that can occur *in vitro* after the culture has been started. Many studies suggest that MCSs have doubling capacity of up to 50 times; after this period the culture is characterized by alteration of a series of cellular characteristics and properties, followed by senescence or even cell transformation. The senescence process is characterized by modifications to morphology and increase of cell volume, reduction in surface marker expression and decrease in differentiation potential. Several molecular mechanisms have already been identified in the senescence process, including DNA injury, accumulation of the cyclin-dependent kinase inhibitor, oxidative stress, telomeric modifications, action of epigenetic factors, and others (Wagner et al, 2010).

Accordingly, the safe and efficient clinical application of stem cells to bone tissue regeneration depends on the elucidation of mechanisms associated with senescence. Moreover, it is essential to understand the mechanisms of action and interaction with other cell types, with different biomaterials, soluble factors, extracellular matrix components (Hidalgo-Bastida et al, 2010) and biochemical and mechanical agents present in the micro-environment *in vitro* and *in vivo*, as well as to keep the proliferation of stem cells restricted to the implanted site and to know the gene control mechanism for safe induction of the desired functions (Gronthos et al., 2000; Discher et al., 2009). The identification of growth factors and the signaling mechanisms involved in the actual control of stem cell renewal and differentiation will allow the design of strategies to block senescence and to safely drive cellular differentiation (Satija et al, 2007).

13. References

Andriano, K.P.; Chandrashekar, B.; McEnery, K.; Dunn, R.L.; Moyer, K.; Balliu, C.M.; Holland, K.M.; Garrett, S.; Huffer, W.E. (2000) Preliminary in vivo studies on the osteogenic potential of bone morphogenetic proteins delivered from an absorbable puttylike polymer matrix. J. Biomed. Mater. Res., Vol. 53, No.1, pp.36-43.

Arinzeh, T.L.; Peter, S.J.; Archambault, M.P.; van den Bos, C.; Gordon, S.; Kraus, K.; Smith, A.; Kadiyala, S. (2003) Allogeneic mesenchymal stem cells regenerate bone in a critical-sized canine segmental defect. J. Bone Joint Surg. Am. Vol., Vol. 85-A, No. 10, pp.1927-1935.

Arrington, E.D.; Smith, W.J.; Chambers, H.G.; Bucknell, A.L.; Davino, N.A. (1996) Complications of iliac crest bone graft harvesting. Clin. Orthop. Rel. Res., Vol. 329, pp.300-309.

Asano, S.; Kaneda, K.; Satoh, S.; Abumi, K.; Hashimoto, T.; Fujiya, M. (1994) Reconstruction of an iliac crest defect with a bioactive ceramic prosthesis. Eur. Spine J., Vol.3, No. 1, pp.39-44.

Augello, A.; De Bari, C. (2010) The regulation of differentiation in mesenchymal stem cells. Human Gene Therapy, Vol. 21, No. 10, pp.1226–1238.

Aust, L.; Devlin, B.; Foster, S.J.; Halvorsen, Y.D.; Hicok, K.; Du Laney, T.; Sen, A.; Willingmyre, G.D.; Gimble, J.M. (2004) Yield of human adipose-derived adult stem cells from liposuction aspirates. Cytotherapy, Vol.6, No. 1, pp.7-14.

Balcik, C.; Tokdemir, T.; Senköylü, A.; Koç, N.; Timuçin, M.; Akin, S, Korkusuz P.; Korkusuz, F. (2007) Early weight bearing of porous HA/TCP (60/40) ceramics in vivo: a longitudinal study in a segmental bone defect model of rabbit. Acta Biomater., Vol. 3, No. 6, pp.985-996.

Barry, F.P.; Murphy, J.M. (2004) Mesenchymal stem cells: clinical applications and biological characterization. Int. J. Biochem. Cell Biol., Vol. 36, No. 4, pp.568-584.

Bauer, T.W.; Muschler, G.F. (2000) Bone graft materials. An overview of the basic science. Clin. Orthop. Relat. Res., Vol. 371, pp.10-27.

Bibbo, C.; Patel, D.V. (2006) The effect of demineralized bone matrix-calcium sulfate with vancomycin on calcaneal fracture healing and infection rates: A prospective study. Foot Ankle Int., Vol. 27, No. 7, pp.487–493.

Bostrom, M.P. (1998) Expression of bone morphogenetic proteins in fracture healing. Clin. Orthop., Vol. 355, pp. S116–S123.

Boyce, T.; Edwards, J.; Scarborough, N. (1999) Allograft bone: the influence of processing on safety and performance. Orthop. Clin. North Am., Vol. 30, No. 4, pp.571-81.

Bruder, S.P.; Kraus, K.H.; Goldberg, V.M.; Kadiyala, S. (1998) The effect of implants loaded with autologous mesenchymal stem cells on the healing of canine segmental bone defects. J. Bone Joint. Surg. Am., Vol. 80, No. 7, pp.985-996.

Bruder, S.P.; Kraus, K.H.; Goldberg, V.M.; Kadiyala, S. (1998) The effect of implants loaded with autologous mesenchymal stem cells on the healing of canine segmental bone defects. J. Bone Joint Surg. Am., Vol. 80, No. 7, pp.985-996.

Bruder, S.P.; Fink, D.J.; Caplan, A.I. (1994) Mesenchymal stem cells in bone development, bone repair, and skeletal regeneration therapy. J. Cell Biol., Vol. 56, No. 3, pp.283–294.

Bucholz, R.W.; Carlton, A.; Holmes, R. (1989) Interporous hydroxyapatite as a bone graft substitute in tibial plateau fractures. Clin. Orthop. Relat. Res., Vol. 240, pp.53-62.

Burg, K.J.L.; Porter, S.; Kellam, J.L. (2000) Biomaterial developments for bone tissue engineering. Biomaterials, Vol. 21, No. 23, pp.2347-2359.

Canalis, E. (1983) The hormonal and local regulation of bone formation. Endocr. Rev., Vol. 4, No. 1, pp.62-77.

Caplan, A. I. (2010) Mesenchymal stem cells: cell-based reconstructive therapy in orthopedics. In: Advances in tissue engineering: stem cells. Vol 2: Stem Cells. Peter C. Johnson & Antonios G. Mikos, eds., pp. 32-45. Mary Ann Liebert, Inc., New Rochelle, NY.

Cavassini, M.M.; Moraes, J.R.E.; Padilha, F.J.G. (2001) Função osteoindutora de fragmentos ósseos conseervados em glicerina a 98%: estudo experimental em ratos. Cienc. Rural, Vol. 31, No. 3, p.445-448.

Chang, W.; Colangeli, M.; Colangeli, S.; Di Bella, C.; Gozzi, E.; Donati, D. (2007) Adult osteomyelitis: Debridement versus debridement plus Osteoset T pellets. Acta Orthop .Belg., Vol. 73, No. 2, pp.238–243.

Chapman, M.W.; Bucholz, R.; Cornell, C. (1997) Treatment of acute fractures with a collagen-calcium phosphate graft material. A randomized clinical trial. J Bone Joint Surg. Am., Vol. 79, No. 4, pp.495-502.

Charbord, P. (2010) Bone marrow mesenchymal stem cells: historical overview and concepts. Human Gene Therapy, Vol.21, No. 9, pp.1045–1056.

Coetzee, A.S. (1980) Regeneration of bone in the presence of calcium sulfate. Arch. Otolaryngol., Vol.106, No. 7, pp.405–409.

Cunha, M.R. Implantes tridimensionais de colágeno polianiônico em falhas ósseas produzidas no fêmur de ratas ovariectomizadas. Tese de doutorado, IB-Unicamp, Campinas, 2006.

Cunha, M.R.; Santos Jr., A.R.; Genari, S.C. (2005) Cultura de osteoblastos sobre membranas de colágeno polianiônico: avaliação preliminar do potencial de indução da formação de tecido ósseo visando reparação tecidual. Bol. Med. Vet., Vol. 1, No. 1, pp.73-85.

Cypher, T.G.; Grossman, J.P. (1996) Biological principles of bone graft healing. J. Foot Ankle Surg., Vol. 35, No. 5, pp.413-417.

De Peppo, G.M.; Sjovall, P.; Lennera, M.; Strehl, R.; Hyllner, J.; Thomsen, P.; Karlsson, C. (2010) Osteogenic potential of human mesenchymal stem cells and human embryonic stem cell-derived mesodermal progenitors: a tissue engineering perspective. Tissue Eng. Part A, Vol. 16, No 11, pp.3413-3423.

De Urgate, D.A.; Alfonso, Z.; Zuk, P.A.; Elbarbary, A.; Zhu, M.; Ashijan, P.; Benhaim, P.; Hedrick, M.H., Fraser, J.K. (2003a) Differential expression of stem cells mobilization-associated molecules on multilineage cells from adipose tissue and bone marrow. Immunol. Lett., Vol. 89, No. 2-3, pp.267-270.

De Urgate, D.A.; Morizono, K.; Elbarbary, A.; Alfonso, Z.; Zuk, P.A.; Zhu, M.; Dragoo, J.L.; Ashijan, P.; Thomas, B.; Benhaim, P.; Chen, I., Hedrick, M.H. (2003b) Comparison of multi-lineage cells from human adipose tissue and bone marrow. Cells tissue organs, Vol. 174, No. 3, pp.101-109.

Dell, P.C.; Burchardt, H.; Glowczewskie Jr, F.P. (1985) A roentgenographic, biomechanical, and histological evaluation of vascularized and non-vascularized segmental fibular canine autografts. J. Bone Joint Surg. Am., Vol. 67, No. 1, pp.105-112.

Derynck, R.; Miyazono, K. (2008) TGF-b and the TGF-b family. In: The TGF-β family, R. Derynck and K. Miyazono, eds., pp. 29-44, Cold Spring, NY.

Discher, D.E.; Mooney, D.J.; Zandstra, P.W. (2009) Growth factors, matrices, and forces combine and control stem cells. Science, Vol. 324, No. 5935, pp.1673-1677.

Docquier, P.L.; Delloye, C. (2005) Treatment of aneurysmal bone cysts by introduction of demineralized bone and autogenous bone marrow. J. Bone Joint Surg. Am., Vol. 87, No. 10, pp.2253-2258.

Doetschman, T.; Eistetter, H.; Katz, M.; Schmit, W.; Kemier, R. (1985) The in vitro development of blastocyst-derived embryonic stem cell lines: formation of visceral yolk sac, blood islands and myocardium. J. Embriol. Exp. Morph., Vol. 87, pp.27-45.

Dong, J.; Uemura, T.; Shirasaki, Y.; Tateishi, T. (2002) Promotion of bone formation using highly pure porous beta-TCP combined with bone marrow-derived osteoprogenitor cells. Biomaterials., Vol. 23, No. 23, pp.4493-4502.

Doornberg, J.N.; Marti, R.K. (2009) Osteotomy and autograft lengthening for intra-articular malunion of the proximal ulna: a case report. Case Report Med., Article: 647126, pp.1-3.

Etheridge, S.L.; Spencer, G.J.; Heath, D.J.; Genever, P.G. (2004). Expression profiling and functional analysis of Wnt signaling mechanisms in mesenchymal stem cells. Stem Cells, Vol. 22, No. 5, pp.849–860.

Fawcett, D.W. (1986) Bone. In: A Textbook of Histology, D.W. Fawcett, ed., pp. 199-238, W.B. Saunders, Igaku-Shoin.

Finkemeier, C.G. (2002) Bone-grafting and bone-graft substitutes. J. Bone Joint Surg. Am., Vol. 84A, No. 3, pp.454-464.

Gaasbeek, R.D.; Toonen, H.G.; van Heerwaarden, R.J.; Buma, P. (2005) Mechanism of bone incorporation of beta-TCP bone substitute in open wedge tibial osteotomy in patients. Biomaterials, Vol. 26, No. 33, pp.6713-6719.

Gamradt, S.C.; Lieberman, J.R. (2004) Genetic modification of stem cells to enhance bone repair. Annals Biom. Engineering, Vol. 32, No. 1, pp.136-147.

Gauthier, O.; Bouler, J.M.; Aguado, E.; Pilet, P.; Daculsi, G. (1998) Macroporous biphasic calcium phosphate ceramics: influence of macropore diameter and macroporosity percentage on bone ingrowth. Biomaterials, Vol. 19, No. 1-3, pp.133-139.

Giannoudis, P.V.; Dinopoulos, H.; Tsiridis, E. (2005) Bone substitutes: an update. Injury. Int. J. Care Injured; Vol. 36, Suppl. 3, pp.S20-S27.

Gogia, J.S.; Meehan, J.P.; Di Cesare, P.E.; Jamali, A.A. (2009) Local antibiotic therapy in osteomyelitis. Semin., Plast. Surg., Vol.23, No. 2, pp.100-107.

Goldberg, V. M.; Stevenson, S. (1992) Biology of autografts and allografts, In: Allografts in orthopaedic practice, A. A. Czitrom and A. E. Gorss, eds, pp. 1-13, Williams & Wilkins, Baltimore.

Groeneveld, E.H.; Burger, E.H. (2000) Bone morphogenetic proteins in human bone regeneration. Eur. J. Endocrinol., Vol. 142, No.1, pp.9-21.

Gronthos, G.; Akitoye, S.O.; Wang, C.Y.; Shi, S. (2006) Bone marrow stromal stem cells for tissue engineering. Periodontology, Vol. 41, No. 41, pp.188–195.

Gupta, M.C.; Theerajunyaporn, T.; Maitra, S.; Schmidt, M.B.; Holy, C.E.; Kadiyala, S.; Bruder, S.P. (2007) Efficacy of mesenchymal stem cell enriched grafts in an ovine posterolateral lumbar spine model. Spine, Vol. 32, No.7, pp.720-726.

Habibovic, P.; de Groot, K. (2007) Osteoinductive biomaterials – properties and relevance in bone repair. J. Tissue Eng. Regen. Med., Vol. 1, No. 1, pp.25-32.

Hadjipavlou, A.G.; Simmons, J.W.; Tzermiadianos, M.N.; Katonis, P.G.; Simmons, D.J. (2001) Plaster of Paris as bone substitute in spinal surgery. Eur. Spine J., Vol. 10, Suppl. 2, pp.S189-S196.

Halbleib, M.; Skurk, T.; De Luca, C.; Von Heimburg, D.; Hauner, H. (2003) Tissue engineering of white adipose tissue using hyaluronic acid-based scaffolds. I: in vitro differentiation of human adipocyte precursor cells on scaffolds. Biomaterials, Vol. 24, No. 18, pp.3125-3132.

Halvorsen, Y-D.C.; Franklin, D.; Bond, A.L.; Hitt, D.; Auchter, C.; Boskey, A.L.; Paschalis, E.P.; Wilkison, W.O.; Gimble, J.M. (2001) Extracellular matrix mineralization and osteoblast gene expression by human adipose tissue-derived stromal cells. Tissue Eng., Vol. 7, No. 6, pp.729–741.

Hench, L.L.; Splinter, R.J.; Allen, W.C.; Greenlee, T.K. Jr. (1971) Bonding mechanisms at the interface of ceramic prosthetic materials. J. Biomed. Mater. Res. Symp., Vol. 2, pp.117-141.

Hench, L.L. (1998) Biomaterials: a forecast for the future. Biomaterials, Vol. 19, No. 16, pp.1419-1423.

Henman, P.; Finlayson, D. (2000) Ordering allograft by weight: suggestions for the efficient use of frozen bone-graft for impaction grafting. J. Arthroplasty, Vol. 15, No. 3, pp.368-371.

Heppenstall, R.B. (1980) Fracture Healing. In: Fracture Treatment and Healing, R.B. Heppenstall, ed., pp. 35-64, WB Saunders, Philadelphia.

Hernigou, P.; Poignard, A.; Beaujean, F.; Rouard, H. (2005) Percutaneous autologous bone-marrow grafting for nonunions: influence of the number and concentration of progenitor cells. J. Bone Joint Surg. Am., Vol. 87, No. 7, pp.1430-1437.

Hicok, K.C.; Du Laney, T.V.; Zhou, Y.S.; Halvorsen, Y.D.; Hitt, D.C.; Cooper, L.F. Gimble, J.M. (2004) Human adipose-derived adult stem cells produce osteoid in vivo. Tissue Eng., Vol. 10, No. 3-4, pp.371-380.

Hidalgo-Bastida, L.A.; Cartmell, S.H. (2010) Mesenchymal stem cells, osteoblasts and extracellular matrix proteins: enhancing cell adhesion and differentiation for bone tissue engineering. Tissue Eng., Part B, Vol. 16, No. 4, pp.405-412.

Hogan, B.L. (1996). Bone morphogenetic proteins: Multifunctional regulators of vertebrate development. Genes Dev., Vol. 10, No. 13, pp.1580–1594.

Hutmacher, D.W.; Schantz, J.T.; Lam, C.X.; Tan, K.C.; Lim, T.C. (2007) State of the art and future directions of scaffold-based bone engineering from a biomaterials perspective. J. Tissue Eng. Reg. Med., Vol. 1, No. 4, pp.245–260.

Ishaug-Riley, S.L.; Crane-Kruger, G.M.; Yaszemski, M.J.; Mikos, A.G. (1998) Three-dimensional culture of rat calvarial osteoblasts in porous biodegradable polymers. Biomaterials, Vol. 19, No. 15, pp.1405-1412.

Jaiswal, N.; Haynesworth, S.E.; Caplan, A.I.; Bruder, S.P. (1997). Osteogenic differentiation of purified, ultureexpanded human mesenchymal stem cells in vitro. J. Cel. Biochem., Vol. 64, No. 2, pp.295–312.

Jayasuriya, A.C.; Kibbe, S. (2010) Rapid biomineralization of chitosan microparticles to apply in bone regeneration. J. Mater. Sci. Mater. Med., Vol. 21, No. 2, pp.393-398.

Junqueira, L.C.; Carneiro, J. (2008) Histologia Básica. 11ª ed., Guanabara Koogan, Rio de Janeiro.

Katz, J.M.; Nataraj, C.; Jaw, R.; Deigl, E.; Bursac, P. (2009) Demineralized bone matrix as an osteoinductive biomaterial and in vitro predictors of its biological potential. J. Biomed. Mater. Res. B Appl. Biomater., Vol. 89, No. 1, pp.127-134.

Keating, J.F.; McQueen, M.M. (2001) Substitutes for autologous bone graft in orthopaedic trauma. J. Bone Jt. Surg. Br., Vol. .83, pp.3-8.

Kelly, C.M.; Wilkins, R.M.; Gitelis, S.; Hartjen, C.; Watson, J.T.; Kim, P.T. (2001) The use of a surgical grade calcium sulfate as a bone graft substitute: Results of a multicenter trial. Clin. Orthop. Relat. Res., Vol. 382, pp.42-50.

Kessel, R.G. (2001) Histologia Médica Básica - A Biologia das Células, Tecidos e Órgãos, Guanabara Koogan, Rio de Janeiro.

Khan, S.N.; Cammisa Jr, F.P.; Sandhu, H.S.; Diwan, A.D.; Girardi, F.P.; Lane, J.M. (2005) The biology of bone grafting. J. Am. Acad. Orthop. Surg., Vol. 13, No. 1, pp.77-86.

Kierszenbaum, A.L. (2004) Histologia e Biologia Celular - Uma Introdução à Patologia, Elsevier, Rio de Janeiro.

Kimelman, N.; Pelled, G, Helm, G.A.; Huard, J.; Schwarz, E.M.; Gazit, D. (2007) Review: Gene and stem cell–based therapeutics for bone regeneration and repair. Tissue Eng., Vol. 13, No. 6, pp.1135-1150.

Kimura, M.; Zhao, M.; Zellin, G.; Linde, A. (2000) Bone-inductive efficacy of recombinant human bone morphogenetic protein- 2 expressed in Escherichia coli: an experimental study in rat mandibular defects. Scand. J. Plast. Reconstr. Surg. Hand Surg., Vol. 34, No. 4, pp.289-299.

Kirschstein, R.; Skirboll, L.R. (2001) The Stem Cell. In: Stem Cells: Scientific Progress and Future Research Directions. Report Prepared by the National Institute of Health, pp. 1-4, Terese Winslow ed. , Virginia.

Kon, E.; Muraglia, A.; Corsi, A.; Bianco P.; Marcacci, M.; Martin, I.; Boyde A.; Ruspantini I.; Chistolini, P.; Rocca M.; Giardino, R.; Cancedda, R.; Quarto, R. (2000) Autologous bone marrow stromal cells loaded onto porous hydroxyapatite ceramic accelerate bone repair in criticalsize defects of sheep long bones. J. Biomed. Mater. Res., Vol. 49, No. 3, pp.328-337.

Le Huec, J.C.; Lesprit, E.; Delavigne, C.; Clement, D.; Chauveaux, D.; Le Rebeller, A. (1997) Tri-calcium phosphate ceramics and allografts as bone substitutes for spinal fusion in idiopathic scoliosis as bone substitutes for spinal fusion in idiopathic scoliosis: comparative clinical results at four years. Acta Orthop. Belg., Vol. 63, No. 3, pp.202-211.

Lee, G.H.; Khoury, J.G.; Bell, J.E.; Buckwalter, J.A. (2002) Adverse reactions to OsteoSet bone graft substitute, the incidence in a consecutive series. Iowa Orthop. J., Vol. 22, pp.35-38.

Lee, C.H.; Singla, A.; Lee, Y. (2001) Biomedical applications of collagen. Int. J. Pharm., Vol. 221, No. 1-2, pp.1-22.

Lee, J.A.; Parret, B.M.; Conejero, J.A.; Laser, J.; Chen, J.; Kogon, A.J.; Nanda, D.; Grant, R., Breitbart, A.S. (2003) Biological Alchemy: Enginnering bone and fat from fat-derived stem cells. Ann. Pastic Surgery, Vol. 50, No. 6, pp.610-617.

Li, J.; Wang, Z.; Pei, G.X.; Guo, Z. (2011) Biological reconstruction using massive bone allograft with intramedullary vascularized fibular flap after intercalary resection of humeral malignancy. J. Surgical Oncol. Vol 104, No.3, pp. 244-249

Lieberman, J.R.; Daluiski, A.; Einhorn, T.A. (2002) The role of growth factors in the repair of bone. Biology and clinical applications. J. Bone Joint Surg. Am., Vol. 84-A, pp.1032-1044.

Macload, L.C.; James, A.W.; Carre, A.L.; Longaker, M.T.; Lorenz, H.P. (2010) A Review of the Therapeutic Potential of Adipose-Derived Stem Cells. Advances In Wound Care, Mary Ann Liebert, Inc., New Rochelle, NY.

Mandracchia, V.J.; Nelson, S.C.; Barp, E.A. (2001) Current concepts of bone healing. Clin. Podiatr. Med. Surg., Vol. 18, pp.55-77.

Marcacci, M.; Kon, E.; Moukhachev, V.; Lavroukov A.; Kutepov, S.; Quarto, R.; Mastrogiacomo, M.; Cancedda, R. (2007) Stem cells associated with macroporous bioceramics for long bone repair: 6- to 7-year outcome of a pilot clinical study. Tissue Eng., Vol. 13, No. 6, pp.947-955.

Marino, J.T.; Ziran, B.H. (2010) Use of solid and cancellous autologous bone graft for fractures and nonunions. Orthop Clin North Am., Vol. 41, No. 1, pp.15-26.

Marsh, J.L. (2006) Principles of bone grafting: non-union, delayed union. Surgery, Vol. 24, No. 6, pp.207-210.

Martin, I.; Muraglia, A.; Campanile, G.; Cancedda, R.; Quarto, R. (1997). Fibroblast growth factor-2 supports ex vivo expansion and maintenance of osteogenic precursors from human bone marrow. Endocrinology, Vol. 138, pp.4456–4462.

Matsuo, A.; Chiba, H.; Takahashi, H.; Toyoda, J.; Abukawa, H. (2010) Clinical application of a custom-made bioresorbable raw particulate hydroxyapatite/poly-L-lactide mesh tray for mandibular reconstruction. Odontology, Vol. 98, No. 1, pp.85-88.

McKee, M.D.; Li-Bland, E.A.; Wild, L.M.; Schemitsch, E.H. (2010) A prospective, randomized clinical trial comparing an antibiotic-impregnated bioabsorbable bone substitute with standard antibiotic-impregnated cement beads in the treatment of chronic osteomyelitis and infected nonunion. J. Orthop. Trauma, Vol. 24, No. 8, pp.483-490.

Meyer, U.; Wiesman, H.; Berr, K.; Kübler, N.; Handschel, J. (2006) Cell-based bone reconstruction therapies - principles of clinical approaches. Inter. J. Oral & Maxillofacial Implant, Vol. 21, No. 6, pp.899-906.

Mizuno, H.; Hyakusoku, H. (2003) Mesengenic potential and future clinical perspective of human processed lipoaspirate cells. J Nippon Med Sch., Vol. 70, No. 4, pp.300-306.

Morone, M.A.; Boden, S.D. (1998) Experimental posterolateral lumbar spinal fusion with a demineralized bone matrix gel. Spine, Vol. 23, No. 2, pp.159-167.

Mosna, F.; Sensebe, L.; Krampera, M. (2010) Human bone marrow and adipose tissue mesenchymal stem cells: a user's guide. Stem Cells Dev., Vol. 19, No. 10, pp.1449-1470.

Mukherjee, D.P.; Tunkle, A.S.; Roberts, R.A.; Clavenna, A.; Rogers, S.; Smith, D. (2003) An animal evaluation of a paste of chitosan glutamate and hydroxyapatite as a synthetic bone graft material. J. Biomed. Mater. Res. B Appl. Biomater., Vol. 67, No. 1, pp.603-609.

Nakagawa, T.; Tagawa, T. (2000) Ultrastructural study of direct bone formation induced by BMPs-collagen complex implanted into an ectopic site. Oral Dis., Vol. 6, No. 3, pp.172-179.

Nandi, S.K.; Roy, S.; Mukherjee, P.; Kundu, B.; De, D.K.; Basu, D. (2010) Orthopaedic applications of bone graft & graft substitutes: a review. Indian J. Med. Res., Vol. 132, No. 1, pp.15-30.

Ogawa, R.; Mizuno, H.; Hyakusoku, H.; Watanabe, A.; Migita, M.; Shimada, T. (2004b) Chondrogenic and osteogenic differentiation of adipose-derived stem cells isolated from GFP transgenic mice. J. Nippon Med. Sch., Vol. 71, No. 4, pp.240-241.

Ogawa, R.; Mizuno, H.; Watanabe, A.; Migita, M.; Shimada, T.; Hyakusoku, H. (2004a). Osteogenic and chondrogenic differentiation by adipose-derived stem cells harvested from GFP transgenic mice. Biopchem. Biophys. Res. Comm., Vol. 313, No. 4, pp.871-877.

Oringer, R.J. (2002) Biological mediators for periodontal and bone regeneration. Compend. Contin. Educ. Dent., Vol. 23, No. 6, pp.501-510.

Parenteau, N.L. (2002). Cells. In: Tissue Engineering Research, pp 19-32. International Technology Research Institute, Baltimore, Maryland.

Park, H.W.; Lee, J.K.; Moon, S.J.; Seo, S.K.; Lee, J.H.; Kim, S.H. (2009) The efficacy of the synthetic interbody cage and Grafton for anterior cervical fusion. Spine, Vol. 34, No. 17, pp. E591-E595.

Passier, R.; Mummery, C. (2003) Origin and use of embryonic and adult stem cells in differentiation and tissue repair. Cardiovascular Res., Vol. 58, No. 2, pp.324-335.

Pelker, R.R.; Friedlaender, G.E. (1987) Biomechanical aspects of bone autografts and allografts. Orthop. Clin. North Am., Vol. 18, No. 2, pp.235-239.

Peltier, L.F.; Bickel, E.Y.; Lillo, R.; Thein, M.S. (1957) The use of plaster of Paris to fill defects in bone. Ann Surg., Vol. 146, No. 1, pp.61-69.

Petite, H.; Viateau, V.; Bensaïd, W.; Meunier, A.; de Pollak, C.; Bourguignon, M.; Oudina, K.; Sedel L.; Guillemin, G. (2000) Tissue-engineered bone regeneration. Nat. Biotechnol., Vol. 18, No. 9, pp.959-963.

Piek, E.; Heldin, C.H.; Ten Dijke, P. (1999). Specificity, diversity, and regulation in TGF-b superfamily signaling. FASEB J., Vol. 13, No. 15, pp.2105–2124.

Pieske, O.; Wittmann, A.; Zaspel, J.; Löffler, T.; Rubenbauer, B.; Trentzsch, H.; Piltz, S. (2009) Autologous bone graft versus demineralized bone matrix in internal fixation of ununited long bones. J. Trauma Management Outcomes., Vol. 3, pp. 11.

Pietrzak, W.S.; Perns, S.V.; Keyes, J.; Woodell-May, J.; McDonald. (2005) Demineralized bone matrix graft: a scientific and clinical case study assessment. J. Foot Ankle Surg., Vol. 44, No. 5, pp.345-353.

Pineda, L.M.; Busing, M.; Meinig, R.P.; Gogolewski, S. (1996) Bone regeneration with resorbable polymeric membranes. III. Effect of poly (L-lactide) membrane pore size on the bone healing process in larger defects. J. Biomed. Mater. Res., Vol. 31, No. 3, p.385-394.

Pittenger, M.F.; Mackay, A.M.; Beck., S.C. Jaiswal, R.K.; Douglas, R.; Mosca, J.D.; Moorman, M.A.; Simonetti, D.W. Craig, S.; Marshak, D.R. (1999) Multilineage potential of adult human mesenchymal stem cells. Science, Vol. 284, No. 5411, pp.143–147.

Porter, J.R.; Ruckh, T.T.; Popat, K.C. (2009) Bone tissue engineering: a review in bone biomimetics and drug delivery strategies. Biotechnol. Prog., Vol. 25, No. 6, pp.1539-1560.

Probst, A.; Spiegel, H.U. (1997) Cellular mechanisms of bone repair. J. Invest. Surg., Vol. 10, No. 3, pp.77-86.

Quarto, R.; Mastrogiacomo, M.; Cancedda, R.; Kutepov, S.M.; Mukhachev, V.; Lavroukov, A.; Kon, E.; Marcacci, M. (2001) Repair of large bone defects with the use of autologous bone marrow stromal cells. N. Engl. J. Med., Vol. 344, No. 5, pp.385-386.

Reddi, A.H. (2001) Bone morphogenetic proteins: From basic science to clinical applications. J. Bone Joint Surg., Vol. 83A, Suppl. A, Part 1, pp.S63-S69.

Reddi, A.H. (1981) Cell Biology and biochemistry of endochondral bone development. Coll. Relat. Res., Vol. 1, No. 2, pp.209-226.

Reddi, A.H. (2000) Morphogenesis and tissue engineering of bone and cartilage: inductive signals, stem cells, and biomimetic biomaterials. Tissue Eng, Vol. 6, No. 4, pp.351-359.

Reed, A.M.; Gilding, D.K. (1981) Biodegradable polymers for use in surgery — poly(glycolic)/poly(Iactic acid) homo and copolymers: In vitro degradation. Polymer, Vol. 22, No. 4, pp.494-498.

Rizzi, S.C.; Heath, D.J.; Coombes, A.G.; Bock, N.; Textor, M.; Downes, S. (2001) Biodegradable polymer/hydroxyapatite composites: surface analysis and initial attachment of human osteoblasts. J. Biomed. Mater. Res., Vol. 55, No. 4, pp.475-486.

Robey, P.G. (2000) Stem cells near the century mark. J. Clin. Invest., Vol. 105, No. 11, pp.1489-1491.

Rodrigues, F.L.; Mercadante, M.T. (2005) Tratamento da falha óssea parcial pelo transporte ósseo parietal. Acta Ortop. Bras., Vol. 13, No. 1, pp.9-12.

Rubin, E.; Farber, J.L. (2002) Patologia. 6.ed., Guanabara Koogan, Rio de Janeiro.

Santos, A.R. Jr. Barbanti, S.H.; Duek, E.A.R.; Dolder, H.; Wada, R.S.; Wada, M.L.F. (2001) Growth and differentiation of Vero cells on poly(L-lactic acid) membranes of different pore diameters. Artif. Organs, Vol. 25, No. 1, pp.7-13.

Santos, A.R. Jr. Ferreira, B.M.P.; Duek, E.A.R.; Dolder, H.; Wada, R.S.; Wada, M.L.F. (2004) Differentiation pattern of Vero Cells cultured on poly(l-lactic acid)/poly(hydroxybutyrate-co-hydroxyvalerate) blends. Artif. Organs, Vol. 28, No. 4, pp.381-89.

Santos, A.R. Jr. Wada, M.L.F. (2007) Polímeros biorreabsorvíveis como arcabouços para cultura de células e engenharia tecidual. Polímeros Cienc Tecnol, Vol. 17, No. 4, pp.308-317.

Santos, A.R. Jr. (2010) Bioresorbable polymers for tissue engineering. In: Tissue engineering, pp. 235-246, Daniel Eberlin, ed., In-Teh, Olajnica.

Satija, N.K.; Gurudutta, G.U.; Sharma, S.; Farhat Afrin, F.; Guptta, P.; Kumar, R.Y.; Singh, V.K.; Tripathi, R.P. (2007) Mesenchymal stem cells: molecular targets for tissue engineering. Stem Cells Dev., Vol. 16, No. 1, pp.7-23.

Seiler, J.G.; Johnson, J. (2000) Iliac crest autogenous bone grafting: donor site complications. J. South Orthop. Assoc., Vol. 9, No. 2, pp.91-97.

Stevenson, S. (1999) Biology of bone grafts. Orthop. Clin. North Am., Vol. 30, No. 4, pp.543-552.

Suominen, E.; Kinnunen, J. (1996) Bioactive glass granules and plates in the reconstruction of defects of the facial bones. Scand. J. Plast. Reconstr. Surg. Hand Surg., Vol. 30, No. 4, pp.281-289.

Tanaka, T.; Kumagae, Y.; Saito, M.; Chazono, M.; Komaki, H.; Kikuchi, T.; Kitasato, S.; Marumo, K. (2008) Bone formation and resorption in patients after implantation of beta-tricalcium phosphate blocks with 60% and 75% porosity in opening-wedge high tibial osteotomy. J. Biomed. Mater. Res. B Appl. Biomater., Vol. 86B, No. 2, pp.453-459.

Tay B.K.; Patel V.V.; Bradford D.S. (1999) Calcium sulfate- and calcium phosphate-based bone substitutes. Mimicry of the mineral phase of bone. Orthop Clin North Am., Vol. 30, No. 4, pp.615-623.

Temenoff, J.S.; Mikos, A.G. (2000a) Review: tissue engineering for regeneration of articular cartilage. Biomaterials, Vol. 21, No. 5, pp. 431-440.

Temenoff, J.S.; Mikos, A.G. (2000b) Injectable biodegradable materials for orthopedic tissue engineering. Biomaterials, Vol. 21, No. 23, pp. 2405-2415.

Thomson J.A.; Itskovitz-Eldor J.; Shapiro S.S.; Waknitz M.A.; Swiergiel J.J.; Marshall V.S.; Jones J.M. (1998) Embryonic stem cell lines derived from human blastocysts. Science, Vol. 282, No. 5391, pp.1145-1147.

Tiedeman, J.J.; Garvin, K.L.; Kile, T.A.; Connolly, J.F. (1995) The role of a composite, demineralized bone matrix and bone marrow in the treatment of osseous defects. Orthopedics, Vol. 18, No. 12, pp.1153-1158.

Törmälä, P.; Pohjonen, T.; Rokkanen, P. (1998) Bioabsorbable polymers: materials technology and surgical applications. Proc. Instn. Mech. Eng. Part H - J. Eng. Med., Vol. 212, No. 2, pp.101-111.

Tsonis, P.A. (2002) Regenerative biology: the emerging field of tissue repair and restoration. Diferentiation, Vol. 70, No. 8, pp.397-409.

Tuli, S.M.; Singh, A.D. (1978) The osteoinductive property of decalcified bone matrix. An experimental study. J. Bone Joint Surg. Br., Vol. 60, No. 1, pp.116-123.

Tuli, R.; Tuli, S.; Nandi, S.; Huang, X.; Manner, P.A.; Hozack, W.J.; Danielson, K.G.; Hall, D.J.; Tuan, R.S. (2003). Transforming growth factor-b-mediated chondrogenesis of human mesenchymal progenitor cells involves N-cadherin and mitogen-activated protein kinase and Wnt signaling cross-talk. J. Biol. Chem., Vol. 278, No. 42, pp.41227–41236.

Undale, A.H.; Westendorf, J.J.; Yaszemski, M.J., Khosla, S. (2009) Mesenchymal stem cells for bone repair and metabolic bone desease. Mayo Clin. Proc., Vol. 84, No. 10, pp.893-902.

Urist, M.R. (1965) Bone: formation by autoinduction. Science, Vol. 150, No. 698, pp.893-899.

Urist, M.R.; Grant, T.T.; Lindholm, T.S.; Mirra, J.M.; Hirano, H.; Finerman, G.A. (1979) Induction of new-bone formation in the host bed by human bone-tumour transplants in athymic nude mice. J. Bone Joint Surg. Am., Vol. 61, No. 8, pp.1207-1216.

Virolainen, P.; Mokka, J.; Seppänen, M.; Mäkelä, K. (2010) Up to 10 years follow up of the use of 71 cortical allografts (strut-grafts) for the treatment of periprosthetic fractures. Scand. J. Surg., Vol. 99, No. 4, pp.240-243.

Wagner, W; Ho, A.D.; Zenke, M. (2010) Different facets of aging in human mesenchymal stem cells. Tissue Eng. Part B., Vol. 16, No. 4, pp.445-453.

Wahl, D.A.; Czernuszka, J.T. (2006) Collagen-hydroxyapatite composites for hard tissue repair. Eur. Cell Mater., Vol. 28, No. 11, pp.43-56.

Weissman, I.L. (2000) Stem cells: units of development, units of regeneration, and units in evolution. Cell, Vol. 100, pp.157-1168.

Yoshikawa, H.; Tamai, N.; Murase, T.; Myoui, A. (2009) Interconnected porous hydroxyapatite ceramics for bone tissue engineering. J. R. Soc. Interface, Vol. 6, Suppl 3, pp.S341-S348.

Zhang Y.; Reddy V.J.; Wong S.Y.; Li X.; Su B.; Ramakrishna S.; Lim C.T. (2010) Enhanced biomineralization in osteoblasts on a novel electrospun biocomposite nanofibrous substrate of hydroxyapatite/collagen/chitosan. Tissue Eng Part A., Vol. 16, No. 6, pp.1949-1960.

Zellin, G.; Hedner, E.; Linde, A. (1996) Bone regeneration by a combination of osteopromotive membranes with different BMP preparations: a review. Connect. Tissue Res., Vol. 35, pp.279-284.

Zuk, P.A.; Zhu, M.; Mizuno, H.; Huang,J.; Futrell, J.W.; Katz, A.J.; Benhaim, P.; Lorenz, H.P., Hedrick, M.H. (2001) Multilineage cells from human adipose tissue: implications for cell-based therapies. Tissue Eng., Vol. 7, No. 2, pp.211-228.

Zuk, P.A.; Zhu, M.; Ashijan, P.; De Urgate, D.A.; Huang, J.I.; Mizuno, H.; Alfonso, Z.C., Frazer, J.I. (2002) Human adipose tissue is a source of multipotent stem cells. Mol. Biol. Cell., Vol. 13, No. 12, pp.4279-4295.

Part 3

Use of Scaffolds

17

Preparation of Deproteinized Human Bone and Its Mixtures with Bio-Glass and Tricalcium Phosphate – Innovative Bioactive Materials for Skeletal Tissue Regeneration

Magdalena Cieslik[1], Jacek Nocoń[2], Jan Rauch[3], Tadeusz Cieslik[4], Anna Ślósarczyk[5], Maria Borczuch-Łączka[6] and Aleksander Owczarek[7]

[1]*Faculty and Institute of Stomatological Materials Science, Medical University of Silesia, Katowice, Bytom,*
[2]*Private Dentistry Practice, Oberhausen,*
[3]*NZOZ – Specialist Dentistry Clinic, Wadowice,*
[4]*Faculty and Clinic of Oral and Maxillofacial Surgery, Medical University of Silesia, Katowice,*
[5]*Faculty of Glass Technology and Amorphous Coatings, AGH - Krakow University of Science and Technology, Kraków,*
[6]*Faculty of Ceramic Technology, AGH - Krakow University of Science and Technology, Kraków,*
[7]*Division of Statistics, Medical University of Silesia, Katowice, Sosnowiec,*
[1,3,4,5,6,7]*Poland*
[2]*Germany*

1. Introduction

Repair of the skeletal system is one of the principal research problems in medical science is closely associated with the field of material engineering. The reasons for using bone implants and grafts include injuries, infections, neoplasms and other hard tissue lesions. Bone replacement materials are predominantly used in medical disciplines such as dentistry, dental surgery, maxillofacial surgery and plastic surgery, as well as in orthopedics and traumatology (Barradas et al., 2011; Kao & Scott, 2007; Precheur, 2007).

From a biological, immunological, and legal point of view, autogenous bone grafting still remains a very popular method in reconstruction following skeletal loss (Block, 2002; Giannoudis et al., 2005). Factors considered in the selection of the source of the bone graft include, among others, the ease of surgical access and the volume of bone mass required (Precheur, 2007). The type of autogenous bone used as a graft (cortical bone vs. cancellous bone) should also be considered. For instance, the higher content of morphogenetic proteins (BMPs) in cortical bone means that grafts of this type induce the process of bone growth more effectively than cancellous bone grafts. Nonetheless, skeletal reconstruction with autogenous bone grafts always requires additional surgical manipulations that constitute an increased burden to the patient and prolong the duration of the procedure. Another very frequent limitation of using autogenous bone is its poor quality, when there is a skeletal system disorder (e.g. osteoporosis) (Bohner, 2010; Giannoudis et al., 2005). Allogenic implants derived from the structures of human bones can be one alternative to autogenous grafts (Ferreira, 2007). Mineralized (FDBA) and demineralized (DFDBA) forms of such implants, additionally subjected to lyophilization, are most frequently used in reconstructive surgery. An advantage of demineralized bone arises from the fact that the organic bone matrix (collagen fibers) has to be exposed in order to remove its mineral components, and therefore so-called matrix proteins (e.g. morphogenetic proteins) can easily diffuse into the implantation site and work osteoinductively (Barradas et al., 2011). Xenogenic implants also play an important role in reconstructive bone surgery. These implants are made of skeletal material obtained from animals (Merkx et al., 2003), in most cases from equine, bovine or porcine bones. Animal material is processed by means of thermal treatment in order to deplete it completely of its organic components (Barakat et al., 2009). As a result, implants lose their immunogenic properties and become neutral to hosts (Liu et al., 2008). Therefore, ready to use preparations for bone replacement (Bio-Oss®, Endobone®) are most frequently available in deproteinized forms (DBBM) (Accorsi-Mendonça et al., 2008). Due to their osteoconductive properties they can serve as an inactive scaffold or platform for the maturation of bone cells present within the defect. They are used in orthopedics, dental and maxillofacial surgery, as well as in periodontology and implantology, etc. (Baldini et al., 2011; Cao et al., 2009; Jian et al., 2008; Merkx et al., 2003; Precheur, 2007; von Wattenwyl et al., 2011). The limited possibilities of modulating the resorption time of such preparations during skeletal tissue reconstruction, however, can be related to the poorer quality of bone at the site of their application.

An alternative solution, eliminating the potential complications associated with the application of materials of autogenous, allogenic or xenogenic origin, is the use of alloplastic implants for the purpose of bone replacement. Such implants can be synthesized from both natural and synthetic materials (Bohner, 2010; Giannoudis et al., 2005). Bioresorbable ceramic based on calcium phosphate plays a distinct role amongst novel synthetic materials used for bone replacement (Barradas et al., 2011). Hydroxyapatite (HAp) is the principal representative of this group, with the widest application in reconstructive surgery. Due to its calcium phosphate content and natural occurrence as an inorganic substance in bones and teeth, hydroxyapatite is characterized by the highest biocompatibility and bioactivity of all currently known implant materials. Additionally, due to the osteoconductive properties of hydroxyapatite, and (to a lesser extent) its osteoinductive properties, hydroxyapatite-based implants can bind directly to bone. Numerous clinical trials, supported by the results of histological observations, have confirmed complete biotolerance to hydroxyapatite ceramic, as well as its positive effects on the process of bone healing and reconstruction. Additionally, hydroxyapatite ceramic can initiate and stimulate various processes that are associated with bone formation (Bellucci et al., 2011; Ravarian et al., 2010; Yuan et al., 2001). Contact between bioactive ceramic and living skeletal tissue induces osteogenesis. As a result, an intermediate binding layer is formed between the living tissue and the implant, serving as a kind of biological glue. The structural similarity of calcium phosphate-based ceramic and the natural mineral components of bone is crucial for infiltration of the implant by the skeletal tissue of the host. This supports the intra-tissue application of hydroxyapatite-based implants whenever long-term remodeling of bone is required. Beta-tricalcium phosphate (betaTCP, beta$Ca_3(PO_4)_2$) is another calcium phosphate that has been used successfully in bone substitution. Its mineralogical analogue is whitlockite (TCP) (Barradas et al., 2011). Similar to HAp, TCP is characterized by high biocompatibility. In comparison to hydroxyapatite materials, it has a higher solubility *in vitro* and a resultant higher susceptibility to resorption and biodegradation in the environment of the living organism (Bohner, 2010; Henkel et al., 2006). TCP is considered an osteoinductive material, which stimulates the processes of bone reconstruction (Wang et al., 2009). Many studies, mostly dealing with the gradual, controlled resorption rate of calcium phosphate ceramic, resulted in the design of the second polymorphic variant of TCP, namely αTCP (Zima et al., 2010). Similar to beta-TCP, it is formed as a result of the non-stoichiometric heating of HAp with a well-defined temperature and defined kinetics of thermal processing. Compared to beta-TCP, αTCP has an approximately five-fold higher susceptibility to resorption in living tissues, along with higher biocompatibility (Oonishi et al., 1999). The high degree of osteointegration of αTCP-based bioceramic is plausibly the result of the higher solubility of α$Ca_3(PO_4)_2$ than of HAp or βTCP. Delivered within the ceramic, calcium and phosphate ions constitute the material for synthesizing the layer composed of non-stoichiometric hydroxyapatite that binds the implant to bone (Kon et al., 1995). Some authors suggest that αTCP can be cytotoxic, probably due to pH changes that are induced *in vitro* (Santos et al., 2002). These suggestions, however, were not confirmed during *in vivo* studies, and αTCP is included in many commercially available bone cements. Since the 1990s, biphasic HAp-βTCP ceramic (BCP biphasic calcium phosphate) has also been used in the reconstruction of skeletal defects (Deculsi, 1998, Deculsi et al., 2003; Fellah et al., 2008; Schwarz et al., 2007).

Bio-glass and apatite-wollastonite glass-ceramic also play an important role amongst the bioactive materials used in the processes of bone synthesis. The biological activity of glass and glass-derived crystalline materials (glass-ceramic materials) is mostly exploited for the manufacture of surface-active implants or implants that can be resorbed in the human body (Giannoudis et al., 2005). The composition of surface-active materials is selected in order to enable interactions between the physiological environment and certain components of the implant. As a result, living tissue is bound to the implant surface. Resorbable glass is characterized by its high content of chemical elements involved in metabolic processes of the human body. In both cases, living tissues can infiltrate the implant and bind it directly to the bone. These processes result not only from the chemical composition of the implants but also from the specific nature of the contained glass substance. The binding abilities of bioactive glass and glass-ceramic implants to skeletal tissue have been the subject of many studies. These studies confirmed the usefulness of such materials in tissue engineering, where they can be used in cell culture (Ferreira, 2007; Xynos et al., 2000). Their application in reconstructive surgery is reflected by their ability to stimulate and supporting bone reconstruction. In view of their confirmed ability to directly bind to skeletal tissue, they have been used successfully in various stomatological disciplines, including implantology, dental surgery and periodontology. Moreover, modern surface engineering has allowed the coating of metal implants with bioactive glass. Such coatings either protect surfaces of metal alloys against corrosion and wear, or stimulate the processes of bone formation in their surroundings. Additionally, resorbable glass can be used as a drug carrier, providing prolonged release of an active substance. Furthermore, some attempts have been made to combine various materials with each other to manufacture glass-containing bioactive composites, used as the components of binders, among others (Bellucci et al., 2011). These experiments were aimed at obtaining biomaterials with better durability parameters and optimal biological characteristics. Various types of bio-glass and glass-ceramics may differ in terms of their biological properties, depending on the technique for their synthesis (Bellucci et al., 2011; Yuan et al. 2001). Amongst various methods of synthesis, the advantages of the chemical sol-gel method are worth noting. This technique enables material whose biological activity is greater than products manufactured using other processing methods to be obtained (Ravarian et al., 2010). Moreover, the sol-gel method does not require high temperature processing. Additionally, this technique allows the production of material with a strictly defined texture and parameters. Due to the possibility of manufacturing porous forms, bio-glass and glass-derived crystalline materials are morphologically similar to bone; this facilitates their infiltration by bone cells at the implantation site. Additionally, the porous structure of the implant enables the supply of fluids and nutrients required for growth to the newly formed bone, as well as the elimination of metabolites (Yuan et al., 2001). Finally, the resorption rate of the implant material is also determined by the degree of its porosity.

In view of the complexity of the biological environment at the implantation site, one principle problem of biomaterial engineering pertains to issues associated with the biodegradation and bioresorption of implanted biomaterials. Determination of degradation time and rate, and the kinetics of this process in the human body, constitute a significant challenge in the design of new implant materials (including bioceramic) used for the purposes of skeletal tissue substitution (Daculsi et al., 2003; Giannoudis et al., 2005). The controlled degradation rate of the implant along with the associated reconstruction of skeletal tissue should result in the formation of bone tissue resembling its natural structure as closely as possible, both at the biological and physico-mechanical levels. One of many methods allowing for the controlled biodegradation of bone replacement materials is the design of implants made of various composites or mixtures. Combining two or more different materials results in the manufacture of an absolutely new composite biomaterial, which is frequently superior in terms of biological and mechanical properties. Due to its specific properties, bioceramic is very frequently used as a basic component of ceramic-ceramic, metal-ceramic and polymer-ceramic systems (Thomas et al., 2005). Hydroxyapatite is very frequently included in bone composites, mostly due to its chemical similarity to the natural components of bone. Some studies have confirmed the positive effects of hydroxyapatite used in combination with metals, ceramic or polymers (Abu Bakar et al., 2003; Bellucci et al., 2011; Choi et al., 2004; Daculsi et al., 2003). The results of research on HAp-containing composites based on resorbable polymers, mostly polylactide (PLA), polyglycolide (PGA), co-polymer of glycolide and lactide (PLGA) or collagen [Cieślik et al., Nagata et al., 2005], seem particularly interesting. These studies confirmed the possible application of such composites as binding and reconstructive elements in reconstructive surgery, as culture media in tissue and genetic engineering, and as drug carriers (Wei & Ma, 2004; Nagata et al., 2003). Although hydroxyapatite is characterized by high biocompatibility, its reactivity with living skeletal tissue is relatively low. In view of these findings, attempts to design composites that combine this biomaterial with markedly more biologically active bio-glass propose an interesting solution. Such a composite can stimulate osteogenesis, leading to the rapid formation of new skeletal tissue around the implant (Yuan et al., 2001; Ravarian et al., 2010). Biological tests of SiO_2-CaO-Mg/natural HAp (bovine bone) composite obtained by means of thermal plasma processing have confirmed its lack of toxicity. Moreover, the glass-ceramic contained in this composite was confirmed to stimulate the growth and proliferation of human fibroblasts (Yoganand et al., 2010). Attempts at bone replacement with materials obtained by combining autogenous bone and deproteinized bovine bone constitute another example of the potential optimization of biological conditions for new skeletal tissue growth within a skeletal defect. The inclusion of autogenous material with osteoinductive properties in such composites results in the enhanced formation of better quality bone (Kim et al., 2009; Pripatnanont et al., 2009; Thorwarth et al., 2006; Thuaksuban et al., 2010). *In vitro* cellular studies have confirmed that allogenic materials of human origin (demineralized bone matrix, deproteinized bone) may also have some osteoinductive activity. This activity was confirmed by an increase in alkaline phosphatase (ALP), osteocalcin (OC) and Ca^{2+} concentrations observed in human bone marrow stromal osteoprogenitor cells (hBMSCs) cultured for three weeks in medium based on human bone components (Zhang et al., 2009).

The increasing demand for bone replacement materials stimulated us to design original composite materials for use in the regeneration of skeletal defects, characterized by both osteoinductive and osteoconductive properties. In designing such composites, we have used well-described biomaterials that have been applied successfully to skeletal defect regeneration, namely bio-glass (BG) and tricalcium phosphate (TCP). Lyophilized human bone obtained from a tissue bank served as a base for designing three types of mixtures: 1) human bone/bio-glass; 2) human bone/TCP; and 3) human bone/bio-glass/TCP. Such variants of material combinations enabled us to analyze the effects of particular components on the process of bone formation, and specifically on its dynamics and on the quality of newly formed bone. Also, the selection of the components included among the analyzed materials (mixtures) was not accidental. We have assumed that both bio-glass and tricalcium phosphate will activate and stimulate osteoblasts to dynamically grow on a biological scaffold of lyophilized human bone, thereby providing optimal biological conditions for the formation of full value bone.

2. Material

The materials used in the study included lyophilized, deproteinized human bone (B) - group B, and its mixtures with: 1) bio-glass (BG) in a proportion of 80:20% weight ratio (B:BG) - group B+BG; 2) tricalcium phosphate (TCP) in a proportion of 80:20% weight ratio (B:TCP) - group B+TCP; 3) bio-glass and tricalcium phosphate in a proportion of 70:15:15% weight ratio (B:BG:TCP) - group B+BG+TCP. The deproteinized human bone and all other components used in the mixtures were in a granulated form with diameters ranging from 0.3 to 0.5 mm.

Human bone used in this study was obtained from the Tissue Bank of the Regional Center of Blood Donation and Treatment in Katowice (Poland). It was cancellous bone subjected to lyophilization, deep freezing and irradiation sterilization with a dose of 35 kGy.

The bio-glass was made from the $CaO\text{-}SiO_2\text{-}P_2O_5$ system with the use of the sol-gel technology, in the laboratory of the Department of Glass and Amorphous Coatings at the AGH Science and Technology University in Cracow, Poland. Its high-calcium A2 variety was used (54% mol. CaO) with a density of 2.9082 g/cm^3, a dominating glassy phase and the beginnings of apatite crystallization. The thermal treatment of the bio-glass was performed at a temperature of 800°C, and its specific surface area (calculated by BET method) amounted to 57.8166 m^2/g.

Resorbable, monophasic βTCP ceramic – a salt of trialkaline orthophosphoric acid $Ca_3(PO_4)_2$ – was synthesized from powdered components obtained by means of wet synthesis in the Bioceramic Laboratory of the Department of Ceramic and Fire Resistant Material Technology at the AGH Science and Technology University in Cracow, Poland (Ślósarczyk & Paszkiewicz, 2005; Zima at al., 2010). The reagents included CaO (obtained by means of calcination of $Ca(OH)_2$ – pure for analysis; MERCK, Poland), and H_3PO_4 (pure for analysis; POCH, Poland).

3. *In vivo* animal experiments

The study was carried out on a group of 48 guinea pigs, with an equal number of both sexes, and weights ranging from 500 to 600 grams. The animals were divided randomly into four groups, which corresponded to 12 animals for each studied composite material (6 males and 6 females). The animal experiments were performed on the 7th, 14th, and 21st days of the study, as well as after the 4th, 8th and 12th experimental weeks. Two guinea pigs (one male and one female) were examined in each experiment. All animal surgical procedures were performed at the Central Experimental Animal Farm at the Medical University of Silesia and were granted permission by the university's Bioethical Board for Experimental Animals.

Before starting any surgical procedures, the animals received general anesthesia with thiopental (0.4 g/kg b.w.). Bone defects (6 mm in diameter and 3 mm in depth) were formed bilaterally on the external surface of the mandibular trunk, 2 mm below its lower edge (between the radices of the incisive and molar teeth) with the aid of a rosette burr placed in the straight hand-piece of a dental machine (Fig.1a). Depending on the experimental group, bone defects on the right side of the mandible were filled with: 1) a preparation of deproteinized human bone (group B), 2) a mixture of deproteinized human bone and bio-glass (group B+BG), 3) a mixture of deproteinized human bone and tricalcium phosphate (group B+TCP), or 4) a mixture of deproteinized human bone with bio-glass and tricalcium phosphate (group B+BG+TCP). Before implantation into the mandibular bone defect, each material was mixed with the animal's blood obtained from the surgical wound (Fig. 1b) in a

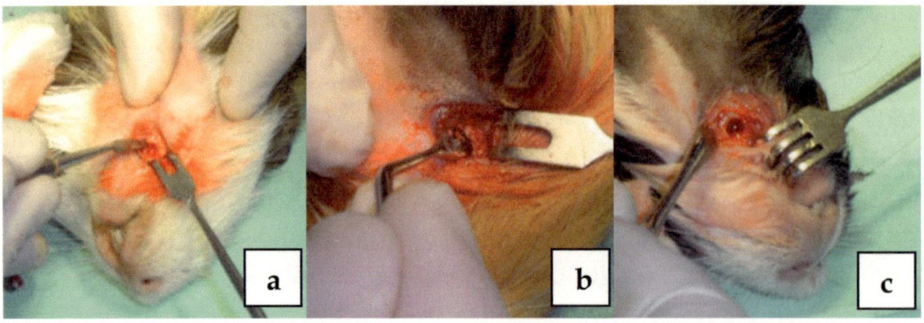

Fig. 1. Bone defect in the mandibular trunk of the experimental animal: a) defect prepared for filling with implant based on deproteinized human bone; b) defect filled with the mixture of deproteinized human bone and bio-glass with animal's blood; c) control defect filled with postoperatively clotted blood

proportion of 25:75% weight ratio (examined material to blood). Blood served as a binder, supplying the implanted material with organic components. Additionally, it prevented the displacement of the implanted granules from within the bone defect, and facilitated their insertion. The defect on the left side of the mandible was left to fill postoperatively with clotted blood and undergo spontaneous healing (Fig.1c), and was considered a control to the right mandibular defect in each animal (control group). The bilateral wounds were closed in multiple layers with Dexon 4.0 sutures.

4. Methods of examination

On each examination day, all guinea pigs were clinically evaluated for surgical wound healing and general condition. Additionally, radiographs were taken in order to assess the regeneration of mandibular bone defects. Newly formed skeletal tissue in the defects was also analyzed quantitatively and qualitatively in terms of bone mineral density at the implantation site (computed tomography and radiographic densitometry). Following euthanasia with Morbital (sodium pentobarbital 133.3 mg/mL and pentobarbital 26.7 mg/mL, 1-2 mL/kg b.w.), the operated area and surrounding tissues were examined macroscopically. Moreover, tissue specimens from the euthanized animals were subjected to histopathologic analysis of the skeletal tissue and bone marrow at the implantation site and at its periphery, including the healing rate of the skeletal tissue and the soft and hard tissue reactions to the implanted material. Histopathologic analysis also included organs involved in detoxification of the body (i.e. liver and kidneys).

Radiographs of the mandibular trunk with the sites of the bone defects were taken with a Heliodent type MD/D - 3195 nr 051692 apparatus (SIEMENS), and AGFA DENTUS M2 CONFORT films for axial pictures (75 mm x 65mm), using the following exposure conditions: 0.16 s, 7 mA and 60 kV.

The tissue specimens taken for the histopathologic examinations were preserved in a 10% solution of buffered formalin. The osseous tissue was decalcified either in a 10% solution of disodium versenate, or electrolytically in Romeis liquid (80 mL of hydrochloric acid + 100 mL of formic acid, diluted to make 1000 mL with water) using PW23 bone decalcifier and an electric current of 0.5A. Next, all the tissues were routinely processed in Technicon's Duo autotechnicon using the sequence of 96% alcohol, acetone and xylene. They were then embedded in paraplast. The obtained cubes were then shaved on a Microm HM335E rotating microtome. The shavings (4-6 microns in thickness) were put onto basic slides, deparaffinized, and stained with hematoxylin and eosin (H&E). They were subsequently mounted with Canada balsam. Histologic slides were analyzed under an Olympus BX50 light microscope equipped with a set for optical (Olympus SC35) and digital microphotography (Olympus Canmedia C-5050 Zoom), at 40 to 400 x magnification. Additionally, the slides were consulted by other histologists using a dual-head Nikon Labophot-2 microscope.

Bone mineral density (BMD) was determined by means of dual-energy X-ray absorptiometry (DXA) with a DPX-L densitometer (LUNAR Radiation Corporation, Madison, USA) using Small Animal Appendicular scanning. Total bone mineral density (BMD; g/cm^2) was determined on the basis of examination of the entire area of the transverse cross-section of the bone defect (28.26 mm^2), analyzed as 0.5 mm^2 to 3 mm^2 sections. One hundred and twenty

BMD measurements were taken in each group (corresponding to 20 measurements per experimental day). In most cases, reproducible results were not included in further analysis; only unique values and the most frequent reproducible values were analyzed, which corresponded to 10 BMD measurements for each day of examination. Additionally, the BMD of the normal mandibular trunk of the guinea pig was determined for the purpose of comparative analysis. Based on 10 consecutive measurements, this value was estimated at 0.51 ± 0.001 g/cm^2. During the measurements, the densitometer was regularly calibrated and controlled according to the manufacturer's recommendations. Both DXA measurements and BMD result analysis were performed by the same investigator.

Additionally, computed tomography (CT) was used to determine the bone mineral density, expressed in Hounsfield units (HU). These measurements were taken using a Somatom Emotion 6 scanner (Siemens; exposure parameters: 13.4 s, 14 mA, 130 kV). Transverse cross-sections of the mandible were visualized in 2 mm slices. Then, cross-sections including the area of the bone defect were selected and analyzed using Volume Viewer software. In each experimental group, six measurements were taken for each examination day. Additionally, six measurements of the normal mandible were taken for the purpose of comparative analysis. Based on these measurements, the normal bone mineral density was determined to be 1218 ± 15.2 HU.

The results of bone density are presented as mean values ± standard deviation. Variables distribution was evaluated by the Shapiro-Wilk test. Homogeneity of variance was assessed by the Levene test. ANOVA for repeated measurements with contrasts analysis were done to assess time and preparation of deproteinized human bone type interaction. The Mauchley test was done to check sphericity. Differences were considered to be statistically significant at $p<0.05$. All calculations were performed using the commercially available statistical package Statistica 9.0.

5. Results

5.1 Clinical observations

Throughout the entire study period, no complications in surgical wound healing were observed in animals of any experimental group. The guinea pigs were calm, which suggested a lack of pain. The animals ate and drank water normally, and neither scraped against the cage nor scratched their wounds during the entire postoperative period. Also, wound dehiscence was not observed in any of the groups.

Tissue edema over the surgical skin wounds resolved 3 to 5 days following surgery and was replaced by protrusions of the tissue. There were no signs of excessive fluid accumulation around the wound or of hematoma formation, but in some animals skin redness was observed around the stitches. In most animals, tissue protrusions persisted until the 21st day after surgery. The stitches were removed 10 to 14 days after surgery. Over the entire study period, the animals gained weight gradually (in a statistically insignificant manner).

5.2 Macroscopic examination

Up to 14th experimental day, the sites of the implanted bone defects were clearly distinguishable from the surrounding tissues as clear, oval protrusions covered with delicate tissue, which could be compressed elastically. Additionally, clearly distinguishable white granulation was visible throughout the superficial tissue in animals of the B+BG and B+BG+TCP groups. Probably, these granules corresponded to the bio-glass particles included in the material implanted in these groups. In the control group, the site of the bone defect was visible as a clear protrusion up to the 7th experimental day. On the next day of examination (the 14th day), however, the site was covered with a tissue whose coloration was darker in comparison to the surrounding tissues.

After the 3rd experimental week, the protrusion visible over the implantation site in groups B+BG, B+TCP and B+BG+TCP was markedly smaller in size, and covered with a hard tissue of more compact texture. Coloration of the tissues covering the implantation site still differed from the color of normal bone. In group BG, white granulation was still visible throughout the superficial tissue. On the 21st day after surgery, the coloration of the bone defect implanted with deproteinized human bone (group B) resembled the color of the surrounding tissues more closely than in previous periods. In both group B and in the control group, clearly distinguishable small areas in the form of a dark-colored spot were visible within the defects. Additionally, tissue protrusion observed over the defects in the control group increased in size when compared to previous periods.

Four weeks following implantation, the area of the bone defect in group B+BG was slightly smaller in size but still clearly distinguishable and differed in color from the surrounding tissues. Similar changes were observed in the B+TCP group; the implantation site in this group was concave, but still appeared hard when compressed. The tissue visible over the implantation site in group B+BG+TCP resembled the surrounding normal tissues the most closely when compared to the other groups. In group BG, white granulation was still visible throughout the superficial tissue covering the bone defect. In the control group, on the other hand, a large protrusion was still visible over the implantation site, with a small dark-colored spot at its center.

After the 8th and 12th weeks of experiment, the implantation sites of the group B animals were still clearly distinguishable on macroscopic examination; they were protruding and differed in color from the surrounding tissues. Protrusions over the implantation sites were also visible in group B+BG, but only up to the 8th week of the study. After this time, the implantation site was hardly distinguishable from the normal tissues, and only the presence of granulation, which was hardly visible throughout the superficial tissue, enabled its visual identification. In group B+BG+TCP, the implantation site could also only be localized due to the subtle appearance of white granulation as early as after 8 weeks of the experiment. In group B+TCP, no concavity was observed over the implantation site beginning from the 8th week of the study, and the tissue covering the defect only slightly differed in color from the surrounding normal bone. After 12 weeks of observation, the tissue over the bone defects in this group had an identical appearance to the surrounding tissues. On the last examination day, the site of the bone defect was distinguishable only in the control group. Although the implantation site was covered with hard tissue, minute concavities and dark spots were still visible on its surface.

5.3 Radiological examinations

Material-related differences in the rates of new skeletal tissue formation were revealed as early as after the 7th day of the study. In group B, spherical translucencies with regular edges were observed on radiographic images, with a size corresponding to the size of the bone defect. The initial process of bone reconstruction, manifested by a foggy appearance of the implantation site, was observed no earlier than after the second experimental week (Fig. 2a). At that same time point, initial signs of bone formation were also observed on radiographic images taken in group B+BG+TCP (Fig. 5a), while in group B+BG this phenomenon was already visible after 7 days (Fig. 3a). Irregular translucencies at the implantation site were seen in this group, but a small shadow was observed in the central zone, whose area and intensity increased with time. Radiologic findings suggested that the beginning of new tissue formation was most delayed in group B+TCP (Fig. 4a) and in the controls (Fig. 6a). In these groups, a distinct shadow was observed no earlier than after the 3rd week of the experiment; this shadow was larger and more intense in B+TCP group.

After the 4th week of the study, radiographic images taken in group B revealed the nearly complete formation of new skeletal tissue. At this point in time, indistinct translucencies were seen only at the periphery of the skeletal defect, suggesting osteogenesis was ongoing in this area. After 8 weeks of the experiment, bone reconstruction was complete in this group. The whole defect was excessively shadowed, suggesting that tissue with greater mineralization was present in this area when compared to normal bone (Fig. 2b). As with group B, in group B+TCP the mineralization of the implantation site was complete after the 8th week of the study, as suggested by a fully shadowed area of the bone defect visible on radiographic images (Fig. 4b). At the same time, a small translucency was visible in the superior-medial aspect of the bone defect in group B+BG. In this group, the process of bone formation was completed no earlier than after the 12th week of the experiment. This was confirmed by the excessive mineralization of skeletal tissue, manifested radiographically as a distinct shadow at the implantation site (Fig. 3b). In the B+BG+TCP group, a homogenous shadow was visible at the implantation site as early as after the 8th week of the experiment, and an incomplete process of skeletal tissue regeneration was suggested only by the presence of spotted translucencies. However, the site of the bone defect could not be distinguished from the surrounding tissues until at least the 12th week of the study (Fig. 5b). In the control group, the process of bone formation was markedly delayed, and after the 8th week of the study a distinct, longitudinal translucency could still be seen at the site of the bone defect. Moreover, non-calcified areas were still visible on the last examination day despite newly formed skeletal tissue present within the entire area of the bone defect (Fig. 6b).

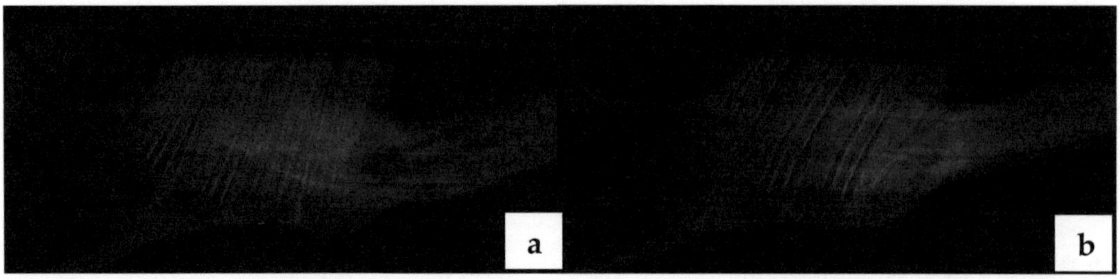

Fig. 2. Radiographic images of the mandible in experimental animals. Bone defect healing in the presence of deproteinized human bone: a) 14th day of the study; b) 8th week of the study

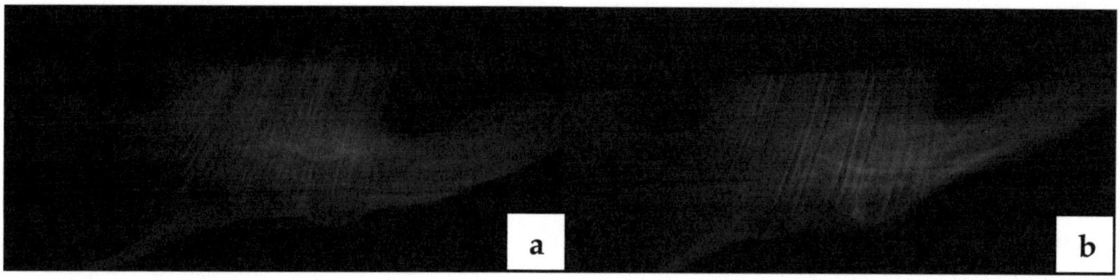

Fig. 3. Radiographic images of the mandible in experimental animals. Bone defect healing in the presence of the mixture of deproteinized human bone with bio-glass: a) 7th day of the study; b) 12th week of the study

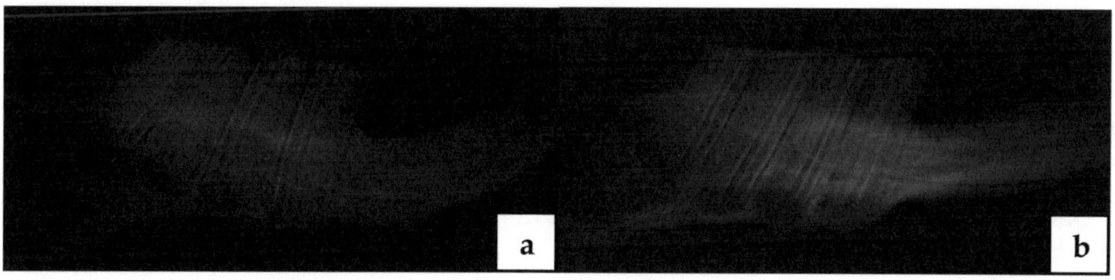

Fig. 4. Radiographic images of the mandible in experimental animals. Bone defect healing in the presence of the mixture of deproteinized human bone with tricalcium phosphate: a) 21st day of the study; b) 8th week of the study

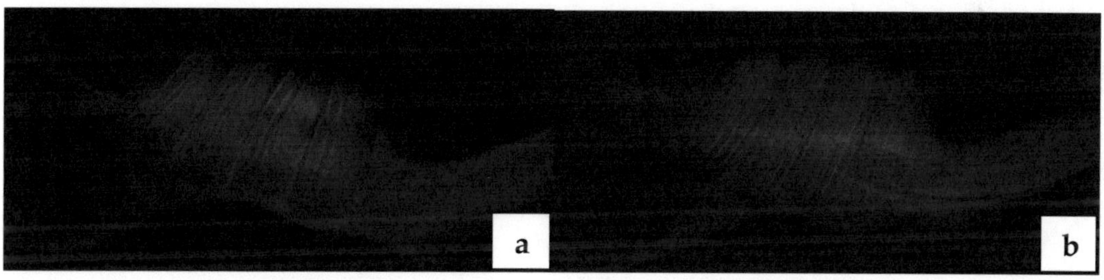

Fig. 5. Radiographic images of the mandible in experimental animals. Bone defect healing in the presence of the mixture of deproteinized human bone with bio-glass and tricalcium phosphate: a) 14th day of the study; b) 12th week of the study

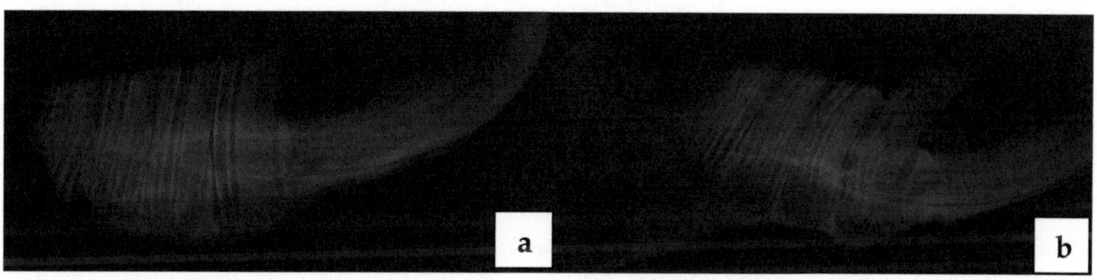

Fig. 6. Radiographic images of the mandible in experimental animals. Bone defect healing on the basis of clotted blood: a) 21st day of the study; b) 12th week of the study

5.4 Bone Mineral Density

5.4.1 Radiographic densitometry

A gradual increase in BMD was observed in all experimental groups and in the controls. The highest BMD value was observed after the 8th week of the study in bone defects implanted with B+TCP (0.40±0.05 g/cm²) and B+BG+TCP (0.39±0.05 g/cm²) (Table 1). However, these values were still lower than the normal BMD of the mandibular trunk determined in guinea pigs (0.51± 0.001 g/cm²).

	BMD [g/cm²]					
Group	**1st week**	**2nd week**	**3th week**	**4th week**	**8th week**	**12th week**
B	0.25 ± 0.01	0.27 ± 0.01	0.28 ± 0.01	0.33 ± 0.02	0.35 ± 0.03	0.35 ± 0.03
B+BG	0.27 ± 0.02	0.30 ± 0.03	0.30 ± 0.03	0.31 ± 0.02	0.31 ± 0.03	0.33 ± 0.01
B+TCP	0.22 ± 0.03	0.24 ± 0.03	0.33 ± 0.01	0.32 ± 0.04	0.40 ± 0.05	0.36 ± 0.02
B+BG+TCP	0.23 ± 0.03	0.28 ± 0.02	0.32 ± 0.03	0.35 ± 0.04	0.39 ± 0.05	0.38 ± 0.02
Control	0.24 ± 0.01	0.25 ± 0.01	0.22 ± 0.02	0.34 ± 0.01	0.33 ± 0.04	0.32 ± 0.02

Table 1. Bone mineral density (BMD) of skeletal defects implanted with various materials and in control defects determined radiographically at various time points in the study

The most regular increase in BMD was observed in defects implanted with B+BG+TCP, followed by B and B+BG. Some irregularities in the time profiles of BMD changes were noted, however, in animals of group B+TCP and in the controls, suggesting inhomogeneous formation of new skeletal tissue (Fig. 7).

Additionally, statistical analysis revealed intergroup differences in the time profiles of BMD changes. The most pronounced differences were observed between groups B and B+BG – 0.98, followed by B+BG vs. B+TCP – 0.79, and B vs. B+TCP – 0.78. Slight differences in the time profiles were noted between the B+BG+TCP group and other material groups, and no significant differences were observed when experimental groups were compared to the control group (Table 2).

	Profile comparison (p-values)			
Group	**Control**	**B**	**B+BG**	**B+TCP**
B	**<0.05**	–	–	–
B+BG	**<0.05**	0.98	–	–
B+TCP	**<0.05**	0.78	0.79	–
B+BG+TCP	**<0.01**	0.20	0.21	0.32

Table 2. The p-values for comparison of bone mineral density (BMD) changes in time (profiles) between the tested material and control groups

Moreover, a relative increase in BMD between the 1st and 12th experimental weeks was calculated for each group (Table 3). The highest relative increase in BMD was observed in bone defects in group B+BG+TCP (25.49) but the increase in the B+TCP group was only slightly lower (24.14). The lowest relative increase in BMD between the 1st and 12th experimental weeks was observed in animals of group B+BG (9.98); the time plot of BMD changes in this group was closest to a vertical line.

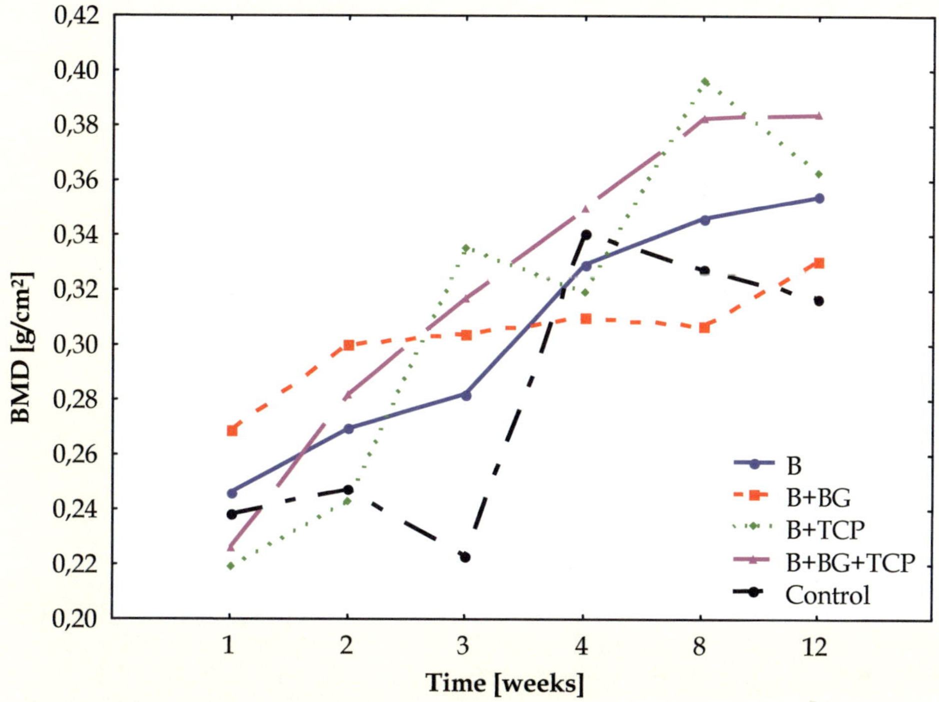

Fig. 7. Time profile of changes in bone mineral density (BMD) determined radiographically in bone defects implanted with various materials and in control defects

Due to non-sphericity (Mauchley test: 0.42; $p<0.001$) only multivariable tests were used. We confirmed that there were statistically significant changes in BMD with time ($F = 894.17$; $p<0.001$). Moreover, the interaction between time and the type of material used (group) was also statistically significant ($F = 35.59$; $p<0.001$).

	12th vs. 1st week	
Group	$\|t\|$	p
B	17.11	**<0.001**
B+BG	9.98	**<0.001**
B+TCP	24.14	**<0.001**
B+BG+TCP	25.49	**<0.001**
Control	13.58	**<0.001**

Table 3. Comparison of bone mineral density (BMD) values between the 12th and 1st weeks with regard to the tested material and control groups

Finally, statistical analysis revealed several significant intergroup differences in BMD values determined in experimental weeks 1 and 12 (Table 4). In the earlier period, the most pronounced difference was observed between groups B+BG and B+TCP (4.33), while the lowest difference pertained to groups B+TCP and B+BG+TCP. After 12 weeks of the experiment, the most pronounced difference in BMD of the bone defect was noted between the control group and the B+BG+TCP group, while the difference between the B and B+TCP groups was the lowest.

	1st week		12th week	
Groups	|t|	P	|t|	p
B vs. Control	0.70	0.48	3.63	**<0.001**
B+BG vs. Control	2.48	**<0.05**	1.41	0.17
B+TCP vs. Control	1.85	0.07	4.54	**<0.05**
B+BG+TCP vs. Control	1.27	0.21	6.40	**<0.001**
B vs. B+BG	1.78	0.08	2.22	**<0.05**
B vs. B+TCP	2.55	**<0.05**	0.91	0.37
B vs. B+BG+TCP	1.98	0.05	2.78	**<0.01**
B+BG vs. B+TCP	4.33	**<0.001**	3.13	**<0.01**
B+BG vs. B+BG+TCP	3.75	**<0.001**	4.99	**<0.001**
B+TCP vs. B+BG+TCP	0.57	0.57	1.87	0.07

Table 4. Comparison of bone mineral density (BMD) values in the 1st and 12th weeks between the tested material and control groups

We have also observed statistically significant differences in the growth of BMD with time between the analyzed groups ($F=8.15$; $p<0.001$). The smallest changes in BMD were yielded by B+BG (0.06±0.01), then the control group (0.08±0.01) and B (0.11±0.02). The largest changes were observed for B+TCP (0.14±0.01) and B+BG+TCP (0.16±0.01). For all paired comparisons, statistically significant differences were noted ($p<0.01$).

5.4.2 Computed tomography

Table 5 summarizes the values of bone density in the experimental groups and in the controls as determined by CT. A gradual increase in the density of healing bone defects was observed in all studied groups. After 12 weeks of the experiment, the highest values of CT bone density were observed in the B+BG (1014.8±53.9 HU) and B+BG+TCP (941.2±28.9 HU) groups, while the lowest values were noted in the controls (812.3±21.8 HU). However, the highest determined values of CT bone density were still lower compared to the bone density of the normal mandible (1218 ± 15.2 HU).

	CT Bone Density [HU]					
Group	1st week	2nd week	3th week	4th week	8th week	12th week
B	647.7 ± 79.1	784.0 ± 82.3	863.2 ± 13.3	763.8 ± 136.1	954.3 ± 56.3	902.8 ± 13.5
B+BG	597.2 ± 48.8	730.5 ± 9.8	814.3 ± 38.1	822.5 ± 99.8	933.3 ± 22.0	1014.8 ± 53.9
B+TCP	410.3 ± 34.0	468.2 ± 59.1	606.7 ± 41.8	804.0 ± 51.7	826.3 ± 39.3	879.2 ± 24.6
B+BG+TCP	454.7 ± 17.1	526.3 ± 35.6	523.2 ± 27.7	725.7 ± 36.3	923.7 ± 31.8	941.2 ± 28.9
Control	256.0 ± 20.4	510.3 ± 28.7	539.0 ± 38.3	597.8 ± 35.3	681.8 ± 19.4	812.3 ± 21.8

Table 5. CT bone density (expressed in Hounsfield units, HU) in bone defects implanted with various materials and in control defects at various time points in the study

The most homogeneous increase in CT bone density values was observed in groups B+BG and B+TCP, and in the controls. Some irregularities in the time profile of bone density were, by contrast, noted in animals of the B and B+BG+TCP groups (Fig. 8).

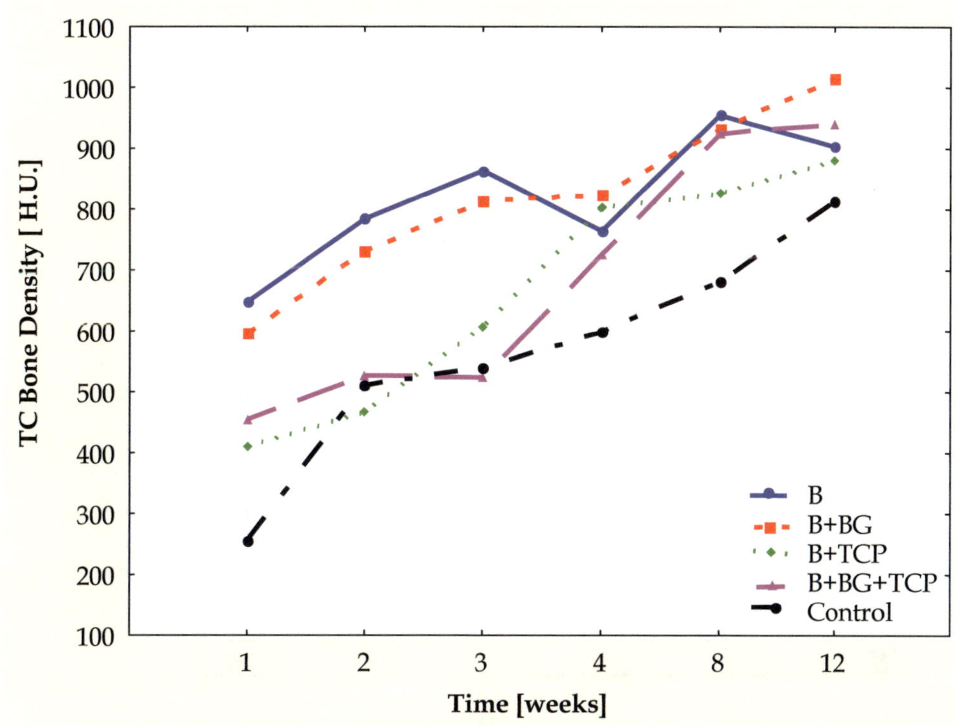

Fig. 8. Time profile of changes in CT bone density (expressed in Hounsfield units, HU) in bone defects implanted with various materials and in control defects

Due to non-sphericity (Mauchley test: 0.0030; $p<0.001$) only multivariable tests were used. We confirmed that there were statistically significant changes in Hounsfield Units with time ($F = 1244.88$; $p<0.001$). Moreover, the interaction between time and the type of material used (group) was also statistically significant ($F = 65.35$; $p<0.001$).

When the time profiles of CT bone density changes were compared between the studied groups, the only significant differences observed were between groups B and B+BG - 0.86, and B+TCP and B+BG+TCP - 0.30 (Table 6).

Profile comparison (p-values)				
Group	**Control**	**B**	**B+BG**	**B+TCP**
B	**<0.001**	–	–	–
B+BG	**<0.001**	0.86	–	–
B+TCP	**<0.001**	**<0.001**	**<0.001**	–
B+BG+TCP	**<0.001**	**<0.001**	**<0.001**	0.30

Table 6. The p-values for comparison of CT bone density (expressed in Hounsfield units, HU) with changes in time (profiles) between the tested material and control groups

The most evident intragroup differences in CT bone density determined in the 1st and 12th experimental weeks were observed in the control group (29.77), but the differences were only slightly less pronounced in groups B+TCP (24.10) and B+BG+TCP (22.94). The least evident differences between the two analyzed time points were noted in the B (10.67) and B+BG (16.77) groups (Table 7).

	12th vs. 1st week	
Group	|t|	p
B	10.67	**<0.001**
B+BG	16.77	**<0.001**
B+TCP	24.10	**<0.001**
B+BG+TCP	22.94	**<0.001**
Control	29.77	**<0.001**

Table 7. Comparison of CT bone density values (expressed in Hounsfield units, HU) between the 12th and 1st weeks with regard to the tested material and control groups

After seven days of the experiment, the most pronounced intergroup differences in CT bone density were observed between the control group and groups B and BG. The least pronounced intergroup differences noted in this period pertained to groups B and B+BG. After 12 weeks of the study, the most evident intergroup differences in CT bone density were again observed between the controls and group B+BG, while the lowest differences were noted between groups B and B+TCP.

	1st week		**12th week**	
Groups	|t|	p	|t|	p
B vs. Control	16.63	**<0.001**	4.91	**<0.001**
B+BG vs. Control	15.23	**<0.001**	10.29	**<0.001**
B+TCP vs. Control	8.48	**<0.001**	3.67	**<0.05**
B+BG+TCP vs. Control	10.36	**<0.001**	6.83	**<0.001**
B vs. B+BG	1.56	0.1325	6.01	**<0.001**
B vs. B+TCP	9.11	**<0.001**	1.39	0.1779
B vs. B+BG+TCP	7.00	**<0.001**	2.14	**<0.05**
B+BG vs. B+TCP	7.55	**<0.001**	7.40	**<0.001**
B+BG vs. B+BG+TCP	5.44	**<0.001**	3.87	**<0.001**
B+TCP vs. B+BG+TCP	2.11	**<0.05**	3.53	**<0.01**

Table 8. Comparison of CT bone density (expressed in Hounsfield units, HU) values in the 1st and 12th weeks between tested material and control groups

We have also observed statistically significant intergroup differences in the relative increase in CT bone density (expressed in Hounsfield units, HU) ($F = 39.22$; $p<0.001$). The smallest changes in CT bone density were yielded by group B (255.2±67.6). Larger changes were observed for B+BG (417.7±25.7), B+TCP (468.8±43.3) and B+BG+TCP (486.5±16.5). The largest change was noted in the control group (556.2±34.4). For all paired comparisons, statistically significant differences (with $p<0.05$) were observed, with the only exception being a comparison between B+TCP and B+BG+TCP.

5.5 Histopathologic analysis

In all analyzed groups, after 7 days of the study, mandibular bone defects had filled with immature fibrous tissue. Numerous, minute granules of deproteinized human bone (with no signs of activity) were seen in this tissue (group B). Additionally, depending on the implant composition, foggy fractions of bio-glass or linearly cracked deposits of tricalcium phosphate could be seen (in groups B+BG, B+TCP and B+BG+TCP). At the same time, active reconstruction of skeletal tissue was observed at the entire periphery of the bone defects implanted with B, B+TCP and B+BG+TCP, as suggested by the presence of ground substance (osteoid) containing numerous immature bone trabeculae covered with osteoblasts. In controls, as well as in groups B and B+BG, blood clots along with the remnants of necrotic bone trabeculae could be seen in the bone marrow at the base of the bone defect. In groups where bio-glass was implanted into the bone defect (B+BG and B+BG+TCP), fragments of this material formed pseudocystic structures that were covered with a thin connective tissue capsule comprised of fibroblasts, fibrocytes and single giant polynuclear cells.

After two weeks of the experiment, the implanted bone defects were filled with mature fibrous connective tissue that contained collagen fibers. Only in group B could immature connective tissue with numerous blood vessels be seen at the periphery of the defect. Immature bone trabeculae were visible in this tissue, surrounded with osteoblasts. Depending on the implant composition, fragments of deproteinized bone granules, bio-glass and/or tricalcium phosphate were seen in the fibrous connective tissue. Around BG and TCP particles, distinct, thick-walled pseudocystic structures could be observed, containing numerous giant polynuclear cells. Additionally, giant polynuclear cells were frequently visible on the surface of deproteinized bone but they did not form distinct capsule-like linear structures in this location. Fragmentation of some TCP particles could be observed due to infiltration by cells composing the previously mentioned cystic structures. Active osteogenesis was evident in all analyzed groups, as manifested by the pronounced growth of numerous immature bone trabeculae covered with osteoblasts. This process was particularly intensive around the particles of implanted material.

In group B, advanced reconstruction of bone defects was observed after three weeks of the study. Most bone trabeculae filling the defect were mature with either no or very little osteoblastic activity. Linearly placed osteoblasts, or even osteoclasts, could be seen on the surface of remaining trabeculae (Fig. 9). A similar advancement in the bone formation process was observed in B+TCP specimens: osteoid was formed on the base of connective fibrous tissue along with the intensive growth of numerous bone trabeculae. Some of these trabeculae were mature already and showed no signs of cellular activity. Skeletal tissue regeneration in this group was particularly enhanced around the fragments of deproteinized bone and tricalcium phosphate (Fig. 10). In contrast to B particles, TCP particles showed signs of dilution and structural fragmentation. Only a few giant polynuclear cells could be seen around the implants, and the previously observed cystic structures of TCP had only residual character. After three weeks of the study, slightly less advanced processes of skeletal defect healing were observed in histologic specimens from groups B+BG and B+BG+TCP. Growth of immature skeletal tissue was observed in these groups on the "scaffold" of deproteinized bone, giving the impression of the implant being "incorporated" into the growing bone trabeculae. Only single giant polynuclear cells could be seen on the surface of specimens from group B+BG. Additionally, thin, linearly placed cells could be observed around some bio-glass particles, forming pseudocystic structures. These structures were visible mostly in areas directly adjacent to fibrous connective tissue. Numerous bone trabeculae were observed around the B+BG+TCP mixture particles, as well as in the fibrous connective tissue between the particles (Fig. 11). Some trabeculae were mature and showed no signs of osteoblastic activity on their surfaces. Some implant particles in this group were covered with giant polynuclear cells forming structures resembling foreign body granulomas. In the control group, all osteoid trabeculae were surrounded by osteoblasts still showing signs of osteoblastic activity.

After four weeks of the study, only the bone defects in group B were nearly completely filled with mature, compact skeletal tissue. This tissue contained numerous "incorporated" granules of deproteinized human bone. Growing bone trabeculae with surface signs of osteoblastic activity could only be seen in a narrow layer of fibrous connective tissue located between the newly formed bone and the bottom of the defect. At the same time, mature cancellous skeletal tissue could only be observed at the periphery of bone defects in groups B+TCP and B+BG+TCP. "Incorporated" fragments of the implanted material, mostly human bone, were visible in this tissue. The central part of the defects was still filled with fibrous connective tissue, showing signs of the ongoing process of bone formation. This area also contained B particles and a few particles of TCP ceramic. These fragments of tricalcium phosphate were covered with a pseudo-capsule comprised of giant polynuclear cells, and were gradually fragmented and resorbed. After four weeks of the experiment, numerous mature bone trabeculae with no signs of cellular activity were observed in specimens from group B+BG (Fig. 12). Some of these trabeculae developed around B and BG particles, giving the impression of "incorporating" implanted material into the structure of reconstructed bone. Some bone defects were still partially filled with mature fibrous connective tissue containing numerous collagen fibers along with particles of implanted material. Additionally, pseudocystic structures could be observed around the BG particles, comprised of linearly placed giant polynuclear cells.

Histopathologic examination performed after eight weeks of the study revealed the completed process of bone healing in group B. Bone structure was fully regenerated, and mature compact bone and cancellous bone could be observed at the defect site, along with normal bone marrow (Fig. 13). Observations made in the B+BG+TCP group after the 8th and 12th weeks of the experiment gave similar findings. In this group bone defects were also filled with mature compact and cancellous skeletal tissue, with "incorporated" particles of bio-glass and deproteinized bone still visible (Fig. 14). These particles showed no signs of activity, and were covered with thin fibrous capsules. In the 8th experimental week, these particles could also be observed in fibrous connective tissue. After the 8th week of the study, the process of osteogenesis was still incomplete in B+BG and B+TCP specimens. In defects implanted with bio-glass, some areas, usually peripheral ones, were still filled with fibrous connective tissue containing numerous collagen fibers. This connective tissue showed signs of ongoing bone formation: maturating or mature bone trabeculae, along with bio-glass particles forming pseudocystic structures (Fig. 15). Particles of deproteinized bone were more rarely evidenced. At the same time, continued bone formation was observed in the central part of bone defects in group B+TCP (Fig. 16). This process was particularly intensive around the B particles, giving the characteristic impression of "incorporating" this material into newly formed bone. After eight weeks of the study, only some bone trabeculae in the control group showed signs of osteoblastic activity, while other areas of the defect contained mature trabeculae.

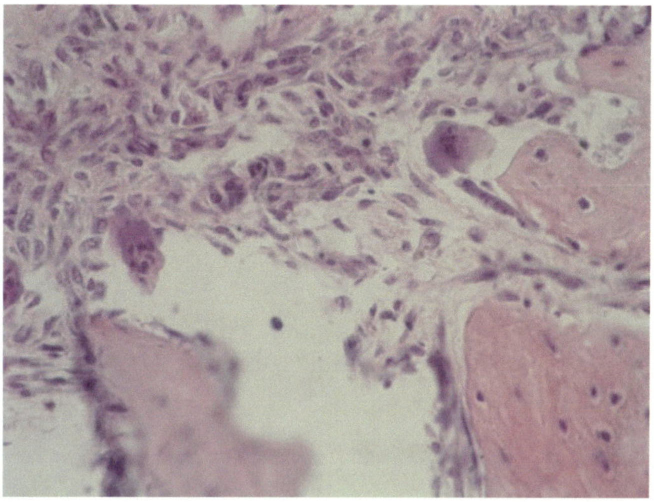

Fig. 9. Histopathologic specimen - 3rd week, group B - single osteoclasts on the surface of bone trabeculae (H&E staining, magnification 400 x)

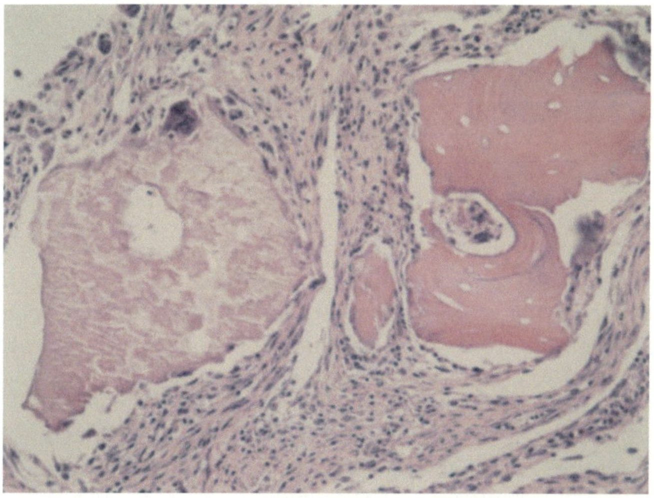

Fig. 10. Histopathologic specimen - 3rd week, group B+TCP - regeneration of skeletal tissue around fragments of deproteinized human bone and tricalcium phosphate (H&E staining, magnification 200 x)

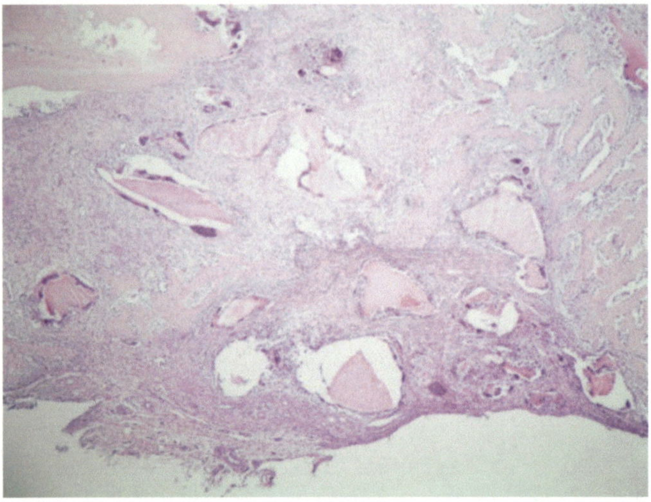

Fig. 11. Histopathologic specimen - 3rd week, group B+BG+TCP - numerous bone trabeculae around the implanted material visible within fibrous connective tissue (H&E staining, magnification 40 x)

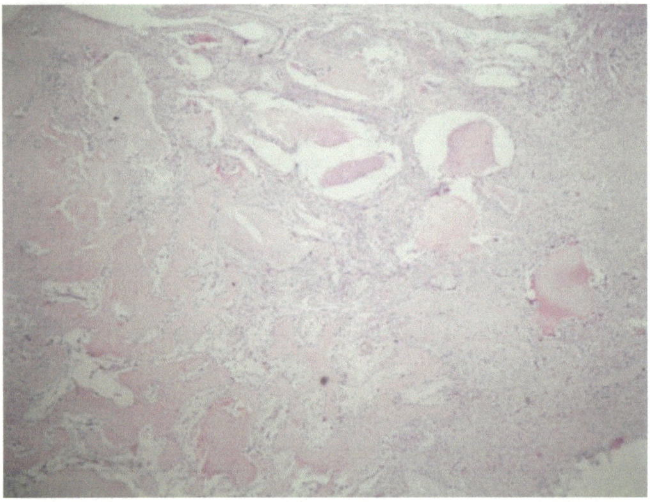

Fig. 12. Histopathologic specimen - 4th week, group B+BG - mature bone trabeculae with no signs of cellular activity, along with particles of deproteinized human bone and bio-glass visible within fibrous connective tissue (H&E staining, magnification 40 x)

After 12 weeks of the study, the process of skeletal tissue regeneration in specimens from groups B+BG and B+TCP was still not complete. Although the defects were nearly filled in their entirety with mature compact and cancellous bone, mature fibrous connective tissue containing numerous collagen fibers and showing signs of ongoing osteogenesis could still be seen in the superficial zone (Fig. 17). This superficial layer contained remnants of deproteinized bone, covered with capsules comprised of giant polynuclear cells, which formed inactive, fibrous foreign body granulomas. Additionally, cystic structures containing foggy remnants of incompletely resorbed TCP (group B+TCP) or BG particles (group B+BG) were revealed in fibrous connective tissue (Fig. 18). In the control group, specimens obtained in the 12th week of the study contained completely matured and fully mineralized bone.

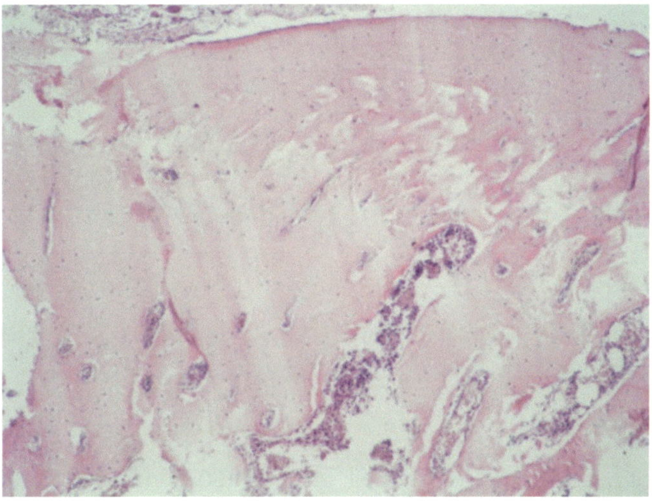

Fig. 13. Histopathologic specimen - 8th week, group B - mature skeletal tissue and bone marrow (H&E staining, magnification 100 x)

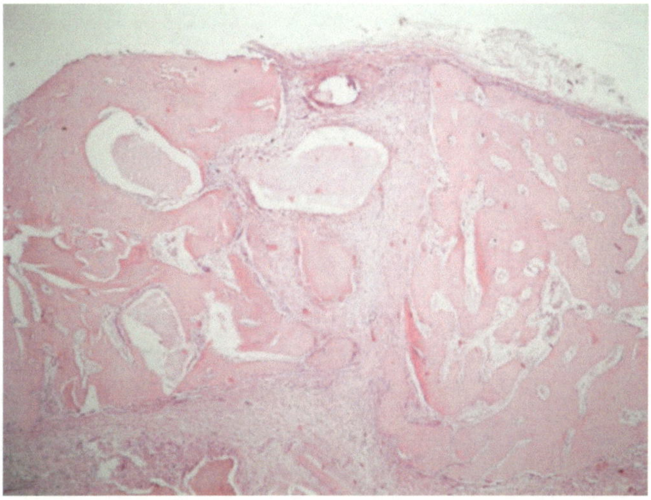

Fig. 14. Histopathologic specimen - 8th week, group B+BG+TCP - mature compact and cancellous skeletal tissue with "incorporated" particles of bio-glass and human bone (H&E staining, magnification 100 x)

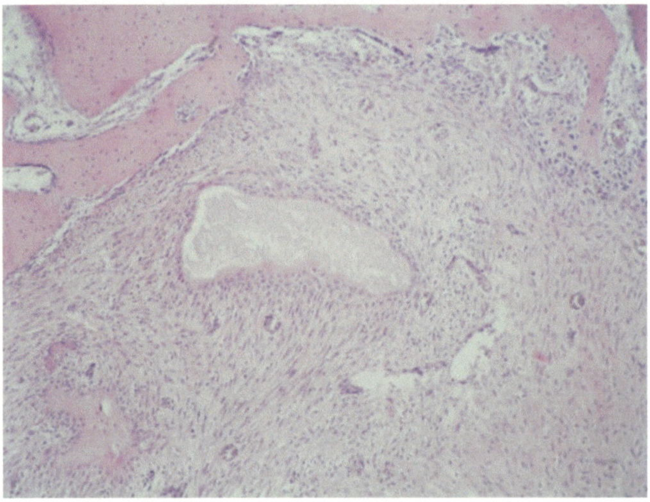

Fig. 15. Histopathologic specimen - 8th week, group B+BG - bio-glass particle forming pseudocystic structure within the mature fibrous connective tissue (H&E staining, magnification 100 x)

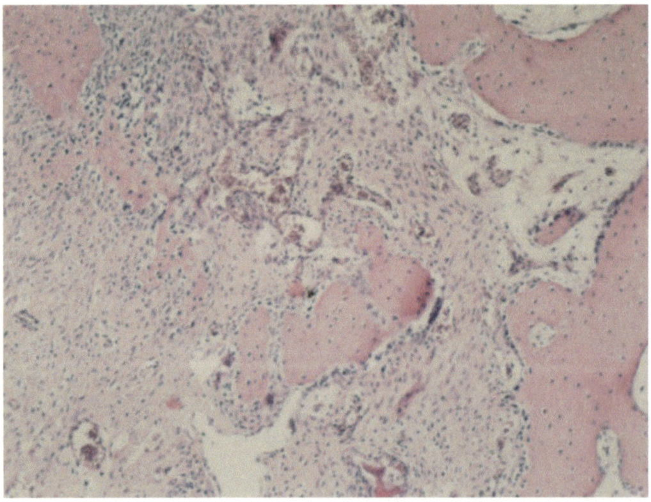

Fig. 16. Histopathologic specimen - 8th week, group B+TCP - fibrous connective tissue with massive osteogenesis (H&E staining, magnification 100 x)

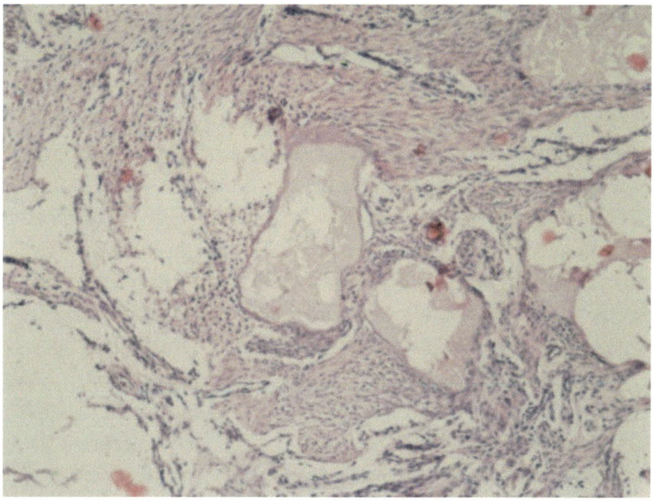

Fig. 17. Histopathologic specimen - 12th week, group B+BG - mature fibrous connective tissue with numerous cystic spaces filled with foggy deposits of bio-glass (H&E staining, 100 x)

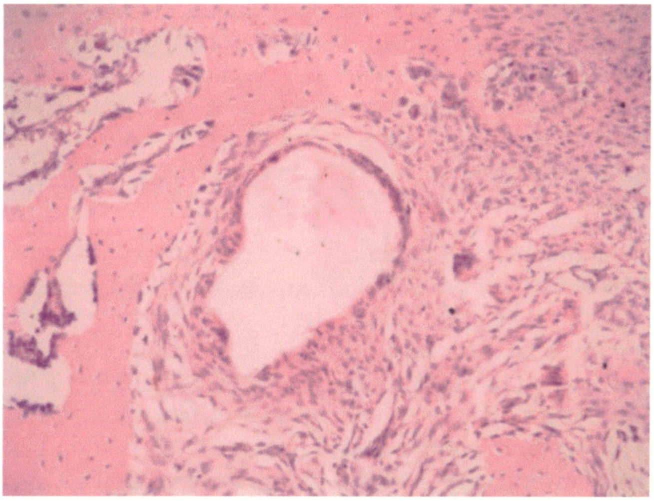

Fig. 18. Histopathologic specimen - 12th week, group B+TCP - cystic structures within fibrous connective tissue containing foggy remnants of tricalcium phosphate (H&E staining, 200 x)

6. Discussion

Due to the limited regenerative ability of skeletal tissue, bone grafting or the implantation of bone derivatives or bone replacement materials is required for the complete healing of large bone defects, whether the result of surgical removal of skeletal cysts or tumors, or caused by other skeletal disorders (Precheur, 2007). The most satisfactory results in stimulating skeletal tissue regeneration have been reported after using autogenous grafts. After being implanted into the bone defect, autogenous grafts can induce all the basic mechanisms responsible for bone reconstruction, i.e. osteogenesis, osteoinduction and osteoconduction (Giannoudis et al., 2005; Merkx et al., 2003). Osteogenic activity is associated with the presence of osteoprogenitor cells in the periosteum, endosteum and bone marrow. In the case of free bone grafts, some of the cells located most superficially may survive and are involved in regeneration processes. The results of bone healing stimulation are definitely most satisfactory when autogenous cancellous bone chips are implanted, since this type of bone contains a high number of osteoprogenitor cells. Osteogenic activity can only be observed in fresh bone grafts. Osteoinduction is associated with the presence of so-called bone morphogenetic proteins (BMP) in the bone matrix. These proteins are released during bone remodeling, and can stimulate minimally differentiated connective tissue cells surrounding the graft to transform into osteoblasts (Barradas et al., 2011). Both fresh autogenic bone grafts and the allogenic grafts obtained from tissue bank (especially when frozen and partly decalcified) have osteoinductive properties. Additionally, BMP preparations can be obtained by extraction from bones, and as of recent, also by biotechnological synthesis in recombinant form. Allogenic bones are very frequently lyophilized, which depletes them of BMP. Additionally, such grafts lose their immunogenic properties due to irradiation sterilization and deep freezing (Bohner, 2010; Liu et al., 2008). This process of sterilization results in decreased durability of the material, and the preserved bone matrix has only osteoconductive properties. It is degraded by osteoclasts with the simultaneous formation of woven bone, which is further transformed into lamellar bone through the process of osteoclasia. Such grafts have been shown to undergo revascularization and remodeling - similar to autogenous grafts, but at a slower rate. Apart from bone implants, organic and inorganic alloplastic bone replacement materials also have osteoconductive properties. Combined with growth factors and autogenous barrier membranes, they are frequently used as basic elements in the process of guided bone regeneration (GBR) (Kao & Scott, 2007; Schwarz et al., 2007).

For various reasons, different bone materials are frequently combined with each other or with alloplastic biomaterials. As a result, biologically improved material compositions are obtained, some of which positively influence the bone formation processes. Combination of natural hydroxyapatite with chitosane resulted in a composite with osteoconductive properties. In the presence of this composite, tibial consolidation in rabbits was observed as early as 12 weeks after implantation, and complete healing was observed after 16 weeks of the study (Yuan et al., 2008). In another study, a composite based on bovine bone with the addition of bio-glass showed no cytotoxicity to human fibroblasts. Moreover, a crystalline carbonated apatite phase was developed on the sample surface as early as 12 days after immersion in simulated body fluid (Yoganad e al., 2010). Another example of the positive effects of combining deproteinized bovine bone with autogenous bone comes from a study in which such a material was used for the regeneration of bone defects in the frontal part of the porcine skull. The presence of autogenous bone in the mixture was the basis for the osteoinductive properties of the material and the more favorable biological conditions for bone growth when compared to deproteinized bovine bone alone (Thorwarth et al., 2006). Similarly, more satisfactory clinical results were reported when deproteinized bovine bone was used in combination with autogenous bone in the management of alveoschisis in humans, instead of bone autograft alone (Thuaksuban et al., 2010). Experiments on rabbits have also given interesting results. It was revealed that the addition of deproteinized bovine bone to autogenous grafts increased the mean optical density of newly formed skeletal tissue, with a simultaneous decrease in its content in bone defects (skullcap). The opposite effects were observed when autogenous bone graft was used alone (Pripatnanont et al., 2009). Using a mixture of allogenic bone and deproteinized bovine bone (BioOss®/Orthoblast II®) for the purposes of maxillary sinus lift did not have results as satisfactory as with the application of deproteinized animal bone or synthetic bone (Osteon®) alone. The individual use of one of these two implant materials was associated with a higher percentage of newly formed osseous fraction collected from the lateral sinus at 4 and 6 months post-operatively (Kim et al., 2009).

In this study we have combined allogenic bone with artificially obtained biomaterials (bio-glass - BG, and/or beta-tricalcium phosphate - TCP) in order to form bone replacement material with improved biological characteristics. Reference materials for comparative analysis of the studied mixtures (B+BG, B+TCP, B+BG+TCP) included lyophilized human bone (B) and the clotted blood of experimental animals. Both clinical observations and further macroscopic, radiographic and histopathologic examinations confirmed that bone defects healed normally in the presence of all studied biomaterials. However, the type of implanted mixture modulated the kinetics of bone formation and the quality of newly formed bone. Bone regeneration was induced markedly earlier whenever biologically active bio-glass was included in the implanted mixture (B+BG, B+BG+TCP). In groups where bio-glass was implanted, irregular shadows were observed on radiographic images of the bone defect sites as early as after two weeks of the study. Probably, these radiographic changes resulted from ongoing reparative processes within the bone. This was additionally confirmed on histopathologic analysis, which revealed intense bone formation processes as early as three weeks after the implantation of BG-containing material. However, bone density measurements (BMD and CT bone density) taken in the early period of this study confirmed the superior quality of newly formed bone only in case of the B+BG mixture. It is plausible that the lower bone densities determined for B+BG+TCP implants resulted from the low content of bio-glass in this mixture. Moreover, as confirmed by histopathologic analysis, resorption of beta-tricalcium phosphate contained in B+BG+TCP already began in the early period of this study. Nonetheless, in the later period of this study, increases in BMD and CT bone density of B+BG implanted bone were markedly lower. As a result, after 12 weeks of the experiment, the defects filled with this mixture were characterized by the lowest BMD values, and histopathologic examination confirmed ongoing bone formation. The final result of bone regeneration was markedly better in the case of defects implanted with B+BG+TCP. In the 12th week, histopathologic analysis revealed mature skeletal tissue (both compact and cancellous bone) at the implantation sites, and this finding was confirmed on radiographic examination. Additionally, new bone formed using B+BG+TCP implantation was characterized by the highest BMD and relatively high CT bone density. Therefore, this regenerated bone most closely resembled the normal skeletal tissue of experimental animals of all mixtures examined. In the 12th week of this study, bone formation processes were still observed in B+TCP implanted defects. Although the BMD of tissue formed on the basis of this implant was higher than in the B+BG implanted bone, it was still lower than in the B+BG+TCP group. Notably, in both the 1st and 12th experimental weeks, only slight differences in BMD and CT bone density were observed between the B+TCP and B+BG+TCP mixtures. Undoubtedly, the process of bone defect regeneration was completed the earliest in group B. Histopathologic studies confirmed that bone defects in this group were filled with mature skeletal tissue with no signs of osteoblastic activity as early as after eight weeks of the study. Early completion of skeletal healing was also confirmed by the radiographic images taken in this group. In the 12th experimental week, BMD of bone defects implanted with human bone alone was higher in comparison to defects filled with B+TCP and B+BG+TCP, in contrast to the early period of this study when BMD and CT bone density in group B were among the highest.

Results of this study suggest that the quality of bone formed in the late period of the experiment was poorest in the control group, as suggested by the low bone density of defects healing on the basis of clotted blood. Since histopathologic studies confirmed that bone formation in controls was complete in week 12, one should not expect further increases in bone density in this group. It is likely that bone density would increase further, however, with prolonged observation of the B+BG and B+TCP groups, since the last densitometric measurements in these groups were taken on incompletely matured bone.

7. Conclusion

This *in vivo* animal study revealed that both lyophilized human bone (B) and mixtures formed on its basis (B+BG, B+TCP, B+BG+TCP) are fully biocompatible materials. We have confirmed that, in the presence of these materials, there is a possibility of forming normal, mature skeletal tissue in mandibular bone defects of guinea pigs. This was possible mostly thanks to the inclusion of deproteinized human bone in analyzed mixtures. Due to its osteoconductive properties, the presence of human bone resulted in favorable biological conditions that promoted skeletal regeneration. Therefore, deproteinized bone formed a scaffold to support the growth of osteogenic cells within the defects. Furthermore, the addition of alloplastic materials, bio-glass and beta-tricalcium phosphate, markedly influenced the rate of bone formation and the quality of newly formed bone. The most satisfactory results were observed in the case of lyophilized human bone mixed with bio-glass and beta-tricalcium phosphate (B+BG+TCP). The group implanted with this material was the only one in which fully matured compact and cancellous bone was observed on histopathologic examination performed after 12 weeks of the study. The particles of bio-glass and beta-tricalcium phosphate included in the mixture induced the processes of bone formation and stimulated the growth of osteogenic cells. As a result of these initiated biological processes, defragmented and resorbed β-TCP was gradually replaced with newly formed skeletal structures. Such a course of bone formation process modulated the quality of newly formed bone, as confirmed by high BMD and CT bone density values determined after 12 weeks in B+BG+TCP implanted defects. As confirmed on histopathologic examination, throughout the three-month period of this study, skeletal regeneration was not completed in defects implanted with B+BG and B+TCP mixtures. This incomplete regeneration was reflected by the lower bone density values observed in these groups when compared to the B+BG+TCP group. In conclusion, the results of this study confirmed our initial assumptions. We have revealed that, in addition to a proper course of bone formation processes, another important outcome in the presence of various bone replacement materials is the high quality of newly formed bone. In our opinion, these two aforementioned results were positively achieved by the mixture of lyophilized human bone with bio-glass and tricalcium phosphate.

8. References

Abu Bakar, M.S., Cheng, M.H.W., Tang, S.M., Yu, S.C., Liao, K., Tan, C.T., Khor, K.A. & Cheang, P. (2003). Tensile Properties, Tension–tension Fatigue and Biological

Response of Polyetheretherketone–hydroxyapatite Composites for Load-bearing Orthopedic Implants. *Biomaterials,* Vol.24, No.13, (July 2003), pp. 2245-2250, ISSN 0142-9612

Accorsi-Mendonça, T., Conz, M.B., Barros, T.C., de Sena, L.Á., de Almeida Soares, G. & Granjeiro, J.M. (2008). Physicochemical Characterization of Two Deproteinized Bovine Xenografts. *Brazilian Oral Research,* Vol.22, No.1, (January-March 2008), pp. 5-10, ISSN 1806-8324

Baldini, N., De Sanctis, M. & Ferrari, M. (2011). Deproteinized Bovine Bone in Periodontal and Implant Surgery. *Dental,* Vol.27, No.1, (January 2011), pp. 61-70, ISSN 1879-0097

Barakat, N.A.M., Khil, M.S., Omran, A.M., Sheikh, F.A. & Kim, H.Y. (2009). Extraction of Pure Natural Hydroxyapatite from the Bovine Bones Bio Waste by Three Different Methods. *Journal of Materials Processing Technology,* Vol.209, No.7, (April 2009), pp. 3408-3415, ISSN 0924-0136

Barradas, A.M.C., Yuan, H., van Blitterswijk, C.A. & Habibovic, P. (2011). Osteoinductive Biomaterials: Current Knowledge of Properties, Experimental Models and Biological Mechanisms. *European Cell and Materials,* Vol.21, (2011), pp. 407-429, ISSN 1473-2262

Bellucci, D., Cannillo, V. & Sola, A. (2011). A New Highly Bioactive Composite for Scaffold Applications: A Feasibility Study. *Materials,* Vol.4, (January 2011), pp. 339-354, ISSN 1996-1944

Block, M.S., Finger, I. & Lytle, R. (2002). Human Mineralized Bone in Extraction Sites Before Implant Placement. *The Journal of the American Dental Association,* Vol.133, No.12, (December 2002), pp. 1631-1638, ISSN 0002-8177

Bohner, M. (2010). Resorbable Biomaterials as Bone Graft Substitutes. *Materials Today,* Vol.13, No.1-2, (January-February 2010), pp. 24-30, ISSN 1369-7021

Cao, K., Huang, W., An, H., Jiang, D., Shu, Y. & Han, Z. (2009). Deproteinized Bone with VEGF Gene Transfer to Facilitate the Repair of Early Avascular Necrosis of Femoral Head of Rabbit. *Chinese Journal of Traumatology (English Edition),* Vol.12, No.5, (October 2009), pp. 269-274, ISSN 1008-1275

Choi, J.W., Cho, H.M., Kwak, E.K., Kwon, T.G., Ryoo, H.M., Jeong, Y.K., Oh, K.S. & Shih, H.I. (2004). Effect of Ag-doped Hydroxyapatite as a Bone Filler for Inflamed Bone Defects. *Key Engineering Materials,* Vol.254-256, (2004), pp. 47-50, ISSN 1013-9826

Cieślik, M., Mertas, A., Morawska-Chochół, A., Sabat, D., Orlicki, R., Owczarek, A., Król, W. & Cieślik, T. (2009). The Evaluation of the Possibilities of Using PLGA co-polymer and Its Composites with Carbon Fibers or Hydroxyapatite in Bone Tissue Regeneration Process – *In Vitro* and *In Vivo* Examinations. *International Journal of Molecular Science,* Vol.10, No.7, (July 2009), pp. 3224-3234, ISSN 1422-0067

Daculsi, G., Laboux, O., Malard, O. & Weiss, P. (2003). Current State of the Art of Biphasic Calcium Phosphate Bioceramics. *Journal of Materials Science Materials in Medicine,* Vol.14, No.3, (March 2003), pp. 195-200, ISSN 0957-4530

Deculsi, G. (1998). Biphasic Calcium Phosphate Concept Applied to Artificial Bone, Implant Coating and Injectable Bone Substitute. *Biomaterials,* Vol.19, No.16, (August 1998), pp. 1473-1478, ISSN 0142-9612

Fellah, B.H., Gauthier, O., Weiss, P., Chappard, D. & Layrolle, P. (2008). Osteogenicity of Biphasic Calcium Phosphate Ceramics and Bone Autograft in a Goat Model. *Biomaterials,* Vol.29, No.9, (March 2008), pp. 1177-1188, ISSN 0142-9612

Ferreira, J.M.F. (2007). Development and In Vitro Characterization of Sol–gel Derived CaO–P_2O_5–SiO_2–ZnO Bioglass. *Acta Biomaterialia,* Vol.3, No.2, (March 2007), pp. 255-262, ISSN 1742-706

Giannoudis, P.V., Dinopoulos, H. & Tsiridis, E. (2005). Bone Substitutes: An Update. Injury, Vol.36, No.3, Supp.1, (November 2005), pp. 20-27,ISSN 0020-1383

Henkel, K.O., Gerber, T., Gundlach, K.K.H. & Bienengräber, V. (2006). Macroscopical, Histological, and Morphometric Studies of Porous Bone-Replacement Materials in Minipigs 8 Months After Implantation. *Oral Surgery, Oral Medicine, Oral Pathology, Oral Radiology, and Endodontology,* Vol.102, No.5, (November 2006), pp. 606-613, ISSN 1079-2104

Jian, Y., Tian, X., Li, B., Qiu, B., Zhou, Z., Yang, Z. & Li, Q. (2008). Properties of Deproteinized Bone for Reparation of Big Segmental Defect in Long Bone. *Chinese Journal of Traumatology (English Edition),* Vol.11, No.3, (June 2008), pp. 152-156, ISSN 1008-1275

Kao, S.T. & Scott, D.D. (2007). A Review of Bone Substitutes. *Oral and Maxillofacial Surgery Clinics of North America,* Vol.19, No.4, (November 2007), pp. 513-521, ISSN 1042-3699

Kim, Y.K., Yun, P.Y., Kim, S.G. & Lim, S.C. (2009). Analysis of the Healing Process in Sinus Bone Grafting Using Various Grafting Materials. *Oral Surgery, Oral Medicine, Oral Pathology, Oral Radiology, and Endodontology,* Vol.107, No.2, (February 2009), pp. 204-211, ISSN 1076-2104

Kon, M., Ishikawa, K., Miyamoto, Y. & Asaoka, K. (1995). Development of Calcium Phosphate Based Functional Gradient Bioceramics. *Biomaterials,* Vol.16, No.9, (June 1995), pp. 709-714, ISSN 2079-4983

Liu, L., Pei, F., Tu, C., Zhou, Z. & Li, Q. (2008). Immunological Study on the Transplantation of an Improved Deproteinized Heterogeneous Bone Scaffold Material in Tissue Engineering. *Chinese Journal of Traumatology (English Edition),* Vol.11, No.3, (June 2008), pp. 141-147, ISSN 1008-1275

Merkx, M.A.V., Maltha, J.C. & Stoelinga, P.J.W. (2003). Assessment of the Value of Anorganic Bone Additives in Sinus Floor Augmentation: A Review of Clinical Reports. *International Journal of Oral and Maxillofacial Surgery,* Vol.32, No.1, (February 2003), pp. 1-6, ISSN 1548-1336

Nagata, F., Miyajima, T. & Yokagawa, Y. (2003). Preparation of Porous Composites Consisting of Apatite and Poly(D,L-lactide). *Key Engineering Materials,* Vol.240-242, (2003), pp. 167-170, ISSN 1013-9826

Nagata, F., Miyajima, T., Teraoka, K. & Yokogawa, Y. (2005). Preparation of Porous Poly(lactic acid)/Hydroxyapatite Microspheres Intended for Injectable Bone Substitutes. *Key Engineering Materials,* Vol.284-286, (2005), pp. 819-822, ISSN 1013-9826

Oonishi, H., Hench, L.L., Wilson, J., Sugihara, F., Tsuji, E., Kushitani, S. & Iwaki, H. (1999). Comparative bone growth behavior in granules of bioceramic materials of various sizes. *Journal of Biomedical Material Research,* Vol.44, No.1, (January 1999), pp. 31-43, ISSN 0021-9304

Precheur, H.V. (2007). Bone Grafts Materials. *Dental Clinics of North America,* Vol.51, No.3, (July 2007), pp. 729-746, ISSN 0011-8532

Pripatnanont, P., Nuntanaranont, T. & Vongvatcharanont, S. (2009). Proportion of Deproteinized Bovine Bone and Autogenous Bone Affects Bone Formation in the Treatment of Calvarial Defects in Rabbits. *International Journal of Oral and Maxillofacial Surgery,* Vol.38, No.4, (April 2009), pp. 356-362, ISSN 1399-0020

Ravarian, R., Moztarzadeh, F., Solati Hashjin, M., Rabiee, S.M., Khoshakhlagh, P. & Tahriri, M. (2010). Synthesis, Characterization and Bioactivity Investigation of Bioglass/Hydroxyapatite Composite. *Ceramics International,* Vol.36, No.1, (January 2010), pp. 291-297, ISSN 0272-8842

Santos, L.A., Carrodeguas, R.G., Rogero, S.O., Higa, O.Z., Boschi, A.O. & Arruda, A.C.F. (2002). α-Tricalcium phosphate cement: "in vitro" cytotoxicity. *Biomaterials,* Vol.23, No.9, (May 2002), pp. 2035-2042, ISSN 2079-4983

Schwarz, F., Herten, M., Ferrari, D., Wieland, M., Schmitz, L., Engelhardt, E. & Becker, J. (2007). Guided Bone Regeneration at Dehiscence-type Defects Using Biphasic Hydroxyapatite + Beta Tricalcium Phosphate (Bone Ceramic®) or a Collagen-coated Natural Bone Mineral (BioOss Collagen®): An Immunohistochemical Study in Dogs. *International Journal of Oral and Maxillofacial Surgery,* Vol.36, No.12, (December 2007), pp. 1198-1206, ISSN 0901-5027

Ślósarczyk, A. & Paszkiewicz, Z. (2005). Method of obtaining highly reactive calcium phosphate powder. Pl Patent 1900486 B1, App. No. 331907, Int. Cl.{7} : C01B 25/32, (December 2005)

Thomas, M.V., Puleo, D.A. & Al-Sabbagh, M. (2005). Bioactive Glass Three Decades on. *Journal of Long-Term Effects of Medical Implants,* Vol.15, No.6, (February 2005), pp. 585-597, ISSN 1050-6934

Thorwarth, M., Schlegel, K.A., Wehrhan, F., Srour, S. & Schultze-Mosgau, S. (2006). Acceleration of de Novo Bone Formation Following Application of Autogenous Bone to Particulated Anorganic Bovine Material *In Vivo. Oral Surgery, Oral Medicine, Oral Pathology, Oral Radiology, and Endodontology,* Vol.101, No.3, (March 2006), pp. 309-316, ISSN 1076-2104

Thuaksuban, N., Nuntanaranont, T. & Pripatnanont, P. (2010). A comparison of Autogenous Bone Graft Combined with Deproteinized Bovine Bone and Autogenous Bone Graft Alone for Treatment of Alveolar Cleft. *International Journal of Oral and Maxillofacial Surgery,* Vol.39, No.12, (December 2010), pp. 1175-1180, ISSN 1399-0020

Von Wattenwyl, R., Siepe, M., Arnold, R., Beyersdorf, F. & Schlensak, C. (2011). Sternum Augmentation with Bovine Bone Substitute in the Neonate. *The Annals of Thoracic Surgery,* Vol.91, No.1, (January 2011), pp. 311-313, ISSN 0003-4975

Wang, S., Zhang, Z., Zhao, J., Zhang, X., Sun, X., Xia, L., Chang, Q., Ye, D. & Jiang, X. (2009). Vertical Alveolar Ridge Augmentation with β-tricalcium phosphate and Aautologous Osteoblasts in Canine Mandible. *Biomaterials,* Vol.30, No.13, (May 2009), pp. 2489-2498, ISSN 0142-9612

Wei G. & Ma P.X. (2004). Structure and Properties of Nano-Hydroxyapatite/Polymer Composite Scaffolds for Bone Tissue Engineering. *Biomaterials,* Vol.25, No.19, (August 2004), pp. 4749-4757, ISSN 0142-9612

Xynos, I.D., Hukkanen, M.V., Batten, J.J., Buttery, L.D., Hench, L.L. & Polak. J.M. (2000). Bioglass 45S5 Stimulates Osteoblast Turnover and Enhances Bone Formation *In*

Vitro: Implications and Applications for Bone Tissue Engineering. *Calcified Tissue International,* Vol.67, No.4, (October 2000), pp. 321-329, ISSN 1432-0827

Yoganand, C.P., Selvarajan, V., Cannillo, V., Sola, A., Roumeli, E., Goudouri, O.M., Paraskevopoulos, K.M. & Rouabhia, M. (2010). Characterization and In vitro-bioactivity of Natural Hydroxyapatite Based Bio-glass–ceramics Synthesized by Thermal Plasma Processing. *Ceramics,* Vol.36, No.6, (August 2010), pp.1757-1766, ISSN 0272-8842

Yuan, H., Chen, N., Lü, X. & Zheng, B. (2008). Experimental Study of Natural Hydroxyapatite/Chitosan Composite on Reconstructing Bone Defects. *Journal of Nanjing Medical University,* Vol.22, No.6, (November 2008), pp. 372-375, ISSN 1007-4376

Yuan, H., de Bruijn, J.D., Zhang, X., van Blitterswijk, C.A. & de Groot, K. (2001). Bone Induction by Porous Glass Ceramic Made from Bioglass (45S5). *Journal of Biomedical Materials Research,* Vol.58, No.3, (May 2001), pp. 270-276, ISSN 1549-3296

Zhang, B., Zhao, J. & Liu, M. (2009). Osteoinductive Activity of Demineralized Bone Matrix and Deprotenized Bone Derived from Human Avascular Necrotic Femoral Head. *Chinese Journal of Traumatology (English Edition),* Vol.12, No.6, (December 2009), pp. 379-383, ISSN 1008-1275

Zima, A., Paszkiewicz, Z. & Ślósarczyk, A. (2010). TCP Bioceramics (αTCP, βTCP, BTCP) for Orthopaedic and Stomatological Applications - Preparation and *In Vitro* Evaluation. *Ceramic Materials,* Vol.62, No.1, (2010), pp. 51-55, ISSN 1644-3470

18

Endochondral Bone Formation as Blueprint for Regenerative Medicine

Peter J. Emans, Marjolein M.J. Caron,
Lodewijk W. van Rhijn and Tim J.M. Welting
*Department of Orthopaedic Surgery, Maastricht University Medical Center,
The Netherlands*

1. Introduction

During our life moving, walking, sport, etc., are essential for our health and quality of life. Both bones and cartilage enable us to do so. Bones support us, allow muscles to move them, and protect vital internal organs. At the end of most bones articular joints are situated. The side where 2 bones form an articular joint, the ends of these bones are covered with hyaline cartilage. This articular cartilage is able to withstand very high mechanical forces with very low friction and thereby enables easy movement. A large number of bones are formed by a process called endochondral ossification. During this process a cartilage template is replaced by bone, in contrast with the cartilage in newly formed joints which remains cartilage. Both articular cartilage and bone mature and this leads to a well organised architecture and specialisation. The arcade-like architecture of cartilage is capable to withstand an enormous amount of intensive and repetitive forces during life. However, the British surgeon William Hunter made the now famous statement that "*From Hippocrates to the present age it is universally allowed that ulcerated cartilage is a troublesome thing and that once destroyed it is not repaired*" (Hunter 1743). In contrast, bone has a very high regenerative capacity. This difference in self-healing capacity may partially be explained by the access to progenitor cells which contribute to tissue repair. For bone repair, progenitor cells of three different sources have been identified. These sources are: (i) progenitor cells form the blood stream since bone is a highly vascularised tissue, (ii) progenitor cells from the overlying periosteum and (iii) progenitor cells from the bone marrow. Cartilage is not vascularised, is not covered by periosteum, nor has a specialized tissue such as bone marrow and this might be part of the explanation for the limited self-repair capacity of cartilage. Although both tissues start from the same mesenchymal cell condensations, the contrast in self-repair is striking (Hunziker, Kapfinger et al. 2007).

From a clinical point of view there is a need for repair of both bone and cartilage. Bone and cartilage were both identified as tissues for which it was thought to be possible to recreate them in a laboratory setting, using the combination of cell isolation culture techniques and carrier materials. The science of combining cells with carrier materials to reproduce tissues in the laboratory is called Tissue Engineering (TE). The collaboration of scientists of different disciplines such as cell biology, biomaterials, biomechanics, engineering and translational medicine has already led to fruitful scientific achievements. However, initial expectations of tissue engineering have not been reached completely. Although some treatments which apply to the principles of TE have reached clinical practice, TE-created tissues are not generated on a large scale (Brittberg, Lindahl et al. 1994; Oberpenning, Meng et al. 1999; Macchiarini, Jungebluth et al. 2008). In addition, time consuming and expensive culture procedures and logistics, multiple operations and quality of the repair that is initiated by TE constructs remain important drawbacks.

Upon implantation of a TE-created construct the introduction of cells, biomaterials, growth factors, etc. in the body will have an effect on the local environment and natural repair mechanisms at the implant site. Since it is largely unknown what this local effect is and how these factors contribute to it, a clear shift is observed in the attempts to repair tissue. This shift includes more specific natural stimuli which trigger and enhance the regenerative capacity of the tissue itself. Injection of stem cells or progenitor cells (cell therapies), and the induction of regeneration by biologically active molecules can all be regarded as an example of Regenerative Medicine (RM). For both TE and RM it becomes more and more evident that studying the underlying natural and developmental processes of cartilage and bone can serve as a blueprint to identify important cell sources, biochemical, biomechanical, structural stimuli and timing thereof. It is expected that insight in these biological mechanisms and the process of endochondral ossification will enhance the progress in the field of both TE and RM.

This chapter describes the first phases of endochondral ossification, bone and cartilage (defects) and current approaches in TE and RM. Parallels with RM and endochondral ossification are identified from where endochondral ossification can serve as a blueprint for future RM approaches.

2. Endochondral ossification

Endochondral ossification is a multistage process that determines the major part of mammalian skeletal development and starts in embryogenesis with condensation of mesenchymal stem cells. The formation of cartilage, a process called chondrogenesis, is a key event in developing limb buds beginning in the center of the condensed mesenchyme. The earliest form of cartilage development is suggested to be 300 million years ago (Urist 1976). In humans, the first skeletal rudiments develop during the 5th week of gestation. In the eight week of the embryological life relatively cell-poor intermediate zones begins to develop, which will form the joint cavities (Gray and Gardner 1950; Anderson 1962; Aydelotte and Kuettner 1992). The diaphyseal cartilage, which is located at the center of the shaft of future long bones, is replaced by bone before birth (primary ossification). However most of the cartilaginous epiphysis at the end of long bones turns into bone after birth (secondary ossification). The remaining cartilage between the primary and secondary ossification centers is called the epiphyseal plate, more commonly known as the growth plate, and it continues to form new cartilage, which is replaced by bone, a process that results in increased length of the bones. Eventually all the cartilage in the growth plate will be converted into bone leaving cartilage only at the articulating surfaces of joints. Although bone and cartilage develop from the same mesenchyme, they have completely different structures, compositions and functions.

Chondrogenesis in both the primary and secondary ossification center and growth plates is characterized by highly proliferative chondrocytes, vectorially dictated to differentiate into hypertrophic chondrocytes before dying from apoptosis. The remaining mineralized extra cellular matrix provides a scaffold for infiltrating blood vessels and for bone cells to adhere to and remodel, setting the stage for *de novo* bone deposition (Kronenberg 2003) (Figure 1). The bone forming cells, osteoblasts, arise from progenitor cells from the overlying periosteal tissue and will form the bone collar (later the cortex) and primary spongiosa (later trabecular bone). In the adult, bone and overlying articular cartilage are attached by an interface of calcified cartilage (Schenk, Eggli et al. 1986). This interface distributes forces and stresses applied during load bearing and acts as a barrier to nutrients. Nutrients for the growing epiphyseal cartilage are supplied by two sources: (i) the synovial cavity and (ii) the vascularized cartilage canals (McKibbin and Maroudas 1979; Kuettner and Pauli 1983). Cartilage and synovium merge at a transitional zone which persists in the adult and is the site of osteophyte formation (Blaney Davidson, Vitters et al. 2007). This osteophyte formation is one of the first examples of endochondral ossification which takes place after growth. Another example is endochondral ossification during fracture healing where a cartilage callus is formed which will be remodelled into new bone. Studying endochondral ossification in normal growth and in healing processes will improve our understanding of both chondrogenesis and osteogenesis and as such may serve as a blueprint for Regenerative Medicine purposes of these tissues.

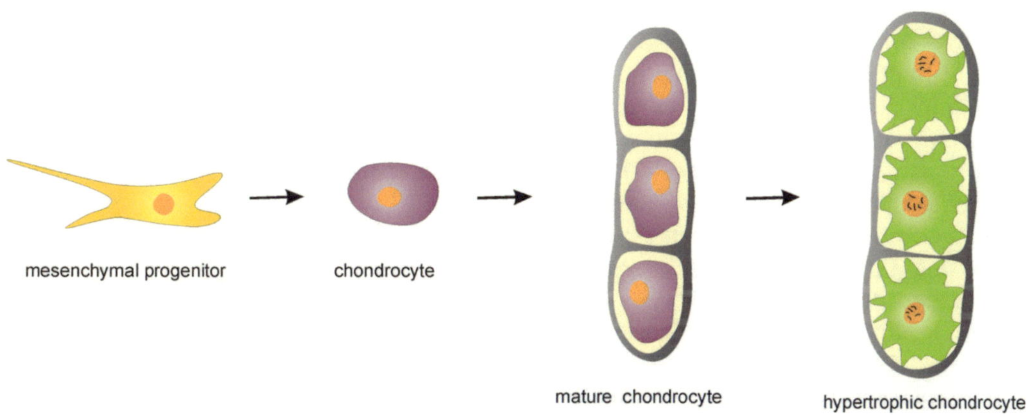

Fig. 1. The different steps of endochondral ossification; mesenchymal progenitor cells condense and undergo chondrogenesis. After maturation these chondrocytes undergo hypertrophy and die by apoptosis leaving a scaffold as a template for bone formation (these last steps are not illustrated nor discussed in this chapter).

2.1 Bone and bone defects

Bone can be formed by 2 different processes, while endochondral bone formation drives most of the skeletal bone formation, bone can also be formed by another process called intramembranous bone formation. During intramembranous bone formation, no cartilage phase is found and progenitor cells directly differentiate into bone. Intramembranous bone formation is largely responsible for the formation of flat bones as can be found in the skull and pelvis. Endochondral bone formation is largely responsible for the formation of bones of the axial skeleton. While in cartilage only one type of cell (chondrocyte) can be found, multiple cell types can be found in bone. Generally the bone forming cells are called osteoblasts and the cells which resorb bone are called osteoclasts. Osteoblasts produce the bone matrix (osteoid) which consists mainly of the organic collagen type I which is mineralized by inorganic hydroxyapatite (calcium phosphate). This gives bones a high compressive strength combined with significant elasticity. When osteoblasts become entrapped in their matrix they become osteocytes; the mature bone cells (Harada and Rodan 2003). Osteoclasts, on the other hand, are multinucleated cells that arise from the monocyte stem-cell lineage and are located at bone surfaces in Howship's lacunae. The cells are equipped with phagocytic-like mechanisms and are characterized by high expression of tartrate resistant acid phosphatase (TRAP) and cathepsin K which are able to break down bone matrix (Boyle, Simonet et al. 2003). The process of bone formation and bone resorption is able to adapt to mechanical forces and as such remodel into the desired architecture (Wolff's law). This process is mostly found in trabecular bone and while no evidence has been found that cartilage adapts/remodels after growth, bone is replaced constantly (Hunziker, Kapfinger et al. 2007). Another important function of bone resorption and formation is controlling homeostasis of important minerals such as calcium and phosphate.

Different specialized structures can be identified in bone; the bone attached to the joint cartilage is called subchondral bone. The zone directly beneath the subchondral bone is called the metaphysis. The metaphysis is characterized by a thin cortex and a highly vascularised trabecular bone. Within this trabecular bone bone marrow can be found. Bone marrow is also present at the inside of long bones where it enables hematopoiesis. In the center of long bones lies the diaphysial bone. Here, trabeculi become more sparse and the cortex thickens. The outer site of all bones is covered by periosteum. This periosteum is largely responsible for appositional growth of long bones as it contains a lining of osteoprogenitor cells. **Bone defects.** In the field of Orthopaedic and Trauma Surgery a large demand exists for autologous or allogenic bone. Clinical problems which fuel this demand are; large segmental bone defects (after infection, trauma or tumor resection), fracture non-unions (e.g. tibia, femur, humerus, carpal bones, and talus), bone defects in the increasing field of prosthesis related revision surgery, and spinal fusions (e.g. spondylolisthesis, discopaty, etc)(Glowacki 1998; Stevenson 1998; Huitema, van Rhijn et al. 2006). Although bone from the iliac crest is the golden standard, it is limited in source and donor site morbidity is a major concern. Alternatively, allografts are expensive and pose the risk of viral infection. While the inorganic part of bone (e.g. TriCalcium Phosphate (TCP), Hydroxyapatite (HA)) is widely explored as ceramics and combined with cells in the field of TE, this approach is not successful in generating a satisfying bone substitute (Petite, Viateau et al. 2000; Kim, Park et al. 2006; Zhao, Grayson et al. 2006). Main drawbacks are mechanical features and handling properties of these ceramics. Combining ceramic with polymers may overcome this problem, but toxic degradation products often affect healing and remodeling of the bone defect (Martin, Shastri et al. 2001; Kim, Park et al. 2006; Zhao, Grayson et al. 2006). In addition, these materials are often inert for Matrix Metallo-Proteins (MMPs) and often interfere with biomechanical signaling which is essential for repair and remodeling of loaded structures such as bone (Wolff's law). Furthermore, increased infection risk in implanted tissue-engineered devices is recently described (Kuijer, Jansen et al. 2007) and supply of oxygen and nutrients is the final aspect of concern when treating bone defects (Shastri 2006). Cells in autologous bone are transplanted from a highly vascularized environment to a hypoxic environment while cells residing in allografts are frozen and stored before transplantation, it is therefore likely that these cells do not contribute to the repair process (Emans, Pieper et al. 2006). Seeding of these materials with bone marrow cells is promising, however also costly, time consuming and infection prone during isolation and expansion (Shastri 2006). Among the disadvantages listed here lies the reason why this topic is currently studied extensively by many groups worldwide.

2.2 Cartilage and cartilage defects

Joint motion is possible by a both structurally and functionally truly remarkable material called hyaline cartilage (Buckwalter and Mankin 1998; Hasler, Herzog et al. 1999; Poole, Kojima et al. 2001). Hyaline cartilage is predominantly found in articular cartilage. Next to hyaline cartilage, two other types of cartilage can be found in the human body; elastic and fibrocartilage. Elastic cartilage is found in the ear, nose-tip and respiratory tract, whereas the menisci and intervertebral discs contain fibrocartilage.

The only cell type found in articular cartilage is the chondrocyte. In contrast to other tissues, the chondrocyte contributes to a relatively low percentage of the cartilage volume in human (1-5 percent). Articular chondrocytes are formed by chondrogenic differentiation of chondroprogenitor cells as described above and in Figure 1, however these cells arrest in the mature chondrocyte phase and normally do not become hypertrophic cells. Each chondrocyte is a metabolically active unit which expands and maintains the extracellular matrix (ECM) in its immediate vicinity (Aydelotte, Greenhill et al. 1988). In adults chondrocytes lack cell-cell contact; therefore communication between cells has to occur via ECM. Furthermore, cartilage is characterized by the absence of blood vessels, lymphatics and nerve fibers. Due to the lack of vascularisation in cartilage the environment is dominated by low oxygen levels and therefore the chondrocytes have an anaerobic metabolism (Schenk, Eggli et al. 1986). This also implicates that chondrocytes have to obtain their nutrients and oxygen via diffusion from the synovial fluid, through the ECM and from the underlying bone.

Structure. In articular cartilage four zones can be distinguished (see Figure 2), based on collagen type II orientation and chondrocyte shape and distribution (Buckwalter and Mankin 1998; Mankin, Mow et al. 2000; Poole, Kojima et al. 2001). In the superficial or tangential zone, chondrocytes are disc shaped and form a layer of several cells thick. The long axis of the cells are parallel to the joint surface and the cells are surrounded by a thin layer of ECM. Thin collagen fibers are oriented parallel with the articular surface. This orientation and the relatively low content of proteoglycans results in high tensile stiffness and the ability to distribute load over the surface. The cells in the transitional or middle zone are more spherical and appear dispersed randomly (Aydelotte and Kuettner 1992; Hunziker 1992), also collagen fibers in this zone are organized randomly. At this zone and at the deep zone, high concentrations of proteoglycans enable the tissue to bear compressive forces. In the radial or deep zone, chondrocytes are ellipsoid, grouped radially in columns of 2-6 cells with their long axes perpendicular to the joint surface. The thicker collagen fibres are also arranged perpendicular to the articular surface. In the calcified zone, chondrocytes are distributed sparsely and remain surrounded by a calcified matrix. The calcified cartilage is less stiff than the subchondral bone. At this calcified zone shear stresses are converted into compressive forces which are in turn transmitted to the subchondral bone (Radin, Martin et al. 1984). The junction between uncalcified and calcified cartilage is called the "tidemark", a line which can be seen on histology (Figure 2). Therefore mechanical forces also change at the tidemark which provides a definite boundary for the uncalcified layer (Donohue, Buss et al. 1983; Aydelotte and Kuettner 1992).

Cartilage defects can arise due to trauma or cartilage degeneration. Although patient's history may differentiate between traumatic and degenerative lesions, the exact cause of cartilage defects often remains difficult to diagnose. Since cartilage has no nerve fibers, cartilage lesions often present with only (minor) effusion of the affected joint or without symptoms. Diagnosis of structures likely to be damaged upon trauma (e.g. subchondral bone, ligaments or menisci), may reveal a cartilage lesion. An X-ray indicates a cartilage lesion in the minority of the cases and Magnetic Resonance Imaging (MRI) is the best non-invasive technique available for diagnosis of cartilage lesions. Important developments are new protocols such as delayed Gadolinium Enhanced MRI of Cartilage (dGEMRIC) and sodium MRI which can visualize cartilage on the Collagen and GAG content level (Gold, Burstein et al. 2006). Overall the MRI is expected to diagnose cartilage lesions in an early stage and will become more important in evaluation of progression of cartilage degeneration and cartilage repair techniques.

As early as 1743 it was recognized that articular cartilage, once destroyed, does not heal spontaneously (Hunter 1995; Hunziker 1999). Whereas the progenitor cells of bone marrow and periosteum contribute to bone formation during fracture healing, articular cartilage is deprived of these progenitors. Although it has been shown that the superficial layer of cartilage and the synovium contain progenitor cells (Dowthwaite, Bishop et al. 2004; Park, Sugimoto et al. 2005), cartilage has a limited ability for self repair (Mankin, Mow et al. 2000; Emans, Surtel et al. 2005). Therefore cartilage and tissue engineering approaches are studied in an attempt to overcome the inability of cartilage to repair itself.

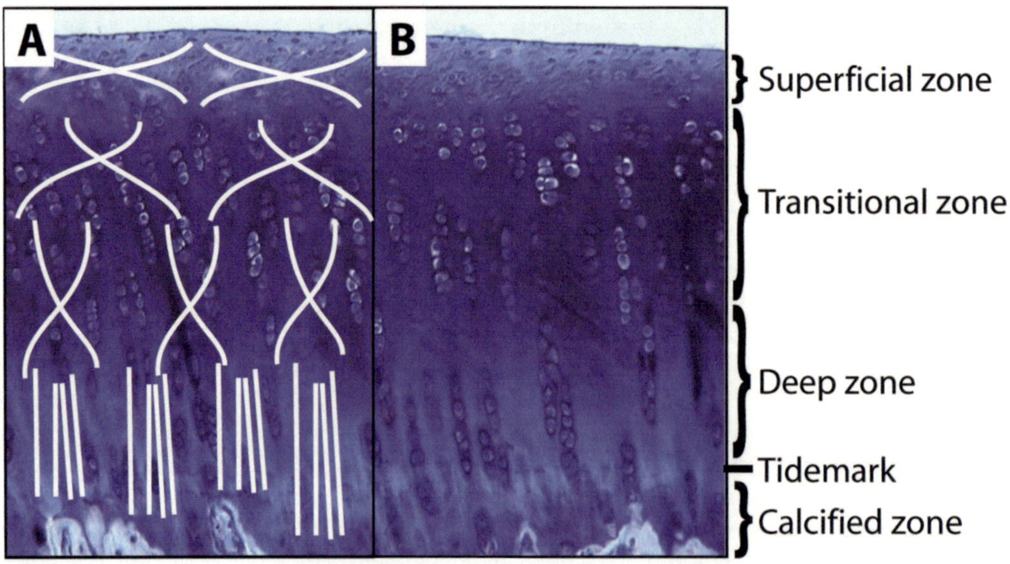

Fig. 2. Architecture of articular cartilage. Four zones can be distinguished with respect to (A) orientation of collagen fibers and (B) cell shape and orientation

3. Tissue engineering and regenerative medicine

Combining technologies from material science, cell biology, and clinical needs has led to the rise of the field of TE and RM. In the 1960's researches proposed the idea of creating tissues in a laboratory which may replace damaged or diseased tissues and cell biologists observed that cells could sort themselves *in vitro* to populations with tissue-like characteristics (Steinberg 1962). Adding a structure (material) such as a collagen gel to fibroblast cultures was shown to further resemble structural characteristics of skin. Later the work of Brittberg and co-workers showed that chondrocytes could be cultured and successfully be transplanted for the repair of cartilage defects (Brittberg, Lindahl et al. 1994). This technique is entitled Autologous Chondrocyte Transplantation or Implantation (ACT or ACI). The combination of specific tissue features and the early findings of culturing and transplanting chondrocytes and fibroblasts, skin, cartilage and bone were identified as tissues which potentially could be repaired by engineering these tissues in the laboratory by combining cells and supporting scaffolds. In the beginning of ACT no artificial structures were used to keep the chondrocytes in the cartilage defect. Optimization of ACT has led to the introduction of collagen meshes to support and maintain chondrocytes which were transplanted into the defect. Already earlier, in the mid-1980s, Langer and co-workers proposed that biodegradable polymers could serve as a scaffold for the organisation and maturation of cells into the desired tissues. As such it was proposed that this approach would enable engineering of thicker and hard tissues such as cartilage. Although cell therapies based on TE for skin are commercially available, which apply to the definition of TE such as Carticel® and Epicel® of Genzyme, the initial expectations of TE and RM have not been met. Although some examples of successful treatment by engineered tissues such as bladder and trachea can be found in the clinic, engineering tissues is not performed on a large scale (Oberpenning, Meng et al. 1999; Macchiarini, Jungebluth et al. 2008).

In the approach to engineer tissues in a laboratory setting and subsequently transplanting them into the body lies the key question; "*until what level should we engineer tissue and when should nature take over?*". It is often the aim of many researchers to engineer a mature tissue which is directly able to take over the function of the diseased tissue or organ. Per example it is often a goal that engineered cartilage and bone should be able to bear mechanical forces directly after implantation. In contrast, in nature a cascade of interactions occur during the process of tissue repair. During this process both the environment as well as the reparative tissue adapt to each other and the biomechanical requirements. In such a manner both integration of repair tissue and tissue remodelling is achieved. The capacity of a mature TE tissue to adapt to the local needs such as integration, remodelling, etc. is lower than a relatively less mature tissue. In addition, in order to create a robust and thicker tissue, the use of scaffolds, growth factors and more differentiated cells may be inevitable. However the question remains whether the local environment is able to adapt in an appropriate manner to all non-physiological stimuli which are introduced. Per example how does the normal tissue remodelling, repair and integration respond to a scaffold which alters local biomechanical stimuli which are known to be essential for tissue remodelling? How do transplanted and environmental cells respond to material properties such as material surface, breakdown products, architecture etc? How does the normal fine-tuned orchestra of tissue repair respond to transplanted cells which are normally not present at a certain phase of tissue repair? Finally, the use of cells in RM and TE approaches often implies the use of two surgical procedures as well as costly and time consuming culture procedures and logistics.

3.1 Bone repair

Natural bone healing. As described above, endochondral ossification drives skeletal growth. Similar sequential steps of endochondral ossification are largely responsible for fracture healing of long bones (Bostrom, Lane et al. 1995; Einhorn 2005). Periosteum is the main source of progenitor cells capable of creating large volumes of non-vascularized cartilage surrounding a fracture (Hall and Jacobson 1975). This first phase of endochondral bone formation is called soft callus. During the second phase chondrocytes become hypertrophic, mineralize (hard callus) (Figure 1), secrete pro-angiogenic factors such as VEGF and finally bone is deposited. In the final phase the newly formed bone is vascularized and will remodel under influence of mechanical forces. Bone healing by endochondral ossification is influenced by many regulatory mechanisms. However, while interaction of Indian hedgehog (Ihh) and Parathyroid hormone related protein (PTHrP) is one of the best known regulatory mechanism in the growth plate, such an interplay is yet unknown for fracture healing (Wu, Ishikawa et al. 1995; Vortkamp, Lee et al. 1996; Volk and Leboy 1999). The role of growth factors during bone healing processes is better studied. Chondrocytes at different stages of maturation release cytokines and growth factors such as Fibroblast Growth Factor (FGF), Transforming Growth Factor (TGF)-β, Bone Morphogenetic Proteins (BMPs) and Vascular Endothelial Growth Factor (VEGF) (Gibson 1998; Gerber, Vu et al. 1999; Blunk, Sieminski et al. 2002). For instance FGF-2 and TGF-β control endochondral ossification by inhibition of chondrocyte proliferation, hypertrophy and apoptosis (Gibson 1998) and in addition from our own findings we know that TGF-β is important for osteo- and chondrogenesis both in *ex vivo* and *in vivo* models (Kuijer, Emans et al. 2003). *In vivo,* in the Osteoarthritic (OA) joint TGF-β is produced. Under the influence of this TGF-β osteophytes are formed which are derived from periosteum adjacent to the joint via endochondral bone formation (van der Kraan and van den Berg 2007). BMPs are also positively involved in ectopic cartilage and bone formation, partly by opposing the actions of the FGF pathways (Yoon and Lyons 2004; Miyazono, Maeda et al. 2005; Yoon, Pogue et al. 2006). Neo-vascularization under influence of VEGF ensures blood vessel formation which supply oxygen and nutrients to osteoblast and osteoclasts. The latter produce MMP-9 and -13 which degrade the matrix surrounding terminally hypertrophic chondrocytes (Gerber, Vu et al. 1999). Blocking VEGF in the hypertrophic zone of the growth plate prevents degradation of this zone which in turn enlarges (Gerber, Vu et al. 1999).

Current approaches for bone repair. Multiple causes may lead to impaired healing of large bone defects. As mentioned before, nature has a good regenerative capacity for fractures, however from a clinical perspective the need for bone is not in fracture repair but mostly for the filling of large bone defects after revision arthroplasty and spondylodesis. These bone defects can be regarded as "non-natural" occurring bone defects and bone healing or filling is impaired at these sites because endochondral ossification does not occur. To deal with this problem a scaffold is introduced as a template for bone ongrowth, ingrowth and remodelling. Currently many bone fillers (scaffolds) and growth factors are available for treatment of bone defects. Taking the scaffold which is formed during endochondral ossification (also see chapter "scaffolds") as a blueprint, bone fillers need to be further optimized; next to being expensive, these aids only address one or a few aspects of the cascade of tissue responses which are necessary for bone repair. Most bone fillers are osteoconductive (supportive) and they lack the timing and onset of essential growth factors to be osteoinductive (stimulating bone growth). Growth factors by themselves have been shown to be osteoinductive but addition of one of the essential growth factors does not necessarily recapitulate the physiological, initial tissue response which leads to fracture/bone repair.

Inflammation is the first and essential phase of tissue repair in general and bone repair in particular. Mimicking this inflammatory response may be a method to enhance bone fracture healing. Several clinical examples such as spondylodesis after infection of the intervertebral disc (e.g. after discography) and the method described by Masquelet confirm that inflammatory responses contribute to osteogenesis (Guyer, Collier et al. 1988; Masquelet and Begue 2010). However, in contrast, from an engineering perspective, the aim is often to create bone which has comparable mechanical features as native bone. The initial mechanical properties of currently used bone chip auto or allografts are incapable of withstanding the mechanical forces to which they are exposed. During impaction of these chips the mechanical properties of the impacted bone as a whole are capable to withstand mechanical forces in a non-loadbearing environment. After vascularisation, bone ingrowth and remodelling of the repaired bone defect adapts to finally bear full loading. As such surgical handling properties, osteoconduction, and most important osteoinduction are features one should aim for rather than engineering mature bone with biomechanical properties comparable to native bone. As mentioned before during endochondral ossification large amounts of cartilage are generated. This cartilage does not have the required mechanical features of the bone it should repair, but does have strong osteoconductive and osteoinductive features. Another challenge when aiming for creation of bone is the scale to which should be generated. During fracture healing bone defects can be repaired by deposition of large amounts of bone which is formed by endochondral ossification. In the pre-remodelling phase of endochondral ossification, the generated bone histologically resembles the metaphysial bone chips which are used on a large scale for bone impaction grafting. In conclusion, regarding endochondral ossification as a blueprint for engineering or regeneration of bone, it has the potential to generate vast amounts of bone, with good handling properties, and is osteoinductive and osteoconductive.

3.2 Cartilage repair

Treatment of damaged cartilage can be grouped to four concepts of principle: the four R's (O'Driscoll 1998). The joint surface can be: (i) resected, (ii) relieved, (iii) replaced or (iv) restored. A joint prosthesis is an example of joint replacement; joint distraction and osteotomies can induce joint relieve. Osteotomies are used to re-align the axis of loading in patients with a malalignment of the leg. By transferring the load to the less affected cartilage (e.g. previously less loaded/damaged cartilage) the damaged part is relieved. Arthodesis is an example of joint resection. For TE and RM techniques the focus is on cartilage restoration.

Restoration implies methods to heal or regenerate the joint surface with or without the subchondral bone into healthy hyaline articular cartilage. Three strategies can be considered when attempts are made to heal or restore cartilage.

i. Subchondral Drilling, Abrasion or Microfracture are techniques to allow penetration of bone marrow through the subchondral bone into the damaged cartilage (Meachim and Roberts 1971; Insall 1974; Mitchell and Shepard 1976; Furukawa, Eyre et al. 1980; Vachon, Bramlage et al. 1986; Bradley and Dandy 1989; Rae and Noble 1989; Kim, Moran et al. 1991; Altman, Kates et al. 1992; Aglietti, Buzzi et al. 1994). These techniques improve the clinical well being of the patient and the joint surface defect may be healed to some extent. However the healing process is inadequate since no functional hyaline cartilage but fibrocartilage is formed (Vachon, Bramlage et al. 1986; Altman, Kates et al. 1992). Nonetheless, these methods are cheap and easy to perform and are therefore seen as the currently best option to relieve the complaints. Other clinical studies have suggested that any beneficial effect is related to the arthroscopic procedure itself. A nonspecific effect might be related to joint lavage rather than the penetration of the subchondral bone (Jackson 1986; Ogilvie-Harris and Fitsialos 1991). In conclusion, these techniques may have some benefit with regard to small defects but no effect has been proven in relation to large defects, osteoarthritic joints or older patients (Kim, Moran et al. 1991).
ii. Implants vary from non-degradable and degradable, cells, periosteum or perichondrium, Osteochondral Autograft Transfer System (OATS or Mosaicplasty) and Osteochondral Allografts (Elford, Graeber et al. 1992; Freed, Vunjak-Novakovic et al. 1993; Nixon, Sams et al. 1993; Hendrickson, Nixon et al. 1994; Reddi 1994; Chu, Coutts et al. 1995; Grande, Halberstadt et al. 1997). The biomaterials and periosteum can be combined with cells or growth factors. Periosteal Arthroplasty is an interesting way of treating cartilage defects since many have reported the chondrogenic potential of periosteum (O'Driscoll, Keeley et al. 1986; O'Driscoll, Keeley et al. 1988; Zarnett and Salter 1989; Nakahara, Bruder et al. 1990; Nakahara, Dennis et al. 1991; Nakahara, Goldberg et al. 1991; Nakata, Nakahara et al. 1992; Iwasaki, Nakata et al. 1993; Gallay, Miura et al. 1994; Iwasaki, Nakahara et al. 1994; Iwasaki, Nakahara et al. 1995; O'Driscoll, Saris et al. 2001; Emans, Surtel et al. 2005). Over 90 percent of collagen type II in the hyaline cartilage formed in the cartilage defects treated with periosteal grafts has been reported (O'Driscoll, Keeley et al. 1986; O'Driscoll, Keeley et al. 1988). Perichondrial Arthroplasty used for human cartilage repair was first described by Skoog et al. (Skoog and Johansson 1976). This technique has been reported to give an initial cartilage repair (Homminga, Bulstra et al. 1990; Homminga, Bulstra et al. 1991). On the long term poor results related to overgrowth of the graft and calcification are reported by Bouwmeester *et al.* (Bouwmeester, Beckers et al. 1997). These authors concluded that a better fixation of the graft might improve the results. In a study comparing periosteum with perichondrium, chondrogenesis was observed significantly more using periosteal grafts (Vachon, McIlwraith et al. 1989). This finding and the accessibility make periosteum to be preferred over perichondrium.
iii. Osteochondral Grafts can be divided in autologous and allogenic. Mosaicplasty or OATS involves harvesting one or more osteochondral plugs from a relatively less weight-bearing region of the joint and subsequent implantation of these plugs into an articular defect. Possible donor site morbidity is bypassed if osteochondral allografts are used (Gross, McKee et al. 1983; Garrett 1986; Czitrom, Keating et al. 1990; Convery, Meyers et al. 1991; Garrett 1994; Ghazavi, Pritzker et al. 1997; Garrett 1998; Horas, Schnettler et al. 2000; Gross, Aubin et al. 2002).

The role of endochondral ossification in cartilage repair. When using progenitor cells for cartilage repair, ossification of the repaired tissue may impair clinical results. Examples hereof are ossification and formation of interlesional osteophytes when applying techniques such as microfracture and periosteum or perichondrium plasty (Bouwmeester, Beckers et al. 1997; Cole, Farr et al. 2011). These findings illustrate that maintaining differentiated progenitor cells in their chondrogenic state remains challenging in cartilage repair. It appears that in contrast to chondrocytes, progenitor cells have the tendency to follow the different phases of endochondral ossification towards hypertrophy and mineralisation when triggered to differentiate into cartilage. As such, locking cells in their desired differentiation state is of the utmost importance when applying these cells for RM purposes. Findings of Hendriks and co-workers showed that chondrocytes stimulate progenitor cells towards chondrogenesis when both cell types are co-cultured (Hendriks, Riesle et al. 2007). These findings were later bolstered by Fisher and co-workers showing that human articular cartilage-derived soluble factors and direct co-culture are potent means of improving chondrogenesis and suppressing the hypertrophic development of mesenchymal stem cells (Fischer, Dickhut et al. 2010). In this study and other work of the group of Richter the PTHrP is an important candidate soluble factor involved in this effect. PTHrP is primarily known as a key regulator in the process of endochondral ossification. Furthermore, we have recently shown that cyclooxygenase (COX) inhibitors are also able to decrease hypertrophy of chondrocytes (unpublished data). Thus studying the process of endochondral ossification and further unravelling how and why articular chondrocytes maintain their phenotype as well as prevention of hypertrophy may enhance cartilage repair techniques by generating stable cartilage which does not lead to intra-lesional osteophytes. Finally, cartilage defects lead to early OA, also in the process of OA more evidence is found that articular chondrocytes loose their capacity to maintain their phenotype and seem to undergo endochondrogenesis since they become hypertrophic and express collagen type X (Saito, Fukai et al. 2010). As such understanding and controlling the process of endochondrogenesis may be of relevance for future insight and treatment of OA.

3.3 Scaffolds

Scaffold or carrier material refers to a wide variety of artificial 2D or 3D structures that are designed for the purpose of tissue engineering. Scaffolds may be seeded with cells before implantation or are designed to recruit or retain cells at the desired place. (Bentley and Greer 1971; Wakitani, Kimura et al. 1989). For bone and cartilage regeneration, relevant cells are (mesenchymal) stem cells of different origins (bone marrow, adipose tissue, dental pulp, iPS etc.) as well as differentiated cells like chondrocytes. Different variables are important parameters for scaffold design: pore diameter, shape, kind of material, (bio)degradability, implantation site, functionalization, mechanical stability and others. Several materials have been and are being explored for this purpose. Generally scaffold materials can be divided in natural or synthetic. Examples of natural material-based scaffolds for cartilage and bone regeneration are: fibrin, hyaluronan, alginate, agarose, demineralized bone matrix, collagen etc. Synthetic scaffold materials include ceramics and copolymers PolyGlycolic Lactic acid (PGLA) and PolyethyleneGlycol-terephthalate/PolyButylene Terephthalate (PEGT/PBT) (Figure 3) etc. When applying a collagen as carrier material, some authors find enhanced cartilage healing, while others conclude that collagen scaffolds have a limited usefulness for chondrocyte grafting in large defects (Wakitani, Kimura et al. 1989; Nixon, Sams et al. 1993;

Sams, Minor et al. 1995; Sams and Nixon 1995). The use of fibrin as carrier material was reported to give superior cartilage healing compared to controls (empty defect) (Hendrickson, Nixon et al. 1994). Within time an ideal scaffold should degrade or allow the populated cells to take over functionality of the artificial tissue implant. Breakdown products and biomechanical features of the scaffold should not negatively interfere with differentiation towards this tissue. It is therefore challenging to design a scaffold with all the optimal characteristics; proper initial mechanical stability, timed release of required growth factors, timed degradation which allows biomechanical stimuli to remodel the formed tissue, no release of degradation products which interfere with tissue repair, good handling properties, etc. Next generation scaffolds will be so called "smart scaffolds". These scaffolds will be loaded with bioactive factors (e.g. TGF-β1 and members of its superfamily such as BMPs) that can directly influence the differentiation pathways (Sellers, Peluso et al. 1997; Sellers, Zhang et al. 2000; Huang, Goh et al. 2002). Effort is being put in e.g. functionalized scaffolds with specific affinity peptides to retain cells (Dong, Wei et al. 2009). Also, the release of e.g. growth factors may be regulated by "on demand" smart systems that depend on incorporated microspheres or proteolytic degradation of linker-peptides. Unfortunately, an ideal material for artificial scaffolds for cartilage and bone regeneration has not been identified yet, as the biological processes involved are far more complex than anticipated.

Fig. 3. A PolyethyleneGlycol-terephthalate/PolyButylene Terephthalate (PEGT/PBT) scaffold produced by using a three dimensional rapid prototyping technique.

The endochondral scaffold. During endochondral ossification nature creates its own scaffold. Hypertrophic chondrocytes die and leave a large scaffold. During this process multiple growth factors are released in an orchestrated manner. The scaffold itself is used as a template for invading cells to deposite bone and provide vascularisation. The scaffold itself is resorbed by osteoclasts which in turn respond to biomechanical and biochemical stimuli. As such the scaffold degrades and simultaneously the proper factors are released. The repair tissue remodels to the appropriate architecture as defined by Wolff's law. Studying this process in detail reveals the challenge when "artificial" scaffolds are designed from a material science point of view, and so far no scaffolds have been created with the same properties capable to dictate the same processes of endochondral remodelling.

3.4 Cells

Cells for "orthopaedic" tissues, such as bone and cartilage, originate from the mesenchymal cell lineage and may be derived from different autologous or allogenic sources. Interestingly, cells used for bone regeneration are almost always progenitor cells, whereas for cartilage regeneration also differentiated cells are used, next to progenitor cells. Some authors prefer the use of chondrocytes for transplantation while others prefer the use of undifferentiated multipotent cells (Skoog and Johansson 1976; O'Driscoll, Keeley et al. 1986; Homminga, Bulstra et al. 1991; Brittberg, Lindahl et al. 1994; Lindahl, Brittberg et al. 2003; Nathan, Das De et al. 2003; Emans, Surtel et al. 2005; Park, Sugimoto et al. 2005).Mature chondrocytes can be released from their cartilaginous matrix, selected and expanded *in vitro*. In this way a relatively small amount of autologous tissue can be used as an appropriate cell source. Both chondrocytes and progenitor cells originating from different cell sources have been studied in combination with various biomaterials (Bentley and Greer 1971; Haynesworth, Baber et al. 1992; Freed, Marquis et al. 1993; Freed, Vunjak-Novakovic et al. 1993; Iwasaki, Nakata et al. 1993; Brittberg, Lindahl et al. 1994; Bruder, Fink et al. 1994; Freed, Grande et al. 1994; Gallay, Miura et al. 1994; Wakitani, Goto et al. 1994; Iwasaki, Nakahara et al. 1995). Bone marrow, adipose tissue, synovium, dental pulp, perichondrium and periosteum can serve as a source for multipotent cells (Skoog and Johansson 1976; Homminga, Bulstra et al. 1991; Bouwmeester, Beckers et al. 1997; Chu, Dounchis et al. 1997; O'Driscoll, Saris et al. 2001; Nathan, Das De et al. 2003; Emans, Surtel et al. 2005; Park, Sugimoto et al. 2005). Numerous publications described subpopulations of progenitor cells in these donor tissues that might be more optimal cell sources than e.g. whole cell pool isolates. However, lots of these studies were only performed *in vitro* and one can question whether selection of subtypes based on cell surface markers may bias the outcome of the intervention in a difficult to predict way. The involvement of subchondral bone may play a role in cell source selection as chondrocytes are capable of producing cartilage under the appropriate conditions, but in a situation where simultaneous bone formation is required (involvement of the subchondral bone), multipotent cells might be a better cell source. After selection and expansion, the main challenge is to keep these cells in the damaged area of the joint and this challenge becomes even bigger in larger defects (Bentley and Greer 1971). Grande *et al.* reported that only 8 percent of the total number of cells in the healing tissue originated from transplanted chondrocytes (Grande, Pitman et al. 1989). Chondrocytes can be maintained in the defect by suturing a periosteal flap or a collagen mesh over the defect (Grande, Pitman et al. 1989; Brittberg, Lindahl et al. 1994; Bartlett, Skinner et al. 2005). As discussed above, chondrocytes can also be seeded in a matrix or scaffold. This matrix can be implanted in a cartilage defect. This Matrix Assisted Chondrocyte Transplantation (MACT) is technically less demanding and has shown identical results compared to Autologous Chondrocyte Transplantation on the short term (Bartlett, Skinner et al. 2005) The use or allogenic chondrocytes has been reported to be successful in rabbits but experiments in horses do not support this finding (Wakitani, Kimura et al. 1989; Freed, Grande et al. 1994; Sams, Minor et al. 1995; Sams and Nixon 1995). Immunological rejection or allogenic chondrocytes upon implantation in rabbits has been reported and this remains a major concern when applying allogenic cells. The use of periosteal tissue or cells has been suggested by several authors and is one of the most clear examples of the use of endochondral ossification as a blueprint for cartilage and bone regenerative medicine. The periosteum is populated with mesenchymal progenitor cells that normally contribute to endochondral bone fracture healing. The differentiation capacity of these cells can also be used to create cartilage or bone for regenerative purposes. This principle is further explained below (see paragraph 3.6).

When using differentiated or undifferentiated cells for cartilage, bone or osteochondral repair it is a challenge to differentiate these cells into their desired state and maintain their desired phenotype. As cells from the mesenchymal lineage, once differentiated into chondrocytes, have a natural tendency for terminal differentiation via the endochondral pathway. This is a big concern for regenerative applications. Much effort is being put in technologies that prevent hypertrophic differentiation of transplanted chondrocytes, while on the other hand for bone regeneration hypertrophic differentiation may be a prerequisite for success (Fischer, Dickhut et al. 2010; Scotti, Tonnarelli et al. 2010).

3.5 Biochemical signaling pathways

Growth factors. In growth plate development, homeostasis of articular cartilage as well as bone formation and maintenance, several signaling pathways are interacting or shared between the different tissues. Indian hedgehog (Ihh) and Parathyroid hormone related peptide (PTHrP) coordinate chondrocyte proliferation and differentiation in the so-called PTHrP-Ihh feedback loop (Kronenberg 2003). This coordination influences the length of proliferative chondrocyte columns as well as chondrocyte hypertrophy. Next to the Ihh and PTHrP loop, fibroblast growth factor crucially regulates chondrocyte proliferation and differentiation. Many of the 22 distinct FGF genes and their four receptor genes are expressed at every stage of endochondral bone formation (Ornitz and Marie 2002). Also Bone Morphogenic Proteins (BMPs) have multiple roles during bone and cartilage formation, as well as growth plate development. Interestingly, BMPs were discovered because of their remarkable ability to induce endochondral bone formation when injected subcutaneously in mice. In a cartilage context, BMPs are involved in early chondrogenesis, cartilage maintenance and hypertrophic differentiation. In a bone context they drive differentiation of progenitor cells to osteocytes and induce alkaline phosphatase activity in osteocytes. TGF-β isoforms are also involved in similar processes and interestingly were found to trigger the formation of osteophytes upon intra-articular injection and during OA (Elford, Graeber et al. 1992; van Beuningen, van der Kraan et al. 1993; van Beuningen, van der Kraan et al. 1994; Hunziker 2001). As osteophyte formation itself is an example of endochondral ossification, the role of TGF-β isoforms in endochondral ossification is supported by this finding. Remarkably, some characteristics of OA resemble chondrocyte differentiation processes during skeletal development by endochondral ossification. Resemblances are: chondrocyte proliferation, chondrocyte hypertrophic marker expression (e.g. Collagen type X and MMP-13), vascularisation and focal calcification of joint cartilage. This suggests that during OA the articular cartilage is terminally differentiating via "normal" endochondral pathways. However, how the mature articular cartilage is kept in its cartilaginous state and why it starts a terminal differentiation program in OA is currently poorly understood. In the final stage of endochondral bone formation secretion of pro-angiogenic factors such as VEGF is essential. Sox9 and RunX2 are important transcription factors. Sox9 is the master regulator of chondrogenesis and acts as a negative regulator for chondrocyte hypertrophy, cartilage vascularisation and bone marrow formation (Hattori, Muller et al. 2010). Amongst others it does this via negatively regulating expression of RunX2 via Nkx3.2 (also known as BapX1) (Yamashita, Andoh 2009). RunX2 is a central regulator for the transition from proliferating to hypertrophic chondrocytes, as it drives the transcription of Collagen type X. Interestingly, RunX2 also drives multiple osteogenic developmental programs. Inflammatory pathways are other key players in endochondral ossification (Einhorn, Majeska et al. 1995; Mountziaris and Mikos 2008). Bone fracture healing by endochondral ossification depends on a haematoma-induced inflammatory environment (Grundnes and Reikeras 1993) and several inflammatory molecules (e.g. IL-6, TNFα, COX-2 and iNOS) are involved in bone fracture repair (Einhorn, Majeska et al. 1995; Mountziaris and Mikos 2008) by influencing chondrocyte maturation and osteogenic development. An important chondrogenic growth factor is Insulin Growth Factor 1 (IGF-1). Together with its receptors and several IGF binding proteins it determines chondrocyte proliferation and differentiation. Importantly IGF-1 appears to play a role in preventing chondrocyte apoptosis. Hence, it determines the pace of hypertrophic differentiation and thus growth plate development and fracture callus maturation. It was shown that IGF-1 exerts its action via NF-κB/p65 signaling (Wu, Gong et al. 2008). Furthermore, IGF-1 also directly influences osteocyte biology. It has been reported that IGF-1 stimulates cancellous bone formation and increases the activity of resident osteoblasts (Zhao, Monier-Faugere et al. 2000). RANK is crucially important for bone homeostasis and remodelling. Activation of RANKL on monocytic cells by RANK on osteoblasts induces osteoclastogenesis of committed monocytic cells. Multinucleation is induced, ultimately leading to the generation of mature bone resorbing osteoclasts (Novack and Faccio 2011). This process is counterbalanced by the soluble factor osteoprotegerin (OPG), thereby preventing bone loss due to osteoclast activation. Activation of the RANKL system is potentiated by prostaglandins. PGE_2, one of the main cyclooxygenase metabolites is reported to increase bone resorption.

In conclusion, the process of endochondral ossification is dictated by spatiotemporal expression and use of variable interacting growth factors and other molecules. It is clear that mimicking this complex, yet incompletely known, tissue formation in an *in vitro* setting on the same scale as TE was expected to do is quite challenging. Several findings such as endochondral ossification after subcutaneous injection of BMPs show that, *in vivo*, this process may be triggered using stimuli which trigger and enhance the regenerative capacity of the tissue itself. In such an approach the amount of unknown stimuli is expected to be limited and the body's own regenerative capacity is used to generate cartilage or bone, which in turn can be transplanted into the damaged site. As such, this approach applies more to the principles of RM than to the principles of TE. The application of a specific *in vivo* trigger to stimulate endochondral bone formation has many advantages; no expensive culture procedures, no more harvesting of cells, and no introduction of factors which possibly conflict with the natural tissue repair and integration. Table 1 summarises the differences in tissue features, currently applied (TE) techniques for restoration, and remaining challenges.

3.6 Examples of endochondral ossification as blueprint for regenerative medicine

Currently for TE purposes cells are harvested during the first operation and the implantation of the graft/cells is performed during the second procedure. A question that remains is the amount of cells that survives the transplantation. It has been shown that periosteal cells show a much poorer survival compared to chondrocytes after transplantation into the hostile environment of a fresh osteochondral defect (Emans, Pieper et al. 2006). However, the disadvantage of using chondrocytes is the fact that the joint is further damaged. It would be ideal to generate cartilage in an ectopical place which does not further interfere with the joint homeostasis, survives the transplantation and is capable to adapt and repair the defect. In line with this, an interesting variation for cartilage repair is a reported by Takahashi *et al.* who used the early fracture callus, induced at the iliac crest (Takahashi, Oka et al. 1995). The early fracture callus was implanted into osteochondral defects of rabbit knees with excellent results. A paper of our group also reported excellent results after transplantation of periosteum derived cartilage callus into osteochondral defects (Emans, van Rhijn et al. 2010). Stevens *et al.* published an interesting paper on inducing chondrogenesis by subperiosteal injection of a hyaluronan-based gel containing the antiangiogenic factor Suramin. The resulting tissue also resembled cartilage of early fracture callus (Stevens, Marini et al. 2005). The main advantage of this approach is that the body is used as its own "*in situ* incubator"; cells provide their own matrix and complex and costly isolation, selection and culturing procedures are bypassed. After this first report focussing on bone, we aimed to control the local environment by injecting a gel into the space between bone and periosteum which would initiate endochondrogenesis. Both agarose and a gel loaded with TGF-β1 were successful to trigger endochondrogenesis. This tissue was harvested during its first chondrogenic phase and successfully implanted into an osteochondral defect where an excellent lateral integration and no calcification of the cartilage adjacent to the joint was observed (Emans, van Rhijn et al. 2010).

It was recognised by the group of Martin that TE and RM attempts to create bone using the intramembranous pathway (Scotti, Tonnarelli et al. 2010). In contrast, during development most bones are formed by endochondral ossification and the parts that do not ossify forms articular cartilage. In addition, during fracture healing bone gaps and defects are often repaired by endochondral bone formation, during which large amounts of callus can be formed. Depending on the phase in which specific tissue is generated by endochondrogenesis, this tissue can be harvested for different purposes. If tissue in the early chondrogenic phase is harvested this may be ideal to heal both bone and cartilage. If this tissue is harvested at a later stage it resembles trabecular bone which has the potential to be used for bone impaction grafting. Compared to the frequently used TE approach to create bone directly (intra-membranous), it seems more logical that endochondral bone formation which is capable to produce large amounts of cartilage and bone, even in an ectopic site, may fuel further research

<table>
<tr><td colspan="6">Challenges in Tissue Engineering</td></tr>
<tr><td colspan="6">Tissue characteristics</td></tr>
<tr><td colspan="2"></td><td>Bone</td><td colspan="3">Cartilage</td></tr>
<tr><td colspan="2">Function</td><td>Weight bearing</td><td colspan="3">Weight bearing and joint articulation</td></tr>
<tr><td colspan="2">Cells</td><td>Osteoblasts/osteocytes and osteoclasts</td><td colspan="3">Chondrocytes</td></tr>
<tr><td colspan="2">Origin</td><td>Mesenchymal and monocyte lineage</td><td colspan="3">Mesenchymal</td></tr>
<tr><td colspan="2">ECM</td><td>Collagen I and Calcium phosphate</td><td colspan="3">Collagen II, Proteoglycans, GAGs and Hyaluronic acid</td></tr>
<tr><td colspan="2">Functional ECM water content</td><td>No</td><td colspan="3">Yes, important</td></tr>
<tr><td colspan="2">Cell-cell contact</td><td>Yes and important</td><td colspan="3">No, 'communication' via ECM</td></tr>
<tr><td colspan="2">Vascularisation</td><td>Yes</td><td colspan="3">No, hypoxic tissue</td></tr>
<tr><td colspan="2">Nutrient/oxygen supply</td><td>Via vascularisation</td><td colspan="3">Via diffusion</td></tr>
<tr><td colspan="2">Remodelling</td><td>Constant (Wolff's law)</td><td colspan="3">Low grade of remodelling?</td></tr>
<tr><td colspan="2">Regenerative capacity</td><td>High</td><td colspan="3">Low</td></tr>
<tr><td colspan="2">Access to progenitor cells</td><td>Bloodstream, Periosteum, Bone marrow</td><td colspan="3">Superficial layer cartilage? Synovium?</td></tr>
<tr><td colspan="2">Endochondral ossification</td><td>Complete</td><td colspan="3">Has to stop at chondrocyte-phase</td></tr>
<tr><td colspan="6">Current approaches in TE</td></tr>
<tr><td colspan="3">Bone</td><td colspan="3">Cartilage</td></tr>
<tr><td></td><td>Advantages</td><td>Disadvantages</td><td></td><td>Advantages</td><td>Disadvantages</td></tr>
<tr><td>Auto- and Allografts</td><td>Osteoconductive
Osteoinductive?
Native bone</td><td>Expensive
Host-vs-graft reaction Infection
Freezing cells?
Donor site morbidity</td><td>Osteochondral grafts</td><td>Native cartilage</td><td>Expensive
Host-vs-graft reaction (allografts)
Infection
Donor site morbidity
Fixation
Cell morbidity during storage</td></tr>
<tr><td>Decellularized bone</td><td>Osteoconductive
Resembles native bone</td><td>Not osteoinductive
Host-vs-graft reaction
Infection</td><td>Subchondral drilling, abrasion, microfracture</td><td>Activation of bone marrow
Very effective
Cheap and easy</td><td>Fibrocartilage</td></tr>
<tr><td>HA/TCP/ Bonefillers</td><td>Osteoconductive
Resembles native bone properties
Can be loaded with cells/growth factors</td><td>Not osteoinductive
Mechanical features
Handling properties
Interference with biomechanical signalling</td><td>ACI/ACT
MACT</td><td>Good integration
Native cartilage</td><td>Infection
Time consuming
Expansion of cells
Fixation
Expensive</td></tr>
<tr><td>Growth factors</td><td>Osteoinductive</td><td>Not osteoconductive
Expensive
Overload of growth factor
Ectopic bone formation</td><td>Implants/ biomaterials</td><td>Can be loaded with cells/growth factors
Initial cartilage repair</td><td>Calcification of cartilage
Fixation
Toxic degradation products</td></tr>
</table>

Challenges		
Current advantages		**Remaining challenges**
Scaffolds	Conductive Can be loaded with cells and/or growth factors to recruit, retain and/or differentiate the cells Immediate initial mechanical stability	Scaffold design: pore diameter and shape (bio)Material: natural or synthetic Biodegradability and degradation at right time Breakdown products Integration / fixation Release of growth factors at right time Interference with tissue environment **Cartilage:** Nutrient supply in scaffold Allow ECM formation, but prevent mineralisation Allow articulation, repetitive mechanical loading and load damping **Bone:** Nutrient and oxygen supply Stimulate vascularisation Allow bone mineralisation Support high mechanical loading **Endochondral:** progenitors differentiate to hypertrophic chondrocytes which leave a natural 'scaffold' for bone cells to adhere and remodel and provides in essential growth factors and vascularisation at appropriate time points.
Growth factors (BMPs, TGF-βs, PTHrP, VEGF, etc)	Inductive Can regulate differentiation of cells Easy	Does not recapitulate total physiological repair response Keep growth factors at damaged area Effect on tissue *in vivo* incompletely known Still expensive
Progenitor cells	High potential to differentiate into required tissue Various origins (bone marrow, dental pulp, adipose tissue, periosteum, blood etc.)	Have to differentiate and remain differentiated into required tissue Infection If allografts: host-vs-graft reaction Keep cells at damaged area **Cartilage:** Nutrient and oxygen supply Have to stop differentiating at chondrocyte phase **Bone:** Nutrient and oxygen supply Vascularisation Cell isolation and culturing is still time consuming and expensive
Adult cells	Inductive of natural tissue Only for cartilage cells?	Donor site morbidity Cells are out of natural environment, can lead to cell death or dedifferentiation If allografts: allogenic reaction Keep cells at damaged area **Cartilage:** Nutrient supply Prevent further differentiation towards hypertrophic chondrocytes

Table 1. Differences in: tissue characteristics, currently applied Tissue Engineering, and remaining challenges for bone and cartilage.

for generating both bone and cartilage. Creating cartilage or bone by triggering endochondrogenesis in an ectopic site bypasses expensive and time consuming culture techniques, logistics, and when triggered by injection of a specific stimulus may even limit the total approach to one operation.

4. Conclusion

From nature it is known that vast amounts of cartilage are formed in the process of endochondrogenesis. Chondrocytes in this cartilage tissue are replaced by a matrix deposited by hypertrophic chondrocytes which die by apoptosis. This matrix is used as an active scaffold for cells that contribute to bone formation. Following embryonic joint formation and post natal growth, the adult skeleton maintains the cellularity and phenotype of articular cartilage, whereas growth plate cartilage completely disappears. This process entitled endochondral ossification can be recapitulated in other places than growth plates. Examples hereof are fracture healing, osteophyte formation and peri-articular ossifications. Even in the process of OA endochondrogenesis plays a role. Next to the formation of osteophytes in OA, evidence has been reported that during the process of OA, articular chondrocytes are triggered to follow the final phase of endochondral ossification (Saito, Fukai et al. 2010).

A scaffold which serves as a template for tissue generation has also been introduced in the field of TE. Thusfar TE has not met initial expectations. Materials used as a scaffold to engineer bone are often engineered to be biocompatible and have good initial biomechanical properties. These properties may interfere with biomechanical stimuli needed for tissue organisation and degradation products from these artificial scaffolds may interfere with the natural healing response. In contrast to a natural endochondral scaffold, artificial scaffolds do not orchestrate ingress of progenitor cells, vascularisation etc.

Periosteum seems to play an important role in postnatal endochondrogenesis. However subcutaneous injection of growth factors leads to generation of bone via the endochondral pathway. The first examples of successful generation of bone and cartilage by triggering the progenitor cells of periosteum are found in literature (Emans, Surtel et al. 2005; Emans, van Rhijn et al. 2010). Also repair of cartilage and bone has been reported to be successful in animal studies using this approach. Using the postnatal endochondrogenic capacity for generation of cartilage and bone has many advantages: expensive culture procedures and logistics are bypassed and sufficient amounts of tissue are likely to be generated. Depending on the stage in which endochondral tissue is harvested, different clinical needs could be treated varying from (osteo)chondral defects to bone defects (Scotti, Tonnarelli et al. 2010). Finally, studying the process of endochondrogenesis may not only be a logical direction for tissue generation, but is also expected to provide useful information how to lock progenitors in the desired phase and will contribute to our understanding of diseases like OA.

5. Acknowledgment

The authors wish to acknowledge the Dutch Arthritis Association and Dutch Stichting Annafonds, as well as Prof. V. Prasad Shastri for the collaboration which has led to further insight into periosteal chondrogenesis.

6. References

Aglietti, P., Buzzi, R., Bassi, P. B. (1994). "Arthroscopic drilling in juvenile osteochondritis dissecans of the medial femoral condyle." *Arthroscopy* 10(3): 286-91.

Altman, R. D., Kates, J., Chun, L. E. (1992). "Preliminary observations of chondral abrasion in a canine model." *Ann Rheum Dis* 51(9): 1056-62.

Anderson, H. (1962). "Histochemical studies of the human hip joint." *Acta Anat* 48: 258-292.

Aydelotte, M.&Kuettner, K. (1992). Heterogeneity of articular chondrocytes and cartilage matrix. New York, Marcel Dekker.

Aydelotte, M. B., Greenhill, R. R.&Kuettner, K. E. (1988). "Differences between sub-populations of cultured bovine articular chondrocytes. II. Proteoglycan metabolism." *Connect Tissue Res* 18(3): 223-34.

Bartlett, W., Skinner, J. A., Gooding, C. R. (2005). "Autologous chondrocyte implantation versus matrix-induced autologous chondrocyte implantation for osteochondral defects of the knee: a prospective, randomised study." *J Bone Joint Surg Br* 87(5): 640-5.

Bentley, G.&Greer, R. B., 3rd (1971). "Homotransplantation of isolated epiphyseal and articular cartilage chondrocytes into joint surfaces of rabbits." *Nature* 230(5293): 385-8.

Blaney Davidson, E. N., Vitters, E. L., van Beuningen, H. M. (2007). "Resemblance of osteophytes in experimental osteoarthritis to transforming growth factor beta-induced osteophytes: limited role of bone morphogenetic protein in early osteoarthritic osteophyte formation." *Arthritis Rheum* 56(12): 4065-73.

Blunk, T., Sieminski, A. L., Gooch, K. J. (2002). "Differential effects of growth factors on tissue-engineered cartilage." *Tissue Eng* 8(1): 73-84.

Bostrom, M. P., Lane, J. M., Berberian, W. S. (1995). "Immunolocalization and expression of bone morphogenetic proteins 2 and 4 in fracture healing." *J Orthop Res* 13(3): 357-67.

Bouwmeester, S. J., Beckers, J. M., Kuijer, R. (1997). "Long-term results of rib perichondrial grafts for repair of cartilage defects in the human knee." *Int Orthop* 21(5): 313-7.

Boyle, W. J., Simonet, W. S.&Lacey, D. L. (2003). "Osteoclast differentiation and activation." *Nature* 423(6937): 337-42.

Bradley, J.&Dandy, D. J. (1989). "Results of drilling osteochondritis dissecans before skeletal maturity." *J Bone Joint Surg Br* 71(4): 642-4.

Brittberg, M., Lindahl, A., Nilsson, A. (1994). "Treatment of deep cartilage defects in the knee with autologous chondrocyte transplantation." *N Engl J Med* 331(14): 889-95.

Buckwalter, J.&Mankin, H. (1998). "Articular cartilage: tissue design and chondrocyte matrix interactions." *AAOS Instr Cours Lect* 47: 487-504.

Chu, C. R., Coutts, R. D., Yoshioka, M. (1995). "Articular cartilage repair using allogeneic perichondrocyte-seeded biodegradable porous polylactic acid (PLA): a tissue-engineering study." *J Biomed Mater Res* 29(9): 1147-54.

Cole, B. J., Farr, J., Winalski, C. S. (2011). "Outcomes After a Single-Stage Procedure for Cell-Based Cartilage Repair: A Prospective Clinical Safety Trial With 2-year Follow-up." *Am J Sports Med* 39(6): 1170-9.

Convery, F. R., Meyers, M. H.&Akeson, W. H. (1991). "Fresh osteochondral allografting of the femoral condyle." *Clin Orthop Relat Res*(273): 139-45.

Czitrom, A. A., Keating, S.&Gross, A. E. (1990). "The viability of articular cartilage in fresh osteochondral allografts after clinical transplantation." *J Bone Joint Surg Am* 72(4): 574-81.

Dong, X., Wei, X., Yi, W. (2009). "RGD-modified acellular bovine pericardium as a bioprosthetic scaffold for tissue engineering." *J Mater Sci Mater Med.*

Donohue, J. M., Buss, D., Oegema, T. R., Jr. (1983). "The effects of indirect blunt trauma on adult canine articular cartilage." *J Bone Joint Surg Am* 65(7): 948-57.

Dowthwaite, G. P., Bishop, J. C., Redman, S. N. (2004). "The surface of articular cartilage contains a progenitor cell population." *J Cell Sci* 117(Pt 6): 889-97.

Einhorn, T. A. (2005). "The science of fracture healing." *J Orthop Trauma* 19(10 Suppl): S4-6.

Elford, P. R., Graeber, M., Ohtsu, H. (1992). "Induction of swelling, synovial hyperplasia and cartilage proteoglycan loss upon intra-articular injection of transforming growth factor beta-2 in the rabbit." *Cytokine* 4(3): 232-8.

Emans, P. J., Pieper, J., Hulsbosch, M. M. (2006). "Differential cell viability of chondrocytes and progenitor cells in tissue-engineered constructs following implantation into osteochondral defects." *Tissue Eng* 12(6): 1699-709.

Emans, P. J., Surtel, D. A., Frings, E. J. (2005). "In vivo generation of cartilage from periosteum." *Tissue Eng* 11(3-4): 369-77.

Emans, P. J., van Rhijn, L. W., Welting, T. J. (2010). "Autologous engineering of cartilage." *Proc Natl Acad Sci U S A* 107(8): 3418-23.

Fischer, J., Dickhut, A., Rickert, M. (2010). "Human articular chondrocytes secrete parathyroid hormone-related protein and inhibit hypertrophy of mesenchymal stem cells in coculture during chondrogenesis." *Arthritis Rheum* 62(9): 2696-706.

Freed, L. E., Grande, D. A., Lingbin, Z. (1994). "Joint resurfacing using allograft chondrocytes and synthetic biodegradable polymer scaffolds." *J Biomed Mater Res* 28(8): 891-9.

Freed, L. E., Vunjak-Novakovic, G.&Langer, R. (1993). "Cultivation of cell-polymer cartilage implants in bioreactors." *J Cell Biochem* 51(3): 257-64.

Furukawa, T., Eyre, D. R., Koide, S. (1980). "Biochemical studies on repair cartilage resurfacing experimental defects in the rabbit knee." *J Bone Joint Surg Am* 62(1): 79-89.

Gallay, S. H., Miura, Y., Commisso, C. N. (1994). "Relationship of donor site to chondrogenic potential of periosteum in vitro." *J Orthop Res* 12(4): 515-25.

Garrett, J. C. (1986). "Treatment of osteochondral defects of the distal femur with fresh osteochondral allografts: a preliminary report." *Arthroscopy* 2(4): 222-6.

Garrett, J. C. (1994). "Fresh osteochondral allografts for treatment of articular defects in osteochondritis dissecans of the lateral femoral condyle in adults." *Clin Orthop Relat Res*(303): 33-7.

Garrett, J. C. (1998). "Osteochondral allografts for reconstruction of articular defects of the knee." *Instr Course Lect* 47: 517-22.

Gerber, H. P., Vu, T. H., Ryan, A. M. (1999). "VEGF couples hypertrophic cartilage remodeling, ossification and angiogenesis during endochondral bone formation." *Nat Med* 5(6): 623-8.

Ghazavi, M. T., Pritzker, K. P., Davis, A. M. (1997). "Fresh osteochondral allografts for post-traumatic osteochondral defects of the knee." *J Bone Joint Surg Br* 79(6): 1008-13.

Gibson, G. (1998). "Active role of chondrocyte apoptosis in endochondral ossification." *Microsc Res Tech* 43(2): 191-204.

Glowacki, J. (1998). "Angiogenesis in fracture repair." *Clin Orthop Relat Res*(355 Suppl): S82-9.

Gold, G. E., Burstein, D., Dardzinski, B. (2006). "MRI of articular cartilage in OA: novel pulse sequences and compositional/functional markers." *Osteoarthritis Cartilage* 14 Suppl 1: 76-86.

Grande, D. A., Halberstadt, C., Naughton, G. (1997). "Evaluation of matrix scaffolds for tissue engineering of articular cartilage grafts." *J Biomed Mater Res* 34(2): 211-20.

Grande, D. A., Pitman, M. I., Peterson, L. (1989). "The repair of experimentally produced defects in rabbit articular cartilage by autologous chondrocyte transplantation." *J Orthop Res* 7(2): 208-18.

Gray, D. J.&Gardner, E. (1950). "Prenatal development of the human knee and superior tibiofibular joints." *Am J Anat* 86(2): 235-87.

Gross, A. E., Aubin, P., Cheah, H. K. (2002). "A fresh osteochondral allograft alternative." *J Arthroplasty* 17(4 Suppl 1): 50-3.

Gross, A. E., McKee, N. H., Pritzker, K. P. (1983). "Reconstruction of skeletal deficits at the knee. A comprehensive osteochondral transplant program." *Clin Orthop Relat Res*(174): 96-106.

Grundnes, O.&Reikeras, O. (1993). "The importance of the hematoma for fracture healing in rats." *Acta Orthop Scand* 64(3): 340-2.

Guyer, R. D., Collier, R., Stith, W. J. (1988). "Discitis after discography." *Spine (Phila Pa 1976)* 13(12): 1352-4.

Hall, B. K.&Jacobson, H. N. (1975). "The repair of fractured membrane bones in the newly hatched chick." *Anat Rec* 181(1): 55-69.

Harada, S.&Rodan, G. A. (2003). "Control of osteoblast function and regulation of bone mass." *Nature* 423(6937): 349-55.

Hasler, E. M., Herzog, W., Wu, J. Z. (1999). "Articular cartilage biomechanics: theoretical models, material properties, and biosynthetic response." *Crit Rev Biomed Eng* 27(6): 415-88.

Hendrickson, D. A., Nixon, A. J., Grande, D. A. (1994). "Chondrocyte-fibrin matrix transplants for resurfacing extensive articular cartilage defects." *J Orthop Res* 12(4): 485-97.

Hendriks, J., Riesle, J.&van Blitterswijk, C. A. (2007). "Co-culture in cartilage tissue engineering." *J Tissue Eng Regen Med* 1(3): 170-8.

Homminga, G. N., Bulstra, S. K., Bouwmeester, P. S. (1990). "Perichondral grafting for cartilage lesions of the knee." *J Bone Joint Surg Br* 72(6): 1003-7.

Homminga, G. N., Bulstra, S. K., Kuijer, R. (1991). "Repair of sheep articular cartilage defects with a rabbit costal perichondrial graft." *Acta Orthop Scand* 62(5): 415-8.

Horas, U., Schnettler, R., Pelinkovic, D. (2000). "[Osteochondral transplantation versus autogenous chondrocyte transplantation. A prospective comparative clinical study]." *Chirurg* 71(9): 1090-7.

Huang, Q., Goh, J. C., Hutmacher, D. W. (2002). "In vivo mesenchymal cell recruitment by a scaffold loaded with transforming growth factor beta1 and the potential for in situ chondrogenesis." *Tissue Eng* 8(3): 469-82.

Huitema, G. C., van Rhijn, L. W.&van Ooij, A. (2006). "Screw position after double-rod anterior spinal fusion in idiopathic scoliosis: an evaluation using computerized tomography." *Spine* 31(15): 1734-9.

Hunter, W. (1995). "Of the structure and disease of articulating cartilages. 1743." *Clin Orthop Relat Res*(317): 3-6.

Hunziker, E. (1992). Articular cartilage structure in humans and experimental animals. New York, Raven Press.

Hunziker, E. B. (1999). "Articular cartilage repair: are the intrinsic biological constraints undermining this process insuperable?" *Osteoarthritis Cartilage* 7(1): 15-28.

Hunziker, E. B. (2001). "Growth-factor-induced healing of partial-thickness defects in adult articular cartilage." *Osteoarthritis Cartilage* 9(1): 22-32.

Hunziker, E. B., Kapfinger, E.&Geiss, J. (2007). "The structural architecture of adult mammalian articular cartilage evolves by a synchronized process of tissue resorption and neoformation during postnatal development." *Osteoarthritis Cartilage* 15(4): 403-13.

Insall, J. (1974). "The Pridie debridement operation for osteoarthritis of the knee." *Clin Orthop Relat Res*(101): 61-7.

Iwasaki, M., Nakahara, H., Nakase, T. (1994). "Bone morphogenetic protein 2 stimulates osteogenesis but does not affect chondrogenesis in osteochondrogenic differentiation of periosteum-derived cells." *J Bone Miner Res* 9(8): 1195-204.

Iwasaki, M., Nakahara, H., Nakata, K. (1995). "Regulation of proliferation and osteochondrogenic differentiation of periosteum-derived cells by transforming growth factor-beta and basic fibroblast growth factor." *J Bone Joint Surg Am* 77(4): 543-54.

Iwasaki, M., Nakata, K., Nakahara, H. (1993). "Transforming growth factor-beta 1 stimulates chondrogenesis and inhibits osteogenesis in high density culture of periosteum-derived cells." *Endocrinology* 132(4): 1603-8.

Jackson, R. W. (1986). "The scope of arthroscopy." *Clin Orthop Relat Res*(208): 69-71.

Kim, H. K., Moran, M. E.&Salter, R. B. (1991). "The potential for regeneration of articular cartilage in defects created by chondral shaving and subchondral abrasion. An experimental investigation in rabbits." *J Bone Joint Surg Am* 73(9): 1301-15.

Kim, S. S., Park, M. S., Gwak, S. J. (2006). "Accelerated bonelike apatite growth on porous polymer/ceramic composite scaffolds in vitro." *Tissue Eng* 12(10): 2997-3006.

Kronenberg, H. M. (2003). "Developmental regulation of the growth plate." *Nature* 423(6937): 332-6.

Kuettner, K.&Pauli, B. (1983). Vascularity of cartilage. New York, Academic Press.

Kuijer, R., Emans, P. J., Jansen, E. (2003). "Isolation and Cultivation of Chondrogenic Precursors from Aged Human Periosteum." *49th Trans ORS* 28: 101.

Kuijer, R., Jansen, E. J., Emans, P. J. (2007). "Assessing infection risk in implanted tissue-engineered devices." *Biomaterials*.

Macchiarini, P., Jungebluth, P., Go, T. (2008). "Clinical transplantation of a tissue-engineered airway." *Lancet* 372(9655): 2023-30.

Mankin, H., Mow, V.&Buckwalter, J. (2000). Articular cartilage repair and osteoarthritis. Rosemont, American Academy of Orthopaedic Surgeons.

Mankin, H., Mow, V.&Buckwalter, J. (2000). Articular cartilage structure, composition, and function. Rosemont, AAOS.

Martin, I., Shastri, V. P., Padera, R. F. (2001). "Selective differentiation of mammalian bone marrow stromal cells cultured on three-dimensional polymer foams." *J Biomed Mater Res* 55(2): 229-35.

Masquelet, A. C.&Begue, T. (2010). "The concept of induced membrane for reconstruction of long bone defects." *Orthop Clin North Am* 41(1): 27-37; table of contents.

McKibbin, B.&Maroudas, A. (1979). "Adult articular cartilage." *Piman Medical* 2E: 461 - 486.

Meachim, G.&Roberts, C. (1971). "Repair of the joint surface from subarticular tissue in the rabbit knee." *J Anat* 109(2): 317-27.

Mitchell, N.&Shepard, N. (1976). "The resurfacing of adult rabbit articular cartilage by multiple perforations through the subchondral bone." *J Bone Joint Surg Am* 58(2): 230-3.

Miyazono, K., Maeda, S.&Imamura, T. (2005). "BMP receptor signaling: transcriptional targets, regulation of signals, and signaling cross-talk." *Cytokine Growth Factor Rev* 16(3): 251-63.

Nakahara, H., Bruder, S. P., Goldberg, V. M. (1990). "In vivo osteochondrogenic potential of cultured cells derived from the periosteum." *Clin Orthop Relat Res*(259): 223-32.

Nakahara, H., Dennis, J. E., Bruder, S. P. (1991). "In vitro differentiation of bone and hypertrophic cartilage from periosteal-derived cells." *Exp Cell Res* 195(2): 492-503.

Nakahara, H., Goldberg, V. M.&Caplan, A. I. (1991). "Culture-expanded human periosteal-derived cells exhibit osteochondral potential in vivo." *J Orthop Res* 9(4): 465-76.

Nakata, K., Nakahara, H., Kimura, T. (1992). "Collagen gene expression during chondrogenesis from chick periosteum-derived cells." *FEBS Lett* 299(3): 278-82.

Nixon, A. J., Sams, A. E., Lust, G. (1993). "Temporal matrix synthesis and histologic features of a chondrocyte-laden porous collagen cartilage analogue." *Am J Vet Res* 54(2): 349-56.

Novack, D. V.&Faccio, R. (2011). "Osteoclast motility: putting the brakes on bone resorption." *Ageing Res Rev* 10(1): 54-61.

O'Driscoll, S. W. (1998). "The healing and regeneration of articular cartilage." *J Bone Joint Surg Am* 80(12): 1795-812.

O'Driscoll, S. W., Keeley, F. W.&Salter, R. B. (1986). "The chondrogenic potential of free autogenous periosteal grafts for biological resurfacing of major full-thickness defects in joint surfaces under the influence of continuous passive motion. An experimental investigation in the rabbit." *J Bone Joint Surg Am* 68(7): 1017-35.

O'Driscoll, S. W., Keeley, F. W.&Salter, R. B. (1988). "Durability of regenerated articular cartilage produced by free autogenous periosteal grafts in major full-thickness defects in joint surfaces under the influence of continuous passive motion. A follow-up report at one year." *J Bone Joint Surg Am* 70(4): 595-606.

O'Driscoll, S. W., Saris, D. B., Ito, Y. (2001). "The chondrogenic potential of periosteum decreases with age." *J Orthop Res* 19(1): 95-103.

Oberpenning, F., Meng, J., Yoo, J. J. (1999). "De novo reconstitution of a functional mammalian urinary bladder by tissue engineering." *Nat Biotechnol* 17(2): 149-55.

Ogilvie-Harris, D. J.&Fitsialos, D. P. (1991). "Arthroscopic management of the degenerative knee." *Arthroscopy* 7(2): 151-7.

Park, Y., Sugimoto, M., Watrin, A. (2005). "BMP-2 induces the expression of chondrocyte-specific genes in bovine synovium-derived progenitor cells cultured in three-dimensional alginate hydrogel." *Osteoarthritis Cartilage* 13(6): 527-36.

Petite, H., Viateau, V., Bensaid, W. (2000). "Tissue-engineered bone regeneration." *Nat Biotechnol* 18(9): 959-63.

Poole, A. R., Kojima, T., Yasuda, T. (2001). "Composition and structure of articular cartilage: a template for tissue repair." *Clin Orthop Relat Res*(391 Suppl): S26-33.

Radin, E. L., Martin, R. B., Burr, D. B. (1984). "Effects of mechanical loading on the tissues of the rabbit knee." *J Orthop Res* 2(3): 221-34.

Rae, P. J.&Noble, J. (1989). "Arthroscopic drilling of osteochondral lesions of the knee." *J Bone Joint Surg Br* 71(3): 534.

Reddi, A. H. (1994). "Symbiosis of biotechnology and biomaterials: applications in tissue engineering of bone and cartilage." *J Cell Biochem* 56(2): 192-5.

Saito, T., Fukai, A., Mabuchi, A. (2010). "Transcriptional regulation of endochondral ossification by HIF-2alpha during skeletal growth and osteoarthritis development." *Nat Med* 16(6): 678-86.

Sams, A. E., Minor, R. R., Wootton, J. A. (1995). "Local and remote matrix responses to chondrocyte-laden collagen scaffold implantation in extensive articular cartilage defects." *Osteoarthritis Cartilage* 3(1): 61-70.

Sams, A. E.&Nixon, A. J. (1995). "Chondrocyte-laden collagen scaffolds for resurfacing extensive articular cartilage defects." *Osteoarthritis Cartilage* 3(1): 47-59.

Schenk, R., Eggli, P.&Hunziker, E. (1986). Articular cartilage morphology. New York, Raven Press.

Sellers, R. S., Peluso, D.&Morris, E. A. (1997). "The effect of recombinant human bone morphogenetic protein-2 (rhBMP-2) on the healing of full-thickness defects of articular cartilage." *J Bone Joint Surg Am* 79(10): 1452-63.

Sellers, R. S., Zhang, R., Glasson, S. S. (2000). "Repair of articular cartilage defects one year after treatment with recombinant human bone morphogenetic protein-2 (rhBMP-2)." *J Bone Joint Surg Am* 82(2): 151-60.

Shastri, V. P. (2006). "Future of regenerative medicine: challenges and hurdles." *Artif Organs* 30(10): 828-34.

Skoog, T.&Johansson, S. H. (1976). "The formation of articular cartilage from free perichondrial grafts." *Plast Reconstr Surg* 57(1): 1-6.

Steinberg, M. S. (1962). "Mechanism of tissue reconstruction by dissociated cells. II. Time-course of events." *Science* 137: 762-3.

Stevens, M. M., Marini, R. P., Schaefer, D. (2005). "In vivo engineering of organs: the bone bioreactor." *Proc Natl Acad Sci U S A* 102(32): 11450-5.

Stevenson, S. (1998). "Enhancement of fracture healing with autogenous and allogeneic bone grafts." *Clin Orthop Relat Res*(355 Suppl): S239-46.

Takahashi, S., Oka, M., Kotoura, Y. (1995). "Autogenous callo-osseous grafts for the repair of osteochondral defects." *J Bone Joint Surg Br* 77(2): 194-204.

Urist, M. (1976). Biogenesis of bone: Calcium and phophorus in the skeleton and blood in vertebrate evolution. Washington DC, Am. Phys. Soc.

Vachon, A., Bramlage, L. R., Gabel, A. A. (1986). "Evaluation of the repair process of cartilage defects of the equine third carpal bone with and without subchondral bone perforation." *Am J Vet Res* 47(12): 2637-45.

Vachon, A., McIlwraith, C. W., Trotter, G. W. (1989). "Neochondrogenesis in free intra-articular, periosteal, and perichondrial autografts in horses." *Am J Vet Res* 50(10): 1787-94.

van Beuningen, H. M., van der Kraan, P. M., Arntz, O. J. (1993). "Does TGF-beta protect articular cartilage in vivo?" *Agents Actions Suppl* 39: 127-31.

van Beuningen, H. M., van der Kraan, P. M., Arntz, O. J. (1994). "Transforming growth factor-beta 1 stimulates articular chondrocyte proteoglycan synthesis and induces osteophyte formation in the murine knee joint." *Lab Invest* 71(2): 279-90.

van der Kraan, P. M.&van den Berg, W. B. (2007). "Osteophytes: relevance and biology." *Osteoarthritis Cartilage* 15(3): 237-44.

Volk, S. W.&Leboy, P. S. (1999). "Regulating the regulators of chondrocyte hypertrophy." *J Bone Miner Res* 14(4): 483-6.

Vortkamp, A., Lee, K., Lanske, B. (1996). "Regulation of rate of cartilage differentiation by Indian hedgehog and PTH-related protein." *Science* 273(5275): 613-22.

Wakitani, S., Kimura, T., Hirooka, A. (1989). "Repair of rabbit articular surfaces with allograft chondrocytes embedded in collagen gel." *J Bone Joint Surg Br* 71(1): 74-80.

Wu, L. N., Ishikawa, Y., Sauer, G. R. (1995). "Morphological and biochemical characterization of mineralizing primary cultures of avian growth plate chondrocytes: evidence for cellular processing of Ca2+ and Pi prior to matrix mineralization." *J Cell Biochem* 57(2): 218-37.

Yoon, B. S.&Lyons, K. M. (2004). "Multiple functions of BMPs in chondrogenesis." *J Cell Biochem* 93(1): 93-103.

Yoon, B. S., Pogue, R., Ovchinnikov, D. A. (2006). "BMPs regulate multiple aspects of growth-plate chondrogenesis through opposing actions on FGF pathways." *Development* 133(23): 4667-78.

Zarnett, R.&Salter, R. B. (1989). "Periosteal neochondrogenesis for biologically resurfacing joints: its cellular origin." *Can J Surg* 32(3): 171-4.

Zhao, F., Grayson, W. L., Ma, T. (2006). "Effects of hydroxyapatite in 3-D chitosan-gelatin polymer network on human mesenchymal stem cell construct development." *Biomaterials* 27(9): 1859-67.

19

Tissue Engineering in Low Urinary Tract Reconstruction

Chao Feng and Yue-min Xu
Department of Urology, Shanghai Jiaotong University-Affiliated 6th People's Hospital, Shanghai, China

1. Introduction

Acquired and congenital abnormalities of the lower urinary tract often require eventual reconstruction. Traditionally, different types of autologous tissue can be chosen for surgery, depending on which organ requires reconstruction. Bladder reconstruction, for example, is usually performed with intestinal tissue while urethral reconstruction can us buccal mucosa, lingual mucosa, colonic mucosa or prepuce skin. However, the problems of a shortage of patients' own tissues, and of nmany complications related to surgery, have not yet been resolved. There is therefore an effort to obtain sufficient tissue resources, to involve fewer complications, to reduce surgery to relatively minor invasion and to achieve better surgical outcomes. These goals may be attainable by the use of tissue engineering techniques.

Over the last 50 years, tissue engineering techniques for low urinary tract regeneration have been applied successfully in a variety of animal models and clinical patients. Rapid advancement has been made in this field, which has broadened the theoretical options for the future of low urinary tract reconstruction. These developments include improvements in cell culture techniques, such as the development of cell resources and identification of markers to isolate and characterize specific cell types. Many new types of natural and synthetic biomaterials for use as scaffold components have been created (1). In addition to these, the applications of nanotechnology and bioreactors have been strengthened within recent decades. Here, we review the literature on the basic principles and latest developments of tissue engineering technologies in lower urinary tract reconstruction.

2. Basic knowledge of tissue engineering in low urinary tract

2.1 Cell sources

2.1.1 Autologous stromal cells

Because epithelial cells are one of the most important components of the lower urinary tract, optimizing sources for them have always been a popular focus of investigators. Traditionally, urothelial cells obtained from bladder or urethra have often been used in previous studies (Fig 1a) (2,3). Although this technique exploits homotypy between the graft cells and host, it involves injury to the genitourinary tract and the operation is complicated.

Fu, *et al.* chose epidermal cells as graft cells because of its abundant resources; they can be obtained by a less invasive method than the traditional method of bladder or urethral biopsy followed by dissection of transitional cells. The results suggest that the epidermal cells can transform to transitional epithelial cells under the influence of the urethral or bladder environment (4). From our experience, we suggest using the oral keratinocytes, such as buccal keratinocytes and lingual keratinocytes, as a source of epithelial cells, (Fig 1b,c). Such cells express the β-defensin, IL-8, which can mediate an innate immune response against microbes (5). Therefore, compound grafts were easily resisted infection both in vitro and in vivo. In addition, these oral keratinocytes expressed AE1/AE3, which is similar to epidermal cells or urothelial cells in a previous report (6). However, 3T3 cells are usually needed as a feeder layer when culturing oral keratinocytes. The purification of oral keratinocytes therefore needs to be should be improved before clinical application.

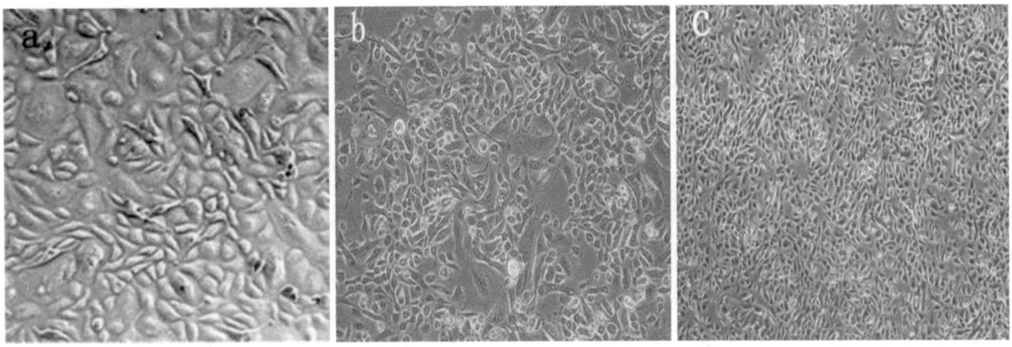

Fig. 1. Morphology the different kinds of epithelial seeding cells.
a. Bladder urethelial cells; b. buccal keratinocytes; c. lingual keratinocytes

To construct 3D bladder or urethral tissue, smooth muscle cell is also necessary. Previously, bladder smooth muscle cells were used for tissue engineering bladder reconstruction (Fig2a). The corpora cavernosa smooth muscle cells were used for constructing the corpora spongiosum, which is one of the most important components of the penile urethra (Fig2b). The advantage of using those cells is that some angiogenic growth factors and their receptors, such as Flk-1 and VEGF, are present in smooth muscle cells. They might contribute to the angiogenesis of bladder or urethral tissue (7). Since contamination by fibroblast cells is a

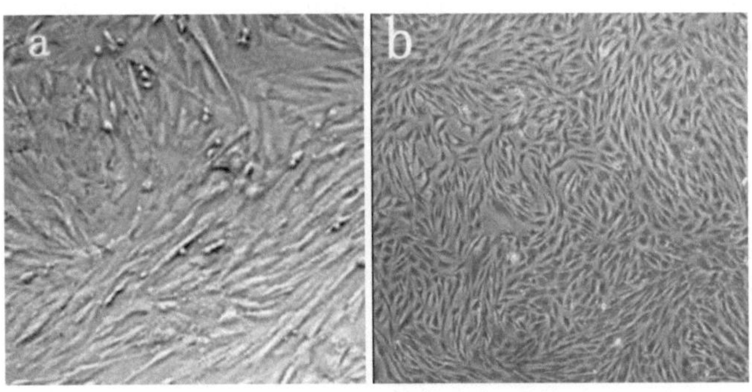

Fig. 2. Morphology the different kinds of smooth muscle cells.
a. Bladder smooth muscle cells; b. corpora cavernosa smooth muscle cells

problem during culturing of these, we advise that a velocity sedimentation method be used to evaluate the purification of smooth muscle cells. Of course, obvious trauma after the procedure is also the shortcoming of this method. As a result of these problems, smooth muscle cells are being replaced by muscle stem cells in urinary reconstruction.

2.1.2 Bone marrow and adipose derived stem cells

Stem cells from bone marrow (BMSC) have been characterized as being either hematopoietic or mesenchymal. They are easily isolated due to their affinity with and adherence to plastic dishes. Their ability to proliferate ensures that even a small number of BMSC multiply into millions of cells under the right culture conditions. Another merit of these cells is that they do not express MHC II, rendering them nonimmunogenic and thereby eliminating possible graft rejection (8). Previous studies showed that BMSCs contained higher concentrations of α-SM actin than did bladder SMC. Meanwhile, BMSCs showed strong response to the Ca^{2+}-ionophore, whereas fibroblasts did not contract their baseline even in the presence of calcium. Those results indicate that BMSCs and smooth muscle cells from low urinary tract are very similar (9). Therefore, BMSCs may serve as an alternative cell source in lower urinary tract tissue engineering.

Stem cells from adipose tissue (ADSCs) have also been popular in tissue engineering research. Adipose tissue is derived from embryonic mesodermal precursors and it contains multipotent progenitor cells that are capable of differentiating into mesenchymal tissue. Since adipose tissue contains 100-1,000 times more pluripotent cells per cubic centimeter

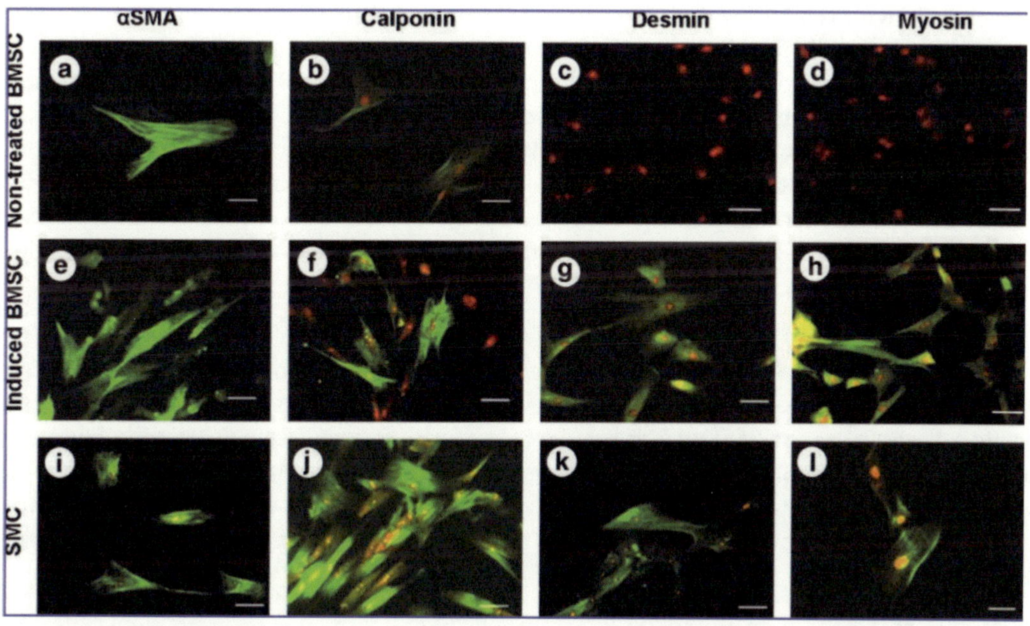

Fig. 3. Myogenic differentiation of human BMSCs using SMC-derived CM. Human BMSCs (p4) were stained with a-SMA (a, e, i), calponin (b, f, j), desmin (c, g, k), and myosin (d, h, l) antibodies without induction as negative control (a–d) and with induction for 14 days (e–h). SMCs were also stained with the same antibodies as a positive control (i–l). (Picture from ref 11)

than does bone marrow, it is easier to obtain ADCSs than other kinds of adult stem cells. Of course, immunoprivilege is also the advantage of this kind of cell. For these reasons, many investigators have selected ADSCs as an ideal source of seeding cells in lower urinary reconstruction, such as repair of bladder and urethra (10).

In many studies, mesenchymal cells have been found to differentiate into many different lineages, such as chondrocytes, osteoblasts, adipocytes, neurons and myoblasts. To urologists, the most interesting thing is the possibility for differentiation of BMSCs or ADSCs into smooth muscle cells and keratinocytes. According to previous reports, these stem cells can acquire a smooth muscle cell phenotype, staining positively for α-SMA, myosin and calponin after being cultured in conditioned medium. Also, culturing in the presence of other myogenic growth factors, such as PDFF-BB,HGF,TGF-β, can also lead to a phenotypic profile of smooth muscle cells (Fig 3) (8,11). Another group has demonstrated the differentiation of marked BMSC into urothelial cells on a seeded scaffold in porcine bladder augmentation, suggesting that mesenchymal stem cells can be made into urothelial cells. However, few additional reports support this result. Since the BMSCs and ADSCs are derived from the mesodermal lineage, more evidence is needed to support that ectodermal lineage cells can be induced from mesenchymal stem cells, such as BMSCs and ADSCs.

2.1.3 Other seeding cells

As well as the autologous stromal cells, BMSCs and ADSCs, other kinds of seeding cells have also shown possibilities for lower urinary tract reconstruction. Drewa (12) *et al.* used hair follicle stem cells for bladder regeneration in rats. This type of cell is CD34 positive, which facilitates the isolation of live epithelial cells with stem cell characterisitcs. In their study, Drewa *et al.* concluded that pluripotent stem cells within rodent hair follicle can differentiate into neurons, glia, keratinocytes and smooth muscle cells. They used an acellular matrix seeded with those cells and achieved a successful bladder wall reconstruction. Further research should be focused on better characterization of these cell populations and on the exact mechanism by which these cells enhance bladder regeneration.

Zhang's study focused on a subpopulation of cells isolated from naturally voided urine (13). This kind of cell demonstrated features typical of progenitor/stem cells, including expression of MSC and pericyte cell surface markers and clonogenic, multipotential, and plastic adhensive capacity. Furthermore, recent study showed that these cells have the capability to differentiate into the urothelial and smooth muscle cells (Fig 4)(14). The latest study has demonstrated the feasibility of forming a tissue-engineered conduit for use in urinary diversion by generating scaffolds seeded with human urine-derived stem cells.

Other cells, such as human amniotic fluid stem cells (AFS), human embryonic stem cells (ES) and human induced pluripotent stem cells (iPS), have also shown a potential for application in lower urinary reconstruction. However, most reports have been rather preliminary investigations. Several key points still need to be studied in depth before the cells can be used in patients.

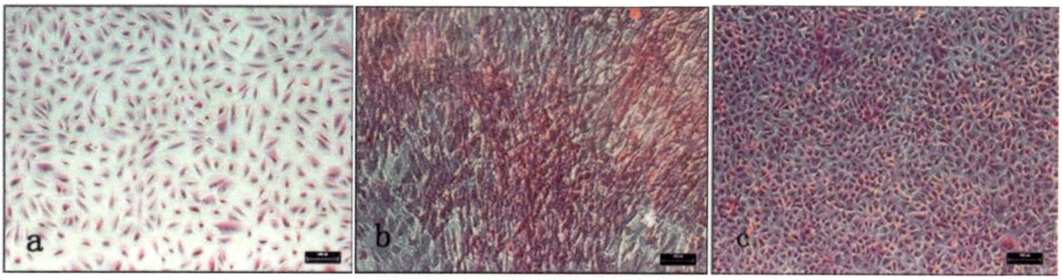

Fig. 4. Morphology of urine-derived stem cells obtained from upper urinary tract (USC-UUT) with differentiation. a. non-treated USC; b. The shape of -UUT changed from an oval to a spindle shape with the addition of myogenic medium; c. a cuboidal shape with the addition of uro-epithelial medium. Scale bar shown is 100 mm (Picture from ref 14).

2.2 Biomaterials

Creating an ideal biomaterial for lower urinary tract reconstruction has been an aspiration of urologists for over a century. An excellent biomaterial for tissue engineering should possess optimal mechanical properties, good biocompatibility, suitable three dimensional structures and degradation rates (15).

2.2.1 Traditional biomaterials

Traditionally, biomaterials can be classified into naturally derived materials, including chitosan, collagen; acellular matrix , such as small intestine submucosa (SIS), bladder acellular matrix (BAMG), acellular corpous spongious matrix (ACSM) and urethral extra matrix (UEM) , as well as synthetic materials, such as PGA and PLGA. Most of them have been used in animal models and human subjects, which will be discussed in the later section. Brehmer provides a useful classification of scaffolds into carrier-, fleece- and sponge-types, according to the structure of biomaterials (16) Carrier-type scaffolds are fiber meshes with very small pore sizes (<15 μm). The pore size of the sponge-type scaffolds is greater than 15 μm. Fleece-type scaffolds have huge interfilamentary spaces (200 μm). In our previous study, we compared the dimensional structures of SIS, BAMG, handmade PGA mesh and ACSM. SEM demonstrated that the pore size of the PGA (>200 μm) was the largest among all biomaterials. The surface pore sizes in SIS were significantly larger than BAMG (58.32 ± 10.31 μm vs 6.77 ± 0.49 μm; $P < 0.05$). Although a looser structure of BAMG could be seen with H&E staining, its pore sizes in surface views were smaller than those of ACSM (6.77 ± 0.49 μm vs 11.12 ± 1.43 μm; $P < 0.05$). An obvious difference of pore diameters in ACSM could be distinguished between urethral surface and cavernosal surface. (2.04 ± 0.32 μm vs 11.12 ± 1.43 μm; $P < 0.05$)(Fig 5)(17). This data can guide the following cell seeding procedure, since cellular growth and infiltration are strongly related to the scaffold's pore sizes. Of course, it should be noted that the structure of PGA or PLGA can be controlled now with the development of electrospin techniques. Therefore, the dimensional structure of synthetic materials is becoming more similar to the naturally derived scaffolds, and even to the original organs.

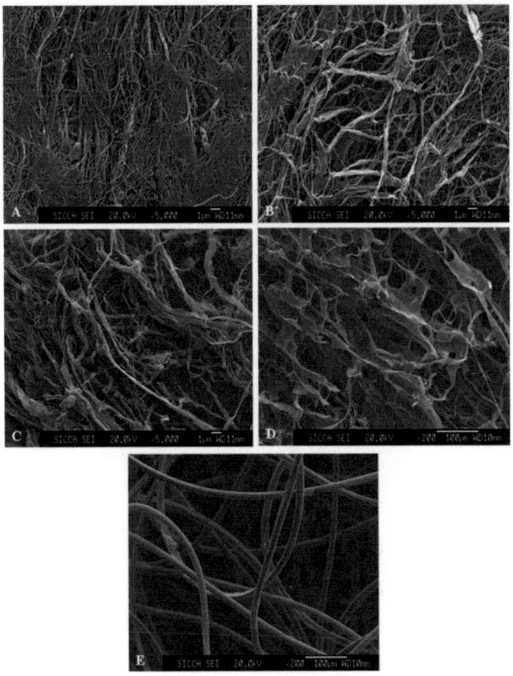

Fig. 5. EMS examination of different materials' surface. (A) urethral surface of ACSM, EMS×5,000, (B) cavernosal surface of ACSM,EMS×5,000 (C) surface of BAMG, EMS×5,000. (D) surface of SIS, EMS_200. (E) surface of PGA, EMS×200.

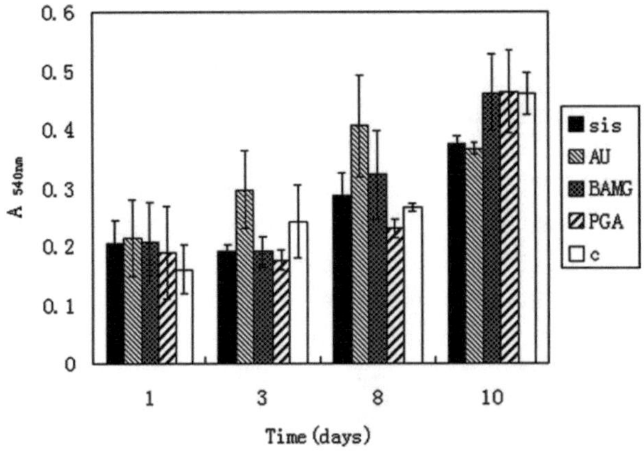

Fig. 6. Metabolic activity of CCSMCs cultured with extracts of various biomaterials or cultured directly in normal medium at 1, 3, 8, and 10 days, as determined by MTT assays. The difference between biomaterials and negative controls was not statistically significant

To address the issue of biocompatibility, our study used the MTT assay technique to evaluate cytotoxicity of different kinds of biomaterials. There were no statistically significant differences in MTT results between the cells cultured with biomaterial extracts and with controls (Fig 6). Thus, we may suggest that all scaffolds could be used safely for lower urinary tract reconstruction.

The mechanical properties of biomaterials are also key to successful reconstruction of the lower urinary tract. For urethral reconstruction, a uniaxial mechanical test is necessary to evaluate the scaffold. In our previous study, all biomaterials exhibited the classic biological nonlinear stress - strain response (Fig. 7) in a mechanical test. The ACSM showed good response in Young's modulus and breaking stress, these being better than in other scaffolds, even the normal rabbit urethra. For the bladder, physiological loading of the tissue involves compressive loads perpendicular to the bladder surface, induced by urine and surrounding pelvic tissues, so biaxial mechanical testing is more realistic. In addition, a burst experiment should also be considered (18) (Fig 8)

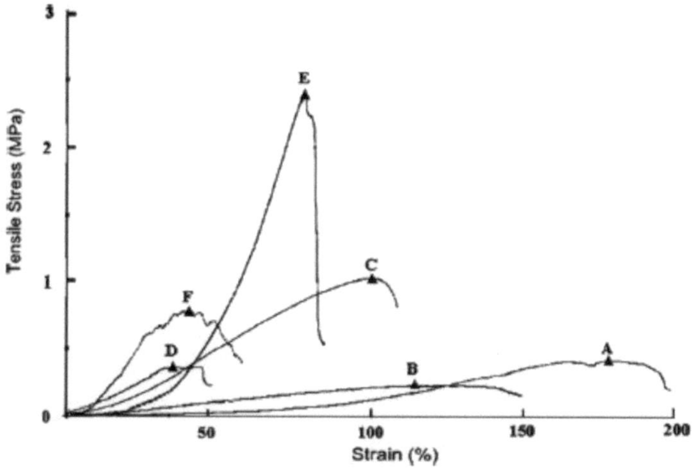

Fig. 7. Stress–Strain curves of various biomaterials. (A) Normal rabbit urethra; (B) SIS; (C) 4-layer SIS; (D) BAMG; (E) ACSM; (F) PGA

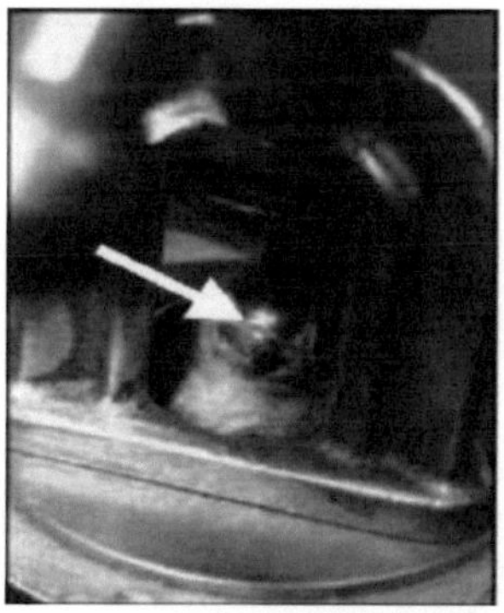

Fig. 8. Picture of ball-burst test with a ruptured test material. Arrow points at the rupture site from (ref 18)

2.2.2 Modified & advanced biomaterials

Since inherit weaknesses always exist in traditional biomaterials, many modified biomaterials have been studied to avoid them.

To enhance angiogenesis, some investigators have modified traditional matrices by incorporating heparin and subsequently loading the heparinized matrices with VEGF. Preliminary studies have shown that this loading of the matrices with VEGF increases the induction of microvessels in both heparinized and non-heparinized matrices, the effect being largest in the case of the heparinized matrices (19).

In order to control the three dimensional structure and degradation rate of synthetic scaffolds such as PGA or PLGA, electrospin techniques are often considered for tissue engineering in lower urinary tract reconstruction. Various materials have been examined for their ability to support cellular adhension, proliferation and formation of a multilayerd urothelium. The results provide the evidence that electrospinning scaffolds show significant benefitsa over commonly used acellular materials in vitro, and suggest that they should be further examined in vivo (20).

2.3 Advanced technique

2.3.1 Bioreactor

The bioreactor is a device that provides a fluid environment for the growth of cells for various applications, such as industrial fermentation and cell culturing. Bioreactors should be introduced in tissue engineering to optimize, through fluid shear, oxygenation and the supply of nutrients, the growth of cells on a 3D scaffold. This approach has been shown to result in better tissue-like constructs than do conventional static culture conditions (21). It is possible to use in vivo graft sites as 'bioreactors' that feature flowing fluids (blood). An example that is commonly used in tissue engineering for lower urinary reconstruction is the greater omentum. Baumert et al (21) used urothelial and smooth muscle cells to seed a sphere-shaped small intestinal submucosa matrix, which was transferred into the omentum after 3wk of cell growth. By this approach, they obtained tissue engineered bladder with a wall thickness was 4 mm. The construct presented a multilayer urothelium on the lumial aspect and deeper fascicles of organized tissue composed of differentiated smooth muscle cells and mature fibroblasts. There was no evidence of inflammation or necrosis (Fig 9). Gu et al (22) implanted 8Fr silastic tubes into the peritoneal cavity of a rabbit. Those tubes were harvested and the tubular tissue covering the tubes was reverted. A pendulous urethral segment of 1.5 cm long was totally excised and urethroplasty was performed with the reverted tubular tissue in an end-to-end fashion. Finally, the results of study showed that the recipients' peritoneal cavity can be used as bioreactor for tissue engineering urethral reconstruction.

More manufactured bioreactors have been designed for tissue engineering bladder. In order to mimic the dynamics of the urinary bladder, bioreactors that imitate the filling and emptying of a normal bladder have been suggested. A bladder bioreactor built this way should be able to recapitulate those dynamics while providing a cellular environment that facilitates cell-cell and cell-matrix interactions. Under the mechanical stimulation from bioreactor, the physiological and mechanical properties of the bladder can be improved. The growth behavior of urothelial cells and bladder smooth cells can be changed, resulting in the cells undergoing adaptive changes in mechanically-stimulated environment (23, 24).

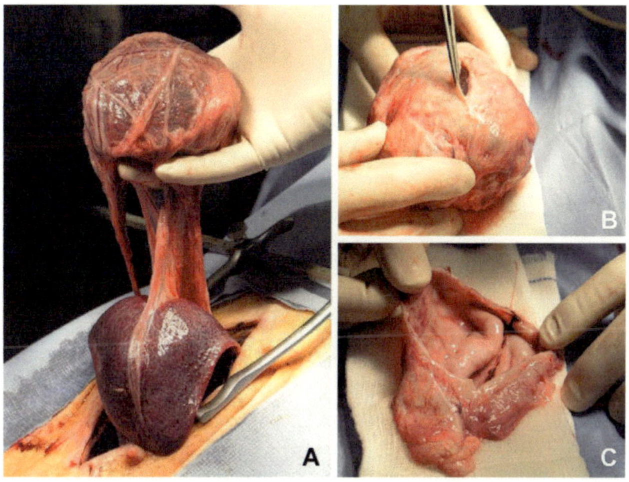

Fig. 9. Harvesting of the matured construct 3 wk after implantation in the omentum (ref 21)

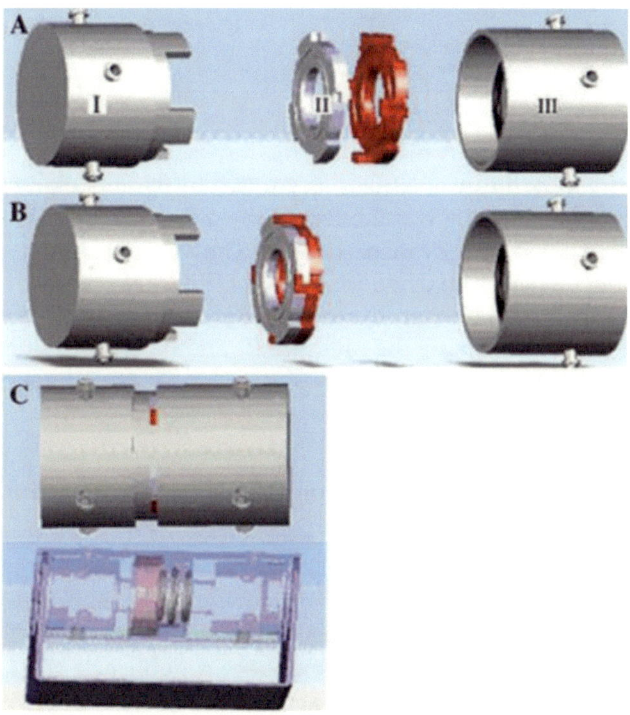

Fig. 10. a. Disassembled urinary bladder bioreactor. I. This chamber will be subjected to controlled pressure and hence would mimic in vitro the urinary bladder chamber. II. Tissue engineered construct ring. III. Compliance chamber (cell culture medium will be re-circulated to accommodate the expansion of the scaffold upon pressure generation). b Interlocking discs for cell-seeded scaffold. c Assembled bioreactor (ref 23)

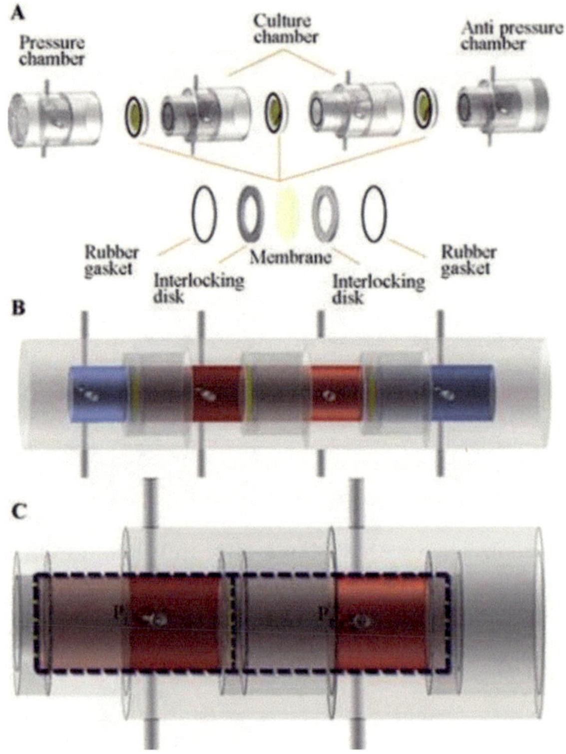

Fig. 11. The bioreactor system. A: A diagram of the disassembled bioreactor, showing the 2 pressure chambers and 2 culture chambers, separated by 3 interlocking rings with elastic membrane. Every interlocking ring was 2 interlocking disks that hold the cell-seeded membrane, which was glued by α-cyanoacrylate. B: The assembled bioreactor with the ports, to which tubing would be attached for medium flow and pressure monitoring. The red parts are culture chambers, blue parts are pressure chambers. C: The assembled culture chambers, the dashed frame showing the pressure P1 and P2 on both sides of cell-seeded membrane. the elastic membrane deformation was driven by pressure difference (P1-P2). (ref 24)

2.3.2 Nanotechnology

Nanotechnology has largely emerged in the last decade of the 20th century as a potential new enabling technology for medicine. For bladder reconstruction, this technology provides a new set of tools to solve many problems that may encountered during the reconstructive procedure. Especially, the incorporation of nanotechnology into bladder tissue engineering materials provides for better bladder materials. Recent published work has demonstrated that increasing of material surface roughness at the nanoscale can improve the adsorption of select proteins important for bladder cell functions (25). Furthermore, some reports showed more bladder smooth muscle cell attachment and growth on polystyrene nanofiber scaffolds fabricated, using an electrospinning technique, to possess surface features at the nanoscale. Cellular adhesive and proliferative ability of keratinocytes were also improved in the nanoscaled scaffold (26).(Table1)

As well as the electrospinning technique mentioned above, another useful nanotechnique is the use of nanoparticles as a delivery system. Mondalek et al (27) investigated the use of PLGA nanoparticles to alter the permeability of SIS scaffolds. Preliminary results indicated that particles ranging from 200 to 500 nm would become imbedded in the SIS scaffold. Particles below this size range would pass through the graft and not become entrapped, and particles above this size range could not penetrate the scaffold. Those results provided the possibility of using nanoparticles to deliver growth factors into seeded cells and scaffolds, to enhance the regeneration of lower urinary tract.

	Bladder Cell Adhesion	Bladder Cell Proliferation	Bladder Cell Synthesis of Proteins	Other Important Functions
Nanostructured PLGA	Highly increased (SMC)[27]	Increased (SMC)[45]	Increased synthesis of elastin and collagen[46]	Decreased formation of calcium stones[29]
	Highly increased (UC)[29]	Highly increased (UC)[29]		
Nanostructured PU	Increased (SMC)[27]	Small increased (SMC)[45]		Decreased formation of calcium stones[29]
	Increased (UC)[29]	Increased (UC)[29]		
Nanostructured PCL	Highly increased (SMC)[27]	Increased (SMC)[45]		
Polymer nanofibers		Increased (SMC)[51]	Production of collagen[51]	Increased cell migration[51]

JC, urothelial bladder cells; SMC, smooth muscle bladder cells.

Table 1. Nanotechnology Approaches to Increase Bladder Cell Functions (ref 26)

2.3.3 Oxygen generating scaffolds

The limitation of oxygen diffusion has led to the general concept that cell or tissue components may not be implanted in large volumes. Many efforts have been made to overcome this limitation. Recently, implantable oxygen releasing biomaterials have been developed in order to provide a sustained release of oxygen to cells and tissues with the goal of prolonging tissue survival and decreasing necrosis (28). In those studies, an oxygen rich compound of sodium percarbonate or calcium peroxide was incorporated into films or 3D constructs of PLGA and used for in situ production of oxygen. In vitro, release of oxygen could be observed from the film more than 24h. Furthermore, these biomaterials were able to extend cell viability growth under hypoxic conditions. Those findings indicate that the use of oxygen generating biomaterials may enhance the scaffold neovascularization after implantation (29). All results suggested that oxygen generating scaffolds can be used for lower urinary tract reconstruction in the future.

3. Applications of tissue engineering in the lower urinary tract

3.1 Bladder reconstruction

Congenital disorders, cancer or trauma can lead to obvious bladder damage. For patients with these problems, bladder reconstructive procedures may be considered. Although gastrointestinal segments are commonly used for bladder augmentation or replacement, multiple complications cannot yet be completely avoided; they include infection, metabolic disturbance and ureolithiasis. A number of animal studies and even clinical experiences have, however, shown the possibility of using tissue engineering techniques to reconstruct bladder tissue. In the laboratory, tissue could be engineered to have function equivalent to the original tissue. In the clinic, patients provided with engineered bladder tissue have obtained satisfactory results.

3.1.1 Animal experiments

Since 1955, many investigators have tried to use different kinds of scaffold for bladder reconstruction in animal models; these have included polyvinyl sponges, polyethylene moulds, Teflon, gelatin sponges, and decellularized pericardial tissue. The outcomes in most studies were unsatisfactory (30). One really successful experiment was reported by Kropp BP et al. (31) in 1995. In this study, the rat underwent partial cystectomy with immediate bladder augmentation with SIS. Host cellular infiltration into the scaffold could be seen 2 weeks after operation. By the end of 48 weeks, the SIS graft presented the three-layered structure of normal bladder, which was indistinguishable from the original bladder. This preliminary study demonstrated the feasibility of using an optimal tissue engineering scaffold for bladder reconstruction. Further study has shown that muscarinic, purinergic and functional cholinergic innervation occurred in rats (32). More recently, bladder regeneration has shown to be more reliable when the SIS was derived from the distal ileum (33). However, graft contraction could be observed in large animal models after using SIS for bladder augmentation, which means that pre-seeding cells may be necessary for tissue engineering-based bladder reconstruction in humans.

Zhang et al. first seeded human bladder urothelial cells and smooth muscle cells onto the SIS by a sandwich culture method (34). This kind of seeding method resulted in organized cell sorting, formation of a well-defined pseudostratified urothelium and multilayered smooth muscle cells with enhanced matrix penetration (Fig 12). The initial study demonstrated that using SIS combined with cell culture could be a valuable model for the study of tissue engineering in bladder reconstruction. To address the problem of graft contraction, Brown et al. showed that opposite-side co-culture of smooth muscle cells and epithelial cells produced a less pronounced matrix contraction than same-side co-culture (20) (Fig 13). The other problem to be addressed in tissue engineering bladder in vitro is cellular infiltration. Recently, Liu et al used preacetic acid (PAA) and Triton X-100

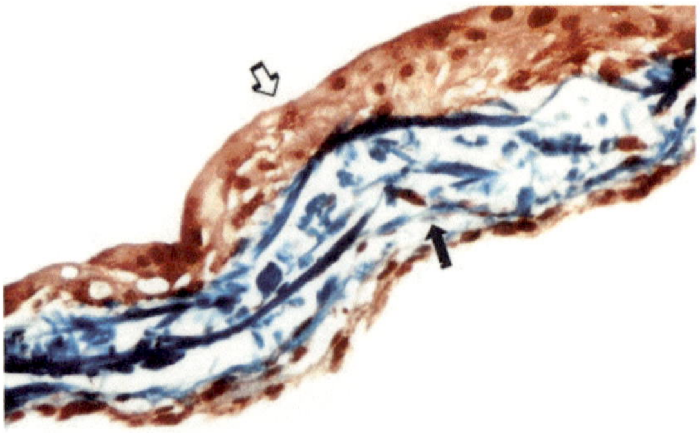

Fig. 12. Sandwich coculture at 28 days shows similar growth pattern to layered coculture technique except that urothelial cells and smooth muscle cells are on opposite sides of small intestinal submucosa membrane. Pseudostratified layer of urothelium is on mucosal surface (open arrow) while multiple layers of smooth muscle cells are on serosal surface and are penetrating into matrix of small intestinal submucosa membrane (solid arrow)(ref 34)

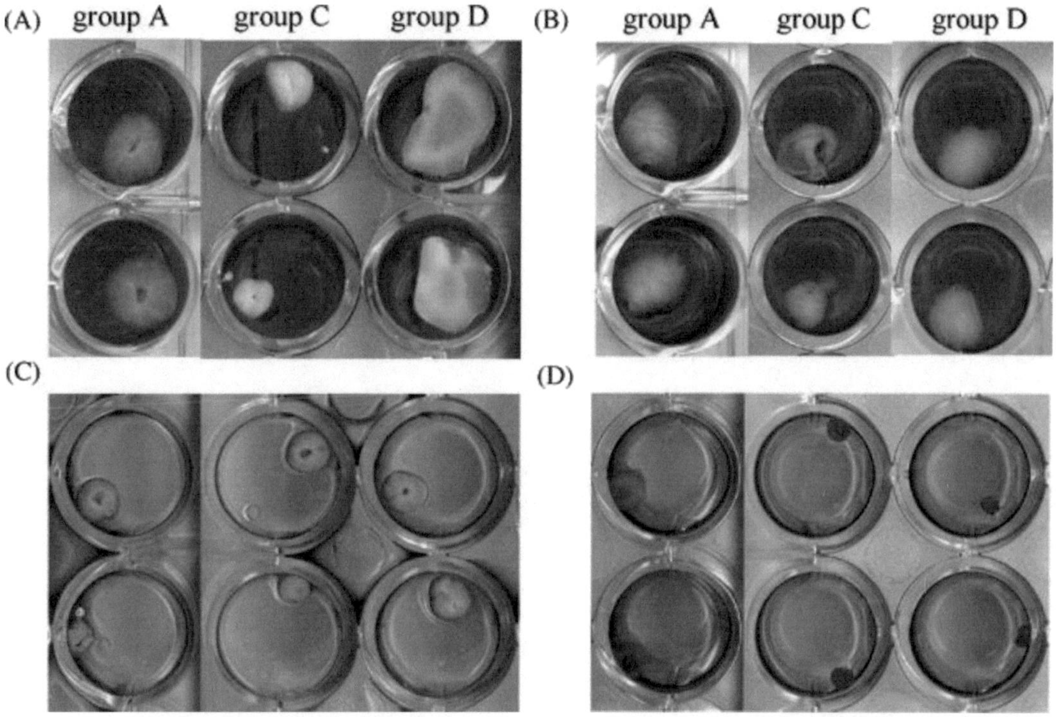

Fig. 13. Gross appearance of cell seeded constructs at 7 and 28 days post seeding. (A) Collagen Gels, 7 days. (B) BAM, 7days. (C) Collagen Gels, 28 days. (D) BAM, 28 days. Group A: SMCs only. Group C: same side co-culture. Group D: opposite side co-culture (ref 20)

to treat the acelluar matrix for bladder tissue engineering reconstruction in vitro (35). This method led to high porosity on the surface of the matrix with about 75% of normal strength. After 3-D dynamic culture, cells could penetrate deeper into the lamina propria of the matrix compared to untreated matrix. (Fig 14). The authors believe that treated scaffold might be more suitable for bladder tissue engineering reconstruction. In order to enhance the vascularization of tissue engineered bladder scaffold in vitro, Baumert et al. transferred the compound matrix into the greater omentum, which has been mentioned above (21).

After meticulous investigation in vitro, Yoo, et al. used scaffold seeded with multiple cell types to reconstruct bladder tissue in 10 beagle dogs, which on which a partial cystectomy had been performed. As a result, 99% a increased in capacity was achieved in the reconstructed bladder. Immunocytochemical analyses confirmed the urothelial and muscle cell phenotypes and showed the presence of nerve fibers (36). Compared to the the technique of seeding with stromal cells, mesenchymal stem cell-seeded scaffold is becoming much more popular in bladder reconstruction. Chung et al. first performed bladder reconstruction using a BMSCs-seeded SIS in rats. At the level of gene expression, regenerated bladder was similar to the control bladder (37). Compared with the BMSCs, ADSCs can be procured more easily. Therefore, we chose ADSCs-seeded scaffolds for bladder reconstruction. At the end of 24 weeks after the operation, the reconstructed bladders reached a mean volume of 94.68±3.31% of the pre-cystectomy bladder capacity in our study. Smooth muscle cells, urothelium and nerve bundles could be detected by

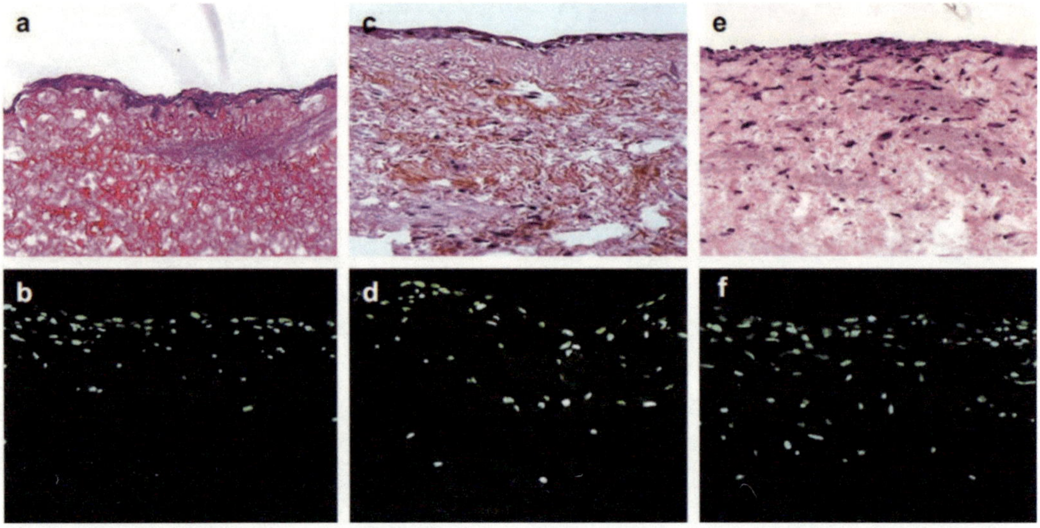

Fig. 14. Cell penetration in 5% PAA-treated BSM in different speeds (0, 10 and 40 rpm) in three-dimensional dynamic culture. Smooth muscle cells and urothelial cell were seeded as layers and co-cultured on the submucosa side of 5% PAA-treated BSM using static culture (left column; a,b) and 3-D rotation culture conditions at 10 (middle column; c,d) and 40 rpm (right column; e,f). H&E staining (a,c,e) and DAPI staining (b,d,f) are shown at 200. Compared with static culture (left column), the cells grew uniformly with deeper penetration in the matrix using 3-D dynamic culture (middle and right column). The cells grown at 40 rpm had deeper penetration of cells within the matrix (e,f) compared to cells cultured at 10 rpm rotation speed (c,d) (ref 35)

immunohistochemical assays. On the contrary, the mean bladder volume was 69.33±5.05% in the control group, made using unseeded scaffolds, and in these there was no evidence of organized muscle or nerve tissue (10) (Fig 15,16). In our study, we also noted that the optimal area for bladder regeneration using seeded scaffold is more than 40-60%, since smaller areas can be regenerated by native bladder tissue. These data provide a useful reference for further clinical application. Other stem cell-seeded scaffolds have also been reported for bladder reconstruction, such as hair-follicle stem cell-seeded scaffolds and urine-derived stem cell-seeded scaffolds (12). However, the number of reports is limited and the actual effectiveness of those scaffold need to be further studied.

As well as traditional cell seeded scaffolds, many modification techniques have been used for tissue engineered bladder reconstructions in animal models. Gregory et al. seeded human adipose stem cells onto PLGA (85:15) bladder dome composites and grafted the result into rat hosts. Results showed that bladder capacity and compliance were maintained in the cell-seeded group throughout the 12 weeks (38) (Fig 17). SIS, modified by hyaluronic acid nanoparticles, has been used for bladder reconstruction. Urinary bladder augmentation has been performed in beagle dogs following hemi-cystectomy using nanoparticle-modified SIS. The results showed that the modified scaffold had significantly higher vascularity compared to unmodified one. This report demonstrated that the nanotechnology can represent a new approach for modifying biomaterials in bladder reconstruction (39). Wei et al.

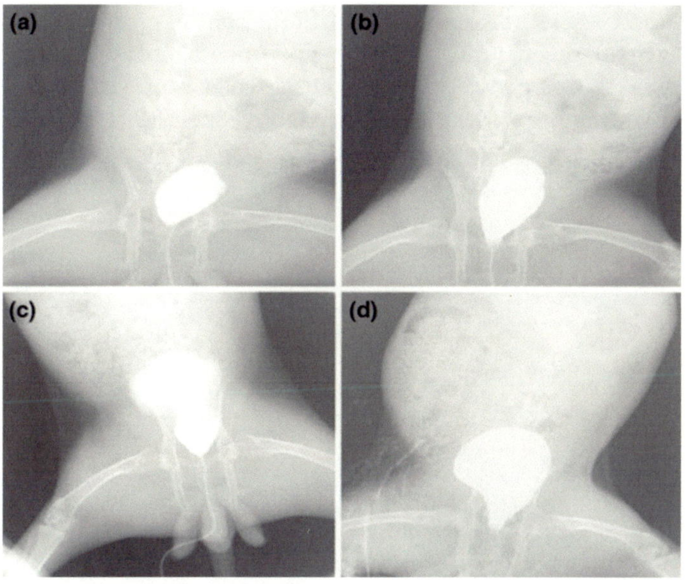

Fig. 15. Cystographies of bladders reconstructed 4 weeks postoperatively. a The control group, b the experimental group 24 weeks postoperatively, c the experimental group and d the experimental group. Cystography demonstrated an improvement in both the shape and capacity of bladders reconstructed with seeded matrices

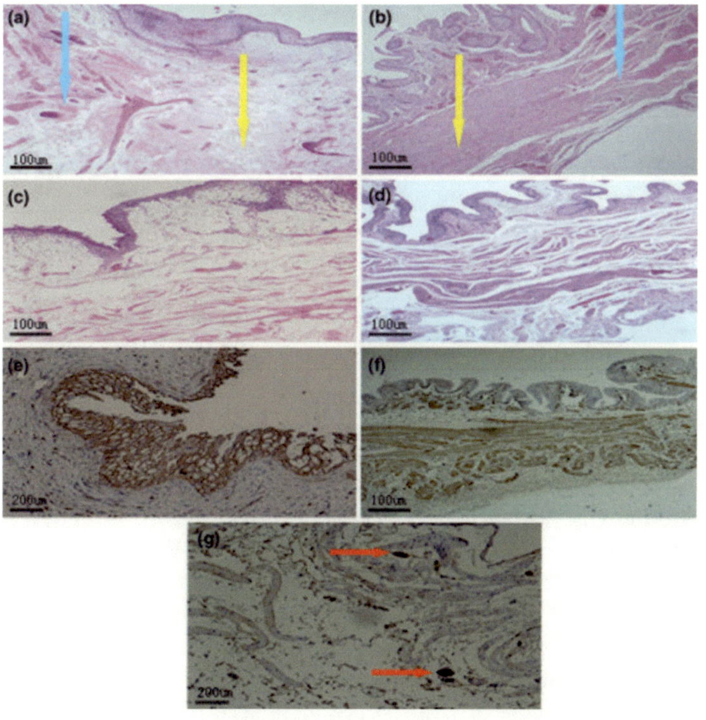

Fig. 16. Histological features of the transplanted grafts. Four weeks postoperatively, native bladder tissue (blue arrow) and in the graft (yellow arrow). a The control group, b the

experimental group 24 weeks postoperatively, c in the control group there is no evidence of organized bladder tissue regeneration, d in the experimental group, the grafts had formed a multilayer epithelium with organized smooth muscle cells. Immunohistochemistry of the transplanted grafts. e Staining with cytokeratin AE1/AE3. α-SM actin. g S-100 (arrows).

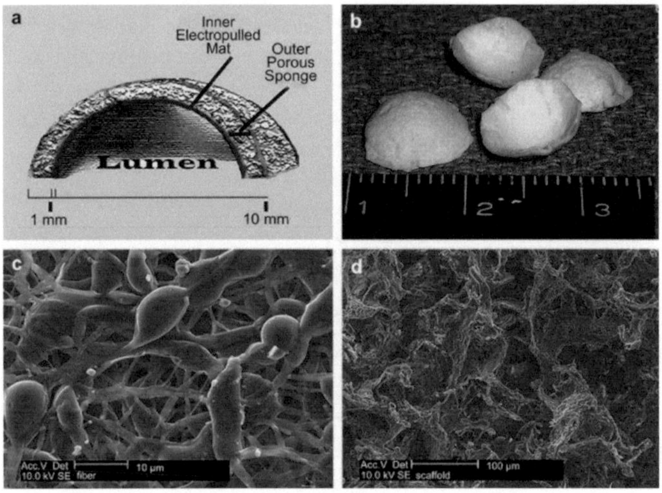

Fig. 17. Construction of the three dimensional synthetic bladder composite. a: Schematic and b: gross micrograph of the three dimensional bladder composite. c:PLGA electropulled microfibers comprising the luminal layer. d: PLGA porous sponge was used as the outer layer (ref 38).

designed a bioreactor to simulate the mechanical properties of bladder. This system successfully generated appropriate pressure waveforms. The viability of cells and tissue structures observed after culture in simulated conditions showed that mechanical stimulation improved the arrangement of cells on scaffold (24).

3.1.2 Clinical application

Although the reports about bladder reconstruction using tissue engineering techniques are few, they are the landmarks in tissue engineered lower urinary tract reconstruction. In Atala's famous study, seven patients with myelomeningocele with high-pressure or poorly compliant bladders enrolled. Urothelial and muscle cells were seeded on a biodegradable bladder-shaped scaffold made of collagen and PGA. Then the biomaterial was used for reconstruction with an omental wrap (40). (Fig 18) After the operation, none of the ultrasounds showed any abnormalities. The cystogram showed the regular shape of bladder after the reconstruction. Urodynamic studies demonstrated significant improvement in volume and compliance in the composite engineered bladders (Fig 19). Postoperatively, it is difficult to distinguish the margin between the composite matrix and the native bladders grossly. All biopsies showed a trilayered structure, consisting of a urothelial cell-lined lumen surrounded by submucosa and muscle. During the post-operative follow-up period all patients had a stable renal function in which serum creatinine was similar to the preoperative status. No metabolic abnormalities were noted. There was no evidence of urinary calculi during the study.

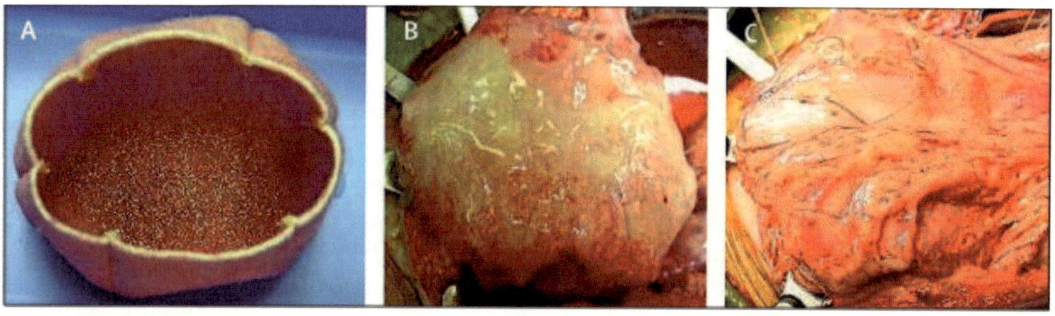

Fig. 18. Construction of engineered bladder scaffold seeded with cells (A) and engineered bladder anastamosed to native bladder with running 4–0 polyglycolic sutures (B). Implant covered with fibrin glue and omentum (C) (Ref 40)

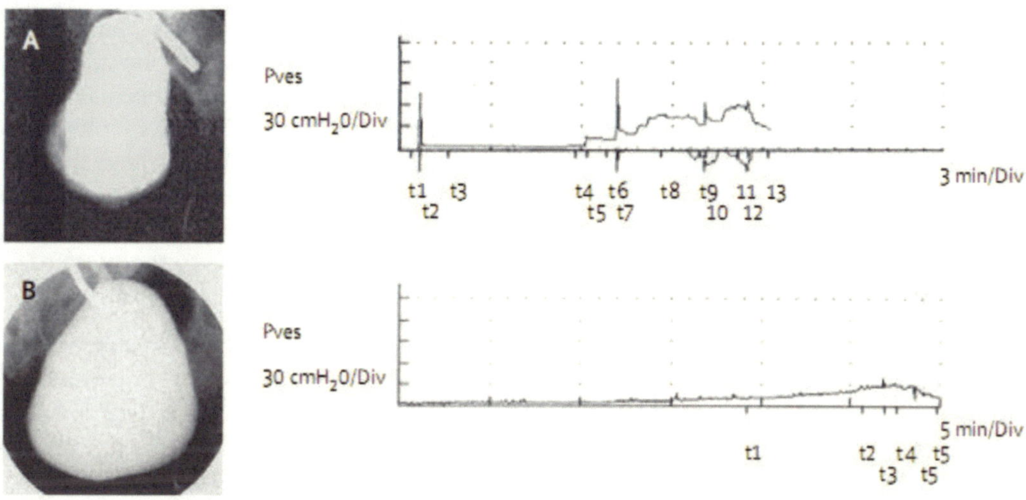

Fig. 19. Preoperative (A) and 10-months postoperative (B) cystograms and urodynamic fi ndings in patient with a collagen-PGA scaff old engineered bladder (Ref 40)

In another clinical study reported in the 2008 AUA, patients, who had received tissue engineered bladder showed increasing capacity and reduced intravesicular pressure. According to these reports, there is a clear reason to hope that the tissue engineered bladder can be utilized for a fully functioning neurogenic bladder. More indications about using this kind of biomaterial might be obtained in the near future.

3.2 Urethral reconstruction

The application of tissue engineering techniques for urethral reconstruction has been developed in recent years and the potential market for a tissue engineered solution for urethral stricture and abnormality will continue to increase in the near future.

3.2.1 Animal experiments

In animal experiments addressing urethral reconstruction, the first attempts used biomaterials alone. Among these experiments, most papers reported the application of SIS in animal urethroplasty. Many results were encouraging. Regenerated urethra contained a well-differentiated epithelium, underneath which was circular smooth muscle and abundant collagen and fibrous connective tissue. The only difference between the SIS-reconstructed urethra and normal urethra was the amount and size of the circular bundles of smooth muscle. However, several key points should be considered before using this kind of biomaterial. First, El-Assmy mentioned that locally prepared SIS and commercially available SIS may lead to the different results (41).This might be related to different pore sizes, which limit the infiltration and migration of cells. Second, the feasibility of using tabularized matrix for urethroplasty is still controversy although SIS has been proven to be useful for onlay urethroplasty. In Shokeir's study, a 3cm segment of the whole urethral circumference was excised and replaced by a tube matrix of the same length and width in 14 dogs. However, all dogs suffered a urethral fistula and/or stricture after the stent removal. This result demonstrated that a tube formed of matrix without seeded cells was not able to replace the long segment including the whole circumference of the canine urethra (42). Third, the length of urethral defect is another key point that should be considered during the urethroplasty. In order to investigate the maximum distance for normal tissue regeneration, Dorin et al. performed the tabularized urethroplasty in 12 male rabbits using acellular scaffold at varying lengths (0.5,1,2 and 3cm). The final result indicated that the maximal defect distance suitable for normal tissue formation using acellular grafts that rely on the native cells for tissue regeneration appears to be 0.5cm (43). Although other reports showed that the synthetic scaffold alone could be used for urethroplasty, the need ror a move to using cell-seeded scaffold is obvious.

In 2003, Bhargava has developed tissue-engineered buccal mucosa for use in substitution urethroplasty. Histologically, the matrix closely resembled the native oral mucosa after culturing for 2 weeks. A gradually increasing thickness of the epidermis and remodeling of the dermis could also be seen (44) (Fig 20). Subsequently, more cell-seeded scaffolds were used for urethral reconstruction in our center. Li et al. replaced urinary epithelial cells with oral keratinocytes seeded on BAMG to reconstruct a tissue-engineered urethra. Histological results showed that multiple layers of keratinocytes had formed at 2 and 6 months after the operation. Obvious margins between graft oral keratinocytes and host epithelium could be noticed in H&E sections (Fig 21b). Fu et al. used foreskin epidermal cell-seeded scaffolds to repair a urethra defect in a rabbit model. During following up, several layers of epidermal cells with abundant vessels in the submucosa were noticed. Moreover, immunofluorescence confirmed the survival of implanted epidermal cells at 1 month after procedure (45,46) (Fig 21a).

Recently, we have investigated the feasibility of constructing 3D structure urethra using multiple seeding cell types. It has also been hypothesized that building three-dimensional constructs in vitro prior to implantation would facilitate matrix vascularization in vivo and minimize the inflammatory response towards the matrix. Therefore, we seeded autologus corporal smooth muscle cells (CSMCs) and lingual keratinocytes into ACSM, using a static-dynamic seeding method. After being cultured 14 days, 6 scaffolds with two kind of cells

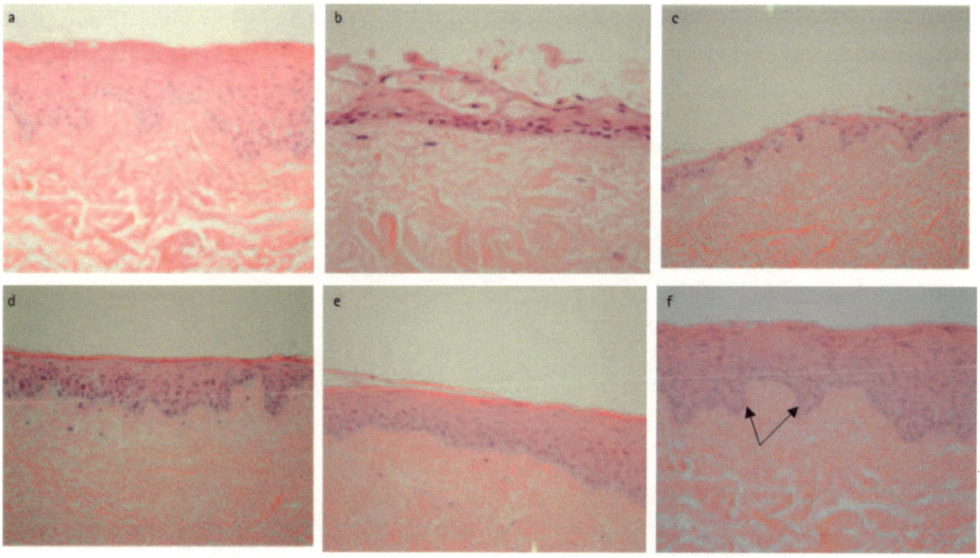

Fig. 20. buccal mucosa culture at a, the air-liquid interface; b, submerged; c, at day 1 ALI; d, at day 5 ALI; e, at day 8 ALI and with f, Protocol 2 (cells on same surface) (Ref 44)

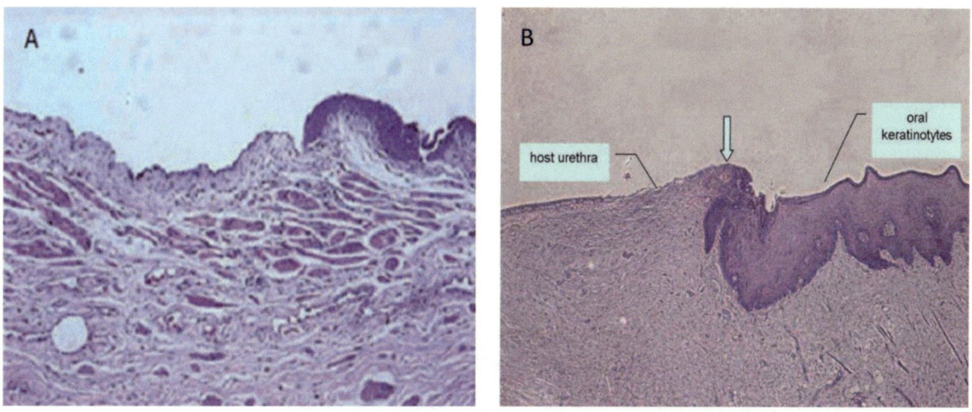

Fig. 21. Histological results of tissue engineering urethral after reconstruction. a. tissue engineering urethra using foreskin seeded scaffold; B, tissue engineering urethra using oral keratinocytes seeded scaffold. (Ref 44,45)

(Group C), 6 scaffolds with only lingual keratinocytes (Group B) and 6 matrices without cells (Group A) were used to repair a rabbit urethral defect. H&E staining of seeded ACSM showed several epithelial layers and well distributed CSMCs in the matrix. The maintenance of wide urethral caliber could be observed in Group C, while strictures were observed in groups A and B (Fig 22). Histologically, the retrieved urethra in group A showed fibrosis and inflammation during 6 months. A simple epithelial layer regenerated in group B but there was still no evidence of CSMCs growing into grafts during study period. A stratified epithelial layer and organized muscle fiber bundles were evident 6 months after implantation in group C (Fig 23). Our results demonstrated that lingual keratinocytes and CSMCs could be used as a source of seed cells for urethral tissue engineering. Using the dynamic-static seeding method, a 3-D urethra could be constructed in vivo. It can provide us an alternative method to treat the urethral disease using tissue engineering techniquea.

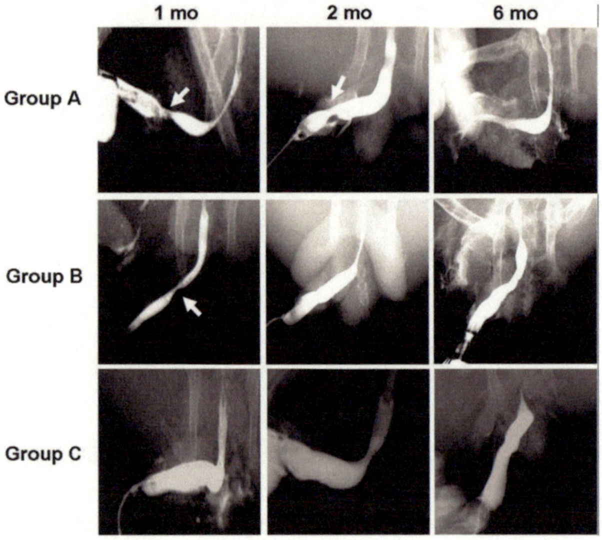

Fig. 22. Comparision of urethrography image in each group at 1,2,6 month after operation. The arrow indicates the stricture site of urethra

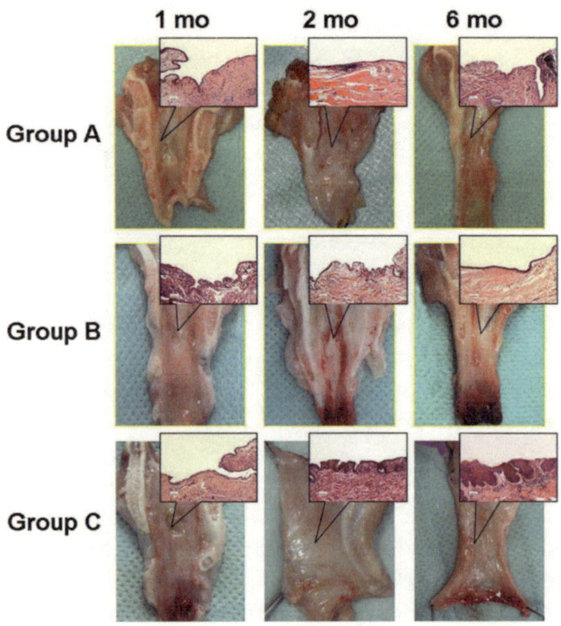

Fig. 23. Macroscopic inspection and H&E staining (inset) of retrieved urethrae in each group at 1,2,6 month after operation. In group A, a urethral stricture existed at every study time point. H&E staining did not show continuous epithelial layers but did show severe inflammation. In group B, strictures could be noticed by gross inspection. Only 1-2 epithelial layers were formed at 6 months after implantation. In group C, patent lumens without

strictures could be observed by the end of 6 months. Meanwhile, multilayer squamous epithelial layers covered the surface of the urethra.

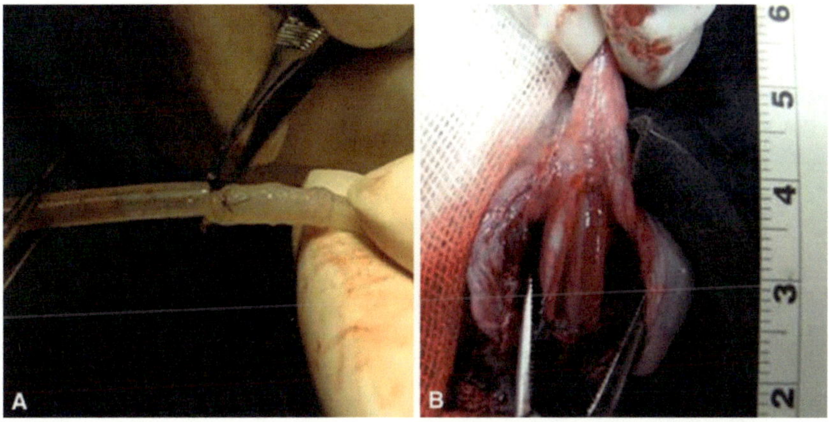

Fig. 24. a. Tubular tissue was gently everted so that the mesothelium lined the lumen. b The everted tubular tissue was interposed and anastomosed as urethral graft (Ref 48)

According to this result, the authors concluded that autologous tissue grown within the recipients' peritoneal cavity can be used successfully for tabularized urethral reconstruction (48). In addition, synthetic matrix combined with seeded cells has also been used for urethroplasty. In Selim's study, the optimal sterilization and cell seeding method for synthetic biomaterials in urethral reconstruction has been investigated. In their study, both PAA and γ-irradiation appear to be suitable methods for sterilizing PLGA scaffolds. And the sterilized PLGA 85:15 is a promising material for tissue engineering urethral reconstruction (49).

3.2.2 Clinical application

Up till now, many urologists have reported successful outcomes of urethral reconstruction using tissue engineering techniques. Most reports have been focused on treating urethral strictures using SIS. Among them, most results were satisfactory (Table 2) (50-56). In the report of Fiala et al., fifty patients with urethral strictures received urethroplasty using SIS. During post-operative follow-up, clinical, radiological, and cosmetic findings were excellent in 80% patients. No complications, such as fistulae, wound infections, or rejection were observed. This is so far the largest reports about using SIS for urethral reconstruction, in terms of numbers of patients. Their results were more satisfacory even than traditional urethral reconstruction using buccal or lingual mucosa for such low complication rate. In our center, we have also used SIS patch to undergo in 16 male patients with urethral strictures. The average length of strictures was 4.6 cm, ranging from 3.5-6 cm (Fig 25). After the operation, urethrography showed a wide patent urethra in all patients. The mean Qmax increased obviously from 3.8ml/s to 25ml/s. Only one patient needed urethral dilation due to the decreasing of Qmax at the end of 5 months. During follow-up, routine urethroscopy was performed in all patients. At the end of 4 weeks after operation, SIS could be easily noticed in the urethral lumen. However, the implanted graft could not be identified from the normal urethra 38 weeks after operation (Fig 26). The HE staining of biopsy showed that stratified squamous epithelial layers had grown on the SIS implanted site, which was similar to normal urethral mucosa (Fig 27). According to these clinical experiences, the use of an acellular matrix SIS for urethroplasty should only be done when the length of urethral stricture is short. Patients with a bulbar urethral stricture are more suitable than those with a urethral stricture in other sites. Of course, the condition of urethral plate should also be considered before using SIS. We believe that urethroplasty using tissue engineered scaffold can achieve a satisfactory outcome that is similar to the gold-standard procedure *provided optimal patients are selected.*

Author	Date	Patient number	SIS type	SIS layer	Stricture length	Recurrence
Mantovani F, et al	2003	1	patch	1	>10cm	none
Le Roux JP	2005	9	tube	1	2-4cm	66.7%(6/9)
Hauser D, et al.	2005	5	patch	4	3.5-10cm	80%(4/5)
Sievert KD, et al.	2005	13	patch	4	4-10cm	30.7%(4/13)
Donkov II,et al	2006	9	patch	4	4-6cm	11.1%(1/9)
Palminteri E,et al	2007	20	patch	4	3-7.7cm	15%(3/20)
Fiala R,et al	2007	50	patch	4	4-14cm	20%(10/50)
Farahat,YA	2009	10	patch	1	0.5-2cm	20%(2/10)

Table 2. Recoder of using SIS for urethral reconstruction in clinic

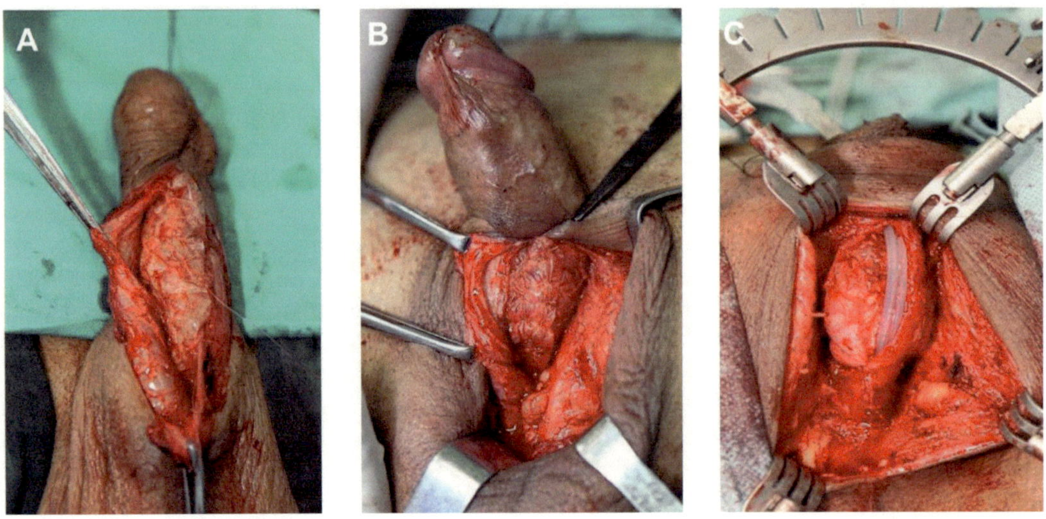

Fig. 25. Application of SIS in urethroplasty.
a. penile urethral stricture; b.bulbopenile urethral stricture; c.bulbar urethral stricture.

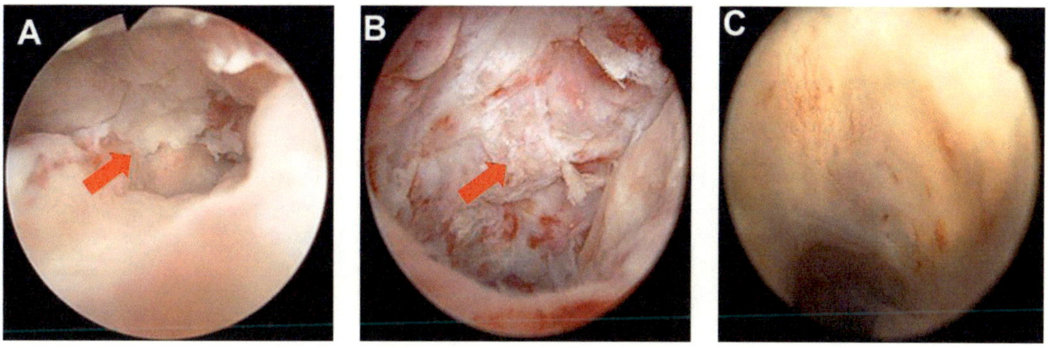

Fig. 26. Urethroscopy after the urethral reconstruction using SIS graft.Arrow headed the implanted site. a:4 weeks after op; b: 6 weeks after op; c: 38 weeks after operation.

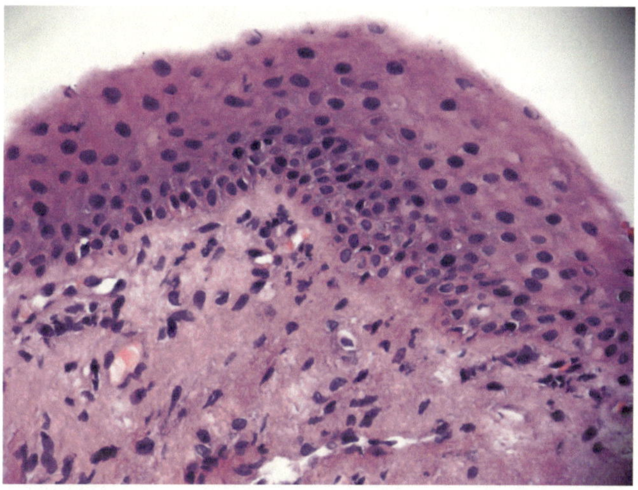

Fig. 27. The HE staining of biopsy showed that stratified squamous epithelial layers have grown on the SIS implanted site.

Cell-seeded scaffolds have also been used for urethral reconstruction in some patients. Based on the previous reports mentioned above, Bhargava et al. useed autologous tissue-engineered buccal mucosa to treat five patients with urethral strictures secondary to lichen sclerosis. After the intimal operation, one patient had complete excision of the grafted urethra and one required partial graft excision. The other three patients required some form of instrumentation although endoscopic appearance showed a patent urethra with the implanted graft in situ (57). Recently, Atala et al. reported that using cell-seeded synthetic tubularized scaffolds to repair urethral defects in five boys. At the end of follow-up, some satisfactory results were obtained. The median end Qmax was 27.1 ml/s, and serial urethrographic and endoscopic studies showed the maintenance of wide calibres without strictures (Fig 28) (58).

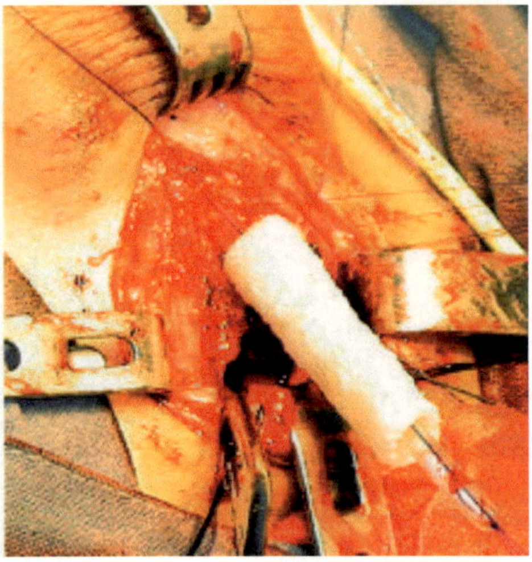

Fig. 28. A cell-seeded graft sutured to the normal urethral margins(Ref 59)

4. Challenges and risks

Based on previous studies, the potential market for a tissue-engineered solution for dysfunctional bladders and small contracted or inflamed bladders is probably far too small for commercial exploitation. The only two potential indications for commercial-scale tissue engineering lower urinary tract tissue are bladder carcinoma and urethral stricture. Cell-seeded scaffolds will probably be further investigated and applied in clinics. hree-dimensional structures and the use of bioreactors will also be more and more popular in tissue engineering research for lower urinary tract reconstruction.

However, several problems need to be solved. For example, the ethical problems about the implanted matrix (and where it is obtained) needs to be further discussed. The potential for carcinogenic problems arising form the use of stem cells is not clear. Optimal methods of cell labeling (for research) still needs to be improved.

Nevertheless, there is no doubt that tissue engineering techniques for lower urinary tract reconstruction will themselves become the gold-standard in the near future. A substantial commercial market will continue to grow and more patients will obtain benefit from this technique.

5. References

[1] Atala A. Tissue engineering for the replacement of organ function in the genitourinary system. Am J Transplant. 2004;4:58-73.

[2] Brown AL, Brook-Allred TT, Waddell JE, White J, Werkmeister JA, Ramshaw JA, et al. Bladder acellular matrix as a substrate for studying in vitr bladder smooth muscle–urothelial cell interactions. Biomaterials. 2005;26:529-543.

[3] De Filippo RE, Yoo JJ, Atala A. Urethra replacement using cell seeded tabularize collagen matrices. J Urol 2002; 168: 1789–1792; discussion 1792–1793

[4] Fu Q, Deng CL, Song XF, Xu YM. Long-term study of male rabbit urethral mucosa reconstruction. Asian J Androl. 2008;10:719-722.
[5] Kimball JR, Nittayananta W, Klausner M, Chung WO, Dale BA. Antimicrobial barrier of an in vitro oral epithelial model. Arch Oral Biol. 2006;51:775-783
[6] Li C, Xu Y, Song L, Fu Q, Cui L, Yin S. Preliminary experimental study of tissue-engineered urethral reconstruction using oral keratinocytes seeded on BAMG. Urol Int. 2008;81:290-295.
[7] Rajasekaran M, Kasyan A, Allilain W, Monga M. Ex vivo expression of angiogenic growth factors and their receptors in human penile cavernosal cells. J Androl. 2003;24:85-90
[8] Drzewiecki BA, Thomas JC, Tanaka ST. Bone marrow-derived mesenchymal stem cells: current and future applications in the urinary bladder. Stem Cells Int. 2011,3; 765-767
[9] Zhang Y, Lin HK, Frimberger D, Epstein RB, Kropp BP. Growth of bone marrow stromal cells on small intestinal submucosa an alternative cell source for tissue engineered bladder. BJU Int. 2005;96:1120-1125.
[10] Zhu WD, Xu YM, Feng C, Fu Q, Song LJ, Cui L. Bladder reconstruction with adipose-derived stem cell-seeded bladder acellular matrix grafts improve morphology composition. World J Urol. 2010;28:493-498.
[11] Tian H, Bharadwaj S, Liu Y, Ma PX, Atala A, Zhang Y. Differentiation of human bone marrow mesenchymal stem cells into bladder cells: potential for urological tissue engineering. Tissue Eng Part A. 2010;16:1769-1779.
[12] Drewa T, Joachimiak R, Kaznica A, Sarafian V, Pokrywczynska M. Hair stem cells for bladder regeneration in rats: preliminary results. Transplant Proc. 2009;41:4345-4351.
[13] Zhang Y, McNeill E, Tian H, Soker S, Andersson KE, Yoo JJ, et al. Urine derived cells are a potential source for urological tissue reconstruction. J Urol 2008;180:2226-2233
[14] Bodin A, Bharadwaj S, Wu S, Gatenholm P, Atala A, Zhang Y. Tissue-engineered conduit using urine-derived stem cells seeded bacterial cellulose polymer in urinary reconstruction and diversion. Biomaterials. 2010;31:8889-8901.
[15] Nakanishi Y, Chen G, Komuro H, Ushida T, Kaneko S, Tateishi T,Kaneko M. Tissue-engineered urinary bladder wall using PLGA mesh-collagen hybrid scaffolds: A comparison study of collagen sponge and gel as a scaffold. J Pediatr Surg 2003;38:1781–1784
[16] Brehmer B, Rohrmann D, Becker C, Rau G, Jakse G. Different types of scaffolds for reconstruction of the urinary tract by tissue engineering. Urol Int 2007;78:23–29.
[17] Feng C, Xu YM, Fu Q, Zhu WD, Cui L, Chen J. Evaluation of the biocompatibility and mechanical properties of naturally derived and synthetic scaffolds for urethral reconstruction. J Biomed Mater Res A. 2010;94:317-325.
[18] Freytes DO, Badylak SF, Webster TJ, Geddes LA, Rundell AE. Biaxial strength of multilaminated extracellular matrix scaffolds. Biomaterials. 2004 May;25:2353-2361.

[19] Yao C, Prével P, Koch S, Schenck P, Noah EM, Pallua N, et al. Modification of collagen matrices for enhancing angiogenesis. Cells Tissues Organs. 2004;178:189-196.

[20] Kundu AK, Gelman J, Tyson DR. Composite thin film and electrospun biomaterials for urologic tissue reconstruction. Biotechnol Bioeng. 2011;108:207-215.

[21] Baumert H, Simon P, Hekmati M, Fromont G, Levy M, Balaton A, et al. Development of a seeded scaffold in the great omentum: feasibility of an in vivo bioreactor for bladder tissue engineering. Eur Urol. 2007;52:884-890.

[22] Gu GL, Zhu YJ, Xia SJ, Zhang J, Jiang JT, Hong Y, et al.Peritoneal cavity as bioreactor to grow autologous tubular urethral grafts in a rabbit model. World J Urol. 2010;28: 227-232.

[23] Farhat WA, Yeger H. Does mechanical stimulation have any role in urinary bladder tissue engineering. World J Urol. 2008;26:301-305.

[24] A novel bioreactor to simulate urinary bladder mechanical properties and compliance for bladder functional tissue engineering

[25] Khang D, Lu J, Yao C, Haberstroh KM, Webster TJ.The role of nanometer and sub-micron surface features on vascular and bone cell adhesion on titanium.Biomaterials 2008, 29:970–983.

[26] Chun YW, Lim H, Webster TJ, Haberstroh KM. Nanostructured bladder tissue replacements.Wiley Interdiscip Rev Nanomed Nanobiotechnol. 2010 Aug 20.

[27] Mondalek FG, Lawrence BJ, Kropp BP, Grady BP, Fung KM, Madihally SV,et al. The incorporation of poly(lactic-co-glycolic) acid nanoparticles into porcine small intestinal submucosa biomaterials. Biomaterials 2008,29:1159–1166

[28] Harrison BS, Eberli D, Lee SJ, Atala A, Yoo JJ. Oxygen producing biomaterials for tissue regeneration. Biomaterials. 2007;28:4628-4634.

[29] Oh SH, Ward CL, Atala A, Yoo JJ, Harrison BS. Oxygen generating scaffolds for enhancing engineered tissue survival. Biomaterials. 2009;30:757-762.

[30] Kropp BP.Small-intestinal submucosa for bladder augmentation:a review of preclinical studies. World J Urol. 1998;16:262-267.

[31] Kropp BP, Eppley BL, Prevel CD, Rippy MK, Harruff RC, Badylak SF, et al. Experimental assessment of small intestinal submucosa as bladder wall substitute. Urology. 1995;46:396-400.

[32] Kropp BP, Sawyer BD, Shannon HE, Rippy MK, Badylak SF, Adams MC, et al. Characterization of small intestinal submucosa regenerated canine detrusor: assessment of reinnervation, in vitro compliance and contractility. J Urol. 1996;156:599-607.

[33] Kropp BP, Cheng EY, Lin HK, Zhang Y. Reliable and reproducible bladder regeneration using unseeded distal small intestinal submucosa. J Urol. 2004;172:1710-1713.

[34] Zhang Y, Kropp BP, Moore P, Cowan R, Furness PD 3rd, Kolligian ME, et al. Coculture of bladder urothelial and smooth muscle cells on small intestinal submucosa: potential application for tissue engineering technology. J Urol. 2000;164:928-934; discussion 934-935

[35] Liu Y, Bharadwaj S, Lee SJ, Atala A, Zhang Y. Optimization of a natural collagen scaffold to aid cell-matrix penetration for urologic tissue engineering. Biomaterials. 2009;30:3865-3873.

[36] Yoo JJ, Meng J, Oberpenning F, Atala A. Bladder augmentation using allogenic bladder submucosa seeded with cells. Urology. 1998;51:221-225.

[37] Chung SY, Krivorov NP, Rausei V, Thomas L, Frantzen M, Landsittel D, et al. Bladder reconstruction with bone marrow derived stem cells seeded on small intestinal submucosa improves morphological and molecular composition. J Urol. 2005;174:353-359.

[38] Jack GS, Zhang R, Lee M, Xu Y, Wu BM, Rodríguez LV.Urinary bladder smooth muscle engineered from adipose stem cells and a three dimensional synthetic composite. Biomaterials. 2009;30:3259-3270.

[39] Mondalek FG, Ashley RA, Roth CC, Kibar Y, Shakir N, Ihnat MA, et al. Enhanced angiogenesis of modified porcine small intestinal submucosa with hyaluronic acid-poly(lactide-co-glycolide) nanoparticles: From fabrication to preclinical validation. J Biomed Mater Res A. 2010 ;94:712-719.

[40] Atala A, Bauer SB, Soker S, Yoo JJ, Retik AB. Tissue-engineered autologous bladders for patients needing cystoplasty. Lancet. 2006;367:1241-1246.

[41] El-Assmy A, El-Hamid MA, Hafez AT.Urethral replacement: a comparison between small intestinal submucosa grafts and spontaneous regeneration. BJU Int. 2004;94:1132-1135.

[42] Shokeir A, Osman Y, Gabr M, Mohsen T, Dawaba M, el-Baz M.Acellular matrix tube for canine urethral replacement:Is it fact or fiction. J Urol. 2004;171:453-456.

[43] Dorin RP, Pohl HG, De Filippo RE, Yoo JJ, Atala A. Tubularized urethral replacement with unseeded matrices: what is the maximum distance for normal tissue regeneration. World J Urol. 2008;26:323-326.

[44] Bhargava S, Chapple CR, Bullock AJ, Layton C, MacNeil S.Tissue-engineered buccal mucosa for substitution urethroplasty. BJU Int. 2004;93:807-811.

[45] Fu Q, Deng CL, Liu W, Cao YL.Urethral replacement using epidermal cell-seeded tubular acellular bladder collagen matrix. BJU Int. 2007;99:1162-1165.

[46] Li C, Xu YM, Song LJ, Fu Q, Cui L, Yin S.Urethral reconstruction using oral keratinocyte seeded bladder acellular matrix grafts. J Urol. 2008 ;180:1538-1542.

[47] Guan Y, Ou L, Hu G, Wang H, Xu Y, Chen J, et al. Tissue engineering of urethra using human vascular endothelial growth factor gene-modified bladder urothelial cells. Artif Organs. 2008;32:91-99.

[48] Gu GL, Zhu YJ, Xia SJ, Zhang J, Jiang JT, Hong Y, et al. Peritoneal cavity as bioreactor to grow autologous tubular urethral grafts in a rabbit model. World J Urol. 2010;28:227-232.

[49] Selim M, Bullock AJ, Blackwood KA, Chapple CR, MacNeil S. Developing biodegradable scaffolds for tissue engineering of the urethra. BJU Int. 2011;107:296-302.

[50] Mantovani F, Trinchieri A, Castelnuovo C, Romanò AL, Pisani E. Reconstructive urethroplasty using porcine acellular matrix. Eur Urol. 2003;44:600-602.

[51] Le Roux JP. Endoscopic urethroplasty with unseeded small intestinal submucosa collagen matrix grafts: a pilot study. J Urol. 2005 ;173:140-143.

[52] Schlote N, Wefer J, Sievert KD. Acellular matrix for functional reconstruction of the urogenital tract. Special form of "tissue engineering"?. Urologe A. 2004;43:1209-1212

[53] Donkov II, Bashir A, Elenkov CH, Panchev PK. Dorsal onlay augmentation urethroplasty with small intestinal submucosa: modified Barbagli technique for strictures of the bulbar urethra. Int J Urol. 2006;13:1415-1417.

[54] Palminteri E, Berdondini E, Colombo F, Austoni E. Small intestinal submucosa (SIS) graft urethroplasty: short-term results. Eur Urol. 2007;51:1695-1701

[55] Fiala R, Vidlar A, Vrtal R, Belej K, Student V. Porcine small intestinal submucosa graft for repair of anterior urethral strictures. Eur Urol. 2007;51:1702-1708;

[56] Farahat YA, Elbahnasy AM, El-Gamal OM, Ramadan AR, El-Abd SA, Taha MR. Endoscopic urethroplasty using small intestinal submucosal patch in cases of recurrent urethral stricture: a preliminary study. J Endourol. 2009;23:2001-2005

[57] Bhargava S, Patterson JM, Inman RD, MacNeil S, Chapple CR. Tissue-Engineered Buccal Mucosa Urethroplasty—Clinical Outcomes. Eur Urol. 2008;53:1263-1269.

[58] Raya-Rivera A, Esquiliano DR, Yoo JJ, Lopez-Bayghen E, Soker S, Atala A. Tissue-engineered autologous urethras for patients who need reconstruction: an observational study. Lancet. 2011;377:1175-1182

20

Novel Promises of Nanotechnology for Tissue Regeneration

Abir El-Sadik
Anatomy and Embryology, Basic Sciences Department,
King Saud Bin Abdulaziz University for Health Sciences, Riyadh
Kingdom of Saudi Arabia

1. Introduction

The term 'nanotechnology' refers to technology that deals with structures and devices of nanometer (10^{-9} meter) size. It involves the design, fabrication and utilization of materials of nanoscale dimensions (Gao & Xu, 2009). The resulting nanomaterials exhibit chemical, physical and biological properties that can differ significantly from those of bulk material. These products can be categorized into metals, ceramics, polymers or composite materials that have nanoscale features. The limited size of their particles leads to a high surface area to volume ratio, improved solubility, multifunctionality, high electrical and heat conductivity and improved surface catalytic activity. (El-Sadik et al., 2010). All these phenomena allow give nanoparticles to interact with biological systems at cellular and molecular levels. These interactions enhance the biomedical applications of nanotechnology giving great promise for improving disease prevention, diagnosis, treatment and in particular tissue regeneration (Murthy, 2007).

Since natural human tissues include nano-scale subcellular and extracellular components, artificial nanomaterials mimic the scales of tissue components (Zhang & Webster, 2009). Cells make contact with other cells and with the extracellular matrix with membrnes that have nanoscale features. It has been shown that nanomaterials, with their biomimetic features, can accelerate the rate of cell growth and proliferation and promote tissue acceptance due to reduced immune response (Oh et al., 2009). One of the most useful properties of nanomaterials, which have been extensively investigated, is their ability to interact with proteins that control cell functions. This may make nanomaterials very useful, and perhaps even necessary, tools for regenerating various tissues such as those of the bone, cartilage, blood vessels and nervous system (Liu & Webster, 2007).

Although a series of technological improvements in tissue regeneration have been acheived using conventional methods, a variety of problems still faces current implants. Nanotechnology could provide several solutions to these problems. A wide range of nanomaterials have been made from organic and inorganic composites, just like conventional materials. However, nanotehnology has the ability to control material properties more closely by assembling components at the nanoscale. These nanomaterials (nanoparticles, nanotubes, nanofibers, nanoclusters, nanocrystals, nanowires, nanorods and nanofilms) can be fabricated by multiple and available nanotechnologies. Electrospining, self assembly, phase separation, photolithography, thin film deposition, chemical etching, chemical vapor deposition and electron beam lithography are all techniques currently used to synthesize nanomaterials with ordered or random nanotopographies (Chen & Ma, 2004).

Conventional tissue replacement, using allografts and autografts, cannot satisfy high performance demands and improvements are necessary. Nanotechnology has been used to fabricate cytocompatible biomimetic nanomaterials that provide biological substitutes useful in restoring and improving tissue functions. Moreover, 2-dimensional tissue cell culture systems on flat glass, coated petri dishes or plastic substrates cannot simulate the natural tissue microenvironments. Normal tissue cells are located in a complex network of 3 dimensional extracellular matrix with nanoscale fibers. Nanomaterials could be fabricated that accurately simulate the dimensions and architecture of natural human tissue, allowing significantly improved performance of the cultured cells (Gelain et al., 2006). The composition and topography of a tissue engineered material could even produce cell-environment interactions that determine the implant fate. Nanomaterials need to be designed to be biocompatible and to function without interrupting other physiological processes. In principle, they can promote normal cell growth and differentiation without any adverse tissue reaction. These nanomaterials must be biodegradable either to be removed via degradation or absorption to leave only native tissue. In addition, nanomaterials used in tissue regeneration should possess biomimetic features that allow cells to react normally to internal and external stimuli and to exchange the signals between those cells and the external environment.

This chapter reviews recent progress in the synthesis of nanomaterials for improving stem cell behavior and tissue regeneration. In addition, it highlights potentially valuable applications of nanotechnology in specific tissue regeneration.

2. Effects of nanomaterials on stem cell behaviour and development of tissue regeneration

Nanotechnology is an extremely promising advancement in synthetic methodologies used to functionalize nanomaterials with biomolecules. Nanomolecules could be modified to desired sizes, shapes, compositions and properties producing different types applied in tissue regeneration such as nanoparticles, nanosurfaces and nanoscaffolds.

2.1 Nanoparticles

Several studies have investigated the influences of different types of nanoparticles on the behaviour of stem cells applied in tissue regeneration. The effects of mesoporous silica nanoparticles conjugated with fluorescein isothiocyanate on human bone marrow mesenchymal stem cells has been investigated by several researchers. Internalization of silica nanoparticles into stem cells is mediated by both clathrin and actin-dependent endocytosis. Once inside the cell, the nanoparticles escaped the endolysosomal vesicles and did not affect stem cell viability or proliferation. They enhanced actin polymerization in mesenchymal stem cells. Moreover, regular osteogenic differentiation was successfully induced in the mesenchymal stem cells after the uptake of mesoporous silica nanoparticles in highly chondrogenic synovium (Huang et al., 2008 & Shi et al., 2009). Fibrin polylactide caprolactone nanoparticles have been designed to induce chondrogenic differentiation in mesenchymal stem cells. These complex nanoparticles facilitated the upregulation of chondrogenesis marker genes. In addition, they effectively sustained chondrogenic differentiation and enhanced chondral extracellular matrix deposition by human adipogenic stem cells. Fibrin polylactide caprolactone nanoparticle complexes could be effectively used for in situ cartilage tissue regeneration from human stem cells (Jung et al., 2009).

The application of nanotechnology to stem cell biology might help to maximize therapeutic benefits and minimize possible undesired effects of stem cell therapy, through delivery of sufficient stem cells to the regions of interest with the smallest number of cells to untargeted regions. Tracking the fate, distribution, proliferation, differentiation of engulfed stem cells employed in tissue regeneration is essential to understand the mechanisms of participation of the cells in tissue repair. Nanotechnology can improve several techniques that would enable non-invasive detection of transplanted stem cells within the desired organs. Iron oxide nanoparticles are inorganic nanoparticles that can be synthesized easily in large quantities and different sizes using simple methods. Several studies reported that when iron oxide nanoparticles bind to the external cell membrane, they do not affect cell viability, although they may detach from the cell membrane or interfere with cell surface interactions (Bulte & Kraitchman, 2004). Superparamagnetic iron oxide nanoparticles are successfully internalized via endocytosis in human mesenchymal stem cells. After their uptake, they are located inside cytoplasmic vesicles. Then, they are transferred to lysosomes in which degradation of the nanoparticles occurs, releasing free iron into the cytoplasm (Jing et al., 2008). Coating the surface of iron oxide nanoparticles modifies the surface of the particles for efficient uptake with minimum side effects on the cells. Coating superparamagnetic iron oxide nanoparticles with dextran improves their stability and solubility and prevents their aggregation. Another example of coating the surface of nanoparticles is provided by coating the superparamagnetic iron oxide nanoparticles with gold. The gold provides an inert shell around the nanoparticles and protects them from rapid dissolution within cytoplasmic endosomes and enhances magnetic resonance imaging (MRI) contrast. It has been shown, however, that dissolved iron oxide nanoparticles may produce free hydroxyl radicals which increase the rate of apoptosis and alterations in cellular metabolism (Emerit et al., 2001).

Concerning the effects of iron oxide nanoparticles on stem cell behaviour, magnetite iron oxide cationic liposomes can be applied efficiently to mesenchymal stem cell techniques. Mesenchymal stem cells incubated in osteogenic medium with these nanoparticles changed their shape from fibroblastic to polygonal, formed calcium nodules and increased in number five-fold compared with controls (Ito et al., 2004). In addition, superparamagnetic iron oxide nanoparticles have been shown to enhance the survival rate of stem cells up to 99%, indicating that these nanoparticles improve stem cell viability (Delcroix et al., 2009). Moreover, superparamagnetic iron oxide nanoparticles did not influence the morphology, cell cycle, telomerase activity, proliferation or differentiation ability of labelled neural stem cells (Kea et al., 2009). Superparamagnetic iron oxide nanoparticles have been successfully applied to tracking the fate of several types of stem cells. For example, the migration of embryonic stem cells and bone marrow mesenchymal stem cells labelled with iron oxide nanoparticles towards a lesion site has been tracked using MRI. This labelling technique offers high resolution, speed, easy access and 3-dimensional capabilities and provides information not only for the transplanted cells, but also for the surrounding tissues, reporting edema or inflammation that may affect the fate of the grafted cells and reduce the recovery of damaged tissue (Sykova & Jendelova, 2007). Another example is tracking human mesenchymal stem cells labelled with superparamagnetic iron oxide nanoparticles after transplantation for articular cartilage repair using MRI (Au et al., 2009). These observations have demonstrated the ability of using iron oxide nanoparticles to be useful in monitoring and tracking the fate of transplanted stem cells apparently without affecting their behaviour, although, selection of the type and concentration of nanoparticles is critically important.

In addition to iron oxide nanoparticles, quantum dots have been much used for cell tracking in in tissue regeneration. Quantum dots are fluorescent semiconducting nanocrystals that overcome the limitations of conventional labelling methods. Several researchers have studied the application of quantum dots to monitoring physiological changes inside living cells by labelling the intracellular organelles or specific proteins with quantum dots. They could monitor cellular migration, track cell lineage and investigate stem cell behaviour. Quantum dots can bind to individual molecules on the cell surface and serve in tracking the motion of those molecules. For example, quantum dots have been applied to demonstrate changes in integrin dynamics during osteogenic differentiation of human bone marrow cells (Chen et al., 2007). Numerous studies have demonstrated a variety of techniques of the cellular uptake of quantum dots. These nanoparticles can be delivered into cells by microinjection, endocytosis, liposome-mediated transfection and special peptide delivery (Chang et al., 2008). Delivered quantum dots were found to escaping lysosomal degradation at the beginning of the uptake. Thereafter, lysosome expression was enhanced and all cellular quantum dots were shown in lysosome vesicles. After the uptake of quantum dots, into several types of stem cells, such as mesenchymal stem cells, cytoskeletal reorganization took place. This action revealed the formation of wide and flat leading lamellipodia filled with a dense actin network (Chang et al., 2009). Human mesenchymal stem cells labelled with quantum dots represented the same viability comparing with the unlabelled human mesenchymal stem cells from the same subpopulation (Shah et al., 2007), suggesting that quantum dots could be used safely for long term labelling of stem cells. Moreover, embryonic stem cells could be labelled with quantum dots for cellular tracking in vivo without affecting the viability, proliferation or differentiation of the embryonic stem cells (Lin et al., 2007). These studies demonstrated that quantum dots could enable cellular and molecular imaging and tracking the fate of stem and progenitor cells used in tissue regeneration with high sensitivity and high spatial resolution. These applications are supported by an extensive number of advanced imaging techniques, giving a great impact on tissue regeneration studies.

2.2 Nanosurfaces

Mammalian cells are surrounded by nanostructures formed by biomolecules arranged geometrically in different configurations. These arrangements affect cell behaviour by producing chemical signals such as growth factors or physical signals such as tensile forces caused by interactions with the surrounding nanostructured extracellular matrix. Nanotechnology provides nanotopographical surfaces that can guide cellular adhesion, spreading, morphology, proliferation and differentiation. Cells react differently according to the nanotopography of their environment, which influences their cytoskeletal organization, attachment, and migration. Nanofabrication techniques provide several types of nanosurfaces for tissue regeneration.

Nanosurfaces of different materials with structural modification, such as the presence of large, medium and small nanoscale grooves, pores, pits, ridges and nodules can be recognized by cultured cells. A wide range of cell types, such as fibroblasts (Dalby et al., 2003b), osteoblasts (Lenhert et al., 2005) and mesenchymal stem cells (Biggs et al., 2008), are influenced by nanoscale grooves with dimensions that mimic those in vivo. Cellular morphology depends on cell type and on groove depth and width. Mesenchymal stem cells seeded on nanogrooves respond by aligning their shape and elongation in the direction of the grooves (Dalby et al., 2003b). Human osteoblasts cultured on ordered nanoscale groove/ridge arrays, fabricated by photolithography, were affected significantly (Biggs et al., 2008). The authors seeded human osteoblasts on grooves of 330 nm depth and different widths (10, 25 and 100 µm in width). They concluded that adhesion formation was not affected in 100 µm wide groove/ridge arrays, although upregulation of genes involved in skeletal development was induced. In addition, increased osteospecific functions were observed. 25 µm wide grooves/ ridges were shown to be associated with a reduction in supermature adhesions and an increase in focal complex formation. However, that osteoblast adhesion was significantly reduced in 10 µm wide groove/ridge arrays. Moreover, grooves manufactured on nanosurfaces promoted the elongation and nuclear polarization in the cultured cells (Charest et al., 2004). Cell membranes stopped at the largest grooves but bridged over the narrowest and deepest ones (Matsuzaka et al., 2003). Electron beam lithography has also been used to generate nanoscale patterns for culturing mesenchymal stem cells (Dalby, 2009). The patterns ranged from highly ordered through controlled disorder, to total randomness. The authors concluded that nanoscale change in surface topography altered mesenchymal stem cell differentiation. Successful osteoconversion of the cultured cells using ± 50 nm level of disorder was demonstrated. The cells focal adhesions interacted with the material surface and affected by several signalling pathways, such as G protein and cytoskeletal signalling. These signalling factors modulated cell sensing, morphology, contractility, proliferation and differentiation. Altering the nanotopography of the surface material influenced the cytoskeletal arrangements (Curtis et al., 2006). Mechanical changes were transmitted from the cytoskeleton to the nucleus, affecting the genomic expression patterns and cell phenotype (Dalby et al., 2007).

Among hard carbon coatings, nanocrystalline diamond has been applied successfully to cultured osteogenic and endothelial cells. Nanocrystalline diamond possesses promising electrical and optical properties, high hardness, low friction coefficient and good compatibility (Bacakova et al., 2007). Nanocrystalline diamond has been used in the form of films to improve the mechanical and physical properties of body implants. In addition, it has been shown to attract cell colonization, its surface nanostructure simulating the architecture of extracellular matrix molecules. Nanocrystalline diamond layers deposited on silicon substrates improves the adhesion and growth of osteogenic and endothelial cells (Grausova et al., 2008). The authors concluded that these nanostructured surfaces gave good support for cellular viability and proliferation and could be applied usefully in tissue regeneration. Furthermore, ultrasmooth nanostructured diamond has been used in orthopaedic implants. Several studies were performed on this material, the authors describing their surface modification techniques and cytocompatability (Clem et al., 2008). The studies demonstrated that hydrogen-terminated ultrasmooth nanostructured diamond surfaces supported robust mesenchymal stem cell adhesion and survival. However oxygen and fluorine terminated surfaces resisted cell adhesion. It was concluded that chemical and physical modifications of ultrasmooth nanostructured diamond could promote or prevent cell/biomaterial interactions. Moreover, mesenchymal stem cell adhesion and proliferation were significantly improved on ultrasmooth nanostructured diamond compared with the commonly used and biocompatible cobalt-chrome. There was also osteoblastic differentiation and deposition of mineralized matrix in mesenchymal stem cells. Ultrasmooth nanostructured diamond was found to reduce debris particle release from orthopaedic implants without influencing osseointegration.

Controllable self-assembly of nanonodules has been demonstrated to occur during chemical depositioning of materials on specifically conditioned microtopographical surfaces (Ogawa et al., 2008). The substrate could be a nonmetalic material such a as biodegradable polymer. The biological potential of the nanonodular surfaces affecting the behaviour of cultured cells using titanium dioxide has been investigated (Kubo et al., 2009). Titanium as a substrate material was proven to be non cytotoxic and was applied in therapeutic and implantable devices used in tissue regeneration. Micro-nano-hybrid surfaces, consisting of nanoscale nodules within microscale pits, were created by applying nanonodular self assembly techniques. These surfaces mimicked the biomineralized matrices with greater surface area and roughness. Changing the assembly time controlled the size of the nanonodules. The addition of nanonodules of different sizes (100 - 300 - 500 nm) to micropits selectively promoted osteoblast functions. In addition, these nanonodular topographies enhanced osteoblastic proliferation and differentiation. These advantages were 3 times greater in the nanonodules with a diameter of 300 nm within the micropits, when implanted in a rat femur model. Cell spread was enhanced on the micro- nano-hybrid surfaces. After 3 hours incubation, osteoblasts were shown to be larger and their cell processes and cytoskeletons started to develop on the nanonodular surfaces, while they remained small and circular on the micropit surface alone. Meanwhile, marked cytoplasmic localization of the focal adhesion protein vinculin was shown on the micro-nano-hybrid surfaces, compared with those on the micropit surface which had faint expression.

Another application of titanium in tissue regeneration is the use of nanocrystalline titanium surfaces. This type of nanometer surface roughness promotes osteoblasted adhesion. This nanosurface enhances cell growth and demonstrates extensive wear resistance due to high hardness and strength (Wang & Li, 2003). Cell compatability studies on nanosized titanium particles showed enhanced osteoblast function and largeer deposition of calcium minerals (Webster et al., 2000). One of the most effective nanostructured titanium surfaces for enhancing the attachment, proliferation and spreading of mesenchymal stem cells is layer-by-layer assembled titanium dioxide nanoparticle thin films. This technique depends on electrostatic attraction between oppositely charged species such as titanium dioxide nanoparticles. The advantage of layer-by-layer assembly is that the adsorption of material cn be controlled with nanometer precision. Titanium dioxide thin films have been proved to be an optimal surface for rapid attachment and spreading of cells (Kommireddy et al., 2005). Increasing the number of layers in titanium dioxide thin films has been shown to increase surface roughness. Higher numbers of attached cells were observed on 4-layer titanium dioxide thin film than on a 1-layer thin film, with a faster rate of spreading on the rougher surface (Kommireddy et al., 2006). Moreover, multilayered and functionalized titanium films composed of chitosan and plasmid DNA demonstrated significant high transfection efficiency in mesenchymal stem cells (Hu et al., 2009). The authors reported high production levels of alkaline phosphatase and osteocalcin. They concluded that multilayered titanium films with chitosan and plasmid DNA promoted the differentiation of osteoprogenitor cells into mature osteoblasts over long time.

2.3 Nanoscaffolds

Nanotechnology provides the tissue regeneration field with nanostructures that might accurately simulate the natural 3-dimensional microenvironment of cells. This approach provides a complex network of nanoscale fibers and extracellular ligands, such as many types of collagens, laminin and fibronectin, that are poorly reproduced in the conventional 2-dimensional systems. Growth of cells in 2-dimensional cultures has been shown to reduce the production of particular extracellular matrix proteins, with consequent morphological changes and increase in spreading. The advancement in the technology of nanostructures enhances the scope of fabricating 3-dimensional nanoscaffolds that could potentially mimic the architecture of natural human tissue. These nanostructured scaffolds could control and direct cellular behaviour and interactions with the extracellular matrix. Scaffolds have been designed in the form of nanofibers, nanotubes, nanowires, nanorods, nanocrystals and nanofilms. These nanostructured scaffolds with their biomimetic features and excellent physicochemical properties, stimulated cellular adhesion, growth, morphology, proliferation, altered gene expression and promoted cellular differentiation. The structural features of these nanoscaffolds were engineered according to the nature of cell response which was desired. The scaffolds were designed in a manner that provided a surface to promote cell attachment, spreading and growth while encouraging the formation of a porous network that offered a suitable path for nutrient transmission and tissue ingrowth (Chen & Ma, 2004). These novel nanoscaffolds had excellent mechanical properties that offered structural support until the new tissue would be formed, as they degraded at a rate matching the new tissue formation and provided substrate for cell migration and survival. They were biocompatible and the products of their degradation were also biocompatible (Smith et al., 2010). These nanostructured scaffolds provided the functional role of the native extracellular matrix with growth factors that regulated the cell fate and bioactive peptide sequences that could bind receptors and activate intracellular signalling pathways (Boudreau & Jones, 1999).

Several techniques have been designed for the fabrication of nanofibrous scaffolds to be employed in tissue regeneration. Electrospinning techniques have been the most commonly used. An electric field is applied to draw a polymer solution from an orifice to a collector, producing polymer fibers with diameters ranging in size from 50 nm to several microns. These resulted lengths mimicked that of native collagen fibrils (Baker et al., 2009). Several types of synthetic and natural biomaterials have been used to form nanofibrous scaffolds such as poly (caprolactone) (PCL), poly (lactic-co-glycolic acid) (PLGA) poly (L-lactic acid) (PLLA), collagen, gelatine and fibrinogen; molecules that have been applied extensively in tissue regeneration. Another technique for nanofibrous fabrication is self-assembly. Molecular self-assembly has been applied to produce supramolecular architectures (Silva et al., 2004). This technique produces nanofiber diameters much smaller than those produced using electrospinning. Molecular self-assembly has been less effective in producing macropores for mass transport and cell accommodation. Phase separation techniques have also been also employed to fabricate nanofibers with diameters ranging from 50 - 500 nm and much higher surface -to-volume ratios than produced by other techniques (Chen et al., 2006).

3. Applications of nanotechnology in specific tissue regeneration

Recent studies have been conducted on the promises and applications of nanotechnology in the regeneration of specific tissues, such as bone, cartilage, vascular and neural tissues.

3.1 Bone and cartilage regeneration

Various types of traumatic bone and cartilage damage - bone fractures, osteoarthritis, osteoporosis or bone tumours - represent common and significant clinical problems. However, the treatment of such problems with traditional implant materials only lasts 10 - 15 years on average and implant failures originating from implant loosening, inflammation, infection, osteolysis and wear debris frequently occur. There is a very urgent need to develop a new generation of cytocompatible bone and cartilage substitutes to regenerate bone and cartilage tissues at diseased sites that could last the life time of the patient (Zhang & Webster, 2009).

Bone is effectively a nanocomposite that consists of a protein-based soft hydrogel template formed of collagen, non-collagenous proteins such as laminin, fibronectin and vitronectin, water, and hard inorganic components such as hydroxyapatite, calcium and phosphate. Specifically, 70% of the bone matrix is composed of nanocrystalline hydroxyapatite which is typically 20-80 nm long and 2-5 nm thick. Nanostructured bone extracellular matrix closely surrounds and affects adhesion, proliferation and differentiation of mesenchymal stem cells, osteoblasts, osteoclasts and fibroblasts. Moreover, cartilage is a poorly regenerating tissue composed of a small percentage of chondrocytes but dense nanostructured extracellular matrix rich in collagen fibers, proteoglycans and elastin fibers. The limited regenerative properties of cartilage originate from a lack of chondrocyte mobility in the dense extracellular matrix as well as an absence of progenitor cells and the vascular network necessary for efficient tissue repair (Vasita & Katti, 2006). Development of nanotechnology might provide clinical medicine with new prospects in bone and cartilage reconstruction. Nanotechnology employs engineered materials with the smallest functional organization called nanomaterials that are able to interact with biological systems at a nanoscale (El-Sadik et al., 2010). Nanomaterials could be grown or self-assembled to stimulate the dimensions of natural entities, such as collagen fibers. After decreasing material size into nanoscale, dramatically increased surface area, surface roughness and surface area to volume ratios could be created, leading to superior physiochemical properties such as mechanical, electrical, optical, catalytic, magnetic properties. These biomimetic features with the nanostructured extracellular matrix of bone and cartilage played a key role in stimulating cell growth as well as guided tissue regeneration (Jang et al., 2009). Numerous researchers fabricated cytocompatible biomimetic nanomaterial scaffolds encapsulating cells, such as stem cells, chondrocytes and osteoblasts. In addition, to the dimensional similarity to bone/cartilage tissue, nanomaterials also exhibited unique surface properties, such as surface topography, surface chemistry, surface wettabilty and surface energy, due to their significantly increased surface area and roughness compared to conventional or micron structured materials. As is known, material surface properties mediate specific protein adsorption and bioactivity, such as fibronectin, vitronectin and laminin, before cells adhere on implants, further, they regulate cell behaviour and dictate tissue regeneration. Furthermore, an important criterion for designing orthopaedic implant materials is the formation of sufficient osseointegration between synthetic materials and bone tissue. Studies have demonstrated that nanostructured materials with cell-favourable surface properties could promote greater amounts of specific protein interactions to more efficiently stimulate new bone growth compared to conventional materials (Webster et al., 2001). This is one of the underlying reasons that nanomaterials are superior to conventional materials for bone growth. Therefore, by controlling surface properties, various nanophase ceramic, polymer, metal and composite scaffolds have been designed for bone/cartilage tissue engineering applications (Zhang & Webster, 2009).

There have been significant advances in the development of bone scaffolds with various compositions and 3 dimensional configurations using a variety of techniques such as the electrospining process for the fabrication of nanofibrous matrices. Several studies have reported the performance of nanofibrous materials in guiding cells to initially adhere to, and spread over, the nanostructures, as well as triggering them to secrete appropriate extracellular matrix molecules targeted to the bone and cartilage tissues. The bone-associated cells and the progenitor/stem cells showed initial responses which were anchorage-dependant. The nanofibrous substratum provided favourable conditions for cell anchorage and growth. Further osteoblastic differentiation and mineralization have also been reported to be regulated in a positive manner on nanofibrous surfaces (Woo et al., 2007). One particular requirement of bone tissue regeneration was that the scaffold should be porous, to incorporate large number of cells. The 3-dimensional scaffolds provided the necessary support for bone cells to attach, grow and differentiate and defined the overall shape of a bone tissue cultured transplant (Jang et al., 2009). Nanofibrous and nanotubular scaffolds were fabricated to mimic collagen fibers in bone and cartilage. Natural collagen is a triple helix self assembled into nanofibers of 300 nm in length and 1.5 nm in diameter. A new nanofiber composite was designed with the same self-assembly pattern as collagen and hydroxyapatite crystals in bone by directly nucleating and aligning the hydroxyapatite on the long axis of a nanofiber. Mesenchymal stem cell behaviour on self-assembled peptide amphiphile nanofiber scaffolds was investigated. Significantly enhanced osteogenic differentiation of mesenchymal stem cells was recorded in the 3-dimensional scaffolds compared to 2-dimensional static conventional tissue cultures.

Other types of nanofibers used in bone regeneration include the natural polymers. Natural polymeric nanofibers, such as poly(caprolactone) (PCL), poly(lactic-co-glycolic acid) (PLGA) poly(L-lactic acid) (PLLA), collagen, gelatine and fibrinogen, are excellent candidates for bone and cartilage tissue engineering applications. These biomaterials possess properties that are useful for bone regeneration, such as biodegradability, flexibility, shape availability and ease of fabrication. Nanoporous polymer matrices can be fabricated via electrospinning, phase separation, particulate leaching, chemical etching and 3 dimensional printing techniques (Zhang & Webster, 2009). Poly(caprolactone) (PCL) was first suggested to be a degradable nanofiber matrix for bone regeneration, and it demonstrated good support of the rat bone marrow stromal cells and *in vitro* matrix formation at 4 weeks, including collagen I and calcium phosphate (Yoshimoto et al., 2003). A cell-nanofiber construct was implanted in rat omenta for 4 weeks (Shin et al., 2004). It revealed the formation of collagen I and mineralization similar to bone like extracellular matrix, highlighting its usefulness in bone tissue regeneration. A combination of degradable polymeric nanofibers with bioactive inorganic metals was proved to enhance osteogenic differentiation and calcification of bone matrix. The inorganic phase improved the biological properties of polymers in the bone forming process. Gelatin-hydroxyapatite nanofibers was fabricated (Kim et al., 2005). Hydroxyapatite nanocrystals were distributed in the gelatin matrix and produced an organized hybrid matrix. This composite enhanced osteoblastic differentiation and could be applied usefully in dentistry. In a similar way, collagen-hydroxyapatite (Song et al., 2008) and chitosan-hydroxyapatite (Zhang et al., 2008) nanofibers were generated mimicking the extracellular matrices.

An additional excellent choice of nanomaterials for the reconstruction of bone tissue was the bone-bioactive inorganics such as bioactive glass, ceramics and calcium phosphates. Silica based sol-gel glass mixed with a polymer binder was generated into a nanofibrous mesh by an electrospining technique. Fibers ranging from 84 nm to 640 nm in size were produced (Kim et al., 2006). The large surface area of the nanofibers, and the consequent ionic reaction with the surrounding medium, induced the formation of a bone mineral-like apatite phase on their surfaces. Osteogenic proliferation and differentiation of rat mesenchymal stem cells were found to be enhanced on the bioactive glass nanofiber substrates more than on conventional bioactive glass. Nanophase metals were investigated for orthopaedic tissue regeneration. They are characterized by the presence of more particle boundaries at their surfaces than the conventional micron metals. Linear patterns of nano-features of titanium were created via electron beam evaporation. These patterns induced greater osteoblast adhesion than the micron-rough regions and guided osteoblast morphology and alignment. Highly porous titanium dioxide nanotube layers were fabricated on titanium by anodization. Titanium was anodized electrochemically in dilute hydrofluoric acid electrolyte solutions to produce nanotubes with diameters of 100 nm and lengths of 500 nm into the titanium dioxide layers of titanium. Nanotubular anodized titanium greatly improved osteoblastic function and significantly increased chondrocytic adhesion, promoting bone and cartilage cellular growth (Zhang & Webster, 2009).

3.2 Vascular tissue regeneration

Researchers have come a long way to develop vascular grafts of great efficacy to replace damaged blood vessels, using materials that produce minimal interactions with the inflowing blood and adjacent tissues. Nanomaterials have been found to improve vascular endothelial and smooth muscle functions. Aligned biodegradable poly(L-lactid-co-epsilon-caprolactone) PLLA-CL (75:25) nanofibrous scaffolds have been tested for their ability to fabricate tubular scaffolds for vessels. These nanofibers demonstrated the mechanical strength needed to sustain high pressure of the human circulatory system and the necessary properties that mimic the dimensions of natural extracellular matrix of human coronary artery. They provided an excellent architecture for endothelial and smooth muscle cell adhesion and proliferation. The aligned fibers affected the behaviour of the smooth muscle cells, and the cytoskeleton is organized to follow the direction of the nanofibers (Xu et al., 2004). Electrospun nanofibers fabricated from natural polymers have been established to develop constructs for vascular tissue regeneration. Electrospun collagen and elastin nanofibers were shown to be good scaffolding systems for the engineering of artificial blood vessels (Boland et al., 2004). Another polymer that promoted the endothelial and vascular smooth muscle cell proliferation was the biodegradable poly(lactic-co-glycolic acid) (PLGA), which produced vascular grafts with nanometer surface features. These nanostructures enhanced fibronectin and vitronectin adsorption from serum leading to better vascular cell responses (Miller et al., 2007). Moreover, self-assembled peptides have been fabricated into scaffolds that mimic the vascular basement membrane with excellent cytocompatability. These peptide scaffolds promote endothelialisation and enhance nitric oxide release and laminin and collagen IV deposition by the endothelial cell monolayer (Genove et al., 2005). Titanium nanostructures have been reported to enhance vascular cell adhesion and proliferation greatly. Competitive endothelial cell functions were promoted over that of vascular smooth muscle cells, solving the problem of the overgrowth of smooth muscle cells in vascular stents (Choudhary et al., 2007).

3.3 Neural tissue regeneration

Nanostructure designs have been shown to promote the functional performance of neuronal cells and neural tissue repair. They possess the necessary cytocompatibilty properties for improved neuronal growth, mechanical properties that last long enough to physically support neural tissue regeneration, and electrical properties that stimulate and control neuron behaviour and guide neural tissue repair. Biodegradable and biocompatible novel nanofibers and nanotubes have been fabricated with controlled architecture and components and efficient topography; they promoted neural tissue regeneration. Nanofibrous poly (L-lactic acid) (PLLA) and poly (caprolactone) (PCL) scaffolds designed via electrospining and phase separation demonstrated significant cytocompatibility properties useful for neural tissue regeneration. Incorporation of laminin into the nanofibers created a biomimetic scaffolds for peripheral nerve repair as laminin is an extracellular protein that promotes neurite outgrowth (Koh et al., 2008). Another example for the addition of laminin onto the poly (L-lactic acid) (PLLA) nanofibers was investigated for the culture of the tissues of rat dorsal root ganglia (Patel et al., 2007). Cultures revealed significant longer neurite length more than those cultured on poly (L-lactic acid) (PLLA) nanofibers without laminin. These findings demonstrated the advantages of biosynthetic nanomaterials over the synthetic ones. Moreover, the topography of the electrospun nanofibers scaffolds affected the behaviour of the cultured dorsal root ganglia. Significant extension and elongation of neurites were shown on aligned fibers compared with cultured on randomly oriented nanofibers. The neurites grew in a radial manner on the aligned nanofibers. Those that grew in the direction of the fibers had a faster growth rate than the others indicating that the aligned nanofibrous scaffolds served in guiding neurite orientation and cell alignment (Chow et al., 2007).

Electrospun Chitosan on poly(caprolactone) (PCL) nanofibrous scaffolds provided excellent mechanical properties that enhanced Schwann cell proliferation (Zhang & Webster, 2009).

Chitosan micro and nanofiber mesh tubes have also been investigated for nerve reconstruction (Wang et al., 2008). The authors observed early recovery of sensory functions and elongation of the regenerating axons in 10 mm rat sciatic nerve gap after implantation of the nanofiber mesh tubes. Covalent binding of synthetic and natural materials have been demonstrated in the conjugation of collagen onto a copolymer of methyl methacrylate and acrylic acid electrospun nanofibers (Cao et al., 2009). Increased neurite length of cortical neural stem cells, in proportion to collagen content, was found, indicating that this combination improved the attachment and viability of the cultured neural stem cells. Peptide nanofibrous scaffolds fabricated by self-assembly induced favourable neural cell responses and enhanced neuronal cell functions, outgrowth and functional synapse formation (Zhang & Webster, 2009). Other types of scaffolds are the carbon nanotubes and nanofibers. They were found to guide axon regeneration and improve neural activity as a result of good electrical conductivity, strong mechanical properties and their similar nanoscale dimensions to neurites. Multiwalled carbon nanotubes have been applied for the growth of neurons: a 200 % increase in total neurite length and a 300 % increase in the number of branches and neurites have been demonstrated. In addition, decreased astrocyte proliferation, and consequent decreased glial scar tissue formation, was shown on carbon nanofibers with a polymer composite. Moreover, it was found that astrocytes attached and proliferated less on carbon nanofibers with the smallest nanometer diameter and the highest surface energy (Mckenzie et al., 2004). Carbon nanofibers were shown to limit astrocyte functions, leading to decreased glial scar tissue formation which is essential for increased neuronal implant efficacy.

4. Safety issues involved in the use of nanotechnology

Despite the wide range of applications of nanotechnology in the tissue regeneration studies, still there is a lack of information concerning the influence of nanomaterials on human health. Data available for the safety of nanomaterials, particularly in the field of tissue regeneration, are limited and the mechanisms of their toxicity are still poorly understood. Several studies indicated that a small size, a large surface area and the ability to generate reactive oxygen species increase the potential of nanomaterials to induce cell injury. However, other studies have indicated that, for example, ceramic nanoparticles were safer to osteoblasts than conventional ceramic microparticles. On the other hand, cellular uptake of nanoparticles and their effects on the physiological processes of the cells and their organelles should be deeply investigated before such materials are applied to human tissues. It has been shown, for example, that degradation of nanomaterials used in artificially engineered joints produced toxic responses due to the use of heavy metals such as iron, nickel and cobalt catalysts (Zang & Webster, 2009).

Recent researches in the field of tracking the engrafted stem cells have demonstrated that the safety of quantum dots depends on their physiochemical properties, dose and exposure. Cytotoxicity of quantum dots has been observed owing to the presence of heavy metals such as cadmium and selenium in their cores. Coating the core of quantum dots was recorded to effectively reduce their toxicity to a significant level. Several strategies have been applied to decrease the toxicity of quantum dots. Coating the core with a shell of zinc sulphide reduces the toxicity by blocking the oxidation of the core by air, making them biologically inert. Another technique uses large protein molecules such as bovine serum albumin to could slow the photo-oxidation of the core. Moreover, labelling quantum dots with biomolecules such as arginine-glycine-aspartic acid removed all the toxic effects on cultured stem cells (Solanki et al., 2008). It is recommended to study the appropriate properties and concentrations of different nanoparticles used in cultured and transplanted cells and their safety limits and to deeply understand the physicochemical, molecular and physiological processes of nanomaterials before introducing them into the human bodies.

5. Conclusion

Nanotechnology has shown great potential for numerous tissue regeneration applications. Nanomaterials have achieved one of the major challenges of tissue regeneration which is mimicking the architecture of natural extracellular matrix. Designed nanostructures such as nanoparticles, nanosurfaces and nanoscaffolds have been used to promote stem cell cultures which will speed up understanding, controlling and guiding tissue regeneration studies of different tissues, such as bone, cartilage, vascular and neural tissues. It is suggested that the creation of such nanostructures would advance greatly the field of tissue regeneration. However, nanomaterials require more testing and investigations before full use in human tissue repair. Further understanding of their interactions with biological systems is still needed.

6. References

Au, K.; Liao, S.; Lee, Y. et al. (2009). Effects of Iron Oxide Nanoparticles on Cardiac Differentiation of Embryonic Stem Cells. *Biochem Biophys Res Commun,* Vol.379, (2009), pp. 898-903

Bacakova, L.; Grausova, L.; Vacik, J. et al. (2007). Improved Adhesion and Growth of Human Osteoblast-Like MG 63 Cells on Biomaterials Modified with Carbon Nanoparticles. *Diamond Relat Mater,* Vol.16, No.12 (December 2007), pp. 2133-2140

Baker, B.M.; Handorf, A.M.; Ionescu, L.C.; Li, W. & Mauck, R.L. (2009). New Directions in Nanofibrous Scaffolds for Soft Tissue Engineering and Regeneration. *Expert Rev Med Devices,* Vol.6, No.5, (September 2009), pp. 515-532

Biggs, M.J.P.; Richards, R.G.; McFarlane, S.; Wilkinson, C.D.W.; Oreffo, R.O.C. & Dalby M. J. (2008). Adhesion Formation of Primary Human Osteoblasts and The Functional Response of Mesenchymal Stem Cells to 330? nm Deep Microgrooves. *J R Soc Interface,* Vol.5, (2008), pp. 1231-1242

Boland, E.D.; Matthews, J.A.; Pawlowski, K.J. et al. (2004). Electrospinning Collagen and Elastin: Preliminary Vascular Tissue Engineering. *Front Biosci,* Vol.9, (2004), pp. 1422-1432

Boudreau, N.J. & Jones, P.L. (1999). Extracellular Matrix and Integrin Signalling: The Shape of Things to Come. *Biochem J,* Vol.339, No.3, (1999), pp. 481-488

Bulte, J.W.M. & Kraitchman, D.L. (2004). Monitoring Cell Therapy Using Iron Oxide MR Contrast Agents. *Curr Pharm Biotechnol,* Vol.5, No.6, (2004), pp. 567-584

Cao, H.; Liu, T. & Chew, S.Y. (2009). The Application of Nanofibrous Scaffolds in Neural Tissue Engineering. *Advanced Drug Delivery Reviews,* Vol.61, No.12, (October 2009), pp. 1055-1064

Chang, J.; Su, H. & Hsu, S. (2008). The Use of Peptide- Delivery to Protect Human Adipose-Derived Adult Stem Cells from Damage Caused by The Internalization of Quantum Dots. *Biomaterials,* Vol.29, (2008), pp. 925-936

Chang, J.; Hsu, S. & Su, H. (2009). The Regulation of The Gap Junction of Human Mesenchymal Stem Cells Through The Internalization of Quantum Dots. *Biomaterials,* Vol.30, (2009), pp. 1937-1946

Charest, J.L.; Bryants, L.E.; Garcia, A.J. & King, W.P. (2004). Hot Embossing for Micropatterned Cell Substrates. *Biomaterials,* Vol.25, (2004), pp. 4767-4775

Chen, V.J. & Ma, P.X. (2004). Nano-Fibrous Poly(L-Lactic Acid) Scaffolds with Interconnected Spherical Macropores. *Biomaterials,* Vol.25, No.11, (2004), pp. 2065-2073

Chen, V.J.; Smith, L.A. & Ma, P.X. (2006). Bone Regeneration on Computer-Designed Nano-Fibrous Scaffolds. *Biomaterials,* Vol.27, (2006), pp. 3973-3979

Chen, H.; Titushkin, I.; Stroscio, M. & Cho, M. (2007). Altered Membrane Dynamics of Quantum Dot-Conjugated Integrins During Osteogenic Differentiation of Human Bone Marrow Derived Progenitor Cells. *Biophys J,* Vol.92, (2007), pp. 1399-1408

Choudhary, S.; Haberstroh, K.M. & Webster, T.J. (2007). Enhanced Functions of Vascular Cells on Nanostructured Ti for Improved Stent Applications. *Tissue Engineering,* Vol.13, (2007), pp. 1421-1430

Chow, W.N.; Simpson, D.G.; Bigbee, J.W. & Colello, R.J. (2007). Evaluating Neuronal and Glial Growth on Electrospun Polarized Matrices: Bridging the Gap in Percussive Spinal Cord Injuries. *Neuron Glia Biol,* Vol.3, (2007), pp. 119-126

Clem, W.C.; Chowdhury, S.; Catledge, S.A. et al., (2008). Mesenchymal Stem Cell Interaction with Ultra-Smooth Nanosrtuctured Diamond for Wear-Resistant Orthopaedic Implants. *Biomaterials,* Vol.29, No.24-25, (August-September 2008), pp. 3461-3468

Curtis, A.S.G.; Dalby, M.J. & Gadegaard, N. (2006). Cell Signaling Arising from Nanotopography: Implications for Nanomedical Devices. *Nanomedicine,* Vol.1, (2006), pp. 67-72

Dalby, M.J.; Riehle, M.O.; Yarwood, S.J.; Wilkinson, C.D. & Curtis, A.S. (2003b). Nucleus Alignment and Cell Signaling in Fibroblasts: Response to a Micro-Grooved Topography. *Exp Cell Res,* Vol.284, (2003b), pp. 274-282

Dalby, M.J.; Biggs, M. J.; Gadegaard, N.; Kalna, G.; Wilkinson, C.D. & Curtis, A.S. (2007). Nanotopographical Stimulation of Mechanotransduction and Changes in Interphase Centromere Positioning. *J Cell Biochem,* Vol.100, (2007), pp. 326-338

Dalby, M.J. (2009). Nanostructured Surfaces: Cell Engineering and Cell Biology. *Nanomedicine,* Vol.4, No.3, (April 2009), pp. 247-248

Delcroix, G.J.; Jacquart, M.; Lemaire, L. et al., (2009). Mesenchymal and Neural Stem Cells Labelled with HEDP-coated SPIO Nanoparticles: In Vitro Characterization and Migration Potential in Rat Brain. *Brain Res,* Vol.1255, (2009), pp. 18-31

El-Sadik, A.O.; El-Ansary, A. & Sabry, S.M. (2010). Nanoparticle-Labeled Stem Cells: A Novel Therapeutic Vehicle. *Clinical Pharmacology: Advances and Applications,* Vol.2, (2010), pp. 9-16

Emerit, J.C.; Beaumount; C. & Trivin, F. (2001). Iron Metabolism, Free Radicals, and Oxidative Injury. *Biomed Pharmacother,* Vol.55, No.6, (2001), pp. 333-339

Gao, J. & Xu, B. (2009). Applications of Nanomaterials Inside Cells. *Nanotoday,* Vol.4, (2009), pp. 37-51

Gelain, F.; Bottai, D.; Vescovi, A. & Zhang, S. (2006). Designer Self-Assembling Peptide Nanofiber Scaffolds for Adult Mouse Neural Stem Cell 3-Dimensional Cultures. *PLoS ONE,* Vol.1, No.1, (2006), e119

Genove, E.;Shen, C.; Zhang, S. & Semino, C.E. (2005). The Effect of Functionalized Self-Assembling Peptide Scaffolds on Human Aortic Endothelial Cell Function. *Biomaterials,* Vol.26, No.16, (June 2005), pp. 3341-3351

Grausova, L.; Kromka, A.; Bacakova, L.; Potocky, S.; Vanecek, M. & Lisa, V. (2008). Bone and Vascular Endothelial Cells in Cultures on Nanocrystalline Diamond Films. *Diamond and Related Materials,* Vol.17, No.7-10, (July-October 2008), pp. 1405-1409

Hu, Y.; Cai, K.; Luo, Z. et al. (2009). Surface Mediated in Situ Differentiation of Mesenchymal Stem Cells on Gene-Functionalized Titanium Films Fabricated by Layer-by- Layer Technique. *Biomaterials,* Vol.30, No.21, (July 2009), pp. 3626-3635

Huang, D.; Chung, T.; Hung, Y. et al. (2008). Internalization of Mesoporous Silica Nanoparticles Induces Transient but not Sufficient Osteogenic Signals in Human Mesenchymal Stem Cells. *Toxicol Appl Pharmacol,* Vol.231, (2008), pp. 208-215

Ito, A.; Hibino, E.; Honda, H. et al. (2004). A New Methodology of Mesenchymal Stem Cell Expansion Using Magnetic Nanoparticle. *Biomechemical and Engineering J,* Vol.20, (2004), pp. 119-125

Jang, J.; Castano, O. & Kim, H. (2009). Electrospun Materials as Potential Platforms for Bone Tissue Engineering. *Advanced Delivery Reviews,* Vol.61, (2009), pp. 1065-1083

Jing, X.; Yang, L.; Duan, X. et al. (2008). Invivo MR Imaging Tracking of Magnetic Iron Oxide Nanoparticle Labelled, Engineered, Autologus Bone Marrow Mesenchymal Stem Cells Following Intra-Articular Injection. *Joint Bone Spine,* Vol.75, (2008), pp. 432-438

Jung, Y.; Chung, Y.; Kim, S.H. et al. (2009). In situ Chondrogenic Differentiation of Human Adipose Tissue-Derived Stem Cells in a TGF-b1 Loaded Fibrin-Poly(Lactide-Caprolactone) Nanoparticulate. *Biomaterials,* Vol.30, (2009), pp. 4657-4664

Kea, Y.; Hu, C.; Jianga, X. et al. (2009). In vivo Magnetic Resonance Tracking of Feridex-Labelled Bone MarroDerived Neural Stem Cells after Autologous Transplantation in Rhesus Monkey. *J Neurosci Methods,* Vol.179, (2009), pp. 45-50

Kim, H.W.; Song, J.H. & Kim, H.E. (2005). Nanofiber Generation of Gelatin-Hydroxyapatite Biomimetics for Guided Tissue Regeneration. *Adv Funct Matr,* Vol.15, (2005), pp. 1988-1994

Kim, H.W.; Kim, H.E. & Knowles, J.C. (2006). Production and Potential of Bioactive Glass Nanofibers as a Next-Generation Biomaterial. *Adv Funct Mater,* Vol.16, (2006), pp. 1529-1535

Koh, H.S.; Yong, T.; Chan, C.K. & Ramakrishna, S. (2008). Enhancement of Neurite Outgrowth Using Nanostructured Scaffolds Coupled with Laminin. *Biomaterials,* Vol.29, No.26, (September 2008), pp. 3574-3582

Kommireddy, D.S.; Ichinose, I.; Lvov, Y.M. & Mills, D.K. (2005). Nanoparticle Thin Films: Surface Modification for Cell Attachment and Growth. *J Biomed Nanotechnol,* Vol.3, (2005), pp. 286-290

Kommireddy, D.S.; Sriram, S.M.; Lvov, Y.M. & Mills, D.K. (2006). Stem Cell Attachment to Layer-by-Layer Assembled TiO2 Nanoparticle Thin Films. *Biomaterials,* Vol.27, No.24, (August 2006), pp. 4296-4303

Kubo, K.; Tsukimura, N.; Iwasa, F. et al. (2009). Cellular Behavior on TiO2 Nanonodular Structures in a Micro-to-Nanoscale Hierarchy Model. *Biomaterials,* Vol.30, No.29, (October 2009), pp. 5319-5329

Lenhert, S.; Meier, M.B.; Meyer, U.; Chi, L. & Wiesmann, H.P. (2005). Osteoblast Alignment, Elongation and Migration on Grooved Polystrene Surfaces Patterned by Langmuir-Blodgett Lithography. *Biomaterials,* Vol.26, (2004), pp. 563-570

Lin, S.; Xie, X. & Patel, M.R. (2007). Quantum Dot Imaging for Embryonic Stem Cells. *BMC Biotechnol,* Vol.7, (2007), pp. 67

Liu, H. & Webster, T.J. (2007). Nanomedicine for Implants: A Review of Studies and Necessary Experimental Tools. *Biomaterials,* Vol.28, No.2, (January 2007), pp. 354-369

Mckenzie, J.L.; Waid, M.C.; Shi, R. & Webster, T.J. (2004). Decreased Functions of Astrocytes on Carbon Nanofiber Materials. *Biomaterials,* Vol.25, No.7-8, (March-April 2004), pp. 1309-1317

Matsuzaka, K.; Walboomers, X.F.; Yoshinari, M.; Inoue, T. & Jansen, J.A. (2003). The Attachment and Growth Behavior of Osteoblast-like Cells on Microtextured Surfaces. *Biomaterials,* Vol.24, (2003), pp. 2711-2719

Miller, D.C.; Haberstroh, K.M. & Webster, T.J. (2007). PLGA Nanometer Surface Features Manipulate Fibronectin Interactions for Improved Vascular Cell Adhesion. *J Biomed Mater Res A,* Vol.81, No.3, (June 2007), pp. 678-684

Murthy, S.K. (2007). Nanoparticles in Modern Medicine: State of the Art and Future Challenges. *International Journal of Nanomedicine,* Vol.2, (2007), pp. 129-141

Ogawa, T.; Saruwatari, L.; Takeuchi, K.; Aita, H. & Ohno, N. (2008). Ti Nano-Nodular Structuring for Bone Integration and Regeneration. *J Dent Res,* Vol.87, (2008), pp. 751-756

Oh, S.; Brammer, K.S.; Julie Li, Y.S.; et al. (2009). Stem Cell Fate Dictated Solely by Altered Nanotube Dimension. *Proc Natl Acad Sci USA,* Vol.106, No.7, (February 2009), pp. 2130-2135, ISSN 1729-8806

Patel, S.; Kurpinski, K.; Quigley, R. et al., (2007). Bioactive Nanofibers: Synergistic Effects of Nanotopography and Chemical Signaling on Cell Guidance. *Nano Lett,* Vol.7, (2007), pp. 2122-2128

Shah, B.S.; Clark, P.A.; Moioli, E.K.; Stroscio, M.A. & Mao, J.J. (2007). Labelling of Mesenchymal Stem Cells by Bioconjugated Quantum Dots. *Nano Lett,* Vol.7, (2007), pp. 3071-3079

Shi, X.; Wang, Y.; Varshney, R.R. et al., (2009). In-vitro Osteogenesis of Synovium Stem Cells Induced by Controlled Release of Bisphosphate Additives from Microspherical Mesoporous Silica Composite. *Biomaterials,* Vol.30, (2009), pp. 3996-4005

Shin, M.; Yoshimoto, H. & Vacanti, J.P. (2004). In Vivo Bone Tissue Engineering Using Mesenchymal Stem Cells on a Novel Electrospun Nanofibrous Scaffold. *Tissue Eng,* Vol.10, (2004), pp. 33-41

Silva, G.A.; Czeisler, C.; Niece, K.L. et al. (2004). Selective Differentiation of Neural Progenitor Cells by High-Epitope Density Nanofibers. *Science,* Vol.303, (2004), pp. 1352-1355

Smith, I.O.; Liu, X.H.; Smith, L.A. & Ma, P.X. (2010). Nano-Structured Polymer Scaffolds for Tissue Engineering and Regenerative Medicine. *Wiley Interdiscip Rev Nanomed Nanobiotechnol,* Vol.1, No.2, (March 2010), pp. 226-236

Song, J.H.; Kim, H.E. & Kim, H.W. (2008). Electrospun Fibrous Web of Collagen-Apatite Precipitated Nanocomposite for Bone Regeneration. *J Mater Sci Mater Med,* Vol.19, (2008), pp. 2925-2932

Sykova, E. & Jendelova, P. (2007). Migration, Fate and In vivo Imaging of Adult Stem Cells in the CNS. *Cell Death Differ,* Vol.14, (2007), pp. 1336-1342

Vasita, R. & Katti, D.S. (2006). Nanofibers and Their Applications in Tissue Engineering. *International Journal of Nanomedicine,* Vol.1, No.1, (2006), pp. 15-30, ISSN 11769114

Wang, L. & Li, D. Y. (2003). Mechanical, Electrochemical and Tribological Properties of Nanocrystalline Surface of Brass Produced by Sandblasting and Annealing. *Surface and Coatings Technology,* Vol.167, No.2-3, (April 2003), pp. 188-196

Wang, W.; Itoh, S.; Matsuda, A. et al. (2008). Influences of Mechanical Properties and Permeability on Chitosan Nano/Microfiber Mesh Tubes as a Scaffold for Nerve Regeneration. *J Biomed Mater Res A,* Vol.84, No.A, (2008), pp. 557-566

Webster, T.J.; Ergun, C.; Doremus, R.H.; Siegel, R.W. & Bizios, R. (2000). Enhanced Functions of Osteoblasts on Nanophase Ceramics. *Biomaterials,* Vol.21, No.17, (September 2000), pp. 1803-1810

Webster, T.J.; Schadler, L.S.; Siegel, R.W. & Bizios, R. (2001). Mechanisms of Enhanced Osteoblast Adhesion on Nanophase Alumina Involve Vitronectin. *Tissue Engineering,* Vol.7, No.3, (2001), pp. 291-301, ISSN 10763279

Woo, K.M.; Jun, J.H.; Chen, J.H.; et al. (2007). Nano-Fibrous Scaffolding Promotes Osteoblast Differentiation and Biomineralization. *Biomaterials,* Vol.28, (2007), pp. 335-343

Xu, C.Y.; Inai, R.; Kotaki, M. & Ramakrishna, S. (2004). Aligned Biodegradable Nanofibrous Structure: A Potential Scaffold for Blood Vessel Engineering. *Biomaterials,* Vol.25, No.5, (February 2004), pp. 877-886

Yoshimoto, H.; Shin, H.; Terai, H. & Vacanti, A. (2003). Biodegradable Nanofiber Scaffolds by Electrospinning and its Potential for Bone Tissue Engineering. *Biomaterials,* Vol.24, (2003), pp. 2077-2082

Zhang, Y.; Venugopal, J.R.; El-Turki, A.; Ramakrishna, S.; Su, B. & Lim, C.T. (2008). Electrospun Biomimetic Nanocomposite Nanofibers of Hydroxyapatite/Chitosan for Bone Tissue Engineering. *Biomaterials,* Vol.29, (2008), pp. 4314-4322

Zhang, L. & Webster, T.J. (2009). Nanotechnology and Nanomaterials: Promises for Improved Tissue Regeneration. *Nanotoday*, Vol.4, No.1, (February 2009), pp. 66-80

Part 4

Modeling and Assessment of Regeneration

21

Non-Invasive Evaluation Method for Cartilage Tissue Regeneration Using Quantitative-MRI

Shogo Miyata
KEIO University
Japan

1. Introduction

Articular cartilage is an avascular tissue covering articulating surfaces of bones and it functions to bear loads and reduce friction in diarthrodial joints. The cartilage can be regarded as a porous gel, mainly composed of large proteoglycan (PG) aggregates having a negative fixed-charge density (nFCD), a water-swollen network of collagen fibrils, and interstitial water, all of which play important roles in load-bearing properties (Lee et al., 1981; Mow et al., 1980).

Although articular cartilage may function well over a lifetime, traumatic injury or the degenerative changes associated with osteoarthritis (OA) can significantly erode the articular layer, leading to joint pain and instability. Because of its avascular nature, articular cartilage has a very limited capacity to regenerate and repair. It is well-known that the natural response of articular cartilage to damage is variable and, at best, unsatisfactory.

Therefore, numerous studies have reported tissue-engineering approaches to restore degenerated cartilage and to repair defects; these approaches involve culturing autologous chondrocytes *in vitro* to create three-dimensional tissue that is subsequently implanted. In these tissue engineering approaches, it is important to assess the biomechanical and biochemical properties of the engineered cartilage. These material properties of the engineered constructs are detectable only via direct measurements that are invasive and require destructive treatments such as histological analysis, biochemical quantification, and mechanical testing. The application and utilization of these tissue-engineering approaches in a clinical setting requires a non-invasive method of evaluating biomechanical and biochemical properties of the actual regenerated cartilage for transplantation. Moreover, the method should be applicable to various aspects of cartilage regenerative medicine, including the characterization of the regenerated tissue during *in vitro* culture and *in vivo* evaluation after transplantation.

Magnetic resonance imaging (MRI) of articular cartilage is well accepted and has become common in recent years. Quantitative MRI techniques have been successfully developed to measure the macromolecular state within cartilage tissue. For example, the relationship between the water content of the degenerated cartilage and water self-diffusion has been reported (Shapiro et al., 2001), while the transverse relaxation time T2 has been related to collagen concentration (Fragonas et al., 1998) and the spatial distribution of collagen, including both fibril orientation and organization (Nieminen et al., 2001; Xia et al., 2002).

The gadolinium-diethylene triamine pentaacetic acid (Gd-$DTPA^{2-}$)-enhanced T1 imaging technique has been used to predict PG content (Bashir et al., 1996) and spatial distribution (Bashir et al., 1999). Furthermore, nFCD can be estimated from consecutive T1 relaxation time measurements using Gd-$DTPA^{2-}$-enhanced MRI and related to PG concentration. This MRI technique is already well-known as the "delayed Gadolinium Enhanced Magnetic Resonance Imaging of Cartilage" (dGEMRIC) technique. This technique is based on the utilization of the two-negative charge of the MRI contrast agent (i.e., Gd-$DTPA^{2-}$). Sulfated glycosaminoglycans (sGAG) in the PGs are negatively charged in the cartilage, giving rise to nFCD; the electric exclusion force between this nFCD and the negatively charged contrast agent result in the inverse distribution of the contrast agent to the PG distribution in the cartilage. Consequently, relaxation time (T1) and nFCD—as determined by dGEMRIC technique —correlate with PG concentration.

Previous studies have reported that, in tissue-engineered cartilage, MR measurements of regenerated cartilage showed correlations with biochemical properties (Potter et al., 2000) and biomechanical properties (Chen et al., 2003). Additionally, the sGAG content and the compressive modulus—the latter of which was determined by unconfined compression tests—showed a trend toward correlation with the nFCD, as determined by the Gd-$DTPA^{2-}$-enhanced MRI technique (Chen et al., 2003; Ramaswamy et al., 2008). In our earlier study, we reported that the nFCD of tissue-engineered cartilage determined by GD-$DTPA^{2-}$-enhanced MRI has been found to correlate with sGAG content (Miyata et al., 2006).

Although the non-invasive assessment of tissue integration and the non-destructive evaluation of molecular structure of the engineered cartilage are important, we believe no previous study has fully evaluated the relationships between the biomechanical properties and MRI measurements of regenerated cartilage consisting of articular chondrocytes. Previous study has indicated that MR images of autologous chondrocyte transplants may show clinically significant variations; neither biochemical properties nor the FCD of regenerated articular cartilage has been evaluated.

In this chapter, we introduce our evaluation technique for tissue-engineered cartilage using quantitative-MRI. We tested the hypothesis that MRI measurements of tissue-engineered cartilage correlate with biomechanical and biochemical properties and that these novel approaches can be used to evaluate cartilaginous matrix material properties during tissue regeneration.

2. Quantitative Magnetic Resonance Imaging (MRI) of tissue engineered cartilage

2.1 Isolation of chondrocytes and preparation of chondrocyte-seeded agarose constructs

We used agarose gel culture for tissue-engineered cartilage model, because agarose is a biocompatible, thermosensitive hydrogel that offers superior homogeneity and stability for assessing both biomechanical and biochemical properties during *in vitro* culture, and has been used widely in cartilage mechanobiology. Chondrocyte-seeded agarose gels were prepared as described previously (Miyata et al., 2006; Miyata et al., 2004).

Articular chondrocytes were obtained from the glenohumeral joints of freshly slaughtered 4- to 6-week-old calves, from a local abattoir. Articular cartilage was excised from the humeral head, diced into ~1 mm^3 pieces, then shaken gently in Dulbecco's Modified Eagle's Medium/Ham's F12 (DMEM/F12) supplemented with 5% fetal bovine serum (FBS), 0.2% collagenase type II, and antibiotics-antimycotics, for 8–10 h at 37°C. Cells were then isolated from the digest by centrifugation and rinsed twice with phosphate buffered saline (PBS). Finally, after the isolated cells were resuspended with feed medium (DMEM/F12 supplemented with 20% FBS, 50 μg/mL L-ascorbic acid, and antibiotics-antimycotics), and the total number of cells was counted with a hemocytometer.

The isolated chondrocytes in the feed medium were mixed with an equal volume of PBS containing agarose with a low melting temperature (Agarose type VII, Sigma, MO) at 37°C, to prepare 1.5 × 10^7 cells/mL in 2% (wt/vol) agarose gel; it was then cast in a custom-made mold to make a large gel plate. After gelling at 4°C for 25 minutes, approximately 50 disks of 8-mm diameter, 1.5-mm thickness were cored out from the large gel plate with a biopsy punch. The chondrocyte-seeded agarose disks were fed 2.5 mL feed medium/disk, every other day and maintained in a 5% CO_2 atmosphere at 37°C.

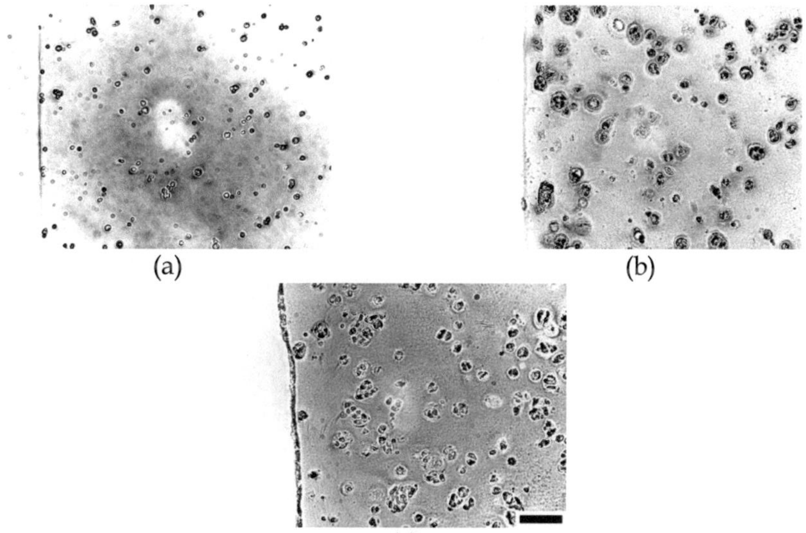

(a) (b) (c)

Fig. 1. Histological appearance of tissue-engineered cartilage at day 3 (a), day 10 (b), and day 28 (c), stained with alcian blue (Miyata et al., 2010). Scale bar = 100 μm.

Alcian blue-stained sections of the cultured specimens are shown in Fig. 1. Over the culture period, the chondrocytes in the agarose gel appeared rounded in shape, similar to those in the "native" articular cartilage. As shown in Fig. 1, the chondrocytes synthesized a thin shell of pericellular matrix (~day 10) and expanded the volume of the cartilaginous matrix (~day 28).

2.2 Magnetic Resonance Imaging (MRI) of cultured chondrocyte-seeded agarose gel

Quantitative MRI evaluations were performed on a 2.0-T Biospec 20/30 System with a B-GA20 Gradient System (Bruker, Karlsruhe, Germany) with a maximum gradient strength of 100 mT/m. The MRI data acquisition and reconstruction were performed using the ParaVision (Bruker) software system. In all MRI experiments, three or four sheets of the disks were stacked in layers and placed into glass tubes containing phosphate buffered saline (PBS) (Fig. 2). The measured parameters included longitudinal (T1) and transverse (T2) relaxation time and water self-diffusion coefficient (Diff). A longitudinal relaxation time map (T1-map) was obtained with a short echo time (TE: 15 ms) spin-echo sequence with different repetition time values (TR: 100 ms to 15 s, 16 steps). A transverse relaxation time map (T2-map) was obtained with a long repetition time value (TR: 15 s) spin-echo sequence with different echo time values (TE: 30 ms to 450 ms, 29 steps). A diffusion coefficient map (Diff-map) was calculated from the images obtained using a conventional diffusion weighted spin-echo (SE-DWI, TR: 15 s, TE: 35 ms) sequence with different *b* values (0, 74, 275, 603, 1059 s/mm^2). All sequences were performed with a field of view (FOV) of 50 × 50 mm^2, matrix size 64 × 64, and slice thickness 3 mm. The values of the relaxation time (T1 and T2) and the relative diffusion coefficient (Diff*) were calculated as the average of the specimen from the obtained T1-, T2-, and Diff-maps. The value of Diff* (= $Diff_S/Diff_P$) was calculated by normalizing the diffusion coefficient of the sample ($Diff_S$) by the diffusion coefficient of PBS ($Diff_P$) around the sample. All MRI measurements were carried out with no contrast agent at room temperature (23°C).

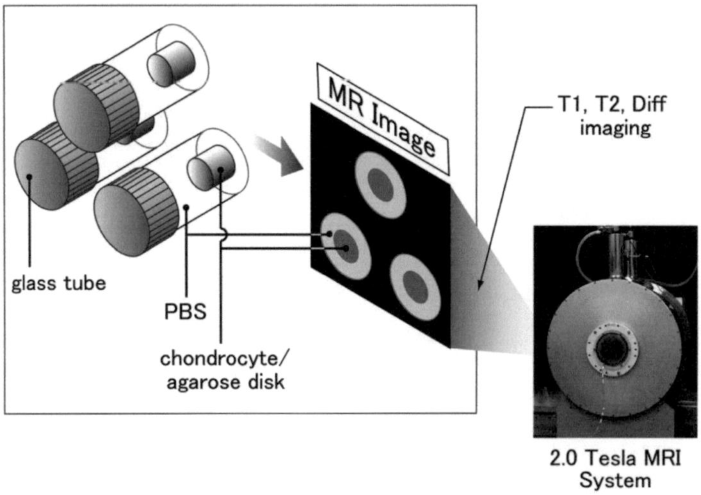

Fig. 2. Schematic diagram of MR Imaging (Miyata et al., 2007).

Figure 3–5 shows the MRI maps of the engineered cartilage. At the first stage of the culture (day 1), T1 and Diff of the engineered cartilage showed values similar to those of the PBS around the cartilage; hence, it was difficult to distinguish the boundaries between the engineered cartilage and the bath solution (PBS) in the MRI maps (Fig. 3a and 5a). By the end of the culture (day 28), the boundaries were distinguished in both T1- and Diff-maps (Fig. 3a–3c and 5a–5c). In contrast, the boundary between the specimen and the PBS remained clear in the T2-map during the culture time (Fig. 4a–4c). The T1, T2, and Diff values of the engineered cartilage were averaged, and the results are summarized in Figure 6. T1 and Diff* of the tissue-engineered cartilage had decreased with an increase in the culture time (Fig. 6a and 6c). On the other hand, T2 of the engineered cartilage showed considerably lower values than those of the PBS in the glass tube throughout the culture time, and these values tended to increase slightly with the culture time (Fig. 6b).

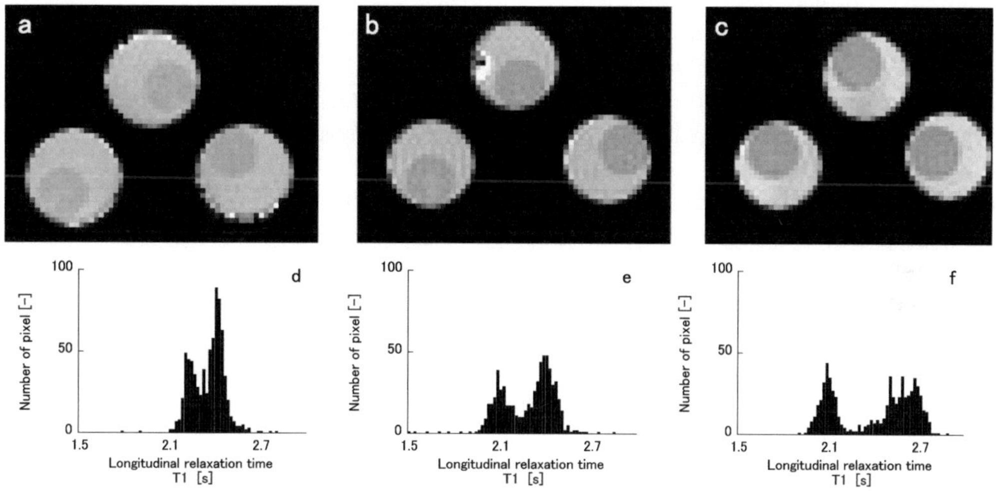

Fig. 3. T1-maps of day 1 (a), day 7 (b), and day 28 (c) post-inoculation specimens, and histograms of the T1 values derived from the MR images on day 1 (d), day 7 (e), and day 28 (f) (Miyata et al., 2007).

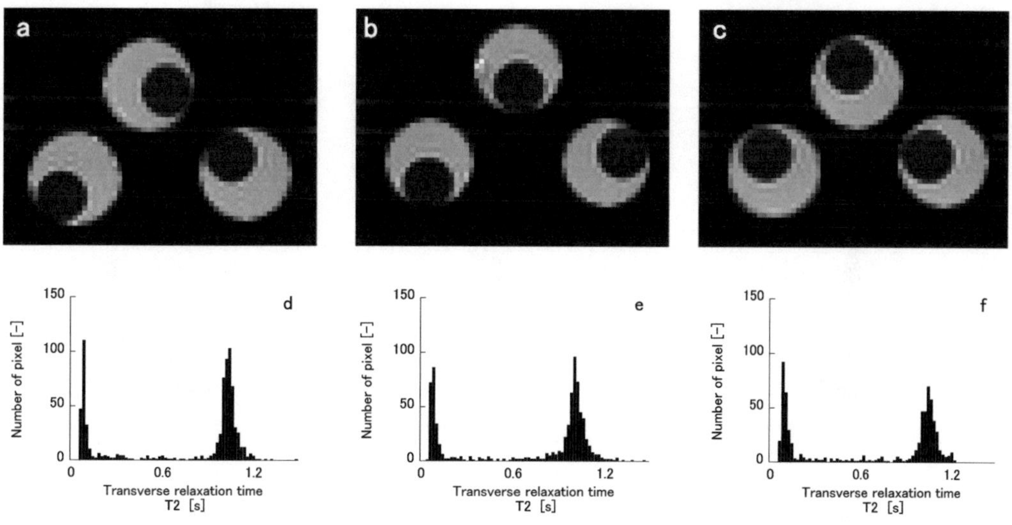

Fig. 4. T2-maps of day 1 (a), day 7 (b), and day 28 (c) post-inoculation specimens, and histograms of the T2 values derived from the MR images on day 1 (d), day 7 (e), and day 28 (f) (Miyata et al., 2007).

Consistent with the results of previous studies (Chen et al., 2003; Potter et al., 1998), our results showed that both T1 and Diff* decreased with an increase in the culture time. On the other hand, T2 tended to increase slightly by the end of the culture time. To understand the MRI properties of the tissue water protons, we have to understand the behavior of water molecules in the tissue at different stages of tissue maturity. With tissue growth and development, proteoglycan and collagen molecules accumulate in the agarose gel, resulting in a large fraction of macromolecule-associated water, which is known as "bound" water. Generally, the water molecules in the "bound" condition show short T1 and T2 relaxation times due to a reduced mobility as compared to "free" water. Thus, water proton relaxation curves, which were described by a single exponential, are derived from the weighted sum of the relaxation behavior of the "free" and "bound" water molecules in the engineered cartilage. This is consistent with our results that the T1 relaxation time and Diff* decreased with an increase in the content of cartilaginous matrix in the agarose gel. In the case of transverse relaxation, the T2 relaxation time of the engineered cartilage showed a value similar to that of the "native" articular cartilage (75–90 ms measured by our MRI system) from the early phase of the culture; further, T2 tended to increase slightly with tissue maturation. Based on this result, we speculate that the transverse relaxation of the water molecules in the engineered construct might be mainly affected by its association with the agarose molecules.

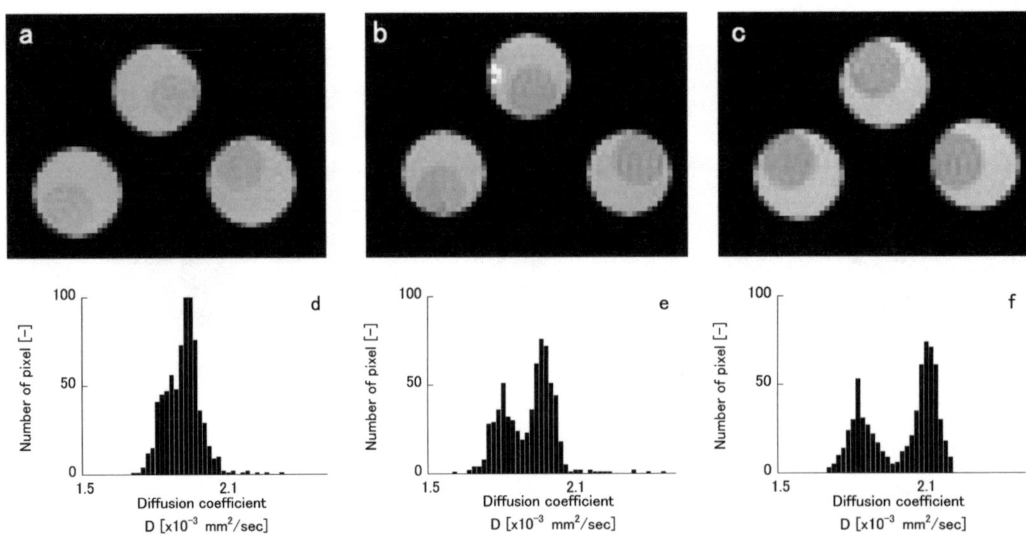

Fig. 5. Diff-maps of day 1 (a), day 7 (b), and day 28 (c) post-inoculation specimens, and histograms of the Diff values derived from the MR images on day 1 (d), day 7 (e), and day 28 (f) (Miyata et al., 2007).

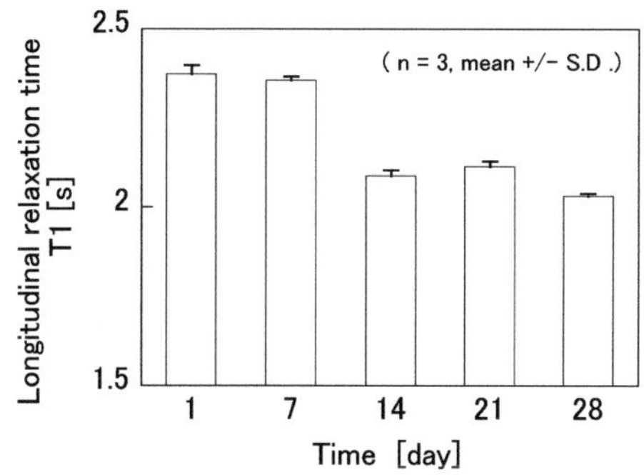

(a)

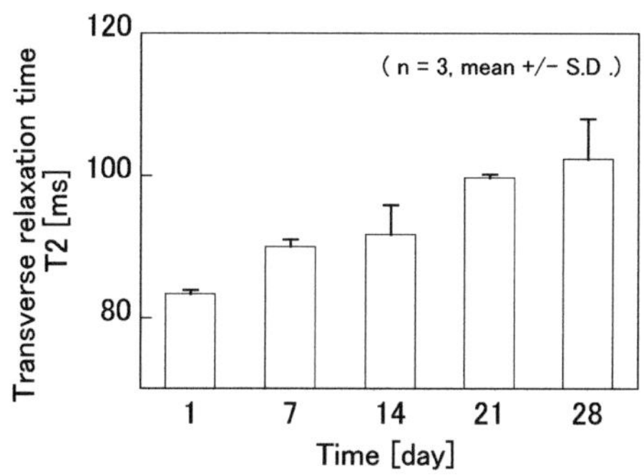

(b)

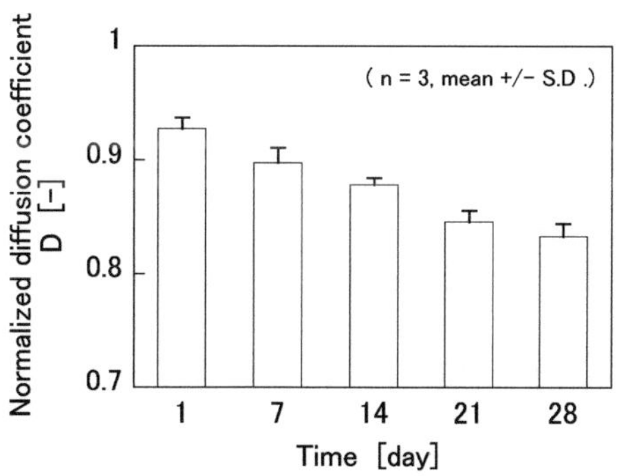

(c)

Fig. 6. Longitudinal relaxation time (a), transverse relaxation time (b), and relative diffusion coefficient (c) of the tissue-engineered cartilage during the culture time (Miyata et al., 2007). The values represent mean +/- S.D. (n = 3).

2.3 Evaluation of fixed charge density of tissue-engineered cartilage

For 'native' articular cartilage, the gadolinium-diethylene triamine pentaacetic acid (Gd-DTPA^{2-}) -enhanced T1 imaging technique has been used to predict the PG content (Bashir et al., 1996) and spatial distribution (Bashir et al., 1999). Furthermore, the negative fixed charge density (nFCD) can be estimated from consecutive T1 relaxation time measurement using $Gd-DTPA^{2-}$-enhanced MRI and be related to the PG concentration. In this study, we used this dGEMRIC technique to monitor and evaluate tissue integration of the engineered cartilage.

The MRI measurements were performed with a 2.0-Tesla Bruker Biospec 20/30 system using $Gd-DTPA^{2-}$ contrast agent. In all MRI measurements, the specimens were put into glass tubes filled with PBS (Fig. 7). The longitudinal relaxation time map, T1-map, was obtained with a short-echo time (TE: 15 ms), spin-echo sequence with different repetition time values (TR: 100 ms to 15 s, 16 steps). Subsequently, the specimens were balanced in PBS containing 1 mM $Gd-DTPA^{2-}$ (Magnevist®, Nihon Schering, Osaka, Japan) for 10−12 hours; the longitudinal relaxation time map in the contrast agent, $T1_{Gd}$-map, was obtained again with a short-echo time (TE: 15 ms), spin-echo sequence with different repetition time values (TR: 30 ms to 5 s, 13 steps). Finally, using the relaxivity (R) value of $Gd-DTPA^{2-}$ in saline (5.24 in our MRI system), the concentration of the contrast agent was estimated using the formula $[Gd-DTPA^{2-}] = 1/R(1/T1_{Gd} - 1/T1)$. The negative fixed charge density (FCD) was calculated as follows

$$\mathrm{nFCD} = \frac{[\mathrm{Na}^+]_b\sqrt{[\mathrm{Gd-DTPA}^{2-}]_t}}{\sqrt{[\mathrm{Gd-DTPA}^{2-}]_b}} - \frac{[\mathrm{Na}^+]_b\sqrt{[\mathrm{Gd-DTPA}^{2-}]_b}}{\sqrt{[\mathrm{Gd-DTPA}^{2-}]_t}} \quad (1)$$

where subscript b stands for bath solution and subscript t stands for cartilaginous tissue(Bashir et al., 1996). All MRI measurements were performed at room temperature 23°C.

In the gadolinium-enhanced MR imaging measurements, longitudinal relaxation time of the bulk PBS containing Gd-DTPA reagent showed 0.179 ± 0.06 seconds in our MRI system. The $T1_{Gd}$ of the cultured specimen increased as a function of tissue maturation (0.197 ± 0.001 to 0.222 ± 0.003 seconds). At the first stage of the culture (day 3), $T1_{Gd}$ of the tissue-engineered cartilage showed values proximate to those of the PBS containing the $Gd-DTPA^{2-}$ agent around the engineered cartilage; hence, it was difficult to distinguish the boundaries between the engineered cartilage and the bath solution in the $T1_{Gd}$-maps (Fig. 8a). By the end of the culture (day 28), the boundaries had become distinct in the T1Gd-maps (Fig. 8). The $[Gd-DTPA^{2-}]$ in the engineered cartilage decreased with increases in culture time. The nFCD, as determined from the $[Gd-DTPA^{2-}]$ in the specimen and bath solution, increased with culture time (Fig. 9).

As time in culture lengthened, the gross appearance of the cultured disk became increasingly opaque. The DMMB assay (Farndale et al., 1986) revealed that the sGAG content of the chondrocyte/agarose disks increased as a function of tissue maturation (0.19 ± 0.27 to 13.2 ± 1.9 mg/mL-disk-vol). Finally, the sGAG content of the reconstructed cartilaginous disk reached approximately 20% of the "native" articular cartilage (data not shown).

To correlate gadolinium-enhanced MRI and biochemical properties, the sGAG content of the tissue was plotted as a function of the FCD. From the linear regression analysis, the FCD correlated significantly with the sGAG content ($r = 0.95$, $n = 30$, $P < 0.001$) (Fig. 10), and the tissue $[Gd-DTPA^{2-}]$ correlated with the sGAG content by $r = 0.83$, $n = 30$, $P < 0.001$.

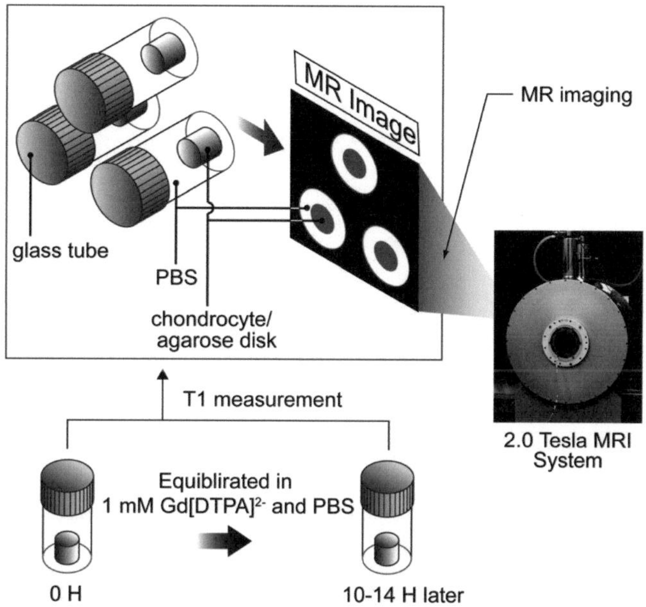

Fig. 7. Schematic diagram of gadolinium-enhanced MRI. In all MRI measurements, the cultured specimens were put into glass tubes filled with phosphate buffered saline (PBS) or 1 mM Gd $DTPA^{2-}$ (Miyata et al., 2010).

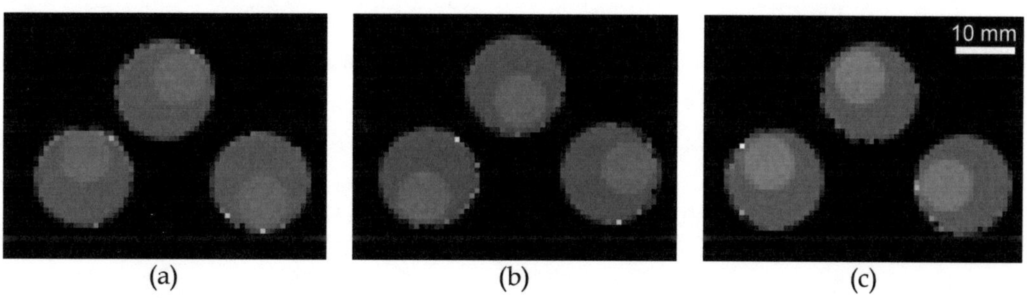

Fig. 8. Quantitative water proton T1 maps in the presence of Gd-DTPA2– at day 3 (a), day 7 (b), and day 28 (c) (Miyata et al., 2010).

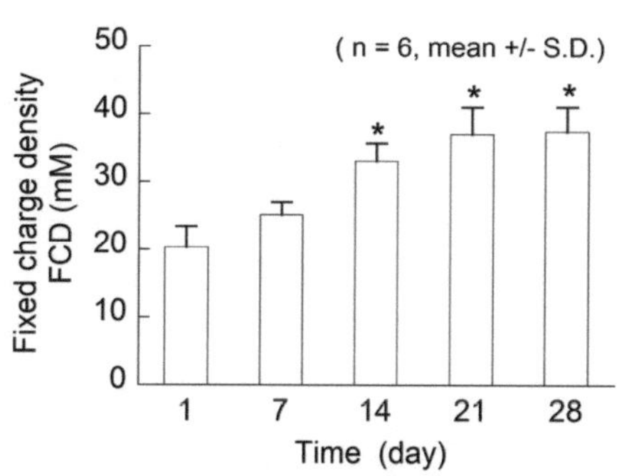

Fig. 9. Tissue fixed-charge density, with time in culture, for tissue-engineered cartilage (Miyata et al., 2010). * indicates significant difference from day 0 ($P < 0.05$).

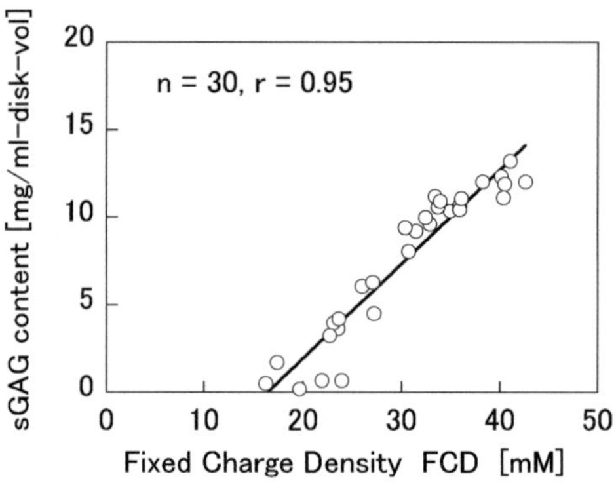

Fig. 10. Scatter plots relating the tissue fixed charge density (FCD) to the sulfated glycosaminoglycan (sGAG) content (Miyata et al., 2006).

2.4 Static and dynamic biomechanical testing of cultured agarose/chondrocyte constructs

Mechanical testing of the disk-shaped specimens was performed with unconfined compression, using impermeable stainless platens in PBS at room temperature. Static compressive properties were measured in a custom-made chamber attached to a material testing device (Autograph 5kNG, Shimadzu, Kyoto, Japan). Stress relaxation tests were performed by applying a ramp displacement at 0.05 mm/min to a 20% static compressive strain, followed by relaxation to equilibrium (2,400 s). The equilibrium compressive modulus (E_{eq}) was calculated from the imposed compressive strain and the equilibrium load, divided by the cross-sectional area of the specimen.

Dynamic compression tests were carried out using a viscoelastic spectrometer (DDV-MF, A&D, Tokyo, Japan) (Miyata et al., 2005). For preconditioning, a 20% static compressive strain was loaded and a sinusoidal displacement of 0.5% compressive strain was then superimposed at a frequency of 1 Hz. After equilibrium had been reached (approximately 20 min), a sinusoidal displacement of 0.5% compressive strain was applied at frequencies ranging from 0.01 to 5.0 Hz. The dynamic compressive modulus (E_{dyn}) was calculated from the ratio of the measured stress amplitude and the applied strain amplitude.

Figure 11 shows the means and standard deviations of the equilibrium compressive modulus E_{eq} and dynamic compressive modulus E_{dyn} versus time in culture, for the tissue-engineered cartilage. With respect to the static compressive property, significant differences were observed in the equilibrium compressive modulus. With increases in culture time, the E_{eq} of the specimens increased and reached approximately 10% of that of "native" cartilage from which the chondrocytes were harvested (0.45 ± 0.12 MPa, n =3). With respect to the dynamic compressive property, significant differences were also observed for testing conditions. The dynamic compressive modulus E_{dyn} of the engineered cartilage depended on both testing frequency and culture time. For each time point, E_{dyn} increased nonlinearly with increases in frequency. The engineered cartilage also exhibited marked stiffening with time in culture. The value of E_{dyn} increased with culture time at each testing frequency (0.01–2.0 Hz).

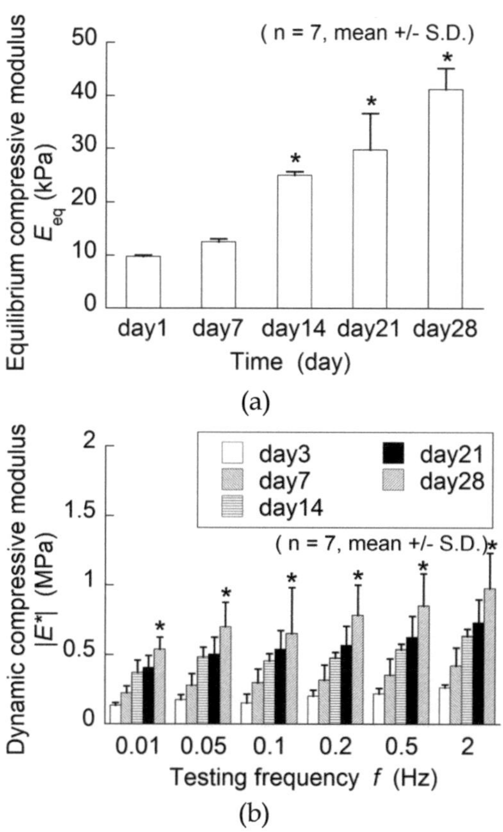

Fig. 11. Equilibrium compressive modulus E_{eq} (a) and dynamic compressive modulus E_{dyn} (b), with time in culture, for the cultured chondrocyte/agarose disks (Miyata et al., 2010). * indicates significant difference from day 0 ($P < 0.05$).

2.5 Relationships between MRI measurements and biomechanical properties of cultured chondrocyte-seeded constructs

To determine the correlations between the quantitative MRI measurements and the biomechanical and biochemical properties of the tissue-engineered cartilage, we performed linear regression analyses among the MRI-derived parameters (T1, T2, Diff, and FCD), the biochemical composition (sGAG content), and the biomechanical properties (E_{eq}, E_{dyn}) of the engineered cartilage.

To confirm the correlation, the E_{eq} of the engineered cartilage were plotted as functions of the T1, T2, and Diff, respectively. The E_{eq} of the engineered cartilage (Fig. 12a) showed a strong correlation with T1 and Diff but a weak correlation with T2 (Fig. 12b and 12c). Similarly, the tissue sGAG concentration (Fig. 13a and 13c) and were found to be strongly correlated with T1 and Diff. Consistent with the results of the previous investigation (Potter et al., 2000), our results showed that T1 relaxation time and Diff showed a significant correlation with the biomechanical properties and the sGAG content of the tissue-engineered cartilage. The results of recent studies have shown that the articular cartilage degeneration induced by collagenase treatment resulted in changes in T2 relaxation time and the equilibrium modulus (Nieminen et al., 2000). In the present study, slight increase in T2 values was observed during the culture.

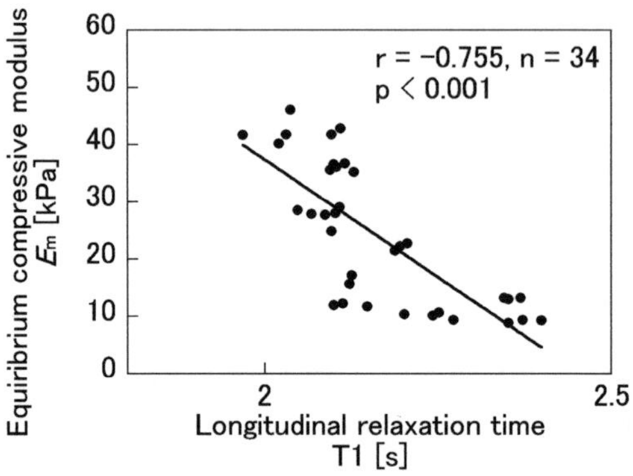

(a)

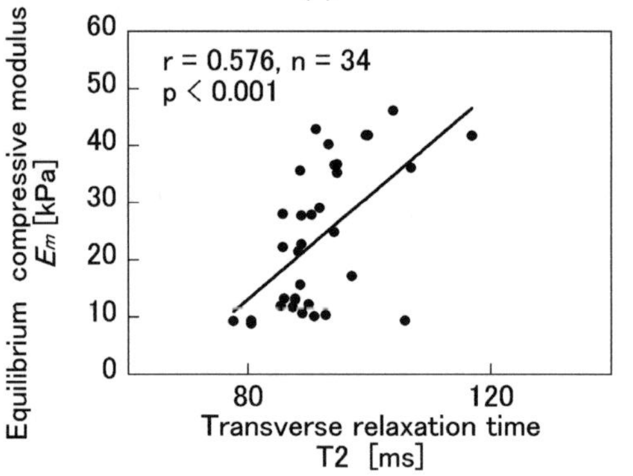

(b)

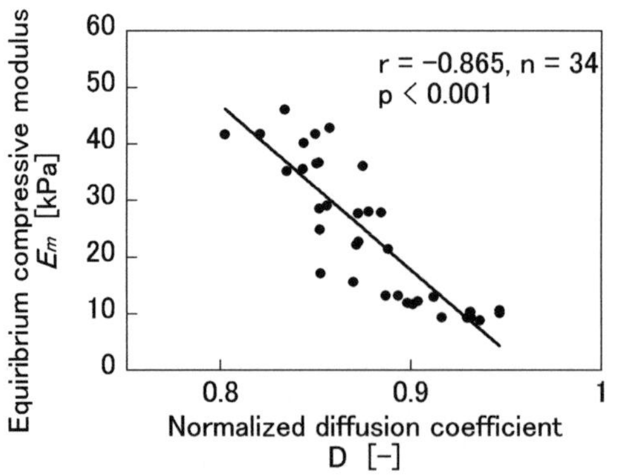

(c)

Fig. 12. Scatter plots for the relationship between the equilibrium compressive modulus E_{eq} and longitudinal relaxation time (a), transverse relaxation time (b), and relative diffusion coefficient (c) (Miyata et al., 2007). Solid line represents the linear regression line.

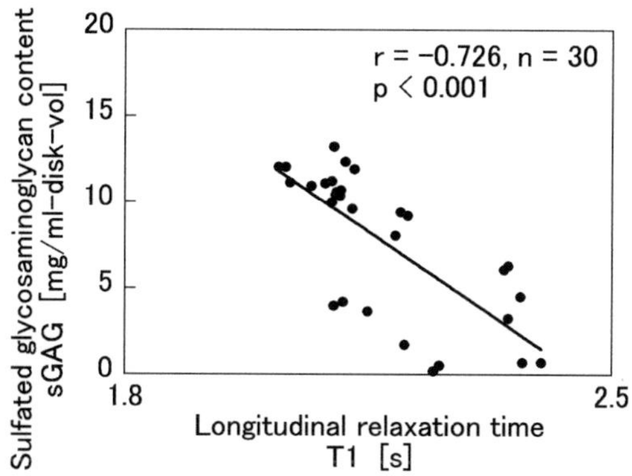

(a)

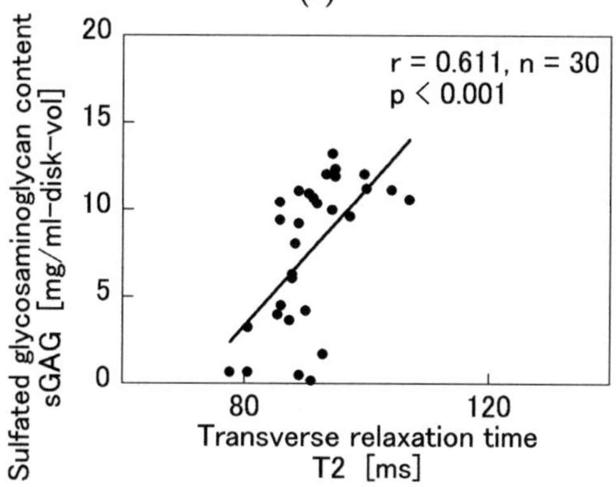

(b)

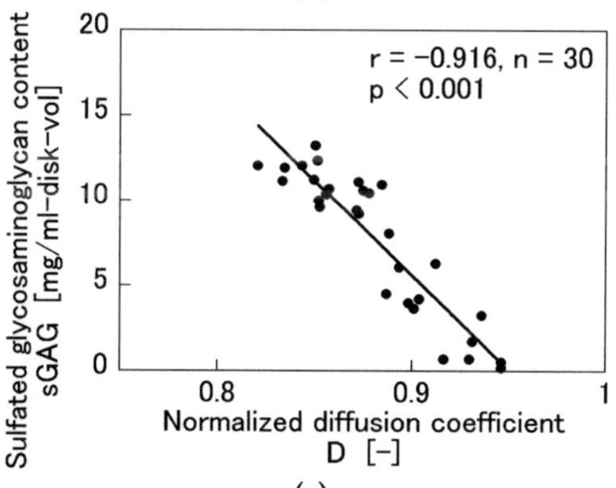

(c)

Fig. 13. Scatter plots for the relationship between the equilibrium compressive modulus E_{eq} and longitudinal relaxation time (a), transverse relaxation time (b), and relative diffusion coefficient (c) (Miyata et al., 2007). Solid line represents the linear regression line.

One possible explanation is that the changes in the biophysical properties might be mainly due to the altered sGAG content, and the synthesis of collagen and the reorganization of collagen network might be insufficient in the agarose gel culture.

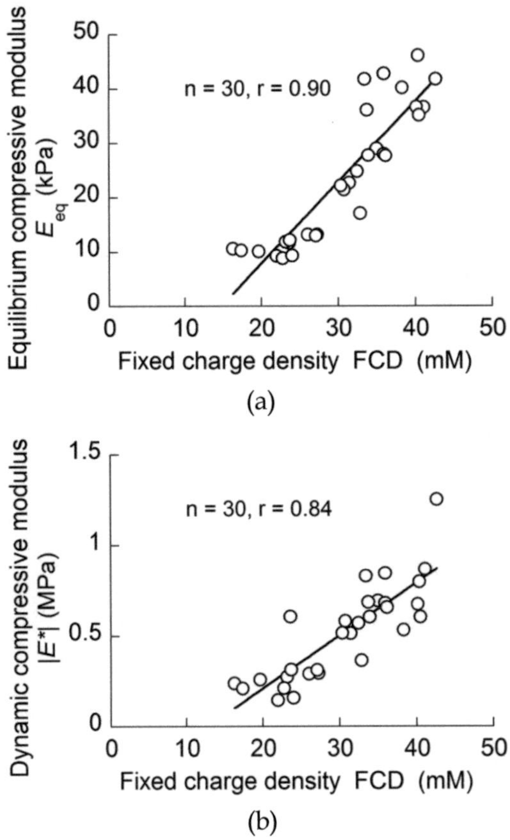

Fig. 14. Typical scatter plots relating the tissue fixed-charge density to equilibrium compressive modulus E_{eq} (a) and dynamic compressive modulus E_{dyn} at 0.5 Hz (b) (Miyata et al., 2010).

	R^2	P
FCD vs. E_{eq}	0.81	< 0.001
FCD vs. E_{dyn}, 0.01 Hz	0.79	< 0.001
FCD vs. E_{dyn}, 0.02 Hz	0.73	< 0.001
FCD vs. E_{dyn}, 0.05 Hz	0.73	< 0.001
FCD vs. E_{dyn}, 0.5 Hz	0.70	< 0.001
FCD vs. E_{dyn}, 2.0 Hz	0.71	< 0.001

Table 1. Linear Pearson correlations between biomechanical and Gd-DTPA^{2-}-enhanced MRI parameters in tissue-engineered cartilage (Miyata et al., 2010).

To evaluate the relationship between Gd-DTPA^{2-}-enhanced MRI parameters and biomechanical properties, the E_{eq} and E_{dyn} of the engineered cartilage were plotted as functions of the nFCD, respectively. From the linear Pearson correlation analysis, it was found that nFCD correlated significantly with E_{eq} and E_{dyn} (Table 1, Fig. 14). The equilibrium compressive modulus showed a higher correlation than the dynamic compressive modulus of all testing frequencies, and the dynamic compressive modulus tended to show a slightly higher correlation at low frequencies (0.01–0.05 Hz). The sGAG of articular cartilage plays a crucial role in static compressive behavior, while collagen bears a dynamic compressive load (Korhonen et al., 2003). Therefore, the nFCD—which is to say, the sGAG content—might show a higher correlation with the equilibrium modulus than with the dynamic modulus. Moreover, the dynamic modulus showed a trend toward correlation with the nFCD at lower frequencies than that of higher frequencies. That might reflect the collagen network levels regenerated in the agarose gel. Nonetheless, the results of recent studies have shown that variations in collagen architecture among varieties of articular cartilage decreased the significance of correlations between Gd-DTPA^{2-}-enhanced MRI and mechanical properties, because the architecture of the collagen network, as well as PGs, plays an important role in the mechanicalproperties of articular cartilage (Nissi et al., 2007). In the present study, the chondrocytes in agarose gel reconstructed the immature collagen network, prompting a low-level effect on the compressive property compared to "native" articular cartilage; therefore, significant correlations might be found between Gd-DTPA^{2-}-enhanced MRI and compressive properties. From these facts, our evaluation methods using Gd-DTPA^{2-}-enhanced MRI could be applicable at the earlier stage of tissue regeneration.

3. Conclusion

In conclusion, we evaluated the changes in the quantitative MRI parameters and matrix FCD of tissue-engineered cartilage that consisted of articular chondrocytes and hydrogels. We found significant linear correlations between the quantitative MRI measurements and the biomechanical and biochemical properties of the engineered cartilage. Finally, we suggest that the quantitative MRI technique can be a useful, non-invasive approach to evaluate the biomechanical properties of regenerated cartilage during *in vitro* culturing process.

4. Acknowledgment

This research was supported in part by a Grant-in-Aid for Young Scientists (B) (No. 18700414) from the Ministry of Education, Science, Sports and Culture of Japan.

5. References

Bashir, A.; Gray, M. L. & Burstein, D. (1996). Gd-DTPA^{2-} as a Measure of Cartilage Degradation. *Magnetic Resonance Medicine*, Vol.36, pp.665-673

Bashir, A.; Gray, M.L.; Hartke, J. & Burstein, D. (1999). Nondestructive Imaging of Human Cartilage Glycosaminoglycan Concentration by MRI. *Magnetic Resonance in Medicine*, Vol.41, pp.857–865

Chen, C. T.; Fishbein, K.W.; Torzilli, P. A.; Hilger, A.; Spencer, R. G. & Horton, W. E., Jr. (2003). Matrix Fixed-Charge Density as Determined by Magnetic Resonance Microscopy of Bioreactor-Derived Hyaline Cartilage Correlates with Biochemical and Biomechanical Properties. *Arthritis and Rheumatism*, Vol.48, pp.1047-1056

Farndale, R. W.; Buttle, D. J. & Barrett, A. J. (1986). Improved Quantitation and Discrimination of Sulphated Glycosaminoglycans by Use of Dimethylmethylene Blue. *Biochimica et Biophysica Acta*, Vol.883, pp.173-177

Fragonas, E.; Mlynárik, V.; Jellús, V.; Micali, F.; Piras, A.; Toffanin, R.; Rizzo, R. & Vittur, F. (1998). Correlation between Biochemical Composition and Magnetic Resonance Appearance of Articular Cartilage. *Osteoarthritis and Cartilage*, Vol.6, pp.24–32

Korhonen, R.K.; Laasanen, M.S.; Töyräs, J.; Lappalainen, R.; Helminen, H.J. & Jurvelin, J.S. (2003). Fibril Reinforced Poroelastic Model Predicts Specifically Mechanical

Behavior of Normal, Proteoglycan Depleted and Collagen Degraded Articular Cartilage. *Journal of Biomechanics*, Vol.36, pp.1373–1379

Lee, R. C.; Frank, E. H.; Grodzinsky, A. J. & Roylance, D. K. (1981). Oscillatory Compressional Behavior of Articular Cartilage and Its Associated Electromechanical Properties. *Journal of Biomechanical Engineering*, Vol.103, pp. 280-292

Miyata, S.; Tateishi, T.; Furukawa, K. & Ushida, T. (2005). Influence of Structure and Composition on Dynamic Visco-Elastic Property of Cartilaginous Tissue: Criteria for Classification between Hyaline Cartilage and Fibrocartilage Based on Mechanical Function. *JSME International Journal: C*, Vol.48, pp.547–554

Miyata, S.; Homma, H.; Numano, T.; Furukawa, K; Tateishi, T. & Ushida, T. (2006). Assessment of Fixed Charge Density in Regenerated Cartilage by Gd-DTPA - Enhanced MR Imaging. *Magnetic Resonance and Medical Science*, Vol.5, No.2, pp. 73-78

Miyata, S.; Numano, T.; Homma, H.; Tateishi, T. & Ushida, T. (2007). Feasibility of Noninvasive Evaluation of Biophysical Properties of Tissue-Engineered Cartilage by Using Quantitative MRI. *Journal of Biomechnaics*, Vol.40, pp. 2990-2998

Miyata, S.; Homma, H.; Numano, T.; Tateishi, T. & Ushida, T. (2010). Evaluation of Negative Fixed-charge Density in Tissue-Engineered Cartilage by Quantitative MRI and Relationship with Biomechanical Properties. *Jounal of Biomechanical Engineering*, Vol.132, No.7, pp.071014

Mow, V.C.; Kuei, S. C.; Lai, W.M. & Armstrong, C.G. (1980). Biphasic Creep and Stress Relaxation of Articular Cartilage in Compression? Theory and Experiments. *Journal of Biomechanical Engineering*, Vol.102, pp.73-84

Nieminen, M.T.; Rieppo, J.; Toyras, J.; Hakumaki, J.M.; Silvennoinen, J.; Hyttinen, M.M.; Helminen, H.J. & Jurvelin, J.S. (2001). T2 Relaxation Reveals Spatial Collagen Architecture in Articular Cartilage: a Comparative Quantitative MRI and Polarized Light Microscopic Study. *Magnetic Resonance in Medicine*, Vol.46, pp.487-493

Nieminen, M.T.; Toyras, J.; Rieppo, J.; Hakumaki, J.M.; Silvennoinen, J.; Helminen, H.J. & Jurvelin, J.S. (2000). Quantitative MR Microscopy of Enzymatically Degraded Articular Cartilage. *Magnetic Resonance in Medicine*, Vol.43, pp.676-681.

Nissi, M.J.; Rieppo, J.; Töyräs, J.; Laasanen, M.S.; Kiviranta, I.; Nieminen, M.T. & Jurvelin, J.S. (2007). Estimation of Mechanical Properties of Articular Cartilage with MRI—dGEMRIC, T2 and T1 Imaging in Different Species with Variable Stages of Maturation. *Osteoarthritis and Cartilage*, Vol.15, pp.24–32

Potter, K.; Butler, J.J.; Adams, C.; Fishbein, K.W.; McFarland, E.W.; Horton, W.E. & Spencer, R.G. (1998). Cartilage Formation in a Hollow Fiber Bioreactor Studied by Proton Magnetic Resonance Microscopy. *Matrix Biology*, Vol.17, pp.513-523

Potter, K.; Butler, J.J.; Horton, W.E. & Spencer, R.G. (2000). Response of Engineered Cartilage Tissue to Biochemical Agents as Studied by Proton Magnetic Resonance Microscopy. *Arthritis and Rheumatism*, Vol.43, pp.1580-1590

Ramaswamy, S.; Uluer, M.C.; Leen, S.; Bajaj, P.; Fishbein, K.W. & Spencer, R.G. (2008). Noninvasive Assessment of Glycosaminoglycan Production in Injectable Tissue-Engineered Cartilage Constructs Using Magnetic Resonance Imaging. *Tissue Engineering Part C: Methods*, Vol.14, pp.243–249

Shapiro, E.M.; Borthakur, A.; Kaufman, J.H.; Leigh, J.S. & Reddy, R. (2001). Water Distribution Patterns Inside Bovine Articular Cartilage as Visualized by 1H Magnetic Resonance Imaging. *Osteoarthritis Cartilage*, Vol.9, pp.533-538

Xia, Y.; Moody, J.B. & Alhadlaq, H. (2002). Orientational Dependence of T2 Relaxation in Articular Cartilage: A Microscopic MRI (microMRI) Study. *Magnetic Resonance in Medicine*, Vol.48, pp.460–469

22

A Mathematical Model for Wound Contraction and Angiogenesis

Fred Vermolen and Olmer van Rijn
Delft Institute of Applied Mathematics, Delft University of Technology
The Netherlands

1. Introduction

Cutaneous wounds, ulcers or burns, as a result of external damage, such as intensive solar exposure or accidents, occur in high numbers and may cause ever lasting traumas. In some cases, the wounds are very painful and impair an individual's daily life in terms of health, and social interaction with his or her surroundings, but also in terms of mobility and ability to work or to perform leisure activities. Furthermore, the aesthetic feeling of patients is highly improved by the treatment of burns and other wounds since otherwise the wounds can make a detrimental impact on the appearance of the patient's skin. Besides aesthetic appearance, also the functioning of the damaged skin is different since infections are possibly more likely to occur near the (scar tissue of the) burn. Furthermore, the conception of temperature and pain by a patient is also altered by the burn, ulcer or cutaneous wound. Therefore, the treatment of these defects, which occurs in very high numbers, is important and hence it is crucial to improve treatments and to make treatments more efficient.

Nowadays many different treatments, such as, in more serious cases, skin transplantation, injection of cell cultures and the implant of artificial skin, are commonly carried out by plastical surgeons in order to relieve the patient and to minimize the aesthetic effects of scar tissue as much as possible. In order to be able to improve medical treatment of scars, it is important to understand the biological mechanism behind scar formation in terms of plastic deformation and a regenerated excess of collageneous matrix. More information about the treatment of scars and the biology behind the scar formation can, among others, be found in Bloemen (2011); van der Veer (2009). The deformations are caused by the contractile mechanism due to the pulling behavior of the (myo-)fibroblasts. Dallon (2008) experimentally found that the nature of contraction is predominantly determined by the following two mechanisms:

- The degree and orientation of motility of fibroblasts. This motility mechanism of the fibroblasts is caused by the attachment and movement of the lamellipodia of the fibroblasts onto the collageneous fibres;
- The attachment of the philapods of the less motile myofibroblasts onto the extra cellular matrix.

These mechanisms were simulated by Dallon (2010) by the use of a spring model. Since Dallon's two-dimensional model tracks the forces exerted by each individual philapod or lamellipodium of the individual (myo)fibroblast, the formalism becomes very expensive from a computational point of view when applied to burns of realistic dimensions with millions of fibroblasts and collageneous fibres. However, his observations and calculations are very useful for a thorough understanding of the implications of the fundamental processes and for the calibration of continuum models that are based on partial differential equations.

The severity of a wound is determined by the thickness of the damaged skin layer of the patient. A minor injury concerns the damage on the corneum, which is replaced with the patient's tissues relatively quickly. This type of injury is referred to as a first degree skin burn or wound. A more severe damage concerns the impairment of the epidermis. In this case reepithialialization has to repair the epidermis by migration and proliferation of keratinocytes. This damage is classified as a second degree burn or wound, but this type of wound will heal without many problems for healthy patients. Contraction does not take place and hence the burn or wound will recover entirely without any plastic deformations or significant scars. Third degree burns or wounds pose a more serious damage in the sense that also (part of) the dermis, and even the subcutis could be disrupted. Here the fibroblast rich dermis, consisting of the collageneous fibrous tissue has to be repaired. This takes place by a sequence of signaling processes to initiate proliferation and movement of fibroblasts and the restoration of the vascular network. As the fibroblasts produce an excess of collageneous fibres, on which they, and the myofibroblasts to a larger extent, exert contractile forces, wound contraction takes place. The process of wound contraction is a useful mechanism for a rapid minimization of exposure of the underlying tissues to hazardous external environments, when the wound is caused by mechanical damage. In particular, in skins of rabbits, percentages of up to 80 percent of the initial wound area have been reported in various experimental studies. In humans, experimental studies evidence that wound contraction is much less significant, being in the order of 5–10 percent. However, in the case of healing of burns, this contraction phenomenon is undesirable, as it gives rise to significant deformations, which can be plastic and furthermore, the patient is left with an excess of collageneous fibres and unpleasantly looking scars. Remodeling is a very slow process which cannot remove all wrinkles as a result of the plastic deformations due to wound contraction.

In order to be able to improve surgical treatments, in terms of effectiveness and minimization of invasion into the patient, it is crucial to know which behavior of the (myo-)fibroblasts cause the plastic deformation of the skin. In the case of a third degree burn or wound, fibroblasts enter the wound area during the proliferative stages of the healing process of the dermis. Subsequently, the fibroblasts start to proliferate and to produce extracellular matrix, referred to as fibrous tissue. Furthermore, as a result of exposure to high strains, fibroblasts differentiate into myofibroblasts, which are considered as weak muscle cells. Myofibroblasts increase the measure of contraction of the wound area of the dermis. This differentiation process also takes place under the influence of a growth factor TF-β, which is produced by the fibroblasts and myofibroblasts depending on the amount of extracellular matrix present. The growth factor also stimulates the production of fibroblasts and its regeneration of collagen. Next to the enhancement of the production of several biological entities, the growth factor increases the chemotactic transport of the fibroblasts towards the wound area. To get more quantitative insight into the influences of the growth factors and other agents on the amount of wound contraction caused by the pulling forces of the (myo)fibroblasts, detailed mathematical models with parameter sensitivity analysis are indispensable. This sensitivity analysis can give insight into the quantification of the influence of all biological parameters involved

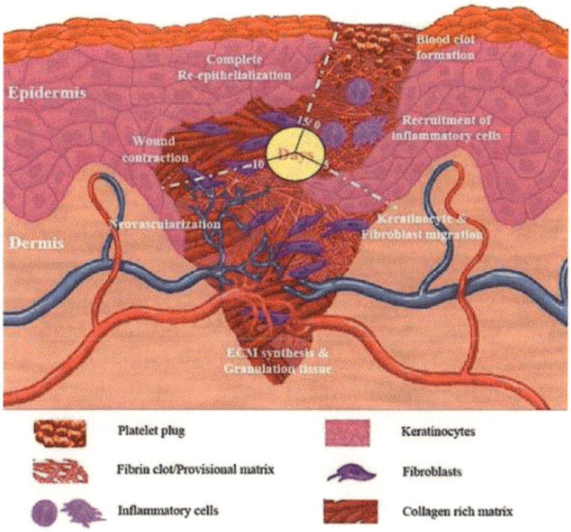

Fig. 1. A schematic of the events during wound healing. The dermis and epidermis are illustrated. The picture was taken with permission from http://www.bioscience.org/2006/v11/af/1843/figures.htm

in this process on contraction and healing, as well as values such that healing of burns is optimized with respect to a minimal level of wound contraction and maximum healing speed. These results are helpful to surgeons to design innovative minimal invasive treatments that are applicable to patients. Furthermore, the mathematical analysis is helpful to make the treatment patient specific.

In this manuscript, we address two partial processes during the proliferative phase: wound contraction and angiogenesis. When the tissue is provided with enough oxygen and nutrients the process of wound closure starts. Cells in the epidermis, which mainly consist of keratinocytes, start regenerating the upperlayer of the wound. Usually the skin can not be replaced fully and some scars are left where the wound was located, due to excess of regenerated collagen. The second and third stage of wound healing do not take place at the same location in the wound. The former is located in the dermis, the latter is limited to the epidermis. The epidermis and the dermis consist of different type of cells and are separated by a so-called basal membrane, see also Figure 1.

In this manuscript, we first present a selection of the currently available mathematical models that seek to describe the biological processes of wound healing as well as possible. The healing process is very complex and many factors contribute to it, therefore simplifications have to be made. The main topics in this thesis are combining two models for angiogenesis and coupling models for the different stages of wound healing. Currently models exist for angiogenesis and dermal regeneration (fibroblasts and collagen) separately, however, only scarcely have there been attempts to couple the models. This is vital as the various stages of wound healing overlap and hence influence each other. The models that we consider in this study are all formulated in terms of partial differential equations (PDEs) that are based on conservation principles. Such coupled models could give more insights into how the process of wound healing evolves. These insights might lead to treatments that reduce healing time, e.g. the use of certain hormones to speed up the healing process. Also scars and other deformations due to incomplete healing might be prevented or reduced. This issue is especially crucial for the treatment of burns, where hypotrophic scars may result after injury.

A lot of simulations have been done in literature to give more insight on how these models behave. Also the dependence of the models on certain parameters is investigated in the present paper. Finally recommendations are made for future research in the topic of mathematically describing wound healing. Many studies are based on PDEs, for instance the work by Murray (2004); Olsen (1995); Maggelakis (2003); Gaffney (2002); Sherratt (1991); Adam (1999); Javierre (2009); Vermolen (2009; 2010; 2011); Schugart (2008); Xue (2009). Many other modeling studies are based on cell-based or Monte-Carlo methods, such as the cellular Potts models used by Glazier (1992); Merks (2009); Plank (2004) to mention a few. This class of model is lattice-based and minimizes a virtual energy functional. Recently, a semi-stochastic continuous cell-based model was formulated by Vermolen-Gefen (2011).

The present manuscript falls within the class of papers, in which one attempts to combine existing mathematical models for various subprocesses occuring during wound healing. Some recent work by this team are Vermolen (2010), and Vermolen (2011), in which the most simplified models for wound contraction, closure and angiogenesis are coupled. The present model is based on a combination of some more advanced models for wound contraction and angiogenesis and gives a more sophisticated formalism for dermal regeneration. The paper is organized as follows. In Section 2, we present several models for processes of angiogenesis and wound contraction. Here, we also present some results of simulations. This is an innovation with respect to other studies. In Section 3, we consider the coupling of the models, including simulations. In Section 4, we describe the numerical methods. Finally, some conclusions are drawn in Section 5.

The character of the current manuscript is rather informal and descriptive about the mathematical concepts used in the present study. The level of abstract mathematics has been reduced tremendously. Further, we note that the simulations shown in this manuscript are still in a preliminary state.

2. Current models

The mathematical model for the proliferative stage of wound healing is usually separated in three distinct parts representing three stages of wound healing. These three stages are wound contraction, angiogenesis and wound closure. Note that the inflammation stage mentioned in Section 1 is not taken into account. This is due to the fact that inflammation only contains the damage and only after the inflammation stage is finished the real healing process starts.

In this chapter we present some of the currently available models on the above mentioned wound healing stages. In Section 2.1, we first present the model on wound contraction. Next in Section 2.2, a model for angiogenesis is presented. This model is a combination of two accepted models that describe a specific feature in the angiogenesis process. The two models take a very different approach on how to model the growth of new blood vessels in the wound. In Section 2.3, a study that attempts to combine models of dermal regeneration and angiogenesis is briefly discussed. In all the models that we deal with, we consider a bounded simply connected domain $\Omega \subset \mathbb{R}^2$. The boundary is denoted by $\partial\Omega$.

2.1 The wound contraction models

During the wound contraction stage fibroblasts (connective tissue cells) invade the wound site and contract the extracellular matrix (ECM). The contraction decreases the area of contact between the wound and its surroundings, thereby reducing the chance of contamination and infection. Furthermore this process is vital in assuring that new blood vessels can be formed in the wound during angiogenesis, since the fibroblasts invading the wound form the tissue in which the new capillaries can grow. The wound contraction stage is limited to the dermis, however, the contraction of the ECM also effects the tissue in the epidermis.

We use the model for contraction due to Javierre (2009), which deals with the presence of myofibroblasts. Myofibroblasts are a kind of weak muscle cells. They are nonmotile cells that differentiate from fibroblasts and transmit and amplify the traction forces generated by the fibroblasts, Vermolen (2009). Secondly, the model incorporates the effects of a growth factor that triggers wound contraction.

The equation concerning the fibroblast concentration u_{fib} becomes

$$\frac{\partial u_{\text{fib}}}{\partial t} + \nabla \cdot \left(\frac{\partial \mathbf{u}}{\partial t} u_{\text{fib}} - D_{\text{fib}} \nabla u_{\text{fib}} + \frac{a_{\text{fib}}}{(b_{\text{fib}} + c_{\text{ecm}})^2} u_{\text{fib}} \nabla c_{\text{ecm}} \right) =$$

$$\left(\lambda_{\text{fib}} + \frac{\lambda^0_{\text{fib}} c_{\text{ecm}}}{C_{1/2} + c_{\text{ecm}}} \right) u_{\text{fib}} \left(1 - \frac{u_{\text{fib}}}{K} \right) - \frac{k_1 c_{\text{ecm}}}{C_k + c_{\text{ecm}}} u_{\text{fib}} + k_2 u_{\text{myo}} - d_{\text{fib}} u_{\text{fib}}, \tag{1}$$

where c_{ecm} and u_{myo} respectively denote the growth factor and myofibroblast concentration. The first term in the left-hand side corresponds to the total accumulation, the second term follows from passive convection due to deformation of the tissue. The third and fourth terms of the left-hand side account for random walk and chemotaxis (movement towards the gradient of the growth factor). In the right-hand side of the above equation, we respectively have the logistic proliferation term up to an equilibrium, in which proliferation is enhanced by the presence of the growth factor, the differentiation term to myofibroblasts under presence of the growth factor, back differiation from myofibroblasts and finally a term dealing with cell death. The cell death rate is denoted by d_{fib}, the myofibroblast to fibroblast differentiation rate by k_2 and the fibroblast to myofibroblast differentiation rate by k_1. Furthermore λ^0_{fib}, $C_{1/2}$ and C_k are known constants that monitor the growth factor's influence on the contraction process and K is a parameter that regulates the equilibrium concentration.

The production term, first on the right hand side, now also incorporates growth factor stimulated proliferation. The other three terms on the right hand side respectively account for differentiation to and from myofibroblasts and cell death.

The PDE for the myofibroblast concentration u_{myo} is similar to equation (1). However, since myofibroblast are nonmotile cells, they will only move due to passive convection. The myofibroblast concentration thus obeys

$$\frac{\partial u_{\text{myo}}}{\partial t} + \nabla \cdot \left(\frac{\partial \mathbf{u}}{\partial t} u_{\text{myo}} \right) = \varepsilon_{\text{myo}} \left(\lambda_{\text{fib}} + \frac{\lambda^0_{\text{fib}} c_{\text{ecm}}}{C_{1/2} + c_{\text{ecm}}} \right) u_{\text{myo}} \left(1 - \frac{u_{\text{myo}}}{K} \right)$$

$$+ \frac{k_1 c_{\text{ecm}}}{C_k + c_{\text{ecm}}} u_{\text{fib}} - k_2 u_{\text{myo}} - d_{\text{myo}} u_{\text{myo}}, \tag{2}$$

where ε_{ecm} is a proportionality constant, further d_{myo} and u^0_{myo} denote the myofibroblasts death rate and the myofibroblast equilibrium concentration respectively. The terms on the left-hand side account for accumulation and passive convection due to deformation of the tissue. The right-hand side contains proliferation of myofibroblasts under presence of the growth factor, differentiation of fibroblasts to myofibroblasts, back differentiation to fibroblasts, and (programmed) cell death (apoptosis).

Both fibroblasts and myofibroblasts contribute to the production of the ECM, furthermore the production is chemically enhanced by the growth factor, Vermolen (2009). The PDE for the ECM density ρ then is given by

$$\frac{\partial \rho}{\partial t} + \nabla \cdot \left(\frac{\partial \mathbf{u}}{\partial t} \rho \right) = \left(\lambda_\rho + \frac{\lambda^0_\rho c_{\text{ecm}}}{C_\rho + c_{\text{ecm}}} \right) \frac{u_{\text{fib}} + \eta_b u_{\text{myo}}}{R^2_\rho + \rho^2} - d_\rho (u_{\text{fib}} + \eta_d u_{\text{myo}}) \rho, \tag{3}$$

where λ_ρ and d_ρ are the ECM production and death rate respectively. The terms on the left-hand side describe accumulation and passive convection, and the right-hand side describes growth factor enhanced production of collagen, as well as decay of collagen. Furthermore λ^0_ρ and C_ρ are known constants that monitor the growth factor's influence on the contraction process. Further, R_ρ is a parameter that quantifies how the ECM production rate depends on the ECM density itself and η_b and η_d are proportionality constants.

The dynamics of the growth factor concentration c_{ecm} are mainly determined by the fibroblasts and myofibroblasts as they produce the growth factor. Also the growth factor is motile, so it is subject to active convection. This leads to the following PDE for the growth factor concentration

$$\frac{\partial c_{\text{ecm}}}{\partial t} + \nabla \cdot \left(\frac{\partial \mathbf{u}}{\partial t} c_{\text{ecm}} - D_c \nabla c_{\text{ecm}} \right) = \frac{k_c (u_{\text{fib}} + \zeta u_{\text{myo}}) c_{\text{ecm}}}{\Gamma + c_{\text{ecm}}} - d_c c_{\text{ecm}}. \tag{4}$$

Here the second and third term on the left hand side account for passive convection and ordinary diffusion. The terms on the right hand side account for growth factor production and growth factor decay respectively. Furthermore D_c is the growth factor diffusion coefficient, k_c denotes the growth factor production rate and d_c the natural decay rate. Also Γ is a parameter that quantifies how the growth factor production rate depends on the growth factor concentration itself and ζ is a proportionality constant.

The initial and boundary conditions for the fibroblast concentration and the ECM density remain the same as in the model due to Tranquillo. For the myofibroblast and growth factor concentration the following initial conditions are imposed

$$u_{\text{myo}}(\mathbf{x}, 0) = 0, c_{\text{ecm}}(\mathbf{x}, 0) = c^0_{\text{ecm}}$$

for $\mathbf{x} \in \Omega_w$ and

$$u_{\text{myo}}(\mathbf{x}, 0) = 0, c_{\text{ecm}}(\mathbf{x}, 0) = 0$$

for $\mathbf{x} \in \Omega_u$. Here c^0_{ecm} denotes the growth factor equilibrium concentration. Furthermore the growth factor concentration satisfy a no flux boundary condition on all boundaries. For the myofibroblast concentration we can apply the same reasoning as for the ECM density ρ and hence no boundary conditions have to be imposed.

The mechanical part of the wound contraction model is based on the linear viscoelastic equations, i.e.

$$-\nabla \cdot \sigma = \mathbf{f}_{\text{ext}}. \tag{5}$$

Here $\sigma = \sigma_{\text{ecm}} + \sigma_{\text{cell}}$, the stresstensor, accounts for the ECM related stress, σ_{ecm}, and the cell stress, σ_{cell}. Furthermore $\mathbf{f}_{\text{ext}}$ represents the external forces acting on the tissue.

In all the below discussed models the ECM related stress tensor σ_{ecm} is given as

$$\sigma_{\text{ecm}} = \mu_1 \frac{\partial \epsilon}{\partial t} + \mu_2 \frac{\partial \theta}{\partial t}\mathbf{I} + \frac{E}{1+\nu}\left(\epsilon + \frac{\nu}{1-2\nu}\theta\mathbf{I}\right). \tag{6}$$

Here the first two terms on the right hand side respresent the viscous effects and the last term the elastic effects. If we let $\mathbf{u} = \mathbf{u}(\mathbf{x}, t)$ denote the displacement of the ECM, then the strain tensor ϵ and the dilation θ in equation (6) are respectively given by

$$\epsilon = \frac{1}{2}\left(\nabla\mathbf{u} + (\nabla\mathbf{u})^T\right) \tag{7}$$

and

$$\theta = \nabla \cdot \mathbf{u}. \tag{8}$$

Furthermore, in relation (6), $\mathbf{I}$ denotes the identity tensor and μ_1, μ_2, E and ν respectively represent the dynamic and kinematic viscosity, Young's modulus and Poisson's ratio.

The external forces acting on the tissue, $\mathbf{f}_{\text{ext}}$, are modelled similarly in all three models, i.e.

$$\mathbf{f}_{\text{ext}} = -s\rho\mathbf{u}. \tag{9}$$

Here $\rho = \rho(\mathbf{x}, t)$ denotes the ECM density and s is the tethering elasticity coefficient.

At time $t = 0$ it is assumed that there is no displacement of the ECM, i.e. $\mathbf{u}(\mathbf{x}, 0) = \mathbf{0}$. Also we assume that $\mathbf{u}$ vanishes at the boundary far away from the wound, i.e. $\mathbf{u}(\mathbf{x}, t) = \mathbf{0}$ for $\mathbf{x} \in \partial\Omega$. This can be justified by taking the computational domain Ω sufficiently large, so that the boundary effects can be ignored.

Since the myofibroblasts transmit and amplify the traction forces generated by the fibroblasts this is also visible in the cell traction term σ_{cell}. For the model due to Olsen et al. this term is given by

$$\sigma_{\text{cell}} = \frac{\tau u_{\text{fib}}\left(1 + \xi u_{\text{myo}}\right)\rho}{R_\tau^2 + \rho^2}\mathbf{I}, \tag{10}$$

where ξ is a constant of proportionality and R_τ quantifies how the cell traction depends on the ECM density.

In Javierre (2009) an extension of the model due to Olsen et al. is presented. Javierre et al. propose the mechanical stress to act as a factor that effects the differentiation from fibroblasts to myofibroblasts. They introduce an estimation of the mechanical stimulus that depends on the dilation $\theta = \nabla \cdot \mathbf{u}$ as

$$p_{\text{cell}}(\theta) = \frac{K_{\text{act}} p_{\max}}{K_{\text{act}}\theta_1 - p_{\max}}\left(\theta_1 - \theta\right)\chi_{[\theta_1, \theta^*]}(\theta) + \frac{K_{\text{act}} p_{\max}}{K_{\text{act}}\theta_2 - p_{\max}}\left(\theta_2 - \theta\right)\chi_{[\theta_2, \theta^*]}(\theta) + K_{\text{pas}}\theta. \tag{11}$$

Here χ denotes the indicator function, i.e.

$$\chi_I(\theta) = \begin{cases} 1, \text{ if } \theta \in I, \\ 0, \text{ else.} \end{cases}$$

and the first two terms on the right hand side account for the contractile stress generated internally by the myosin machinery and transmitted through the actin bundles, Javierre (2009). The third term on the right hand side establishes the contractile stress supported by the passive resistance of the cell. See also Figure 2 for a plot of $p_{\text{cell}}(\theta)$.

Furthermore, in equation (11), the compression and traction strain limits are respectively denoted by θ_1 and θ_2, p_{max} respresents the maximal contractile force exerted by the actomyosin machinery and K_{max} and K_{pas} the volumetric stiffness moduli of the active and passive components of the cell. Also the parameter θ^* can be computed from K_{act} and p_{max} as $\theta^* = \frac{p_{\text{max}}}{K_{\text{act}}}$, Javierre (2009).

To incorporate the effects of the mechanical stimulus on the fibroblast to myofibroblast differentiation an extra factor is found before the differentiation term (the second term on the right hand side) in equation (1). The fibroblast to myofibroblast differentiation term changes to

$$\frac{p_{\text{cell}}(\theta)}{\tau_d + p_{\text{cell}}(\theta)} \frac{k_1 c_{\text{ecm}}}{C_k + c_{\text{ecm}}} u_{\text{fib}}, \qquad (12)$$

where τ_d is a parameter that quantifies how the differentiation rate depends on the mechanical stimulus. Note that this term is also present in the PDE for the myofibroblast concentration and that thus the second term on the right hand side of equation (2) also changes to the above.

The mechanical stimulus also effects the the cell stresses and thus the cell traction term $\boldsymbol{\sigma}_{\text{ccll}}$. In Javierre (2009) it is assumed that $\boldsymbol{\sigma}_{\text{cell}}$ depends linearly on $p_{\text{cell}}(\theta)$ and thus that

$$\sigma_{\text{cell}} = p_{\text{cell}}(\theta) \frac{u_{\text{fib}}\left(1 + \xi u_{\text{myo}}\right)\rho}{R_\tau^2 + \rho^2} \mathbf{I}. \qquad (13)$$

In order to give more insight in the process of wound contraction we did some simulations with the model due to Javierre (2009). We will describe the numerical techniques in Section 4. Further, the input data can be found in the appendix. As a computational domain we use the unit square, i.e. $\Omega = \{\mathbf{x} = (x, y) |\, 0 \le x, y \le 1\}$, and the initial wound is given by $\Omega_w = \left\{\mathbf{x} |\, |\mathbf{x}| \le \frac{1}{2}\right\}$. The results are given in Figures 3 to 5, where we show the solution two days after injury. The computations have been done using parameter values taken from Javierre (2009), see also the appendix.

In Figure 3 we show the fibroblast and myofibroblast concentration two days after injury. The myofibroblast concentration is highly concentrated around the wound edge, whereas the fibroblast have invaded the wound. Figure 4 shows the ECM density and the growth factor concentration two days after injury. We see that the ECM density is slightly elevated at the wound edge, which is to be expected since the wound is healing there. Furthermore the growth factor concentration has spread throughout the computational domain, but is still concentrated inside the wound. In Figure 5, the displacement pattern due to contractile forces exerted by fibroblasts and myofibroblasts is plotted. If we compare Figure 5 with Figure 3 we see that the ECM displacement is largest at places of high myofibroblast concentration. Here

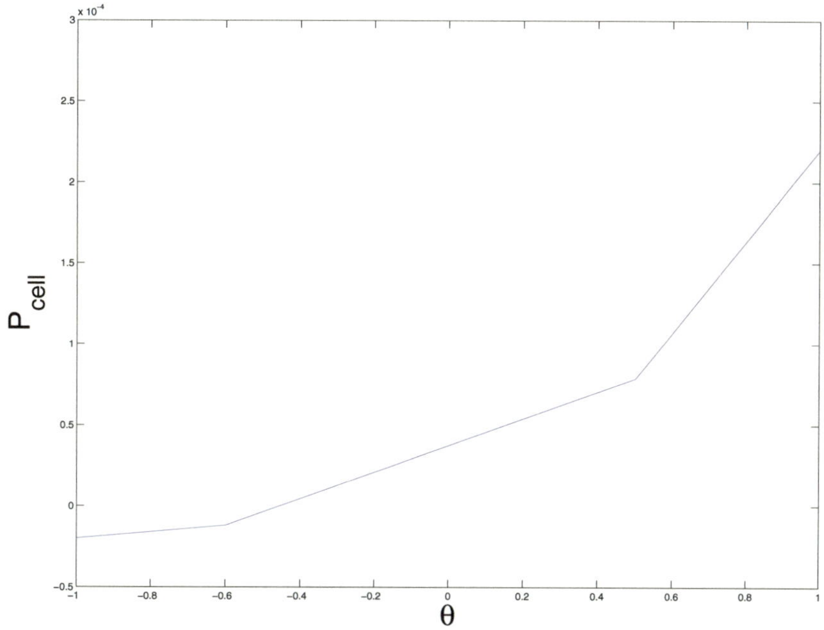

Fig. 2. The mechanical stimulus p_{cell} as a function of the dilation θ, computed with values from Javierre (2009).

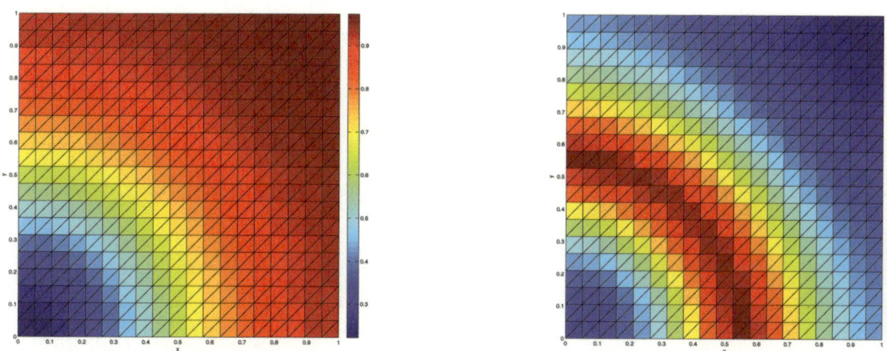

Fig. 3. Normalized fibroblast (left) and myofibroblast (right) densities two days after injury. All input data have been given in the appendix.

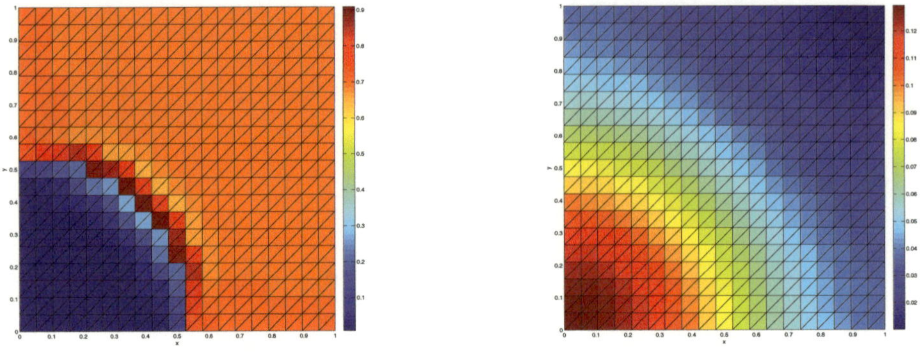

Fig. 4. Normalized ECM density (left) and growth factor concentration (right) two days after injury.

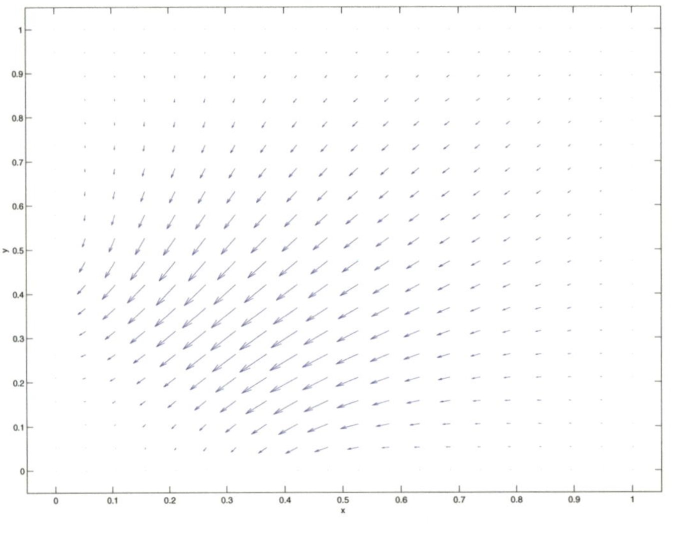

Fig. 5. Displacement of the ECM two days after injury.

the effect of the myofibroblasts, as they transmit and amplify the traction forces generated by the fibroblasts, can clearly be seen.

The normalized solutions in time at $\mathbf{x} = \mathbf{0}$, furthest in the wound, are shown in Figure 6. This gives an indication of the development of the solutions in time. We see that the fibroblast concentration increases towards it equilibrium, whereas the myofibroblast concentration first rises and then falls again. This is due to the fact that, as the wound heals, the myofibroblasts either differentiate back to fibroblasts again or undergo apoptosis and hence eventually will disappear completely. Also the growth factor concentration eventually goes to zero as the wound is healed. The ECM density slowly rises in time and will eventually reach its equilibrium, allthough it can take some time before the ECM is completely restored. We finally

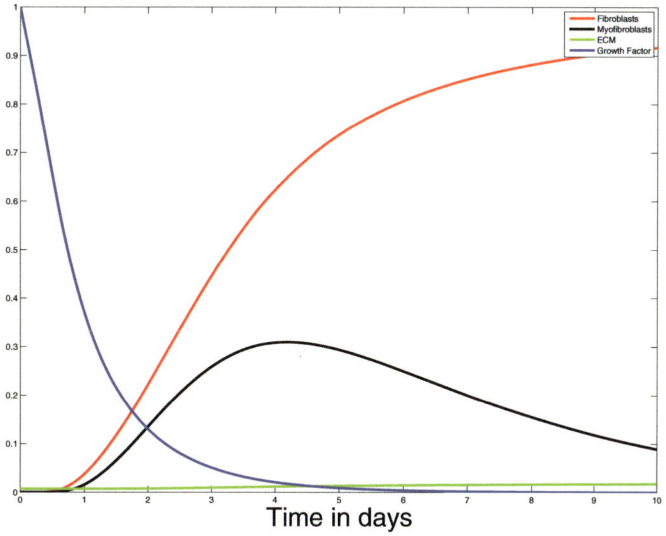

Fig. 6. Normalized solutions furthest in the wound, at $\mathbf{x} = \mathbf{0}$, in time.

remark that all the simulations in this subsection were obtained without any coupling with the model for angiogenesis.

2.2 Angiogenesis

In this section, we present the model for angiogenesis. These two models each take a very different approach on how to model this process. Where the model due to Maggelakis (2003) focusses on the relation between a lack of oxygen and capillary growth, the model due to Gaffney (2002), attempts to model the migration of endothelial cells into the wound. Both models adress an important aspect of angiogenesis, but on the other hand both also miss an important aspect.

In this section we present a novel angiogenesis model based on the models of Maggelakis and Gaffney et al. In this model we attempt to unite the two approaches and thus create a model that is suitable for dealing with both aspects of angiogenesis. This gives a new model for angiogenesis. By combining these two aspects in one model we hope to create a more accurate model of the process of angiogenesis. As a basis we take the model of Gaffney et al. and extend it with a negative feedback mechanism for oxygen. This assures that both the endothelial cell migration and the oxygen shortage aspects are covered in this novel angiogenesis model.

The model due to Gaffney et al. is given by

$$\begin{aligned} &\frac{\partial u_{\text{tip}}}{\partial t} - \nabla \cdot (D_1 \nabla u_{\text{tip}} + D_2 \nabla u_{\text{end}}) = f(u_{\text{tip}}, u_{\text{end}}), \\ &\frac{\partial u_{\text{end}}}{\partial t} - \nabla \cdot \lambda_5 (D_1 \nabla u_{\text{tip}} + D_2 \nabla u_{\text{end}}) = g(u_{\text{tip}}, u_{\text{end}}) \end{aligned} \tag{14}$$

where the left-hand sides account for accumulation and transport of tips and endothelial cells by means of a biased random walk process. Further, the functions f and g are given by

$$f(u_{\mathrm{tip}}, u_{\mathrm{end}}) = \lambda_2 u_{\mathrm{tip}} - \lambda_3 u_{\mathrm{tip}}^2 - \lambda_4 u_{\mathrm{tip}} u_{\mathrm{end}},$$

$$g(u_{\mathrm{tip}}, u_{\mathrm{end}}) = \lambda_6 a u_{\mathrm{end}}(u_{\mathrm{end}}^0 - u_{\mathrm{end}}) + \lambda_6 \chi u_{\mathrm{tip}} u_{\mathrm{end}}(u_{\mathrm{end}}^1 - u_{\mathrm{end}})$$
$$+ \lambda_5(\lambda_3 u_{\mathrm{tip}}^2 + \lambda_4 u_{\mathrm{tip}} u_{\mathrm{end}}).$$

These functions were taken from Gaffney (2002). The first term of the function f models tip-branching, as each tip has a likelihood to bifurcate. The second term of the function f accounts for tip-tip anastomosis, which is the process that two tips have a probability to join and hence form a loop, by which a tip is no longer a tip, and hence the number of tips decreases then. The third term of the function f stands for tip-sprout anastomosis, which is the process that a tip may converge into a branch, which closes the network as well and thereby also decreasing the number of tips. For the function g, we distinguish the following terms: The first one contains ordinary logistic proliferation under standard conditions, the second term accounts for an increased logistic growth due to the presence of tips. Finally, the third term accounts for the extent of anastomosis. To incorporate the effects of the macrophage derived growth factor (MDGF) concentration c_{md} on the growth of capillary tips and the proliferation of endothelial cells we assume that λ_2 and λ_6 are functions of the macrophage derived growth factor c_{md}, which is released as a result of a shortage of oxygen. These functions must be zero if there is no MDGF present and rise as the MDGF concentration rises. Furthermore the effects of additional MDGF must be lower if there is already a lot of MDGF present. Therefore we let

$$\lambda_2(c_{\mathrm{md}}) = \lambda_2^0 \frac{c_{\mathrm{md}}}{c_{\mathrm{md}} + \tau_{\mathrm{tip}}},$$

$$\lambda_6(c_{\mathrm{md}}) = \lambda_6^0 \frac{c_{\mathrm{md}}}{c_{\mathrm{md}} + \tau_{\mathrm{end}}},$$

where τ_{tip} and τ_{end} qualify how respectively λ_2 and λ_6, accounting for tip regeneration and logistic proliferation of endothelial cells respectively, depend on the MDGF concentration. The functions $\lambda_2(c_{\mathrm{md}})$ and $\lambda_6(c_{\mathrm{md}})$ are given in Figure 7 for $0 \leq c_{\mathrm{md}} \leq 1$.

This settles the trigger mechanism that the MDGF concentration fullfills in the growth of new capillaries. To get a good overview of what the model looks like we give the PDEs that drive the model, i.e.

$$\frac{\partial u_{\mathrm{oxy}}}{\partial t} = D_{\mathrm{oxy}} \Delta u_{\mathrm{oxy}} - \lambda_{\mathrm{oxy}} u_{\mathrm{oxy}} + \lambda_{13}\left(u_{\mathrm{tip}} + \frac{\lambda_{\mathrm{oxy}} u_\theta u_{\mathrm{end}}}{\lambda_{13} u_{\mathrm{end}}^0}\right), \tag{15}$$

$$\frac{\partial c_{\mathrm{md}}}{\partial t} = D_{\mathrm{md}} \Delta c_{\mathrm{md}} - \lambda_{\mathrm{md}} c_{\mathrm{md}} + \lambda_{21} Q(u_{\mathrm{oxy}}), \tag{16}$$

$$\frac{\partial u_{\mathrm{tip}}}{\partial t} = \nabla \cdot \{D_1 \nabla u_{\mathrm{tip}} + D_2 u_{\mathrm{tip}} \nabla u_{\mathrm{end}}\} + f(u_{\mathrm{tip}}, u_{\mathrm{end}}), \tag{17}$$

$$\frac{\partial u_{\mathrm{end}}}{\partial t} = \lambda_1 \nabla \cdot \{D_1 \nabla u_{\mathrm{tip}} + D_2 u_{\mathrm{tip}} \nabla u_{\mathrm{end}}\} + g(u_{\mathrm{tip}}, u_{\mathrm{end}}), \tag{18}$$

where we explain the terms occuring in the above PDEs.

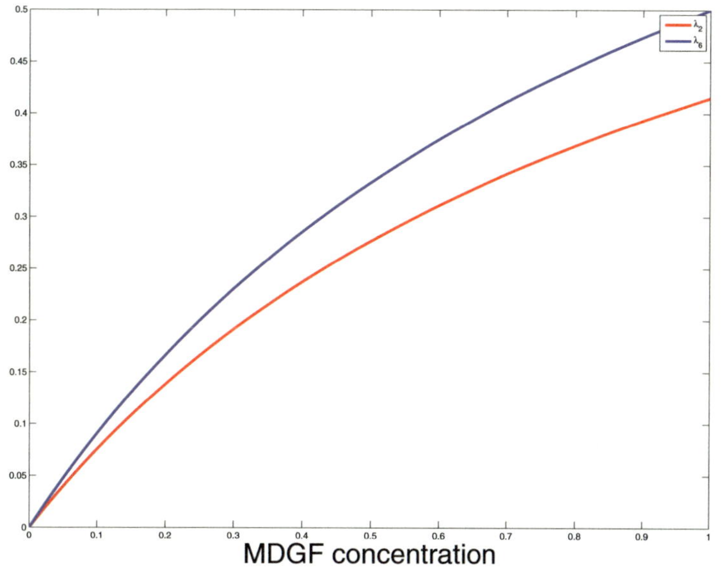

Fig. 7. λ_2 and λ_6 as functions of c_{md}. Here we used $\tau_{\text{tip}} = \tau_{\text{end}} = 1$, $\lambda_2^0 = 0.83$ and $\lambda_6^0 = 1$.

- the first equation: the first, second and third term in the right-hand side, respectively, stand for diffusion of oxygen (u_{oxy}), consumption of oxygen, and supply of oxygen due to the blood flow in the tips and capillaries (via diffusion through the capillary walls);
- the second equation: the first, second and third term in the right-hand side, respectively, model diffusion of macrophage derived growth factor (VEGF), natural decay of VEGF, and secretion by the macrophages as a result of a shortage of oxygen;

The last two equations were described earlier.

Further, the functions f and g are given by

$$f(u_{\text{tip}}, u_{\text{end}}) = \lambda_2^0 \frac{c_{\text{md}}}{c_{\text{md}} + \tau_{\text{tip}}} u_{\text{tip}} - \lambda_3 u_{\text{tip}}^2 - \lambda_4 u_{\text{tip}} u_{\text{end}},$$

$$g(u_{\text{tip}}, u_{\text{end}}) = \lambda_6^0 \frac{c_{\text{md}}}{c_{\text{md}} + \tau_{\text{end}}} \left(a u_{\text{end}} (u_{\text{end}}^0 - u_{\text{end}}) + \chi u_{\text{tip}} u_{\text{end}} (u_{\text{end}}^1 - u_{\text{end}}) \right) + \lambda_5 (\lambda_3 u_{\text{tip}}^2 + \lambda_4 u_{\text{tip}} u_{\text{end}}).$$

To illustrate how this new model behaves we show some results in Figures 8 to 10. The input-data can be found in the appendix. As a computational domain we use the unit square, i.e. $\Omega = \{\mathbf{x} = (x, y) |\ 0 \le x, y \le 1\}$, and the initial wound is given by $\Omega_w = \left\{\mathbf{x} |\ |\mathbf{x}| \le \frac{1}{2}\right\}$.

In Figure 8 we see the oxygen and MDGF concentration seven days after injury. The oxygen tension decreases as one moves into the center of the wound. The relation between a lack of oxygen and the production of the macrophage derived growth factor can clearly be seen. In

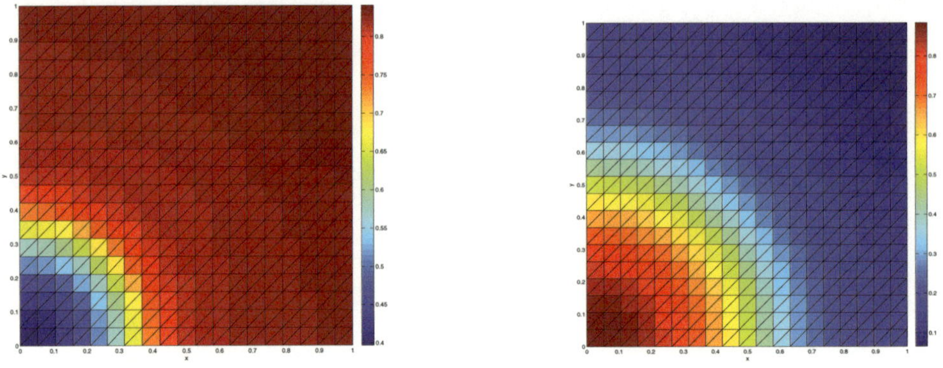

Fig. 8. Normalized oxygen (left) and macrophage derived growth factor (right) concentration seven days after injury.

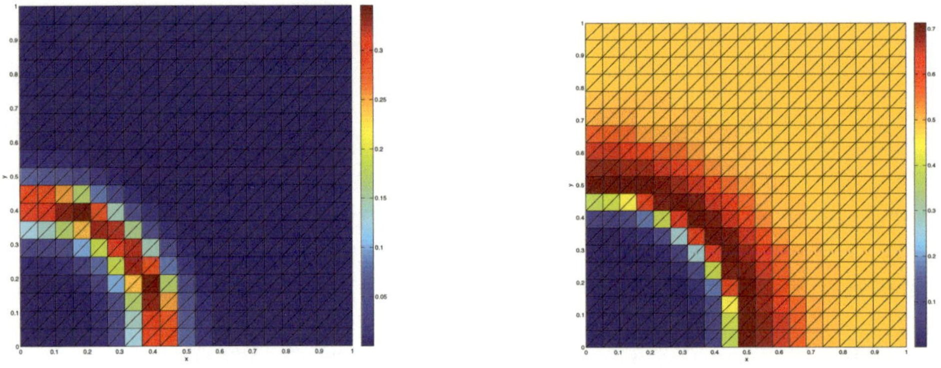

Fig. 9. Normalized capillary tip concentration (left) and endothelial cell density (right) seven days after injury.

areas where the oxygen concentration is low, i.e. inside the wound, the MDGF concentration is at its peak and vice versa. Hence, the MDGF concentration is maximal at the center of the wound. The time for ingress of macrophages has been assumed to be negligible. Also in Figure 10 we see that as the oxygen concentration rises the MDGF concentration drops at approximately the same rate.

Furthermore in Figure 9 we see a front of capillary tips, which slowly moves towards the center of the wound. Also the endothelial cell density is slighty elevated at the wound edge, since this is where the new capillaries are formed. This can also be seen in Figure 10, where the endothelial cell density first peaks and then drops again towards an equilibrium. The capillary tip concentration also peaks, but then drops to zero again. This is to be expected since eventually all capillary tips join together in the newly formed capillary network. The local maximum of the oxygen concentration at $t \approx 18$ days, follows from the peak in capillary tips. As the shortage of oxygen decreases, the concentration of macrophage derived growth factors decreases down to zero.

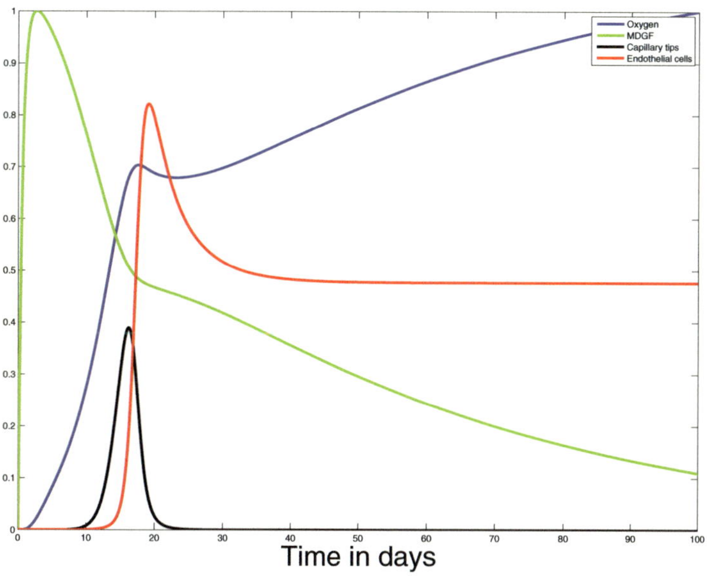

Fig. 10. Normalized solutions furthest in the wound, at $\mathbf{x} = \mathbf{0}$, in time.

3. The coupled model

For all the stages during the proliferative phase of wound healing mathematical models have been presented that attempt to describe the processes involved in these stages. These models only focus on one wound healing stage particularly and do not take into account the interactions between these stages. One example of interaction is the need of oxygen for the growth of cells and thus the dependency of the profileration terms on the oxygen concentration. But there are several more interactions that we can think of, including some that will not be taken into account here either.

In Vermolen (2010) and Vermolen (2011) an attempt is made to combine models of the three wound healing stages into one mathematical model. In this chapter we take the novel model on angiogenesis, presented the previous subsection, and make an attempt to couple it with the wound contraction model due to Javierre (2009), given in Section 2.1. We will investigate both the influence of the oxygen concentration on the wound contraction mechanism and the influence of the fibroblast concentration on capillary growth. Furthermore the displacement of the extracellular matrix (ECM) will also effect the angiogenesis process, this will also be taken into account. For now we will not consider wound closure.

In the wound contraction stage several processes require energy to work. The growth of new tissue consumes energy as cells divide and form new cells, i.e. mitosis. Also the active movement of the fibroblasts requires energy since the cells crawl over each other. This energy needed for wound contraction comes primarily from the food that an individual consumes, however, oxygen is needed to put the energy to use, i.e. the cells require oxygen to get the most out of the energy. This means that the oxygen concentration is good indicator for the amount of energy that is available.

If we look at the partial differential equation (PDE) for the fibroblast concentration in the Javierre model,

$$\frac{\partial u_{\text{fib}}}{\partial t} + \nabla \cdot \left(\frac{\partial \mathbf{u}}{\partial t} u_{\text{fib}} - D_{\text{fib}} \nabla u_{\text{fib}} + \frac{a_{\text{fib}}}{(b_{\text{fib}} + c_{\text{ecm}})^2} u_{\text{fib}} \nabla c_{\text{ecm}} \right) =$$
$$\left(\lambda_{\text{fib}} + \frac{\lambda^0_{\text{fib}} c_{\text{ecm}}}{C_{1/2} + c_{\text{ecm}}} \right) u_{\text{fib}} \left(1 - \frac{u_{\text{fib}}}{K} \right) - \frac{p_{\text{cell}}(\theta)}{\tau_d + p_{\text{cell}}(\theta)} \frac{k_1 c_{\text{ecm}}}{C_k + c_{\text{ecm}}} u_{\text{fib}} + k_2 u_{\text{myo}} - d_{\text{fib}} u_{\text{fib}}, \quad (19)$$

the third term on the left hand side denotes the active mobility due to random walk and the first term on the right hand side represents fibroblast proliferation. To incorporate the effects of the oxygen level on these two processes we replace these terms with the following,

$$\frac{u_{\text{oxy}}}{\tau_{\text{oxy}} + u_{\text{oxy}}} \left(-D_{\text{fib}} \nabla u_{\text{fib}} + \frac{a_{\text{fib}}}{(b_{\text{fib}} + c_{\text{ecm}})^2} u_{\text{fib}} \nabla c_{\text{ecm}} \right)$$

for the mobility and

$$\frac{u_{\text{oxy}}}{\tau_{\text{oxy}} + u_{\text{oxy}}} \left(\lambda_{\text{fib}} + \frac{\lambda^0_{\text{fib}} c_{\text{ecm}}}{C_{1/2} + c_{\text{ecm}}} \right) u_{\text{fib}} \left(1 - \frac{u_{\text{fib}}}{K} \right)$$

for the proliferation of fibroblasts. The extra term in front, $\frac{u_{\text{oxy}}}{\tau_{\text{oxy}} + u_{\text{oxy}}}$, assures that if there is no oxygen present, i.e. no energy can be consumed, the two processes are stopped. When the tissue is saturated with oxygen the processes continue at their normal rate (the term then approaches one). Note that due to the PDE for the oxygen concentration the oxygen level will never rise to dangerous levels (the concentration moves towards an equilibrium), i.e. oxygen poisoning will never take place.

The proliferation of myofibroblasts also consumes energy and thus the same dependency on the oxygen concentration can be found in the PDE for the myofibroblasts. This concludes the influence of the oxygen level on wound contraction covered in this model.

The other way around wound contraction also effects the growth of new capillaries. First the displacement of the ECM causes passive convection of the variables of the angiogenesis model. This means that in each of the four equations an extra term is found that describes this passive convection, i.e.

$$\nabla \cdot \left(\frac{\partial \mathbf{u}}{\partial t} u_i \right),$$

where u_i denote the four variables of the novel angiogenesis model.

Furthermore the new capillaries need tissue to grow in. At first all tissue in the wound has been destroyed by the injury and so the capillaries can not grow. Gradually fibroblasts invade the wound and new tissue is formed. Only then can the recovery of the capillary network start. This is why it is reasonable to let the growth of capillaries depend on the fibroblast concentration.

The growth terms in (17) and (18) are given by

$$\lambda^0_2 \frac{c_{\text{md}}}{c_{\text{md}} + \tau_{\text{tip}}} u_{\text{tip}}$$

and

$$\lambda_6^0 \frac{c_{\text{md}}}{c_{\text{md}} + \tau_{\text{end}}} \left(a u_{\text{end}} (u_{\text{end}}^0 - u_{\text{end}}) + \chi u_{\text{tip}} u_{\text{end}} (u_{\text{end}}^1 - u_{\text{end}}) \right)$$

respectively. Similarly to how the oxygen concentration is coupled with the fibroblast PDE we couple the fibroblasts to the capillary tip and endothelial cell density PDEs. Therefor we introduce the factor

$$\frac{u_{\text{fib}} / u_{\text{fib}}^0}{\tau_{\text{fib}} + u_{\text{fib}} / u_{\text{fib}}^0}$$

in front of both production terms. This results in no capillary growth if there are no fibroblasts present, i.e. there is no appropriate dermal tissue for them to grow in. As the fibroblast concentration rises the rate of capillary growth also rises towards its normal rate, since the term approaches one.

This concludes the coupling between the wound contraction model due to Javierre and the novel angiogenesis model covered in this thesis. Of course there are several other interaction between them. One can for instance think about the differentation from fibroblasts to myofibroblasts and back, which surely also consumes energy. Alternatively, we could think of also linking the chemotaxis term to the oxygen content. This could be a subject for further study.

Next we present some results of the computation done on the coupled model. The values of the parameters used can be found in appendix. We vary the diffusion speeds of the fibroblasts and the oxygen, since we consider these to be of great importance (the problem seems to be diffusion dominated) and we would like to study the effect of these parameters on the solution.

In Figures 11 to 13 the results of the simulations can be found. The figures show the normalized solutions in time furthest in the wound, i.e. at $\mathbf{x} = \mathbf{0}$. The difference in diffusion speeds can clearly be seen in the graphs.

We see that with slow fibroblast diffusion the growth of capillaries start off at a later point in time. This is to be expected since the fibroblasts provide the tissue in which the capillaries can grow. The growth of the capillaries does follow a similar course. First the capillary tips find their way to the center of the wound and after that the capillary network is restored, which decreases the number of tips. As the number of tips decreases down to zero, the oxygen content decreases for a short while due to the lack of this tip-source. Later, the oxygen tension increases to its undamaged equilibrium.

Further, having slower oxygen diffusion, the oxygen concentration depends more on the growth of new capillaries. In Figure 13 we see that the oxygen concentration rises far slower than 11 and 12. The oxygen concentration grows due to capillaries transporting oxygen to the wound side and not primarily due to diffusion.

We see that the varying the diffusion speeds has a major effect on the solutions. This confirms our idea that the problem is diffusion dominated. Although the solutions follow similar patterns in all cases, the differences can also be seen clearly.

Furthermore the coupling between the two models can also be seen. Especially with low diffusion speeds we see that the fibroblast concentration clearly depends on the oxygen concentration. Also the growth of new capillaries starts off later than in the uncoupled model, see Section 2.2. This can also be explained by the coupling, since the growth of capillaries now

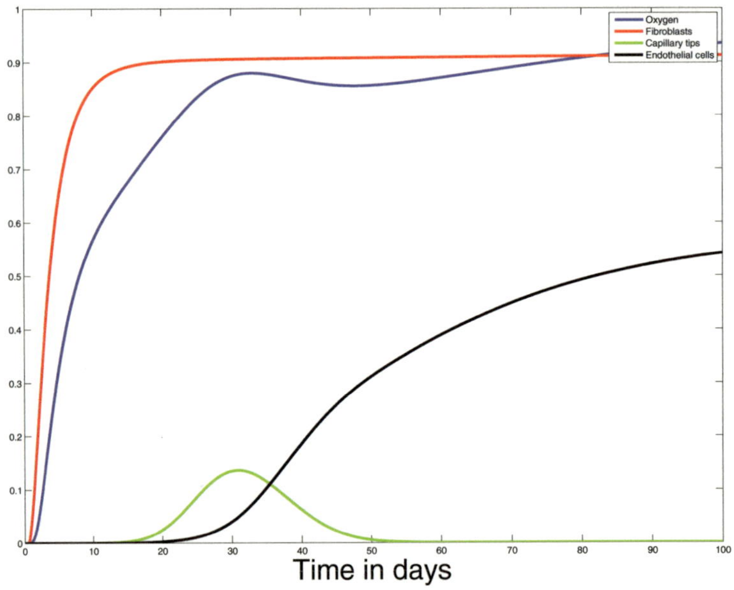

Fig. 11. Normalized solutions furthest in the wound, at $\mathbf{x} = \mathbf{0}$, in time for $D_{\text{fib}} = 0.02$ cm^2/days and $D_{\text{oxy}} = 0.01$ cm^2/days.

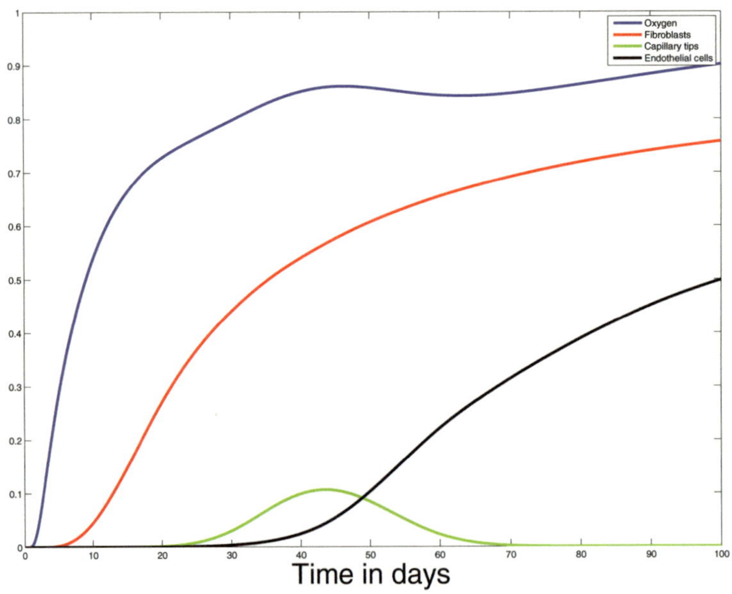

Fig. 12. Normalized solutions furthest in the wound, at $\mathbf{x} = \mathbf{0}$, in time for $D_{\text{fib}} = 0.002$ cm^2/days and $D_{\text{oxy}} = 0.01$ cm^2/days.

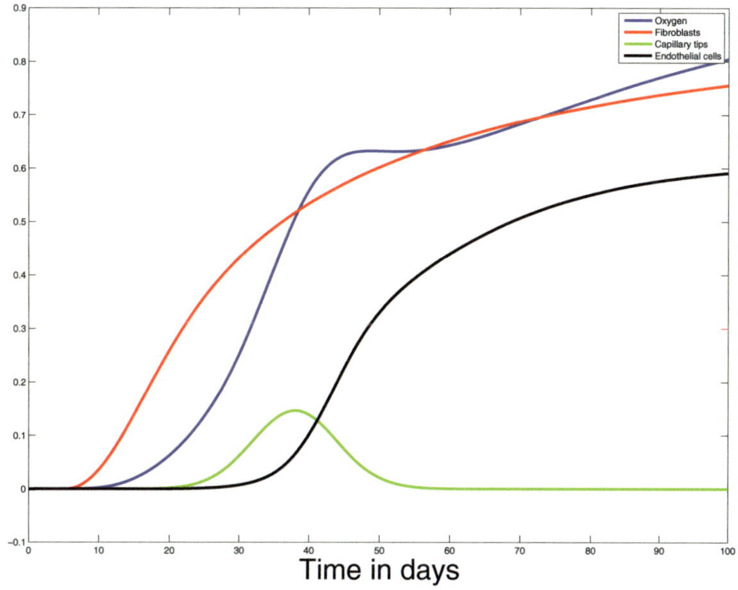

Fig. 13. Normalized solutions furthest in the wound, at $\mathbf{x} = \mathbf{0}$, in time for $D_{\text{fib}} = 0.002$ cm^2/days and $D_{\text{oxy}} = 0.001$ cm^2/days.

depends on the fibroblast concentration. So the growth can not begin untill there is enough tissue available, i.e. the fibroblast concentration is non-zero.

4. Numerical methods

We use a Galerkin finite-element method with linear triangular elements to solve the resulting system of PDEs. The reaction-transport equations are solved using a second-order Trapezoidal Time Numerical Integration Method. The nonlinear problem at each time step is solved using a Newton's method, which gives quadratic convergence. For the numerical integration of the spatial integrals in the weak formulation, we use the second order accurate Newton-Cotes integration rule. The numerical solution will thereby have a quadratic accuracy in both element size and time step. For an efficient implementation, the reactive-transport equations are solved as a fully coupled problem in which the degrees of freedom are numbered in a nodewise fashion. The mechanical parameters were taken at the current time step, since at each time-step the mechanical problem is solved before the biological reactive-transport equations are solved numerically.

The visco-elastic equations are solved using a Galerkin finite-element method with linear triangular elements. The system is solved as a fully coupled problem with a nodewise arrangement of the degrees of freedom, being the vertical and horizontal displacement. The concentrations and cellular densities from the previous time step are used in the mechanical balance equations from visco-elasticity. For the time-integration, a Trapezoidal Numerical Integration Method is used as well to have second order accuracy in time as well. The linear elements give a second-order accuracy for the element size.

5. Discussion

In this paper, we described and coupled some mathematical models for wound contraction and angiogenesis as subprocesses for dermal regeneration post-wounding. It can be seen in Figure 10 that in the course of healing, the endothelial cell density, which is the most important measure for the vascular network density, first increases up to value above the equilibrium value, which corresponds to the fully healed state. This increase coincides with the temporary high capillary tip density at these times. After this intermediate stage, the capillary tip density decreases down to zero, and the endothelial cell density decreases to the equilibrium as time proceeds. This qualitatively agrees with the rabbit ear chamber images from the experiments that were carried out by Komori (2005), among others.

Furthermore, the numerical solutions converge to the undamaged state at the longest scales. This is consistent with clinical situations if the initial wound is very small. However, scar formation due to an excessive regeneration of collagen is not taken into account in the present model. Hence, the present model is incapable of predicting the occurrance of ulcers or hypertrophic scars that can result for large wounds or burns. This issue will be dealt with in future, as well as the extension of the present model by the incorporation of the epidermal layer which starts healing as a result of triggering of keratinocytes by the signaling agents that are secreted by the fibroblasts in the dermis. We are also interested in the development of the basal membrane, which separates the dermis from epidermis, in the course of time.

We also note that the model is entirely based on the continuum hypothesis, and that all densities and concentrations should be considered as averaged quantities over a *representative volume element*. At the smaller scale, one has to use a more cellular approach, such as the cellular automata (cellular Potts) models or (semi-) stochastic approaches. A translation between these classes of models is highly desirable and hardly found in literature.

Finally, we note that modeling of wound healing is very complicated, due to the enormous complexity of the healing process. The present model can already help in obtaining insight into the relation between several parameters, such as the influence of various parameters on the speed and quality of wound healing. In future studies, we intend to include the effects of several treatment of wounds, such as pressure or drug treatments, on the kinetics of wound healing. The models can then be used as a tool for medical doctors to evaluate the influence of certain treatments on wound healing, as well as to determine the circumstances for optimal healing with respect to minimization of hypertrophic scars in the case of burns. Here, we also remark that all data are patient-dependent and hence all parameters in the model are exposed to a stochastic nature. In order to deal with the issues of uncertainties, we intend to employ stochastic finite element methods in near future.

6. Conclusions

We presented an advanced model for the coupling of angiogenesis and wound contraction. The model gives a more complete picture than most of the presently existing formalisms in literature. However, still some room for improvement remains. The current model could also be extended with the healing of the epidermis located on top of the dermis. This will be done in a future study.

The first important innovations in the present study are the formulation of a new angiogenesis model, which consists of a driving force due to lack of oxygen (Maggelakis) and a balance of both the capillary tips and capillaries by tracking these quantities as individual parameters. The second important innovation concerns the combination of the wound contraction model due to Javierre (2009) with the newly developed angiogenesis model. As an important factor of communication between both processes, we use the oxygen tension (on fibroblast mobility and proliferation), the presence of macrophage derived growth factor influencing the logistic proliferation rate of endothelial cells and the fibroblast density on the production of endothelial cells. Of course, these coupling terms have been assumed from educated guesses. An experimental validation could be decisive on the validity of the assumptions used in the present model. The calculations in the present manuscript are preliminary and a parameter sensitivity analysis is to be carried out.

We also note that there is a large uncertainty for the values of all the parameters used. We intend to use stochastic methods to better deal with the occurrance of patient dependencies and other uncertainties in the data. This will be tackled in future studies. The present paper was more descriptive, rather than mathematical.

7. Appendix: Parameter values

In this Appendix, we show the data that we used for the calculations. The first two tables show the parameter values for the skin regeneration models ((myo)fibroblasts and collagen). The third table gives the data of the visco-elastic model for wound contraction. The fourth table gives the input for the angiogenesis model. Further, the values of the coupling parameters are scattered throughout the tables as well.

Parameter	Description	Value	Dimension
u^0_{fib}	Fibroblast density in healthy tissue	10^4	cells/cm^3
D^0_{fib}	Fibroblast diffusion rate	$2\times10^{-2}/2\times10^{-3}$	cm^2/day
a_{fib}	Determines, with b_{fib}, the maximal chemotaxis rate per unit GF conc.	4×10^{-10}	g/cm day
b_{fib}	GF conc. that produces 25% of the maximal chemotatic response	2×10^{-9}	g/cm^3
λ_{fib}	Fibroblast proliferation rate	0.832	day^{-1}
λ^0_{fib}	Maximal GF induced proliferation rate	0.3	day^{-1}
K	Determines the fibroblast equilibrium density	1×10^7	cells/cm^3
$C_{1/2}$	Half-maximal GF enhancement of fibroblasts proliferation	1×10^{-8}	g/cm^3
k_1	Maximal fibroblast to myofibroblast differentation rate	0.8	day^{-1}
k_2	Myofibroblast to fibroblast differentation rate	0.693	day^{-1}
C_k	Half-maximal GF enhancement of fibroblast to myofibroblast differentation	10^{-8}	g/cm^3
d_{fib}	Fibroblast death rate	0.831	day^{-1}
ε_{myo}	Myofibroblast to fibroblast logistic growth rate proportionality factor	0.5	-
d_{myo}	Myofibroblasts death rate	2.1×10^{-2}	day^{-1}
τ_{oxy}	Determines oxygen enhancement of (myo)fibroblasts proliferation	2×10^{-4}	

Table 1. Parameters related to the fibroblast and myofibroblast equations.

Parameter	Description	Value	Dimension
ρ^0	Collagen concentration in healthy tissue	0.1	g/cm^3
ρ_{ini}	Initial collagen concentration in the wound	10^{-3}	g/cm^3
λ_ρ	Collagen production rate	7.59×10^{-10}	g^3/cm^6 cell day
λ_ρ^0	Maximal rate of GF induced collagen production	7.59×10^{-9}	g^3/cm^6 cell day
C_ρ	Half-maximal GF enhancement of collagen production	10^{-8}	g/cm^3
R_ρ	Half-maximal collagen enhancement of ECM deposition	0.3	g/cm^3
η_b	Myofibroblast to fibroblast collagen production rate proportionality factor	2	-
d_ρ	Collagen degradation rate per unit of cell density	7.59×10^{-8}	cm^3/cell day
η_d	Myofibroblast to fibroblast collagen degradation rate proportionality factor	2	-
c_{ecm}^0	Initial GF concentration in the wound	10^{-8}	g/cm^3
D_c	GF diffusion rate	5×10^{-2}	cm^2/day
k_c	GF production rate per unit of cell density	7.5×10^{-6}	cm^3/cell day
ζ	Myofibroblast to fibroblast chemical production rate proportionality factor	1	-
Γ	Half-maximal enhancement of net GF production	$\times 10^{-8}$	g/cm^3
d_c	GF decay rate	0.693	day^{-1}

Table 2. Parameters related to the collagen and growth factor equations.

Parameter	Description	Value	Dimension
p_{max}	Maximal cellular active stress per unit of ECM	10^{-4}	N g/cm2 cell
K_{pas}	Volumetric stiffness moduli of the passive components of the cell	2×10^{-5}	N g/cm2 cell
K_{act}	Volumetric stiffness moduli of the active filaments of the cell	1.852×10^{-5}	N g/cm2 cell
θ_1	Shortening strain of the contractile element	-0.6	-
θ_2	Lengthening strain of the contractile element	0.5	-
τ_d	Half-maximal mechanical enhancement fo fibroblast to myofibroblast differentation	10^{-5}	N g/cm2 cell
μ_1	Undamaged skin shear viscosity	200	N day/cm2
μ_2	Undamaged skin bluk viscosity	200	N day/cm2
E	Undamaged skin Young's modulus	33.4	N/cm2
ν	Undamaged skin Poisson's ratio	0.3	-
ξ	Myofibroblasts enhancement of traction per unit of fibroblasts density	10^{-3}	cm^3/cell
R_τ	Traction inhibition collagen density	5×10^{-4}	g/cm^3
s	Dermis tethering factor	5×10^2	N/cm g

Table 3. Parameters related to the mechanical behaviour of cells and the ECM.

Parameter	Description	Value	Dimension
u^0_{oxy}	Oxygen concentration in healthy tissue	5	mg/cm^3
D_{oxy}	Oxygen diffusion rate	$10^{-2}/10^{-3}$	day^{-1}
λ_{oxy}	Oxygen decay rate	2×10^{-2}	day^{-1}
λ_{13}	Oxygen transport rate	1	day^{-1}
u_θ	Threshold value for macrophage derived GF	5	mg/cm^3
D_{md}	Macrophage derived GF diffusion rate	0.1	day^{-1}
λ_{md}	Macrophage derived GF decay rate	1	day^{-1}
λ_{21}	Macrophage derived GF production rate	10	day^{-1}
u^0_{tip}	Normalized initial capillary tip concentration in the small strip facing the wound	1	-
D_1	Capillary tip diffusion rate	3.5×10^{-4}	cm^2/day
u^0_{end}	Normalized endothelial cell density in healthy tissue	1	-
D_2	Endothelial cell diffusion rate	3.5×10^{-4}	cm^2/day
λ^0_2	Capillary tip growth rate	0.83	day^{-1}
τ_{tip}	Half-maximal macropahge derived GF enhancement of capillary tip growth	1	-
λ_3	Rate at which two capillary tips meet	0.83	day^{-1}
λ_4	Rate at which a capillary tip meets another capillary	0.85	day^{-1}
λ^0_6	Endothelial cell proliferation rate	1	day^{-1}
τ_{end}	Half-maximal macropahge derived GF enhancement of endothelial cell proliferation	1	-
a	Determines, with χ and u^1_{end}, the equilibrium endothelial cell density	0.25	-
χ	Determines, with a and u^1_{end}, the equilibrium endothelial cell density	0.3	-
u^1_{end}	Determines, with a and χ, the equilibrium endothelial cell density	10	-
λ_5	Capillary tip to capillary proportionality factor	0.25	-
τ_{fib}	Half-maximal fibroblast enhancement of capillary growth and endothelial cell proliferation	1	-

Table 4. Parameters related to angiogenesis.

8. References

Bloemen, M.C.T., van der Veer, W.M. Ulrich, M.M.W., van Zuijlen, F.B., Niessen, F.B., Middelkoop, E.: *Prevention and curative management of hypertrophic scar formation*, to appear in Burns, 2011.

van der Veer, W.M., Bloemen, M.C.T., Ulrich, M.M.W., Molema, G., van Zuijlen, P.P.M., Middelkoop, E., Niessen, F.B.: *Potential cellular and molecular causes of hypertrophic scar formation*, Burns, 35, 15–29, 2009.

Dallon, J.C., Ehrlich, H.P.: *A review of fibroblast populated collagen lattices* Wound repair and regeneration, 16, 472–479, 2008.

Dallon, J.C.: *Multiscale modeling of cellular systems in biology*, Current opinion in colloid and interface science, 15, 24–31, 2010.

urray, J.D.: *Mathematical Biology II: Spatial Models and Biomedical Applications*, New York: Springer-Verlag; 2004.

Olsen, L., Sherratt, J.A. and Maini, P.K.: *A mechanochemical model for adult dermal wound closure and the permanence of contracted tissue displacement*, Journal of Theoretical Biology 1995; 177:113-128.

Maggelakis, A.: *A mathematical model for tissue replacement during epidermal wound healing*, Applied Mathematical Modelling 2003; 27:189-196.

Gaffney, E.A., Pugh, K. and Maini, P.K.: *Investigating a simple model for cutaneous wound healing angiogenesis*, Journal of Theoretical Biology 2002; 45:337-374.

Sherratt, J.A. and Murray, J.D.: *Mathematical analysis of a basic model for epidermal wound healing*, Journal of Theoretical Biology 1991; 29:389-404.

Adam, J.A.: *A simplified model of wound healing (with particular reference to the critical size defect)*, Mathematical and Computer Modelling 1999; 30:23-32.

Javierre, E., Moreo, P., Doblaré, M. and García-Aznar, J.M.: *Numerical modeling of a mechano-chemical theory for wound contraction analysis*, International Journal of Solids and Structures 2009; 46:3597-3606.

Vermolen, F.J. and Javierre, E.: *A suite of continuum models for different aspects in wound healing*, In: Bioengineering research of chronic wounds, studies in mechanobiology, tissue engineering and biomaterials, New York: Springer-Verlag; 2009.

Vermolen, F.J. and Javierre, E.: *Computer simulations from a finite-element model for wound contraction and closure*, Journal of Tissue Viability 2010; 19:43-53.

Vermolen, F.J. and Javierre, E.: *A finite-element model for healing of cutaneous wounds combining contraction, angiogenesis and closure*, submitted, 2011.

Schugart, R.C., riedman, A., Zhao, R., Sen, C.K.: *Wound angiogenesis as a function of tissue oxygen tension: a mathematical model*, Proceedings of the National Academy of Sciences USA, 105(7), 2628–2633, 2008.

Xue, C., Friedman, A., Sen, C.K.: *A mathematical model of ischemic cutaneous wounds*, Proceedings of the National Academy of Sciences USA, 106(39), 16783–16787, 2009.

Graner, F., Glazier, J.: *Simulation of biological cell sorting using a two-dimensional extended Potts model*, Phys. Review Letters, *69*, 2013–2016, 1992.

Merks, M.H., Koolwijk, P.: *Modeling morphogenesis in silico and in vitro: towards quantitative, predictive, cell-based modeling*, Mathematical modeling of natural phenomena, *4(4)*, 149–171, 2009.

Plank, M.J. and Sleeman, B.D.: *Lattice and non-lattice models of tumour angiogenesis*, Bull. Mathem. Biol., *66* (2004), 1785–1819.

Vermolen, F.J. and Gefen, A.: *A semi-stochastic cell-based formalism to model the dynamics of migration of cells in colonies*, to appear, 2011.

Kumar, V., Abbas A. and Fausto, N.: *Robbins and Cotran Pathologic Basis of Disease*, Philadelphia: Elsevier Saunders; 2005.

Komori, M., Tomizawa, Y., Takada, K. and Ozaki, M.: *A single local application of recombinant human basic fibroblast growth factor accelerates initial angiogenesis during wound healing in rabbit ear chamber*, Anesth Analg, *100* (2005), 830–834.